Elonka, Stephen Michael.

Standard heating and power boiler plant questions and answers.

$62.95

DATE			

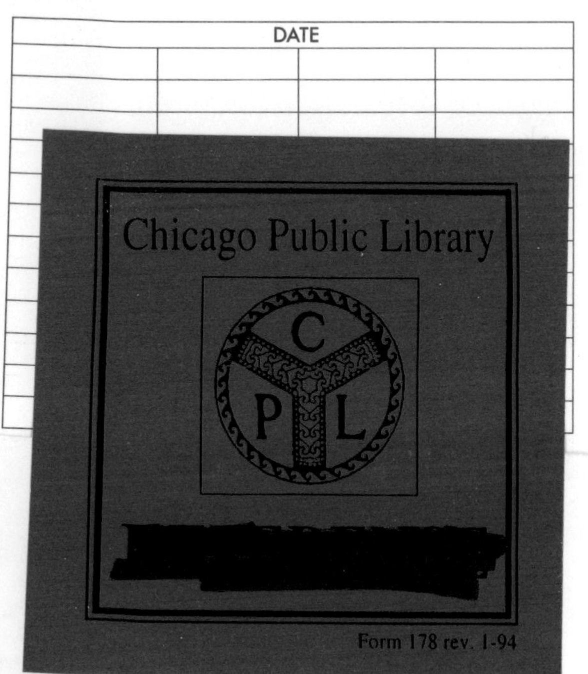

Standard Heating and Power Boiler Plant Questions and Answers

RELATED BOOKS OF INTEREST

Standard Basic Math and Applied Plant Calculations (1978)
Stephen M. Elonka
Standard Plant Operators' Manual (3d ed., 1980)
Stephen M. Elonka
Standard Refrigeration and Air Conditioning Questions and Answers (2d ed., 1973; reprint 1980)
Stephen M. Elonka and Quaid W. Minich
Standard Boiler Operators' Questions and Answers (1969; reprint 1978)
Stephen M. Elonka and Anthony L. Kohan
Standard Industrial Hydraulics Questions and Answers (1967; reprint 1979)
Stephen M. Elonka and Orville H. Johnson
Standard Electronics Questions and Answers (1964)
Stephen M. Elonka and Julian L. Bernstein
Standard Instrumentation Questions and Answers (1962; reprint 1979)
Stephen M. Elonka and Alonzo R. Parsons
Standard Plant Operator's Questions and Answers (2d ed., 1981)
Stephen M. Elonka and Joseph F. Robinson
Standard Boiler Room Questions and Answers (3d ed., 1982)
Stephen M. Elonka and Alex Higgins
Electrical Systems and Equipment for Industry (1978)
Arthur H. Moore and Stephen M. Elonka
Marmaduke Surfaceblow's Salty Technical Romances (1979)
Stephen M. Elonka
Plant Energy Systems: Energy Systems Engineering (1967)
The Editors of Power
Mechanical Packing Handbook (1960)
Stephen M. Elonka and Fred H. Luhrs
Boiler Operator's Guide (2d ed., 1981)
Harry M. Spring and Anthony L. Kohan

Standard Heating and Power Boiler Plant Questions and Answers

A revision of *Standard Boiler Operator's Questions and Answers*, first published in 1969

Stephen Michael Elonka

Contributing Editor, Power *magazine; Licensed Chief Marine Steam Engineer, Oceans, Unlimited Horsepower; Licensed as Regular Instructor of Vocational High School, New York State; Member: Instrument Society of America, National Association of Power Engineers (life, honorary); Author:* The Marmaduke Surfaceblow Story; Plant Operators' Manual; *Coauthor:* Standard Plant Operator's Questions and Answers, *Volumes I and II;* Standard Refrigeration and Air Conditioning Questions and Answers; Standard Instrumentation Questions and Answers, *Volumes I and II;* Standard Electronics Questions and Answers, *Volumes I and II;* Standard Industrial Hydraulics Questions and Answers; Handbook of Mechanical Packings.

Anthony Lawrence Kohan

Manager, Boiler and Machinery Technical Specialists, Boiler and Machinery Department, Royal Insurance Company; National Board Commissioned Inspector (various state boiler inspector commissions); Certified Safety Professional; Member: American Society of Mechanical Engineers, National Society of Professional Engineers, American Welding Society.

McGraw-Hill Book Company

New York St. Louis San Francisco Auckland
Bogotá Hamburg Johannesburg London Madrid
Mexico Montreal New Delhi Panama Paris
São Paulo Singapore Sydney Tokyo Toronto

Library of Congress Cataloging in Publication Data

Elonka, Stephen Michael.
Standard heating and power boiler plant questions and answers.

Rev. ed. of: Standard boiler operators' questions and answers. [1969]
Bibliography: p.
Includes index.
1. Steam-boilers—Handbooks, manuals, etc.
I. Kohan, Anthony Lawrence. II. Elonka, Stephen Michael. Standard boiler operators' questions and answers.
III. Title.
TJ289.E4 1984 621.1'83 83-19538
ISBN 0-07-019277-4

3 4 5 6 7 8 9 10 11 BKP BKP 9 0 9 8 7 6 5 4 3 2 1 0

ISBN 0-07-019277-4

The editors for this book were Patricia Allen-Browne and Laura Givner, the designer was Mary Ann Felice, and the production supervisor was Sally Fliess. It was set in Caledonia by Byrd Data Imaging Group.

Printed and bound by Book Press.

Dedicated to the numerous people involved with boiler systems operation, inspection, maintenance, and repair, who must have a thorough knowledge of boilers and connected auxiliaries involving construction, instrumentation, and safeguards if we are to keep power and process services flowing efficiently and safely

Contents

Preface

This book is a major expansion of *Standard Boiler Operators' Questions and Answers*, coauthored by the same authors. This volume includes additional material on heating and power boiler systems that are used in the commercial, industrial, and power-generation fields. For example, one chapter is devoted to the problem of integrating boiler and turbogenerator operations and controls. It explains how steam contaminants may affect both the turbine and the boiler and thus why turbine and boiler systems require close coordination between boiler and turbine operating crews.

The question-and-answer format of the first volume has been retained, and the text has been expanded and changed as needed. This volume includes a review of boiler types, methods used in construction, ASME code requirements, NB and jurisdictional inspection, repairs and reporting procedures, fuels, firing, combustion controls and flame-safeguard systems, water treatment, operating and maintenance problems, failures, code calculations to check for the strength of boiler components, safety devices, and many other features related to the operation and maintenance of an energy system. Pollution-control requirements, energy conservation, cogeneration, and combined cycles are also stressed in the text.

The rapid growth of controls and automatic operating devices requires a review of how these may malfunction and cause forced outages. The modular boiler-turbine arrangements in large power plants require all operating people to be more cognizant than ever in analyzing systems from limit readings or printouts. Major damage may develop from a minor cause that used to be corrected by operators who were spread out in the generating plant to watch key components. Too much reliance on sensors and display boards can be misleading at times because there is a limit to the amount of pickups available, and the economic costs involved may be prohibitive. Thus there is still a need for human intelligence in observing and analyzing operations and in diagnosing problems so that the equipment will operate economically and safely.

While this volume is arranged in question-and-answer format, as are the other books in the Elonka "standard" series, the questions are not limited only to those that may be asked at job interviews or on operating-license examinations. It was the wish of the authors to expand the answers in order to stress more engineering fundamentals; thus the reader will better understand the point of each question. This approach should prepare the reader for answering any questions that may arise on the job, as well as questions asked at interviews and on any written examination on heating and power boiler systems.

Plant engineers and energy management people should also find this material useful. It will help them assess their current operating and maintenance practices and determine how they may be improved to assure a more economical and reliable operation.

Stationary-license examiners will find this volume useful in preparing examinations; thus it will help bring the standards for operating engineers up to a uniform level. As it is today, there is little similarity in license requirements and examinations given in various parts of the country. For example, an operator with a chief engineer's license in one state may not even qualify as a fireman in another state. (The Canadian examinations are more uniform and of a higher level.)

Special dedication and appreciation should be offered to the grimy boiler inspectors who, in their work, are so often amazed at the many accidents they investigate which could have been easily avoided if certain fundamentals were understood by the operators and owners of boilers.

The authors wholeheartedly thank the editors of *Power* magazine for offering their valuable help. They also thank the many manufacturers and professional societies that furnished illustrations and technical information. These include the American Society of Mechanical Engineers, the American Boiler Manufacturers Association, the Edison Electric Institute, the National Board of Boiler and Pressure Vessel Inspectors, the National Fire Protection Association, the Uniform Boiler and Pressure Vessel Laws Society, and too many others to list here.

Suggestions for improvement of the text will always be welcome, as will the submission of any corrections or additions for the text.

Stephen Michael Elonka
Anthony Lawrence Kohan

CHAPTER **ONE**

System Application

Today's heating and power-generating systems are complex because of automation and the required pollution-control devices. To burn fuel efficiently and cleanly, the responsible plant operator must know not only heat-generating equipment but also the systems, auxiliaries, and instrumentation tied in with heating boilers and steam generators. In this chapter we look at the various systems, then cover basics, the generation of heat and steam, some of the required boilers, and the ASME boiler code.

BUILDING HEATING

Q Name the five basic mediums for building heating.
A Steam, hot water (hydronic system), hot air, electricity, and solar energy are the five basic heating mediums. Steam and hot water are usually generated at a central point, as in Fig. 1.1, then distributed through piping to the heat-transfer surfaces in the spaces to be heated.

Hot air is circulated by fans or through ducts or by natural means directly from the furnace surfaces or by steam or hot-water coils. Electric boilers, or electric resistance heaters, are light and easy to spot in most locations. And solar energy is the newest medium.

Q Name some advantages of hot-water heating.
A Ease of control is the biggest claim for hot-water heating. And because the heating medium is a liquid, the amount of heat given up depends mostly on the temperature difference between the room air and the hot water. Thus accurate control is possible by varying the quantity or temperature of the water pumped through the spaces to be heated.

Q What are the main advantages of hot-air heating?
A Hot-air heating not only offers positive air circulation and ventila-

Fig. 1.1 Four 40,000 lb steam per hour two-drum boilers modernized with Coen Company gas/oil burners in heating plant of Illinois State University. (*Courtesy of Coen Company.*)

tion but also makes it fairly easy to filter, humidify, or treat the air. Air also forms a natural complement to summer air conditioning and is usually the most rapid system to start.

Q Name the principal advantages of steam heating.
A Because steam gives up most of its heat by condensation, only a small amount needs to be circulated (one-fiftieth by weight, or one-thirtieth by volume) as compared with water. Also, high heat-transfer rates allow small heating surfaces. And finally, steam pressure assures positive circulation without pumps, making steam systems flexible and usually less costly to install.

Q Why is heating with electricity gaining in use today?
A True, heating with electricity has been increasing recently. Two

basic reasons are (1) no pollution and (2) our abundance of coal for generating electricity at a central station. While electric resistance heating is more expensive, it is clean and safer. Also, it can be applied to do most heating jobs.

Q Explain how energy from the sun is today converted into heat for buildings.
A The most common method is capturing sunlight outside in thermal collectors for heating and also for cooling buildings. Thermal collectors are also used for heating liquids for domestic and process use. A liquid (either water or antifreeze) or air is circulated through solar collectors on the outside of the building. Then the heat is transferred to tanks containing water or rocks, which absorb and store the heat and through which water or air is circulated for use in the spaces to be heated.

Q Name the main advantages of infrared for comfort heating.
A Infrared provides efficient, controllable heat where spot, or limited-area, heating is needed.

> EXAMPLE: Large spaces may be heated by a suitably designed infrared system because these systems project from 50 to 80 percent of their total heat input to the floor, similar to a spotlight illuminating a dark area, without inducing air motion. Also, relocation is easy to meet changing needs.

ENERGY

Q What conservation-of-energy factors must be reviewed when thermal (heat) systems are considered?
A The conservation of energy law states that energy can be neither created nor destroyed. But energy can change in form, with definite and invariable amounts occurring in the transformation. Potential energy and kinetic energy are the most familiar forms of energy. Potential energy is the energy that exists when a body (or working medium) is at an elevation above a datum plane on the earth. Potential energy is expressed in foot-pounds.

> EXAMPLE: If a 600-lb body is 30 ft above a datum plane, the potential energy is $600 \times 30 = 1800$ ft · lb.

Kinetic energy is energy associated with a body in motion. Most often cited is a bullet traveling at high speed and having kinetic energy. Physics books express kinetic energy in the following equation:

$$KE = \frac{WV^2}{2g} \quad \text{ft} \cdot \text{lb}$$

where W = weight of moving body, lb
V = velocity of body, ft/s
g = acceleration of gravity, 32.3 ft/s^2

EXAMPLE: If a body weighing 30 lb is 20 ft above the ground, what will be its velocity when it strikes the ground? What is its potential energy?

The potential energy is 20 × 30 = 600 ft · lb.

Substituting in the KE equation above, realizing that

$$\text{KE} = \text{potential energy} = 600 \text{ ft} \cdot \text{lb}$$

$$600 = \frac{30V^2}{2(32.2)}$$

and solving for V,

$$V = \sqrt{\frac{(600)(2)(32.2)}{30}} = 35.9 \text{ ft/s} \qquad \textit{Ans.}$$

Internal energy, the heat or energy stored in a working medium, is thought to be the molecular activity in the molecules composing the working medium. It is commonly called heat.

When a liquid is changed to a gas by heat, the *heat of vaporization*, or energy, dissociates the molecules from the liquid form to the gaseous form. It takes additional energy to make such a transformation well above that required to raise the temperature of a substance. Internal energy is usually expressed in British thermal units per pound of the working substance. The mechanical equivalent of heat is 778 ft · lb = 1 Btu.

Internal-energy Btu content is also dependent on the pressure and temperature of the working medium.

Q What is meant by *pressure*?

A Pressure is that force in pounds per square inch (psi) caused by a fluid (liquids and gases are fluids) tha is confined in a pressure vessel or piping. Pressure as discussed here will be confined to boiler application; it is also used in structure, soil analysis, and process flow calculations, etc. In boiler and pressure vessel strength calculations, pressure is expressed as pounds per square inch gage (psig). In physics and thermodynamic calculations involving heat flow or steam and water characteristics, *absolute pressure* (expressed as pounds per square inch absolute and abbreviated psia) is used instead of gage pressure. ASME boiler code calculations are always in gage pressure.

Q How do *gage* and *absolute* pressure differ?

A Gage pressure, as the name implies, is the pressure noted on a pressure gage when it is installed at any opening into the pressure part of the pressure vessel. It indicates the pressure inside the vessel. But on the outside of the vessel, the atmosphere is also exerting a pressure on the outer surface of the shell. *Absolute pressure* is thus the total pressure of gage pressure *plus* atmospheric pressure. Absolute pressure is influenced by the elevation above sea level. At sea level, for example, the atmospheric pressure is 14.7 psi. Thus a 100-psi-rated boiler would show a 100-psi gage pressure and have 114.7-psi absolute pressure. In determining the heat content of steam for a boiler in Btu per pound, steam tables must be checked, and absolute pressure used.

Q Is there a difference between *force* and *pressure*?
A Yes. *Pressure* is always expressed in pounds per unit area (in Europe, kilograms per unit area). It is the force exerted on a unit area, which in the United States is expressed in pounds per square inch, abbreviated as psi. *Force* is expressed simply by the weight term, *pounds*. But there is a relationship between force and pressure.

EXAMPLE: If a plate of 10 × 10 in area has 100 psi acting on it, the force on the plate is found by the following fundamental equation:

$$F = P \times A$$

where F = force
P = pressure
A = area (upon which pressure acts)

For the plate in this problem, force = 100 × (10 × 10) = 10,000 lb ***Ans.***

Q Give some conversion factors for pressure.
A Pressure is usually expressed in pounds per square inch but can also be expressed as pounds per square foot, inches of mercury for vacuum conditions, feet of water, etc.

EXAMPLE: 1 lb/in^2 = 144 lb/ft^2 = 2.31 ft of water.
1 standard atmosphere = 14.7 lb/in^2.
1 inch of mercury (in Hg) = 0.491 lb/in^2, or 13.6 in of water.
1 ft of water = 0.434 lb/in^2.

The unit pressure exerted by a *fluid* depends on the height of the fluid above a datum and its density, or weight per unit volume. Height times density of one fluid = height times density of another fluid, or

$$h_1 d_1 = h_2 d_2$$

if the fluids are at the same pressure.

EXAMPLE: What would one standard atmosphere at 14.7 lb/in^2 be in (1) feet of water? (2) inches of mercury?

1. To convert to feet of water, the atmospheric pressure must be converted to pounds per square foot.

$$h_1 d_1 = \text{water pressure}$$

$$h_1 d_1 = 14.7 \times 144 = \text{height} \frac{\text{density of water}}{\text{ft}^3}$$

or $14.7 \times 144 = 62.3\ \text{lb/ft}^3\ (h_1)$, where 62.3 = weight of 1 ft^3 of water. Solving:

$$h_1 = 34\ \text{ft} \qquad \textbf{\textit{Ans.}}$$

2. Density of mercury is 0.4893 lb/in^3 at 70°F.

$$h_1 d_1 = \text{pressure} = 14.7 = 0.4893\ (h_1)$$

$$h_1 = \frac{14.7}{0.4893} = 30.03\ \text{in (at 70°F)} \qquad \textbf{\textit{Ans.}}$$

STEAM TABLES

Q What is specific volume?

A Specific volume v is defined as the volume occupied by a unit weight of a fluid or gas under specified conditions of pressure and temperature. Steam tables provide specific volumes for defined pressures and temperatures of water and steam.

Density is the reciprocal of the specific volume or pounds per cubic foot, etc., of a substance.

EXAMPLE: From Table 1.1, what is the specific volume v_f of water at 100°F? The table shows v_f (liquid) is 0.01613 $\text{ft}^3\text{/lb}$. The density would be $1/0.01613 = 61.99\ \text{lb/ft}^3$. ***Ans.***

Q How is the term *enthalpy* defined?

A The law of conservation of energy states that any fluid entering a device, plus any energy added in the device, must equal the energy leaving the device. From the general energy equation, the term internal energy (U) plus the flow work $Pv/778$, or

$$U + \frac{Pv}{778} = \text{enthalpy} = h$$

where P = absolute pressure, lb/ft^2
v = ft^3 specific volume

Enthalpy h is expressed in Btu per pound.

Table 1.1 Temperature

Temperature, t, °F	Absolute pressure, p, psi	Specific volume: Sat. liquid v_f	Specific volume: Sat. vapor v_g	Enthalpy: Sat. liquid h_f	Enthalpy: Evap. h_{fg}	Enthalpy: Sat. vapor h_g	Entropy: Sat. liquid s_f	Entropy: Sat. vapor s_g
32	0.08854	0.01602	3306	0.00	1075.8	1075.8	0.0000	2.1877
35	0.09995	0.01602	2947	3.02	1074.1	1077.1	0.0061	2.1770
40	0.12170	0.01602	2444	8.05	1071.3	1079.3	0.0162	2.1597
45	0.14752	0.01602	2036.4	13.06	1068.4	1081.5	0.0262	2.1429
50	0.17811	0.01603	1703.2	18.07	1065.6	1083.7	0.0361	2.1264
60	0.2563	0.01604	1206.7	28.06	1059.9	1088.0	0.0555	2.0948
70	0.3631	0.01606	867.9	38.04	1054.3	1092.3	0.0745	2.0647
80	0.5069	0.01608	633.1	48.02	1048.6	1096.6	0.0932	2.0360
90	0.6982	0.01610	468.0	57.99	1042.9	1100.9	0.1115	2.0087
100	0.9492	0.01613	350.4	67.97	1037.2	1105.2	0.1295	1.9826
110	1.2748	0.01617	265.4	77.94	1031.6	1109.5	0.1471	1.9577
120	1.6924	0.01620	203.27	87.92	1025.8	1113.7	0.1645	1.9339
130	2.2225	0.01625	157.34	97.90	1020.0	1117.9	0.1816	1.9112
140	2.8886	0.01629	123.01	107.89	1014.1	1122.0	0.1984	1.8894
150	3.718	0.01634	97.07	117.89	1008.2	1126.1	0.2149	1.8685
160	4.741	0.01639	77.29	127.89	1002.3	1130.2	0.2311	1.8485
170	5.992	0.01645	62.06	137.90	996.3	1134.2	0.2472	1.8293
180	7.510	0.01651	50.23	147.92	990.2	1138.1	0.2630	1.8109
190	9.339	0.01657	40.96	157.95	984.1	1142.0	0.2785	1.7932
200	11.526	0.01663	33.64	167.99	977.9	1145.9	0.2938	1.7762
210	14.123	0.01670	27.82	178.05	971.6	1149.7	0.3090	1.7598
212	14.696	0.01672	26.80	180.07	970.3	1150.4	0.3120	1.7566
220	17.186	0.01677	23.15	188.13	965.2	1153.4	0.3239	1.7440
230	20.780	0.01684	19.382	198.23	958.8	1157.0	0.3387	1.7288
240	24.969	0.01692	16.323	208.34	952.2	1160.5	0.3531	1.7140
250	29.825	0.01700	13.821	218.48	945.5	1164.0	0.3675	1.6998
260	35.429	0.01709	11.763	228.64	938.7	1167.3	0.3817	1.6860
270	41.858	0.01717	10.061	238.84	931.8	1170.6	0.3958	1.6727
280	49.203	0.01726	8.645	249.06	924.7	1173.8	0.4096	1.6597
290	57.556	0.01735	7.461	259.31	917.5	1176.8	0.4234	1.6472
300	67.013	0.01745	6.466	269.59	910.1	1179.7	0.4369	1.6350
310	77.68	0.01755	5.626	279.92	902.6	1182.5	0.4504	1.6231
320	89.66	0.01765	4.914	290.28	894.9	1185.2	0.4637	1.6115
330	103.06	0.01776	4.307	300.68	887.0	1187.7	0.4769	1.6002
340	118.01	0.01787	3.788	311.13	879.0	1190.1	0.4900	1.5891
350	134.63	0.01799	3.342	321.63	870.7	1192.3	0.5029	1.5783
360	153.04	0.01811	2.957	332.18	862.2	1194.4	0.5158	1.5677

Table 1.1 Temperature (*Continued*)

Temperature, t, °F	Absolute pressure, p, psi	Specific volume: Sat. liquid v_f	Specific volume: Sat. vapor v_g	Enthalpy: Sat. liquid h_f	Enthalpy: Evap. h_{fg}	Enthalpy: Sat. vapor h_g	Entropy: Sat. liquid s_f	Entropy: Sat. vapor s_g
370	173.37	0.01823	2.625	342.79	853.5	1196.3	0.5286	1.5573
380	195.77	0.01836	2.335	353.45	844.6	1198.1	0.5413	1.5471
390	220.37	0.01850	2.0836	364.17	835.4	1199.6	0.5539	1.5371
400	247.31	0.01864	1.8633	374.97	826.0	1201.0	0.5664	1.5272
410	276.75	0.01878	1.6700	385.83	816.3	1202.1	0.5788	1.5174
420	308.83	0.01894	1.5000	396.77	806.3	1203.1	0.5912	1.5078
430	343.72	0.01910	1.3499	407.79	796.0	1203.8	0.6035	1.4982
440	381.59	0.01926	1.2171	418.90	785.4	1204.3	0.6158	1.4887
450	422.6	0.0194	1.0993	430.1	774.5	1204.6	0.6280	1.4793
460	466.9	0.0196	0.9944	441.4	763.2	1204.6	0.6402	1.4700
470	514.7	0.0198	0.9009	452.8	751.5	1204.3	0.6523	1.4606
480	566.1	0.0200	0.8172	464.4	739.4	1203.7	0.6645	1.4513
490	621.4	0.0202	0.7423	476.0	726.8	1202.8	0.6766	1.4419
500	680.8	0.0204	0.6749	487.8	713.9	1201.7	0.6887	1.4325
520	812.4	0.0209	0.5594	511.9	686.4	1198.2	0.7130	1.4136
540	962.5	0.0215	0.4649	536.6	656.6	1193.2	0.7374	1.3942
560	1133.1	0.0221	0.3868	562.2	624.2	1186.4	0.7621	1.3742
580	1325.8	0.0228	0.3217	588.9	588.4	1177.3	0.7872	1.3532
600	1542.9	0.0236	0.2668	617.0	548.5	1165.5	0.8131	1.3307
620	1786.6	0.0247	0.2201	646.7	503.6	1150.3	0.8398	1.3062
640	2059.7	0.0260	0.1798	678.6	452.0	1130.5	0.8679	1.2789
660	2365.4	0.0278	0.1442	714.2	390.2	1104.4	0.8987	1.2472
680	2708.1	0.0305	0.1115	757.3	309.9	1067.2	0.9351	1.2071
700	3093.7	0.0369	0.0761	823.3	172.1	995.4	0.9905	1.1389
705.4	3206.2	0.0503	0.0503	902.7	0	902.7	1.0580	1.0580

For a boiler, since no mechanical work is done, the enthalpy of the fluid leaving minus the enthalpy of the fluid entering is equal to the heat added per pound of the fluid entering the boiler, or

$$Q \text{ (added)} = h_2 - h_1 \qquad \text{Btu/lb}$$

This equation shows that the heat added is equal to the enthalpy of leaving steam h_g minus the feedwater enthalpy h_f. See Tables 1.1 and 1.2.

EXAMPLE: Water at 250°F enters a boiler and saturated steam at 600 psia leaves the boiler. How many Btu per pound were added?

$$Q = h_2 - h_1$$

Table 1.2 Pressure

Absolute pressure, *p*, psi	Temperature, *t*, °F	Specific volume: Sat. liquid v_f	Specific volume: Sat. vapor v_g	Enthalpy: Sat. liquid h_f	Enthalpy: Evap. h_{fg}	Enthalpy: Sat. vapor h_g	Entropy: Sat. liquid s_f	Entropy: Sat. vapor s_g
1.0	101.74	0.01614	333.6	69.70	1036.3	1106.0	0.1326	1.9782
2.0	126.08	0.01623	173.73	93.99	1022.2	1116.2	0.1749	1.9200
3.0	141.48	0.01630	118.71	109.37	1013.2	1122.6	0.2008	1.8863
4.0	152.97	0.01636	90.63	120.86	1006.4	1127.3	0.2198	1.8625
5.0	162.24	0.01640	73.52	130.13	1001.0	1131.1	0.2347	1.8441
6.0	170.06	0.01645	61.98	137.96	996.2	1134.2	0.2472	1.8292
7.0	176.85	0.01649	53.64	144.76	992.1	1136.9	0.2581	1.8167
8.0	182.86	0.01653	47.34	150.79	988.5	1139.3	0.2674	1.8057
9.0	188.28	0.01656	42.40	156.22	985.2	1141.4	0.2759	1.7962
10	193.21	0.01659	38.42	161.17	952.1	1143.3	0.2835	1.7876
11	197.75	0.01662	35.14	165.73	979.3	1145.0	0.2903	1.7800
12	201.96	0.01665	32.40	169.96	976.6	1146.0	0.2967	1.7730
13	205.88	0.01667	30.06	173.91	974.2	1148.1	0.3027	1.7665
14	209.56	0.01670	28.04	177.61	971.9	1149.5	0.3083	1.7605
14.696	212.00	0.01672	26.80	180.07	970.3	1150.4	0.3120	1.7566
15	213.03	0.01672	26.29	181.11	969.7	1150.8	0.3135	1.7549
16	216.32	0.01674	24.75	184.42	967.6	1152.0	0.3184	1.7497
17	219.44	0.01677	23.39	187.56	965.5	1153.1	0.3231	1.7449
18	222.41	0.01679	22.17	190.56	963.6	1154.2	0.3275	1.7403
19	225.24	0.01681	21.08	193.42	961.9	1155.3	0.3317	1.7360
20	227.96	0.01683	20.089	196.16	960.1	1156.3	0.3356	1.7319
21	230.57	0.01685	19.192	198.79	958.4	1157.2	0.3395	1.7280
22	233.07	0.01687	18.375	201.33	956.8	1158.1	0.3431	1.7242
23	235.49	0.01689	17.627	203.78	955.2	1159.0	0.3466	1.7206
24	237.82	0.01691	16.938	206.14	953.7	1159.8	0.3500	1.7172
25	240.07	0.01692	16.303	208.42	952.1	1160.6	0.3533	1.7139
26	242.25	0.01694	15.715	210.62	950.7	1161.3	0.3564	1.7108
27	244.36	0.01696	15.170	212.75	949.3	1162.0	0.3594	1.7078
28	246.41	0.01698	14.663	214.83	947.9	1162.7	0.3623	1.7048
29	248.40	0.01699	14.189	216.86	946.5	1163.4	0.3652	1.7020
30	250.33	0.01701	13.746	218.82	945.3	1164.1	0.3680	1.6993
35	259.28	0.01708	11.898	227.91	939.2	1167.1	0.3807	1.6870
40	267.25	0.01715	10.498	236.03	933.7	1169.7	0.3919	1.6763
45	274.44	0.01721	9.401	243.36	928.6	1172.0	0.4019	1.6669
50	281.01	0.01727	8.515	250.09	924.0	1174.1	0.4110	1.6585
55	287.07	0.01732	7.787	256.30	919.6	1175.9	0.4193	1.6509
60	292.71	0.01738	7.175	262.09	915.5	1177.6	0.4270	1.6438
65	297.97	0.01743	6.655	267.50	911.6	1179.1	0.4342	1.6374

Table 1.2 Pressure (*Continued*)

		Specific volume		Enthalpy			Entropy	
Absolute pressure, p, psi	Temperature, t, °F	Sat. liquid v_f	Sat. vapor v_g	Sat. liquid h_f	Evap. h_{fg}	Sat. vapor h_g	Sat. liquid s_f	Sat. vapor s_g
70	302.92	0.01748	6.206	272.61	907.9	1180.6	0.4409	1.6315
75	307.60	0.01753	5.816	277.43	904.5	1181.9	0.4472	1.6259
80	312.03	0.01757	5.472	282.02	901.1	1183.1	0.4531	1.6207
85	316.25	0.01761	5.168	286.39	897.8	1184.2	0.4587	1.6158
90	320.27	0.01766	4.896	290.56	894.7	1185.3	0.4641	1.6112
95	324.12	0.01770	4.652	294.56	891.7	1186.2	0.4692	1.6068
100	327.81	0.01774	4.432	298.40	888.8	1187.2	0.4740	1.6026
110	334.77	0.01782	4.049	305.66	883.2	1188.9	0.4832	1.5948
120	341.25	0.01789	3.728	312.44	877.9	1190.4	0.4916	1.5878
130	347.32	0.01796	3.455	318.81	872.9	1191.7	0.4995	1.5812
140	353.02	0.01802	3.220	324.82	868.2	1193.0	0.5069	1.5751
150	358.42	0.01809	3.015	330.51	863.6	1194.1	0.5138	1.5694
160	363.53	0.01815	2.834	335.93	859.2	1195.1	0.5204	1.5640
170	368.41	0.01822	2.675	341.09	854.9	1196.0	0.5266	1.5590
180	373.06	0.01827	2.532	346.03	850.8	1196.9	0.5325	1.5542
190	377.51	0.01833	2.404	350.79	846.8	1197.6	0.5381	1.5497
200	381.79	0.01839	2.288	355.36	843.0	1198.4	0.5435	1.5453
250	400.95	0.01865	1.8438	376.00	825.1	1201.1	0.5675	1.5263
300	417.33	0.01890	1.5433	393.84	809.0	1202.8	0.5879	1.5104
400	444.59	0.0193	1.1613	424.0	780.5	1204.5	0.6214	1.4844
500	467.01	0.0197	0.9278	449.4	755.0	1204.4	0.6487	1.4684
600	486.21	0.0201	0.7698	471.6	731.6	1203.2	0.6720	1.4454
700	503.10	0.0205	0.6554	491.5	709.7	2101.2	0.6925	1.4296
800	518.23	0.0209	0.5687	509.7	688.9	1198.6	0.7108	1.4153
1000	544.61	0.0216	0.4456	542.4	649.4	1191.8	0.7430	1.3897
2000	635.82	0.0257	0.1878	671.7	463.4	1135.1	0.8619	1.2849
3000	695.36	0.0346	0.0858	802.5	217.8	1020.3	0.9731	1.1615
3206.2	705.40	0.0503	0.0503	902.7	0	902.7	1.0580	1.0580

From the table,

$$Q = h_g\,(600 \text{ psia}) - h_f\,(250°\text{F})$$

$$Q = 1203.2 - 218.48 = 984.72 \text{ Btu/lb} \qquad \textit{Ans.}$$

See Appendix A for other conversion factors.

Q Name the three enthalpies generally used in energy systems.

A At one time, the expressions *heat in the water*, and *heat in the vapor* were used for the modern term *enthalpy*. Thus enthalpy means the heat content of the fluid. In dealing with water and steam, three enthalpies are to be noted as follows:

1. Enthalpy of saturated liquid h_f in Btu, which is the heat content of the water at a certain pressure and temperature under consideration.
2. Enthalpy of evaporation h_{fg} in Btu, which is the heat required to evaporate 1 lb of water to steam at that pressure and temperature.
3. Enthalpy of saturated vapor h_g in Btu, which is the heat content of the saturated steam at the pressure and temperature being considered.

The enthalpy of saturated steam is thus a sum of the enthalpy of saturated liquid plus the enthalpy of evaporation, or the *total* heat content of the saturated steam in Btu per pound, or $hg = h_f + h_{fg}$.

Q Explain briefly how to use the steam tables.

A Tables 1.1 and 1.2 give the properties of water and of saturated steam. The only difference is that in Table 1.1 we enter with the boiler temperature, while in Table 1.2 we enter with the boiler pressure (psia). For example, Table 1.1 shows that for water to boil at 100°F, the absolute pressure must be 0.95 psi. Table 1.2 shows that at 40 psia, water boils at 267°F. It is not necessary to use all the digits given in the table. Most practical work doesn't require it. Engineers rarely need to figure water temperatures closer than the nearest degree, or heats or enthalpies closer than the nearest Btu.

After the first two, the columns are the same in both tables. *Sat. liquid* means liquid water at the saturation or boiling temperature. *Sat. vapor* means steam at the boiling temperature. When water is boiling in a closed container, both the water and the steam over it are in a saturated condition. Steam is saturated when generated by a boiler without a superheater. For steam, *Saturated* means steam that contains no liquid water yet is *not* superheated (still at boiling temperature). Note that the absolute pressure is gage pressure plus about 15 lb. Now, in Table 1.2, try reading across the line for 50 psia (35 psig).

Boiling temperature is 281°F. At this temperature 1 lb of water fills 0.0713 ft^3 and 1 lb of saturated steam fills 8.51 ft^3. Specific volume is in cubic feet per pound of water or steam. Thus it takes 250 Btu to heat the pound of water from 32°F to the boiling point and another 924 Btu to evaporate it, making a total of 1174 Btu. As mentioned, enthalpy used to be called *heat* in the old steam tables and it is given in Btu per pound. The last three columns of the old tables were labeled *heat of the liquid*, *heat of vaporization*, and *total heat*.

Q For what type of practical problems do operating engineers use steam tables?

A For calculating the amount of heat needed to convert feedwater at any given temperature into saturated steam at any given pressure or temperature.

> EXAMPLE: A boiler generates saturated steam at 135 psig (150 psia). The enthalpy, or heat of the final steam, is 1194 Btu/lb. The amount of heat required to produce this steam in an actual boiler will depend on the temperature of the feedwater. Suppose the feedwater temperature is 180°F. Table 1.1 shows that the heat in the water is 148 Btu. Then the heat supplied to turn this water into steam is merely the difference, or 1194 − 148 = 1046 Btu.
>
> It is easy from this to figure the boiler efficiency. Let us say the boiler generates 10 lb steam per pound of coal burned and the coal contains 13,000 Btu/lb. Then, for every 13,000 Btu put in as fuel, there is delivered in steam 10 × 1046 = 10,460 Btu.
>
> The efficiency of any power unit is its output divided by its input; so here 10,460 ÷ 13,000 = 0.805, or 80.5 percent efficiency. *Ans.*

For most purposes Table 1.1 is not needed to get a close value of the heat of the liquid. Just subtract 32 from the water temperature. For example, the enthalpy of water at 180°F is the heat required to raise it from 32 to 180°F, or a difference of 148°F. This takes about 148 Btu. But it will not work out so closely for very high temperatures. Take water at 300°F. Table 1.1 gives 269.7 Btu, while our simple method gives 300 − 32 = 268 Btu, close enough for most purposes.

Below 212°F the rule is extremely accurate, never out by more than 1/10 Btu. Just remember that condensing of steam in a condenser, heater, radiator, or process is nothing but heating the evaporation, worked backward.

Q Explain how to use the steam tables for superheated steam.

A The first column of Table 1.3 gives the absolute pressure and (directly below it in parentheses) the corresponding saturation temperature, or boiling temperature. In the next column v and h stand for volume of 1 lb and its heat content. For example, at 150 psia the volume of 1 lb is 0.018 ft^3 for liquid water and 3.015 ft^3 for the saturated steam. The corresponding heat contents of 1 lb are 330.5 Btu and 1194.1 Btu.

The temperature columns give the volume and heat content per pound for superheated steam at the indicated temperature. Take steam at 150 psia, superheated to a total temperature of 600°F. Look in the 600°F column opposite 150 psia. The volume is 4.113 ft^3, as against 3.015 ft^3 for saturated steam at the same pressure. This is natural because steam expands like a gas when superheated. Also, the heat content is naturally higher, 1325.7 Btu instead of 1194.1 Btu. Note that

Table 1.3 Superheated Steam

Abs. pressure, psi (sat. temp)	*	Sat. liquid	Sat. vapor	Temperature, °F							
				300	400	500	600	700	800	900	1000
15	v	0.016	26.29	29.91	33.97	37.99	41.99	45.98	49.97	53.95	57.93
(213.03)	h	181.1	1150.8	1192.8	1239.9	1287.1	1334.8	1383.1	1432.3	1482.3	1533.1
20	v	0.016	20.09	22.36	25.43	28.46	31.47	34.47	37.46	40.45	43.44
(227.96)	h	196.2	1156.3	1191.6	1239.2	1286.6	1334.4	1382.9	1432.1	1482.1	1533.0
40	v	0.017	10.498	11.040	12.628	14.168	15.688	17.198	18.702	20.20	21.70
(267.25)	h	236.0	1169.7	1186.8	1236.5	1284.8	1333.1	1381.9	1431.3	1481.4	1532.4
60	v	0.017	7.175	7.259	8.357	9.403	10.427	11.441	12.449	13.452	14.454
(292.71)	h	262.1	1177.6	1181.6	1233.6	1283.0	1331.8	1380.9	1430.5	1480.8	1531.9
80	v	0.018	5.472	......	6.220	7.020	7.797	8.562	9.322	10.077	10.830
(312.03)	h	......	282.10	1183.1	1230.7	1281.1	1330.5	1379.9	1429.7	1480.1	1531.3
100	v	0.018	4.432	......	4.937	5.589	6.218	6.835	7.446	8.052	8.656
(327.81)	h	298.4	1187.2	......	1227.6	1279.1	1329.1	1378.9	1428.9	1479.5	1530.8
150	v	0.018	3.015	......	3.223	3.681	4.113	4.532	4.944	5.352	5.758
(358.42)	h	330.5	1194.1	......	1219.4	1274.1	1325.7	1376.3	1426.9	1477.8	1529.4
200	v	0.018	2.288	......	2.361	2.726	3.060	3.380	3.693	4.002	4.309
(381.79)	h	355.4	1198.4	......	1210.3	1268.9	1322.1	1373.6	1424.8	1476.2	1528.0
300	v	0.0189	1.5433	......	......	1.7675	2.005	2.227	2.442	2.652	2.859
(417.33)	h	393.8	1202.8	......	......	1257.6	1314.7	1368.3	1420.6	1472.8	1525.2
400	v	0.0193	1.1613	......	......	1.2851	1.4770	1.6508	1.8161	1.9767	2.134
(444.59)	h	424.0	1204.5	......	......	1245.1	1306.9	1362.7	1416.4	1469.4	1522.4
500	v	0.0197	0.9278	......	......	0.9927	1.1591	1.3044	1.4405	1.5715	1.6996
(467.01)	h	449.4	1204.4	......	......	1231.3	1298.6	1357.0	1412.1	1466.0	1519.6
600	v	0.0201	0.7698	......	......	0.7947	0.9463	1.0732	1.1899	1.3013	1.4096
(486.21)	h	471.6	1203.2	......	......	1215.7	1289.9	1351.1	1407.7	1462.5	1516.7
800	v	0.0209	0.5687	......	......	......	0.6779	0.7833	0.8763	0.9633	1.0470
(518.23)	h	509.7	1198.6	......	......	......	1270.7	1338.6	1398.6	1455.4	1511.0
1000	v	0.0216	0.4456	......	......	......	0.5140	0.6084	0.6878	0.7604	0.8294
(544.61)	h	542.4	1191.8	......	......	......	1248.8	1325.3	1389.2	1448.2	1505.1
1200	v	0.0223	0.3619	......	......	......	0.4016	0.4909	0.5617	0.6250	0.6843
(567.22)	h	571.7	1183.4	......	......	......	1223.5	1311.0	1379.3	1440.7	1499.2
1400	v	0.0231	0.3012	......	......	......	0.3174	0.4062	0.4714	0.5281	0.5805
(587.10)	h	598.7	1173.4	......	......	......	1193.0	1295.5	1369.1	1433.1	1493.2

* v = specific volume, ft^3/lb; h = enthalpy, Btu/lb.

this table gives the actual temperature of the superheated steam rather than the degrees of superheat, which is a different thing. If the steam has been superheated from a saturation temperature of 358 to 600°F, the superheat is

$$600 - 358 = 242°F$$

These superheat tables are used similarly to the saturation tables described in the previous question. Let us take a problem. How much heat does it take to convert 1 lb of feedwater at 205°F into superheated steam at 150 psia and 600°F? The heat in the steam is 1325.7 (1326 Btu). The heat in the water is 205 − 32 = 173 Btu. Then the heat required to convert 1 lb of steam is 1326 − 173 = 1153 Btu.

Q Name two methods of calculating boiler efficiency.
A To calculate boiler efficiency, the method is the same as that for finding the efficiency of practically any other piece of power equipment; namely, efficiency is the useful energy output divided by the energy input. For example, if we get out three-quarters of what we put in, the efficiency is ¾, or 0.75 percent. In the case of a boiler unit, we feed in Btu in the form of coal, oil, or gas, and we get out useful Btu in the form of steam. Thus, the first method states that boiler efficiency can be figured directly from the total fuel burned in a given period and the total water evaporated into steam in the same period, but converting everything to units of heat such as Btu. It is more common to figure, first, the evaporation per pound of fuel fired and then, from this, the efficiency. Another method is from data on heat lost up the stack and through other boiler components. Figuring this way,

$$\text{Boiler efficiency} = \frac{\text{fuel energy input} - \text{energy lost}}{\text{fuel energy input}}$$

Q Calculate boiler efficiency using the steam generated vs. the fuel consumed. Assume that for one calendar month of regular operation, the coal consumed is 682,000 lb and the steam generated is 6,400,000 lb at 179 psig and superheated to a total temperature of 520°F.
A First, the actual evaporation per pound of coal fired is

$$\frac{6{,}400{,}000}{682{,}000} = 9.40 \text{ lb}$$

For rough estimates of efficiency the heat content of the coal may be taken from the statement of the company supplying the coal. For accurate work the operator must collect a good average sample of the coal and send it to a laboratory for testing. Assume that the figure is 13,260 Btu/lb of coal as fired. Remember that this pound of coal produces 9.4 lb of steam.

The absolute steam pressure is

179 + 15 = 194 lb abs

Then, the steam tables show that the total heat of 1 lb of steam at 194 psia and 520°F temp is 1280.4 Btu. Assuming that the feedwater temperature is 208°F, its heat content above water at 32°F is merely 208 − 32 = 176 Btu. Thus the heat put into each pound of steam produced by the boiler is 1280.4 − 176.0 = 1104.4 Btu. The heat put into 9.4 lb of steam will be 10,381 Btu. Then, boiler efficiency equals heat put into 9.4 lb steam divided by the heat in 1 lb of coal, or

$$\frac{10{,}381}{13{,}260} \times 100\% = 78.3\% \text{ efficiency} \qquad \textit{Ans.}$$

Q What are some average small-boiler efficiencies?
A Table 1.4 gives some average boiler efficiencies as found under tests by a consulting engineering firm.

Q How is power defined?
A Power is defined as the rate at which work is performed. There is thus a time factor involved. Some common conversions:

$$1 \text{ hp} = \frac{33{,}000 \text{ ft} \cdot \text{lb}}{\text{minute}} = \frac{550 \text{ ft} \cdot \text{lb}}{\text{second}}$$

$$1 \text{ kw} = 1000 \text{ W} = 1.34 \text{ hp}$$
$$1 \text{ hp} = 0.746 \text{ kW}$$
$$1 \text{ hp} = 33{,}000 \times 60 = 1{,}980{,}000 \text{ ft} \cdot \text{lb/h}$$

If the above is divided by 778 lb/Btu,

$$1 \text{ hp} = 2544 \text{ Btu/h}$$
$$1 \text{ kW} = 2{,}654{,}000 \text{ ft} \cdot \text{lb, or } 3413 \text{ Btu/h}$$

Q How are Fahrenheit (°F) and Celsius (°C) converted?
A Use

$$°\text{F} = \frac{9}{5}(°\text{C}) + 32$$

$$°\text{C} = \frac{5}{9}(°\text{F} - 32)$$

EXAMPLE: Change 20°C to °F. Substituting,

$$°\text{F} = \frac{9}{5}(20) + 32$$

$$°\text{F} = 36 + 32 = 68°\text{F} \qquad \textit{Ans.}$$

Table 1.4 Average Overall Efficiencies of Boilers of 500 hp and Under

Type of boiler	Percent of rating	Gas	Oil	Coal (stoker-fired)	Coal (hand-fired)
SBI steel heating	100	75	78	75	60
	125	73	76	73	58
	150	70	73	70	55
Horizontal return tube, 4-in tubes	100	73	76	73	58
	125	70	73	70	55
	150	65	68	65	50
Horizontal return tube, 3-in tubes	100	75	78	75	60
	125	72	75	72	57
	150	67	70	67	52
Scotch marine	100	76	79	76	61
	125	74	77	74	59
	150	71	74	71	56
Water tube	100	77	80	77	
	125	76	79	76	
	150	75	78	75	
Low-head water tube	100	75	78	75	
	125	74	77	74	
	150	73	76	73	

NOTE: The above efficiencies apply to boilers of 500 hp and under, without preheaters and superheaters. Where the larger-capacity boilers and steam generators are to be considered, efficiencies will increase for all fuels in approximately the same ratios. Where powdered coal is used with preheaters, efficiencies can go as high as 87 percent plus, while gas is limited to 84 percent plus.

HEAT TRANSFER

Q Why is heat transfer important to energy systems?

A Practically all heat-energy systems depend on transferring heat from one body to another, or converting this energy by heat-balance equations. In a boiler (Fig. 1.2) the chemical energy of a fuel is converted to heat, which, by properly designed surfaces, transfers this energy to a working medium such as water and steam. The steam eventually is converted to mechanical, electrical, or process-energy needs. All involve transferring energy from one datum to another.

Boiler-plant managers or operators and designers are constantly striving to keep heat-transfer surfaces at an optimum condition in order to maintain the high efficiency of converting fuel to useful energy. This is in addition to controlling combustion in order to obtain good burning in the furnace by the best fuel/air ratio. Thus, understanding heat transfer will assist energy-system operators in controlling heat losses.

Q Besides heat-transfer considerations, what other factors have determined the way boilers are constructed?

A Early boilers were very simple, usually a shell or cylinder full of

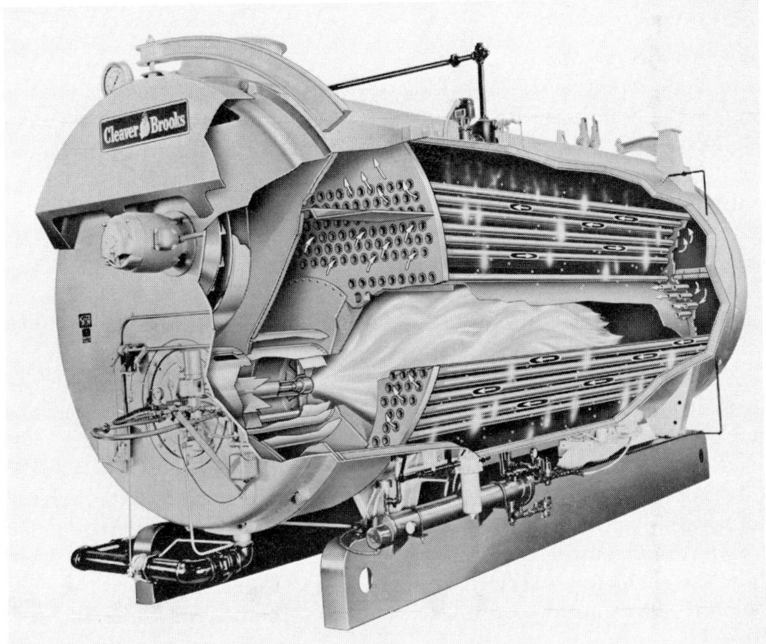

Fig. 1.2 Heat-transfer surfaces of fire-tube boiler designed to convert heat efficiently to water, then into steam. (*Courtesy of Cleaver Brooks.*)

water suspended over a brick-enclosed fire. Heat was applied on the bottom of the shell, and there were no tubes or flues. But as the science of heat transfer became better known, designers began to apply this knowledge to the simple shell vessels so as to produce more efficient and economical boilers. Thus a variety of specialized forms and arrangements are used to extract as much heat as possible from the fuel source.

New designs also had to be evolved for the greater capacity that was required, for new fuels, new materials, rising pressures and temperatures, new methods of construction, and better knowledge of stresses. And the search for better boilers to convert the inherent heat energy of a fuel into a useful heat medium goes on year after year.

Q How is heat transferred from one substance to another?
A The basic laws of heat transfer stipulate that when energy is transferred from one body to another a temperature difference must exist. Another fundamental law is that heat may be transferred from a high-temperature region to one of lower temperature, but *never* from a lower-temperature region to one of higher temperature. And this flow of heat may occur in one of, or a combination of, these three ways: (1) *conduction*, (2) *convection*, and (3) *radiation*. These three methods of heat transfer are utilized in boiler design to convert fuel energy into a useful heat medium.

Q What is *conduction*?
A Conduction is the transfer of heat from one part of a material to another or to a material with which it is in contact. Heat is visualized as molecular activity—crudely speaking, as the vibration of the molecules of a material. When one part of a material is heated, the molecular vibration increases. This excites increased activity in adjacent molecules, and heat flow is set up from the hot part of the material to the cooler parts. In boilers, considerable surface conductance between a fluid and a solid takes place, for example, between water and a tube and gas and a tube, in addition to conductance through the metal of a tube, shell, or furnace.

While surface conductance plays a vital part in boiler efficiency, it can also lead to metal failures when heating surfaces become overheated, as may occur when surfaces become insulated with scale. The surface conductance, when expressed in Btu per hour per square foot of heating surface for a difference of one degree Fahrenheit in temperature of the fluid and the adjacent surface, is known as the *surface coefficient* or *film coefficient*. Figure 1.3 shows stagnant areas near the tube where the film coefficient will reduce heat transfer. Figure 1.4 shows the effect on the temperature gradient as heat flows across the films and tube metal.

Q How is *conductance* expressed?
A As the coefficient of thermal (heat) conductivity, defined further as

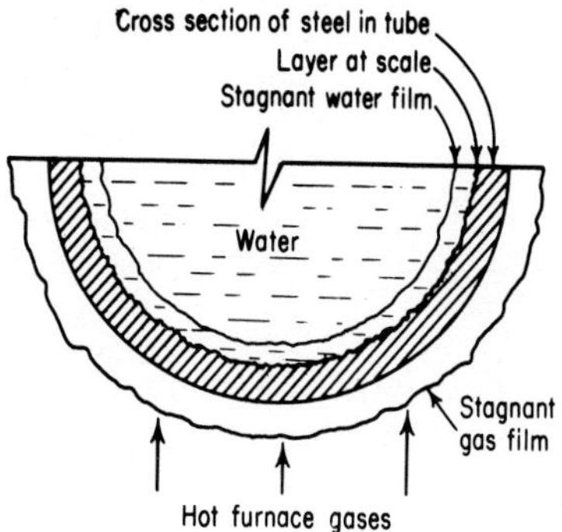

Fig. 1.3 Film coefficients of stagnant water and gas can effect heat transfer through a boiler tube.

the quantity of heat that will flow across a unit area in unit time if the temperature gradient across this area is unity. In physical units it is expressed as *Btu per hour per square foot per degree Fahrenheit per foot*. Expressed mathematically, the rate of heat transfer Q by conduction across an area A, through a temperature gradient of degrees Fahrenheit per foot T/L is

$$Q = kA \frac{T}{L}$$

where k = coefficient of thermal conductivity.

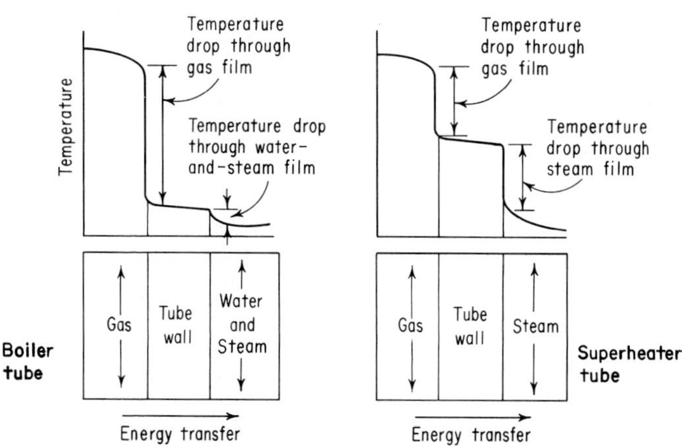

Fig. 1.4 Temperature gradients from hot gas through the tube wall to fluid inside depend greatly on the resistance of thin films adhering to the tube surfaces.

Note that k varies with temperature. For example, mild steel at 32°F has a thermal conductivity of 36 Btu/(h · ft^2 · °F/ft), whereas at 212°F it is 33.

Q How would you define *convection*?
A *Convection* is the transfer of heat to or from a fluid (liquid or gas) flowing over the surface of a body. It is further refined into *free* and *forced convection.* Free convection is *natural* convection causing circulation of the transfer fluid due to a difference in density resulting from temperature changes.

> EXAMPLE: In Fig. 1.5 the heated water and steam rise on the left, are displaced by cooler (heavier) water on the right. This causes free convection of heat transfer between heat on one side of the U tube and cooler water on the other side. Actually, conduction has to take place first between the gas film and metal of the tube, then through the water. But if the water did not circulate, eventually equal temperatures would result. Heat transfer would then cease.

Forced convection results when circulation of the fluid is made positive by some mechanical means, such as a pump for water or a fan for hot gases. The heat transfer by convection is thus aided mechanically.

Q How is heat transfer by *convection* expressed?
A As *Btu per hour per square foot per degree Fahrenheit.* Calculations of free and forced convections involve fluid flow, streamline flow, turbulent flow, critical velocity, film theory, dimensional analysis, and friction. All these are beyond the scope of this book. But no matter what

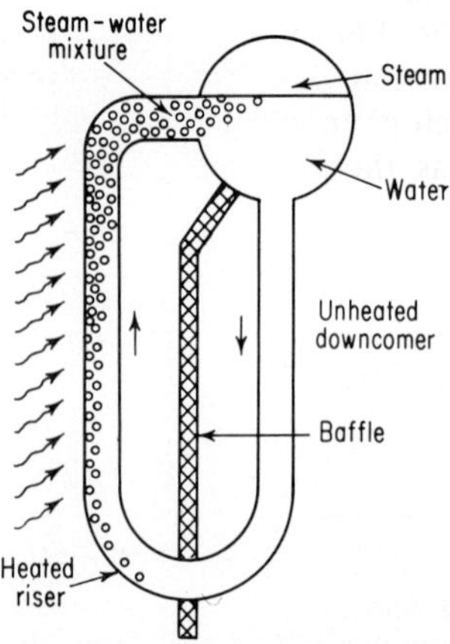

Fig. 1.5 Heated water rises while cooler water drops to replace it.

kind of boiler, remember that heat transfer by convection is involved in water circulation and also, except for atomic plants, in draft or hot-gas circulation.

Q If forced convection increases the rate of heat transfer by convection, why not add more heating surface and thus increase the overall efficiency of energy conversion?
A Adding boiler surface may increase the heat absorption, but as shown in Fig. 1.6, the temperature gradient will drop more and more. Then at some point the gain in efficiency will be far less than the cost of adding heating surface. Further, the mechanical power required for forced circulation will also increase with the addition of heating surface by convection.

Q How does the pressure in a boiler affect water circulation by convection?
A Note that in Fig. 1.7 more tube area is required at lower pressure than higher pressure for the same circulation to exist. But the force producing circulation is less at high pressure than at low pressure. This involves the change in the specific weight of water and steam as pressures increase. The mixture actually weighs less in pounds per cubic foot at higher pressures. For example, in the sketch in Fig. 1.7*b* at the critical pressure (3206.2 psia) water and steam have the same specific weight. Friction losses due to flow are generally less at higher pressure. This is primarily due to more laminar, or streamlined, flow and less turbulent flow in the tubes.

Q How does the boiling of water in a tube affect *heat transfer*?
A When boiling occurs in a tube, bubbles of vapor are formed and liberated from the surface in contact with the liquid. This bubbling action creates voids (Fig. 1.8*a*) of the on-again-off-again type, because of the rapidness of the action. This creates a turbulence near the heat-transfer surfaces, which generally increases the heat-transfer rate. But the loss of wetness as the bubbles are formed may diminish heat transfer.

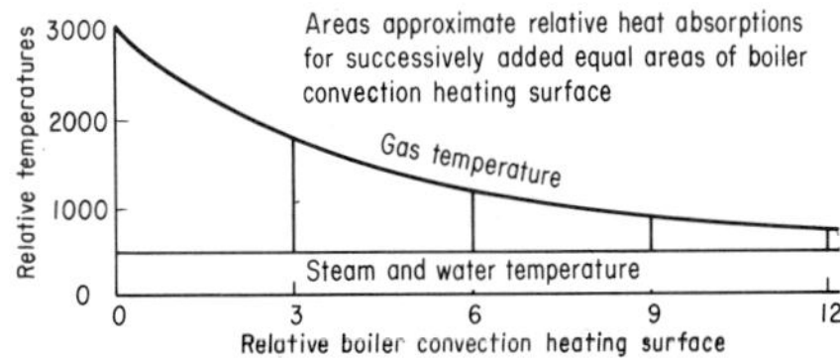

Fig. 1.6 Adding boiler surface increases heat absorption, but at reduced temperature differential per square foot.

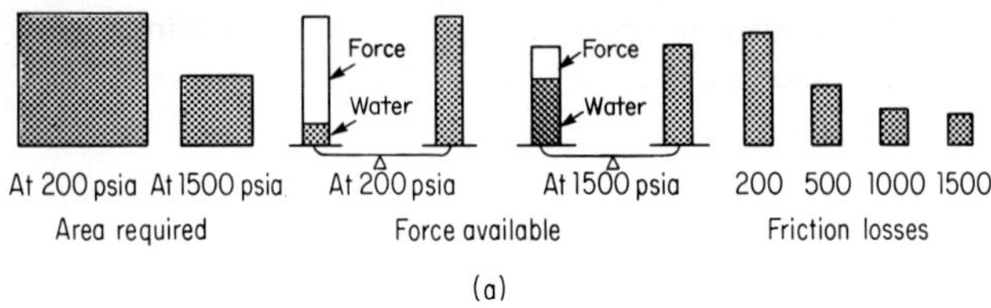

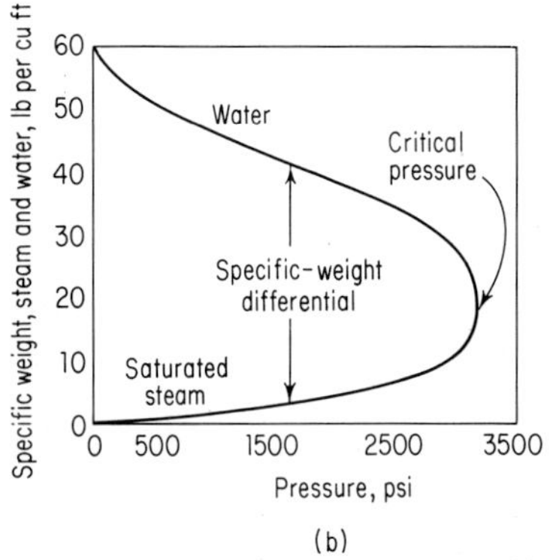

Fig. 1.7 Heat-transfer areas required and forces creating water circulation in a boiler.

Pressure has a marked effect on the boiling and heat-transfer rate. With higher pressures (Fig. 1.8*b*), bubbles tend to give way to what is called film boiling, in which a film of steam covers the heated surface. This phenomenon is very critical in boiler operation, often causing water-tube failures due to starvation, even though a gage glass may

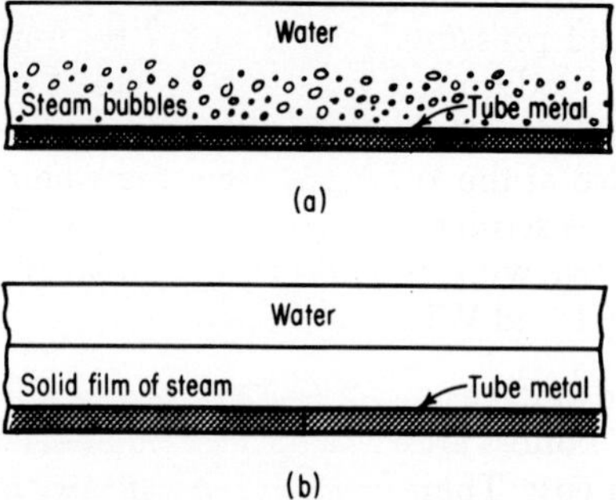

Fig. 1.8 Bubbles of water vapor or a solid film of steam is formed near tube surfaces, depending on pressure.

show water. It is further compounded by the formation of scale and other impurities along the boiling area of a tube.

Q Define *radiation.*
A *Radiation* is a continuous form of interchange of energy by means of electromagnetic waves without a change in the temperature of the medium between the two bodies involved. Radiation is present in all boilers. In fact, all boilers utilize all three means of heat transfer: *conductance*, *convection*, and *radiation.*

Q With heat transfer playing such an important part in boiler design, how are boilers classified on the basis of *heat* transfer?
A There are two broad general classifications, *fire tube* (see Fig. 1.2) and *water tube*, into which most boilers can be grouped, no matter what the arrangement. The exception is cast-iron boilers, which are grouped separately. A fire-tube (FT) boiler is one in which the products of combustion, or flue gases, pass on the inside of the tubes. A water-tube (WT) boiler is one in which water passes on the inside of the tubes. A combination FT and WT boiler is one in which part of the tube arrangement is of the FT type and part is of the WT type.

Q What is the principal difference between WT and FT boilers in addition to water and gas passages?
A The tubes in the FT boiler are contained within the shell or drum. In contrast, the tubes in most WT boilers are located outside the shell or drum. Because smaller sizes of FT boilers can be built into compact factory-assembled units, this design lends itself to good engineering of the entire assembly, including controls. But as the units become larger, the capacity of the FT boiler becomes limited because of the larger size of shell required. It is here that the WT boiler has a distinctive advantage, as tube arrangement can take many different forms to obtain more and more heating surface. Thus the WT units are capable of greater capacity and pressure, which would be impossible in an FT boiler. The largest modern steam generators are of the WT design.

But the WT unit is not limited to only larger sizes. Compact coil-type boilers are also of the WT type. Here the water goes through the coils, while hot gases scrub the outside of the coils. The coil-type boiler is directly competitive with the FT vertical tubular (VT) type boiler. See later chapters on FT and WT boiler types.

Q Name some advantages and disadvantages of FT boilers.
A As a class, FT boilers are of simple and rugged construction and of relatively low first cost. Their larger water capacity makes them somewhat slow in steaming up to operating pressure. But this larger hot-water capacity provides accumulator action (heat storing) that makes it possible to meet steam load changes quickly.

Because a sphere is the ideal shape to resist internal pressure, noncylindrical sections and flat surfaces must be stayed to give added resistance to internal pressure. This is a big disadvantage.

In any shell the force tending to burst it along the length is twice that tending to burst it around the girth. In the critical longitudinal direction, the strength required to resist bursting is proportional to the product of pressure and diameter. That is why high pressures and large diameters would lead to extremely thick shell plates. Hence, there is a definite economical limit on pressure (250 psi) and capacity that can be reached with shell-type boilers of the FT type. In the United States, capacity rarely exceeds 50,000 lb of steam per hour, which is roughly 750 boiler hp. In Europe, where larger FT boilers have always been popular and economical and boiler codes are different, units reach over 50,000 lb/h.

Q Name five advantages of WT boilers.

A Principal advantages are:

1. Greatly higher capacity can be obtained. Thus larger heating surfaces are exposed to the radiant heat of the fire.

2. Because the shell or drum (used for water and steam storage only) is not exposed to the radiant heat of the fire, it is not subjected to overheating. Thus it can be constructed of heavier plate. Accordingly, it can also be designed for much higher pressures.

> EXAMPLE: Fire-tube boilers generally are limited to a maximum design working pressure of 250 psi. Water-tube boilers range from 15 psi to above the supercritical pressure of 5000 psi.

3. Most parts of the boiler are accessible for cleaning, repair, and inspection.

4. The general design permits higher operating efficiencies.

5. The furnace proportions are such that various fuels can be used without making alterations. Thus, during price fluctuations of various fuels, economy can be gained by using the lower-priced level.

USAGE AND APPLICATION

Q Name some common terms used in the study of boilers.

A The study of boilers involves also such terms as boilers, steam generators, critical-pressure boilers, low pressure, high pressure, steam, hot-water-heating boilers, hot-water-supply boilers, and also boiler code requirements.

A number of these terms are directly related to legal requirements drawn up by state and city laws. Thus they are of great importance not only to boiler operators, operating and mechanical engineers, and maintenance and service people, but also to those in management. These are the people in charge of (1) plant safety, (2) fiscal and legal policy affected by plant insurance, costs, and hazards, and (3) the city or state governments having jurisdiction over plant equipment.

Q We know that boilers have a great potential for causing loss of life and damage to property. But in addition to safety, what other knowledge is required when working with boilers?
A The use of fired and unfired pressure vessels is rapidly expanding in industry, commercial buildings, apartment houses, and in every economic activity around the world. Most of us working with machinery sooner or later come in contact with fired and unfired pressure vessels. Thus, we should become familiar with this equipment. Today, control devices, automatic equipment, computers, and other sophisticated items are being applied to fired pressure vessels.

But do not make the mistake of thinking that boilers and other pressure vessels thus automated are perfectly reliable because they operate like a robot. On the contrary, these safety devices make the inherent danger even greater when complete dependence is placed on them.

The use of nuclear reactors is also expanding, but a reactor is also basically a heated pressure vessel. Thus certain boiler fundamentals also apply to the construction and operation of reactors.

State and municipal laws involving boilers and pressure vessels require a knowledge of this subject to comply with legal statutes. And not to be forgotten when working with this equipment are the factors of reliability, service, efficiency, costs, loss of use, and, of course, safety.

Q Do state laws apply only to *high-pressure boilers*?
A Certainly not. At one time this was true, but with the advent of more and more automatic devices on boilers, with less and less reliance on full-time boiler operator attendance, coupled with some serious explosions on *low-pressure boilers*, more and more states have adopted, and *are* adopting, low-pressure boiler laws and even *unfired pressure vessel* laws. The trend for more rigid laws affecting pressure equipment is going to continue because *low pressure* on a fired vessel, under certain conditions, becomes *high pressure* in a matter of minutes.

Q How would you define a boiler?
A A *boiler* is a closed pressure vessel in which a fluid is heated for use external to itself by the direct application of heat resulting from the combustion of fuel (solid, liquid, or gaseous) or by the use of electricity or nuclear energy. These can use steam, hot water, or other working substances.

Q What is a *high-pressure steam boiler*?
A A *high-pressure steam boiler* is one which generates steam or vapor at a pressure of more than 15 pounds per square inch gage (psig). Below this pressure it is classified as a low-pressure steam boiler.

Q Define a *miniature high-pressure boiler*.
A According to Section I of the Boiler and Pressure Vessel Code of the American Society of Mechanical Engineers (ASME), a *miniature boiler* is a high-pressure boiler which does not exceed the following limits: (1) 16 in inside diameter of shell, (2) 5 ft^3 gross volume exclusive of casing and insulation, and (3) 100 psig pressure. If it exceeds any of these limits, it is a *power boiler*. Most states follow this definition.

Q What is a *power boiler*?
A A *power boiler* is a steam or vapor boiler operating above 15 psig and exceeding the miniature boiler size. This *also includes* hot-water-heating or hot-water-supply boilers operating above 160 psi or 250 degrees Fahrenheit (°F).

Q Define a *hot-water-heating boiler*.
A A *hot-water-heating boiler* is a boiler used for space hot-water heating, with the water returned to the boiler. It is further classified as low pressure if it does not exceed 160 psi or 250°F. But if it exceeds any of these, it becomes a high-pressure boiler. See Fig. 1.9.

Q What is a *hot-water-supply boiler*?
A A *hot-water-supply boiler* is a boiler furnishing hot water to be used externally to itself for washing, cleaning, etc. If it exceeds 160 psi or 250°F, it becomes a high-pressure power boiler.

> NOTE: The ordinary domestic-type hot-water-supply heater directly fired with oil, gas or electricity may be classified as a hot-water-supply boiler, depending on the state. The ASME Low-Pressure Heating Boiler Code, Section IV, stipulates that it becomes a hot-water-supply boiler if any of the following is exceeded: (1) heat input over 200,000 Btu/h, (2) water temperature over 200°F, and (3) nominal water-containing capacity of 120 gal. If it is below these limitations, it is designed under Section VIII, Unfired Pressure Vessels, ASME Boiler and Pressure Vessel Code. Some state laws start classifying these hot-water heaters as hot-water-supply boilers if the input exceeds 100,000 Btu. Other states incorporate all fired vessels (including those electrically fired) under the Boiler Code Regulations, excluding only vessels in private residences or apartment houses with six families or less. So always check your state law.

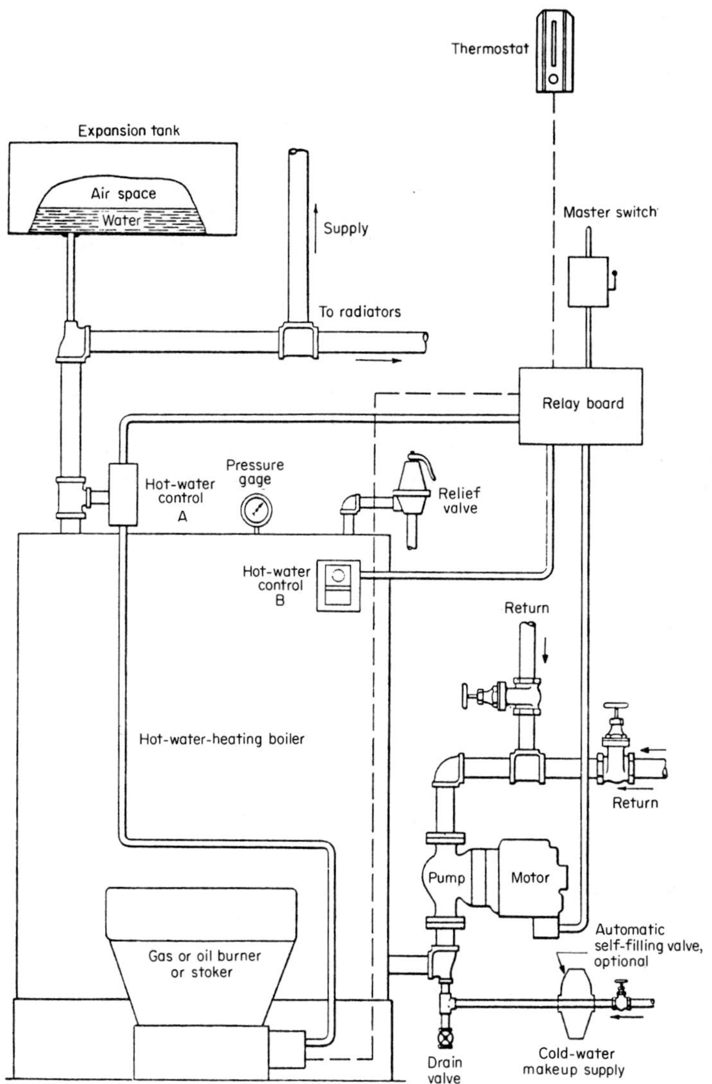

Fig. 1.9 Components needed for a hot-water-heating boiler system.

BOILER CODES

Q The latest ASME Boiler and Pressure Vessel Code is very important when working with boilers. How did this code come to be applied and administered?

A Before the ASME code, manufacturers had their own individual construction techniques. There were no fixed legal standards, but after more and more boiler accidents occurred as the country became more industrialized, the public demanded laws for protection. One of the first states to adopt a state code was Massachusetts, but it took a boiler explosion which killed 58 persons and injured 117. Thus the ASME set up a committee in 1911 to formulate standards of construction of steam boilers and pressure vessels. This committee is now called the Boiler and Pressure Vessel Code Committee, and the codes drawn up are called the ASME Boiler and Pressure Vessel Code, consisting of the following: Section I, Power Boilers; Section II, Material Specifications; Section III, Nuclear Vessels; Section IV, Low-Pressure Heating Boilers; Section VII, Suggested Rules for Care of Power Boilers; Section VIII, Unfired Pressure Vessels; Section IX, Welding Qualifications.

In time, the ASME code became recognized as a standard in the United States and even in foreign countries. States and cities started to adopt one or more sections of the code to make it legal. Many state and city representatives, including some from Canada, are now on the ASME Boiler and Pressure Vessel Code Committee. There are also various technical groups, called subcommittees, covering such subjects as power boilers; fire-tube boilers; steam boilers in service; material specifications; steel plates; steel tubular products; steel castings, forgings, and boltings; nonferrous materials; properties of metals (with subgroups on toughness, fatigue strength, strength properties); strength of weldments; nuclear power; heating boilers; unfired pressure vessels; welding; safety-valve requirements; code symbols and stamps; openings and bolted connections; special design; nondestructive testing; and vessels under external pressure.

Q What is a *supercritical pressure* boiler?
A Steam and water have a critical pressure at 3206.2 psi absolute (psia). At this pressure steam and water are at the same density, which means that the steam is compressed as tightly as the water. When this mixture is heated above the corresponding saturation temperature of 705.4°F for this pressure, dry, superheated steam is produced to do useful high-pressure work. This dry steam is especially well suited for driving turbine-generators. A supercritical boiler is thus one that operates above the supercritical pressure of 3206.2 psia and 705.4°F saturation temperature.

Q What is a *once-through boiler*?
A This refers to a boiler or steam generator which receives feedwater at one end of continuous tubes and discharges steam at the other end. See the chapter on WT boilers.

Q Define a *waste-heat boiler.*

A This is a boiler which uses by-product heat such as from a blast furnace in a steel mill, exhaust from a gas turbine, or by-products from a manufacturing process. The waste heat is passed over heat-exchanger surfaces to produce steam or hot water for conventional use.

Q Do code rules apply to waste-heat boilers?
A Yes. The same basic construction rules still apply, and the usual auxiliaries and safety features normally required on any fired pressure vessel are needed.

Q What is a *steam generator*?
A Engineers prefer to use the term *steam generator* instead of *steam boiler*, as *boiler* refers to the physical change of the contained fluid whereas the term *steam generator* covers the whole apparatus in which this physical change is taking place. But in ordinary usage, both are essentially the same thing. Most state laws are still written under the old, basic *boiler* nomenclature.

Q Are there any other classifications of boilers dependent on the nature of service intended?
A Yes, the traditional classifications are *stationary*, *portable*, *locomotive*, and *marine*, defined as follows. A stationary boiler is one which is installed permanently on a land installation. A portable boiler is a boiler mounted on a truck, barge, small river boat, or any other such mobile-type apparatus. A locomotive boiler is a specially designed boiler, specifically meant for self-propelled traction vehicles on rails (also used for stationary service). A marine boiler is usually a low-head-type special-design boiler meant for ocean cargo and passenger ships with an inherent fast-steaming capacity.

Q Is it permissible to build *cast-iron boilers* for steam usage above 15 psig?
A No. Under the present boiler code, cast-iron boilers for steam usage are limited to an AWP (allowable working pressure) of 15 psig. Cast-iron boilers are specifically restricted by the ASME code, Section IV, to be used exclusively for low-pressure steam *heating*. Process work usually means heavy-duty service of continuous steaming and heavy makeup of fresh cold water. This will cause rapid temperature changes in a cast-iron boiler, resulting in cracking of the cast-iron parts. Thus the code restricts their use to steam-heating service only.

IBR, SBI, AND EDR

Q In heating-load calculations, the terms *IBR rated*, *SBI rated*, and *EDR* are often used. What do these terms mean?
A These terms affect the output rating of a boiler. Thus they are

important in sizing a boiler for heating a certain size space. They also affect the safety valve required on a boiler. The preferred modern trend is to rate a boiler by the Btu-per-hour-output method. The terms mean the following:

IBR stands for the Institute of Boiler and Radiator Manufacturers, which rates cast-iron boilers. IBR-rated boilers usually have a nameplate indicating net and gross output in Btu per hour. Gross output is further defined as the net output plus an allowance for starting, or pickup load, and a piping heat loss. The net output will show the actual useful heating effect produced. The ASME code states it is the gross heat output of the equipment that should be matched in specifying relief-valve capacity.

SBI stands for the Steel Boiler Institute. The nameplate data shown on SBI-rated boilers are not uniform, but the style or product number may be shown. The manufacturer's catalog will often show an SBI rating and an SBI net rating. The SBI rating tends to show the sum of SBI net ratings, plus 20 percent extra for piping loss, not including the pickup allowances noted under IBR ratings. Thus, it is difficult to obtain the true gross output to determine safety relief capacity from these data. But the SBI does require the square feet of heating surface to be stamped on the boiler. With this, the ASME rule of minimum steam safety-valve (SV) capacity in pounds per hour per square foot of heating surface is used.

EDR means *equivalent square feet of steam radiation surface*. It is further defined as a surface which emits 240 Btu/h with a steam temperature of 215°F at a room temperature of 70°F. With hot-water heating, the value of 150 Btu/h is used with a 20°F drop between inlet and outlet water. This term is used by architects and heating engineers in determining the area of heat-transfer equipment required to heat a space. Thus boiler capacity is obtained indirectly from a summation of the EDRs. Table 1.5 shows typical cast-iron boiler ratings from a manufacturer's catalog. Table 1.6 shows ratings of the Steel Boiler Institute. Table 1.7 shows the EDRs for different types of radiators.

Q What other terms are used to indicate boiler output?

A These three terms are often used with pressure and temperature listings.

1. For steam boilers, the actual evaporation in pounds per hour. For hot-water boilers, the Btu-per-hour output for the given pressures and temperatures is stamped on the boiler. Today this is the preferred method. Also used are
2. Square feet of heating surface.
3. Boiler horsepower.

Q What is meant by *heating surface* in a boiler?

Table 1.5 Typical IBR Ratings for Cast-Iron Boilers from Manufacturer's Catalog

Boiler number	Gross boiler hp	Gross IBR output, 1000 lb/h	Net IBR rating*: Ft² steam	Net IBR rating*: 1000 lb/h water	IBR burner capacity based on 150,000 Btu/gal (heavy oil), gal/h	IBR gas input, 1000 lb/h	IBR chimney size for heavy oil†: Size, in	IBR chimney size for heavy oil†: Height, ft	Inside diam of rectang. smoke pipe to fit over smokehood outlet, in	Burner capacity based on 140,000 Btu/gal (light oil), gal/h	Heating surface, ft²
450-8	47.3	1584	5,060	1214.7	13.20	1980	16 × 20	25	15 × 19	14.15	288.7
450-9	53.7	1796	5,795	1391.2	14.90	2238	16 × 20	26	15 × 19	16.00	325.3
450-10	60.0	2008	6,495	1559.0	16.65	2496	20 × 20	27	15 × 19	17.85	361.3
450-11	66.3	2220	7,180	1723.6	18.35	2753	20 × 20	28	15 × 19	19.70	398.0
450-12	72.6	2428	7,855	1885.1	20.10	3011	20 × 20	29	15 × 19	21.50	434.0
450-13	78.9	2640	8,540	2049.7	21.80	3269	20 × 24	30	19 × 19	23.35	470.7
450-14	85.2	2852	9,225	2214.3	23.50	3527	20 × 24	31	19 × 19	25.20	506.7
450-15	91.5	3064	9,910	2378.9	25.25	3785	20 × 24	32	19 × 19	27.05	543.3
450-16	97.7	3270	10,580	2538.8	26.95	4043	24 × 24	33	19 × 19	28.90	579.3
450-17	104.0	3490	11,290	2709.6	28.70	4300	24 × 24	34	24 × 19	30.70	616.0
450-18	110.2	3700	11,970	2872.7	30.40	4558	24 × 24	35	24 × 19	32.55	652.0
450-19	116.8	3910	12,650	3035.7	32.10	4816	24 × 24	36	24 × 19	34.40	688.7
450-20	123.1	4120	13,330	3198.8	33.85	5074	24 × 28	37	24 × 19	36.25	724.7

* Net IBR ratings shown include allowance for piping loss and pickup load.
† If the heavy oil burner being used requires a draft over the fire greater than 0.06 in, the chimney height should be adjusted accordingly.

Table 1.6 Data of SBI-Rated Steel Boilers for Steam or Hot-Water Heating

	Boiler number													
	HM-880	HM-881	HM-882	HM-883	HM-884	HM-885	HM-886	HM-887	HM-888	HM-889	HM-890	HM-891	HM-892	HM-893
SBI gross output, hp	64	75	91	108	134	161	188	215	269	322	376	430	483	537
Steam, ft^2	9,000	10,500	12,750	15,000	18,750	22,500	26,250	30,000	37,500	45,000	52,500	60,000	60,000	75,000
Water, ft^2	14,400	16,800	20,400	24,000	30,000	36,000	42,000	48,000	60,000	72,000	84,000	96,000	108,000	120,000
1000 lb/h	2,160	2,520	3,060	3,600	4,500	5,400	6,300	7,200	9,000	10,800	12,600	14,400	16,200	18,000
Steam (212°F), lb/h	2,208	2,588	3,140	3,726	4,623	5,555	6,486	7,418	9,281	11,109	12,972	14,835	16,664	18,527
Certified output, hp	78	92	111	131	164	195	228	261	326	392	456	521	585	651
Certified output, 1000 lb/h	2,625	3,060	3,718	4,374	5,464	6,559	7,650	8,745	10,929	13,114	15,300	17,485	19,671	21,855
Firing rate:														
Oil, gal/h	22	26	31	37	46	54	64	73	91	110	127	146	167	182
Gas, 1000 lb/h	3,280	3,830	4,650	5,460	6,840	8,200	9,560	10,920	13,670	16,400	19,120	21,900	24,600	27,300
Heating surface (waterside), ft^2	429	500	608	715	893	1,072	1,250	1,429	1,786	2,143	2,500	2,857	3,214	3,571
Heating surface (fireside), ft^2	397	463	563	662	827	993	1,157	1,323	1,654	1,985	3,215	2,645	2,976	3,307
Furnace volume (SBI min), ft^3	52.1	60.8	73.8	86.8	108.5	130.2	151.8	173.5	216.9	260.3	303.6	346.9	390.3	433.6
Net furnace volume, ft^3	62.5	67.7	85.4	94.2	118.5	149.9	164.1	185.1	228.2	271.3	333.5	348.7	393.1	437.6
Safety valve capacity, lb steam/h	3,432	4,000	4,864	5,720	7,144	8,576	10,000	11,432	14,288	17,144	20,000	22,856	25,712	28,568

Table 1.7 EDR per Square Foot of Surface for Different Heaters under Approximate Btu per Hour Capacities

Type of heater	EDR
Thermal circulating air, free-standing radiator; 240 Btu/h per square foot of surface	1.0
Thermal circulating, direct-indirect radiator; 360 Btu/h per square foot of surface	1.5
Thermal circulating, indirect radiator, recirculating; 360 Btu/h per square foot of surface	1.5
Thermal circulating, indirect radiator, outside air; 480 Btu/h per square foot of surface	2.0
Mechanical circulating, indirect radiator, recirculating; 720 Btu/h per square foot of surface	3.0
Mechanical circulating, indirect radiator, outside air from 0 to 130°F at 960 Btu/h per square foot of surface	4.0
Mechanical recirculating, indirect radiator, outside air from 0 to 70°F at 1440 Btu/h per square foot of surface	6.0

NOTE: Normal operating conditions for radiators assume steam at 215°F and space temperature at 70°F.

A This is the fireside area in a boiler exposed to the products of combustion. This area is usually calculated on the basis of areas on the following boiler element surfaces: tubes, fireboxes, shells, tube sheets, and projected area of headers. Heating surface is another method of measuring boiler output.

Q What is meant by *boiler horsepower*?

A This is an old method of rating output, dating back to boiler-steam engine usage. But it is slowly being replaced by the Btu-per-hour-output method or the pounds-per-hour-output method. A boiler horsepower (boiler hp) is defined as the evaporation into dry saturated steam of 34.5 lb of water per hour at a temperature of 212°F. Thus 1 boiler hp by this method is equivalent to an output of 33,475 Btu/h and was commonly taken as 10 ft^2 of boiler heating surface. But 10 ft^2 of boiler heating surface in a modern boiler will generate anywhere from 50 to 500 lb of steam per hour. Today the capacity of larger boilers is stated as so many *pounds of steam per hour.*

STEAM-HEATING SYSTEMS

Q Upon what energy principle is steam heating based?

A See Fig. 1.10. The basic principle involves giving up the heat of vaporization in a radiator or convertor. Practically all heating comes from the steam giving up this heat as it is condensed. For example, at

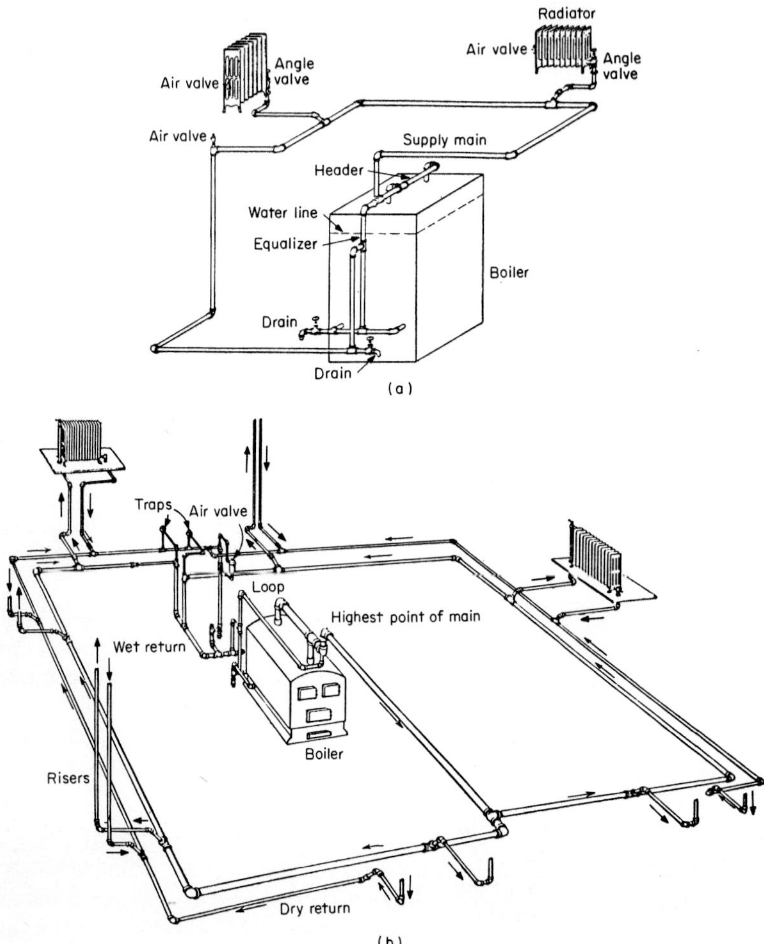

Fig. 1.10 Steam-heating systems. (*a*) One-pipe air vent system; (*b*) two-pipe steam and dry-return system. *Note:* Heavy arrows indicate supply and return flow from load or risers. Fine arrows indicate pitch downward of pipe.

14.7 psi and atmospheric pressure, a dry pound of steam gives up 970.3 Btu. Steam flows in pipes because of pressure differences. When the steam condenses, the pressure and volume are reduced; thus flow takes place from the boiler as the steam flows to the space where steam has condensed.

Q In a steam-heating boiler installation, is it permissible to use the *same pipe* to distribute the steam and also to return the condensate?

A Yes. This is called a one-pipe system, whereas a two-pipe system has the steam going out from the boiler in one pipe and condensate returning in another pipe. But the one-pipe system is susceptible to water hammer (at elbows and dead pockets). Thus the high-velocity steam will drive the returning water against the elbows or dead pockets of the piping with an effect similar to a hammerblow. The one-pipe system is becoming obsolete.

Q What is the difference between a *gravity return* and a *mechanical return* on a steam-heating system?
A When all the heating elements such as radiators, convectors, and steam coils are located above the boiler and no pumps are used, it is called a gravity return, as all the condensate returns to the boiler by gravity. If traps or pumps are installed to aid the return of condensate, the system is called a mechanical return system. In addition to traps, this system usually includes a condensate tank, a condensate pump, or a vacuum tank or vacuum pump.

Q What is the advantage of a *vacuum-pump return system*?
A The supply steam pressure required to force the steam up a riser will be less. Also, greater circulation of steam and condensate through the piping system can be obtained, thus accomplishing faster heat transfer. See Fig. 1.11.

Q How is the size of the condensate tank required for a steam-heating system determined?
A This depends on how fast circulation is obtained on the condensate and return line. For example, in a high-head installation, the condensate will return faster than in a spread-out installation of low head. Generally, a condensate tank should be about one-half the output rating of the boiler. This means that the condensate tank should be sized to handle one-half the pounds-per-hour rating of the boiler. Thus a boiler rated at 5000 lb/h would require the following size condensate tank by this rule:

$$\frac{5000}{2} = 2500 \text{ lb}$$

$$\frac{2500}{8.33 \text{ lb/gal}} = 300 \text{ gal} \qquad \textbf{\textit{Ans.}}$$

But this should be increased by 25 percent, as only 75 percent of the tank will normally be filled. Thus a tank of 375 gal is required.

NOTE: Condensate tank sizes may be governed by local plumbing and piping codes; so always check them. This is not a boiler code item.

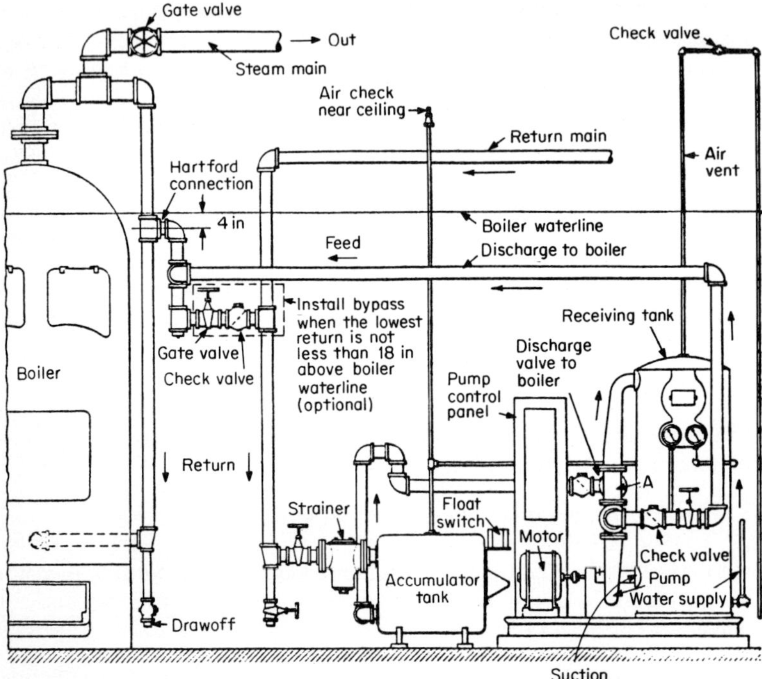

Fig. 1.11 Vacuum-system return used with steam-heating boilers.

Q Some of the *older heating boilers*, of both cast-iron and steel construction, are not stamped as to pressure. What is the allowable pressure on these boilers?
A The maximum allowable pressure is 15 psi steam and 30 psi water. This is the standard, old ASME specification before the states adopted stamping laws for low-pressure heating boilers.

Q Are *stop valves* on the steam supply line required for a single boiler installation that is used for low-pressure heating? How about multiple boilers?
A Not if there are no other restrictions in the steam and condensate line and all condensate is returned to the boiler. But if a stop valve (or trap) is placed in the condensate return line, a valve is required on the steam supply line. For multiple boilers, a stop valve must be used on the steam supply line from each boiler and also on the condensate return line to each boiler.

Q On cast-iron steam-heating boilers, where should the *makeup* water be introduced to the boiler?

A Always to the return line, never to a cast-iron section, as this may thermal-shock the section and cause it to crack.

Q What is a Hartford loop on a steam-heating boiler?
A The Hartford loop (Fig. 1.12) is the piping connecting the steam supply line with the condensate return line at the middle of the loop. The loop then is fed to the bottom of the boiler. This is used in a gravity return system and ensures a water level *above* the lowest safe water level, which could otherwise occur when the boiler is firing (and the steam pressure is higher than the condensate return pressure).

Q What size *expansion tanks* are required on a hot-water heating boiler of the closed type?
A The required minimum capacity of the air-cushion tank (mandatory to have when system is of the closed type) may be determined as follows:

EDR (equivalent direct radiation), ft^2	Boiler output, Btu/h	Required size tank, gal
Up to 350	52,500	18
Up to 450	67,500	21
Up to 650	97,500	24
Up to 900	135,000	30
Up to 1100	165,000	35
Up to 1400	210,000	40
Up to 1600	240,000	20 (2 tanks)
Up to 1800	270,000	30 (2 tanks)
Up to 2000	300,000	35 (2 tanks)
Up to 2400	360,000	40 (2 tanks)

For systems with more than 2400 ft^2 of equivalent direct water radiation (EDR), the required capacity of the cushion tank should be increased on the basis of 1 gal tank capacity per 33 ft^2 of additional EDR. The above table is based on the following formula, which should be used if all information required is available:

$$V_t = \frac{(0.00041T - 0.0466)V_s}{(P_a/P_f) - (P_a/P_o)}$$

where V_t = minimum volume of expansion tanks required, gal
V_s = volume of hot-water system, not including tanks, gal
T = average operating temperature, °F
P_a = atmospheric pressure (absolute, not gage pressure)
P_f = fill pressure (absolute, not gage pressure)
P_o = maximum operating pressure (absolute, not gage pressure)

Q To what does the term *packaged boiler* refer?
A A packaged boiler is a completely factory-assembled boiler, either

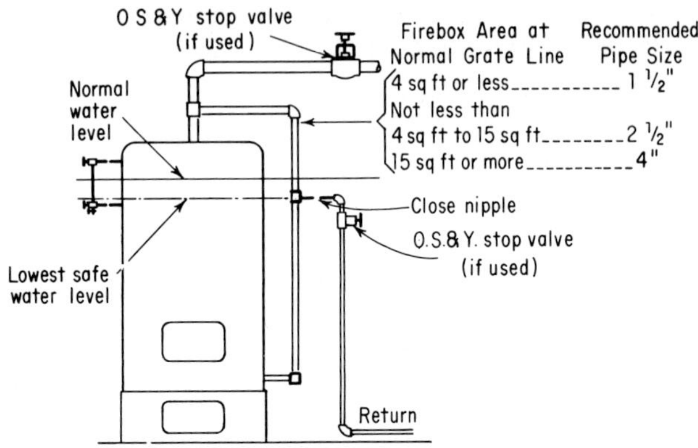

Fig. 1.12 Hartford loop is the recommended return-pipe loop connection.

water tube or fire tube, including boiler firing apparatus, controls, and boiler safety appurtenances. The trend today is to apply these units for industrial use as well as for some power generation. In the low-pressure field, most boilers are packaged units, unless a problem arises at the installation site because of headroom, access opening into the plant, etc.

A shop-assembled boiler is less costly than a field-erected unit of equal steaming capacity. While it is not an off-the-shelf item, it can generally be put together and delivered much faster than a field-erected boiler. Installation and start-up times are substantially shorter. Shop-assembled work can usually be better supervised and done at lower cost. In addition, packaged boilers require much less space, often 50 percent less in overall volume. Pressures are also rising in packaged boilers. Latest trends indicate that 950 psi is available, and higher pressure will be built into these units.

SELECTING BOILERS

Q Name five important factors to consider when selecting a boiler.

A 1. Because most boiler manufacturers design their boilers around the fuel to be burned, carefully consider the available fuel. Fuel selection may be a question of local price conditions, licensing laws for operators (see *Standard Plant Operator's Questions and Answers*, vol. II, for 56 pages of requirements in the United States and Canada), smoke ordinances, auxiliary equipment needed to burn the fuel, availability of a second fuel to supplement the primary fuel during normal operation (or emergency conditions), and experience of operating personnel with a particular fuel.

2. Capacity and pressure required will determine the type of boiler, since some boilers are limited on capacity and pressure, others are not.

3. Space conditions in the building or property will have a marked effect on the capacity of the unit considered. They will also affect the type of firing and even the fuel that can be used.

4. Cost is a very vital element, but a word of caution. First cost may be high on one make of boiler, but it may more than compensate by lower operating and maintenance expenses. So a careful evaluation is required rather than selection of the unit on first cost alone.

5. Individual preference may have a heavy bearing on the type of boiler selected, just as it does on other equipment. Experience with a scotch marine boiler, for example, may cause the user to prefer this type of boiler.

Q What is the most common *size* boiler sold in the United States?
A A capacity range of 50,000 lb/h or less accounts for 60 to 80 percent of unit sales per year.

Q Of what *pressure range* are these boilers?
A Most of these boilers are under 300 psi, indicating the tremendous amount of boilers being used at moderate pressures and temperatures for process, industrial work, incidental power, and heating purposes.

Q What is the trend for pressure and capacity in the public utility boiler field?
A The trend is for single boiler-turbine installations: one boiler supplies one steam turbine-generator. While 5000 psi and 1100°F temperatures are presently limits, capacity is increasing, and the largest units are of capacities of well over 6,000,000 lb/h.

Q Is higher pressure and temperature progress gradual because we cannot calculate the stresses imposed on such units?
A No; although partly true, the main problem is developing materials that will physically and chemically withstand higher temperatures and pressures. The goal of metallurgists in the boiler field and the steel industry is to develop alloys that have the following desirable characteristics: (1) high-temperature strength to resist long-term creep, (2) structural stability to resist intergranular crystal changes from service conditions, (3) surface stability to resist corrosion, erosion, and oxidation on both the waterside and fireside, and (4) alloys that can be reasonably well machined, welded, and otherwise fabricated for the boiler shapes required and that can also be reasonably priced for boiler and pressure-vessel usage.

HIGH-PRESSURE BOILER SYSTEMS

Q What is a high-pressure boiler system?

A This is a system in industry that produces steam for process needs, electric power, and combinations of process needs and power generation. Utility systems have as their main purpose the generation of electric power.

Q What size units are generally used in industrial plants?

A Most boilers are well under 100,000 lb/h. About 10 percent of industrial boilers sold in the United States are over 250,000 lb/h. The biggest purchasers of large industrial boilers are paper manufacturers, with metal and chemical manufacturers following. Oil companies are also large boiler users.

Industrial companies with large process-steam needs will also produce electricity as a by-product. This is done by installing a high-pressure boiler with a corresponding high-pressure turbogenerator. Process steam is extracted from the turbine at various stage pressures per process needs while the turbine drives the generator. The term *cogeneration* defines this simultaneous production of process-steam needs and electricity.

Cogeneration is encouraged to some extent by recent federal legislation which requires utilities to buy excess power that may be generated by industrial plants. The electric rates to be paid by the utilities to industry under this law reflect the energy cost savings the utilities realize by being able to avoid capacity additions of their own.

Q How are utility systems classified?

A The boiler may be of the natural- or forced-circulation type, the boiler and turbogenerator may be arranged in modular form (no common steam header with other boilers), and the pressure may be a subsupercritical unit used for power generation.

Figure 1.13 shows a 1475 psi, 977°F unit with a continuous output rating with coal burning of 518,000 lb/h. Table 1.8 gives further particulars of this unit.

Q What is a supercritical once-through boiler?

A This is a boiler system which operates above the supercritical pressure of 3206.2 psia and 705.4°F saturation temperature. It does not have the fluid recirculation when operating at full pressure and temperature that a natural-circulation boiler has. The fluid is brought up to pressure and temperature in series-connected fluid passes; the name *once-through* is applied to this type of unit. Figure 1.14 shows a supercritical twin-furnace unit delivering 5015 psia and 2,000,000 lb/h of steam at 1200°F to a tandem-arranged steam turbogenerator driving a 325,000-kW generator.

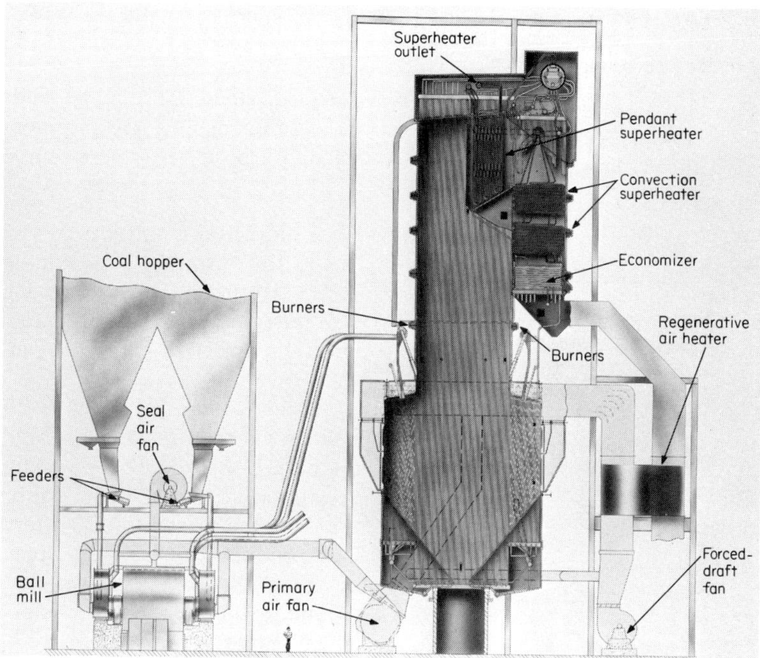

Fig. 1.13 Subsupercritical water-tube boiler rated 1475 psi, 977°F at 518,000 lb/h for power generation. (*Courtesy of Foster Wheeler Corp.*)

Figure 1.15 is another schematic of a once-through unit showing recirculating pumps and boiler feed pump loop in this type of unit.

Q What is a combined cycle?
A See Fig. 1.16. This is a system that uses a gas turbine to generate hot gases and drive a generator with the exhaust from the gas turbine going through a waste-heat boiler. The waste-heat boiler generates steam, which is then passed through a turbogenerator in the conventional manner to also produce electric power.

NUCLEAR-REACTOR SYSTEMS

Q What two nuclear systems are generally used in the United States to generate power?
A The two most common systems used are the boiling-water reactor (BWR), in which the water in the reactor boils, and the pressurized-water reactor (PWR), in which water is heated at high pressure and then passed through a heat exchanger external to the reactor to make the

Table 1.8 Design Details of a Coal-Fired Subsupercritical Boiler

Heating surfaces:	
Boiler	3,840 ft^2
Waterwalls	10,020 ft^2
Convection superheater	19,465 ft^2
Economizer	16,380 ft^2
Total	49,745 ft^2
Air heater	138,400 ft^2
Total furnace volume	57,800 ft^3
Total furnace surface	10,566 ft^2
Steam capacity	518,000 lb/h
Steam pressure	1,400 psi
Superheater steam outlet temperature	977°F
Temperature water feed entering unit	405°F
Temperature water feed leaving economizer	490°F
Temperature air entering unit	80°F
Temperature air leaving air heater	757°F
Temperature flue gas leaving furnace	1980°F
Temperature flue gas leaving economizer	815°F
Temperature flue gas leaving air heater	320°F
Air weight entering unit	638,000 lb/h
Draft in furnace	0.15 in H_2O
Heat losses:	
Dry gas	6.14%
Hydrogen and moisture in fuel	2.92%
Moisture in air	0.14%
Unburned combustibles	2.50%
Radiation	0.26%
Miscellaneous losses	1.50%
Total losses	13.45%
Boiler efficiency	86.54%
Coal proximate analysis:	
Moisture	8.0%
Volatile matter	6.30%
Fixed carbon	60.70%
Ash	25.00%
Ash fusion temperature	2240°F

steam for a turbogenerator. Figure 1.17 shows the two systems with their main components.

PWR and BWR systems are designed with capacities that may range to 17,000,000 lb/h of steam at 1050 psia and 550°F. The primary loop of a PWR system may operate at 2250 psia, providing water for the steam generator in the 540 to 600°F range. Most PWR steam generators are recirculating U-tube units on a per-loop basis. This will consist of an evaporator section with inverted U-shaped tubes and an overhead steam-drum section with a moisture separator.

The once-through steam generators are usually of straight-tube design. Coolant from the reactor enters the upper head, flows down through the tubes and out through the lower head.

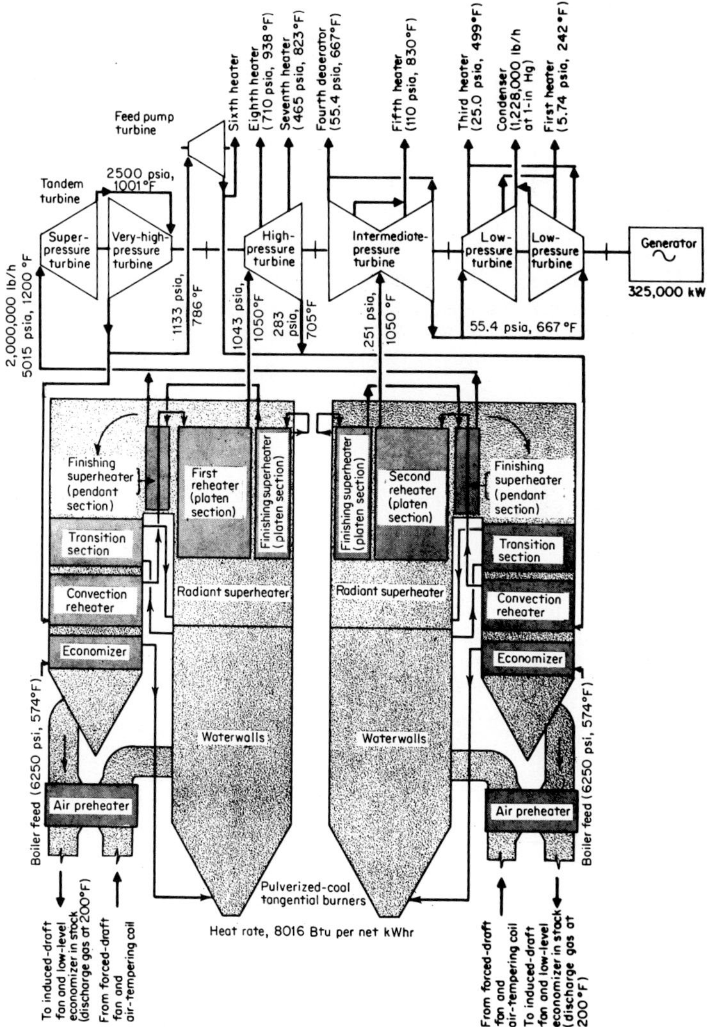

Fig. 1.14 Supercritical unit, 5015 psia, 1200°F supplies tandem 325,000-kW turbogenerator as shown above.

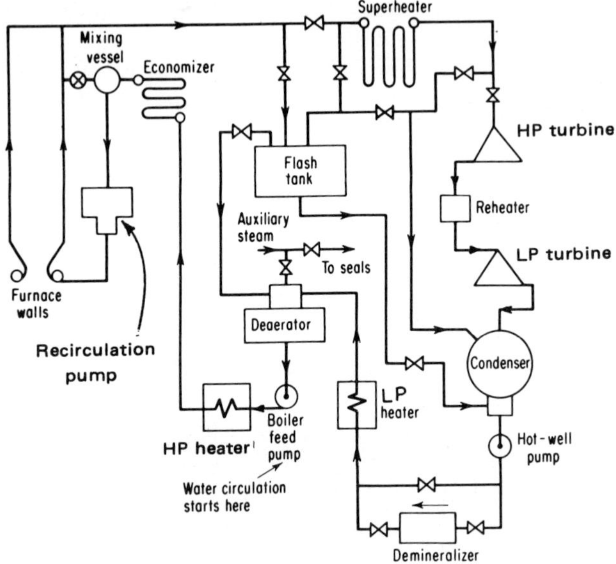

Fig. 1.15 Schematic of once-through boiler and turbogenerator arrangement.

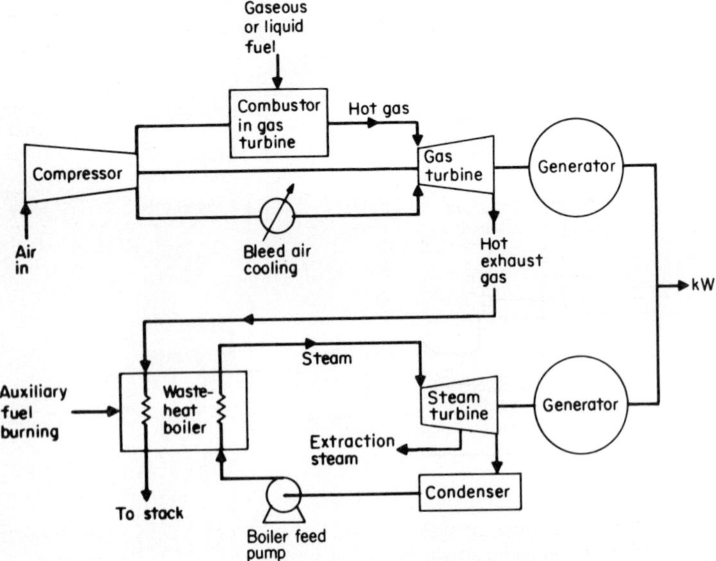

Fig. 1.16 Combination gas and steam-turbine cycle using waste-heat boiler to supply electric-power and process-steam needs.

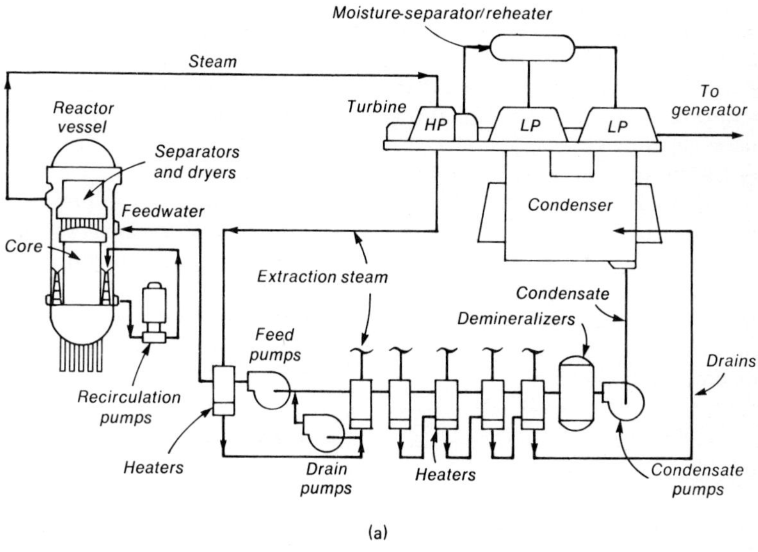

(a)

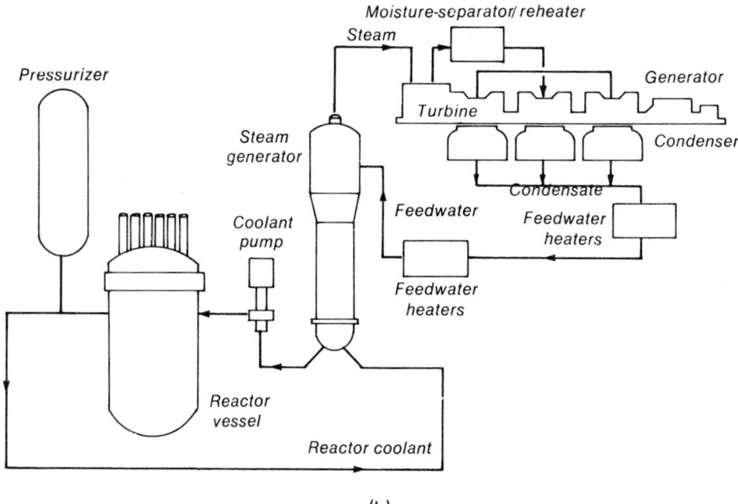

(b)

Fig. 1.17 Two nuclear-reactor systems used for power generation. (*a*) Boiling-water reactor (BWR); (*b*) pressurized-water reactor (PWR).

Q What important factor must be considered in the selection of industrial boilers?
A Industrial processes require a wide range of boiler sizes, pressures, and temperatures as well as operating conditions. A significant consideration today is the fuel to be used—oil, gas, coal, and even waste fuels and process waste heat. Reliability to produce steam in amounts up to rated capacity at designed operating pressure and temperature whenever called upon to do so is another important consideration.

Forced outages can range from simple tube failures, the improper operation of a relay in a burner supervisory system, or a serious furnace explosion, dry firing of a component of the boiler, which can cause extensive property damage, process interruption, and even personnel injury or death. Thus, there is an increased emphasis on reliability as unit size increases and spare or standby equipment may be nonexistent or limited in capacity.

CODES AND REGULATIONS

Q Do most states in the United States have some form of boiler laws?
A Yes, most state jurisdictions have some form of boiler laws. These are usually adopted from the ASME Boiler and Pressure Vessel Code requirements, or the National Board of Boiler and Pressure Vessel Inspectors Inspection Code. Some familiarity with these laws will assist energy managers and operators in meeting legal requirements.

Q Explain the ASME boiler code stamp authorization system.
A The ASME boiler code and the National Board of Boiler and Pressure Vessel Inspectors Inspection Codes are important source documents for legal requirements in the various states and municipalities that have adopted boiler safety laws. In addition to maintaining active boiler and pressure-vessel committees in order to keep the published codes up to date with developing technology, the ASME issues to qualified manufacturers, assemblers, material suppliers, and nuclear power plant owners code symbol stamps indicating that the manufacturer has received authorization from the ASME to build boilers and pressure vessels to the ASME code.

A fundamental principle of the AMSE Boiler and Pressure Vessel Code is that a boiler or pressure vessel, to be stamped ASME code-designed, must receive third-party authorized inspection during construction for compliance with the prevailing code requirements. Most third-party inspections are performed by authorized boiler and pressure-vessel inspectors who have appropriate experience and have passed a written examination in a jurisdiction. They must be employed either by the state or by an insurance company licensed to write boiler and pressure-vessel insurance in the jurisdiction where the boiler or pressure vessel is to be built, and in some cases the installation's location also must be considered. With uniform requirements for inspectors that have been promoted and implemented by the National Board of Boiler and Pressure Vessel Inspectors, a boiler or pressure vessel inspected by a properly credited National Board (NB) inspector will generally be accepted in all jurisdictions.

The manufacturer or contractor who wishes to build or assemble boilers or pressure vessels under an ASME certificate of authorization must first agree with an authorized inspection agency that code inspections will be performed by the agency. This is usually arranged by both parties signing a contract with the inspection work done on a fee basis.

Q What is the National Board of Boiler and Pressure Vessel Inspectors?

A The National Board of Boiler and Pressure Vessel Inspectors is composed of chief inspectors of states and municipalities in the United States and Canadian provinces. This organization has established criteria for boiler inspectors' experience requirements, the promotion and conductance of uniform examinations, and testing that are used by the jurisdictions. The National Board issues commissions to inspectors passing an NB examination, which are accepted on a reciprocal basis by most jurisdictions, thus providing a "portability" feature to a credential.

Most areas of the United States and all jurisdictions in Canada require that high-pressure boilers be subjected to periodic inspection by an authorized inspector. In most jurisdictions, this consists of annual internal inspection of power boilers and biennial inspection of heating boilers and usually of pressure vessels for those states that have adopted laws on low-pressure boilers or unfired pressure vessels. If the results prove satisfactory, the jurisdiction issues an inspection certificate, authorizing use of the vessel for a specific period.

There are three types of inspectors who make the legal inspections and reports to a jurisdiction that a boiler is safe or unsafe to operate or that it requires repairs before it can be operated:

1. State, province, or city inspectors see that all provisions of the boiler and pressure-vessel law, and all the rules and regulations of the jurisdiction, are observed. Any order of these inspectors must be complied with, unless the owner or operator petitions (and is granted) relief or exception.

2. Insurance company inspectors are qualified to make ASME code inspections. If commissioned under the law of the jurisdiction where the unit is located, they can also make the required periodic reinspection. As commissioned inspectors, they require compliance with all the

Table 1.9 States Having Boiler and Pressure-Vessel Reinspection Laws

State	Accept insurance company reports (X = yes)	Require inspection for		
		High-pressure boilers	Low-pressure boilers	Unfired pressure vessels
Alabama	No law	—	—	—
Alaska	X	X	X	X
Arizona	X	X	X	—
Arkansas	X	X	X	X
California	X	X	—	X
Colorado	X	X	X	—
Connecticut	X	X	X	—
Delaware	X	X	X	—
District of Columbia	X	X	X	—
Florida	No law	—	—	—
Georgia	No law	—	—	—
Hawaii	X	X	X	X
Idaho	X	X	X	X
Illinois	X	X	X	—
Indiana	X	X	X	X
Iowa	X	X	X	X
Kansas	X	X	—	—
Kentucky	X	X	X	—
Louisiana	X	X	X	X
Maine	X	X	X	—
Maryland	X	X	X	X
Massachusetts	X	X	X	X
Michigan	X	X	X	—
Minnesota	X	X	X	X
Mississippi	X	X	X	X
Missouri	No law	—	—	—
Montana	X	X	X	—
Nebraska	X	X	X	X
New Mexico	No law	—	—	—
Nevada	X	X	X	X
New Hampshire	X	X	X	X
New Jersey	X	X	X	X
New York	X	X	X	—
North Carolina	X	X	X	X
North Dakota	X	X	X	—
Ohio	X	X	X	—
Oklahoma	X	X	—	—
Oregon	X	X	X	X
Pennsylvania	X	X	X	X
Rhode Island	X	X	X	—
South Carolina	No law	—	—	—
South Dakota	X	X	X	—
Tennessee	X	X	X	X

Table 1.9 States Having Boiler and Pressure-Vessel Reinspection Laws (*Continued*)

State	Accept insurance company reports (X = yes)	Require inspection for		
		High-pressure boilers	Low-pressure boilers	Unfired pressure vessels
Texas	X	X	X	—
Utah	X	X	X	—
Vermont	X	X	X	X
Virginia	X	X	X	X
Washington	X	X	X	X
West Virginia	X	X	—	—
Wisconsin	X	X	X	X
Wyoming	No law	—	—	—

provisions of the law and rules and regulations of the authorities. In addition, they may recommend changes that will prolong the life of the boiler or pressure vessel.

3. Owner-user inspectors are employed by a company to inspect unfired pressure vessels for direct use and not for resale by such a company. They also must be qualified under the rules of any state or municipality which has adopted the code. Most states do not permit this group of inspectors to serve in lieu of state or insurance company inspectors.

Tables 1.9 and 1.10 list the states, cities, and counties in the United States that have some form of installation and periodic reinspection requirements on boilers and some unfired pressure vessels. These laws vary a great deal. For example, on low-pressure boilers, reinspection requirements may be limited to installations located in places of public assembly. Others include all heating boilers, except those located in private residences or in apartment houses with six families or less. Therefore, local or state laws should be checked for more specific requirements.

The same principle applies to operating-engineer licensing laws listed in Table 1.11. Local requirements vary quite a bit on experience needed, grades, and degree of responsibility. It is suggested that the appropriate jurisdiction be contacted for further details.

Q How do environmental regulations affect heat-energy systems?

A Technology plays a central role in maintaining the standard of living to which society has become accustomed and which can affect the daily life of the average citizen. Some of these impacts have been negative—for example, major power blackouts, noise near jetports, pollution of air and water resources, etc. In response to some of these

Table 1.10 Cities and Counties Having Boiler and Pressure-Vessel Reinspection Laws

City or county	Accept insurance company reports (X = yes)	High-pressure boilers	Low-pressure boilers	Unfired pressure vessels (UPV)
Albuquerque, N.Mex.	X	X	X	—
Buffalo, N.Y.	X	X	X	—
Chicago, Ill.	No	X	X	—
Dearborn, Mich.	X	X	X	X
Denver, Colo.	No	X	X	X
Des Moines, Iowa	X	X	X	—
Detroit, Mich.	UPV only	X	X	X
E. St. Louis, Mich.	No	X	X	X
Greensboro, N.C.	X	X	X	X
Kansas City, Mo.	X	X	X	X
Los Angeles, Calif.	X	X	X	X
Memphis, Tenn.	X	X	X	X
Miami, Fla.	X	X	X	X
Milwaukee, Wis.	X	X	X	X
New Orleans, La.	X	X	X	X
New York, N.Y.	X	X	X	—
Oklahoma City, Okla.	X	X	X	—
Omaha, Neb.	X	X	X	—
Phoenix, Ariz.	X	X	X	X
St. Louis, Mo.	X	X	X	X
San Francisco, Calif.	X	X	X	X
San Jose, Calif.	X	X	X	—
Seattle, Wash.	X	X	X	X
Spokane, Wash.	X	X	X	X
Tacoma, Wash.	X	X	X	X
Tampa, Fla.	X	X	X	X
Tucson, Ariz.	X	X	X	X
Tulsa, Okla.	No	X	X	X
University City, Mo.	No	X	X	—
White Plains, N.Y.	X	X	X	—
Arlington County, Va.	X	X	X	—
Dade County, Fla.	X	X	X	X
Fairfax County, Va.	X	X	X	X
Jefferson Parish, La.	X	X	X	X
St. Louis County, Mo.	X	X	X	X

negative aspects, the public attitudes toward the social value of technology have been changing, and government and private groups have become more actively involved in questioning and even suggesting restraining the advancement of technology. For example, safety from chronic effects, such as the long-term effects of radiation or the long-

Table 1.11 Jurisdictions Having Operating-Engineers' Licensing Laws for Boilers

Jurisdiction	High-pressure boilers	Low-pressure boilers
U.S. cities and counties		
Buffalo, N.Y.	X	—
Chicago, Ill.	X	—
Dearborn, Mich.	X	X
Denver, Colo.	X	X
Des Moines, Iowa	X	X
Detroit, Mich.	X	X
E. St. Louis, Ill.	X	X
Kansas City, Mo.	X	X
Los Angeles, Calif.	X	X
Memphis, Tenn.	X	X
Miami, Fla.	X	X
Milwaukee, Wis.	X	X
New Orleans, La.	X	X
New York, N.Y.	X	—
Oklahoma City, Okla.	X	X
Omaha, Neb.	X	X
St. Joseph, Mo.	X	X
St. Louis, Mo.	X	X
San Jose, Calif.	X	—
Spokane, Wash.	X	X
Tacoma, Wash.	X	X
Tampa, Fla.	X	X
Tulsa, Okla.	X	X
University City, Mo.	X	X
White Plains, N.Y.	X	—
Jefferson Parish, La.	X	X
St. Louis County, Mo.	X	X
States		
Alaska	X	X
Arkansas	X	X
District of Columbia	X	X
Massachusetts	X	—
Minnesota	X	X
Montana	X	X
Nebraska	—	X
New Jersey	X	X
New York	X	X
Ohio	X	X
Pennsylvania	X	X

Table 1.11 Jurisdictions Having Operating-Engineers' Licensing Laws for Boilers (*Continued*)

Jurisdiction	High-pressure boilers	Low-pressure boilers
Canadian provinces		
Alberta	X	X
British Columbia	X	X
Manitoba	X	X
New Brunswick	X	X
Newfoundland and Labrador	X	X
N.W. Territory	X	X
Nova Scotia	X	—
Ontario	X	X
Quebec	X	X
Saskatchewan	X	X
Yukon Territory	X	X

NOTE: Because of variations in the laws, it is necessary to check the jurisdiction for specific requirements in licensed operators.

term exposure to potential carcinogen-type products or toxic materials, must be evaluated; and the disposal of both waste and used products that may be potentially harmful to the environment requires attention.

Thus, fuel-burning systems for boilers, and nuclear-energy systems, must be designed and then operated and maintained so that air pollution and waste disposal will have minimal effects on the environment. As a minimum they also must comply with legal requirements on established threshold limits for air and water pollution, as well as radiation levels in nuclear-plant applications.

Legal requirements on boilers and nuclear-power-plant equipment no longer are limited to establishing safe construction codes. They have been expanded into requirements on controls, on devices to prevent furnace explosions, and on measurements to limit air pollution and radioactive contamination. Owners and operators must periodically review their operation and maintenance practices in order to make sure they comply with these additional legal requirements of the jurisdiction in which the equipment is located.

CHAPTER **TWO**

Fire-Tube Boilers for Heat and Process Applications

The earliest boilers built were of the simple shell type, still used in large numbers for heating and high-pressure process needs, where saturated steam up to 250 psig may be needed; capacities range to 50,000 lb/h. Above this pressure and capacity, water-tube boilers are usually used. The reemergence of solid-fuel burning, due to the potential shortage of oil and gas, may reactivate FT boiler design of the type where such fuel can be burned, such as the HRT and economic boiler. This chapter reviews some features of fire-tube (FT) boilers, and the types that may be found from past and even future applications.

HEAT TRANSFER

Q What is the heat-transfer mechanism in shell-type boilers (Fig. 2.1) to form steam efficiently?

A See Fig. 2.2*a*, which shows a simple drum heated from beneath. Steam forms in bubbles at the heated surface. The bubbles and the heated water are displaced by heavier, solid water, and circulation currents bring bubbles to the surface, where the steam is released into the space above. Here, in simplest form, is the three-part cycle common to every steam generator: (1) flow of water to heated areas, (2) flow of steam and heated water to upper areas, and (3) release of steam. It is the designer's task to provide conditions favorable to the effective operation of this circulation cycle.

Figure 2.2*b* and *c* shows how the simple drum may be modified to create better steaming conditions. Putting the heat source inside the shell increases heated surface, and steam formation tends to become more uniformly distributed over the mass of water.

In the sketches of Fig. 2.2, steam formation and circulation flow take place within a single, large mass of water. This condition character-

Fig. 2.1 Three-pass ASME fire-tube boiler has hot flue gases inside tubes and water on outside. (*Courtesy of York Shipley, Inc.*)

izes the broad class known as *fire-tube* boilers. In *water-tube* boilers, water and steam flow in a relatively large number of individual tubular paths which are always heated externally.

Note that flue gases are *inside* the tubes of FT boilers, and water is *inside* the tubes of WT boilers. One way to increase the heating surface

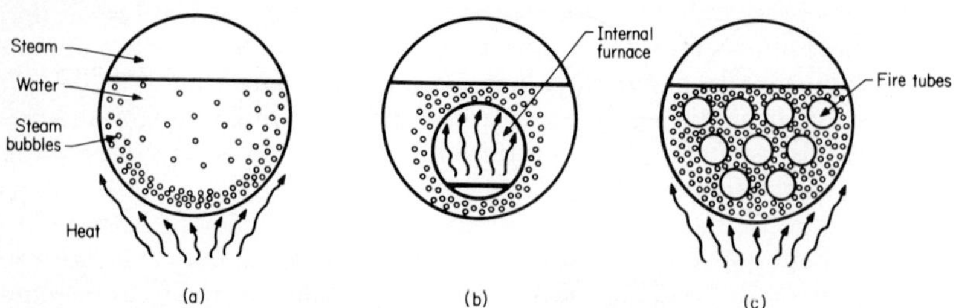

Fig. 2.2 Steam-bubble formation. (*a*) Simple drum is heated from lower shell; (*b*) internal furnace heats water all around furnace peripheral surface; (*c*) tubes inside shell add to heating surface. (*Courtesy of Power magazine.*)

of FT boilers is to direct the hot gases through the tubes in more than one pass, as in Fig. 2.1. Figure 2.3 shows the tube area provided per pass for flue-gas passages. The manufacturer has deliberately decreased progressively the cross-sectional area of each pass in order to maintain high gas velocities throughout the boiler. The hot gases occupy less volume as they pass through the four passes on a per-pass basis, and also have less heat content. When a high velocity is increased or maintained through the unit, a more uniform heat transfer takes place per pass in the boiler.

FIRE-TUBE BOILER HISTORY

Q Briefly discuss the history of fire-tube boiler development.

A The first boiler consisted of a vessel full of water with a fire burning under it. The efficiency and capacity limitations of such an arrangement brought improvements in the general means to increase heating surface without increasing overall boiler dimensions. This may be done by putting the furnace inside the boiler shell, by traveling hot gas through tubes inside the water space, or by a combination of both methods. The result is the *fire-tube* boiler.

In such a boiler, water and steam are contained within a single space formed by the shell, and flow of hot gas is divided into a number of separate streams. Although in some types part of the heat flows through the shell to the water inside, most of the work is done by heat transfer from hot gas inside tubes to water surrounding them—hence the name fire-tube.

Simplicity of construction makes this general class of boilers rugged and relatively low in first cost. They are good steamers, and the large volume of water in the shell produces the equivalent of accumulator

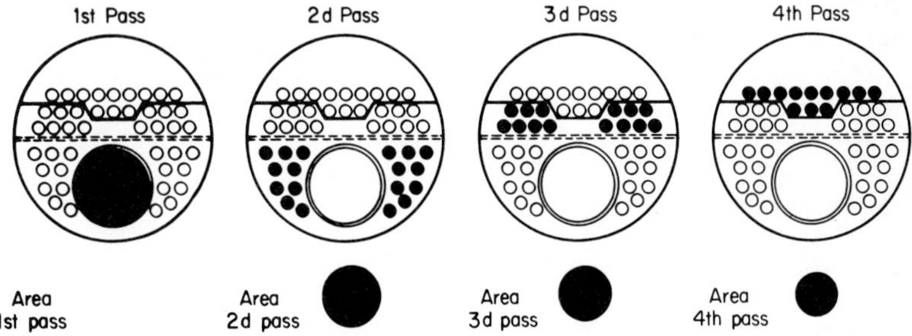

Fig. 2.3 Four passes of flue gas through FT boiler provide added heating surface. Decreasing cross-sectional area of each pass maintains high flue-gas velocity through boiler. (*Courtesy of Cleaver Brooks.*)

action, making it possible to meet load changes quickly. On the other hand, the overload capability of fire-tube boilers is limited by danger of overheating. With suitable design, relatively high efficiency can be obtained.

The fact that the entire steam-making process takes place within a shell limits both boiler size and pressure. Maximum diameter of cylindrical shells runs about 120 in and top pressure about 400 psi, but by far the greatest number of fire-tube boilers operate at pressures below 300 psi. Their general characteristics make fire-tube boilers well suited for small plants, and they are widely used for supplying heating loads.

Later chapters will review the forces acting on boiler shells and thus will assist in understanding the reasons for fire-tube construction and operation. Because pressure of a fluid exerts itself equally in every direction, a sphere represents the strongest shape for resisting internal pressure. For practical reasons, boiler-shell shape departs from the spherical, but most designs utilize cylindrical or curved shapes as far as possible. Internal braces or stays overcome the tendency of flat plates to bulge. Later chapters will show that the force tending to burst the shell along its length is twice that tending to burst it around the girth. This means that a longitudinal joint must be approximately twice as strong as a circumferential one, with resulting differences in construction.

It will also be shown that the strength required to resist bursting is proportional to the product of pressure and diameter. An increase in either means a corresponding increase in required strength, and if both go up, the strength must go up as the product. It can be seen that high pressures and large diameters would lead to extremely thick shell plates, hence the rather definite limit on pressure and size of fire-tube boilers.

There are many types of fire-tube boilers to meet varying requirements. A broad division may be made between those intended for use with an external furnace built of refractory and those with one form or another of internal furnace, often called *firebox* boilers. Of the "brick-set" types, the horizontal-return-tubular boiler, or HRT, was at one time the most common.

CONSTRUCTION DETAILS

Q How are tubes attached to the tube sheets in FT boilers?

A Tubes in all FT boilers must be rolled and beaded (Fig. 2.4*a*) or rolled and welded. If rolled and welded in high-pressure boilers, see Power Boilers, Section I, ASME Boiler and Pressure Vessel Code. Tubes are beaded to prevent the ends from being burned off by the hot

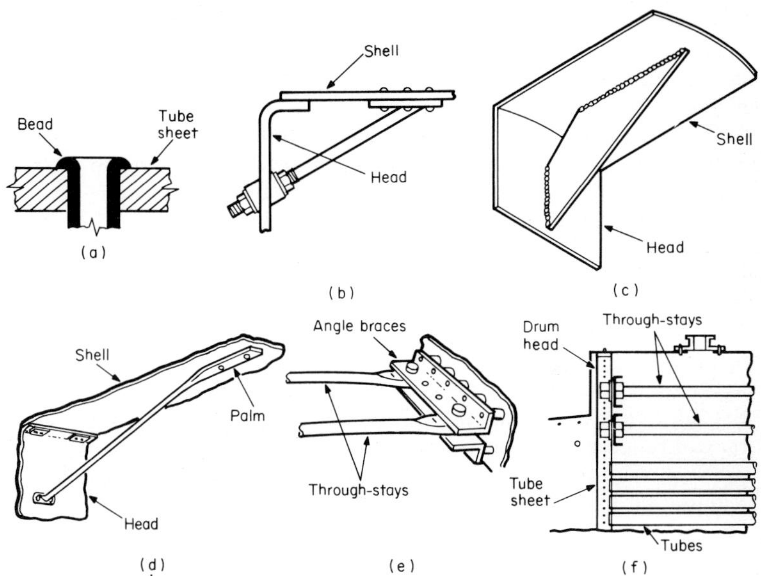

Fig. 2.4 Details of FT boiler construction. (*a*) Rolled and beaded fire tube; (*b*) threaded diagonal stay; (*c*) welded gusset stay; (*d*) riveted diagonal stay; (*e*) through stays using angle steel braces; and (*f*) through stays using outside nuts and washers.

gases in this area. Beading also increases heat transfer near the tube sheet and tube juncture.

Q What would cause a fire tube to burst or explode?
A Fire tubes are normally under external pressure; thus they may collapse but not burst. The biggest problems are loosening of tubes in the tube sheet, cracking, burning, and corrosion of tube ends, waterside pitting and corrosion leading to leakage, fireside corrosion and pitting leading to leakage or pulling out of the tube sheet (due to poor rolling). Another problem is scale buildup on the waterside, leading to overheating and possible sagging and loosening in the tube sheet.

Q What is the code requirement for stay bolts, or stays screwed through plates?
A See Fig. 2.4*b*, *c*, *d*, and *e*. The ends of the stay bolts, or stays screwed through the plate, must extend beyond the plate not less than two threads when installed, after which the ends must be riveted over without excessive scoring of the plate. They may also be fitted with threaded nuts, provided the stay bolts, or stays, extend through the nut.

If stay bolts are solid, 8 in in length or less, and threaded, they must be drilled with telltale holes (Fig. 2.5) at least $^3/_{16}$ in in diameter to a depth extending at least ½ in beyond the inside of the plate. If the stay bolt is reduced in diameter between the ends, the telltale hole must extend ½ in beyond the point where the reduction in diameter begins. Telltale holes are not required for bolts over 8 in in length or if the stay bolt is attached by welding and code rules have been followed. Since stay bolts usually break near the plate supporter, warning is given by water flashing from the telltale hole.

Q What are the requirements for welded-in stay bolts in high-pressure boilers?

A The stay bolts and stays must be inserted in countersunk holes through the plate, except for diagonal stays, which can be fillet-welded to the shell but not the head, provided the weld is not less than ⅜ in and the fillet weld continues the full length of the stay. Stay bolts inserted by welding cannot project more than ⅜ in beyond the plate exposed to products of combustion. The welding must be stress-relieved after it is completed. Radiographing (x-ray photograph) is not required.

HORIZONTAL-RETURN TUBULAR BOILERS

Q What was the first boiler evolved from the simple shell boiler design?

A The horizontal-return tubular (HRT) boiler (Fig. 2.6) is available today in smaller sizes as packaged units. The HRT boiler consists of a cylindrical shell, today usually fusion-welded, with tubes of identical diameter running the length of the shell throughout the water space. The space above the water level serves for steam separation and storage. A baffle plate (or dry pipe) is ordinarily provided near the steam outlet to obtain greater steam dryness.

The HRT boiler is simple in construction, has a fairly low first cost, and is a good steamer. It is more economical than the vertical tubular or locomotive types, but the scotch marine boiler is replacing it. One

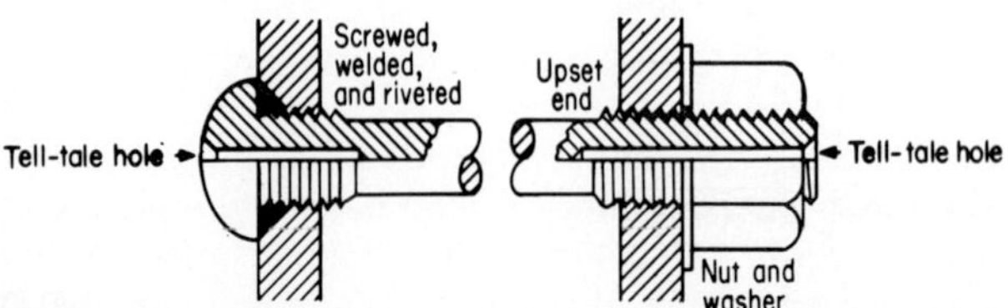

Fig. 2.5 Stay bolts are used to strengthen flat sheets against bulges and possible rupture.

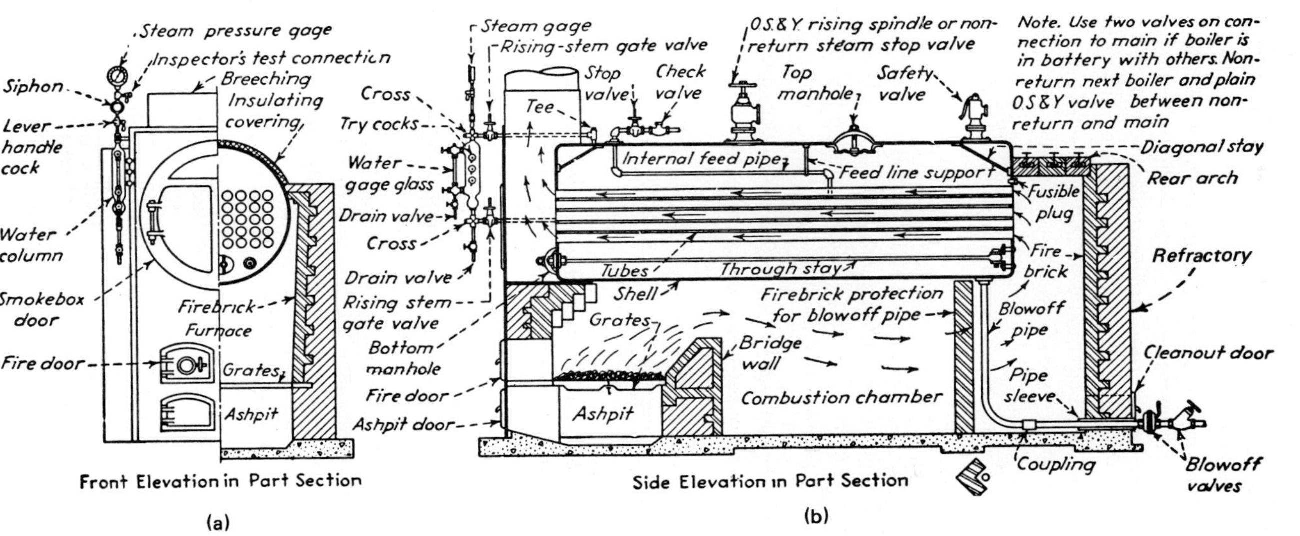

Fig. 2.6 Horizontal-return tubular boiler (HRT) is an older design, but still found in many smaller industrial plants. (*a*) End view; (*b*) side view.

disadvantage is that hard deposits of scale are difficult to remove from water surfaces of the inner rows of tubes. Another disadvantage is the danger of burning the lower shell plates above the fire if thick scale or deposits of mud form on the waterside on these plates. But the difficulty of cleaning scale from the tubes holds true for all other types of FT boilers.

HRT boilers are not very practical in shell sizes over 96 in in diameter or for pressures exceeding 200 psi. Thick plates for higher pressures exposed directly to the flames would deteriorate very rapidly from overheating, because of poor heat transfer to the water. This could lead to bags (bulges) in the shell.

Bags also result when mud or other sediment from water settles on the bottom of the shell. For this reason an HRT boiler should be pitched 1 to 2 in toward the blowdown connection. Also, proper blowdown and water chemical treatment must be used to avoid scale or sediment formation and to keep solids to a minimum.

Q Why is a brick pier (Fig. 2.6) built in front of the blowdown pipe on older HRT boilers, and what other precautions are needed on blowdown pipes?

A On older HRT boilers the blowdown line is in the path of the hot gases, thus must be protected from overheating. A V-shaped pier of firebrick, usually in front of the blowdown pipe, deflects the hot gases.

> CAUTION: If not properly protected, the blowdown pipe may melt under certain conditions and drain the boiler, possibly leading to a serious explosion.

The blowdown connection is usually a steel pad on the shell. On older boilers, the blowdown pipe runs through a section of the furnace and out through a bushing in the rear brick wall. Three points of possible distress are:

1. The blowdown pad and the screwed or welded-in blowdown pipe may develop leakage due to expansion and contraction.
2. The blowdown pipe firebrick protective wall may deteriorate, leading to possible overheating of the blowdown pipe, then rupture, thus draining the boiler.
3. The rear brick wall may settle and bear on the blowdown pipe where it goes through the wall, causing the pipe to break.

Q Describe the outside suspension type of setting for an HRT boiler.

A See Fig. 2.7. This type of steel structural suspension support is required on all HRT boilers 72 in in diameter and over, as brickwork cannot support this heavier weight. On older boilers the pad on the shell was riveted to the shell, with rivets designed with a safety factor of 12½.

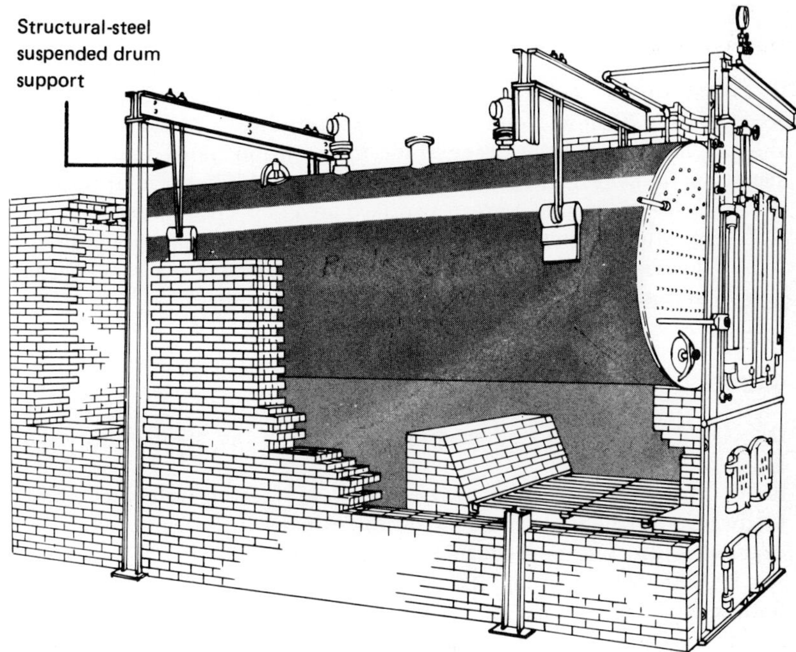

Fig. 2.7 HRT boilers with diameters over 72 in must have outside drum suspension support.

But if the pads are welded, full peripheral fillet welds are required, and these must be stress-relieved on high-pressure boilers. Radiographing is not required.

Q How is feedwater introduced into an HRT boiler?
A Usually through the front head or top part of the shell by means of a bushing. If the shell is 40 in in diameter, the discharge of the feedwater line must be about three-fifths the length of the boiler from the hottest end. This prevents solids from the feedwater settling on the hottest boiler surfaces.

Q What are the most common staying methods used on an HRT boiler?
A Four methods are used:

1. Through stays below the tubes because there is insufficient room for other stays, which would accumulate deposits more readily.
2. Diagonal stays above the tubes to support the flat unstayed tube sheet above the tubes.
3. Gusset stays above the tubes.

4. If not over 36 in in diameter and not over 100 psi, structural shapes can be used if arranged according to ASME code requirements.

Q What causes water to circulate in an HRT boiler?
A As Fig. 2.8 shows, feedwater is introduced near the bottom of the shell, is heated, and thus rises to the top as it lowers in density. The staggered tubes (Fig. 2.8*a*) allow better heat transfer because upcoming water must flow around a greater surface area of each hot tube. As Fig. 2.8*b* shows, baffled passages between parallel tube arrangements allow colder (heavier) water to drop unobstructed to the bottom, where it is heated and must wash against the hot tubes in rising.

CAUSES OF BAGS, BLISTERS

Q Name three causes of bagging (bulging) of the shell of an HRT boiler.
A 1. Oil (from lubricated pump rods, etc.) getting into the boiler feedwater and being carried to the lower part of the shell, which is exposed to fire, and thus causing overheating.

2. Scale or mud (from sediment in the water) deposits on the lower portion of the shell, thus restricting heat transfer.

3. Excessively localized flame on a portion of the shell, thus causing overheating.

Q What is the difference between a bag and a blister on the shell of a boiler?
A A bag is a bulged-out section of a portion of the shell, extending through the full thickness of the shell, caused by overheating and

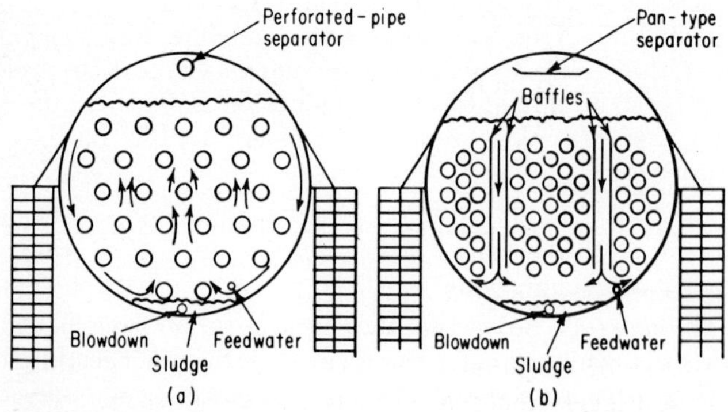

Fig. 2.8 Water circulation in HRT boiler. (*a*) Using staggered tubes; (*b*) using baffled tubes.

pessure. A bag can be driven (hammered) back if the remaining thickness is at least as strong as the longitudinal joint and if the quality of the metal has not been affected. A blister is actually a separation of the metal from the shell plate, caused by impurities rolled into the shell plate when formed. But only the outside layer will blister from the heat because the remaining thickness is not affected. A blister cannot be driven back but must be cut out and the edges trimmed; if the remaining metal is sound, the pressure on the boiler must be reduced to correspond to the thickness of the remaining sound metal. If pressure cannot bc reduced, the entire blistered section, including sound metal, must be cut out, a flush-welded patch formed and butt-welded in, with localized stress relieving required after welding. If the diameter of the blister is over 8 in, the weld must be radiographed. After it is shown to be satisfactory, a hydrostatic test of 1½ times the maximum allowable pressure is required to check the repair. All welding must be done by a certified welder, and the repair must be approved by a commissioned boiler inspector.

> WARNING: If a repair of this kind is made without the approval of a commissioned boiler inspector, the entire boiler can be condemned by the local enforcing authorities, and in a noncode state, the insurance on the boiler may be canceled.

Q Is oil at the waterline of an HRT boiler serious? How is oil removed from inside the boiler?
A Oil on any part of the waterside is serious. First check the fuel-oil heater (if oil-fired) for leaks, then the piping to the oil heater. As soon as the source of contamination is found, take measures to prevent oil from entering the boiler. A check valve installed on the steam line supplying a fuel-oil heater will prevent oil from backing into the boiler. Condensate should go to an open drain. If condensate goes to a condensate tank, install an oil separator and keep it in good condition. Also, there are double-shell oil heaters on the market which confine the leakage of an oil break to the inner shell.

Oil may be removed from a boiler by boiling it out with soda ash and caustic soda, using 1 lb of each per 1000 lb of water in the boiler. Carry the boiler at about 5 psi pressure and continue boiling for 2 or 3 days, depending on the extent of oil contamination. Then empty the boiler, wash it thoroughly with fresh water, and again check the internal surfaces.

Q What is the lowest permissible water level in an HRT boiler?
A The lowest level should never be less than 3 in above the tubes.

Q During an inspection, what are some of the areas to check carefully on an HRT boiler?

A Internally, on the section above the tubes, check for corrosion and pitting. Look for grooving on the knuckles of heads, shells, welds, rivets, and tubes. Check the seams for cracks, broken rivet heads, porosity, and any thinning near the waterline of the shell plate. Check all stays for soundness and proper tension. Examine the internal feed pipe for soundness and support, and see that it is not partially plugged. Check the openings to the water-column connections, safety valve, and pressure gage for scale obstruction. Also check shell and tube surfaces for scale buildup. Follow the same procedure internally below the tubes. Then check the opening to the blowdown connections and make sure that the bottom of the shell is pitched toward blowdown and that it has no blisters or bulges.

Externally, remove the plugs from the crosses of water-column connections and make sure they are free of scale. Examine blowoff piping pad and the blowoff pipe to make sure it is protected from the fire, that the pipe is sound, and that blowoff valves are in good order. Examine tube ends and rivets or welds for cracks and weakening of the tube to the tube-sheet connection. Check for fire cracks around the circumferential seam and for leakage at the calked edge. Then examine the setting and supports for soundness.

USE OF WELDING

Q Are fire-tube boilers still fabricated by riveted joints and connections?

A Fire-tube boilers are now fabricated by welding, instead of by riveting as was previously the practice. In the usual design, all shell joints are butt-welded, connection nozzles are welded in place, and diagonal stays are also fastened by welding. Handling of through stays and stay bolts varies, but one threaded end is usually retained by tension adjustments.

Q When is an access hole required in the front tube sheet *below* the tubes in HRT boilers?

A When the boiler is 48 in in diameter or larger.

Q When is an access hole required *above* the tubes of any HRT boiler?

A If externally fired, for boilers of 40-in diameter or over; if internally fired, for boilers of 48-in diameter or over.

Q How much larger than the fire-tube diameter may the tube holes be? Compare this with a water-tube unit.

A For fire tubes, $\frac{1}{32}$ in larger in diameter at the fire end and $\frac{1}{16}$ in larger at the opposite end. For a water tube, it is $\frac{1}{32}$ in larger at either end. This clearance is needed so that the tubes can be inserted into the holes before rolling.

Q What was the maximum permitted length of a course on the shell of a riveted HRT boiler?
A Twelve feet.

Q Why, instead of through-to-head stays, are diagonal stays not used below the tubes, as above the tubes, to brace the tube sheet?
A Because there is usually insufficient room for the proper number without placing them too close together or using sizes larger than is practical. Also, they would tend to hold loose scale and sludge and prevent its free movement to the blowdown.

Q How much space should there be in front of the boiler in planning installation?
A Sufficient room for replacement of tubes.

Q If, on looking in the furnace, the bottom of the shell is found to be bulged, what should be done?
A The boiler should be shut down immediately and then inspected by an authorized boiler inspector. The inspector's recommendations should be followed before the boiler is returned to service.

Q Why are boilers over 72 in in diameter required to be supported by the outside-suspension type of setting?
A Because the weight of the larger boilers may be in excess of the safe load on brickwork. Crushing or buckling of the walls might result from the load of a large boiler full of water.

Q What are two dangerous conditions offered by a weakened rear arch?
A (1) If the arch collapses wholly or in part, the upper part of the rear tube sheet may become overheated and damaged. (2) Anyone walking on top of the arch may cause it to collapse, and that person will fall through to a horrible death.

Q What precaution is required with a flush-front-set HRT boiler?
A The front arch protects the dry-sheet and front head seam from damage by overheating. This arch should be kept in good condition.

Q Why should the hole in the brick wall through which the blowdown pipe passes be inspected?
A To see that the pipe is not resting on the bottom of the opening. If it does, it is an indication that the boiler is probably settling and that it may not have sufficient clearance to allow free expansion and contraction.

Q Would loose bricks lying in the firebox be of any interest? If so, why?
A They might be from the closing-in line or from some point where they are supposed to protect part of the boiler from overheating. Also, they might be from a point where their loss would weaken the walls.

Q Name three points where brickwork should protect the boiler from overheating.
A The rear arch, the closing-in-line along the sides of the shell, and the front arch in the flush-front-set boiler.

Q How and under what conditions may the segments of the heads above the tube dispense with diagonal or through-stays?
A In boilers not exceeding 36-in diameter or 100 psi maximum allowable pressure, the segments may be braced by stiffening with channel irons or angle irons (riveted or welded back to back) riveted or welded to the tube sheet. The specifications for structural form should comply with code rules.

Q What is the difference between a through-brace and a through-to-head brace? Where is each used? Why?
A The through-brace has washers and nuts on each end. The through-to-head brace has nuts and washers on one end; the other end is forged into an eye that is held clear of the rear head by a pin or other construction. The through-stays may be used above the tubes, for the rear outside nuts are protected from burning off by the rear arch.

The through-to-head braces are used below the tubes where the rear ends have to be protected from the high-temperature gases.

Q If a fusible plug is used, where should it be located?
A Near the centerline of the rear tube sheet not less than 1 in above the top row of tubes.

Q How may the proper height of a water column be checked?
A Fill the boiler with water to a level 3 in above the top row of tubes at their highest end (usually the front). The water column should be at a level where water just shows in the bottom of the gage glass. Or the height of a water column may be checked by measurement from any common point of elevation.

Q Why do the flat segments of the tube sheets or heads require bracing?
A Because internal pressure tends to bulge these areas outward into a semispherical shape.

Q Is it practical to team up older HRT boilers with newer packaged types when modernizing for plant expansion?

A Yes. Figure 2.9 shows at left an older 125-hp riveted HRT boiler and at right a newer two-pass, 300-hp welded scotch dry-back boiler. At the time the new boiler was installed, the older one was converted from burning coal to burning Bunker C oil, as does the new unit. They both supply 135-psi steam to a common header. The type and size of boilers working as a team do not matter, so long as both are designed for supplying steam (or hot water) at the same pressure, which is very important because of the safety-valve settings.

ECONOMIC, OR FIREBOX, BOILER

Q What boiler is an adaptation of the HRT boiler?
A The so-called economic, or firebox, type in which the flue gases make two (Fig. 2.10) or three passes (Fig. 2.11), built for both low and high pressure. The boiler is self-supported in its special casing, and thus requires little brickwork. But this type has the same size and pressure limitations as the HRT boiler. The flat surfaces on each side require staying. The front water legs are stayed to each other by means of stay bolts. In the back, the side sheets are stayed by through-braces passing between the horizontal tubes.

The economic-type boiler is considered to be an externally fired, FT design because its steel-encased combustion chamber is not a pressure part of the boiler. Boilers of this type are usually shipped as a unit and thereby qualify as being among the first so-called compact designs.

Fig. 2.9 Older HRT boiler (left) teams up with newer scotch dry-back unit as steam demands increase with plant growth.

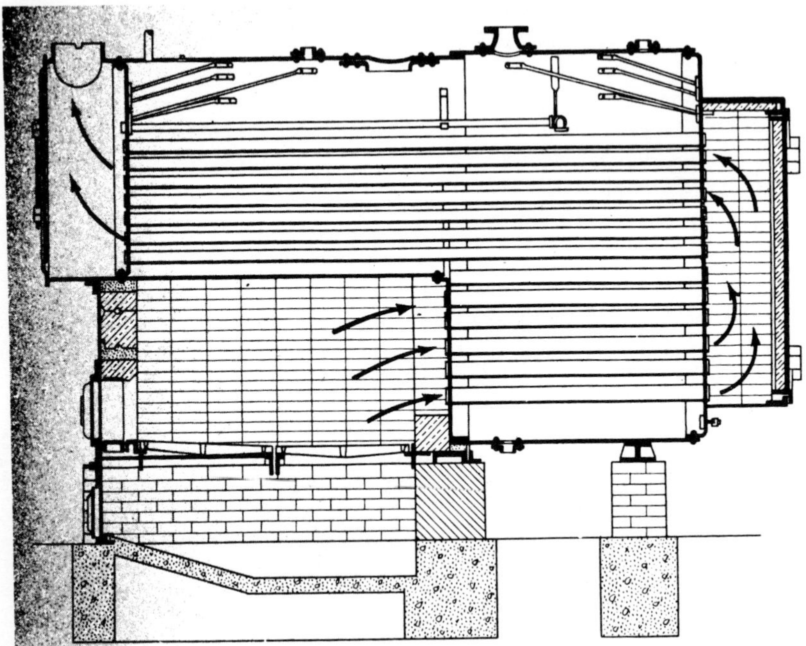

Fig. 2.10 Economic two-pass boiler with short, large-diameter tubes for first pass and longer, smaller-diameter tubes for second pass.

LOCOMOTIVE FIREBOX BOILERS

Q Describe the locomotive firebox (LFB) boiler and name all its parts.
A See Fig. 2.12. Like the VT and scotch marine types, the LFB boiler is an internally fired FT unit. But its shell is horizontal, and the firebox is not contained within the cylindrical portion of the boiler. The firebox is rectangular in shape with a curved top known as a crown sheet. This crown sheet is supported by radial stays screwed into the crown sheet and the outer wrapper sheet. Ends of radial stays are riveted over. The inner sheets of the firebox are connected to the outer side sheets by stay bolts. Space between these sheets is called the water legs. The first tubes are within the barrel and run from the firebox tube sheet to the smokebox head of the barrel (see Fig. 2.12). This head in the smokebox is formed by extending the barrel beyond the tube sheet.

The firebox front sheet above the crown sheet, and the smokebox tube sheet above the tubes, are supported by longitudinal stays. In some cases diagonal stays also are used for this purpose. The steam dome provides additional steam storage space and allows the main steam

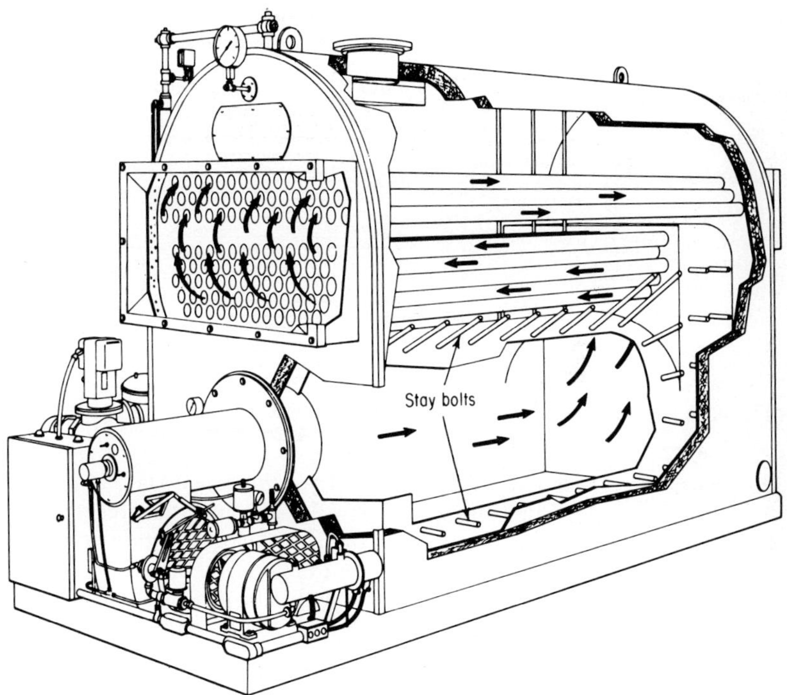

Fig. 2.11 Economic boiler with three passes. Flat sidewalls are water-jacketed and heavily stayed.

outlet to be taken off at a considerable height above the waterline, thus reducing the possibility of water carrying over with steam. The steam space extends over both the furnace and the barrel, which usually has 3-in tubes. All tubes are of one diameter and length.

As is true of most internally fired FT boilers, some water spaces are very difficult to clean, either mechanically or manually. Also, the LFB boiler is limited as to pressure and capacity, just as is the HRT boiler.

Q Where are stay bolts most likely to break in the LFB boiler?
A Usually the top row of stay bolts and the first row of radial stays, with fracture occurring close to the inner surface of the outer sheet (wrapper sheet). This area is a high-heat zone, causing large expansion and contraction movements.

Q Where does internal and external corrosion occur on an LFB boiler?
A For internal corrosion, check: (1) waterline and top row of the tubes because of oxygen and other impurities released when boiling; (2) top of

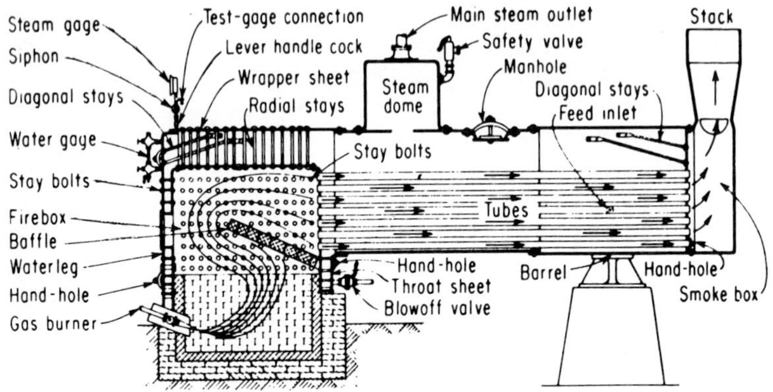

Fig. 2.12 Locomotive-type boiler for stationary use has only one pass for flue gas.

the crown sheet, on and around the ends of the radial stays and stay bolts; (3) water legs because of oxygen release and the presence of corrosive sediment; and (4) bottom of the barrel where pitting occurs if the boiler is improperly laid up when out of service.

External corrosion on an LFB boiler should be especially looked for around handhole plates and at the bottom of the first head. Rainwater wetting the sooty smokebox will corrode its bottom and also the handholes in this area. The bottom of the barrel should be watched for corrosion because of possible dampness or leakage. Corrosion due to sediment deposits forming at the bottom of water legs and their handholes may eat through the plate and gaskets of handholes. Leaking tubes or stay bolts cause corrosion at the furnace end of the tube sheets.

Q In case of low water, what is the most dangerous part of the LFB boiler?

A The crown sheet, because it is first uncovered by low water and therefore will overheat, leading to a possible explosion.

Q What is the lowest permissible water level in an LFB boiler?

A Three inches above the crown sheet if the shell is *more* than 36 in in diameter, but only 2 in above the crown sheet if the shell is *less* than 36 in in diameter.

Q From which end are the tubes removed and replaced in the LFB boiler?

A Because all tubes in the LFB boiler are removed and replaced from the smokebox end, always provide sufficient room at this end. In fact, in any boiler installation, there must be room for removing old tubes and inserting new ones without having to break building walls, remove doors or windows, etc.

Q Explain the meaning of externally and internally fired boilers.
A Externally fired boilers have a separate furnace built outside the boiler shell. The HRT boiler is probably the most widely known example of the externally fired boiler.

In internally fired boilers, the furnace forms an integral part of the boiler structure. The VT, LFB, and scotch marine (SM) are well-known examples of internally fired boilers.

SCOTCH MARINE BOILERS

Q What is a packaged FT boiler?
A These boilers are the lineal descendants of the basic scotch design. They represent the bulk of FT boilers being manufactured today. The American Boiler Manufacturers Association (ABMA) defines a packaged FT boiler as "a modified scotch-type boiler unit, engineered, built, fire-tested before shipment, and guaranteed in material, workmanship, and performance by one firm, with one manufacturer furnishing and assuming responsibility for all components in the assembled unit, such as burner, boiler, controls, and all auxiliaries. . . ."

The SM, or scotch dry-back, type may have two to four circular furnaces when the boiler is of large diameter. Figure 2.13*a* shows an SM-type boiler with three furnaces and wet-back construction. Note the stay bolts in the back, girder stays on the crown sheets in the back furnace, through-stay rods above and below the tubes, and corrugated furnaces.

Q Describe the dry-back SM-type boiler.
A This boiler (Fig. 2.13*b*) is an adaptation to stationary practice of the well-known SM wet-back boiler. It consists of an outer cylindrical shell, a furnace, front and rear tube sheets, and crown sheet. The hot gases from the furnaces pass into a refractory-lined combustion chamber at the back (sometimes built into a hinged or removable plate) and are returned through the fire tubes to the front of the boiler and thence to the uptake. This boiler is suitable for coal, gas, and oil firing.

In the wet-back design (Fig. 2.13*a*) the shell, tube, and furnace construction are similar to the dry-back type, but the combustion chamber, being inside the shell, is surrounded by water. Thus no outside setting or combustion chamber refractory is needed. The dry-back type is a quick steamer because of its large heating surface. It is also compact and easily set up and shows fairly good economy.

The internal furnace is subject to compressive forces and so must be designed to resist them. Furnaces of relatively small diameter and short length may be self-supporting if the wall thickness is adequate. For larger furnaces, one of these four methods of support may be used: (1)

corrugating the furnace walls; (2) dividing the furnace length into sections with a stiffening flange (Adamson ring) between sections; (3) using welded stiffening rings and (4) installing stay bolts between the furnace and the outer shell. If solid fuel (coal, wood, etc.) is to be fired, a bridge wall may be built into the furnace at the end of the grate section.

Q How many passes do the flue gases usually take in an SM boiler?
A Up to four passes (Fig. 2.14*a*) are taken by flue gases, depending on the design. While more passes can be used, today four is the practical limit. The two-pass design is the most popular for average plants, whether for high- or low-pressure service. Designers also give attention to the number and arrangement of the second-pass tubes. For example, waterside inspection and cleaning are easier when the tubes are aligned uniformly both vertically and horizontally. But a staggered layout of tubes tends to give a more circuitous flow of water around these tubes, thus promoting increased heat transfer. Improved transfer from gas to water is also obtained by using slip-in spiral fittings (retarders) at the tube inlet to impart a swirl to the hot gases (Fig. 2.14*b*). Some boilers have tubes finned on the fireside known as extended-surface tubes.

PERFORMANCE AND RATINGS

Q How are performance and ratings established on packaged boilers?
A It was rare for the makers of older conventional field-assembled FT boilers to know what firing setting and draft conditions their units

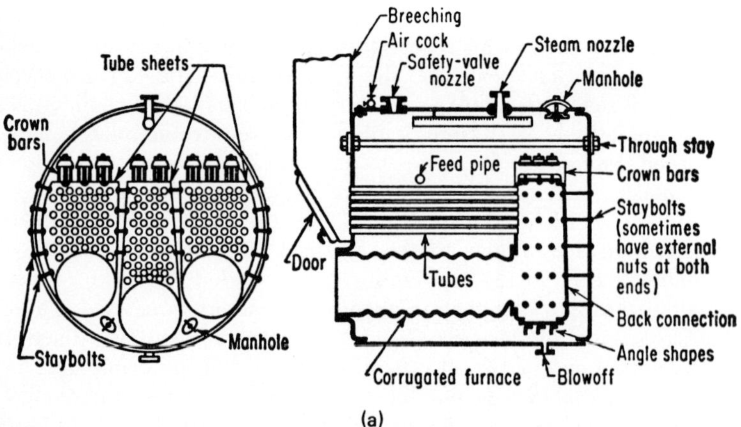

Fig. 2.13a Scotch marine (SM) wet-back boiler with three furnaces.

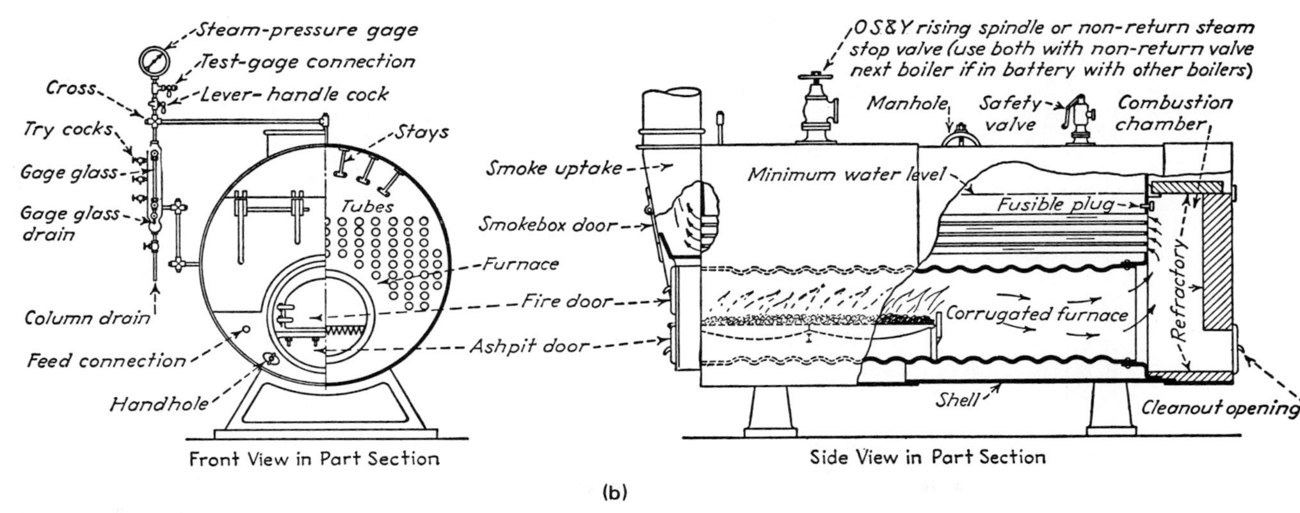
Steam-pressure gage
Test-gage connection
Cross
Lever-handle cock
Try cocks
Stays
Gage glass
Gage glass drain
Tubes
Furnace
Column drain
Fire door
Feed connection
Ashpit door
Handhole
Front View in Part Section
OS&Y rising spindle or non-return steam stop valve (use both with non-return valve next boiler if in battery with other boilers)
Manhole
Safety valve
Combustion chamber
Smoke uptake
Minimum water level
Fusible plug
Smokebox door
Refractory
Corrugated furnace
Shell
Cleanout opening
Side View in Part Section
(b)

Fig. 2.13b SM dry-back boiler with corrugated furnace.

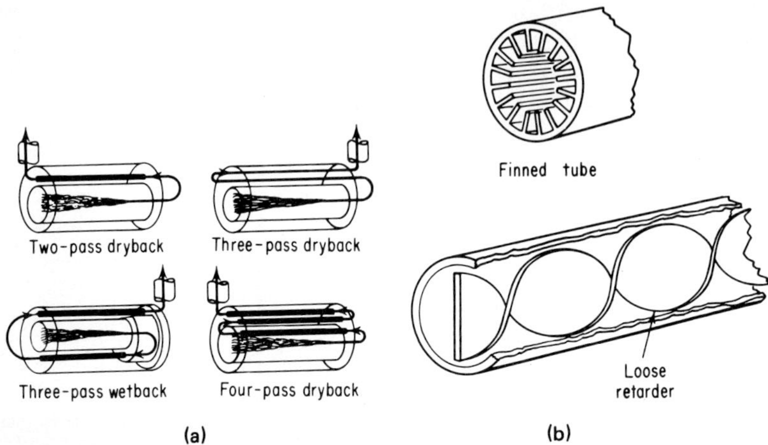

Fig. 2.14 (*a*) Flue gas passes in scotch marine boiler; (*b*) methods to improve heat absorption in fire tubes by use of finned tubes and retarders.

would encounter in actual application. So it became customary to assume that, as a guide, under average conditions, 10 ft^2 of heating surface would produce 34.5 lb of steam per hour, from and at 212°F. This was established as one boiler horsepower. But with such ratings boilers working under favorable conditions would operate at 150 to 200 percent of rating.

In contrast, today ratings of continuous maximum boiler output mean exactly what they say. This is of great importance when an older boiler is replaced by a packaged unit. Thus, instead of using the rating of the old boiler, we must estimate actual steam loads to be met, then select the capacity of the new unit to meet these needs. If we do not, the packaged boiler will be forced beyond its rating, and rapid tube deterioration, tube-sheet cracking, weld deterioration, and general heavy accelerated wear will take place, even though the waterside is kept clean from impurities and operated with care. Under forced conditions, the danger of low water is also increased, as the safety margin has been narrowed down.

Q How does the more efficient use of heating surfaces in packaged boilers affect the boiler operator?

A One result of working heat-exchanger surfaces harder has been the necessity of keeping both fireside and waterside clean. Thus good water treatment and maintenance are important. Compact designs also tend to make surfaces less accessible for inspection and cleaning. Thus today's higher heat-transfer rates can easily cause overheating, especially if forced. This results in loose tubes in tube sheets, cracks between ligaments on tube sheets, weld cracks in high-heat zones, bulged furnaces, and low water. Even more dangerous is the complete reliance on automatic controls to safely cycle a boiler, without periodically checking the controls for (1) conditions of electrical contacts, (2) electrical connections, (3) water column connections, (4) waterside plugging of pressure switches, (5) low-water fuel cutoffs, (6) soot accumulation in tubes, (7) operation of solenoid valves in fuel cutoff lines, (8) firing-equipment timing and operation of flame-failure devices, and (9) operation of safety valves.

But being automatic does not mean that everything has been designed into the unit, including self-maintenance. Remember that more highly skilled operators are required for the packaged, automatic units and that more knowledge is required on how the controls and safety devices function. Many boilers have lost days of service because untrained personnel did not even know what control was malfunctioning, so they could not keep the boiler in operation.

Boiler accidents are frequently caused by someone manipulating the controls without knowing the dangerous effects this can produce. The blocked-in relay, shorted-out control and jumper wires, rapid introduction of fresh cold water into a boiler, all are caused by unskilled operators. In contrast, properly trained operators quickly diagnose the malfunctioning boiler components and thus contribute to boiler safety and dependability on packaged, automatic boilers as they did on older, manual-operated units. By careful maintenance and operation, the efficiency of modern units will also be kept at a peak.

SM BOILER SIZE

Q What is the most prominent boiler used in smaller steam systems firing oil/gas or potentially solid fuels?

A Fire-tube boilers of either the scotch marine or firebox type serve in most industrial plants where saturated-steam demand is less than about 50,000 lb/h and pressure requirements are less than 250 psig. The general arrangement of these boilers has changed little in the past 25 years, although size has increased.

The largest scotch marine boiler offered by a U.S. manufacturer is rated 2000 hp, or 69,000 lb/h, using 34.5 lb/boiler hp. It has two combustion chambers in a single shell, measuring about 13 ft in diameter and more than 30 ft long. Boilers with single combustion chambers come in sizes up to about 1200 hp, or 41,400 lb/h.

Scotch marine boilers generally are specified only when liquid and gaseous fuels are burned, and they usually are cheaper than comparable water-tube boilers without economizers in sizes up to about 40,000 lb/h. Figure 2.15 shows a proposed modified SM boiler design for burning pulverized fuels.

Other advantages over water-tube boilers include easier tube replacement, minimal headroom required, and low susceptibility to cold-end and casing corrosion.

Older firebox-design boilers offer a larger ratio of furnace volume for heat-transfer surface than scotch marine boilers, making them more suitable for burning solid fuels.

> EXAMPLE: The economic boiler has a combustion chamber—or firebox—with a flat bottom and vertical sides going into an arched crown sheet that forms the top.

The furnace, which requires stays for structural support, is cooled at the sides and back by water legs between it and the outer casing. Fire tubes heat water in the boiler while conveying combustion gases from the furnace to the stack, as they do in scotch marine units. Combustion gases normally make three passes through the boiler before being discharged.

SM BOILER SPECIFICATIONS

Q What are typical boiler specifications for a packaged scotch marine boiler?

A Good examples are the specifications of the four-pass unit in Fig. 2.16 offered by the prominent manufacturer Cleaver Brooks. The unit is preassembled for immediate mounting on the floor or on a simple foundation and is ready for attachment to water, steam, fuel, blowoff, electrical, and vent connections. After the boiler-burner test, fittings which cannot remain mounted during shipment are removed at union connections and match-marked for simple reassembly.

The boiler shell is furnished with an adequate number of handholes to faciliate boiler inspection and cleaning. All the necessary tappings for blowoff, boiler feed, injector, and pop valves are provided in accordance with code requirements.

Cleanout openings furnish full access to all fire surfaces. Flue-gas extensions are lined with high-grade refractory and high-temperature insulating material. Front and rear heads are of bolted flange design and sealed with asbestos locomotive head gaskets. The entire tube area is accessible for cleaning from the front or rear.

The vent outlet includes sight doors at the front and rear for observation of stack conditions. Observation ports—heat-resistant ports for the observation of combustion conditions—are provided at each end

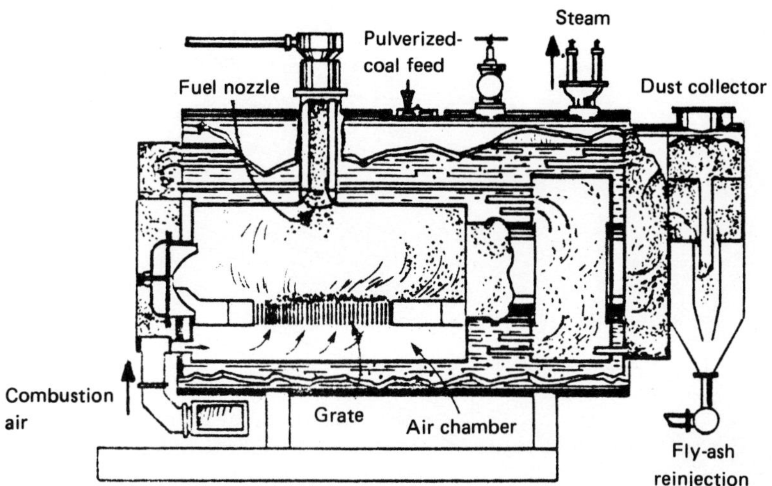

Fig. 2.15 Wet-back SM boiler modified for pulverized solid fuel burning. Note stationary grates above lower furnace surface. These are primarily used to blow air into furnace for combustion and draft. (*Courtesy of Power magazine.*)

of the furnace. The entire boiler assembly and frame receive one coat of machinery paint.

Boiler fittings include the following: (1) Water column—complete with gage, three trycocks, blowdown valve, test valve, and vacuum breaker. (2) Low-water cutoff—mounted on or cross-connected to water column. Control is wired in burner control circuit to prevent burner operation with boiler water below safe level. (3) Boiler feed control—included as an integral part of primary low-water cutoff control mounted on or cross-connected to water column. It automatically actuates a motor-driven boiler feed pump to maintain boiler water level within normal high and low limits. The pump control circuit has a momentary contact switch for manual operation of the boiler feed pump. (The boiler feed pump is *not* included with the boiler unless specifically mentioned in the proposal.) (4) Steam pressure gage—located at the front of the boiler shell, this includes a siphon and cock. (5) Pop safety valves—one or more valves of a type and size to comply with code specifications. (6) Injector—mounted on the side of the boiler within full view of the water column. All necessary valves are provided.

For burning commercial No. 2 oil, or lighter, the following equipment is furnished installed or connected on the steam boiler to form one integral unit unless specifically noted: (1) Burner—pressure-atomizing-type oil burner with full automatic electric ignition and motor-driven blower. The burner is suitable for operation with oil having a viscosity

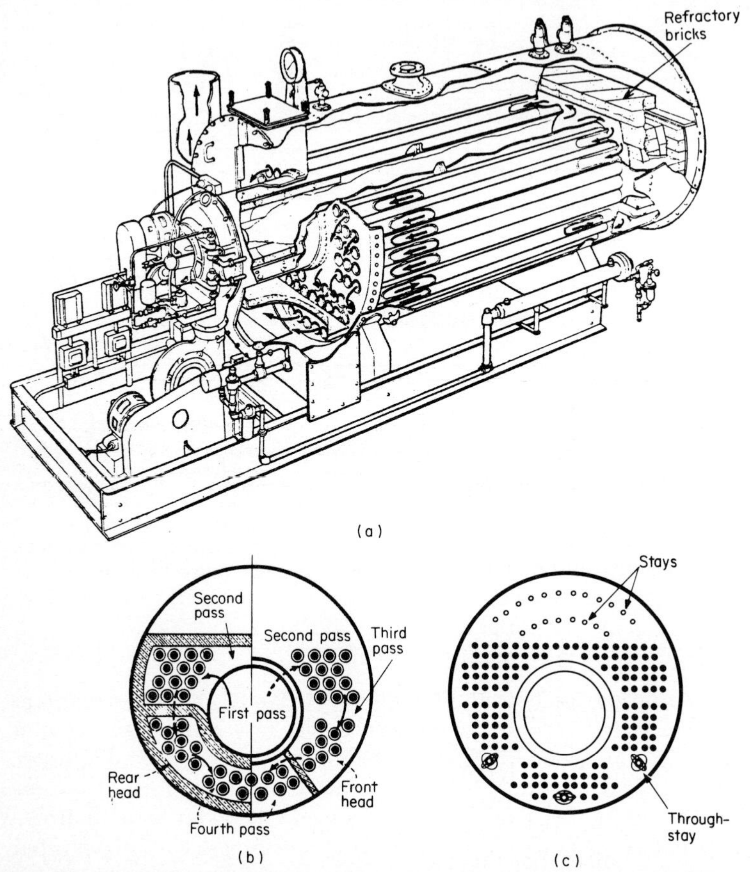

Fig. 2.16 (*a*) Details of four-pass, dry-back, packaged SM boiler; (*b*) left sketch shows gas flow as directed by baffles with right side being front of boiler and left side being rear. Sketch on right side shows tube spacing in passes for cleaning, inspection, and maintenance. (*Courtesy of Power magazine.*)

not exceeding 50 seconds Saybolt Universal at 100°F, operating on the on-off principle. (2) Blower—combustion air supplied by motor-driven blower mounted integrally on the unit, complete with V-belt drive with full safety guard. The air damper is manually adjustable to provide proper air/fuel ratio for full on-and-off firing. (3) Steam-pressure controls—mounted at front of boiler to actuate burner controller. (4) Oil pump—capacity equal to at least twice maximum burning rate at 100 psi pressure. Pump connected to blower shaft by V-belt drive. (5) Burner and flame-failure controller—an electronic programming relay actuated by the steam-pressure controls provides proper cycling of start, stop, ignition, and safety shutdown. A photoelectric flame detector is used in connection with the electronic relay to provide immediate shutdown in the event of flame failure. (6) Ignition transformer—10,000-V secondary with midpoint grounded. (7) Fuel-line piping system—consists of terminal block for suction and return line, check valve, fuel-oil suction strainer, pressure regulator, pressure gage, solenoid oil valve, and fuel-oil shutoff valve all integrally mounted on the unit. (8) Control panel—mounted on boiler frame—includes motor starter, burner sequence controls, and burner control switch fully wired in conformance with national wiring codes and ready for electrical service connections. Special cold-heat-, and moisture-resistant wiring materials are used throughout. Similar automatic features are provided for burning heavier oils and for gas-fired units.

VERTICAL TUBULAR BOILERS

Q What is a vertical tubular boiler?
A The vertical tubular (VT) boiler is an internally fired FT unit. It is a self-contained unit requiring little or no brickwork. Requiring little floor space, it is popular for portable service, such as on cranes, pile drivers, hoisting engines, and similar construction equipment. Vertical tubular boilers are used for stationary service where moderate pressures and capacity are required for process work, such as pressing, drying roll applications in various small laundries, and in the plastics industry.

The coil-type WT boiler (described in later chapters) is a competitor of the VT boiler for small capacities and lower pressures to 150 psi. But the VT boiler is limited in capacity and pressure even more than the horizontal FT boiler. For this reason, most VT boilers of the FT type seldom exceed 300 hp, or about 10,000 lb/h capacity with a maximum pressure of 200 psi.

Q What are the most common vertical FT boilers?
A There are five general classifications:

1. Standard straight-shell type with dry top (Fig. 2.17*a*).
2. Straight-shell type with wet top (Fig. 2.17*b*).
3. Manning boiler with enlarged firebox.
4. Tapered course bottom with enlarged firebox.
5. For even smaller capacity, the vertical tubeless unit (Fig. 2.18).

Q Describe the vertical dry-top boiler.
A The vertical, internally fired, FT boiler (Fig. 2.17*a*) consists of a vertical cylindrical shell which contains a cylindrical firebox and a number of small fire tubes. Heat radiated from the fire passes through

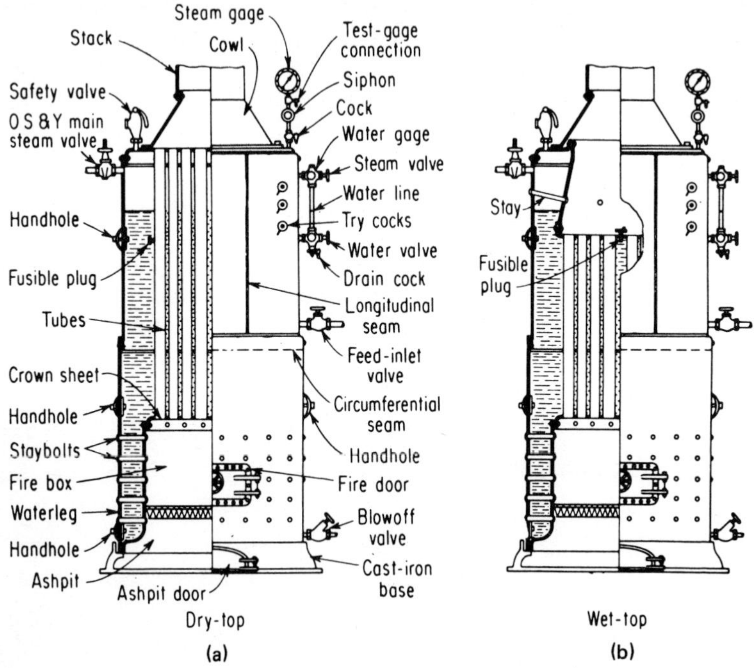

Fig. 2.17 Vertical tubular boiler details. (*a*) Dry top; (*b*) wet top has submerged head.

the firebox plates to the water in the boiler. The hot gases pass upward through the fire tubes to the smoke stack, giving up part of their heat to the metal of the tubes, which in turn transfer the heat to the water inside the boiler.

The upper tube plate forms the top head of the boiler. As the water level is carried about two-thirds of the length of the tubes above the lower tube sheet, or down sheet, the top tube sheet is dry (steam on one side, hot gases on the other side)—hence the name dry top. The main drawback is that tubes may become overheated above the waterline. Then they loosen in the tube sheet. But one advantage is that the upper third of the tubes being dry, they give a slight amount of superheat to the steam.

Handholes and cleanout plugs are provided at convenient points for washing out and inspection. In the smaller sizes, the bottom of the shell rests on a cast-iron base that also forms the ashpit.

The vertical flueless type (Fig. 2.18) is popular for low-capacity steam of up to about 150 psi. There are several designs. Instead of tubes, the water is either in a jacket around the furnace (Fig. 2.18) or in a water-

and-steam cylinder inside the furnace. On one type, reversal of gas flow takes place between openings on the jacket connected to the upper cylinder. Some designs have down-firing. There are no stay bolts or stays, as flat-surface widths are kept narrow or small in area. Thus the boiler plate thickness is adequate for resisting the bending stresses. Numerous solid U-shaped heat conductors are welded to the surface of the shell to provide better heat transfer.

Q Describe the vertical wet-top, or submerged-top boiler.

A This boiler is similar in construction to the dry-top vertical, except for the top head. In the wet-top boiler (Fig. 2.17*b*), a conical-shaped

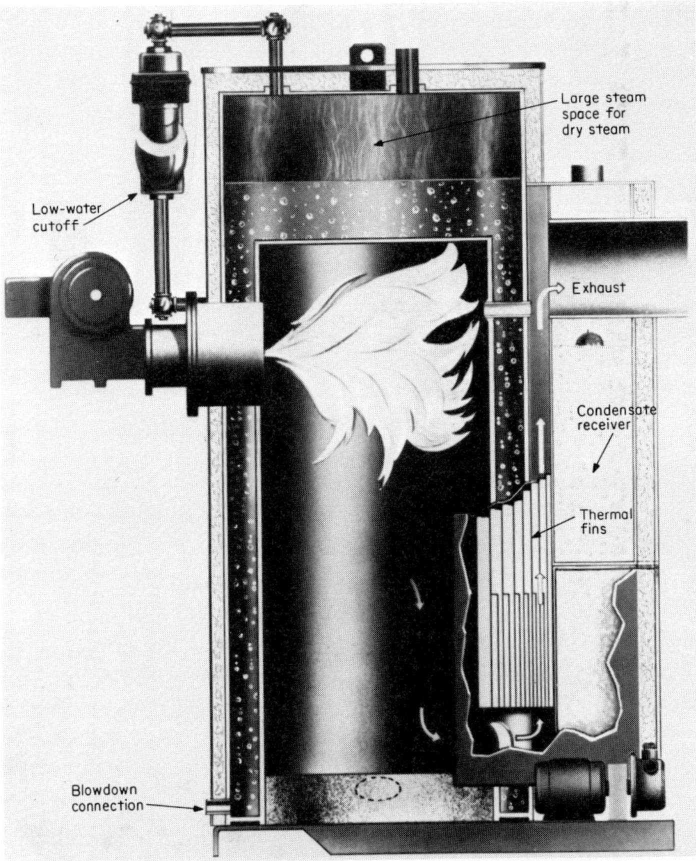

Fig. 2.18 Tubeless boiler with built-in condensate receiver and pump burns oil or gas or is combination fired. (*Courtesy of York-Shipley, Inc.*)

plate is welded (in modern boilers) to the top head, and the upper tube sheet forms the bottom of this cone. Thus the waterline can be carried above the upper tube sheet so that it and all the tubes are below the water level. Other construction details are the same as in the dry-top boiler.

Q What are the advantages and disadvantages of VT FT boilers?
A Advantages are (1) compactness and portability, (2) low first cost, (3) very little floor space required per boiler horsepower, (4) no special setting required, and (5) quick and simpler installation.

Disadvantages are that (1) the interior is not easily accessible for cleaning, inspection, or repair; (2) the water capacity is small, making it difficult to keep a steady steam pressure under varying load; (3) the boiler is liable to prime (carry over with steam) when under heavy load because of the small steam space; and (4) the efficiency is low in smaller sizes because hot gases have a short, direct path to the stack and thus much of the heat goes to waste.

SAFE WATER LEVEL

Q What is the safe water level in the dry-top VT boiler?
A The code indicates no specific water level for this type of boiler, except to state: "It shall be at a level at which there shall be no danger of overheating any part of the boiler when in operation at that level." But the level is generally taken as a minimum at a point two-thirds the height of the shell, above the bottom head, or crown sheet or tube sheet. This is the same minimum requirement as for miniature vertical FT boilers.

For the submerged wet-top type, the minimum water level must be at least 2 in above the top tube ends, except for miniature boilers, where it can be 1 in.

Q Where does sludge usually settle in a VT boiler?
A Deposits settle in the 4-in-wide water legs (maximum per code) because they have restricted circulation. Cleanout openings should be periodically opened around the circumference in the water legs and bottom tube sheet so all areas are accessible for cleaning. Some units have a continuous drain in the bottom of the water legs. When this boiler is opened for cleaning, a length of chain can be pulled around inside to get the sludge and scale to a cleanout opening for removal.

Q In calculating furnaces for external pressure (especially for VT boilers), the terms *tubes*, *flues*, and *furnaces* are often used. Is there a difference?

A Yes. Tubes are hollow cylindrical elements of up to 5 in in diameter. Flues are hollow cylinders exceeding 5 in in outside diameter and used for the conveyance of gases with temperatures of 850°F or less. Furnaces are hollow cylinders exceeding 5 in in outside diameter in which combustion takes place, or which are used for the conveyance of gases having temperatures exceeding 850°F.

Q Where does internal and external corrosion usually take place in a VT boiler?

A Internal corrosion:

1. At the waterline, usually on the tubes. This is due to oxygen and organic materials being released there during the process of boiling.
2. In the vicinity of the feedwater discharge, as a result of oxygen release.
3. On top of the lower tube sheet, because of scale formation.
4. On and around the ends of the stay bolts, as a result of the stresses imposed and the subsequent expansion and contraction (this leads to stress corrosion).
5. In the water legs, especially at the bottom, where grooving may occur in addition to the usual pitting under the scale.

External corrosion:

1. On top of the top tube sheet, bottom tube sheet, and tube ends, because of acid formed by the damp soot in contact with the products of combustion.
2. Around all access hole, handhole, and washout openings. Here corrosion is from leakage due to poor gaskets, improper handhole cover installation, and thermal expansion and contraction which loosens opening closures.
3. At the bottom of shell, water legs, and furnace sheet, as a result of soot attack.
4. Around all openings including gage connections, safety-valve connections, steam-outlet connection, feedwater connection, blowdown connection, and water-column connection, because of leakage.

INSPECTION DETAILS

Q How are designers attempting to improve access for the cleaning of compact, packaged boilers?

A The highly compact design of packaged boilers, as a class, makes access to internal heating surfaces, especially the waterside, very difficult to provide. So designers are trying many variations in design and construction to increase the convenience of access for cleaning, inspection, and maintenance. Typical of these efforts is the hinged-door

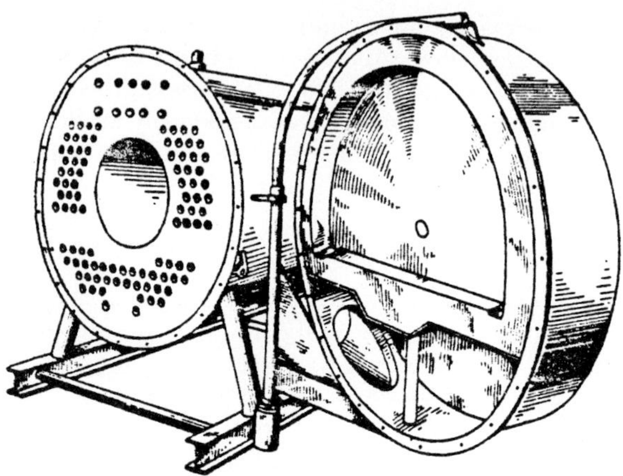

Fig. 2.19 Rear swinging head on dry-type SM boiler permits easier tube cleaning of soot deposits.

back construction (Fig. 2.19) in which the entire rear chamber swings back from the rear tube sheet to give access for inspection and cleaning.

Handholes are also conveniently located in the heads, tube sheets, or shells. But this creates a gasket-leakage maintenance problem. On SM boilers, the ASME code requires the following inspection openings: handholes—at least four, one on the bottom of the shell, one above the furnace, one on each side of the first head or shell below the tubes, and one near the front ends.

Table 2.1 Minimum Thickness for Shell and Tube Sheets for Various Diameters of Boilers in Inches

Inside diameter of shell	High pressure		Low pressure	
	Shell	Tube sheet	Shell	Tube sheet
36 in and under	1/4	3/8	1/4*	5/16
Over 36 in to 42 in	5/16	3/8	1/4	5/16
Over 42 in to 54 in	5/16	7/16	5/16	3/8
Over 54 in to 60 in	3/8	1/2	5/16	3/8
Over 60 in to 72 in	3/8	1/2	3/8	7/16
Over 72 in to 78 in	1/2	9/16	3/8	7/16
Over 78 in	1/2	9/16	7/16	1/2

Note: The above thicknesses are minimum requirements regardless of pressure and are based strictly on shell diameters of the fire-tube boiler involved. Later chapters will show how minimum thicknesses may be determined for different pressures.

* For low-pressure service, with shell diameters 24 in or less, and for a maximum allowable working pressure of 30 psi, shell plate 3/16 in in thickness may be used.

An access hole is mandatory on the shell, or head above the tubes, if the boiler has a diameter 48 in and over. Below this diameter, a handhold may be substituted.

Scotch marine boilers of the wet-back design must also have an inspection opening to the water space at the rear of the combustion chamber.

There are similar inspection and washout opening requirments for HRT boilers, vertical boilers, and similar fire-tube units. The intent of the code requirements is to make accessible for inspection the shell and stayed surfaces exposed to heat from the furnace, areas with sluggish circulation, or pockets where mud and sludge can accumulate, such as on the bottom of water legs. On low-pressure HRT boilers an access hole is required in the front head below the tubes if the boiler shell is 60 in in diameter and over. An access hole is required *above* the tubes on all horizontal fire-tube boilers over 60 in in diameter.

Q What are the minimum code thickness requirements for shell and tube sheets of fire-tube boilers made of ferrous plate materials?

A See Table 2.1.

CHAPTER **THREE**

Industrial and Utility Water-Tube Boilers for Process and Power Generation

To gain additional heating surface outside a shell as in a fire-tube boiler, water-tube boilers were developed in the late 1800s. The first units were simple front to back straight-tube types with headers connecting the tubes to drums. Later, bent-tube designs permitted greater expansion in capacity and pressure rating. This chapter will review the different WT boiler types and the characteristics of construction and also operation.

STEAM FORMATION IN WATER-TUBE DESIGNS

Q Figure 3.1 is a D-type two-drum vertical bent-tube boiler with water-cooled furnace and pressurized firing, widely used in industry. Explain the principles for steam formation in water-tube boilers.

A As shown in Fig. 3.2*a* in simple water-tube circuits, steam forms in the heated side of the tube. The resulting steam-water mixture weighs less than the cold water on the unheated side and is therefore displaced. In the drum at the top, steam bubbles rise to the water's surface, releasing steam.

In Fig. 3.2*b*, circulation is shown in the straight-tube boiler design. The basic idea of water-tube circulation applies in the straight-tube boiler, but the downcomer tubes and rear header are not unheated. Design aims to hold heating to the amount needed to bring water to saturation temperature without steam formation.

In Fig. 3.2*c* the steam formation and circulation are shown for a three-drum bent-tube boiler. Circulation becomes more complex in bent-tube multidrum designs. The rear tube bank acts as a downcomer to the mud drum; most steaming occurs in inclined-tube banks. The steam-water mixture enters the steam drum, both above and below the water's surface.

Fig. 3.1 Water-tube steam generator widely used in industry has water-cooled furnace which is pressurized. (*Courtesy of Combustion Engineering, Inc.*)

Although circulation circuits of bent-tube boilers appear complex, the general principle of the simple water-tube circuit holds. Here two additional drums, in effect, take the place of the headers of the straight-tube design. The top drum serves for steam release and feedwater admission. Blowdown connections for removal of suspended solids are generally made at the system low point, in this case the bottom drum.

Q What is meant by DNB when discussing circulation?
A DNB is an abbreviation of departure from nucleate (having a nucleus, or cluster, as steam bubbles) boiling. The DNB phenomenon is considered to occur when circulation in a bank of tubes is not adequate; it involves not matching properly the flow of combustion energy across the tubes with the water and steam flow in the tubes. As a result, tubes can overheat even though normal water levels are maintained in the boiler.

Q Explain how forced circulation assists in preventing DNB.
A Figure 3.3*a* shows that as pressures increase, pumps are used to promote circulation of water through the tubes and thus avoid the possibility of local tube-water starvation due to a mismatch of mass flows between input heat and absorbent heat surfaces. In Fig. 3.3*b* the

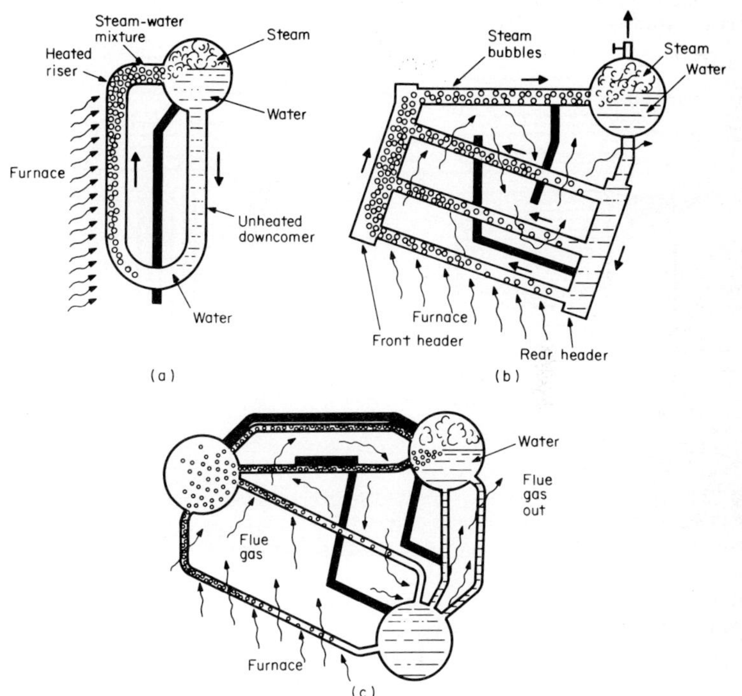

Fig. 3.2 Steam formation and natural circulation in water-tube boilers. (*a*) Simple water-tube circuit depends on gravity and change in water density for circulation. (*b*) Straight-tube header-type boiler has water going down rear header; steam bubbles up on front header. (*c*) Bent-tube design counts on downcomer pipes or tubes to promote circulation.

once-through boiler has no recirculation; a subcritical pressure unit uses a steam separator.

Q What is meant by controlled circulation in a large steam generator?
A Controlled circulation is a design by which water can be carefully apportioned to furnace walls, boiler-tube sections, parallel tubes or tube, in accurate, predetermined amounts. Water can even be changed in total flow or distribution within a certain range at any subsequent time in operation. This is usually done by installing circulating pumps between the boiler drum and water inlet to the heat-absorbing surfaces. The result is positive flow in one direction at all times, regardless of heat application.

With forced circulation, small-diameter, thin-walled tubes can be used where natural circulation design would not be possible because of

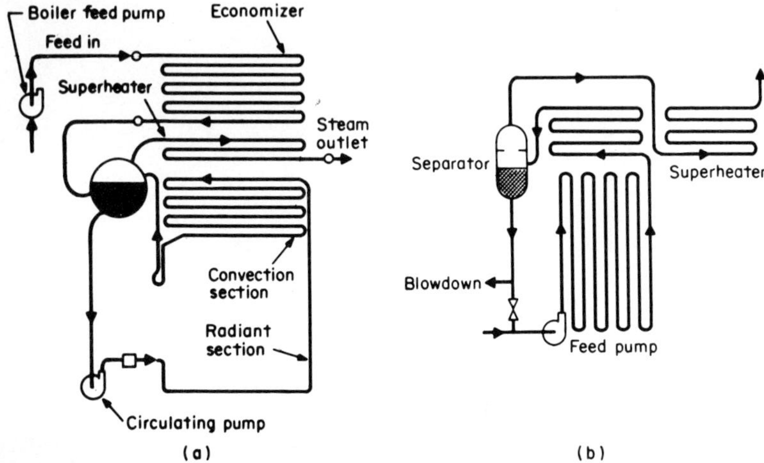

Fig. 3.3 (*a*) Forced-circulation boilers use a circulating pump in furnace-water circulation loop to assure flow and to avoid DNB. (*b*) Once-through boiler has no recirculation; subcritical pressure unit uses steam separator.

the high temperature the natural circulation may have to operate against. This is especially true of large-flow and high-pressure units (supercritical) and once-through designs.

Q What is the trend in natural-circulation boilers to prevent tube-water starvation or DNB?
A Internal ribbed tubes are being used instead of smooth tubes. The internal ribbing creates a centrifugal action that directs the water droplets in the tubes toward the tube surface, thus preventing the formation of a steam film or nucleate-boiling steam barrier.

Q What is meant by once-through design?
A Figure 3.4 is a once-through design; feedwater leaves the tubes as steam. Shown is the circulation for a subcritical pressure unit which employs a separator to remove any moisture in the steam.

NOTE: No recirculation exists in the once-through design.

SUPERCRITICAL DESIGNS

Q Explain the term supercritical boiler.
A A *supercritical boiler* operates above the supercritical pressure of 3206.2 pounds per square inch absolute (psia) and 705.4°F saturation temperature. Steam and water have a critical pressure at 3206.2 psia. At this pressure, steam and water are at the same density, which means that the steam is compressed as tightly as the water. When this mixture is

heated above the corresponding saturation temperature of 705.4°F for this pressure, dry, superheated steam is produced to do useful high-pressure work. This dry steam is especially well suited for driving turbine-generators.

Supercritical-pressure boilers are of two types: once-through and recirculation. Both types operate in the supercritical range above 3206.2 psia and 705.4°F. In this range the properties of the saturated liquid and saturated vapor are identical; there is no change in the liquid-vapor phase, and therefore no water level exists, thus requiring no steam drum as such.

Q What are the chief design and safety differences between the WT boiler and FT boiler?

A In contrast to the FT design, WT-type boilers (Fig. 3.5) feature one or more relatively small drums with a multiplicity of tubes in which a water-steam mixture circulates. Heat flows from outside the tubes in this mixture. This subdivision of pressure parts makes possible large capacities and high pressures.

The WT boiler is safer, largely because most of the water at the hottest part of the furnace is in small components (tubes). Thus if a tube ruptures, only a comparatively small volume of water is instantly released to flash into steam. As a rule, all parts of the WT boiler are more accessible for cleaning, inspection, and repairs. But very long, bent tubes may be difficult to clean on some designs. Water-tube boilers in the larger sizes are faster steamers because of their large heating surface, long gas travel, and rapid and positive water circulation. On small coil-type WT boilers, there may be no difference in steaming time. For the same reasons, larger WT boilers can carry much greater overloads and respond more readily to sudden changes and fluctuations

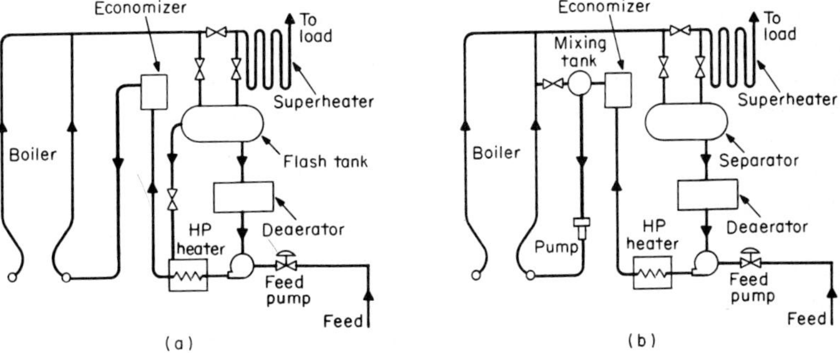

Fig. 3.4 Two supercritical boiler systems. (*a*) Bypass flash tank is used for low-load and start-up operation; (*b*) a circulating pump is used for start-up and low-load operation.

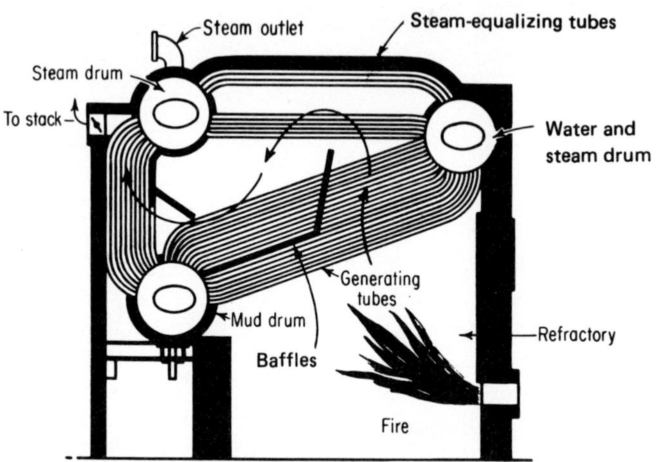

Fig. 3.5 Flue-gas flow in bent-tube water-tube boiler.

in demand. Also, the drum in WT boilers is not exposed to the radiant heat of the fire.

The biggest advantage over FT boilers is the freedom to increase the capacities and pressures. That is impossible with FT boilers because the thick shells and other structural requirements become prohibitive over 50,000 lb/h capacity and over 300 psi. The large capacities and pressures of the WT boiler have made possible the modern large utility-type steam generators.

TUBE DIAMETERS

Q Is it true that for the same diameter and thickness of tube, a WT boiler has more heating surface than an FT type?

A According to the ASME code, the heating surface is always calculated from the side exposed to the products of combustion. And since in a WT boiler the products of combustion are on the outside of the tube, the outside diameter (OD) of the tube is used in calculating the heating surface. In an FT, the products of combustion are on the inside of the tube; thus the inside diameter (ID) is used in calculating the heating surface. Since the OD is larger than the ID, it is assumed that a WT has a larger heating surface area than an FT (Fig. 3.6).

But heating surface depends on other factors such as tube arrangement, length of gas travel, use of efficient burning equipment, and cleanliness and maintainability of heat-exchanger surfaces. So don't assume that WT boilers are always more efficient than FT boilers.

Q What materials are used in boiler tubes?

A Large modern boilers consist of thousands of feet of steel tubes of various diameters and wall thicknesses; so the type of metal used is of great importance. Low-carbon steel is used in most WT boilers operating over a wide temperature range. Convection temperatures run between 500 and 700°F. Medium-carbon steel, with 0.35 percent maximum of carbon, permits higher stress levels than low-carbon steel at temperatures up to 950°F. For superheater tubes, which must resist temperatures above 950°F, alloy steels are required. These may contain chromium, chromium-molybdenum, and chromium-nickel. They may be of ferritic structure, or, for higher furnace temperatures, of an austenitic structure, or stainless-steel type.

Q How are water tubes attached in a WT boiler?

A For high-pressure service, the ASME code stipulates the following attachment of WT tubes to tube sheets, headers, and drains: (1) expanded or rolled and flared (Fig. 3.7); (2) flared not less than ⅛ in, rolled, and beaded; (3) flared, rolled, and welded; (4) rolled and seal-welded, provided the throat of the seal weld is not more than ⅜ in and the tubes

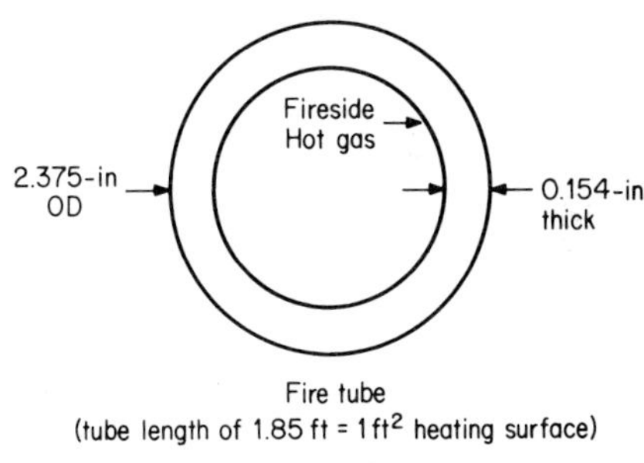

Fire tube
(tube length of 1.85 ft = 1 ft^2 heating surface)

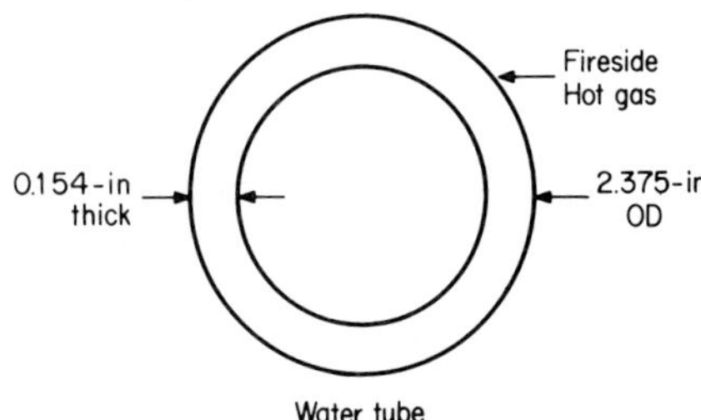

Water tube
(tube length of 1.61 ft = 1 ft^2 heating surface)

Fig. 3.6 Comparing heating surfaces of FT with WT for tubes of same diameter.

are rerolled after welding; (5) superheater, reheater, waterwall, or economizer tubes may be welded without rolling or flaring, provided that the welds are heat-treated after welding and the welding is done according to code requirements.

Q The term *tube sheet* is often applied to a WT boiler equipped with only steam drums and mud drums. Is not this term incorrect as applied to WT boilers?

A No. The term *tube sheet* refers to any sheet of a WT boiler where tubes are inserted, whether in a flat sheet or a drum.

Q How is the term *ligament* defined, and how does this term affect WT boiler design?

A A ligament is the section of solid plate between tube holes. But the drilling of these holes in a pattern weakens the solid plate. Thus in designing or calculating the strength of a boiler drum, the ligament efficiency must be considered in determining the safe working pressure. And the ligament efficiency will always be in reference to tube hole arrangement, whether a steam drum or mud drum. The term *tube sheet* may be used, even though a tube sheet such as on FT boilers does not actually exist. In WT boiler drum design, the weld or rivet and the ligament tube efficiency of the drum are very crucial in establishing

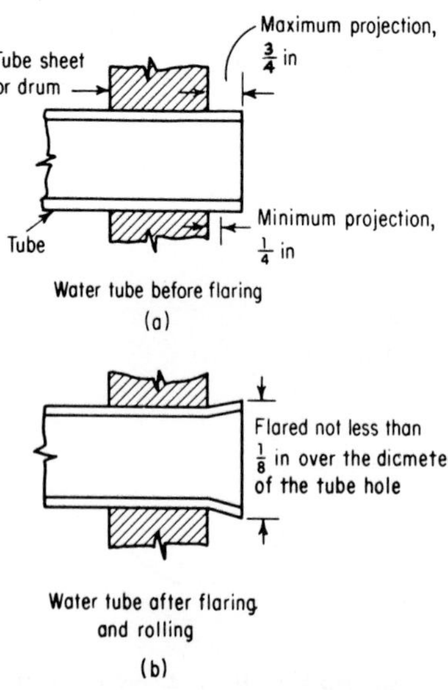

Fig. 3.7 Methods of attaching water tubes.

plate thickness required for a given pressure with a given material (see Chap. 9).

Q Why are tubes flared in a WT boiler?
A To add to the holding power of the tubes after rolling, and also to prevent the tubes from pulling out of the tube holes if the holes should become enlarged from overheating caused by low water or other reasons.

Q What precautions should be used when rolling or expanding tubes in WT boilers?
A Most tubes today are expanded with power-driven expanders instead of by hand rolling. This results in rolling just enough to obtain tight joints. But with power-driven rollers, there is always the danger of overrolling, thus damaging the metal by cold-working, which causes the metal to become harder and more brittle. It may also result in loss of ductility and ability to stretch and contract with temperature changes. Cold-working often causes surface tears and marks which will reduce resistance to corrosion. Thus it is better to underroll, whether on new or repair work. One aid to properly rolling a new tube is to note when the mill scale (or paint) on the tube sheet around the tube begins to flake off. Stop rolling at this time, as the joint is tight enough.

Before tubes are inserted for rolling, all tube ends and tube seats should be cleaned of oil, grease, rust, etc. Rust should be removed with a fine abrasive cloth. Expanders should be kept clean and washed frequently with kerosene. Vegetable oil or other specially prepared expander lubricant should be used to lubricate the expander rolls. One end of the tube should be held in place with a wedge while being expanded so it does not move in the seat. The tubes should enter the holes parallel to the centerline of the holes and have equal projection from the tube sheet or header at each side of the tube sheet.

The bottom tubes should always be rolled in first, then the next row upward so that oil, dirt, etc., do not drip off the ends of the tubes into empty tube seats not yet filled with tubes. Be sure to use the correct expander to suit the size and thickness of the tubes and tube seats.

STRAIGHT-TUBE WT BOILERS

Q What was the first widely used WT boiler?
A Figure 3.8 shows an early straight-tube WT boiler. The tubes are placed in the furnace, and the shell above is used primarily as a storage tank for water and steam. Circulation from the drum is down the back

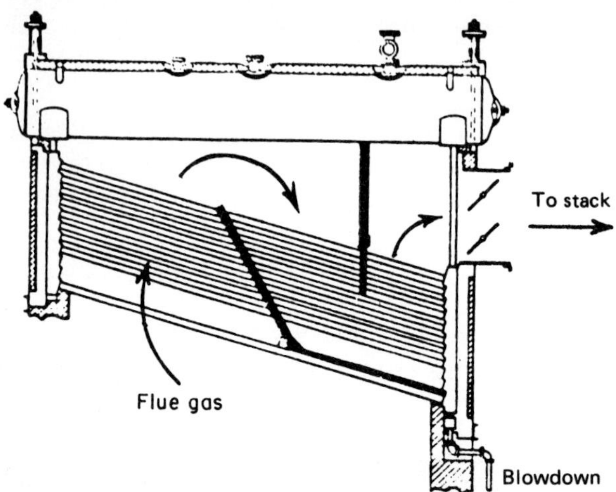

Fig. 3.8 Early sectional-header water-tube boiler.

headers, through the water tubes, and up through the front header. With this arrangement the tubes on boilers began to be separated from the internal shell, in contrast to FT boilers. The design shown has one drum; larger boilers had two or three drums. The drum runs from the front to the rear of the boiler. The inclined straight steel tubes, usually of 4-in OD, are connected with the drum by pressed-steel headers of the sectional type, the tubes being staggered in pairs. The mud drum below the rear headers collects sediment and is blown out from time to time. Tube headers are in one piece for each vertical row of tubes.

Header handholes (for tube cleaning) are closed by bolted covers with machined joints.

Q Name the various types of straight-tube WT boilers.
A The straight-tube design is usually classified by the type of header and by the direction of the drums or the tubes. As a result there are, in general, six types: (1) sinuous- or sectional-header, (2) box-header, (3) longitudinal-drum, (4) cross-drum, (5) horizontal-tube, and (6) vertical-tube types.

Q The terms *sectional header* and *sinuous header* are often used with WT boilers. How do they differ?
A The sectional header, which is also referred to as a sinuous, or serpentine, header (Fig. 3.9*a*), is either a casting or forging. In older low-pressure boilers, the headers were also constructed of cast iron. But for high pressure, they were limited by the ASME code to a maximum

design pressure of 160 psi and 350°F. In contrast, headers constructed of forged steel have been used for pressures up to 1200 psi.

Q What is a box header?

A A box header (Fig. 3.9*b*) is constructed of flat plates, referred to as a tube sheet and tube cap sheet. But these surfaces must be stayed to prevent deformation. The sides, top, and bottom are flanged and riveted (welded on new boilers) to the tube sheets and cap sheets. The staying of sheets limits pressure for a box header to about 600 psi.

Q Describe the longitudinal-drum and the cross-drum WT boilers; name advantages and disadvantages of each.

A The longitudinal-drum type (Fig. 3.10*a*) has the drum, or drums, parallel to the inclined tubes and above the headers. In the longitudinal-drum box-header type, the water leg at the high end of the tubes has a flanged semicircular throat, which is welded directly to the drum (or drums). At the low end of the tubes, the rear box header is connected with the drum (or drums) by tubes expanded into the top of the headers and into a throat connection welded to the drum (or drums).

The cross-drum type (Fig. 3.10*b*) has tubes running from the upper part of the front header (or headers) to the drum, then entering the drum steam space. The rear header (or headers) is connected by tubes rolled into the tube holes in the lower (water space) part of the drum, and also into the tube holes in the top of the headers. The cross-drum type, in addition to being more economical to build, provides for better circulation of water.

Q How are longitudinal-drum WT boilers supported?

A They are suspended from crossbeams attached to the steam drum

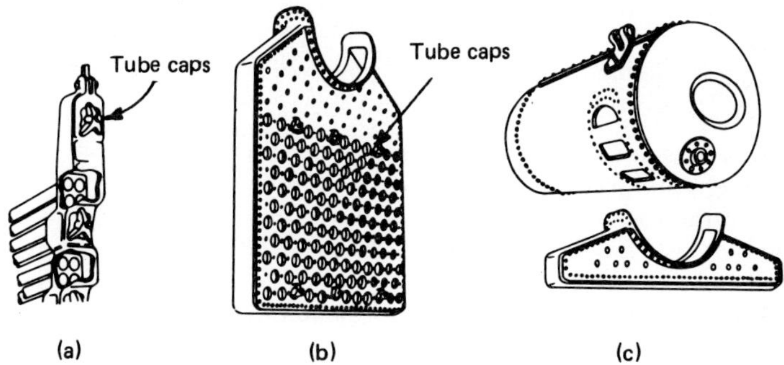

Fig. 3.9 Header-type boiler details. (*a*) Sinuous header; (*b*) box header; (*c*) drum fits into crossbox.

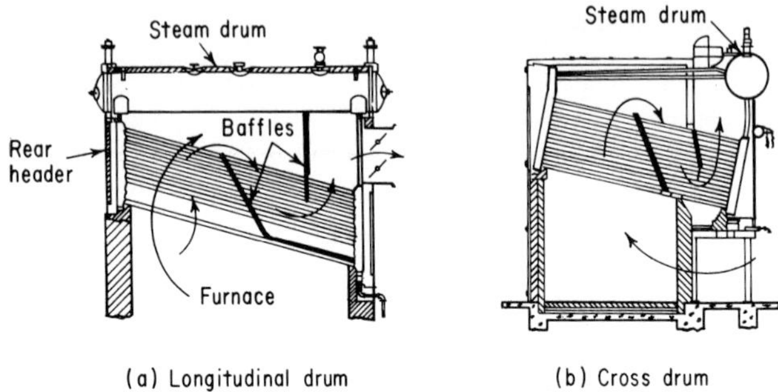

Fig. 3.10 (*a*) Longitudinal-drum and (*b*) cross-drum water-tube boilers.

and supported by steel columns. One method is the U-bolt support shown in Fig. 3.11. The channels are mounted on vertical steel columns, away from the high-temperature zones of the boiler drum setting and thus protected from heat.

CAUTION: If encased in brickwork (older boilers), periodically remove the brickwork to check for corrosion of the U bolts.

Q Where does most of the corrosion take place on box-header boilers?
A Heavy leakage at handhole plates causes corrosion on the wrapper or external sheet where the closing caps are located. Leakage of these handholes finds its way down to the bottom of the headers (Fig. 3.9), which is usually concealed in brickwork. Thus leakage from above causes undetected corrosion. When inspecting box-header boilers having bricked-in bottom headers, always remove the brickwork and check this surface for corrosion.

Q Where do sinuous-header boilers usually corrode?
A On the tube entrance outer plate, from leakage at the handholes, or on caps on the openings. Also between the sinuous headers, which are usually packed or filled with asbestos yarn.

Q Describe a vertical WT boiler and list its advantages.
A One common type of vertical WT boiler (Fig. 3.12) has a steam drum at the top and a mud drum at the bottom. The vertical tubes are rolled into the flat tube sheets of both drums. Sling stays brace the tube sheets. Obviously, the dished heads opposite the tube sheets require no staying. The main advantage of this type of boiler is that it requires less floor space per unit of capacity.

The Wickes boiler has a large number of vertical water tubes. The boiler is enclosed in a brick setting and supported by brackets fastened to the lower drum, which rest on steel or concrete pillars. This design

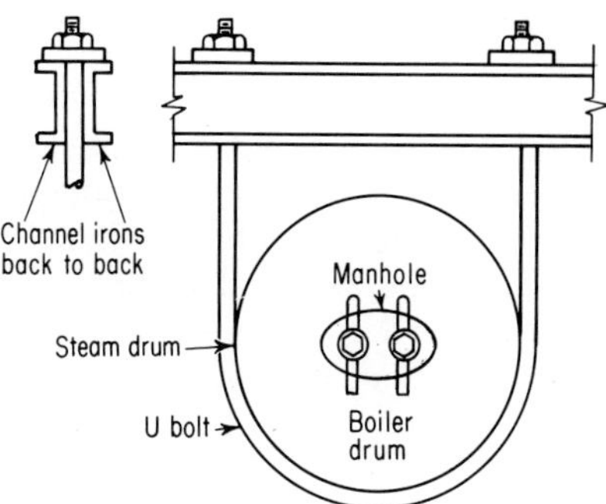

Fig. 3.11 U-bolt method of supporting longitudinal-drum water-tube boiler.

allows the boiler to expand and contract freely without greatly disturbing the setting. A vertical tile baffle in the center of the tube bank extends downward through the rear bank. The water circulation is in the same direction, up the front tubes and down the rear tubes. The primary combustion takes place in the furnace, which is called a dutch oven. Feedwater enters through the top steam drum, while blowdown is from the mud drum. Both top and bottom drums have access holes for tube replacement and for cleaning purposes. The access holes usually are flanged into the dished heads.

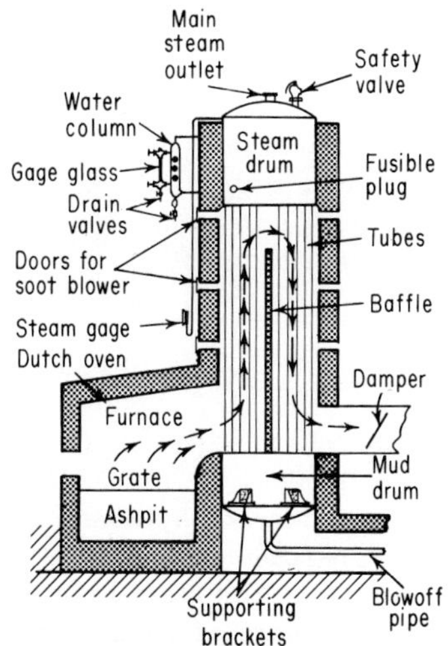

Fig. 3.12 Vertical straight-tube water-tube boiler has steam and mud drum.

CONSTRUCTION DETAILS

Q Why are baffles used in the passages of WT boilers?
A Baffles deflect the hot gases back and forth between the tubes a number of times to enable greater heat absorption by the boiler tubes. They also permit designing for better temperature differences between tubes and gases throughout the boiler. Baffles help maintain gas velocity, eliminate dead pockets, deposit fly ash and soot for proper removal, and prevent high draft losses.

Q What happens when a furnace baffle breaks?
A The gases short-circuit one or more passes, causing excessive flue-gas temperatures and a loss in efficiency and also capacity. Overheating and damage might result in those parts of the boiler designed for low gas temperatures. Thus on any outage inspection, the baffles should always be carefully checked for erosion, breaks, leakage (around tubes), or dislocation, as tube failure may result.

Q What type of firing door is requried on WT boilers?
A The inward-opening type, or a type provided with self-locking door latches which omit springs or friction contact. Reason? So the door cannot be blown open from pressure inside the furnace in case of tube rupture or furnace explosion, and thus possibly burn or scald personnel standing nearby.

> IMPORTANT: Explosion doors, if used, and if located in the setting walls within 7 ft of the firing floor or operating platform, must have deflectors to divert any blast.

Q What are some causes of tube failures in WT boilers?
A (1) Solid deposits, (2) low-water conditions, (3) corrosion, (4) slagging of gas passages which restrict normal heat transfer, (5) high concentration of heat in some tube areas, (6) stress corrosion, (7) flame impingement, (8) poor circulation, and (9) steam cutting and external erosion and corrosion by soot blowers that are improperly located or in poor condition.

BENT-TUBE WT BOILERS

Q Why are bent tubes used in WT boilers?
A Bent tubes are more flexible than straight tubes. Boilers can be made wide and low where headroom is limited, or narrow and high where floor space is at a premium. Also, bent-tube boilers allow more heating surface to be exposed to the radiant heat of the flame. Drums serve as convenient collecting points in the steam-water circuit and for

separation of steam and water. Thus boilers with two, three, and sometimes four drums have been used. As boilers grew in size (made possible by bent-tube design), the demand for more active furnace cooling increased. It was then that waterwalls and other improvements in design were made. A better knowledge of fluid dynamics resulted in simpler and much safer methods for the circulation of waterwall fluids, on both the gas and steam sides.

Two-drum boilers, even boilers with but one drum on top and one or two large headers at the bottom, became commonplace. Thousands of boilers of the vertical-header type and of multidrum bent-tube design are still operating today. The tubes are bent because:

1. Heat-transfer reasons make it impossible to use straight tubes.

2. The bent tube allows for free expansion and contraction of the assembly, usually on the lower mud-drum end, as the upper drum (or drums) is separated or suspended by steel structures.

3. The bent tubes enter the drum radially to allow many banks of tubes to enter the drum.

4. Bent tubes allow greater flexibility in boiler-tube arrangement than is possible in straight-tube boilers.

Q What was one of the first bent-tube boiler designs that is still in use?
A The Stirling boiler (Fig. 3.13), which has three steam drums and one mud drum. Steam-equalizing tubes are between the middle and rear steam drum and between the front and rear drums. Water-circulating tubes connect the front and middle steam drums below the waterline for the boiler. Rear tubes connecting the middle drum to the mud

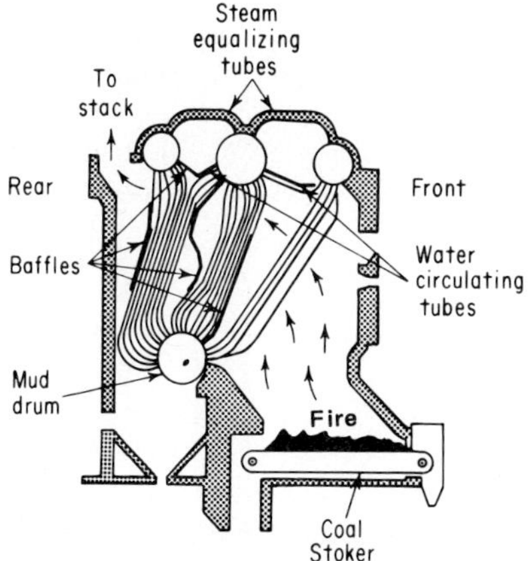

Fig. 3.13 Sterling WT boiler is older type, has bent tubes.

drum deviate from the middle bank to the generating tubes. They join in the rear bank of generating tubes on the back steam drum, to ensure positive circulation throughout the boiler.

Baffles protect the three steam drums the full width of the setting, thus protecting the longitudinal riveted seam, on each drum, from contact with flue gas. Feedwater enters the back top drum through a welded-in feed pipe. The water circulates downward through the back tubes to the mud drum and then up to the middle and front drums. The equalizing tubes keep the water level equal in these two drums.

Q In the three-drum boiler of the Stirling type, what areas have to be watched for possible sources of trouble?

A Figure 3.14 shows a unit with front and middle boiler banks

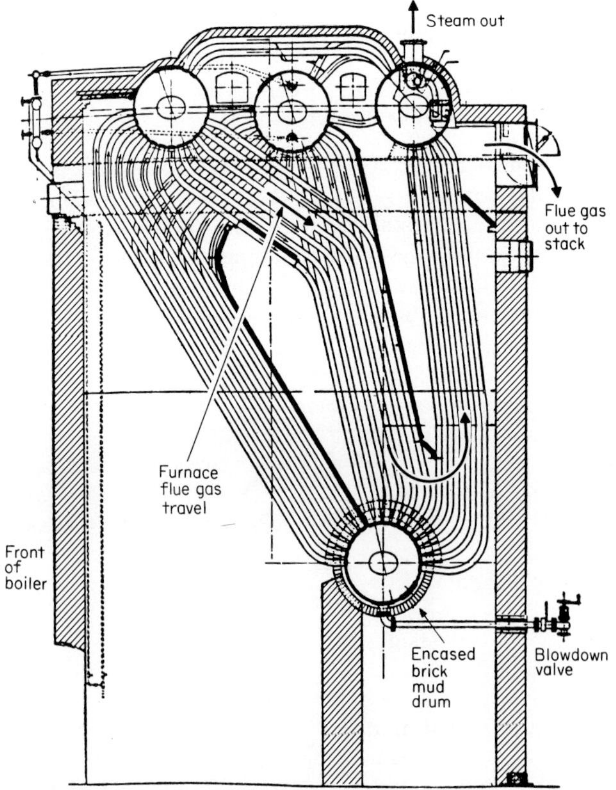

Fig. 3.14 Four-drum Stirling-type boiler with front boiler bank connected to front and middle drum for better circulation.

connected into both the front and middle upper drums. This arrangement equalizes the discharge of the steam-and-water mixture to improve circulation and reduce carryover with the steam. Boilers of this general type were usually designed for pressures from 160 to 1000 psi, and capacity range from 7500 to 350,000 lb/h steam. Both the top and bottom drums have brickwork built partly around them. Because the boiler code required the longitudinal joint to be away from furnace heat, the joint was usually under this brickwork. Even if it means removing some brickwork, always check the condition of longitudinal and circumferential rivets, including the calked edges, and look for caustic embrittlement affecting the drums.

Q What is meant by an integral-furnace boiler?

A The integral-furnace bent-tube WT boiler is so-called because the bottom drum (near the bottom of the furnace) makes it possible to package the furnace integrally with the boiler. Thus on older units only insulation and brickwork had to be erected at the site. Of the bent-tube boilers, the two-drum types are the simplest.

Today these two-drum boilers are available in standard sizes, ranging from 15,000 to 300,000 lb/h of steam with pressures from 160 to 1000 psi to 900°F temperatures. The demand for compact, standardized WT boilers is growing, with the two-drum bent-tube type a leader in the field. They are shop-assembled, equipped with a water-cooled furnace, and arranged for pressure-firing of either oil or gas. And they are ready for shipment by truck or railway car, completely equipped with firing equipment, oil pumps, oil heater, forced-draft fan, boiler trip, feedwater regulator, soot blowers, casing, setting, and automatic control system. Thus with pressure firing, good arrangement of heating surfaces, tight baffles, tight casing construction, and automatic combustion control, more efficient operation is possible.

Steam generators of the integral type arranged for pulverized coal, oil, or gas firing are built in sizes up to 425,000 lb/h steam. While pulverized fuel, oil, or gas are most frequently used, stokers can also be applied.

Modern packaged WT boilers have grown in popularity and size since their inception in the 1940s. Today, most packaged WT boilers follow one of the three designs shown in Fig. 3.15. These are known as A, D, and O types.

Figure 3.16 features some common bent-tube boilers with water and gas circulation shown in each. Note that blowdown is always from the bottom drum, where sediment gathers. Some of these boilers come equipped with automatic surface-blow equipment, which is different from blowdown (bottom-blow). Surface blow refers to blowing the floating scum from the waterline level in the steam drums.

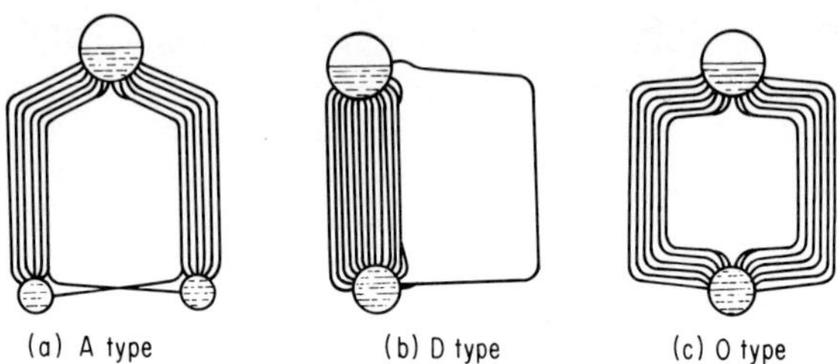

Fig. 3.15 Three typical modern packaged water-tube structural configurations.

WATER CIRCULATION

Q How does water circulate through a two-drum bent-tube boiler?
A Figure 3.17 shows a Keeler steel-encased unit with a full-length steam drum and a half-length mud drum. Feedwater enters (dotted line) the boiler through the front head of the upper drum, to which an internal feed pipe is attached. This pipe discharges the water at the rear end of

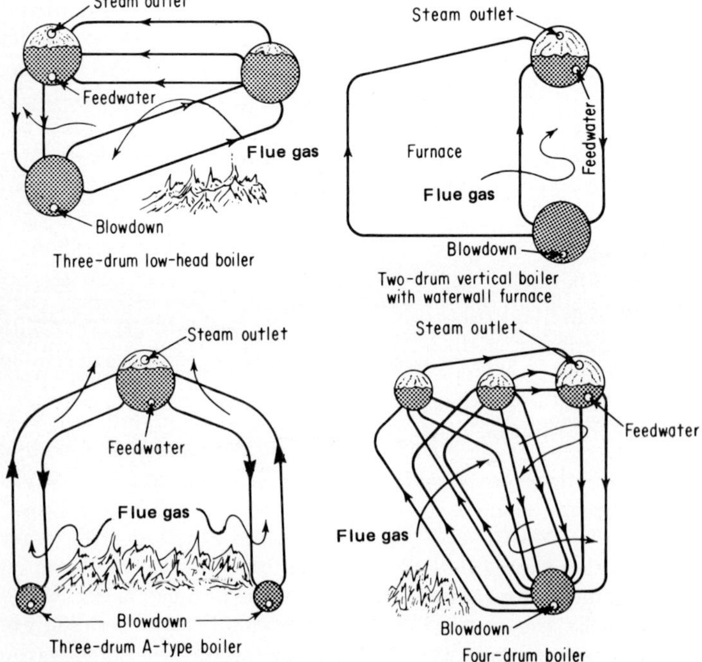

Fig. 3.16 Circulation of water and hot flue gas in water-tube boilers.

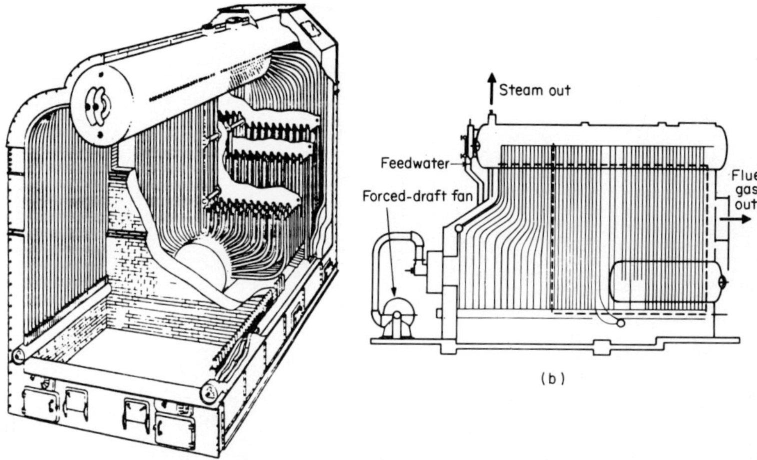

Fig. 3.17 Keeler boilers with short and long drums are of the bent-tube type. (*a*) Cutaway view; (*b*) side view.

the upper drum, from where it flows down the rear bank of tubes into the rear end of the lower drum, then horizontally through the short circulating tubes into the waterwall headers, forward and up through the side waterwall tubes into the upper drum, thus completing the cycle of circulation.

A second cycle of circulation occurs in the main bank of boiler tubes, flowing downward from the rear end of the upper drum to the rear end of the lower drum at the rear of the tube nest, then upward from the forward end of the lower drum directly above. This circulation occurs without the use of internal baffle plates, thus making it possible to operate at high overloads with dry steam and with no disturbance of water level (normally at the center of the upper drum). The furnace waterwall tubes do not depend upon the upper drum for their water supply but receive water directly from the lower drum. Also, the steam generated in the furnace waterwall tubes is freely discharged directly into the upper drum without the use of intermediate headers.

Q What types of steam generators are used in industry and small utilities today?

A Figure 3.18 features a 40,000 lb/h unit operating at 450 psi and 750°F. Note the comparatively large diameter drums and headers designed for easy access. Other features include wide spacing of wall tubes backed by refractory. The air heater on this unit can be bypassed at low ratings. Figure 3.19 is a packaged WT boiler of the two-drum D

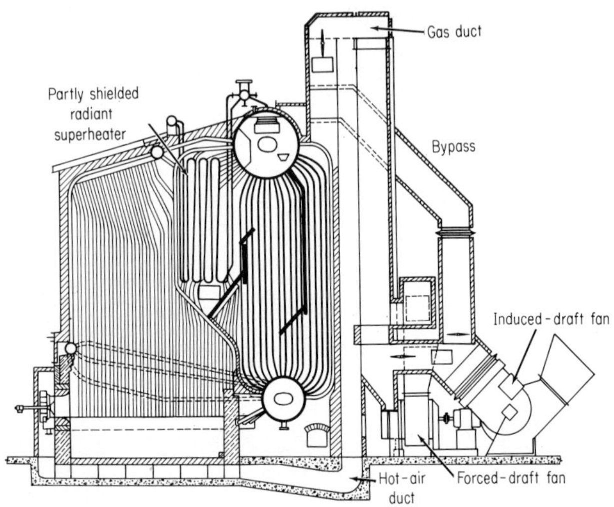

Fig. 3.18 A typical modern steam generator used in industry and in small utilities.

type, of up to 80,000 lb/h of steam. It comes complete with fan, burner, and controls for oil, gas, or combination firing.

TRENDS IN LARGE INDUSTRIAL BOILERS

Q What is the trend in large industrial boilers?
A Large paper mills and large chemical and oil-refinery plants are using boilers designed for higher pressures and capacities. Some units have reached the 2,000,000 lb/h capacity, and 1250-psi pressures are common. A stimulant to higher pressures and temperatures is the interest in *cogeneration,* which is the simultaneous production of electric energy and process-steam needs. Thus any excess power generated can be sold to a utility system at a price equal to the cost the utility can experience by not having to provide the power. This federal legislation is an attempt to improve the utilization of our oil and gas resources by combining power production with process-heat needs.

Today there is a decided trend in larger industries to burn coal and wood wastes of all types, again for the purpose of lowering energy costs or saving fuel purchases.

Q What does the term *circulation ratio* mean?
A It is the amount of water circulated in a boiler in pounds per hour vs. the steam generated.

Fig. 3.19 Packaged water-tube boiler of the two-drum D type for burning oil or gas, comes complete with controls, ready to install. (*Courtesy of Cleaver Brooks.*)

EXAMPLE: If 10 lb of water is circulated in a boiler to make 1 lb of steam, the circulation ratio is 10. This ratio can become important if too much steam begins to form in a steam-water mixture in the generating tubes. Too much steam in the mixture may create *steam blanketing* of the tube, and eventually local tube overheating because the heat from the fireside cannot be absorbed on the water-steam side.

WATERWALLS

Q Explain the importance of waterwalls as a heat-absorption surface in a WT boiler.

A Waterwalls consist of relatively close-spaced vertical tubes forming the four walls of the furnace. They were originally developed to cool and protect the furnace lining. One design of large power-generating boilers has 144-ft-high, 0.340-in-thick tubes at the hottest furnace zone (below 85-ft elevation) but only 0.320-in-thick tubes above.

Depending on the type of boiler, the waterwall heating surface may account for only 10 percent of the boiler's total heating surface, yet represent as much as 50 percent of the total heat absorption. Waterwalls perform three basic functions: (1) protect the insulated walls of the furnace, (2) absorb heat from the furnace to increase the unit's generating capacity, and (3) make the furnace airtight (on pressurized furnaces with tangent welded tubes).

Q How is furnace heat transferred to the waterwalls?
A Heat is transferred to the waterwall tubes as radiant heat from the zone of highest temperature in the furnace.

> NOTE: Because of the great amount of heat absorbed by that part of the boiler, feedwater must be of the best quality. Also, the circulation of water must be rapid and plentiful to ensure positive flow through each tube at all times.

Q Should waterwall headers be blown down while the boiler is under load?
A Positively not.

> WARNING: Under no circumstance should waterwall headers be blown down while the boiler is operating. If blown, the boiler's normal circulation will be upset and the overheated tubes will bulge or rupture.

Q How are waterwalls assembled?
A Figure 3.20 shows some typical arrangements. The waterwall in Fig. 3.20*a* is designed for moderate cooling. This design has the tubes spaced apart and the wall surface composed of part firebrick. The brick is backed with several layers of insulation and strong steel casing. Reinforcing metal lath is often used in wall construction. Figure 3.20*b* shows how tangent tubes are placed in the furnaces of many large and small boilers. The staggered tube arrangement offers a high heat-absorbing surface that is backed by solid block, or plastic, insulation and a strong steel external casing. Figure 3.20*c* shows steel lugs or longitudinal fins welded to nontangent wall tubes. In some designs the lugs protrude from the tube into the furnace and are covered with a chrome-base refractory or slag. To ensure furnace tightness, adjacent fins are often welded.

Figure 3.20*d* shows newer gastight casings, known as membrane walls. Here tightness is obtained by welding a flat strip of metal between the tubes. This eliminates the casing and many of its problems. Insulation is applied directly to the outside of the tubes, while metal lagging is attached to give the outer surface durability and good appearance. Figure 3.20*e* shows how the outer casing, insulation, and steel-skin casing are often constructed.

TYPES OF BOILER TUBES

Q Briefly explain tubes other than waterwalls used on WT boilers.

A Depending on the design for pressure, capacity, and final temperature, a modern, large steam generator will have superheater, economizer, reheater, and air-heater tubes. The basic purpose of these tubes is to extract all possible heat from the fuel input, thus lowering the exit temperature going up the stack.

Figure 3.21*a* shows two typical curves; the upper curve indicates gas temperatures to the exit portion, and the lower curve shows feedwater temperature entering the boiler, then slowly being raised by the hot gas to its final steam temperature. Of course, the various tubes provide the necessary heat transfer to accomplish just this.

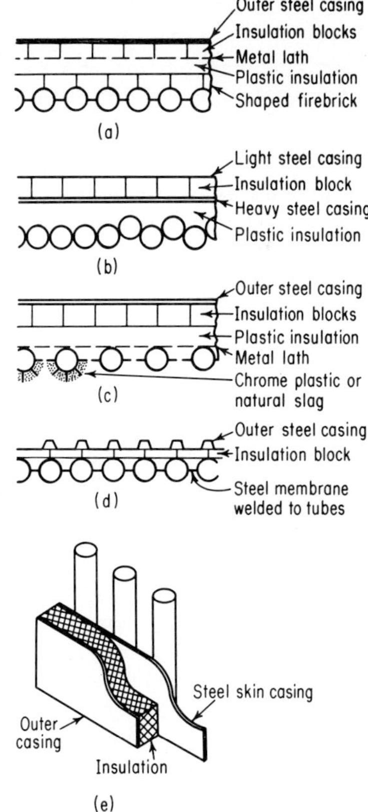

Fig. 3.20 Typical design variations of water-cooled airtight (usually) walls for larger boilers.

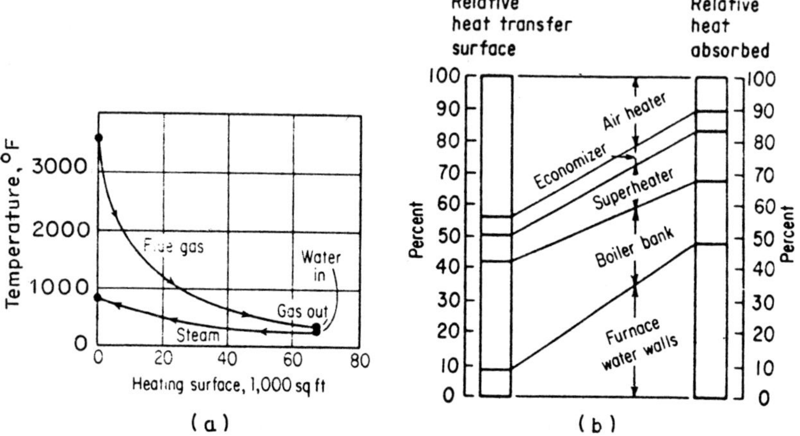

Fig. 3.21 Extraction of heat is dependent on tube surface area provided. (*a*) Curves show flue-gas temperature declining as water and steam are heated; (*b*) boiler components absorb different rates of heat.

Figure 3.21*b* shows how the various tubes perform in a typical boiler as to percentage of total heat absorbed by the furnace-wall tubes, main boiler bank tubes, superheater, economizer, and air heater. Note how the furnace-wall tube heating surface of only 10 percent absorbs 50 percent of total heat-transfer surface. The reason for this is a basic heat-transfer law. As the gas temperatures are cooled in their passage through the boiler, the temperature drop slowly brings into effect the law of diminishing returns as the temperature difference becomes smaller and smaller. Thus to get a certain temperature rise, it takes more heating surface. Because the furnace is the highest-temperature zone where the greatest temperature difference exists, most heat absorption takes place even with a moderate heat-surface area.

Q Name some types of superheaters in general use.

A Figure 3.22 shows some typical superheaters and their locations in the boiler. The general classification includes radiant and convection types, depending on whether they absorb radiant or convection heat. The interdeck type has tubes arranged between banks of primary boiler tubes. The pendant type is a suspended series of coils, usually shielded against radiant heat by a screen of boiler tubes. It is often arranged as the first steam heater before the steam goes to the superheater outlet header. The platen type is similar to the pendant type, but the tubes are in one plane. Usually the steam goes through the platen superheater before it enters the pendant superheater.

Q What are economizers, and how are they classified?

A Economizers serve as traps for removing heat from the flue gases at moderately low temperature, after they have left the steam-generating and superheating sections of the boiler. The general classification is:

1. Horizontal- or vertical-tube, according to the direction of gas flow with respect to the tubes in the bank.
2. Parallel flow or counterflow, with regard to the relative direction of gas and water flow.
3. Steaming or nonsteaming, according to thermal performance.
4. Return-bend or continuous-tube types.
5. Plain-tube or extended-surface types, according to the details of design and the form of the heating surface.

The tube bank may be further designated as of the staggered or in-

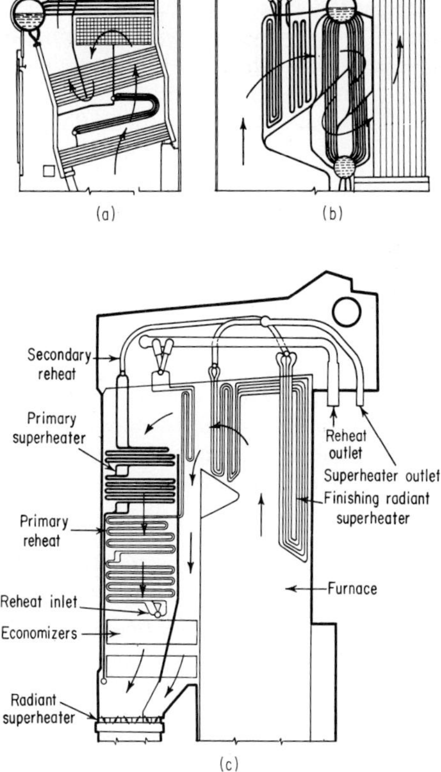

Fig. 3.22 Typical superheaters and their locations in modern steam generators.

line arrangement, with regard to the pattern and spacing of tubes which affect the path of gas flow through the bank, its draft loss, heat-transfer characteristics, and ease of cleaning.

Figure 3.23*a* shows some extended-tube types used for economizers. Theoretically, economizer heating surface could be added to a boiler until the exit temperature neared the outside air temperature. But an abnormally high heat-surface area would be needed. Further, each fuel burned has a dew-point temperature which can cause moisture accumulation on the economizer and corrode the surface in a short time. Figure 3.23*b* shows an economizer as installed in a stack for furnace exhaust to heat feedwater.

Q Describe the reheater.

A A reheater is essentially another superheater used in modern utility boilers for boosting plant efficiency. While the superheater takes steam from the boiler drum, the reheater obtains used steam from the high-pressure turbine at a pressure below boiler pressure. This lower-pressure steam passing through the reheater is heated to 1000°F and then is introduced into the intermediate or low-pressure turbine. Reheaters, like superheaters, are also classified according to their location in the boiler as convection or radiant. Convection superheaters and reheaters may be of the horizontal or pendant type.

AIR HEATERS

Q What is the purpose of air heaters?

A Air heaters make the final heat recovery from boiler flue gases with which they preheat the incoming furnace air for its combustion with fuel. Thus some fuel is saved which would otherwise be used in heating the air-fuel mixture up to its ignition point. But the temperature of the flue gas must not be reduced below its dew point, since moisture would condense out of the flue gas. That would cause water to combine with sulfur and possibly carbon dioxide, also carbon monoxide, to form highly corrosive sulfurous and carbonic acids. Figure 3.24*a* shows a design with air passing inside the tubes, while hot gas crosses at right angles. The design in Fig. 3.24*b* has hot gases flowing through the tubes, while air is directed across the tubes by baffles.

Figure 3.25 shows typical inlet and outlet temperatures of feed, air, and flue gas for a 518,000 lb/h boiler designed for 1475 psig. Note the great rise in the air temperature at different loadings, which range from an inlet of 95°F to air leaving the heater at 750°F. There is a great pickup of heat for the air to be used in burning pulverized coal.

Q What is a regenerative air heater?

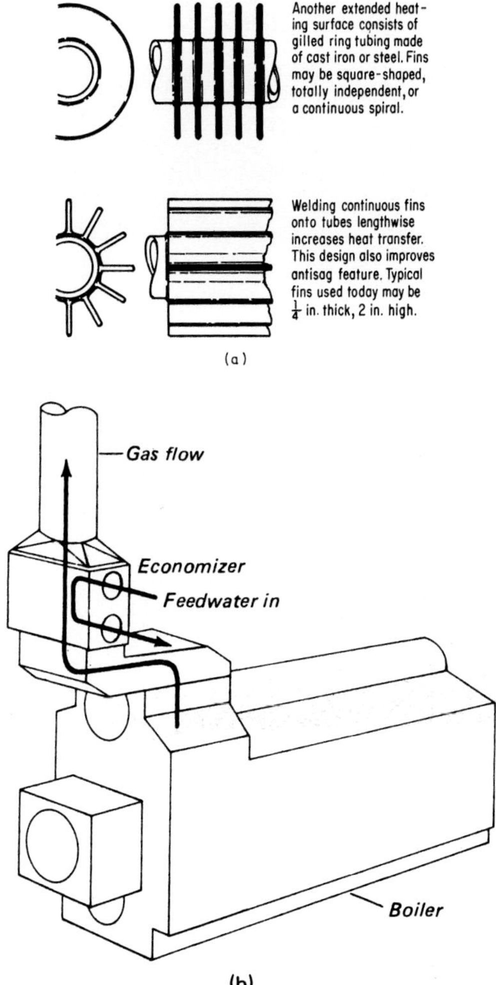

Another extended heating surface consists of gilled ring tubing made of cast iron or steel. Fins may be square-shaped, totally independent, or a continuous spiral.
Welding continuous fins onto tubes lengthwise increases heat transfer. This design also improves antisag feature. Typical fins used today may be $\frac{1}{4}$ in. thick, 2 in. high.
(a)
Gas flow
Economizer
Feedwater in
Boiler
(b)

Fig. 3.23 (*a*) Extended surface designs on economizer tubes transfer more heat; (*b*) economizer installed in stack for furnace exhaust to heat feedwater. (*Courtesy of Zurn Industries, Inc.*)

A This type of air heater (Fig. 3.26) offers a large surface of contact for heat transfer. It usually consists of a rotor which turns at about 2 to 3 r/min and is filled with thin, corrugated metal elements. Hot exhaust gases pass through one half of the heater; incoming air passes to the

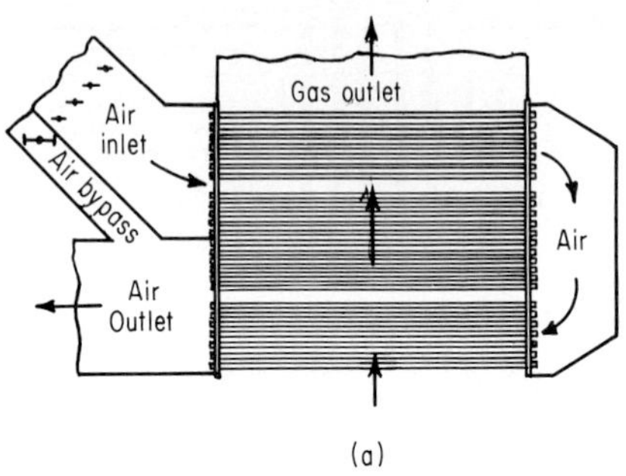

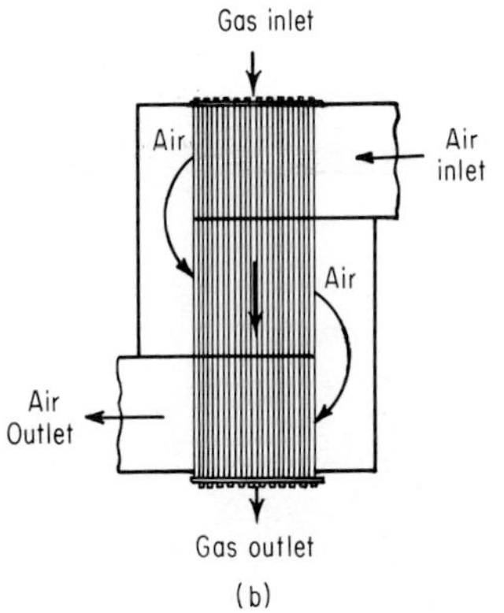

Fig. 3.24 Types of air heaters used to recover low-temperature heat.

furnace through the other half. As the rotor turns, the heat-storage elements transfer the heat picked up from the hot zone to the incoming air zone.

Q What is a downcomer?
A A large vertical tube or pipe for circulating water from the water space of the steam drum to waterwall headers.

> NOTE: The downcomer is always placed outside the boiler casing so that it does not absorb heat from the furnace or boiler proper. And it must not disturb the natural gravity circulation of the cooler water downward, which might occur if it absorbed heat.

Q How are heating surfaces distributed on larger utility-type boilers?
A An example is a natural-circulation, reheat-type steam generator: capacity 2,390,000 lb/h, superheated steam at 2460 psig and 1005°F at the outlet, 2,141,000 lb/h reheated steam at 455 psig and 1000°F at

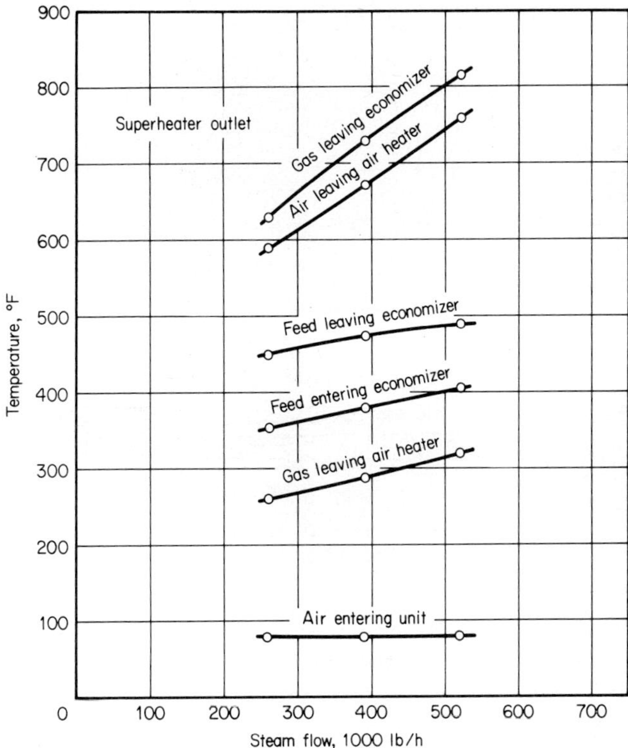

Fig. 3.25 Temperatures of air, feed, and flue gas at different boiler loads for a 518,000 lb/h hour, 1475 psig rating.

outlet, oil-fired. Here are the heating areas:

Boiler, 8080 ft^2
Waterwalls, 20,250 ft^2
Radiant superheater, 23,225 ft^2
Convection superheater, 59,750 ft^2
Economizer, 86,500 ft^2
Air heater, 302,000 ft^2
Reheater, 84,900 ft^2
Total furnace volume, 184,000 ft^3
Total furnace surface, 41,000 ft^2

MODULAR CONSTRUCTION

Q Explain the term *modular construction.*

A Figure 3.27 shows furnace front, roof, and rear walls of a 500,000 lb/h steam generator. The tube panels making up a wall are completely

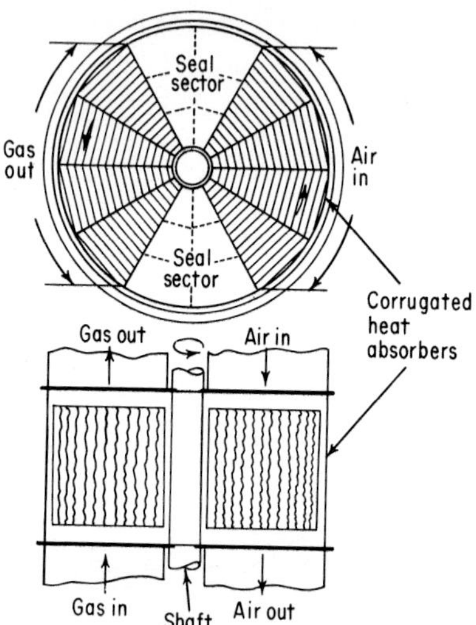

Fig. 3.26 Regenerative air heater has rotating heat-storage elements.

fabricated in the manufacturer's shop and are field-assembled on a panel or modular basis—hence the term modular construction. The panel shown is for a rectangular furnace of 29 ft 2½ in wide throughout and 100 ft high from the lower waterwall header to the junction of the front wall and roof tubes.

Modular construction permits higher quality control in the fabricator's shop in comparison with field erection. It also permits faster field erection of the boiler by joining the shop-assembled components on a segmented, planned basis.

Q Name the different types of fabricated tubes that may be used in water-tube boilers.

A Tubes are made to suit conditions and orders as follows: seamless tubes of the following composition: carbon steel, hot- or cold-drawn alloy steels, hot- or cold-drawn stainless steels. Thicknesses may range from 0.049 to 2.75 in OD (outside diameter).

For welded tubes, size ranges from ¾ to 10¾ in OD with wall thickness from 0.049 to 0.625 in.

For bimetallic tubing, see Fig. 3.28*a*. This tubing consists usually of a stainless-steel outer layer bonded together with a carbon steel. Such tubings are used to provide corrosion resistance on the fireside of the tube, such as in the highly corrosive black-liquor recovery boilers used in large paper mills.

For ribbed tubing, see Fig. 3.28*b*. This tubing has internal, spiral ribbing to promote circulation and thus avoid DNB or departure from nucleate boiling.

For external finned tubing, see Fig. 3.28*c*. These are used in boiler membrane walls. They eliminate welding bars to tubes to make a membrane by being formed as shown. Only the fins are welded together to form a (usually) airtight furnace wall.

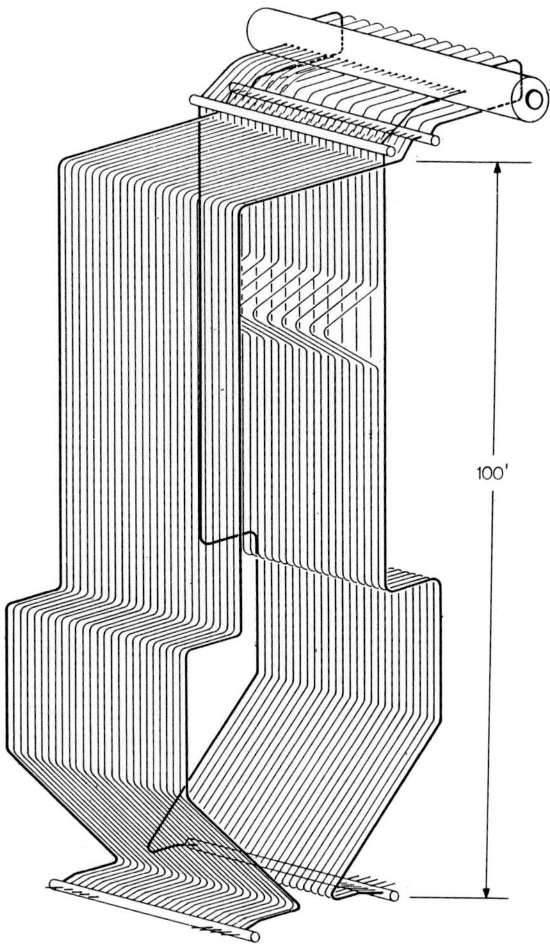

Fig. 3.27 Modular construction is used to erect furnace front, roof, and rear walls of 500,000 lb/h WT steam generator.

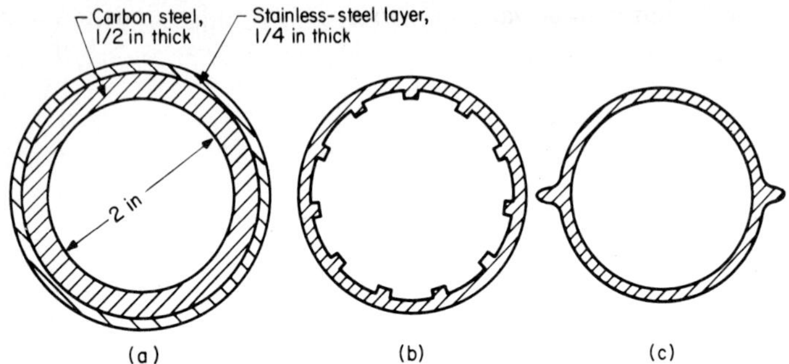

Fig. 3.28 Special WT tubes are used for different services or for ease of erection. (*a*) Bimetallic tube for stress-corrosion resistance; (*b*) ribbed tube for promoting circulation; (*c*) external finned tube for ease of welding membrane wall.

DRUM SUPPORT ARRANGEMENTS

Q What is the advantage of a top-supported boiler drum?

A The top-supported structural design (Fig. 3.29) permits free expansion of all pressure parts such as tubes and drums, and also accommodates or makes easier the installation of platforms. It also permits designing for such load factors as wind and earthquake, which makes this type of boiler popular for outdoor installation in chemical, refinery, and paper-mill plants.

The illustrated unit has a furnace that facilitates burning pulverized coal, oil/gas, and low-heat-value gases such as blast-furnace gas. The boiler is of the single-pass type with no baffles, which facilitates fireside cleaning.

Q Name the chief advantages of the type of bottom structural supported steam generator for industrial use shown in Fig. 3.30.

A These are usually offered as partially shop-assembled units for fast field erection. Thus the boiler is close to being a packaged compact boiler. The unit shown is also adaptable to limited space conditions. Sizes of the unit generally range from 200,000 to 1,000,000 lb/h. A drainable superheater can be provided if needed. Pressure is to 1500 psig and 900°F. Liquid and gaseous fuel is generally fired in the furnace. Bottom-supported units can be designed wide and with low heights.

TREND IN UTILITY BOILERS

Q What is the trend in utility boilers?

A The fuel to be used and the expected present and future service conditions influence selection of design, capacities, pressure, temperature, draft, and similar variables. Pressure considerations may involve selecting a drum-type WT boiler to generate at the subcritical pressure or the once-through supercritical type. Pressures for subcritical types are usually 1900 or 2600 psig, whereas supercritical units are usually designed for 3800 psig and over, well above the critical pressure of 3206.2 psig.

Capacities range from about 1,000,000 to nearly 10,000,000 lb/h; thus energy is provided to drive 125- to 1300-MW turbogenerators.

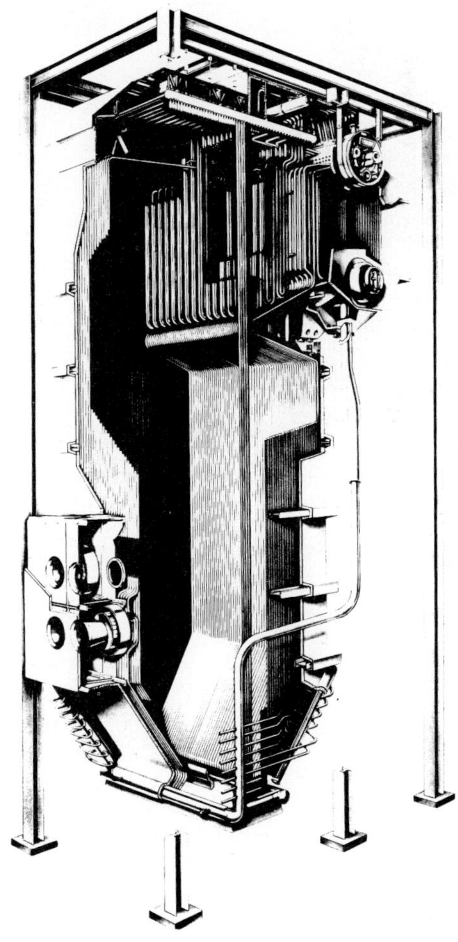

Fig. 3.29 Top-supported structural design, modular erected boiler of 1,000,000 lb/h and 1800 psig. (*Courtesy of Foster Wheeler Corp.*)

Fig. 3.30 Compact, bottom-supported liquid- and gaseous-fuel-fired boiler, 350,000 lb/h, 1500 psig, and 900°F. (*Courtesy of Foster Wheeler Corp.*)

Figure 3.31 shows a gigantic 6,500,000 lb/h supercritical steam generator designed for 3785 psig and 1005°F.

Q Describe a once-through steam generator and system.

A Once-through boilers receive feedwater at one end and discharge steam at the other end. Figure 3.32 shows a typical plant layout of a large utility unit. The entire process of heating, steam formation, and superheating is carried out in a continuous flow; thus a steam drum is not needed.

The design consists of many once-through circuits discharging into a common outlet. The system illustrated handles pressures either below or above the critical pressure of 3206 psia. Water-steam flow is through the furnace walls, the primary horizontal superheater, and, finally, the first and second sections of the secondary (pendant) superheater.

A supercritical-pressure cycle must maintain active furnace-wall circulation at low loads and start up when little or no steam goes to the turbine. In the system shown, the steam generated discharges to a flash tank. The steam separated from this fluid is directed to turbine seals, deaerator, and high-pressure heater. It can also be used for rolling the turbine. But an automatically controlled high-pressure stop valve and a return-line valve between flash tank and superheater return the cycle to normal when quarter load is reached.

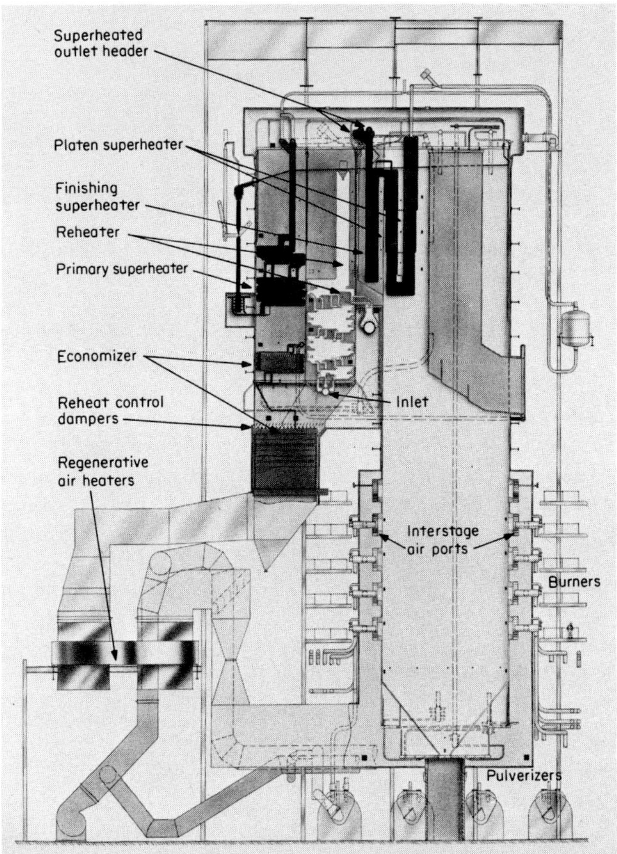

Fig. 3.31 Gigantic coal-fired once-through supercritical steam generator produces nearly 6.5 million lb/h of 3785-psig steam at 1005°F. (*Courtesy of Foster Wheeler Corp.*)

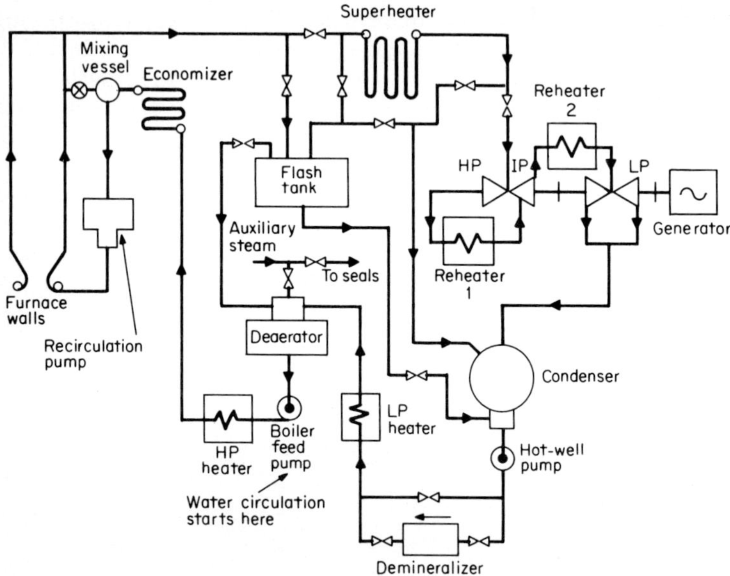

Fig. 3.32 Once-through unit plant layout where entire process of heating, steam formation, and superheating is carried out in a continuous flow.

Q Describe a typical supercritical-pressure steam generator.

A One such generator is a 900-MW unit that produces 6,400,000 lb of steam per hour. The steam generator is of the twin-furnace type with water-cooled center wall in between. Feedwater enters at 550°F and picks up about 100°F in the economizer. A recirculating pump circulates the fluid between successive waterwalls. But when it reaches the first section of the horizontal superheater at a temperature of around 800°F, the fluid is no longer water, but a vapor mixture which must be dried and superheated.

After passing through finishing superheaters, 1003°F steam is ready for the high-pressure (HP) turbine. Temperature is maintained at the exact control point by attemperation (spraying water into the steam) before it enters the finishing superheater. When that area is reached, the steam has absorbed about 83 percent of boiler capacity. Most of the energy is then released in the HP turbine. The exhaust steam from the HP turbine returns to the boiler and circulates through two sets of reheaters, bringing the temperature back up to 1003°F. The reheated steam discharges to an intermediate-pressure turbine, then to a low-pressure (LP) turbine and condenser. The resulting condensate is treated by a filter-demineralizer, and more heat is added in a series of feedwater heaters and a deaerator. The feedwater is pumped through HP heaters into the steam generator and is ready for a new cycle.

DRUM COMPONENTS

Q How thick are the steam drums of large utility-type boilers?

A The thickness depends on the design pressure, the diameter of the drum, and the material being used. For a general idea, consider the drum specifications of a steam generator with natural circulation, reheater section, economizer, waterwall heating surface, combination radiant and convection superheater, convection-type reheater, and regenerative-type air preheaters fired by oil, gas, or coal. Also assume that the unit has a continuous output rating of 2,390,000 lb/h and design pressures of 2574 psi at the drum and 2460 psi at the superheater outlet.

The drum specifications are 60 in ID and $6^{11}/_{16}$-in-thick SA-212 Grade B firebox steel. The overall length of the drum is 100 ft between the feed pipe nozzle walls. The hemispherical heads are 60 in ID and $5^{7}/_{8}$-in-thick SA-212 Grade B firebox steel.

Q On modern large steam generators, what components are placed inside the steam drum? Why?

A Figure 3.33 illustrates the internals of a typical drum, which

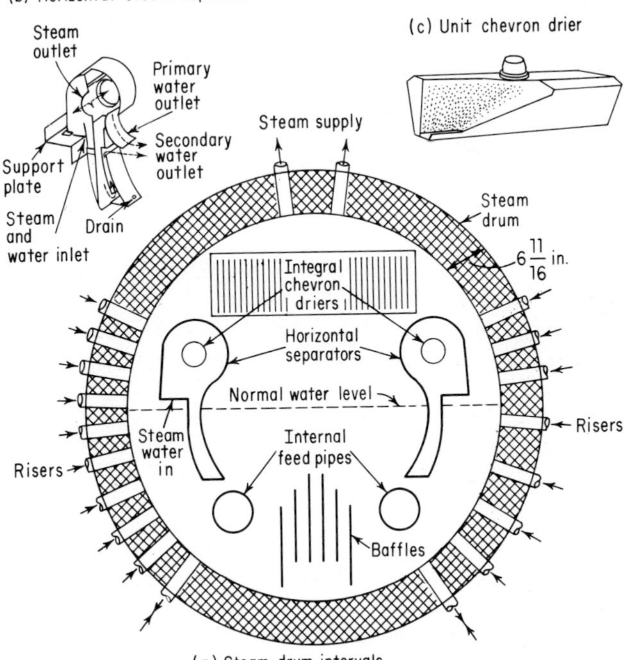

Fig. 3.33 Steam drum has horizontal steam separator and integral chevron driers.

performs two essential functions:

1. Separates steam from water to provide the downcomer system with steam-free water necessary for proper and safe circulation.
2. Separates moisture from steam to provide high steam quality.

The drum internals shown provide both functions by means of two stages of separation. The normal water level is 1½ in below the horizontal centerline of the drum. Vortex eliminators separate the steam and water passages in the drum.

The total circulating steam and water mixture from the steam-generating tubes is directed to the horizontal turboseparators (Fig. 3.33*b*). The steam and water mixture enters the separators and is centrifuged by following the curved contour of the separator dome. Most of the separated water is discharged horizontally at the water level of the drum. The separated steam flows to the chevron driers (Fig. 3.33*c*) at the top of the drum. The steam flows from the chevron through the dry box, then out the top of the drum via steam tubes across the roof to the partition wall superheater.

Q Do not the integral chevron driers, separators, and other hardware restrict the flow of steam outward from any opening in the drum and thus restrict the opening to a safety valve?

A The opening to the main steam line will usually be adequate to handle the design flow capacity. The ASME boiler code has definite rules on the opening to a safety valve. For example, internal collecting steam pipes, splash plates, or pans are permitted to be used near safety valve openings, provided "the total area for inlet of steam thereto is not less than twice the aggregate areas of the inlet connections of the attached safety valves. The holes in such collecting pipes must be at least ¼ in. in diameter."

In the case of steam scrubbers or driers, the ¼-in-diameter opening does not apply, provided "the net free steam inlet area of the scrubber or drier is at least 10 times the total area of the boiler outlets for the safety valves." Therefore, when inspecting the internals of drums, check the condition of the ¼-in-diameter holes in the collecting pipes. They must be free and clear. The same applies to the openings on driers, because plugged driers on collecting pipes could lead to restrictions of the safety-valve openings. That would reduce relieving capacity flow, which could be dangerous.

Q Have special ASME and code exceptions or rules been made in reference to once-through forced-flow steam generators which have no fixed water lines and steam lines?

A Yes, quite a few. For example:

1. It is permissible to design the pressure parts for different pressure levels along the path of water-steam flow.

2. No bottom blowoff pipe is required.

3. If stop valves are installed in the water-steam flow path between any two sections of line, certain safety valves or power-actuated pressure-relief valves, with control impulse interlocks, are required (this is different from the typical boiler with safety valves protecting the entire boiler).

4. Pressure gages required are more numerous at various sections. No water-gage glass or gage cocks are required.

Q Describe the LaMont boiler shown in Fig. 3.34.

A This design is called a controlled-circulation boiler because the quantity of water passing through the boiler is from 3 to 20 times the amount evaporated. Thus two pumps are required, one for circulating the high rate of flow through the tubes (no natural circulation), the other as a conventional boiler feed pump. The feed pump operates on the same principle as most boiler feed pumps by maintaining a constant level of water in the drum.

The function of controlled circulation is to establish a flow through the first section of inlet (Fig. 3.34 shows a flow diagram) to the boiler to prevent the water in the tubes from evaporating to complete dryness. Instead, evaporation is only to the extent that dissolved solids and salts will remain in solution. This solution (a mixture of water and steam) passes to the steam and water drum, where the steam is separated out, while the excess water is removed. The separated water, along with feedwater, is returned to the pumps through downcomers.

By means of continuous or intermittent blowdown, some water and solids in solution are removed. The separated steam is passed through

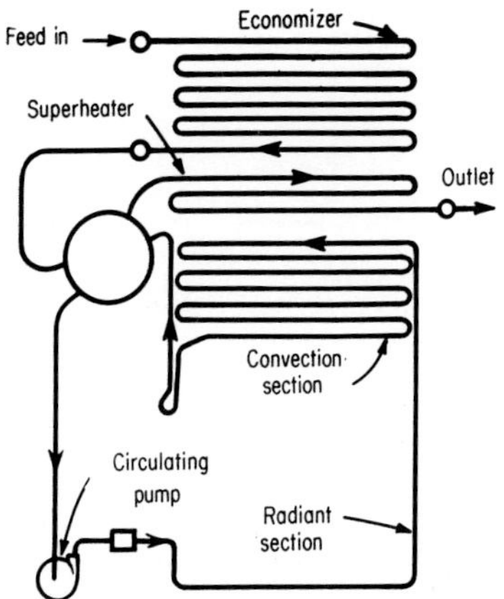

Fig. 3.34 LaMont controlled (forced) circulation boiler schematic diagram.

the superheater for final usage. These boilers have been built for pressures up to the supercritical and for capacities in excess of 100,000 lb/h.

VARIABLE-PRESSURE OPERATION

Q What is the advantage of variable-pressure operation when utilizing once-through boilers?
A At low capacity, once-through units will achieve fuel economy, in comparison with full-pressure operation by the furnace firing rate being adjusted to superheater temperature changes due to the variable or low-load operation.

Q Explain the reheat bypass system schematic in Fig. 3.35.
A This is a system used to continue firing a once-through boiler on a turbogenerator trip even though boiler turbogenerators are usually arranged to trip jointly on load rejection. It is a common European practice on once-through designs.

In Fig. 3.35, assume load rejection occurs by a transmission-line fault opening the main breaker and unloading the turbine-generator valves to prevent overspeeding. The steam produced by the steam generator then bypasses the turbine entirely, going through the reheater to the condenser. Spray is used to control the steam temperature going to the condenser.

The steam-generator firing system does not trip and shut down the unit. Rather, steam output is reduced at a programmed rate to a

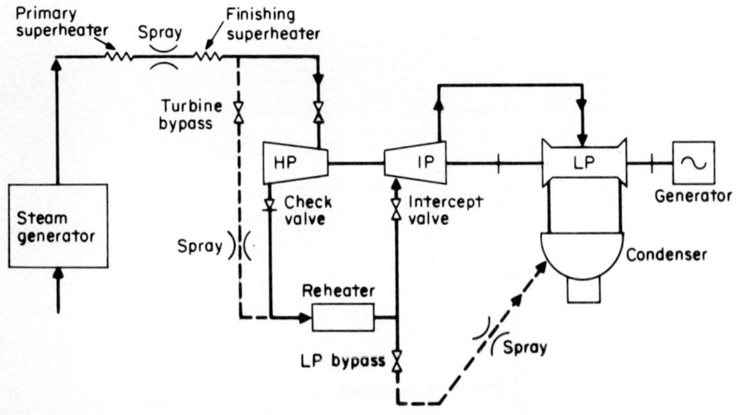

Fig. 3.35 Turbine bypass keeps once-through boiler generating when the turbogenerator trips.

convenient load plateau. The turbine remains on line carrying house load, usually about 5 percent. When the fault is cleared, the turbine is resynchronized and brought again to a desired load in step with the steam generator.

Q What operating factors may affect heat transfer and fluid flow in WT boilers?

A Designers strive to provide the following: (1) sufficient radiant and convection heat-absorbing surface areas, (2) proper air/fuel ratio for good combustion, (3) proper circulation of steam and water within the tubes to match firing rates, (4) coordination of auxiliary equipment such as fans, blowers, and feed pumps to keep flows of air, water, flue gas, and steam in balance.

Problems occur when operation of boilers (in time) affects heat-absorption rates or fluid flow across heat-absorbing surfaces. These variables may not be matched properly with firing rates. The result may be overheated tubes.

When heat-transfer surfaces, including superheaters, are sized, usually some allowance is made for the effect of external, or fireside, deposits on tube surfaces. However, if deposits exceed this allowance, the tubes cannot absorb the heat designed for. Also, fireside deposits can cause tube corrosion and plugging of flue-gas passages. Maldistribution of this flue gas as a result may create high draft losses and tube erosion.

And deposits on the internal surfaces of tubes result in lower heat-transfer rates and possibly tube overheating. Water treatment to prevent deposits from occurring is the chief method of avoiding this trouble.

On air heaters and economizers, it is necessary to avoid operating at too low a temperature level on the fireside. If not controlled, the corrosion effects of acid from the flue gas depositing on these heating surfaces can be very detrimental. Partial load operation can be a problem here. Additional protection is provided for air heaters to avoid "dew point" operation at low load by installing supplementary steam coil or electric heated air heaters.

Economizers' inlet temperatures must be controlled by preheating the water above the flue-gas dew-point temperature. This will prevent condensation on the economizer and thus stop acid formation from sulfur dioxide in the flue gas. Thus, it can be seen, WT boilers must be carefully operated in order to avoid costly breakdowns in tubes. Later chapters will cover additional problem areas.

CHAPTER **FOUR**

Heating Systems and Their Applied Boilers

Today, heating systems utilizing steam and hot-water boilers outnumber industrial boilers at least 10 to 1. There are more heating boilers in use than industrial and utility boilers simply because there are more properties that require heating systems for human habitation and comfort, such as homes, factories, offices, stores, theaters, schools, and similar structures where people live, congregate, and work. This chapter is devoted to reviewing some basics, features of heating systems, and some types of boilers used therein.

BASICS

Q Name three ways that heat is gained or lost in a building.
A As Fig. 4.1 shows: (1) By conduction or transmittal of heat through walls from the air on one side to the air on the other side of the wall. (2) By leakage through cracks and openings in windows, doors, and similar openings. (3) By radiation from two bodies at different temperatures where both emit and absorb impinging radiation. But a hotter body emits more than it receives, or the reverse is true of a colder body, such as a building, which receives more from a hotter body than it emits.

Q How is heat transfer through walls and ceilings shown in Fig. 4.2 reduced? And how about leakage?
A To reduce heat transfer through walls and ceilings, the insulating quality of the wall must be improved by (1) adding insulation and (2) providing other barriers such as air space to the conduction of heat across the wall.

Leakage, of course, is reduced by sealing the cracks or openings through the use of calking compounds, weather strips, storm windows and doors, and similar blocking devices to minimize openings.

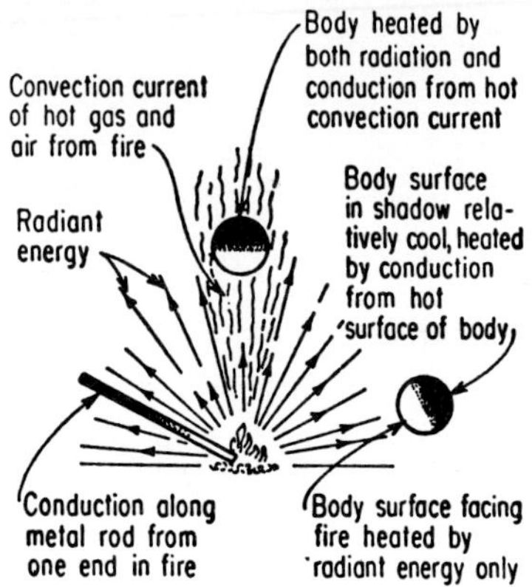

Fig. 4.1 Heat transfer. (1) Conduction heats bar; (2) radiation heats globe at right; (3) both radiation and convection current carry heat to upper globe. (*Courtesy of Power magazine.*)

Q What is the coefficient of conductivity?

A The resistance of an object to heat flow can be predicted. Thus heavy, dense materials transmit heat by conduction faster than do lighter, less dense materials. The heat-transfer property of a material is

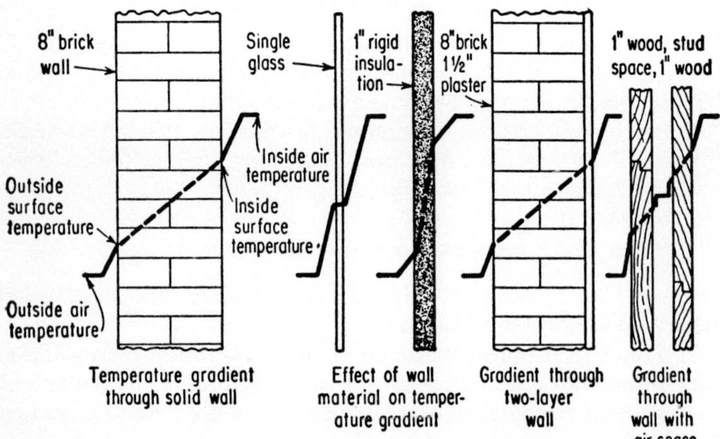

Fig. 4.2 Fundamentals of heat loss through wall. Total temperature difference between inside and outside is sum of temperature drops at each layer of resistance. (*Courtesy of Power magazine.*)

designated by the letter k, which is known as the coefficient of conductivity.

EXAMPLE: Brick has a k of 5.0. This means that 5 Btu/h can pass through a 1-in-thick 1-ft^2 brick for each degree Fahrenheit difference between the outside and inside temperature of the brick. The heat flow through a 10-in brick is only 5.0 divided by 10, or 0.5. So the thicker the insulation, the smaller the heat loss.

RULE: For solid, uniform materials, the rate of flow k is given for a 1-in thickness. To get the rate of flow for any other thickness, divide k by the actual thickness (inches) of the material.

HEAT-TRANSFER CALCULATIONS

Q How is the heat transferred across a wall calculated?
A Generally, the following equation is used:

$$Q = UA\,(t_i - t_o)$$

where Q = heat transferred, Btu/h
A = area of wall, ft^2
U = overall coefficient of heat transfer, Btu/(h · ft^2 · °F)
t_i = inside temperature, °F
t_o = outside temperature, °F

U is usually found in heat-transfer or building material supplier's handbooks, depending on the wall construction.

EXAMPLE: The wall of a room facing the outside of a house is 20 ft long by 8 ft high. The wall construction is such that heating and ventilating reference tables indicate the overall conductivity of the wall, or $U = 0.349$ Btu/(h · ft^2 · °F). If the outside design temperature for the room is −15°F and the inside design room temperature is 70°F, what will be the maximum heat loss through this wall?

Substituting in the equation,

$$Q = 0.349\,(8 \times 20)\,(70 + 15)$$
$$Q = 4746.4 \text{ Btu/h is lost} \qquad \textit{Ans.}$$

Q What is a degree-day?
A Architects and engineers use degree-days in calculating heating loads for a structure. Usually a 65°F inside temperature is taken as the standard. The difference between this 65°F inside temperature and the average outside temperature for the particular locality is used as an index of the heating requirements for the structure.

One degree-day is assumed to occur for every degree the average outside temperature is below 65°F during a 24-h (or single-day) period.

EXAMPLE: If for 30 days the outside temperature averages 35°F, how many degree-days were accumulated in the 30 days?

$$(65 - 35)\,30 = 900 \text{ degree-days} \qquad \textit{Ans.}$$

NOTE: Use Table 4.1 for typical degree-days for United States and Canadian cities.

Q Give an illustration of the use of degree-days in determining the heat lost for a building in Newark, N.J., with a heating load of 600,000 Btu/h operating between 70°F inside temperature and 5°F outside temperature.

A From Table 4.1, degree-days for Newark are 5383. And the temperature difference for heat being lost is 70 − 5 = 65°F.

$$\text{Heat load per °F} = \frac{600{,}000}{65} = 9231 \text{ Btu/(h} \cdot \text{°F)}$$

This is the temperature difference per hour. Then for 24 h,

$$\text{Heat loss per °F} = 24 \times 9231 = 221{,}544 \text{ Btu per degree-day}$$

For the season, the heat lost would thus be

$$221{,}544 \times 5383 = 1{,}192{,}561{,}352 \text{ Btu heat lost} \qquad \textit{Ans.}$$

Q In the above problem, if oil with a heating value of 144,000 Btu/gal is used, and a heating system with an average efficiency of 75 percent is assumed, how many gallons of oil would have to be used per heating season?

A Heat lost per season = heat input per season × efficiency

$$\text{Heat lost} = \text{gallons used} \times \frac{\text{Btu}}{\text{gal}} \times \text{efficiency}$$

Therefore,

$$\frac{\text{Heat lost per season}}{\text{Btu/gal} \times \text{efficiency}} = \text{gallons used}$$

Substituting,

$$\frac{1{,}192{,}561{,}352}{144{,}000 \times 0.75} = 11{,}042 \text{ gallons used per season} \qquad \textit{Ans.}$$

Q On steam-heating systems, where does most of the useful heating come from when supplying radiators or convectors?

A The useful heat comes mostly from the latent heat given off by the steam as it condenses. Remember that for atmospheric pressure, one pound of dry steam gives up 970.3 Btu in condensing.

Table 4.1 Average Design Temperatures, Degree-Days, and Wind Velocities for United States and Canadian Cities

State and city	Design temp, °F	Avg. winter wind velocity, mi/h and direction	Degree-days
Alabama			
Birmingham	5	9N	2352
Mobile	15	8N	1471
Montgomery	10	7NW	1884
Arizona			
Flagstaff	−15	7SW	7145
Phoenix	25	4E	1405
Tucson	20		1845
Arkansas			
Fort Smith	10	8E	3213
Hot Springs	5		2665
Little Rock	5	10NW	2811
Texarkana	10		2219
California			
Los Angeles	30	6NE	1504
Pasadena	25		2067
Sacramento	25	8SE	2653
San Diego	35	5NW	1645
San Francisco	35	8N	3264
Colorado			
Colorado Springs	−25		6518
Denver	−20	8S	5874
Fort Collins	−30		6877
Grand Junction	−15	5NW	5570
Connecticut			
Bridgeport	−5		
Hartford	−5	8NW	6036
New Haven	−5	10N	5895
Waterbury	−15		5661
Delaware			
Wilmington	5	NW	4789
District of Columbia			
Washington	0	7NW	4626
Florida			
Jacksonville	25	9NE	890
Miami	35	9NW	None
Pensacola	20	14N	1249
Tallahassee	25	N	1228
Tampa	35	7NE	None
Georgia			
Atlanta	10	12NW	2890
Augusta	15	6NW	2161
Macon	15	6NW	2201
Savannah	15	9NW	1490
Idaho			
Boise	0	5SE	5552
Lewiston	5	5E	4924
Pocatello	−5	10SE	6459
Illinois			
Chicago	−100	13SW	6290
Peoria	−10	8S	6109
Rockford	−10		6847
Springfield	−10	10NW	5373
Indiana			
Evansville	0	10S	4244
Fort Wayne	−10	11SW	5925
Indianapolis	−5	12SW	5298
Iowa			
Davenport	−10	9NW	6389
Des Moines	−15	8NW	6384
Dubuque	−10	7NW	6790
Iowa–Con't.			
Sioux City	−20	12NW	6898
Kansas			
Dodge City	−10	10NW	5035
Topeka	−10	10S	5307
Wichita	−5	12S	4673
Kentucky			
Frankfort	0	S	4211
Lexington	0	13SW	4618
Louisville	0	10SW	4180
Louisiana			
Baton Rouge	20	E	1349
New Orleans	25	9N	1024
Shreveport	15	9SE	1938
Maine			
Eastport	−10	13W	8520
Lewiston	−15	NW	7502
Portland	−10	9NW	7012
Maryland			
Baltimore	5	8NW	4533
Cambridge	5	NW	4245
Frostburg	−5	W	5613
Massachusetts			
Boston	0	12W	6045
Fitchburg	−10		6732
Lowell	−15		6504
Springfield	−10		6464
Michigan			
Ann Arbor	−5		6877
Detroit	0	13SW	6490
Flint	−10		7179
Grand Rapids	−10	9NW	6535
Lansing	−10	7SW	7048
Marquette	−10	11NW	8693
Saginaw	−10	11SW	7063
Minnesota			
Duluth	−25	13SW	9480
Minneapolis	−20	12NW	7850
St. Paul	−20	12NW	7926
Mississippi			
Jackson	15	SE	1959
Meridan	10	6N	2160
Vicksburg	15	8SE	1823
Missouri			
Hannibal	−10	10SW	5248
Kansas City	−5	11SW	4852
St. Louis	−5	12S	4585
Springfield	−10	11SE	4428
Montana			
Billings	−25	W	7119
Butte	−20	NW	8272
Havre	−30	10SW	8700
Helena	−25	7SW	8054
Nebraska			
Lincoln	−10	11S	5999
North Platte	−20	8W	6131
Omaha	−10	9NW	6128
Nevada			
Las Vegas	20	S	2844
Reno	−5	5W	5892

Table 4.1 Average Design Temperatures, Degree-Days, and Wind Velocities for United States and Canadian Cities (*continued*)

State and city	Design temp, °F	Avg. winter wind velocity, mi/h and direction	Degree-days
New Hampshire			
Berlin	−25		8867
Concord	−20	7NW	7353
New Jersey			
Atlantic City	10	16NW	5176
Dover	0	SW	6270
Elizabeth	0	NW	5302
Jersey City	5	NW	5193
Newark	5	NW	5383
Phillipsburg	−5	W	5694
Trenton	5	13SW	4933
New Mexico			
Albuquerque	5	N	4298
Roswell	−10	7S	3399
Santa Fe	0	8NE	6063
New York			
Albany	−5	8S	6580
Buffalo	−5	17W	6822
Elmira	0	NW	6412
New York City	0	17NW	5347
Rochester	−5	11W	6732
Syracuse	−15	14SW	6893
Watertown	−15	SW	7298
North Carolina			
Asheville	5	10NW	4232
Charlotte	10	7SW	3153
Raleigh	15	7SW	3234
Wilmington	20	9SW	2302
Winston Salem	10	W	3904
North Dakota			
Bismarck	−30	9NW	9192
Grand Forks	−25	NW	9764
Ohio			
Cincinnati	−5	8SW	4703
Cleveland	−5	15SW	6155
Columbus	−10	12SW	5398
Toledo	−5	12SW	6077
Oklahoma			
Ardmore	10	N	2374
Oklahoma City	−5	13N	3613
Tulsa	−5	N	3559
Oregon			
Pendleton	−15	SE	5119
Portland	15	7S	4469
Salem	15	S	4618
Pennsylvania			
Bethlehem	−5	11NW	5230
Erie	0		6273
Harrisburg	5	8W	5375
Philadelphia	5	11NW	4855
Pittsburgh	−5	12W	5235
Reading	−5	8NW	5389
Scranton	−10	7SW	6129
York	−5	W	5449
Rhode Island			
Providence	0	14NW	6015
South Carolina			
Charleston	20	11SW	1769
Columbia	15	8NE	2364
S. Car.–Con't.			
Greenville	15	9NE	3380
South Dakota			
Aberdeen	−30	NW	8709
Rapid City	−20	8W	7070
Tennessee			
Chattanooga	15	8NE	3118
Knoxville	10	7SW	3670
Memphis	5		2950
Nashville	5	10NW	3507
Texas			
Amarillo	−5	13SW	4335
Austin	10	N	1585
Brownsville	30	S	342
Dallas	10	11SE	2256
El Paso	10	10NW	2428
Fort Worth	10	11NW	2148
Galveston	25	12SE	1016
Houston	25	9SE	1157
San Antonio	20	8N	1202
Utah			
Salt Lake City	−5	7SE	5555
Vermont			
Burlington	−10	12S	7514
Virginia			
Lynchburg	10	5NW	4020
Norfolk	15	13N	3350
Richmond	10	8SW	3727
Roanoke	5	W	4068
Washington			
Seattle	10	11SE	4966
Spokane	−10	7SW	6355
Tacoma	10	7SW	5181
Walla Walla	−10	6S	4808
West Virginia			
Charleston	0	W	3789
Fairmont	−10		5047
Huntington	−5	W	4734
Parkersburg	−10	7SW	4948
Wisconsin			
Ashland	−20	SW	9066
Eau Claire	−20	NW	7970
La Crosse	−25	7S	7322
Madison	−10	10NW	7429
Milwaukee	−10	12W	7245
Wyoming			
Cheyenne	−20	14NW	7463
Sheridan	−30	6NW	8113
Canada			
Vancouver, B.C.	10	5E	5976
Victoria, B.C.	10	9N	5777
Winnipeg, Man.	−35	12SW	11,166
Fredericton, N.B.	−25	9NW	9099
Yarmouth, N.S.	0	13NW	7694
Port Arthur, Ont.	−40		10,803
Toronto, Ont.	−15	14SW	7732
Montreal, Que.	−20	15SW	8717
Quebec, Que.	−25	15SW	8628

Source: American Society of Heating and Ventilating Engineers.

EXAMPLE: Assume that a tank holds 700,000 lb of water that is to be heated from 60 to 75°F by 212°F steam. How many pounds of steam would be required?

The steam will give up latent heat of 970.3 Btu/lb and also 1 Btu/(°F · lb).

Heat required to raise the water to 75°F is

$$700{,}000\,(75 - 60) = 10{,}500{,}000 \text{ Btu}$$

The steam will give up $970.3W$ of latent heat, where W = pounds of steam required.

The steam will also give up heat after it is condensed and as it drops from 212 to 75°F, or

$$\text{Heat given up by condensed steam} = (212 - 75)W, \text{ or } 137W$$

Equating the Btu needed to raise the 700,000 lb of water to the steam heat given up, we have

$$10{,}500{,}000 = 970.3W + 137W$$

$$\frac{10{,}500{,}000}{1107.3} = W$$

$$W = 9483 \text{ lb of steam} \qquad \textit{Ans.}$$

NOTE: Keep in mind that the most Btu given up by steam is in condensing, when it gives up its latent heat.

Q Why are pockets of water undesirable in a steam-pipe system?

A Water hammer may occur. This is noticeable by intermittent clicking and knocking in a steam-heating system with improper drainage. Lack of drainage of condensate causes water pockets which act as pools for passing steam to condense. This brings local reduced pressures which the moving steam tries to fill. The result is that water is carried with the fast-moving steam, and as the water hits elbows and walls of the piping system, the impact of the water in the piping can create high water-hammer forces.

Q Referring to Table 4.1, if it is known that 25,000 gal of oil is used in one year with a heating value of 144,000 Btu/gal, how do we calculate the heating load per hour for a building with a heating-system efficiency of 78 percent? Assume the building is located in Cleveland and is to maintain 68°F inside temperature.

A From Table 4.1, Cleveland is considered to have an average of 6155 degree-days with an average low temperature of −5°F.

Let HL = heating load per hour, Btu/h.

$$\frac{\text{HL}}{68 + 5}\,24 = \text{Btu heat load per degree-day}$$

or

$$24\frac{HL}{73} = \text{Btu heat load per degree-day}$$

For the season,

$$\text{Btu load} = \text{degree-day} \times \text{Btu load/degree-day}$$

or

$$\text{Seasonal heat load} = \frac{6155(24)(HL)}{73}$$

Now,

$$\text{Seasonal heat load} = \text{heat input (on seasonal basis)} \times \text{efficiency}$$
$$\text{Seasonal heat load} = 25{,}000(144{,}000)(0.78)$$

Equating the two heat-load equations:

$$\frac{6155\ (24)(HL)}{73} = 25{,}000(144{,}000)(0.78)$$

and solving for HL;

$$HL = \frac{25{,}000(144{,}000)(0.78)(73)}{6155(24)}$$

$$HL = 1{,}387{,}652 \text{ Btu/h} \qquad \textit{Ans.}$$

If we assume that 970.3 Btu is given up per pound of steam needed to deliver the heating load per hour indicated, a low-pressure boiler of the following capacity is needed:

$$\frac{1{,}387{,}652}{970.3} = 1430 \text{ lb steam/h}$$

PIPING SYSTEMS

Q What is the purpose of air-venting steam-piping systems?

A Pipes and radiators contain air when steam is first generated in a heating system. The air will accumulate in radiators as steam enters the piping system and thus will retard heat transfer. Acceptable air-vent valves allow the air to pass out but close when steam starts to emerge from the opening.

In Fig. 4.3, when water enters the eliminator, the float rises and closes the outlet which remains closed until enough air collects in the chamber to drop the float so air is vented out at the top. If steam instead

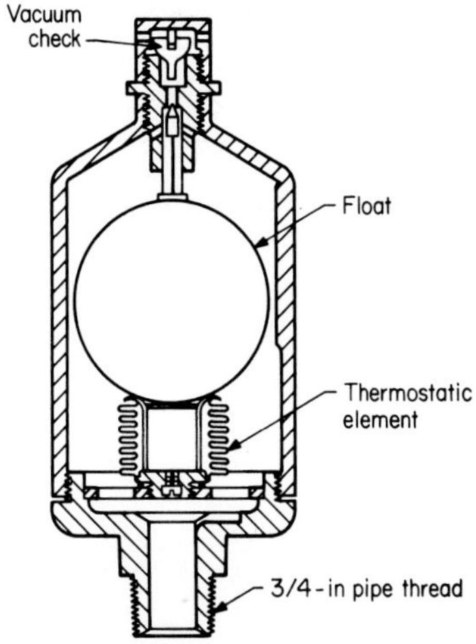

Fig. 4.3 Radiator or convector automatic air eliminator. (*Courtesy of Sarco Co.*)

of cooled condensate enters the eliminator, the thermostatic element, which is filled with a volatile liquid, expands and forces the outlet shut, preventing the escape of steam. A check valve on top prevents air from entering if the system is under a vacuum. Eliminators of this type are also used with hot-water systems, but for this use the thermostatic element is not required, nor is it always needed with steam.

Q Why should there be supply and return shutoff valves on a heating boiler?
A So that the boiler can be isolated from the piping system for testing, cleaning, or repairs.

Q What is a *dry-* and *wet-return main* on a steam-heating piping system?
A A *dry-return* main is one that has the return piping above the waterline of the boiler, while a *wet-return* main has the return piping below the waterline of the boiler. Wet returns are preferred over dry returns except where the return piping may be subject to freezing.

Q What is meant by equivalent length of pipe?
A Valves, elbows, and similar fittings offer resistance to flow as does straight-run piping. Values of the resistance to flow of steam through valves, elbows, and fittings have been experimentally developed, and these are expressed in equivalent length of straight pipe. Table 4.2 shows such values. By adding the equivalent length of pipe to straight-run pipes, total resistance to flow is easily calculated.

Table 4.2 Equivalent Length of Straight Pipe (in Feet) for Elbows, Valves, and Other Fittings with Same Resistance to Flow as Straight Pipe of Same Diameter

Size of pipe, in	Kind of fitting					
	Standard elbow	Side outlet tee	Gate valve	Globe valve	Angle valve	Reducing coupling
½	1.3	3	0.3	14	7	1.5
¾	1.8	4	0.4	18	10	2.0
1	2.2	5	0.5	23	12	2.5
1¼	3.0	6	0.6	29	15	3.0
1½	3.5	7	0.8	34	18	3.5
2	4.3	8	1.0	46	22	5
2½	5.0	11	1.1	54	27	6
3	6.5	13	1.4	66	34	7
3½	8.0	15	1.6	80	40	9
4	9.0	18	1.9	92	45	10
5	11.0	22	2.2	112	56	13
6	13	27	2.8	136	67	15
8	17	35	3.7	180	92	20
10	21	45	4.6	230	112	25
12	27	53	5.5	270	132	30
14	30	63	6.4	310	152	35

EXAMPLE: A 3-in steam main is 124 ft long and contains three elbows, two globe valves, and one tee. What is its equivalent length to resist steam flow?

For three elbows,

Equivalent length is $3 \times 6.5 = 19.5$ ft

For two globe valves,

Equivalent length is $2 \times 66 = 132.0$ ft

For one tee,

Equivalent length is $1 \times 13 = 13.0$ ft

Total equivalent length	164.5 ft
Plus straight-run pipe	124.0 ft
Total length of pipe is	188.5 ft *Ans.*

Q What is a return and steam trap?

A Most steam-heating systems will have a return trap to prevent the possibility of boiler water's backing into the return side. It is a form of check valve. Steam traps should be installed in lines wherever condensate must be drained as rapidly as it accumulates and wherever condensate must be recovered for heating, for hot-water needs, or for return to boilers. They are a *must* for steam piping, separators, and all steam-heated or steam-operated equipment.

Inverted open-float steam traps can be used to return condensate to regions where the pressure is less than that in the trap. They can lift condensate against a head not exceeding the pressure in the trap. Steam traps should be selected on the basis of their discharge capacity, and not by pipe size. Capacity is determined by the design and construction of the trap, the size of the orifice, and the effective seat pressure.

Nonpumping traps are used to drain condensate from steam lines, separators, and steam chambers of a wide variety of machinery. An efficient trap should expel condensate but prevent the wasteful blowing through of stream. Many types are available. Figure 4.4 shows a thermostatic radiator trap used on steam-heating systems. When water or steam enters this trap from the radiator at too high a temperature, the metallic bellows, which is partly filled with a volatile liquid, expands, closing the port opening. This prevents steam from the radiator from flowing out before it gives up more latent heat. Water or condensate at lower temperature will travel along the bottom and pass through to the condensate return-line system.

VACUUM HEATING SYSTEM

Q How does a mechanical heating system differ from a vacuum heating system?

A Both permit establishing a greater pressure differential between

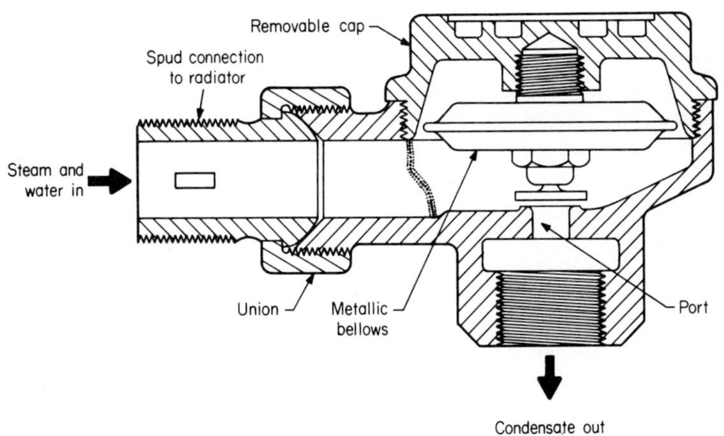

Fig. 4.4 Radiator trap with thermostatic element.

supply and return lines, thus permitting faster circulation. Both use pumps to promote circulation. The greater differential pressure makes it possible to use much smaller piping. The largest use of vacuum steam-heating systems is found in very large office buildings, hotels, apartments, and similar occupancies, where small pipes are desirable to promote rapid and well-balanced steam circulation throughout the building.

Q Are traps essential in vacuum steam-heating systems?
A Yes, more so than in gravity-return systems. Under the differential pressures which act in vacuum systems, every connection between the supply main and the return main must be trapped by a thermostatic or mechanical device.

> NOTE: One untrapped connection can defeat the operation of the entire scheme by passing enough vapor or air to prevent the pump from maintaining a pressure *below* atmospheric.

Vacuum traps are almost exclusively of the thermostatic type, with ports which open a trifle to pass water or air but which close tightly under steam or vapor temperatures. Thus the return pipes of a vacuum system when in proper condition do not transport anything warmer than the condensate, which under vacuum is much cooler than steam at atmospheric pressure. A typical vacuum trap is shown in Fig. 4.4.

Float traps may also be used for passing condensate from the supply side to the return side of vacuum heating systems. They are often used for this purpose, but separate provision must also be made in such cases for venting the air.

Q Describe a typical vacuum-pump return used in steam-heating systems.
A Figure 1.11 shows the return arrangements of one type of vacuum system. Condensate and air flow into the accumulator tank, which is placed at the low point to which the system can drain. A float-operated switch on the accumulator tank operates to start and stop the pump motor.

As centrifugal pumps cannot effectively pump air, the pump itself is not connected to the accumulator but receives suction water from the receiving tank. The pump delivers this water at high velocity through jets at *A* (in sketch), where the kinetic effect of the water jets in a combining tube aspirates (pulls) water and air from the accumulator. In a diffuser tube the kinetic energy of the jets is sufficient to compress the air-water mixture to the receiving-tank pressure, which is about atmospheric. The mixture in the receiving tank is separated, the air passing to the outside and the water dropping down into the tank to recirculate through the pump.

Only a portion of the water discharge from the pump is required by the aspirating jets, and the remainder is sent directly into the boiler as feedwater. An automatic control maintains the proper level of water in the receiving tank by closing down on the discharge valve to the boiler feed line until the level is reestablished by additional water entering from the jet circuit.

BOILERS FOR HEATING SYSTEMS

Q What are the two most prominent heating systems used today?
A Steam and hot-water systems are the most frequently used systems. They are further divided into low- and high-pressure systems. Both systems use either steel boilers of the fire-tube or water-tube types or cast-iron boilers. Steel boilers were discussed in the previous two chapters, with the smaller sizes generally used for low-capacity process needs or for space-heating purposes.

Q To what pressure can hot-water heating or hot-water supply boilers be built and still qualify as low-pressure boilers?
A Pressure of 160 psi and 250°F. If either is exceeded, the boiler qualifies as a high-pressure boiler and must be built under Section I of the ASME code, Power Boilers.

CAST-IRON HEATING BOILERS

Q In a broad, general sense, is the cast-iron boiler featured in Fig. 4.5 an FT or a WT type?
A The cast-iron (CI) boiler is basically a WT type because the water is inside the cast sections (no tubes) and the products of combustion are on the outside. But because of the limitations of cast iron, the boiler code treats CI boilers as a special type, without considering the heat-transfer method. Many CI boilers are stamped by the manufacturer as CI WT boilers, which should not be misinterpreted as a code classification.

Q Why are boilers built of cast iron?
A Primarily because of the high resistance of cast iron to corrosion. Also CI boilers retain heat longer because of the more massive construction of the components. On early boilers, cast parts eliminated the need for riveted, and later for welded, joints at seams of drum and headers. Cast iron also eliminated the costlier labor involved in assembling a steel boiler. Sections were cast to a pattern and then assembled without too much trouble, compared with the many details needed on earlier steel boilers. Then with the growth of central heating, CI boilers, along with room radiators, became immediately acceptable.

Q What grade of cast iron is used for CI boilers?

A *Cast iron* is a term applied to many iron-carbon alloys which can be cast in a mold to make a particular shape. But for CI boilers, gray cast iron is generally used. When the casting is cooled slowly in the molds, part of the carbon separates out as graphite. This makes the gray cast iron less brittle and easier to machine. Also, alloyed with nickel, chromium, molybdenum, vanadium, or copper, considerable tensile-strength properties can be achieved. The general practice is to classify cast iron by class:

Class No.	Ultimate tensile strength, psi
20	20,000
25	25,000
30	30,000
35	35,000
40	40,000

TESTING CAST-IRON BOILERS; ASME STAMPING

Q What *hydrostatic test* is required on HW (hot-water) heating or HW supply boilers built of cast iron and operating over a pressure of 30 psi?

A Each section of a cast-iron boiler must be subjected to a hydrostatic test of 2½ times the maximum allowable pressure *at the shop* where it is built. Cast-iron boilers marked for working pressures over 40 psi must be subjected to a hydrostatic test of 1½ times the maximum allowable pressure in the field (when erected and ready for service). After the boiler is in service and a hydrostatic test is required, the test should be at 1½ times the maximum allowable pressure.

Q In testing a steam or hot-water CI boiler that has been in service, what hydrostatic test pressure should be used?

A Cast-iron boilers are limited to a maximum working pressure of 15 psi for steam usage. The code allows field testing at a maximum of 1½ times the allowable pressure for either steam or hot-water boilers. There is a further stipulation that the test pressure must be controlled so the hydrostatic pressure imposed cannot be exceeded by more than 10 psi. Thus, for a steam boiler of CI construction, 22.5 psi would be the hydrostatic pressure imposed in the field.

For hot-water boilers, the test pressure would be 1½ times the allowable pressure for hot-water service. Therefore, a 30-psi hot-water-heating boiler can be subjected to a 45-psi hydrostatic test, a 60-psi hot-water boiler would require a 90-psi hydrostatic test, etc.

Fig. 4.5 Cast-iron vertical-flue design with controls mounted on front of boiler. (*Courtesy of Burnham Corporation.*)

Q What *stamping* does the ASME require on cast-iron boilers?
A The marking should consist of the following: (1) manufacturer's name, (2) maximum allowable pressure in psi, (3) capacity in pounds per hour for steam or Btu per hour for water service.

Q The symbol NB is often noted on boilers, with a number following it. What does this stand for?
A NB stands for National Board of Boiler and Pressure Vessel Inspectors. It means that the boiler's design and fabrication were followed in the shop by an NB-commissioned inspector, including the witnessing of the hydrostatic test and signing of data sheets required by the ASME.

Q What stamping should be noted on a CI boiler to indicate that it is built under ASME rules?
A The first stamping to note is the ASME Heating Boiler Code symbol shown in Fig. 4.6. The other stamping required is one indicating allowable pressures and capacities, and whether the boiler is for water or steam usage.

Q What is the minimum-size safety valve (SV) allowed on a CI steam boiler? Is the same true for a hot-water CI boiler?

Fig. 4.6 Official symbol for stamps to denote ASME Standard Heating Boilers, Section IV.

A ASME requires a minimum ¾-in SV for both steam and hot-water CI boilers.

Q What is meant by an officially rated ASME pressure-relief valve?
A An officially rated ASME pressure-relief valve is meant for hot-water boilers. It must be stamped for its pressure setting and its Btu-per-hour relieving capacity. Also, it must be equipped with a hand test lever, must be spring-loaded, and must not be of the adjustable screw-down type. A typical SV ASME symbol is shown in Fig. 4.7.

TYPES OF CAST-IRON BOILERS

Q Name the types of CI boilers built today.
A Cast-iron boilers are built to various shapes and sizes but can be grouped into the three following broad classifications:

1. Round CI boilers (Fig. 4.8) consist of a firepot (furnace) section with base, a crown sheet section, one or two intermediate sections, and a top or dome section. The sections are held together by tie rods or bolts with push nipples (Fig. 4.9) interconnecting the waterside of the sections. Thus water circulates freely through the nipples from section to section. Fuel is burned in the center furnace, with the flue gases rising and flowing through the various gas passages of the water-filled sections, then out to the stack. The round CI boiler used to be popular for HW-supply service, and was often stamped "hy-test" to indicate an allowable pressure of 100 psi for domestic hot-water service. But it is rapidly being replaced by fired steel-welded water heaters or by electrically heated water heaters.

Fig. 4.7 ASME standard symbol for safety relief valves.

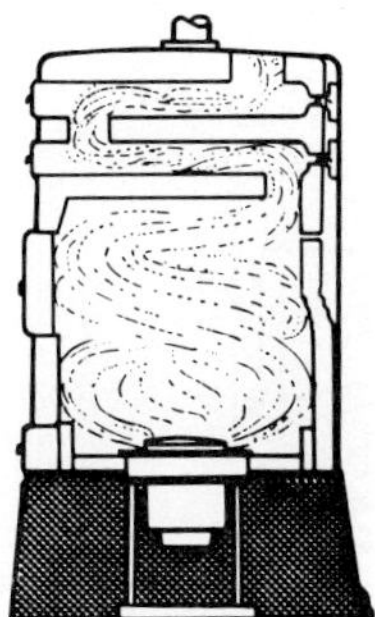

Fig. 4.8 Round cast-iron boiler with wall-flame rotary oil burner.

2. Sectional boilers (vertical) consisting of sections assembled front-to-back, with sections standing vertically and assembled by means of push nipples or screwed nipples. See Figs. 4.10 and 4.11.

3. Sectional boilers (horizontal) consisting of assembled sections stacked like pancakes. Here each section is laid flat in relation to the base. This type of vertical stacking may be supplemented by having three vertically stacked boilers side by side and interconnected to gain additional capacity. In this arrangement a common supply and return header is used with no intervening valves between the vertically stacked boilers. These units are usually gas-fired, with a burner for each vertical stacking.

Q Name the two design features used to aid in assembly of vertical CI boilers.

A 1. Internal, tapered push nipples inserted into holes of the vertical section. Then by means of through tie rods, or short tie rods (Fig. 4.9), the sections are pulled together by tightening nuts against washers on

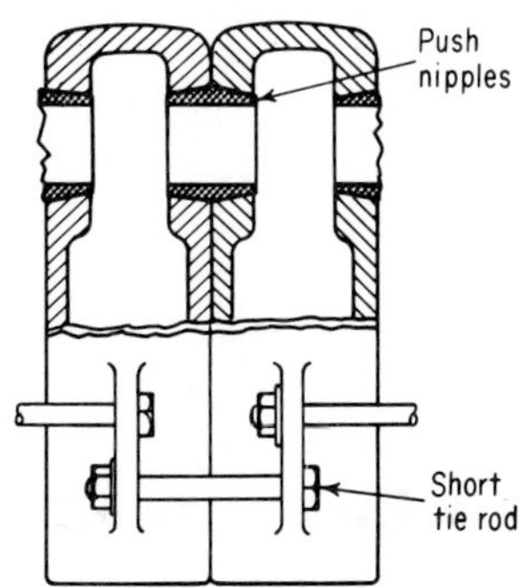

Fig. 4.9 Details of push nipple and short tie-rod construction.

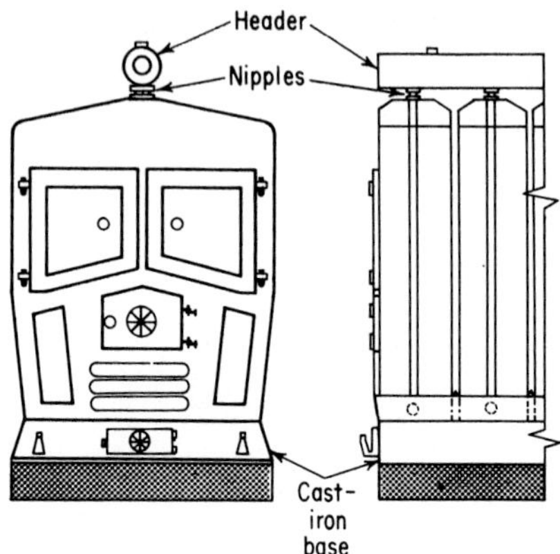

Fig. 4.10 Vertical-sectional horizontal-stacked cast-iron boiler of the screwed-nipple type.

the tie rod. That interconnects the sections on the waterside, enabling them to withstand pressure as assembled sections.

2. External header type (Fig. 4.12) where the sections are indivdually assembled to headers (supply drum and return drums) by means of threaded nipples, locknuts, and gaskets. This type of assembly allows replacing an intermediate section, because only the locknuts, gaskets, and threaded nipple have to be removed on each header to slide a section out. In contrast, with through tie-rod construction, all the sections in front of the intermediate section to be replaced have to be removed first to get at the affected section.

GAS TRAVEL

Q How does the flue gas travel through a CI boiler?

A In the design shown in Fig. 4.13, flue gases rise from the combustion area into uptakes between each boiler section at the right and left

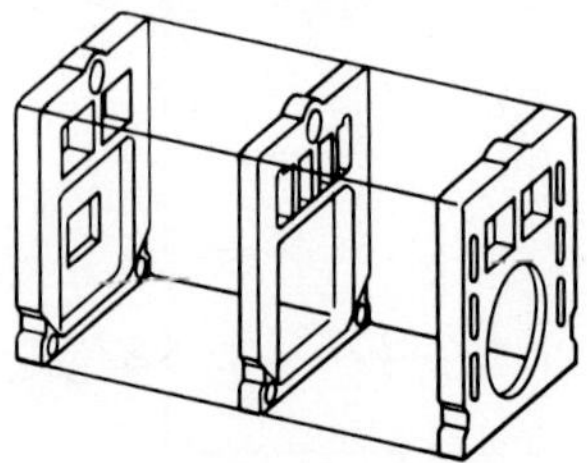

Fig. 4.11 Assembling vertical sectional cast-iron boiler of push-nipple type.

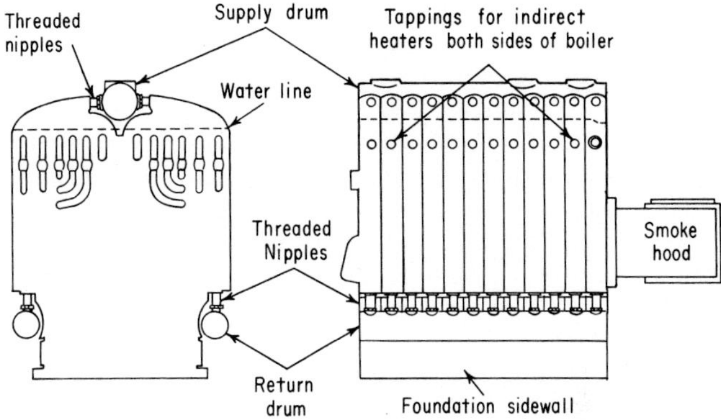

Fig. 4.12 Header-drum and screwed-nipple construction of cast-iron boilers.

side. The gases are directed through the two outside flue passages to the front of the boiler and then back through the center flue ways to the smoke collar. The uptakes between each section expose more radiant-heat-absorbing surfaces and divide the hot gas volume into small gas streams. These gas streams are directed to the cooler sides of the firebox. Multiple uptakes combined with three-pass design prevent gases taking shortcuts to the chimney. Thus longer flue-gas travel and higher velocities increase heat absorption of the secondary heating surfaces.

Today CI boilers are also built for forced-draft firing. This is possible by sealing the space between sections with asbestos rope seals. The advantages claimed are that the pressurized furnace can use a shorter chimney and less boiler-room space because the unit is smaller and more efficient. Flame-retention-type burners are used which hold the flame in front of the nozzle, guiding it in a shape-controlled pattern.

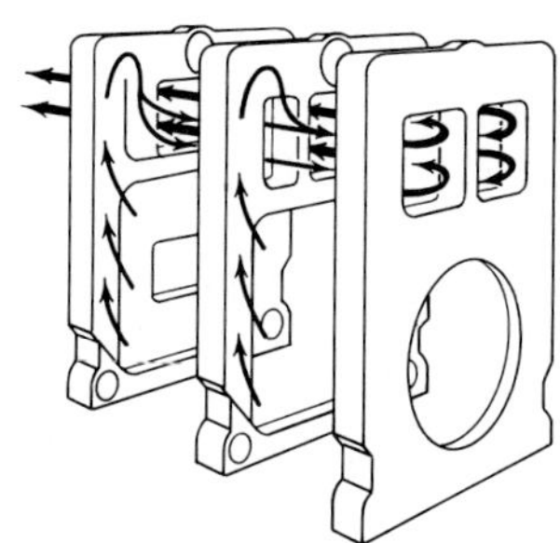

Fig. 4.13 Gas travel through tie-rod assembled-type cast-iron boiler.

The sections of these boilers have a newly designed, grooved, seal strip which receives the asbestos rope. When installed, the outer edge of the rope is accessible between sections so that the boiler can be visually checked for furnace tightness. The rope is compressible, thus allowing for contraction and expansion of the boiler. The assembled boiler, partially illustrated in Fig. 4.13, requires no combustion chamber. The bottom and sides of the assembled sections form the furnace. Note this arrangement's similarity to a WT boiler with waterwalls and bottom furnace tubes. Here also, flame-retention burners limit the flame travel and avoid flame impingement on the sections.

Figure 4.14*a* shows the external-header boiler and its gas travel. This unit has two half-sections for each vertical row, sometimes referred to as pork chop sections. The design shown may also be provided with a manifold on both sides (Fig. 4.14*b*) to obtain domestic hot water.

Q Does a CI boiler require a bottom blowoff pipe and valve?

A Yes. The ASME code requires each boiler to have a blowoff pipe connection fitted with a valve or cock, of not less than ¾-in pipe size. It must be connected with the lowest water space practicable.

Q What is the minimum size of pipe required for connecting a water column to a steam-heating boiler?

A The minimum size of ferrous or nonferrous pipe must be of 1-in diameter.

Q Does the ASME code require a low-water fuel cutoff on a CI steam-heating boiler?

A Yes. Each automatically fired steam- or vapor-system CI boiler (or low-pressure steel steam boiler) must be equipped with an automatic

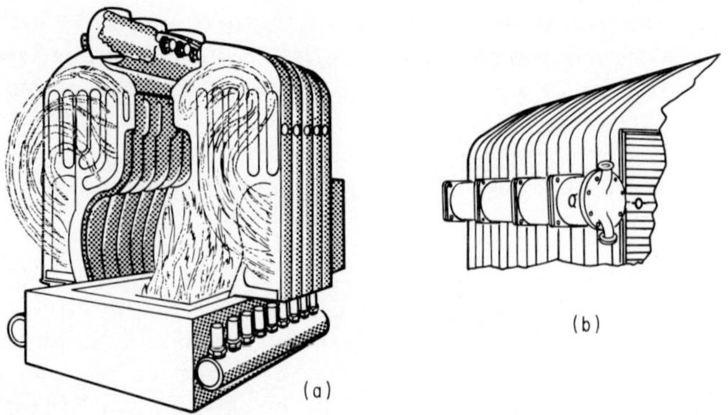

Fig. 4.14 (*a*) Gas travel in header-type cast-iron boiler. (*b*) Manifold for heating domestic water.

low-water fuel cutoff. It must be so located as to automatically cut off the fuel supply when the water level drops to the lowest safe waterline. The safe waterline cannot be lower than the lowest visible part of the water gage glass.

CAST-IRON BOILER FAILURES

Q Can CI boilers explode, or do they merely crack?
A When controls are not operating properly, or if the safety relief valve is stuck or of the wrong size, a CI boiler will explode. Many have. The big difference between a shell-type steel boiler explosion and a CI boiler explosion is that a CI boiler will usually fragmentize into smaller pieces of the affected sections. A steel boiler, on the other hand, rips and tears along the sheet of a drum or shell. Then it flies apart in the form of curved panels of steel. But in each explosion, the danger to life and property is very great.

Q Name the most frequent failure on a CI boiler.
A Cracking of a section or sections, permitting steam and water to gush out of the cracks, and thus making the boiler inoperable.

Q Does a cracked section require a complete replacement of all sections on a CI boiler?
A Only if the model and type of boiler section are obsolete and no longer available. The reason is that cast sections made from a pattern to be built and then cast in a foundry are too expensive a repair. Therefore, the obsolescence of a CI boiler can be very rapid if the model owned or operated is no longer stocked by the manufacturer with replacement sections. On steel boilers, especially welded boilers, repairs can be made more readily by replacing defective areas of sheets, veeing out and rewelding cracks, etc.

If sections are available, and one or two sections are cracked, the defective section or sections can be removed and the replacement installed without too much difficulty (see Fig. 4.15). On older through-tie-rod construction, this may mean first removing good sections in front of the defective section. But on the external header type with screwed nipples, only the defective section has to be removed.

Q Cannot cracks in cast-iron sections be repaired by welding?
A Yes, but the work must be done by a firm, or welder, very experienced in welding cast iron. At times, depending on the location and length of the crack, economics will dictate a cheaper repair. Thus a more reliable repair may be to replace the cracked section. For example, if of through-tie-rod construction and the section is cracked in an intermediate inaccessible area, the unit will have to be dismantled. Then the

Fig. 4.15 Replacement section of cast-iron boilers.

area of the crack will have to be thoroughly cleaned on the fireside (also waterside, if possible), followed by slow and careful welding.

But if the crack is small and in an accessible area, it may be possible to repair the crack in position, if done properly by an experienced welder. Fusion welding of the *cold-welding* type is recommended, using iron, steel, stainless steel, nickel, or bronze electrodes. But the experience of the welder may dictate the electrode.

Q What is meant by cold welding?
A Cold welding is a method of fusion welding that has been used for years in the repair of castings such as CI boilers, cylinder blocks, and stationary CI objects. Basically, cold welding is done by keeping the heat input to the casting at a minimum while keeping the casting and repair area at a low temperature until welding is completed. While no preheating is used, it is best to have the casting at room (ambient) temperature. Of course, any foreign matter should be removed before welding.

Q Name at least seven causes of CI boilers cracking.
A Cast-iron boilers of different poured shapes, with unpredictable stress-concentration areas, geometry, service factors, waterside fouling, soot and carbon accumulation on the fireside, and rapid temperature changes are all causes for cracking. Thus they may never fail twice in an identical manner. Careful investigation will usually point to one or more of the following causes:

1. Rapid introduction of cold water into a hot boiler, as may occur with poorly operated manual makeup or an automatic feed device.

2. Makeup line connected to a section instead of to the return line to temper the cold water with the returning condensate or hot water (the code requires makeup to *enter return lines*).

3. Controls not functioning and thus overstressing the boiler by either pressure or temperature.

4. Insufficient water in the boiler.

5. Internal concentrated deposits, blanking off proper heat transfer or obstructing circulation from section to section.

6. Defective material or casting which does not become noticeable until after several years of service.

7. Poor assembly or workmanship, or improper installation of the boiler as to pitch, alignment, tie-rod tightening, screwed-nipple fitting, etc.

SECTION CONNECTIONS TO BE CHECKED

Q What should be checked on section connections?

A When checking new boilers, always check the type of nuts securing the tie rods. If solid steel or brass nuts are used, make sure they are only hand-tight, or backed off a few threads. The first choice on new boilers is collapsible washers, with shallow split brass nuts. The second choice is split shallow nuts backed off hand-tight (or cloverleaf-type nuts that fail when a slight expansion takes place). On an older unit idle during summer months and located in a damp basement, make sure the holding nuts are not tight. Also determine whether the tie rods are rusted into their holes, which may have the same binding effect as tight securing nuts. Obviously, if the rods are free and nuts slacked off, there should be no problem of expansion cracking. If rust growth plus tight tie rods are the cause of cracking, the boiler must be dismantled and the rust buildup removed by chipping.

Header-type boilers have no tie rods or tapered nipples; so nothing can be adjusted to allow for abnormal expansion. In addition to rust depositing between the sections, rapid start-up can cause serious damage to these units. Make sure older boilers without good blowdown facilities do not develop scale buildup, because it results in cracking. Restriction of water supply and circulation can also be caused by scale buildup in these units, resulting in overheating.

ASME RECOMMENDATIONS

Q What are the ASME recommendations for connecting steam-heating boilers in battery?

A See Fig. 4.16 and note the following:

1. Return-loop connection shown eliminates the need of check

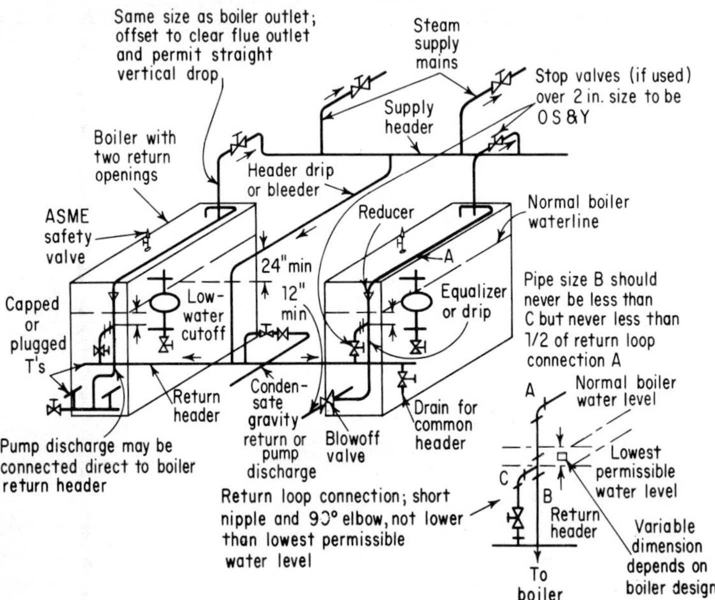

Fig. 4.16 ASME recommendations for connecting steam-heating boilers in battery.

valves on gravity-return systems, but in some localities a check valve is a legal requirement.

2. When pump discharge piping exceeds 25 ft, install a swing check valve as shown and at the pump discharge.

3. If pump discharge is looped above normal boiler waterline, install a spring-loaded check valve at the return header and at the pump discharge.

4. Where supply pressures are adequate, feedwater may be introduced directly to a boiler through an independent connection (see the latest ASME code).

5. The return connections shown for a multiple-boiler installation may not always ensure that the system will operate properly. Thus in order to maintain proper water levels in multiple-boiler installations, it may be necessary to install supplementary controls or other suitable devices.

Q Name the four ASME recommendations for return-pipe connections to steam-heating boilers.

A 1. The return-pipe connections of each boiler supplying a gravity-return steam-heating system should be arranged to form a loop as shown in Fig. 4.16. Then the water in each boiler cannot be forced out below the safe water level.

2. For hand-fired boilers with a normal grate line, the recommended pipe sizes detailed as A in Fig. 4.16 are 1½ in for 4 ft^2 or less of firebox area at the normal grate line, 2½ in for areas more than 4 ft^2 and up to 14.9 ft^2, and 4 in for 15 ft^2 or more.

3. For automatically fired boilers which do not have a normal grate line, the recommended pipe sizes detailed as A in Fig. 4.16 are 1½ in for boilers with a minimum SV relieving capacity of 250 lb/h or less, 2½ in for boilers with minimum SV relieving capacity from 251 to 2000 lb/h, inclusive, and 4 in for boilers with more than 2000 lb/h minimum SV relieving capacity.

4. Provision should be made for cleaning the interior of the return piping at, or close to, the boiler.

Q What are the ASME recommendations for connecting hot-water-heating boilers in battery.

A See Fig. 4.17. The pipe size can usually be obtained from the boiler manufacturer if the model and type are identified. Also see the ASME

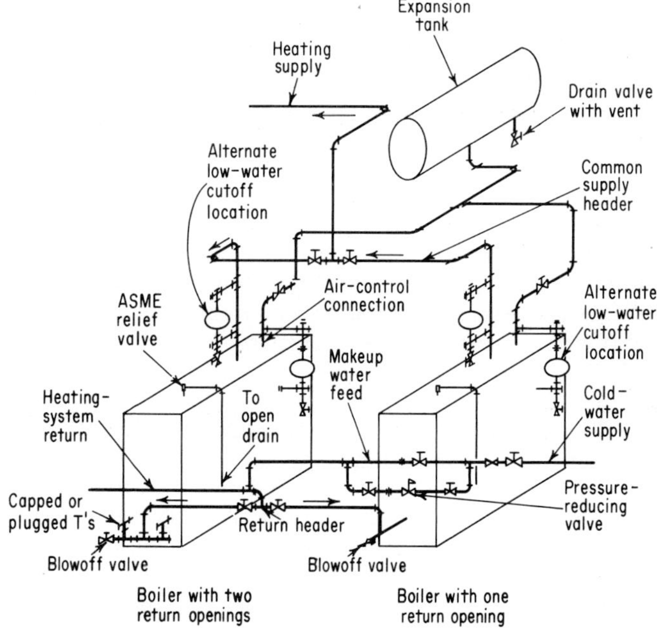

Fig. 4.17 ASME recommendations for connecting hot-water-heating boilers in battery.

and the local code. Note that acceptable shutoff valves, or cocks, in the connecting piping may be installed for convenience of control testing and service.

USE OF HIGH-PRESSURE BOILERS FOR LOW-PRESSURE HEATING SYSTEMS

Q If a high-pressure boiler is used for heating low-pressure apparatus through a *reducing valve*, what should be guarded against on the low-pressure side of the system?
A Overpressure, due to lack of proper pressure setting and capacity of a safety valve (SV) or to no SV should the reducing valve fail or stick in the open position. To guard against overpressure, install a safety relief valve on the low-pressure side of the reducing valve, with adequate capacity in pounds per hour to match the reducing-valve capacity.

The pressure setting of the SV should always be based on the maximum allowable pressure of the weakest equipment on the low-pressure side of the reducing valve. The SV should be as near the reducing valve as possible. In the event of multiple vessels on the low-pressure side of the reducing valve, do *not* combine total SV on each vessel. *Always assume* only one vessel may be operating, and, therefore, total relieving capacity of all SV may be *no protection.*

For example, suppose there are six kettles, each with an SV rated for 200 lb/h, supplied by a common steam line off a reducing valve. If the reducing valve can pass 900 lb/h, and if five kettles are shut off, the vessel operating will not be properly protected. Why? Because it will have only 200 lb/h of protection against 900 lb/h of steam flow. In this type of situation, an SV on the common line, right after the reducing valve, is required, with a minimum relieving capacity of 900 lb/h. Proper *overpressure* protection requires both pressure and capacity to be matched with the reducing-valve capacity to avoid a serious explosion. There have been numerous explosions for this very reason, *lack of overpressure protection on the low-pressure side of reducing valves.*

The NB has drawn up the following method to calculate the required relief-valve capacity in pounds per hour on the low-pressure side of a reducing station. These rules are drawn from thermodynamic principles, involving flow through a nozzle. A coefficient of discharge had to be assumed (around 70 percent). Check with the reducing-valve manufacturer for the maximum flow and then install the recommended capacity. In lieu of data from the manufacturer, use the following formula:

$$\text{RVC} = \tfrac{1}{3} \times \text{OC} \times \text{VSPA}$$

where RVC = relief valve capacity required, lb steam/h
OC = orifice capacity, lb steam/(h · in^2) (Table 4.3)
VSPA = valve size pipe areas, in^2 (Table 4.4)

Where a pressure-reducing valve is supplied by steam from the boiler, the capacity of the SV or valves on the low-pressure side of the system need not exceed the capacity of the boiler for obvious reasons.

NOTE: Most pressure-reducing valves are arranged with a valved bypass, which also acts as a potential steam source hazard in case the bypass is left open. Where such a valved bypass is used, the following formula should be used to determine the steam flow rate through the *bypass:*

$$RVC = ½ \times OC \times BPA$$

where RVC = relief valve capacity, lb steam/h
OC = orfice capacity, lb steam/(h · in^2) (Table 4.3)
BPA = bypass pipe area, in^2 (Table 4.4)

The *larger* of the relief-valve capacities calculated by the above two formulas should be used for selecting the relief valve for the vessel.

EXAMPLE: Suppose a high-pressure boiler operating at 125 psi distributes steam to a series of 40-psi ASME-constructed retorts through a 1½-in pressure-reducing valve provided with a 1-in globe valve bypass. Determine the proper ASME relief-valve protection for the retorts. Using data in Tables 4.3 and 4.4 and the first of the two formulas above:

$$W = ⅓ \times 7200 \times 2.04 = 4896 \text{ lb steam/h} \qquad \textit{Ans.}$$

Checking the bypass steam flow according to the second formula gives

$$W = ½ \times 7200 \times 0.86 = 3100 \text{ lb steam/h} \qquad \textit{Ans.}$$

The potential steam flow through the pressure-reducing valve is 4896 lb/h rated capacity, or 4896 × 1000, or 4,896,000 Btu/h. This is the minimum capacity required of the SV to protect the low-pressure side against overpressure.

HEATING BOILER SAFETY VALVES

Q What is the maximum size safety valve permitted on a heating boiler?

A Section IV of the ASME code stipulates 4½-in diameter.

Table 4.3 Orifice Relieving Capacities [lb/(h · in^2)] for Determining the Proper Size of Relief Valves Used on Low-Pressure Side of Reducing Valves

Outlet pressure,	Pressure-reducing valve inlet pressure, psi								
psi	125	100	85	75	60	50	40	30	25
110	4550								
100	5630								
85	6640	4070							
75	7050	4980	3150						
60	7200	5750	4540	3520					
50	7200	5920	5000	4230	2680				
40	7200	5920	5140	4630	3480	2470			
30	7200	5920	5140	4630	3860	3140	2210		
25	7200	5920	5140	4630	3860	3340	2580	1485	
15	7200	5920	5140	4630	3860	3340	2830	2320	1800
10	7200	5920	5140	4630	3860	3340	2830	2320	2060
5	7200	5920	5140	4630	3860	3340	2830	2320	2060

Table 4.4 Valve Size Pipe Areas

	Standard		
Nominal pipe size, in	Actual ext diam, in	Approx int diam, in	Approx int area, in²
3/8	0.675	0.49	0.19
1/2	0.840	0.62	0.30
3/4	1.050	0.82	0.53
1	1.315	1.05	0.86
1 1/4	1.660	1.38	1.50
1 1/2	1.900	1.61	2.04
2	2.375	2.07	3.36
2 1/2	2.875	2.47	4.78
3	3.5	3.07	7.39
3 1/2	4.0	3.55	9.89
4	4.5	4.03	12.73
5	5.563	5.05	19.99
6	6.625	6.07	28.89
8	8.625	8.07	51.15
10	10.750	10.19	81.55
12	12.750	12.09	114.80

NOTE: In applying these rules, the area of the pipe is always based upon standard weight pipe and the inlet size of the pressure-reducing valve (high pressure inlet side to reducing valve).

Q What pressure rise is permitted on a boiler before the safety-valve capacity may be considered inadequate for heating boilers?
A For steam-heating boilers, the safety-valve capacity should be sufficient at maximum firing rate that the pressure does not rise more than 5 psi. For hot-water-heating boilers, the relief-valve capacity must be sufficient to prevent the pressure from rising more than 10 psi with the burners operating at the maximum firing rate.

Q Name the two pressure controls required on a steam-heating boiler and the temperature controls required on a hot-water-heating boiler.
A Automatically fired steam-heating boilers require an operating-pressure cutout switch set lower than the allowable pressure and an upper-limit cutoff switch also set lower than the allowable pressure but never at a pressure higher than 15 psi. For hot-water-heating boilers, an operating temperature cutout switch set less than the maximum allowable temperature is required as well as an upper-limit temperature cutout switch set no higher than 250°F.

CHAPTER **FIVE**

Boilers for Special Applications

Concern about fuel costs and fuel availability is generating great interest in waste-heat boilers, waste-fuel burning in boilers, cogeneration, and similar methods of extracting all the energy out of sources that in the past may have been wasted. Here we review some of these methods that aim to accomplish a better utilization of all fuel resources with special boiler systems that extract as much energy as possible from fuel. The return to solid-fuel burning created possible air-pollution problems in the form of acid rain due to high SO_2 emissions. Fluidized-bed burning is a promising technique to overcome the SO_2 emission problem, also reviewed here.

COIL-TYPE BOILERS

Q What are special boiler systems?
A Any of the better-known designs of boilers that have unusual applications and therefore cannot be labeled as a conventional fired fire-tube, water-tube, or cast-iron design.

Q Are controlled-circulation boilers built in smaller sizes?
A Yes, they are popular in sizes of 15 to 300 hp (500 lb/h to about 10,000 lb/h). One design has steam generated in one continuous spiral helical-wound coil (Fig. 5.1), with no headers in between. Rapid steaming ability is a feature of several coil-type WT boilers. They have forced circulation and intense heat release from a gas or oil burner. Most come to full capacity and pressures within 5 min after start-up. Operation is automatic. They may generate as little as 1500 lb/h, although some produce 15,000 lb/h at 900 psi.

The boiler illustrated is shipped complete, ready to operate, fully automatic at pressures to 300 psi. In addition to several units often

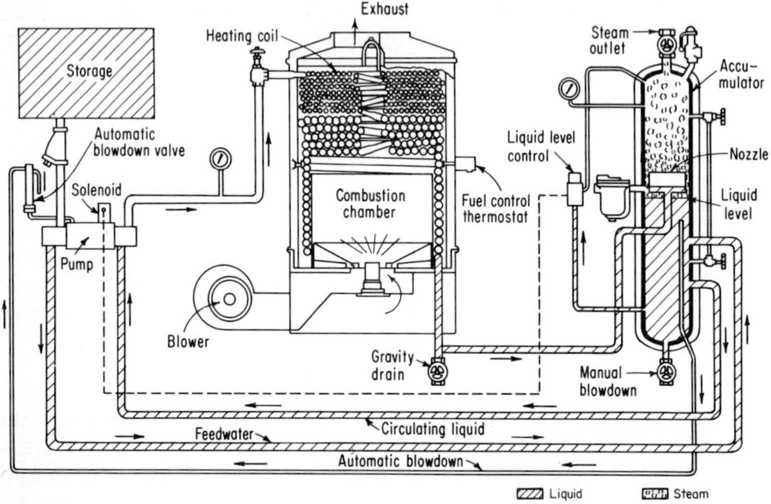

Fig. 5.1 Coil-type unit of once-through design has accumulator. (*Courtesy of Clayton Corp.*)

installed in batteries, they are also used for emergency steam service, available on a rental basis, delivered where needed.

An unusual feature is the ring thermostat tube (Figs. 5.1 and 5.2). This tube is an integral part of the heating coil, actuated directly by combustion heat. In case of water shortage or any excessive heat condition, the thermostat control will expand beyond the set point and thus shut off the fuel supply.

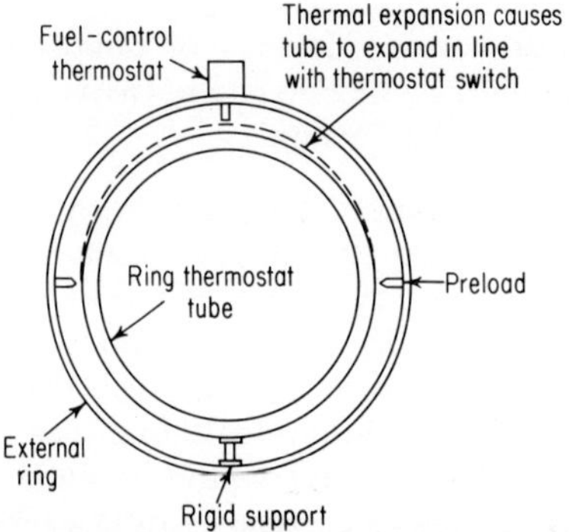

Fig. 5.2 Expanding overheated ring thermostat shuts off the fuel to prevent dry firing.

Another design has three nests of coils and inlet and outlet headers. It is widely used in small sizes up to 300 hp, and pressures to 250 psig.

Q Name some disadvantages of coil-type boilers.
A Any malfunction in the loop may cause a coil failure from overheating. This could come about from pump failure, partial blockage of inlet and outlet lines, blocked tubes (scale), fireside soot accumulation in concentrated heat zones, malfunction of controls, etc. Thus it is essential to keep both the waterside and fireside of the coils clean; proper feedwater treatment is vital. A pressure differential chart is often used to show whether the difference between suction and discharge pressure on the recirculation pump exceeds a certain pressure differential. If so, it is a sign of tube or flow obstruction. In a conventional boiler of multitube design, a leaky tube can be plugged, then the boiler operated until it is convenient to replace the affected tube. With a coil-type boiler the entire coil must usually be replaced.

WASTE-HEAT BOILERS

Q Where are waste-heat boilers used besides in utilizing the exhaust heat from gas turbines and internal-combustion engines?
A Where by-product heat of sufficient Btu content is available from a manufacturing process. With air-pollution laws becoming more stringent, more waste-heat boilers will probably be installed.

Q What is meant by a combined cycle?
A This is a cycle which uses a combination of gas-turbine exhaust, steam generators, and steam-turbine drives (generator or other) in an attempt to improve cycle efficiency by better extraction of useful energy from fuel.

Combined-cycle installations may be categorized into four broad classifications, each of which is primarily dependent on how the steam generator is used in conjunction with the gas turbine:

1. Gas turbine plus unfired steam generator.
2. Gas turbine plus supplementary-fired steam generator (Fig. 5.3).
3. Gas turbine plus furnace-fired steam generator.
4. Supercharged furnace-fired steam generator plus gas turbine.

Simple gas turbines without regenerative heating of incoming air have poor thermal efficiencies or heat rates. Depending on the particular manufacturer and fuel utilized, heat rates ranging from 13,000 to 16,000 Btu/kW · h may be obtained with current designs.

This heat rate may be improved by a number of major design modifications, such as increased compressor pressure ratios and higher gas-turbine inlet-gas temperatures. However, these create problems

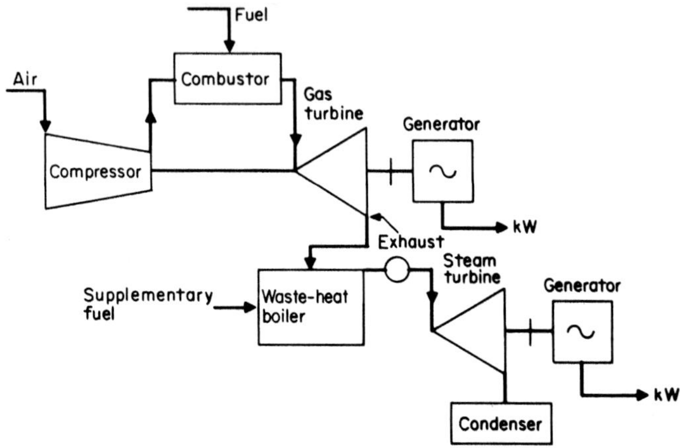

Fig. 5.3 Simple combined cycle uses exhaust heat from gas turbine to generate steam for steam turbine.

with materials of construction that are still under development. The station heat rate associated with present gas-turbine designs may be further improved by utilizing the energy contained in the turbine exhaust gases.

EXAMPLE: A 50-MW gas turbine might discharge about 2,200,000 lb/h of exhaust gases at about 900°F, with an equivalent value of 450 $\times 10^6$ Btu/h of energy, or approximately 75 percent of the fuel input. This can be recovered partially in a waste-heat boiler.

Where supplementary fuel is used, gas-turbine exhaust contains from 16 to 18 percent oxygen and may be used as an oxygen source to support further combustion. Therefore, a modification of the simple waste-heat application is the use of a supplementary firing system located in the connecting duct between the gas turbine and the steam generator. See Fig. 5.3. The firing system will utilize a portion of the oxygen contained in the gas-turbine exhaust and be selected to limit the maximum gas temperature entering the steam generator to approximately 1200 to 1300°F. By maintaining this maximum gas temperature level, the steam generator will retain its relatively simple arrangement.

With a given gas turbine size and this gas-temperature limit, the steam generation will double that of a simple waste-heat application, and the steam turbine will supply a greater proportion of the plant load. For most current designs, the steam turbine will supply approximately 50 percent of the total electrical output. The higher steam-generator inlet-gas temperature will allow steam conditions to be increased to levels of 1500 psig and 950°F. The steam-turbine designs are nonreheat and may be either condensing or noncondensing.

Q What must be considered in integrating a gas-turbine steam generator and steam turbine?
A The selection of the gas turbines and steam generators must be carefully coordinated in order to supply sufficient combustion air to the boiler throughout the load range to allow safe operation. The general arrangement of the major steam-generator components, such as the furnace, superheat, and reheat sections, utilizes the same concepts as large utility units, and while sufficiently challenging, no new problems result because of the combined-cycle application.

One important factor in the selection of the boiler is the degree of independence which the plant will require of the gas turbine and the steam turbine. If the two systems must always operate in series, the plant design is considerably simplified, particularly in the selection of the fans and connecting ductwork. However, in practice, most plants require that each of the prime power suppliers be capable of operating either in series or independently, in order to improve plant reliability and flexibility.

Q What is meant by topping and bottoming cycles in combined-cycle systems?
A In a topping cycle the fuel is burned to produce electricity, and the exhaust heat is used to make process steam in a waste-heat boiler. In a bottoming cycle, the fuel is burned for process use, such as in a lime kiln, smelting, and similar process applications, and the exhaust heat from the process is used to generate electricity.

Q What is cogeneration?
A Cogeneration is the coincident generation of electricity and process-steam needs for a manufacturing complex. Usually high-pressure steam is produced in on-site steam generators. This high-pressure steam is passed through a turbogenerator to obtain electric power and process steam, by extracting the steam from the turbine at a pressure needed for the process. Several extraction pressures can be used. The overall heat balance for a cogeneration plant can thus be attractive in comparison with just buying power and generating only process steam.

Q What kinds of boilers are used in waste-heat applications?
A Depending on conditions, these types:

1. Fire-tube boilers, both the vertical and horizontal types, if waste gas is relatively clean.

2. Straight-tube WT boilers, for clean or moderately dust-laden waste gas.

3. Water-tube of the bent tube (Stirling) type, for very heavy dust loadings.

4. Positive-circulation boilers, for clean, low-temperature gases.

5. Pressurized or supercharged boilers, for gas-turbine exhaust (Velox type).

Q What is meant by the term *total energy system*?

A Total energy refers to the reappearance of generating plants in large apartment-house projects, office-building complexes, shopping centers, and other locations where year-round comfort cooling and heating are required. By proper design, it is possible to generate electricity by means of diesel engines or gas turbines and then use the exhaust heat (previously wasted) to heat the premises in winter. The warm engine-cooling water discharge under pressure is also used to generate hot water or steam. By means of large absorption refrigeration machines, or centrifugal compressor machines utilizing exhaust steam, the premises are cooled in summer. Other by-products of exhaust heat include hot-water supply for washing purposes. The electricity generated can be used for lighting or for driving pumps and fan motors for the air-conditioning and heating systems. Thus a very efficient heat rate can be obtained (up to 70 percent), and the energy of the fuel is spread in three directions: (1) electricity for light and power, (2) heating, and (3) cooling. Therefore, the term total energy system is used, because all energy required is supplied.

INCINERATOR PLANTS

Q What types of boilers are used in incinerator plants?

A Four-drum Stirling-type boilers were used in older installations, whereas today two-drum types are used. The two-drum types have a furnace designed to utilize the waste heat from a municipal incinerator. Usually the gases from the incinerator enter through an opening in the side wall of the furnace near the floor. The rear wall of the furnace may have an oil burner to supplement the waste heat when more steam is required. A convection-type superheater is used behind the water-cooled bridge wall. The garbage is not burned inside some boilers, but rather in outside ovens connected to the furnace. The flue gas is led to these boilers by means of a bricked-in tunnel. Still another type of unit has a pendant superheater suspended over the waste-gas inlet. It is a three-drum boiler, two drums on top and one on bottom. While used in large petrochemical plants for pressures up to 450 psig at 550°F, it can also be designed for incinerator service. And capacity is up to 112,000 lb/h.

With tougher air-pollution laws, many cities are turning to more efficient European incinerators. Figure 5.4 is one example. The boiler in the plant shown has a superheater and economizer. The refuse is burned on a barrel-grate stoker, and the steam generated is sold for district heating.

Q Assume a process of a continuous nature providing waste heat to a waste-heat boiler, thus making the securing, or killing, of a fire or process extremely expensive, if not impossible. How can such a boiler be secured in case of low water to avoid burning the tubes?

A This is an extremely important design consideration. And if designed for, it is also a very important operating check that should be made periodically to ensure that the unit is in working condition. This test should be the same as when testing low-water fuel cutoffs on a suspended fuel-fired standard boiler. Basically the design should include a mechanism, either a heavy damper or other quick-closing mechanism. In this way in case of low water, the waste gases can be cut off from the boiler and bypassed to the stack. Then the basic process will not be interrupted, and the boiler will also be saved from serious damage due to overheating.

A preferred method is a device that automatically diverts the gases as soon as the water level drops to a predetermined dangerous level. Remember that *time* is of extreme importance in a low-water condition; so the heat input must be quickly removed.

IMPORTANT: On waste-heat boilers, the testing and duplication of a low-water condition, and actions to be taken, are often neglected by

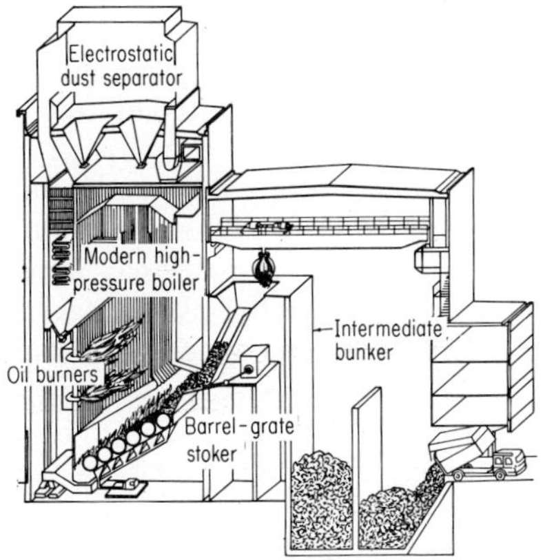

Fig. 5.4 Incinerator plant has modern high-pressure boiler and barrel-grate stoker.

the operators. They do not realize that low water can and does happen on these units. Presumably they also fear to interrupt a process.

WASTE-FUEL-FIRED BOILERS

Q From where do some gaseous waste fuels come?

A The steel industry has large quantities of gaseous by-product energy available. Heat content varies from less than 100 Btu/ft^3 for blast-furnace gas to 525 to 600 Btu/ft^3 for coke-oven gas. The main problem is getting this gas clean enough to avoid fouling the burners.

Oil refineries (catalytic cracking of crude petroleum) produce large volumes of gas as a by-product of catalyst regeneration. This gas contains 5 to 8 percent CO (carbon monoxide), about twice that much CO_2 (carbon dioxide), and air. The gas temperature is around 500°F, with a heat content of about 145 Btu/lb. Increasingly, refineries reclaim this energy by burning the gas, together with oil or gas and additional air, in a CO steam generator.

> NOTE: A CO unit is also needed in any area where air pollution is a problem because it not only saves heat but burns up the CO and any unburned hydrocarbons that escape from the regenerator of a catalytic cracking unit.

Q From where do some liquid waste fuels come?

A Liquid by-products include residue from chemical processes such as tar and pitch. These can be handled in conventional oil burners, but for satisfactory results, they must be heated to maintain viscosity at the proper level. Filtration is also required to remove any solid contaminants. High moisture content can be poured off (decantation) or emulsified.

Recovery steam generators in the pulp and paper industry recover chemicals used for pulping from either the black-liquor alkaline process or the red-liquor acid process. In addition, they generate a high proportion of the steam needed for the plant. Black and red liquors do not burn alike; thus the steam generators used are not alike. Black liquor in the kraft process is very difficult to burn. Large furnaces are needed to keep the temperature relatively low because the liquor has a high content of low-fusion-temperature ash. Smelt collects on the refractory sloping hearth, and a reducing atmosphere must be maintained in the lower part of the furnace for chemical conversion. Also, since superheater and boiler surfaces have a tendency to coat with slag, they operate at low absorption rates. Thus frequent soot blowing and shot cleaning of heating surfaces are necessary.

Red liquor in the MgO (magnesia oxide) process, on the other hand, burns completely in suspension, making little or no slag. Thus a smaller steam generator can be used for an equivalent amount of steam production.

Q What is meant by black liquor?
A In the papermaking industry, wood chips are loaded into large pressure vessels (digesters). After water, sodium sulfide, and sodium hydroxide are added, live steam is introduced. Cooking turns the resultant solution black, hence the name black liquor. This strong liquor is withdrawn from the digester (the remaining pulpy mass is used for making paper) and is stored in tanks where it is joined by a weaker liquor solution used to wash the pulpy mass from the digester. In this state the liquor from the digesters cannot be burned; so it is concentrated by evaporating some of the water. Crushed salt is also added until it becomes over 50 percent solid. The resultant liquor is so viscous that it must be heated to around 220°F before it can be pumped. Heating is by direct steam, or the heat exchangers would soon plug from the viscous liquor. The high solid concentration also makes the liquor abrasive.

Q Why is the boiler called a black-liquor recovery boiler?
A Because it performs three functions: (1) recovers the soda ash used in the cooking of the wood, (2) makes steam for process and power generation, and (3) disposes of the black liquor from the digesters, thus serving the environment. The black liquor could be harmful if dumped into streams, and this is not permitted by environmental laws.

Q How is the capacity of black-liquor boilers expressed?
A The capacity of a black-liquor recovery boiler is expressed in the amount of tons per day of solids it can burn in 24 h.

> EXAMPLE: A unit designed for 1600 psi with a capacity of 392,000 lb/h of steam is rated at the same time as being able to burn 800 tons/day of black liquor.

Q Describe the slanting-bottom furnace and the decanting furnace used in black-liquor recovery boilers.
A The two major manufacturers in the United States are B&W and Combustion Engineering (CE). The B&W units use a slanting-bottom furnace (Fig. 5.5). The oscillating burners spray the black liquor on the furnace walls. The deposited liquor dries, forms char, and falls to the hearth, where sodium chemicals are smelted and the organics in the char are turned to gas which burns in the furnace. The smelted chemicals drain down the sloping floor through the water-cooled spout and then into the dissolving tank.

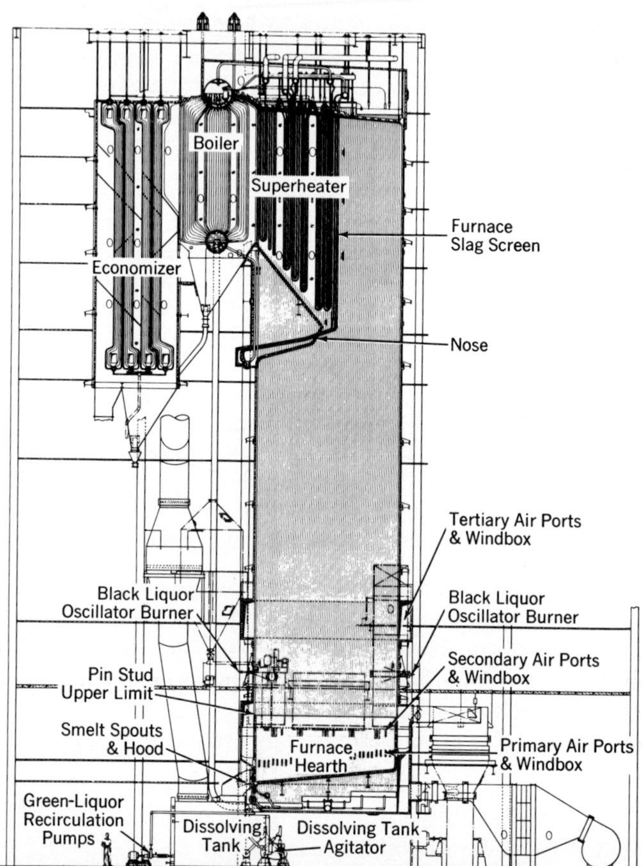

Fig. 5.5 Black-liquor recovery boiler has slanted-bottom furnace for smelt removal to dissolving tank. (*Courtesy of Babcock & Wilcox Boiler Co.*)

The CE unit sprays the black liquor into the center of the furnace. The bottom of the furnace in the CE unit is flat and is called a decanting furnace. Most other boiler details are similar for the two manufacturers' designs.

Q What three possible furnace explosions can occur in a black-liquor recovery boiler?

A Serious furnace explosions on recovery boilers have instituted research as to the cause. It is generally believed that the smelt in the furnace can chemically combine with water to produce a chain-reaction type of detonation in the furnace. This has produced rapid pressure buildup in the furnace with shock-wave results to the structural components of the furnace. The paper industry, insurance companies, and the boiler manufacturers have drawn up guidelines, including emergency procedures, in order to prevent smelt-water explosions. Tube integrity is stressed, and NDT (nondestructive testing) inspections of welds in the furnace area are recommended, as is black-liquor-concentration monitoring. Among the novel procedures recommended is rapid draining of the boiler to a level 8 ft above the low point of the furnace floor at any time when water is suspected of entering the furnace area.

Thus three possible sources for an explosion are: (1) tube failure leaking water into the furnace to cause a smelt-water reaction, (2) weak liquor, usually below 50 percent concentration, being sprayed into the furnace to also cause a possible smelt-water reaction, and (3) sudden ignition of unburned fuel from the auxiliary burners when supplementary fuel such as oil or gas is used. A furnace explosion can result if unburned fuel from the auxiliary burners is suddenly ignited.

Q How is the concentration of black liquor being sprayed into the furnace monitored?

A Refractometers proved suitable for measuring the concentration of solids are used to monitor the black liquor. Two refractometers in series are specified by the Black Liquor Recovery Boiler Advisory Committee. Dissolved solids content below 58 percent requires automatic diversion of the black liquor away from the firing guns.

Q What is meant by the emergency shutdown procedure on black-liquor recovery boilers?

A This procedure must be initiated whenever water in any amount is known to be entering the furnace and cannot be stopped immediately, or when any water leak develops in a pressure part (except for leaks external to the furnace, such as handhole gaskets and gage glasses).

1. Sound an alarm to clear the recovery area of unnecessary personnel.

2. Immediately stop firing all fuel. Secure the unit's auxiliary fuel system at a remote location.

3. Immediately shut off feedwater and all other water sources to the boiler except for smelt spouts.

4. Shut down air supply to primary air ports immediately. Continue operation of the forced-draft fan; supply as much air flow as possible to the secondary and tertiary air ports (if present). Regulate the induced-draft speed or damper to maintain a balanced draft in the furnace.

5. Drain the boiler as rapidly as possible and in accordance with the manufacturer's recommendations to a level 8 ft (2.4 m) above the low point of the furnace floor.

6. Reduce steam pressure as rapidly as possible after the boiler has been drained to the 8-ft (2.4-m) level.

These emergency shutdown procedures should be signed by policy-making management and posted at each recovery-unit control station.

In addition to the above, the Black Liquor Recovery Boiler Advisory Committee had drawn up logic diagrams and sequential operation steps in the firing of black liquor or auxiliary fuels in these special boilers. These should be referred to for details on requirements.

SOLID-WASTE FUEL

Q From where do some solid-waste fuels come?

A Solid by-product fuels are many. Among them are wood chips, sawdust, hulls from coffee and nuts, corn cobs, bagasse (waste product from sugar cane), coal char (residue from low-temperature carbonization of coal), and petroleum coke (final solid residue from a refinery). Each product must be handled in a special manner because of differences in moisture content, consistency, specific weight, and heat content.

The furnace rather than the steam generator is affected when these special fuels are used. Products like bagasse, which has about 50 percent moisture, require a dutch oven. The Ward furnace is a popular design for bagasse, both in the United States and abroad. Here bagasse fuel is partly dried and burned in refractory cells below a radiant arch. The combustion of gases is completed above the arch. Spreader stokers can also be used. See Fig. 5.6.

In refining sugar from cane, the juice squeezed out of the cane eventually is processed into sugar. The remaining fibrous, tenacious, and bulky crushed cane is called bagasse. It is also moist. Depending on where it is grown and the efficiency of the juice extractor, bagasse contains 30 to 50 percent wood fiber and 40 to 60 percent water. Heating value is 8000 to 8700 Btu/lb as a dry solid, with a yield of about 4500 Btu/lb at around 45 percent moisture content.

The art of burning wood refuse (ranging from hogged wood to sawdust) has progressed to a point where a spreader stoker or a cyclone burner does an efficient job, generally in combination with some coal.

Q How is wood bark burned?

A Figure 5.7 shows a bark-burning system. Changing fuel economics have enhanced the attraction of wood as a fuel, and sophisticated wood-fired steam-generating equipment is now available to accommodate a growing number of users, especially in the wood and paper industry. In addition to refinement in design and extension of application, boiler

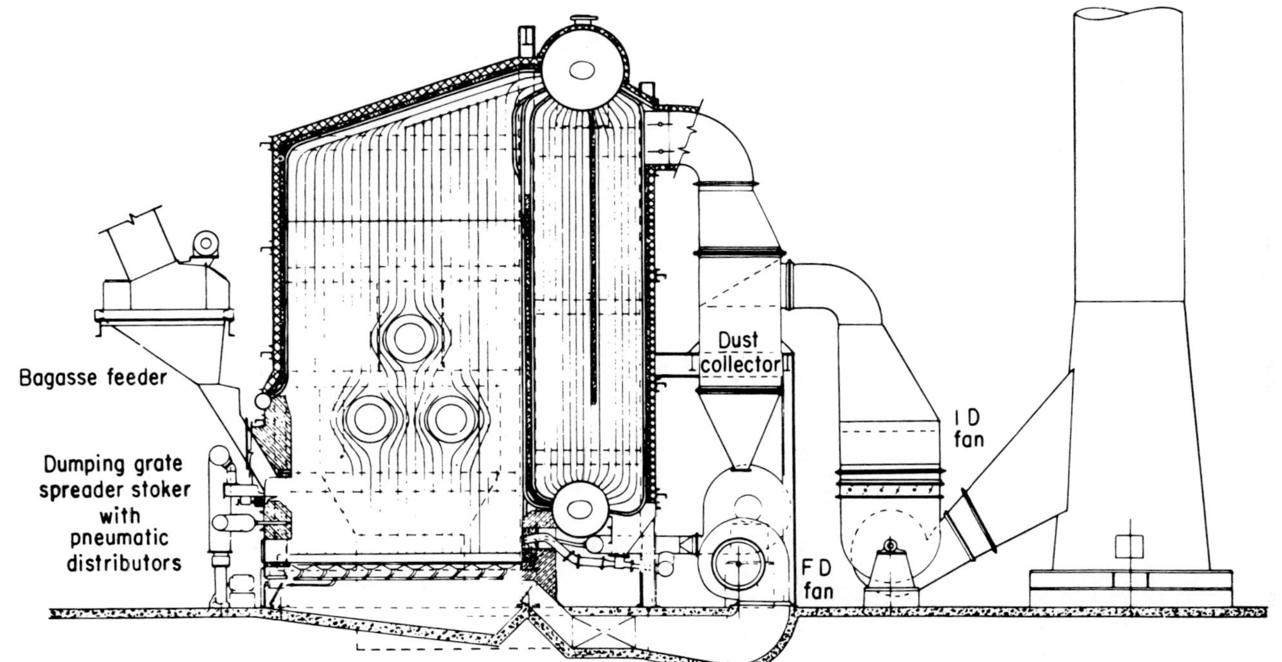

Fig. 5.6 Bagasse-burning water-tube boiler with dump-type spreader stoker and auxiliary oil burners; capacity 100,000 lb/h, 250 psi. (*Courtesy of Riley Corp.*)

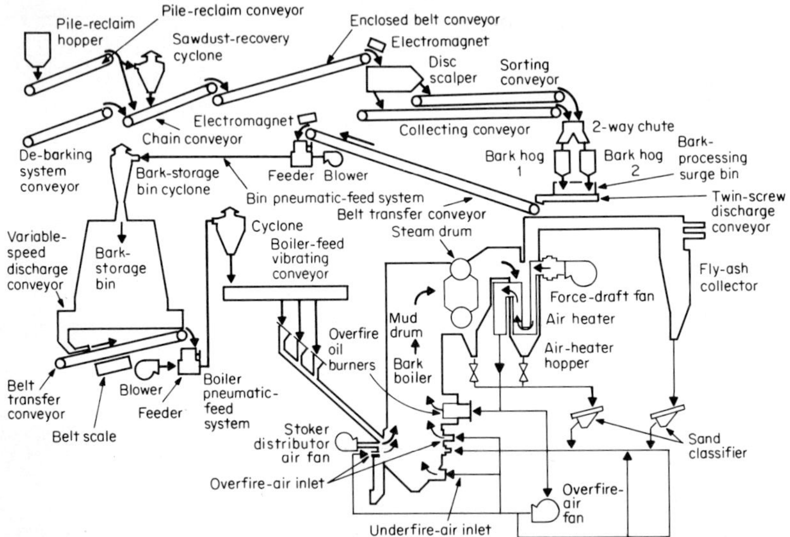

Fig. 5.7 Bark-burning boiler process flow. (*Courtesy of Power magazine.*)

manufacturers have made innovations in methods employed to utilize this fuel more efficiently and also to meet environmental needs.

The burning of wood refuse requires the evaporation of moisture, the distillation and combustion of volatile compounds, and the combustion of remaining carbon materials. These three steps are accomplished either separately or simultaneously depending on the type of equipment installed. Boiler manufacturers offer a variety of wood-feeding, grate-burning, and ash-removal techniques that may be used in a combination best suited for any particular installation.

Large waste-wood boilers may have:

1. Pneumatic spreader distributors with flat water-cooled grates.

2. Pneumatic spreader distributors with continuous ash-discharge grates. The type of grate selected for a specific installation is a function of unit size, stack-emission criteria, fuel sizing, provision for auxiliary fuels, and economic considerations.

3. Inclined water-cooled grate. Steam generators for wood-refuse firing may also employ an inclined water-cooled grate which permits the wood refuse to slide into the furnace by gravity, and such units are in widespread use. Wood refuse is dried as it slides down the grate and is burned near the bottom of the slope and on the horizontal portion. Ash is transported intermittently by steam jets and removed through discharge doors or dumping grate sections.

Large WT-type wood-refuse firing units have been developed along the same lines as those burning fossil fuels. Refractory furnaces have been replaced by welded fin-tube, water-cooled furnaces, resulting in low maintenance requirements and the elimination of furnace deposits. Baffled boiler banks have been superseded by open-pass boiler-bank arrangements that reduce the possibility of fly-ash erosion and permit effective cleaning by soot blowers.

Provision for combustion of oil or coal as auxiliary fuels is often incorporated in these designs. The capacity of large WT-type units and their steam pressures and temperatures have increased steadily. Units fueled by wood refuse currently produce up to 575,000 lb of steam per hour at 1350 psig and temperatures of 900°F. They are an important source of steam for large paper mills and also help dispose of wood wastes of all types.

CORROSION-RESISTANT TUBES

Q What is meant by a composite tube as used in waste-fuel boilers?
A These are tubes with a stainless-steel rolled overlay for corrosion resistance on the fireside of the tubes. The main alloyed-steel tube is designed to carry the full steam pressure. The stainless-steel overlay protecting the alloyed stainless-steel tube cannot be used in calculating the safe pressure on the tube. They are used in boilers with known corrosive firesides such as the black-liquor boiler.

Q What other methods exist to improve the corrosion resistance of tubes on the fireside in waste-fuel boilers?
A Flame spraying, plasma spray, and even hard-face welding have been used, mostly in Finland, Japan, and France.

Q What metals are used in the above procedures?
A For flame spraying, nickel aluminide and nickel titanide are used by specialty companies active in this work. For plasma spray coating, nickel titanium and chromium/nickel-titanium combinations are used. For hard-face welding, austenitic alloys containing 18 percent chromium or more are used. One specification shows:

Cr	Ni	C	Mo
21.9	14.8	0.01	2.6

The above shows that fireside corrosion can be a serious inspection and maintenance problem on some waste-heat and -fuel boiler applications. Special tube material must be considered to withstand the possible corrosion attacks.

Q Describe the Loddby furnace (Fig. 5.8).

A The Loddby furnace, which is a separate refractory-lined furnace at the front of a shop-assembled boiler, is designed to sustain combustion of high-moisture sulfite liquor alone or with a minimum of auxiliary fuel, depending on the liquor's Btu value. In operation, liquid wastes are sprayed into the combustion chamber through the firing gun, and air is admitted tangentially along the length of the furnace, creating a cyclone effect.

The pressure between the wall and the center of the furnace causes the hot gases to recirculate back to the burner area. This way, liquid wastes are dried before reaching the wall area. The boiler is generally arranged with a single-pass, in-line, vertical-tube bank for minimum draft loss. Firing assemblies can be arranged for any combination of low-ash liquid fuels.

HIGH-TEMPERATURE HOT-WATER BOILERS

Q When does a boiler become a high-temperature hot-water (HTHW) unit?

A According to the ASME code, when the temperature exceeds 250°F or the pressure exceeds 160 psi. In practice, 350 to 450°F is considered high-temperature water, 250 to 350°F is medium temperature.

Q What are some of the advantages claimed for HTHW systems?

A A basic advantage is that the heat-storage capacity of water per cubic foot is considerably greater than that of steam at equivalent

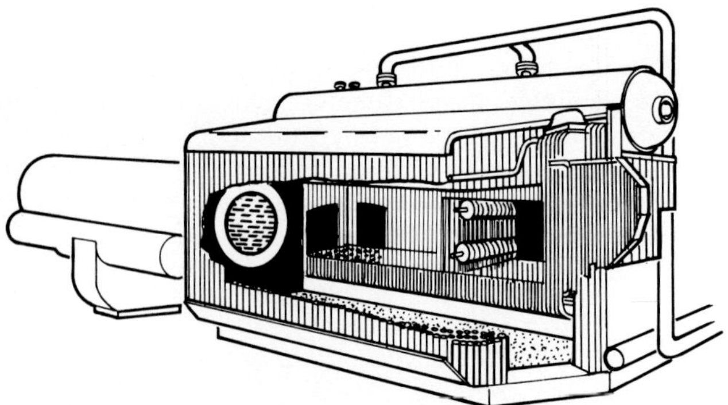

Fig. 5.8 Loddby furnace is a refractory-lined furnace combined with a packaged boiler. Liquid wastes are fired into combustion chamber and air is admitted tangentially, creating cyclone effect. (*Courtesy of Power magazine, November 1981.*)

saturation pressures (see Table 5.1). This inherent reserve (flywheel effect) permits closer temperature control and more rapid response to changing load demands. The claim is also that substantial savings in capital investment and operating and maintenance costs are possible. For example, steam traps, valves, condensate tanks, and expansion tanks are not always needed. But in an HTHW system of the forced-circulation type, a pump is required to circulate the hot water. However, the pump can be located near the boiler and not elsewhere in the condensate system.

Since a closed hot-water system is under constant pressure, very little makeup is required because of flashing. Losses occur chiefly at the pumps and amount to 0.5 percent of the total system content per day. This compares with up to 15 percent per day makeup needed for steam systems. The heat-storage capacity of an HTHW unit and the elimination of flashing and losses due to leaks permit a smaller heat generator. And smaller-diameter distribution lines can be used, thus exposing less circumference for heat loss. The heat storage of an HTHW pipeline is 18.4 times greater than the Btu content of a comparable steam line.

Q What kinds of boilers are generally used in HTHW applications?

A Many types, burning coal, oil, gas, or other fuels. Forced-circulation boilers, such as the LaMont type (Fig. 5.9), are very popular. But regardless of design, they produce a mixture of water and steam which flows from the heater exit to the expansion drum (Fig. 5.10). Not all boilers have a steam drum. The basic idea is that for a given output a

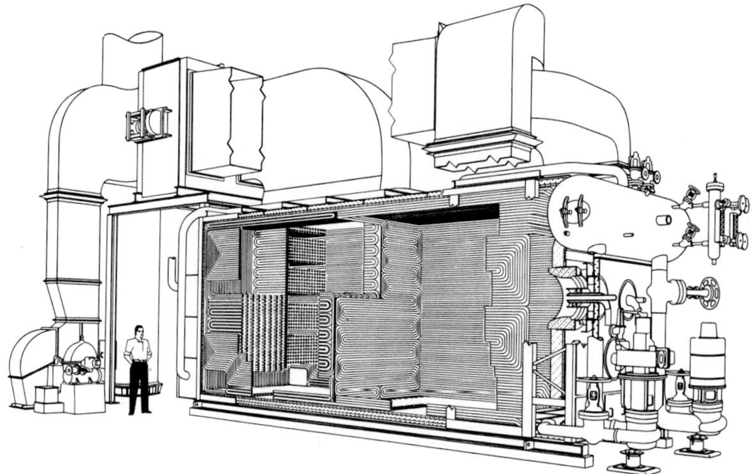

Fig. 5.9 High-temperature hot-water boiler of forced-circulation LaMont type.

Table 5.1 Comparison of Relative Heat Content of Water and Steam

			Density		Heat content		Heat content ratio
Abs pressure, psi	Saturated temp, °F	Return temp, °F	Lb/ft³ water	Lb/ft³ steam	Btu/ft³ water	Btu/ft³ steam	Water/steam
14.7	212	180	59.8	0.0373	1,923	37.4	51.5/1
29.8	250	200	58.9	0.0724	2,972	72.1	41.2/1
67.0	300	200	57.3	0.1547	5,820	156.7	37.2/1
134.6	350	200	55.6	0.299	8,550	306.5	27.9/1
247.3	400	200	53.5	0.537	11,090	555	20.0/1
343.7	430	200	52.4	0.741	12,570	768	16.4/1
566.1	480	200	50	1.222	14,820	1268	11.7/1

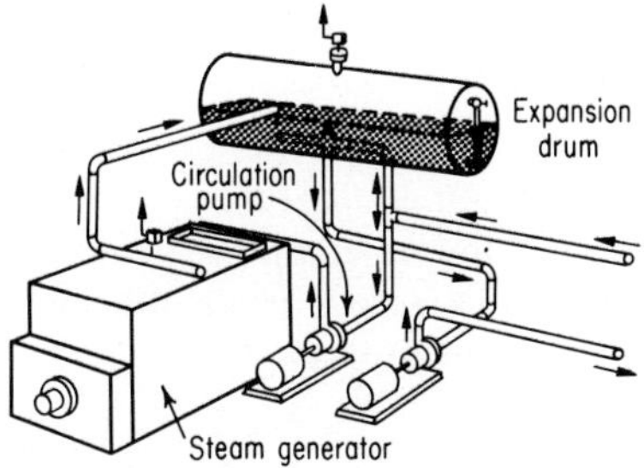

Fig. 5.10 Steam-cushioned method for high-temperature hot-water system.

smaller combustion space and thus a smaller boiler can be used, if the waterwall subjected to high gas temperature can absorb larger heat quantities per unit area. (See Fig. 3.3*a* for a flow diagram.)

Q Name one method used for pressurizing HTHW systems.
A A steam-cushioned system (Fig. 5.10) is generally used because of its simplicity. Here the HTHW unit discharges directly from one or more generators into a common expansion drum. Steam is flashed to maintain constant pressure within the drum. To permit free vapor release, the expansion drum is located above the generator, with enough height to furnish a reasonable net positive suction head (NPSH) for the circulation pump.

Q How is circulation through the boiler tubes controlled in forced-circulation boilers of HTHW systems?
A The pump-circulation direction must not buck the natural circulation of the heated water in the boiler. In the LaMont design, each tube or bank of parallel tubes has an orifice. These create an artificial pressure drop to induce flow from higher to lower pressure through the tubes. A carefully designed natural recirculation system is also used to support the forced circulation, especially in high furnace-absorption zones. This is done by proportioning the flow through the tubes to match the varying furnace-heat-absorption zones. Thus downcomer tubes are used outside the heated zone toward the lower, or water, section. This affects, or restores, upward circulation to the hotter-water zone.

Q What kinds of valves should be used for HTHW applications?
A Valve seats, plugs, and bodies should be made only of cast steel, forged steel, or steel alloys. Valve seats should be stainless chrome-nickel steel, to avoid corrosion and erosion by flow of water. Pressure ratings should follow the power boiler code rating on stop valves, feed

valves, and blowdown valves, which is 125 percent of boiler AWP (allowable working pressure).

Q Do HTHW generators require a water and gage glass?
A Only if a natural-circulation boiler has a drum, which is used as an expansion tank. If the boiler is completely filled with water under pressure and has an external expansion tank, the power boiler code does not require a water gage glass or gage cocks.

Q On forced-circulation boilers for HTHW, what provision should be made to prevent hot spots, or lack of circulation, in case the pump fails?
A A pressure differential or flow switch should be installed to shut off the fuel-burning equipment in case no water is flowing.

WARNING: This is very important for automatically operated boilers.

Q Should a low-water fuel cutout be used on an HTHW generator of the suspended fuel-fired type?
A Some states now require a low-water fuel cutout on low-pressure hot-water heating systems. Thus it follows that this safety device is even more necessary on an HTHW boiler. It should be installed on the boiler to shut off the fuel in case the water level drops below a dangerous level.

THERMAL-LIQUID HEATERS AND VAPORIZERS

Q What is a thermal-liquid heater?
A As heat equipment was developed, high temperatures at moderate pressures were recognized as having potential uses in process work. For example, to obtain 705°F with water, the saturation pressure is 3206 psi (critical pressure). So special fluids were developed to obtain high temperatures at moderate pressures. Thus the term *thermal-liquid heater* means a closed vessel in which a heat-transfer medium other than water is heated without vaporization, and the heated fluid gives up its heat or does useful work outside the closed vessel.

Q What is a fluid-vaporizer generator?
A This also is a closed vessel in which a heat-transfer medium, other than water, is vaporized under pressure by the application of heat. Here also, the heat-transfer medium is used externally to the closed vessel.

Q Why are high-temperature fluids used instead of water?
A These fluids can be used as heat-transfer media at temperatures and pressures that could make water or steam uneconomical. The allowable temperature range varies all the way from −100 to 1000°F, and a uniform temperature is obtained over the entire heat-transfer surface.

Economy is the big advantage because at moderate pressures, higher temperatures can be obtained with these media.

EXAMPLE: Up to the 350 or 360°F temperature level, steam around 150 psi is satisfactory. But if temperatures from 450 to 1000°F are needed, the corresponding steam pressure may be excessive for economic operation.

Q Illustrate a thermal-liquid system for process.

A Figure 5.11 shows an application for the plastics industry. Thermal liquid at 550°F supplies plant process and space heating in this system. A calender temperature-control valving system and recirculating pumps are part of the circuit. Liquid-type heaters are generally of the WT type, whereas vaporizers may be either FT or WT design.

Q Do fire insurance companies and local fire regulations have requirements on liquid-phase heaters or on vapor-phase heaters?

A Yes. Because a distinct fire hazard may exist with some of the organic fluids, including mineral oils, fire regulations are in effect.

Q How is the discharge capacity of SVs determined on liquid- and vapor-phase heaters?

A The pressure setting of the SV should be based on the maximum allowable working pressure of the unit. The capacity of the SV should be based on the maximum Btu-per-hour output of the heat generator, whether for liquid- or vapor-phase heaters.

ELECTRIC BOILERS

Q Is there any other type of boiler in competition with the fired VT boiler?

A Yes, the electric boiler is popular in areas where electric utilities

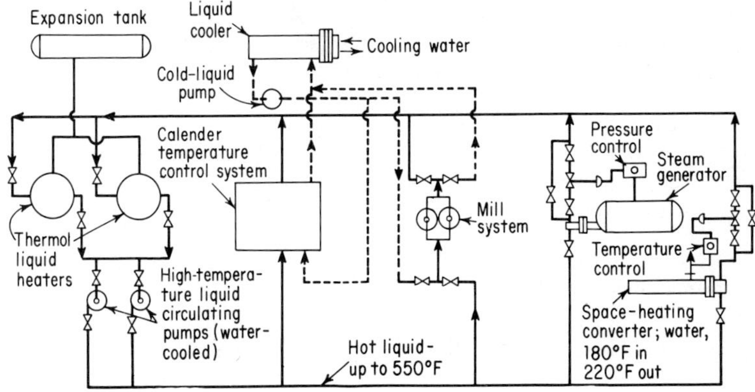

Fig. 5.11 Thermal-liquid system supplies 550°F liquid for plastics industry. Dashed lines denote cold liquid, solid lines hot liquid.

are promoting electric heating. Basically, all electric boilers have the same efficiency, nearly 98 percent. Power consumption, which is proportional to the steam generated, is figured at 10 kilowatthours (kW · h) of 220-V (or higher) current for 1 boiler horsepower (boiler hp) or 34.5 lb of steam per hour. But this is roughly 3 times the cost of steam generated in a small boiler fired with No. 2 oil. Models today have capacities of up to 60 boiler hp, which is about 2000 lb of steam per

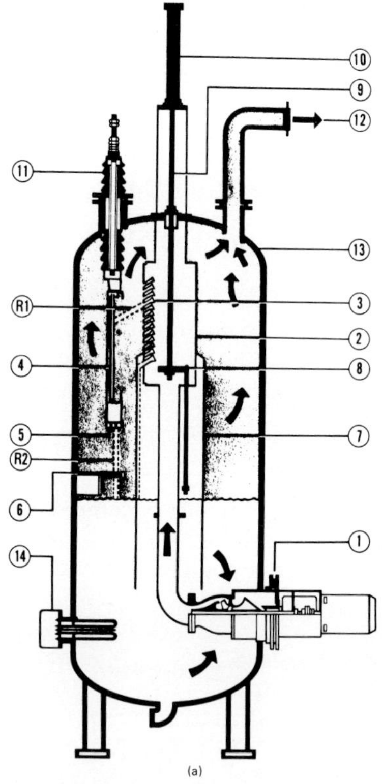

Fig. 5.12a High-voltage steam boiler. 1, circulating pump; 2, nozzle stock; 3, nozzle; 4, electrode plate; 5, nozzle plate; 6, counter electrode; 7, control sleeve; 8, control linkage; 9, control cylinder rod; 10, control cylinder; 11, insulator; 12, steam outlet; 13, boiler shell; 14, stand-by heater; R1, upper water jet; R2, lower water jet. (*Courtesy of Coates, Division of Cam Industries.*)

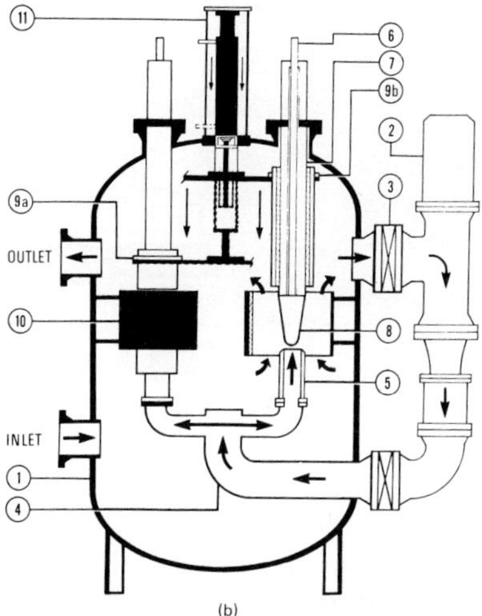

Fig. 5.12b Model BAHW high-voltage-electrode hot-water boiler. 1, boiler shell; 2, circulation pump; 3, pump-isolating valves; 4, distribution manifold; 5, guide tubes; 6, electrode rod; 7, electrode lead-through insulators; 8, electrode; 9*a*, load-control insulator, lowered position; 9*b*, load-control insulator, raised position; 10, neutral shield; 11, hydraulic cylinder assembly. (*Courtesy of Coates, Division of Cam Industries.*)

hour. Small, portable units for cleaning go down to as little as 1 boiler hp. Steam pressures go as high as 600 psi, although 100 to 150 psig is average. See Fig. 5.12.

Q How are electric boilers classified?

A There are two basic types:

1. Units with heating elements constantly submerged in water (Fig. 5.13). These elements do not depend on conductivity or resistance of the water for heating and steam generation.

2. Units with electrical electrodes located in a central generating chamber (Fig. 5.14). The water level recedes as the demand for steam decreases. Thus at no load, the electrodes are entirely out of water, and there is no electrical consumption.

While low-water-level controls are required for the boilers shown in Fig. 5.13, they are not needed for the second type. The reason is that

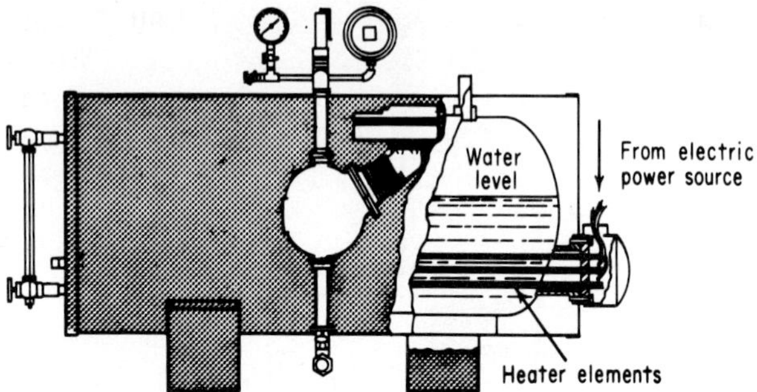

Fig. 5.13 Resistance-type electric steam generator.

they cannot generate steam when the water is low or when no steam is required. To ensure that the boiler carries the proper salinity, a conductivity control is generally supplied. This means that salts are added or the unit is blown down, depending on the condition of the water. All electric boilers should be built in accordance with the ASME boiler code and also approved by Underwriters Laboratories.

Q Explain how the electrode principle works.

A Figure 5.14 shows a cylinder, open at the bottom, welded to the inside of the upper head of a pressure tank. This cylinder divides the tank into two concentric chambers. The outer is the regulating, the inner the generating chamber. Suspended within the generating chamber are the three electrodes with electric power connected to them.

To operate, a prescribed quantity of electrolyte is dissolved in water

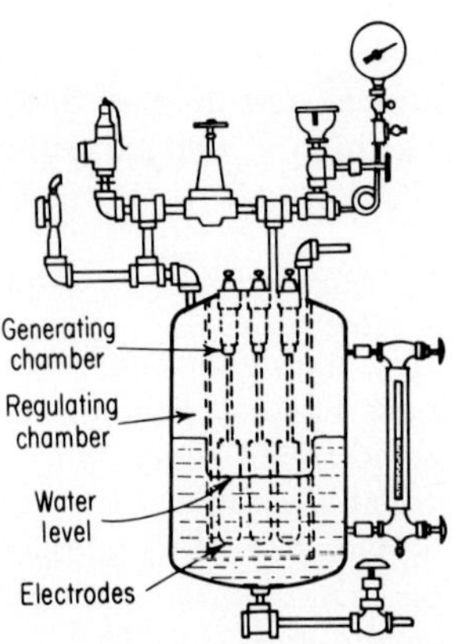

Fig. 5.14 Electrode-type electric steam generator.

and poured into the generator through the hand fill. (In larger steam generators, it is added with the feedwater pump.) This electrolyte remains in the generator until drawn off with the water through the drain valve. Electric power is switched on, and heat is generated by the resistance of the water to the passage of current between the solid electrodes. Thus steam generated in the chamber flows through the outlet pipe and through the steam header, then through the pressure regulator and out the regulating chamber.

Q Do state boiler code laws requiring boilers of over 15 psi pressure to be opened annually for internal inspection also apply to electric boilers?
A Most states include electric-type boilers in their rulings. The ASME boiler code states that electric boilers of a design employing a removable cover which will permit access for inspection and cleaning of the shell, and having a normal water content not exceeding 100 gal, need not be fitted with washout or inspection openings.

The usual access for inspection and cleaning is the electrode cover connection to the inside of the boiler. This must be pulled out to check the internal conditions of the boiler. At the same time, check the condition of the electrical resistance elements.

Q How is the SV capacity determined on an electric boiler if there are no typical heating surfaces?
A The capacity of the SV is determined by the kilowatt input. The minimum SV capacity must be at least 3.5 lb/h per kilowatt input. This is true whether it is a high- or low-pressure boiler. On a Btu-per-hour basis, the requirement is 3500 Btu/h for each kilowatt input.

Q An electric boiler supplies 600 lb/h of steam at 100 psi. If there is no return and feedwater temperature is at 60°F, what would be the cost of electricity for an 8-h period if the kilowatthour rate is \$0.05? Efficiency = 70 percent.
A It is necessary to determine the Btu input and then convert this to kilowatthours.

From steam tables, the enthalpy of saturated steam at 115 psia pressure is 1189.7 Btu.

h sat liq at 60°F is 28.1 Btu

The total heat required to deliver the 600 lb/h is

$$\text{TH} = \frac{(1189.7 - 28.1)600(8)}{0.7} = 7{,}965{,}257 \text{ Btu}$$

$$\text{Cost} = \frac{7{,}965{,}257(0.05)}{3415}$$

where 3415 = Btu/kW · h

Cost = \$116.67 ***Ans.***

FLUIDIZED-BED-BURNING BOILERS

Q Define the term *fluidized-bed combustor* (FBC).

A Fluidized-bed combustors were invented in the 1960s. They use sand-sized materials to form a bed in the boiler, usually on a grid (Fig. 5.15). The flow of combustion air causes these particles, which are

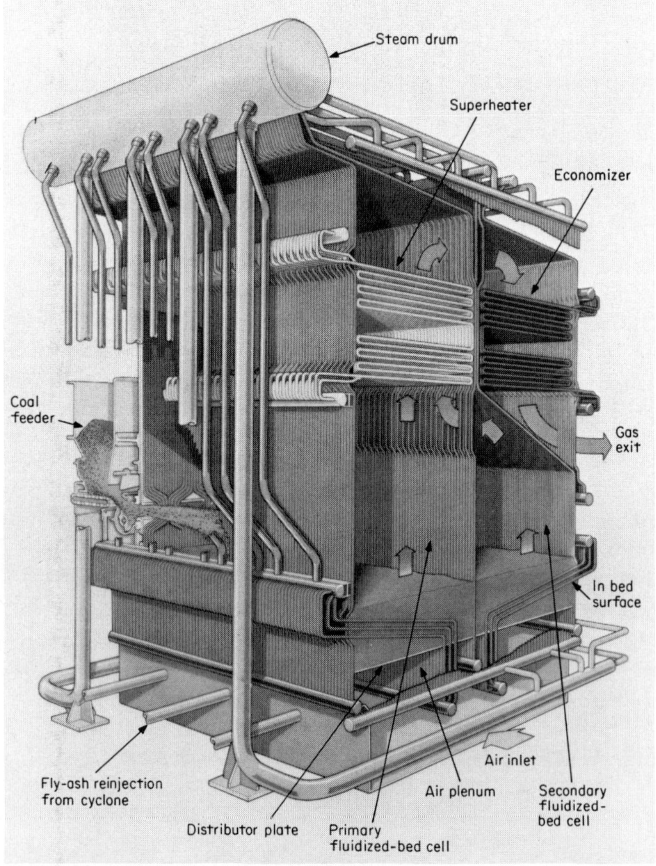

Fig. 5.15 Fluidized-bed-burning boiler uses sand-sized materials to form a bed, usually in the grid as here.

supported by a grate, to bubble, giving the bed the appearance of a boiling fluid. Fluidized-bed advantages are that conventional coal combustors control emissions of fly ash (smoke) and various gases, including SO_2 and NO_x, by means of wet and dry scrubbers, baghouses, and electrostatic precipitators. According to fluidized-bed boiler manufacturers, these systems can account for up to 40 percent of the cost of building and operating conventional coal-combustion systems.

FBCs can produce stack gases as clean as those from using natural gas, and without the need for wet scrubbers. The boiling action of the bed causes coal to burn at a much lower temperature than is possible in a stoker—typically 1500 to 1600°F vs. 3000°F. The lower temperature allows the FBC to operate below the ash fusion point of coal (the temperature at which the ash in the coal melts), avoiding the formation of "clinkers" (chunks of fused ash) and making automatic operation of the bed possible. In addition, this temperature is below the point at which most NO_x pollutants are formed.

Q How is SO_2 removed?

A The low temperature also enables the burner to remove SO_2 without the use of a stack-gas scrubber, by employing ground limestone as the bed material. The limestone reacts with the SO_2 that is emitted as the coal burns, forming calcium sulfate (gypsum). This reaction takes place only at these relatively low temperatures.

Since acid rain is considered a serious environmental problem and is being blamed on coal-firing plants, if the SO_2 emission problem can be solved by fluidized-bed combustors, the United States will be able to burn more coal efficiently and minimize the environmental effect.

Q Describe a typical fluidized-bed boiler installation.

A See Fig. 5.16. The sketch shown includes four external downcomers supplying the lower headers of the unit enclosure and partition wall (separating bed A from bed B) and the sloped tubes within the fluidized bed. Flue gases are cooled by a conventional boiler bank in which the cold-end tubes act as downcomers, with the remaining tubes functioning as steam-generating risers.

Both beds have separate spreader-stoker overbed coal-feed systems. Coal sized to less than 1¼ in is injected into the beds by the overbed spreader feeders. These feeders do not require drying of the coal. Sized to less than ⅛ in, limestone is fed by gravity at a single point in each bed. Because of the long residence time of the limestone in the fluidized bed and the rapid burning rate of the coal, less than 5 percent of the bed consists of fuel.

Since the fluidized bed operates at a temperature of 1600°F, well below the ash-softening temperature of the fuel, no slag formation

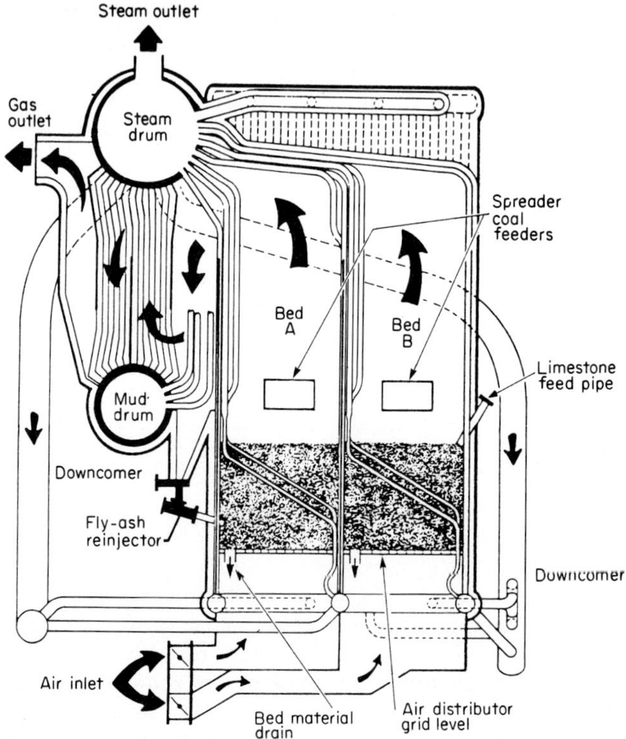

Fig. 5.16 Fluidized-bed combustor boiler, WT type, generates 100,000 lb/h up to 625 psi. (*Courtesy of Foster Wheeler Corp.*)

occurs. This eliminates the need for soot blowers as required on other types of coal-fired steam generators.

Q Can fluidized-bed burning of coal with limestone meet particulate emission standards of the government?

A Manufacturers of fluidized-bed-burning boilers are active in promoting fluidized-bed-burning boilers for this very reason. They claim that combustion of coal in a fluidized bed of limestone makes it possible to reduce SO_2, NO_x, and particulate emission to levels specified by the federal government as well as by most state and local authorities.

Control of SO_2 is achieved by maintaining the fluidized-bed temperature in the range of 1500 to 1600°F and operating with limestone bed material. The raw limestone ($CaCO_3$) will calcine to lime at the normal bed operating temperature. The lime (CaO) will react with SO_2 released during coal combustion and form gypsum ($CaSO_4$) on the surface of the particles. Affinity for sulfur capture is maintained by continuously feeding limestone and maintaining proper bed inventory with the bed-material removal system.

NO_x emissions from fluidized-bed combustion are low because of the temperature of the fluidized beds. Particulate emissions can be kept low by means of a baghouse dust collector designed for 99.75 percent efficiency. Use of the baghouse is claimed to have resulted in excellent stack opacity readings and no visible plume from the stack.

A recent test at a fluidized-bed-burning boiler showed the results in Table 5.2 in meeting stack emissions.

Q How is fluidized-bed burning controlled?

A Most fluidized-bed facilities include a combustion-control and safety interlock system that is typical for industrial coal-fired facilities. Typically, the system provides automatic control when the load is between 25 and 100 percent. Operator action is required only at approximately 50 percent load to light off or shut down part of a bed. The safety interlock system will automatically shut the facility down should any critical operating component fail or if bed temperature or furnace pressure does not remain in the proper range. The controls are also designed to react to input from SO_2 and oxygen gas-analysis systems.

WASTE-HEAT BOILERS IN CHEMICAL PLANTS

Q What are some considerations in using waste-heat boilers in large chemical-manufacturing projects?

A The waste-heat boiler is often used as a process controller involving temperatures, and flows of the process. Thus, chemical-plant operators will often treat a waste-heat boiler as if it were in the same category as other process equipment, without realizing the special hazards of boiler operation.

Table 5.2

	Stack emission limits		
Type of emission	EPA limit	Washington, D.C., limit	Measured
SO_2 (lb/10^6 Btu)	1.20	0.78	0.48
NO_x (lb/10^6 Btu)	0.70	0.70	0.30
Particulates (lb/10^6 Btu)	0.10	0.056	
Stack opacity, %	20	20	17*

Source: Foster Wheeler Corp.

* High reading, opacity range is 4 to 17 percent with an average of approximately 10 percent.

EXAMPLE: Whether a process should be shut down or diverted around the boiler if a low-water condition develops in the boiler is a common problem, quite often with insufficient emergency planning by chemical-plant operators for this situation. The same is true of a tube failure in a waste-heat boiler.

In some processes radiant heat can damage a boiler even if fuel is cut off on low water. Additional pumping facilities may be needed to keep the tubes from overheating. Process upsets may make it difficult to remove a steam generator before securing other equipment. In the interval, the steam generator may be exposed to serious overheating damage because of this delay.

There is a need for process-equipment and steam-generator operators to be trained on the flow of the process, steam, fuel, and water and the effect on the process, its catalysts, and any steam generator if flow is intercepted or if an emergency breakdown situation develops in the train. By constant training, malfunctions will be clearly understood as to their potential effects, and combined actions will be taken to minimize the potential damage or effects to the process and associated equipment in the process train. Suitable alarms and their interpretations on a priority-system basis will then be possible.

CHAPTER **SIX**

Nuclear-Power-Generation Pressure Vessels

Nuclear-power-plant equipment consists of many pressure vessels, including those that generate heat under pressure, called *reactors*, because in its core a nuclear reaction is taking place. Steam or high-pressure hot water is also made in nuclear reactors; by means of heat exchangers, the steam produced then follows a conventional flow to a turbogenerator. This chapter briefly reviews some of these systems. However, the reader is referred to more advanced texts to study nuclear energy more fully.

Q Briefly review the structure of an atom that makes possible nuclear energy.

A A review of the structure of an atom will reveal that the nucleus is made of protons and neutrons defined as follows:

A *proton* is a *positive-charged particle* of matter. There is usually one proton in the nucleus for each negative-charged particle, or electron, circling the nucleus. The positive-charged proton attracts the negative-charged electron whirling around the nucleus, and this prevents the electron from flying off by centrifugal force just as the earth is held in orbit around the sun by gravitational forces.

A *neutron* is matter in the nucleus with no electric charge, and it weighs about the same as a proton. There is a strong binding force holding together protons and neutrons. The number of neutrons is not fixed but varies with atoms, although this number does affect the atomic weight. Neutrons do not affect the chemical properties of the elements, as protons and electrons do. Since neutrons have no charge, they are very useful in nuclear-bombardment applications. They travel in straight lines until they come in contact with other matter. This makes them very useful in splitting the nucleus of an atom.

Electrons are negatively charged particles, quite a bit lighter than protons. In fact, they weigh about $\frac{1}{1800}$ the weight of a proton. Electrons are arranged in rings around the nucleus, with definite numbers of electrons per ring. That is, the first ring holds 2 electrons, the second 8, the third 18, until a maximum of 32 is reached. The total number of electrons in an atom is equal to the number of protons and neutrons in the nucleus, which is also the atomic weight of the atom. Uranium has the highest atomic weight at 92 and also has the greatest number of electrons surrounding the nucleus.

Q What is the relationship between mass and energy in nuclear reactions?

A The fundamental law of physics on which a nuclear reaction is based is Einstein's law of the interrelationship of mass and energy, namely, energy equals mass times velocity squared ($E = mv^2$). One must realize, though, that the velocity term is near the speed of light. It has been established, for example, that 1 lb of uranium 235, when hit by a neutron bullet, will *explode* to form lighter atoms and spare neutrons. Some of the mass, about one-thousandth of a pound, is lost in the form of energy which is equivalent to about 11.4 million kilowatthours (kW · h) of energy. Many other principles of nuclear physics are involved in a nuclear reaction, but they are beyond the scope of this book.

Q Define fissionable material.

A Material that is capable of capturing neutrons, and therefore splitting an atom into two or more particles, is called a *fissionable material.* Uranium 235, uranium 233, and plutonium 239 are fissionable materials. When an atom is split by a neutron traveling at high speed so that two or more fragments are split from the atom, it is called *fission.* During fission, two to three neutrons are released from the split atom. If a chain reaction is to result, one of these free neutrons must be captured by another atom and cause fission of that atom; then in turn it must produce another free neutron to establish the chain reaction of fission. The ratio of the number of neutrons which cause fission to the number of neutrons initially produced is called the *effective multiplication.*

Q Is power generated primarily by nuclear fission?

A The present method of nuclear power generation is by fission, or the splitting of the nuclei of certain heavy atoms by bombarding them with neutrons. Hitting the uranium-isotope ^{235}U nucleus, these neutrons split the atom to form new elements, such as krypton and barium, release energy in the form of heat, and liberate additional neutrons from the nucleus that can bombard another nucleus to keep the reaction going. Other atoms that fission are ^{233}U, thorium, plutonium, and ^{238}U.

Q Does the breeder reactor also employ nuclear fission?

A Under development are breeder reactors for power generation. The breeder reactor is so named because it produces more fissionable fuel material than it consumes. The now wasted uranium 238 and low-grade thorium ores would be converted in a breeder reactor, through neutron bombardment, to a fissionable fuel from the dwindling supply of the present fuel used, uranium 235.

FUSION POWER

Q What is the basic principle in fusion power?

A Fusion power plants are still in the conceptual and development stages. Fusion requires temperatures to 100 million degrees Fahrenheit, which no known material can withstand. It is necessary to develop intense magnetic fields to confine and also to obtain the hydrogen-gas plasma that is the fusion fuel. In the fusion process, hydrogen or its isotopes, deuterium and tritium, are fused to form helium, the next higher element in the atomic scale. The joining or fusion takes place in nuclear ovens which generate intense heat that must be extracted for the production of useful energy. Intense development is going on all over the world in an effort to find an economical method of extracting usable energy from a fusion nuclear reaction.

Q What are some of the difficulties in the use of controlled nuclear fusion?

A The central problem in the use of controlled fusion is in developing safe and economical methods to isolate the fuel at kinetic temperatures of 100 million degrees or more from contacting the "cold" walls of the reaction chamber. It is necessary to use a nonmaterial force field to satisfy such containment requirements. A gaslike state called a plasma is used. At fusion temperatures, all matter is ionized into free negative electrons and positive nuclei; therefore, a gas must be used. This plasma has charged particles so they can be acted on in flight by a strong magnetic field in order not to come in contact with the walls of the containment. The basic theory is that a powerful magnetic field can contain the plasma long enough for it to be heated to the temperatures needed for the nuclei to fuse. To initiate and sustain the fusion process, the nuclei of the fuel atoms must approach each other with sufficient power to overcome their coulomb repulsive forces.

Figure 6.1 shows an open and closed magnetic confinement system. This arrangement, called a tokamak or toroidal system, is receiving intense study and development throughout the world. The tokamak system shown in Fig. 6.1*a* is a closed confinement design.

In the open confinement system of Fig. 6.1*b*, the magnetic field lines do not close on themselves within the plasma-confinement region.

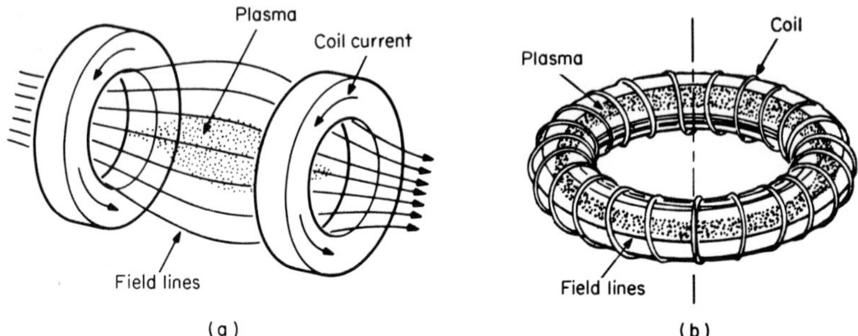

Fig. 6.1 Open and closed magnetic confinement systems are used to develop fusion reaction. (*a*) Open system—simple magnetic mirror; (*b*) closed system—simple torus.

Particles in an open system can escape from the plasma-confining region if their velocity components parallel to the field lines are sufficiently high.

However, increasing the magnetic field strength in a region will decrease the parallel velocity of a particle as it approaches the stronger field. If the field becomes strong enough, and the parallel velocity is not too high, it will decrease to zero and the particle will be "reflected" back along the field line in the opposite direction. This is the principle of mirror confinement, illustrated in Fig. 6.2.

NUCLEAR POWER REACTORS

Q For the generation of steam, how does a nuclear power reactor compare with a steam boiler?

A The reactor also is a pressure vessel. It generates heat produced by the fission(splitting) of atoms. In the boiling-water reactor type (BWR) shown in Fig. 6.3, steam inside the core goes to a conventional steam plant for driving a turbine. In a PWR, or pressurized-water reactor, water is heated under high pressure, and steam is generated in water to steam-heat exchangers (see Fig. 6.4). This system also employs a pressurizer as a form of expansion tank for the reactor water. In another type, liquid sodium metal (sodium graphite type) is circulated through the reactor core, the liquid metal giving up its heat in a heat exchanger where steam is formed. Thus basically, the reactor serves the same purpose as a boiler, but instead of burning fuel the heat is generated by the splitting of atoms. Just as with boilers, there are various types of reactors in use, the pressurized-water type, fast-breeder type, heavy-water type, etc. In the United States, the BWR and PWR are the main installed units.

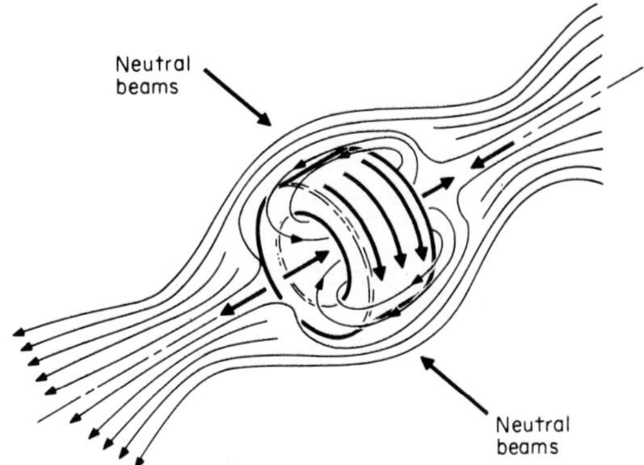

Fig. 6.2 Field-reversed mirror. Mirror confinement of magnetic fields is being studied to develop fusion reactors.

Q Name the types of reactors possible, based on coolant used.

A From the many possible arrangements of fissionable and fertile fuel, moderator, and coolant that can constitute a chain-reacting system, six types have emerged as principal contenders for full-scale electric power generation: (1) pressurized water, or PWR; (2) the closely related boiling water, or BWR; (3) sodium-cooled, graphite-moderated; (4) gas-cooled; (5) heavy-water-cooled; and (6) organic-cooled, heavy-water-moderated reactors.

Q Name the five main components of a PWR nuclear power plant.

A (1) Nuclear reactor; (2) steam generators or heat exchangers; (3) piping to circulate the coolant between the reactor and the steam generators; (4) pumps to circulate the coolant; and (5) a pressurizer to prevent the coolant from boiling and to permit volume changes in the

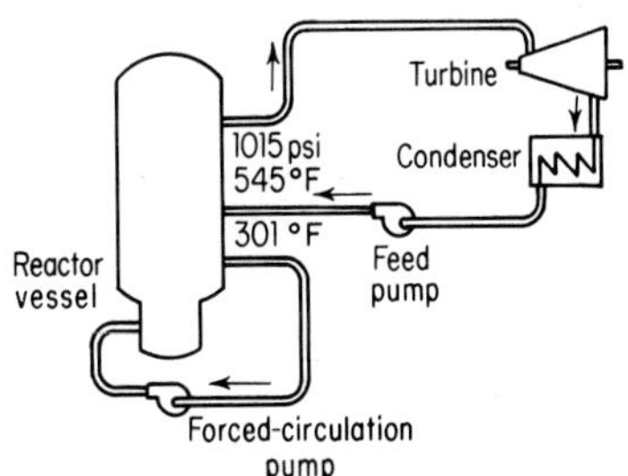

Fig. 6.3 BWR nuclear reactor, instead of a boiler, is used in this otherwise conventional steam-generating plant.

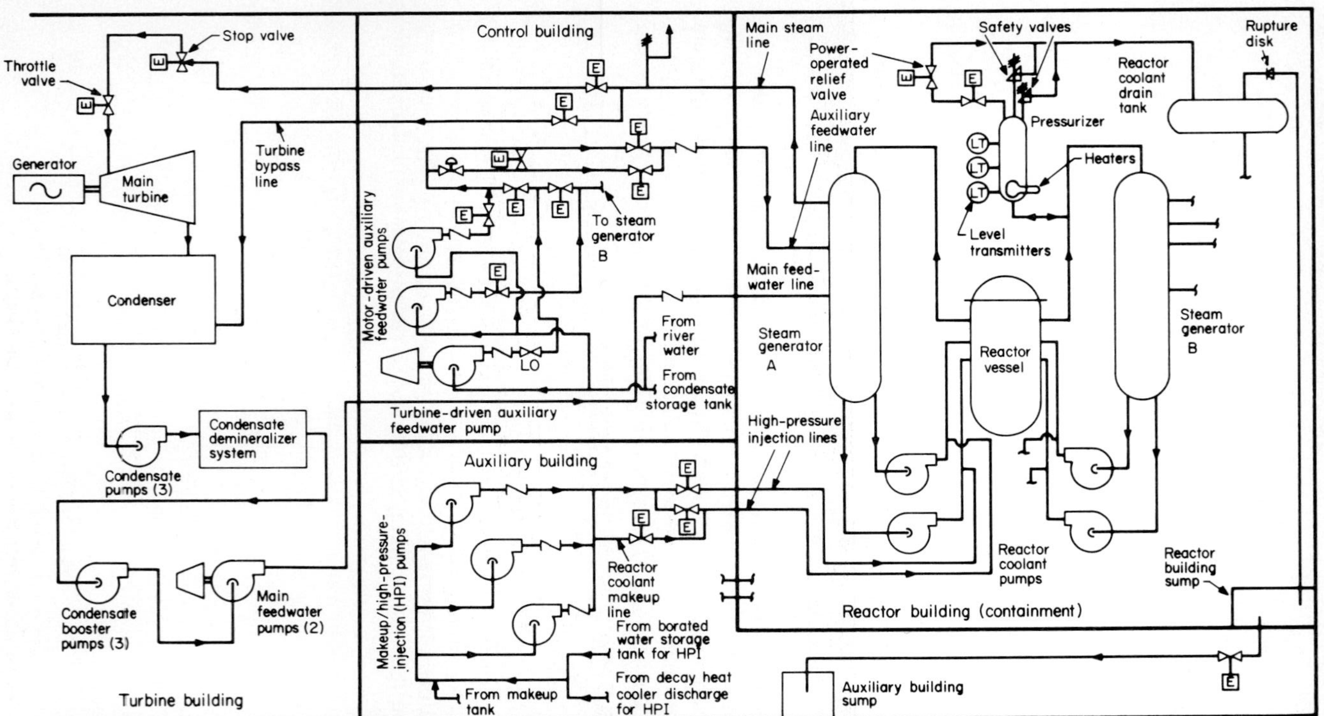

Fig. 6.4 PWR nuclear power plant has high-pressure injection pumps in reserve to cool reactor in an emergency. (*Courtesy of Spring and Kohan, Boiler Operator's Guide, McGraw-Hill.*)

coolant without excessive pressure rise with coolant temperature changes. See Fig. 6.4.

Q What are a shim rod and scram rod as applied to nuclear plants?
A A shim rod is a control rod used for making coarse adjustments in the reactivity of a reactor's chain reaction, whereas a regulating rod makes fine adjustments in the reactivity. Reactor control is also achieved by varying the liquid level for those reactors that use a liquid as a moderator of reactivity.

A scram rod is a safety rod that is capable of shutting down a reactor very quickly in the event the shim or control rods fail to control the reactivity within prescribed limits.

Q What is meant by the term *poison*?
A The term *poison* applies to fission products in the fuel elements of a reactor that absorb neutrons and thus affect the reactivity of the reactor. The two most prominent fission products considered poisonous are xenon 135 and iodine 135. These are produced when a reactor uses uranium. The poison formed as fission products eventually reduces the output of the reactor; as a result, the fuel elements are spent, which eventually requires the reactor to be refueled. Poison can also be injected into a reactor to scram it, or shut it down, under critical emergency conditions.

Q Compared with a conventional boiler, what safety considerations must be given a nuclear power reactor used for the generation of steam?
A The energy potential of the fuel mass in a nuclear reactor and the lethal radioactivity of many fission products demand more stringent safety measures than a conventional power plant. They may be considered under two heads: (1) control and instrumentation for safe reactor operation, and (2) means to prevent escape of radioactivity during normal operation, possible reactor runaway, or other failure.

Reactor instrumentation can be grouped into three classes: control, safety, and monitoring. The major requirement in control instrumentation is measurement and display of the heat rate produced in the core over a complete range from subcritical to full-power operation. This is done by measuring the neutron flux level, or rate of neutron fission, with neutron detectors. Different types of instrumentation are provided for shutdown, start-up, and low-power and high-power operation.

For shutdown and start-up measurements, pulse counters are used, often taking the form of ionization chambers filled with boron trifluoride gas. Operating at a particular voltage, they give a pulse proportional to the incident radiation. Alternatively, fission chambers, coated with a fissionable material, are capable of detecting neutrons by fissions inside the chamber. Each neutron produces an electric pulse. In either case, the output passes through a pulse amplifier to counting rate meters, logarithmically scaled since neutron flux increases exponentially. Start-up and similar low-power detectors retract into the biological shield during high-power operation. For power measurements in the normal operating range, where temperature effects on reactivity become important, instruments are linearly scaled. Ionization chambers act directly on high-impedance potentiometer recorders.

Reactor emergency tripping is based on set limits being executed for such factors as neutron density change, fuel-element temperature, power level, and coolant flow. Then reactor "scram" takes place: control rods are automatically inserted into the core, either by power or by releasing the drive mechanism and allowing them to fall by gravity. "Scram" time is usually 2 to 3 s.

REACTOR CONTAINMENT

Containment of fission products demands several lines of defense. One sodium-cooled reactor vessel breeder plant is surrounded by a thermal and biological shield system. A stainless-steel thermal wall protects the shielding from heat; borated graphite acts as a neutron absorber. Gamma rays emitted by the absorber are captured in the outer concrete shell. A suppression pool (Fig. 6.5) is a safety measure to

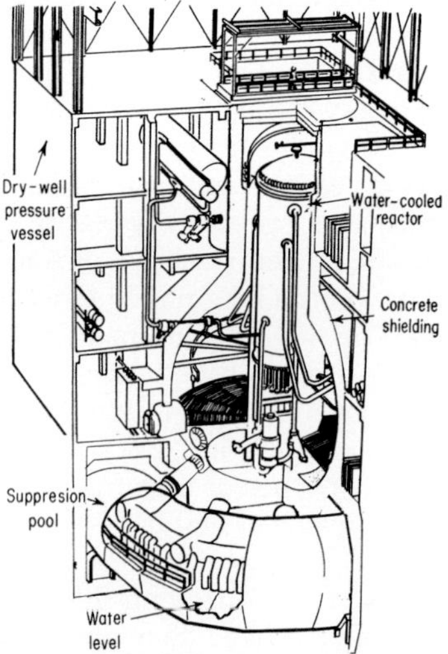

Fig. 6.5 Supression pool condenses steam if reactor or piping should rupture in this sodium-cooled breeder reactor.

condense steam, if the water-cooled reactor pressure vessel or the piping ruptures.

Monitoring the reactor power level is the purpose of the neutron detector (fission counter or ionization chamber) placed within the core (Fig. 6.6). Detectors may be alternatively placed outside the reactor, treating the core as a point source of neutrons, as shown in Fig. 6.7. The extreme range of power level, as much as 10 decades, requires multiple detectors with overlapping ranges. To gain data on fuel, heat transfer, and metallurgical characteristics, extensive in-core sensing instrumentation may be applied. An instrumented fuel assembly (Fig. 6.6) is used in a boiling D_2O reactor to determine fuel power limits.

REACTOR SHIELDING

Q What is the purpose of reactor shielding?

A Three distinct kinds of radiation are emitted from radioactive material. Alpha and beta rays have little penetrating power, but gamma rays can penetrate great thicknesses. Neutrons, too, have great penetrating power and are the primary radiation hazard in an operating reactor.

To safeguard personnel against neutrons, gamma rays, and heat, the reactor, and much of its auxiliary equipment, must be enclosed within thermal and biological shielding. Gamma rays can be absorbed by a

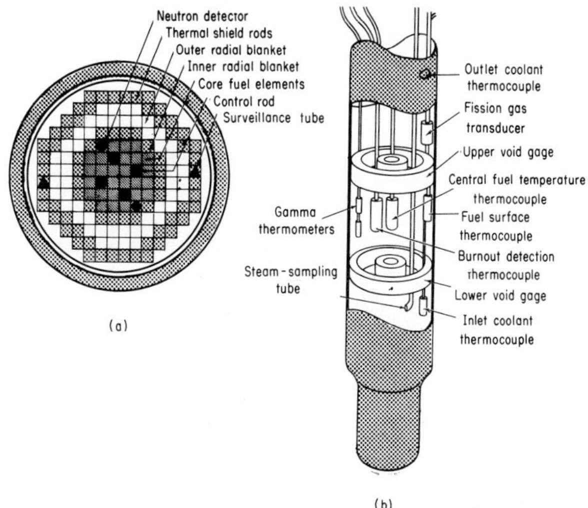

Fig. 6.6 Neutron detector is placed within core to monitor reactor reaction inside core.

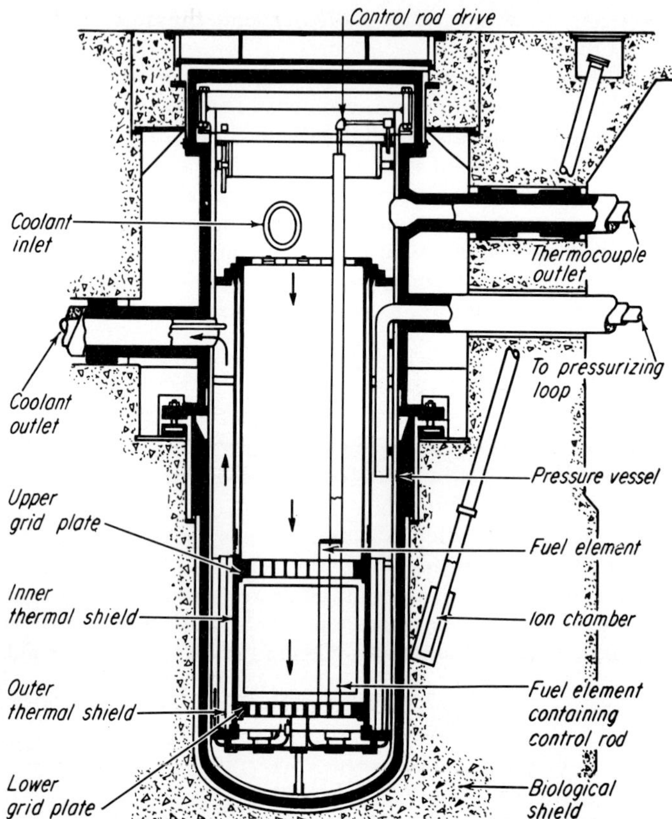

Fig. 6.7 Detectors are placed outside the reactor vessel in this organically cooled and moderated reactor.

number of materials, particularly those with the greatest density, such as lead or steel. The same effect is obtained by using a much greater thickness of water or concrete. For land-based reactors space and weight limitations are not major considerations. The cost of material and construction is usually more important. The cheapest and most widely used shielding material is concrete. When several feet thick, a concrete biological shield is an excellent neutron absorber. The addition of some percentage of denser material to the mix, such as iron or barytes, will improve local gamma shielding. Alternatively, the use of more expensive magnetic concrete affords protection against both gamma rays and neutrons.

The intensity of radiation follows an inverse square law as the distance from the source increases. Therefore, the amount of shielding required can be reduced by building the shield at a greater distance from the core. Since concrete is not able to withstand high heat, a thermal shield is constructed between the main shield and the reactor. This may be steel or a separate thin concrete vessel. Coolant channels may be provided in the concrete vessel to carry away heat. The major problem is preventing radiation-leaking paths where coolant inlet and outlet pipes, control rods, and other hardware leave the reactor. Any such annular paths must be stepped to avoid a direct "line of sight" path.

Q How is the reactor designed to withstand a pressure explosion?
A The reactor containment shell is a thin cylindrical or spherical steel pressure vessel surrounding the shielding. Its purpose is to accommodate energy released by sudden, uncontrolled fission in the event of a reactor accident. Since personnel work inside, the shell must be provided with access locks. The pressure within is maintained below atmospheric pressure so that any leakage of radioactive gas from the reactor is retained in the shell and passes through specially filtered and monitored ducts.

The pressure-suppression containment system eliminates the need for a large containment shell, at least in the case of water-cooled reactors. Instead, the reactor pressure vessel and its associated pumps and pipework are enclosed in a second pressure vessel sized to accommodate the steam formed in the event of vessel or pipe rupture. Ducts are connected to this outer vessel. They are led to an annular suppression pool in which the steam is condensed. Entrained fission products are then removed by absorption.

Q What emergency systems are usually provided to prevent core meltdown?
A Core meltdown from a loss of coolant has received great attention by designers and regulatory agencies. To prevent core meltdowns and consequently possible release of radioactive material to the biosphere, reactor systems are equipped with numerous safety devices to forewarn of a developing incident and also to initiate backup emergency core-cooling systems if needed. This system is intended to replenish cooling water that might be lost through a rupture of the primary cooling system. All emergency core-cooling systems are designed to inject water into the pressure vessel rapidly enough to keep the core cooled. Boiling-water reactors employ a core spray and an independent low-pressure coolant injection system.

Figure 6.4 shows the makeup of high-pressure injection pumps that are designed to inject cooling water into the reactor of a PWR unit in the event of an emergency condition.

CODE CLASSIFICATION OF NUCLEAR PRESSURE VESSELS

Q Since a nuclear reactor is basically a pressure vessel heated by nuclear energy, why did the ASME draw up a separate Nuclear Vessel Code and not use the existing Power Boiler Code, Section I, and the Unfired Pressure Vessel Code, Section VIII?

A At the beginning of nuclear-reactor-vessel development, all parties involved did go to the ASME boiler code committees for guidance on nuclear-reactor-vessel construction. Special rules were passed, called *cases*, as each problem of the nuclear-reactor vessel was carefully reviewed. When enough cases developed, a separate code had to be adopted for these reasons:

1. Nuclear vessels have many unusual design considerations which require more stringent rules than are needed for boilers and other pressure vessels.

2. There are three classes of nuclear vessels defined as follows: The ASME construction code provides for three levels of quality for design and construction, depending on the element of radioactive risk that may be involved with the component under consideration. The choice of quality level for a given component requires a knowledge of how the component functions in the process or system. *Class 1* components are those that are a part of the reactor-coolant pressure boundary, i.e., those components containing coolant for the core and piping and similar components that cannot be isolated from the core and its radioactivity. *Class 2* components are those involved with reactor auxiliary systems which are not part of the reactor-coolant pressure boundary but which are in direct communication with it, such as a residual heat-removal system. *Class 3* systems are those components that support Class 2 systems without being part of them.

3. Since a nuclear-reactor vessel may be radioactive for years, periodic internal inspections in the usual sense as applied to boilers and pressure vessels are impossible. Sophisticated testing by nondestructive means has to be employed.

4. The hazard in a nuclear-reactor vessel is not only a pressure explosion, but a radioactive contamination hazard far more serious to life than a steam pressure explosion only. A whole area may be affected by radioactive fallout. Thus more stringent design and fabrication rules are required than for boilers or pressure vessels.

5. Nuclear-reactor vessels can be subjected to very sudden heating or cooling temperature changes. This creates abnormal thermal stresses and cycling fatigue stresses. The power boiler code does not have enough provisions for calculating these stresses, but the Nuclear Vessel Code requires these to be calculated.

6. The material to be used on a nuclear vessel may have to

withstand radiation effects which may affect its properties. Materials for nuclear vessels thus require this consideration, whereas materials for boilers and pressure vessels may not.

7. Inspection during fabrication of a nuclear-reactor pressure vessel has to be far more thorough than inspection of a boiler because of the hazards involved. Rigid quality control must also be exercised far above the usual practice in boiler and pressure-vessel construction.

Q What special considerations are made for Class 1 vessels because they are part of the reactor coolant pressure boundary?

A The design of Class 1 vessels is far more exacting in terms of code requirements than for conventional boilers. Complete stress analysis is required on all major components of Class 1 vessels, and this must be done by a registered professional engineer. All forms of stress must be considered and combined to obtain the actual stress on each critical section of the vessel. These include normal, shear, discontinuity, bending, thermal, cycling, and fatigue stresses.

Corrosion allowance must be provided if analysis shows that it is required. All loading conditions must be considered, including: (1) internal and external pressure or combinations thereof; (2) the weight of the vessel and its contents; (3) superimposed loads, such as other vessels, insulation, piping, and cladding; (4) wind loads, snow loads, and earthquake loads; (5) reactions of supporting lugs, rings, saddles, and other supports; (6) temperature effects; and (7) environment effects due to radiation.

The allowable design stress is based on the yield strength, not on the ultimate strength, as is the case for boilers and unfired pressure vessels.

NDT Requirements Welding requirements are *stiffer*. No backing strips are allowed to remain. Full radiographing is required on all welds subject to stress caused by pressure. Some welds require ultrasonic examination also. The material must be certified by the material manufacturer to comply with all sections of the Nuclear Vessel Code. Heat-treatment test coupons are required. Ultrasonic examination is required for steel plates on reactor vessels and all other plates 4 in and over in thickness. Standards are established for the ultrasonic technique employed. The ultrasonic test is by means of a pulse-echo instrument.

Forgings must be inspected ultrasonically and their surface inspected by the magnetic-particle method. Forgings of nonmagnetic material must be examined on the surface by the liquid-penetrant method. Castings must be examined by radiographing, ultrasonic, magnetic-particle, and liquid-penetrant methods.

Pipes, tubes, and fittings must be examined over their full length by radiographic, ultrasonic, magnetic-particle, liquid-penetrant, or eddy-current methods, based on the material to be examined. Bolts and bolting material must be examined by the wet magnetic-particle method or by the liquid-penetrant method. Carbon steel, alloy steel, and chromium-alloy steels must be subjected to impact tests to determine ductility and brittleness.

Quality Control and Documentation During the fabrication phase, it is important to keep complete and detailed records of all the inspections, to be sure that the inspections were made at the proper time, in the proper sequence and not after the fact, and on all material released in fabrication.

Testing Vessel test plates must be made. The welding procedure and the nondestructive examination methods used should be the same as those used in the fabrication of the vessel. Important test specimens must be made. A final hydrostatic test of 1¼ times the design pressure is stipulated. Pneumatic tests may be used in lieu of the hydrostatic test.

Marking, Stamping, and Reports Design calculations and specifications must be filed with the state enforcement authority responsible, at the point of the installation, for the vessel. The design specifications must be certified as to compliance on vessel classification, detailed report on operating conditions for a complete evaluation of the design, and construction and inspection according to code rules by a registered professional engineer experienced in pressure-vessel design. All necessary data sheets must be included. The vessels to be marked with the N symbol must be stamped with the following information: (1) class of vessel; (2) manufacturer; (3) design pressure at coincident temperature; (4) manufacturer's serial number; and (5) year built.

Q Who must have certificates of authorization from the ASME if involved with nuclear-power-plant components?

A The ASME code for nuclear-power-plant components now requires certificates of authorization to be obtained and retained by the following organizations that may be active in nuclear-power-plant work: manufacturers, engineering organizations, fabricators, installers, material manufacturers and suppliers, owners and agents of a proposed plant. The aim and purpose of certification is to have a quality-assurance program from all accredited organizations that have met certain minimum code standards. The ultimate aim is that all work performed will be conducted in rigid compliance with approved and controlled specifications and procedures, that critical activities will be verified by qualified personnel from an organization independent of work performance (third-party inspection program), that measurement and tests results will be completely documented and analyzed for conformance to established specifications and code requirements, and that appropriate actions will be taken to preclude the recurrence of discrepancies and deficiencies.

Most certificate holders are required to have a quality-assurance manual. This document must be kept current and must explain in detail the controlled manufacturing system that is being followed in order to achieve code compliance with specifications.

Q List the ASME documents that are applicable to nuclear-power-plant components.

A The Nuclear Regulatory Commission (NRC) of the federal government has numerous requirements and standards for nuclear power plants, including impact-on-the-environment considerations. Section III of the ASME Boiler and Pressure Vessel Code now consists of eight documents:

- Division 1: Nuclear Power Plant Components, General Requirements
- Division 1: Nuclear Power Plant Components, Class 1 Components
- Division 1: Nuclear Power Plant Components, Class 2 Components
- Division 1: Nuclear Power Plant Components, Class 3 Components
- Division 1:Nuclear Power Plant Components, Class MC Components
- Division 1: Nuclear Power Plant Components, Component Support
- Division 1: Nuclear Power Plant Components, Core Support Structures
- Division 2: Nuclear Power Plant Components, Code for Concrete Vessels and Containment

The ASME also has a Section XI code entitled *Rules for In-Service Inspection of Nuclear Power Plant Components.*

Q What document and organization besides the ASME defines the qualifications of nuclear inspectors?

A Another document that affects the construction of nuclear power plants is American National Standard Institute (ANSI) N626, entitled *Qualification and Duties for Authorized Nuclear Inspection*, which the ASME has also adopted as a requirement. This document defines authorized inspection agencies and authorized nuclear inspectors and supervisors. Great emphasis is placed on knowledge of welding and how it can affect metals in the examinations given to potential nuclear inspectors. Knowledge of nondestructive testing (NDT) methods and interpretation of results are also highlighted in the requirements. Quality-control procedures in monitoring a construction job are also stressed in the duties of a nuclear inspector, as are traceability of documentation on materials, welding procedures, qualification of welders, and similar areas where variables in construction may exist. The nuclear code has been extended to pumps and valves to ensure the integrity of the pressure-containment parts and not only reactors and pressure vessels.

Q What is the primary purpose of requiring in-service inspections of nuclear-power-plant components?

A The primary objective of Section XI of the ASME code is to provide a means of ensuring that the mechanical integrity of the primary coolant system is maintained throughout the operating life of the facility. This objective is accomplished through the requirement for the conductance of minimum periodic inspections of critical nuclear components, such as Class 1 vessels and their weld zones and other highly stressed areas. A preservice inspection is required. Usually ultrasonic inspection techniques are used that can be duplicated later during required reinspections. This permits comparisons to be made to note changes. Automatic inspection devices are utilized as much as possible. The use of remotely operated inspection equipment under properly planned procedures has reduced the radiation hazard to examining and inspection staffs. The frequency of inspections and acceptable criteria are detailed in the ASME code for nuclear-power-plant components. Fracture mechanics is used to analyze the seriousness of flaws where no code standards exist.

NDT PERSONNEL

Q How are NDT personnel qualified and classified to perform NDT work?

A Nondestructive testing personnel must be qualified, and these qualifications are graded as follows:

Level 1 people must have experience or training in the performance of the inspections and tests that they are required to perform. They should be familiar with the tools and equipment to be employed and should have demonstrated proficiency in their use. They must be familiar with inspection and measuring equipment calibration and control methods and be capable of verifying that the equipment is in proper condition for use.

Level 2 people must have experience and training in the performance of required inspections and tests and in the organization and evaluation of the results of the inspections and tests. They must be capable of supervising or maintaining surveillance over the inspections and tests performed by others and of calibrating or establishing the validity of calibration of inspections and measuring equipment. They must have demonstrated proficiency in planning and setting up tests and must be capable of determining the validity of test results.

Level 3 people must have broad experience and formal training in the performance of inspections and tests and should be educated through formal courses of study in the principles and techniques of the inspections and tests that are to be performed. They should be capable of planning and supervising inspections and tests, reviewing and approving procedures, and evaluating the adequacy of activities to accomplish objectives. They must be capable of organizing and reporting results and of certifying the validity of results.

Q What are the requirements for NDT supervisors?
A Personnel involved in the performance, evaluation, or supervision of nondestructive examinations, including radiography, utlrasonic, penetrant, magnetic-particle, or eddy-current methods, must meet the Level 3 qualification specified in SNT-TC-1A of the American Society for Nondestructive Testing and supplements. Those personnel involved in the performance, evaluation, and supervision of gas-leak test methods must meet the qualification requirements specified for a Level 2 person.

Personnel who are assigned the responsibility and authority to perform project functions must have as a minimum the level of capability shown in Table 6.1. When inspections and tests are implemented by teams or groups of individuals, the one responsible must participate and must meet the minimum qualifications indicated.

A file of records of personnel qualifications must be established and maintained by the owner. This file should contain records of past performance, training, initial and periodic evaluations, and certification of the qualifications of each person.

Q What requirements or standards apply to nuclear operators?
A Operator qualification and training are governed and determined by federal requirements. The criteria for the selection and training of nuclear-power-plant personnel are contained in American National Standard 3.1, entitled *Selection and Training of Nuclear Power Plant Personnel.* Most nuclear plants require a minimum presence during operation of at least one senior reactor operator, two reactor operators, and two auxiliary operators who may not be licensed yet. Operator training is receiving increased attention, as is the need for supporting staff such as a safety engineer to supplement the operating staff in case an emergency situation develops in operation. The many valves and subsystems in a nuclear power plant as well as electrically activated

Table 6.1 Minimum Levels of Capability for Project Functions

Project function	Level 1	Level 2	Level 3
Approve inspection and test procedures			X
Implement inspection and test procedures	X		
Evaluate inspection and test results		X	
Reporting of inspection and test results		X	

controls require a broad knowledge of the interplay between systems. See Fig. 6.4. The many pumps and valves shown can lead to errors of valve opening and closing and also increase the possibility of electrical and mechanical failures. Complete reliance on automatic redundancy systems cannot anticipate all the possibilities of malfunction or sensing that a malfunction is taking place; therefore, skilled operators well versed and trained in the design and operation requirements of nuclear power plants are needed.

Q What fire-resistive barriers are required in nuclear power plants?
A Fire protection in nuclear power plants has received major attention. Fire-resistive barriers of not less than 3-h rating are required to separate the following areas in a nuclear plant: administration building, battery rooms, boilers used for starting, cable penetrations, cable shafts, cable tunnels, computer rooms, control building or room, decontamination areas, fire pump rooms, switch gear room, and so on.

Q What is under consideration for nuclear-waste disposal?
A The technology exists to dispose of high-level nuclear waste, but approval by the federal government is required. Under design consideration is a system that converts high-level liquid waste to a solid and immobilizes this solid in glass for burial in deep, underground salt caves in remote areas away from the human environment.

Q Differentiate between surface examination and volumetric examination.
A Surface examination includes the liquid-penetrant method and magnetic-particle examination. It is used to verify surface or near surface cracks or discontinuities. Volumetric examination is for the purpose of detecting subsurface discontinuities to the extent that the entire volume of metal below the metal surface may be examined. The two principal examination methods used are radiographic examination and ultrasonic testing.

BASELINE INSPECTION

Q Explain the term preservice or baseline inspection of nuclear-power-plant components.
A These are examinations and inspections that are performed on nuclear-power-plant components prior to operation. These examinations and inspections must be performed with visual, NDT, and other examination methods which can be duplicated after the plant has been in service. Periodic testing after the plant has been in service is required. Components in the primary radioactive area require as a

minimum one inspection every 10 years. All records generated during the preservice inspection must be retained by the owner for future comparison of the results obtained during the in-service inspections, and thereafter for the life of the components. If this procedure is followed, it is expected that changes in welds, material, and similar considerations will be detected for serious developments. These could include such conditions as crack formations, porosity, corrosion, thinning, embrittlement, and similar possible deteriorating phenomena.

Q Does the ASME code address itself to pump and valve integrity checking at periodic intervals?
A The primary purpose for code-required periodic pump and valve inspections and tests is to make sure they will be able to operate as intended during any operational upset which will require these pumps to keep the reactor cool for a safe shutdown. As a result, there are several requirements. Among these are:

1. Valves which have devices at remote locations to indicate if they are open or closed must be tested by the operators at defined intervals to make sure that the indicating devices are operating correctly.
2. Regular exercising of valves is required in order to make sure that the stems move up and down properly and are not frozen.
3. Pumps are required to be test-run periodically.
4. Diesel generators that provide blackout power must be test-run at stated intervals to test their ability to provide emergency power.

Q What mechanical devices are used in inspecting highly radioactive areas of a nuclear power plant?
A Giant robots are being used to inspect the welds of reactors. The typical reactor pressure vessel may be four or five stories high and about 12 ft in diameter. Since the vessel becomes highly radioactive during service, it is flooded at the start of inspection to provide a protective shield. The robot operates within the water-filled area in order to inspect welds, nozzles, and even the top flange ligaments between the bolt holes used to attach the vessel's cover.

The robot of the Westinghouse Corp. carries an array of 16 ultrasonic inspection transducers and is under the control of a computer. The robot has nine degrees of freedom to position the transducers to the area of welds to be inspected. An operator can follow the progress of the inspection by the robot and can stop it at certain points for closer examination if required.

Robot inspection is quicker and far less expensive than human inspection. It also reduces the need for human beings to be exposed to the radioactivity.

RADIOACTIVITY MONITORING

Q How is radioactivity detected?

A The instruments commonly used are Geiger-Müller counters, scintillation counters, gamma survey meters, and proportional counters. The instrument to be used depends on the type and density of radiation to be measured. Geiger-Müller counters are used to detect beta and gamma radiation and are not effective for measuring alpha radiation.

Q What procedures are generally followed in decontaminating radioactive material?

A Radioactive material connot be destroyed. Therefore, the process of decontamination involves one of the following methods of lessening the hazard of radioactivity: (1) Isolate the area of radioactive contamination until such time as the radioactivity is decreased to a safe level as a result of radioactive decay. The half-life of the contaminant will influence this procedure. (2) Treat the surface so that the radioactive material is absorbed, cleaned off, swept, etc., and then taken to a site where the radioactive substance will not harm people. The most difficult decontamination is where the contaminant is absorbed into porous material such as concrete. Complete removal of walls, floors, and similar contaminated areas may be necessary under this type of contamination.

Q Name some typical methods employed to protect workers from excessive radiation doses.

A Some typical factors considered in the protective scheme are generally the following: (1) Control the length of an exposure; (2) control the distance between the human body and the radiation source; (3) provide shielding between the body and the source of radiation; (4) establish a strict and precise radiation-monitoring program in the work area; and (5) have medical facilities available at all times to handle any accidental exposures above normal stipulated levels.

Q What are the present NRC dose limits on radiation?

A Federal regulations currently limit occupational radiation dose to 1¼ rems in any calendar quarter or specified 3-month period. However, when there is documented evidence that a worker's previous occupational dose is low enough, a licensee may permit a dose of up to 3 rems per quarter or 12 rems per year. The accumulated dose may not exceed $5(N - 18)$ rems, where N is the individual's age in years; i.e., the lifetime occupational dose may not exceed an average of 5 rems for each year above the age of 18.

Q How is radiation dose defined?

A The term *dose* is the quantity of radiation absorbed per unit mass by the body or any portion of the body. The term *rem* means "roentgen-equivalent man or mammal."

Q Define a roentgen.
A A roentgen is a unit exposure dose of gamma radiation (x-ray) such that the associated corpuscular emission per 0.001293 g of air produces, correspondingly in air, ions carrying 1 esu of electricity.

Q Define a rad.
A The rad is a measure of the dose of any ionizing radiation to body tissues in terms of the energy absorbed per unit mass of the tissue. One rad is the dose corresponding to the absorption of 100 ergs per gram of tissue [1 millirad (mrad) = 0.001 rad].

Q What is the relationship between a rem and rad?
A The rem is a measure of the dose of any ionizing radiation to body tissue in terms of its estimated biological effect relative to a dose of 1 roentgen (R) of x-rays. The relationship between a rem and rad is that the dose in rems is equal to the dose in rads multiplied by the appropriate relative biological effectiveness (RBE). The RBE is defined as the ratio of gamma x-rays to the dose that is required to produce the same biological effect by the specific radiation under consideration.

Q Explain the methods used in monitoring the radiation exposure of personnel.
A Monitoring the radiation levels in an area with portable or remote instrumentation allows prediction of the exposures which will be sustained by personnel working in the areas.

The techniques used to assess the personnel exposure in restricted areas involve use of personnel dosimetry equipment, including self-loading pocket dosimeters and personnel film badges.

Another technique of assessing the exposure of various personnel relies on *estimates of the activity*, utilizing monitoring data, from some portable survey equipment and air samples of laboratory and bioassay procedures.

Bioassay procedures are used to determine the amount of a radionuclide present in the body. The various bioassay procedures consist of excretion analysis (feces, breath, and urine), analysis of clinical samples (blood, nose wipes, sputum tissues), and direct analysis utilizing gamma scintillation counting for the entire body or for the individual organs which are known to have highly specific proclivities for nuclides.

Q How is a worker's radiation dose determined?
A A worker may wear two types of radiation-measuring devices. A self-reading pocket dosimeter records the exposure to incident radiation

and can be read out immediately upon finishing a job involving external exposure to radiation. A film badge or TLD (total life dosage) badge records radiation dose, either by the amount of darkening of the film or by storing energy in the TLD crystal. Both these devices require processing to determine the dose and are considered more reliable than the pocket dosimeter. A worker's official report of dose received is normally based on film or TLD badge readings.

Q How much radiation does a nonnuclear worker receive in a year?
A A government study indicates that the average person is exposed to radiation from conception onward. The environment contains naturally occurring radiation. Cosmic radiation from outer space and the sun contributes additional exposure. The use of x-rays and radioisotopes in medicine and dentistry may add more radiation. A government study showed the estimated average individual exposure in millirems from natural background radiation and other sources of radiation (see Table 6.2). Thus, the average individual in the general population receives about 0.2 rem of radiation exposure each year from sources that are a part of our natural and manufactured environment. By the age of 20 years, an individual has accumulated about 4 rems. The most likely target for reduction of population exposure is thus in medical uses.

Q What are the typical radiation doses of workers in nuclear plants?
A A government report which was based on data received from nuclear licensees stressed the following dosage levels. Data were received on the occupational doses in 1977 of approximately 100,000 workers in power reactors, industrial radiography, fuel-processing and fabrication facilities, and manufacturing and distribution facilities. Of this total group, 85 percent received an annual dose of less than 1 rem according to these reports; 95 percent received less than 2 rems; fewer

Table 6.2 U.S. General Population Exposure Estimates (1978)*

Source	Average individual dose, mrem/year
Natural background	100
Release of radioactive material by mining, milling, etc.	5
Medical	90
Nuclear weapons development (primarily fallout)	5–8
Nuclear energy	0.28
Consumer products	0.03
Total	~ 200 mrem/year

* Adapted from a report by the Interagency Task Force on the Health Effects of Ionizing Radiation published by the Department of Health, Education, and Welfare.

than 1 percent exceeded 5 rems in any 1 year. The average annual dose of these workers who were monitored and had measurable exposures is about 0.65 rem. A study completed by the EPA, using 1975 exposure data for 1,260,000 workers, indicated that the average annual dose for all workers who received a measurable dose was 0.34 rem.

Q Who is responsible for monitoring radiation levels in a nuclear power plant?

A The owner of the facility is responsible for monitoring all areas for radiation. Radiation officers are usually employed who are qualified by training and experience in radiological health so that they can properly evaluate radiation hazards. It is their responsibility to establish and administer a radiological protection program. They coordinate the program with station technical and supervisory personnel by:

1. Planning the activities of personnel who must enter radiation areas to limit dosages.
2. Monitoring the actions and procedures of all individuals working in such areas.
3. Conducting preentry and postentry briefings on procedures that could improve or reduce exposures for the task on hand or to be faced in the future.
4. Studying the numerous U.S. Nuclear Regulatory Guides on nuclear safety requirements in order to advise staff of the requirements and the procedures needed to implement them.

Q What are the major issues facing the continuing development of nuclear power?

A The negative impact to nuclear power has been the question of nuclear safety, especially from malicious mischief or sabotage causes. Waste disposal and the rising costs of nuclear power and any unit outage costs have also been cited by opponents of nuclear power.

Improvements are being made in monitoring reactor performance by the addition of sensors at critical parts of the reactor loop, such as:

1. Relief- and safety-valve indicators so that the operators know if the valves are open or closed in the control room.
2. Increasing the readout ranges on core temperatures.
3. Automatic closure of relief valves on pressurizers when pressure returns to normal. This is to prevent loss of water from a stuck relief valve.
4. Increased emphasis on operator training for every conceivable abnormal event in the nuclear plant.

Maximizing the availability of operating plants has an economic impetus, since unit outages require purchasing more expensive fossil-fuel power. The trend is developing in reassuring the public that there

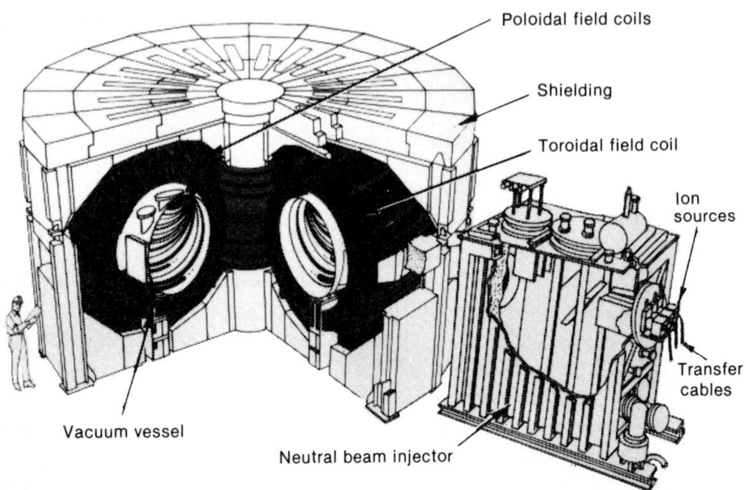

Fig. 6.8 Sketch of Princeton University's tokamak fusion reactor, which will use deuterium-tritium as a fuel.

is a need for nuclear power and that safety of nuclear power has been adequately designed for. In the final analysis, however, it will be the depletion of presently known fossil fuels that will require the continued use of nuclear power to maintain our present standard of living.

Q Can fusion power solve the earth's energy problems in the decades ahead?

A Scientists are stressing that the controlled fusion reaction of two or more isotopes of hydrogen into helium would solve the world's energy problems for millennia, because we would be duplicating the sun's process. Nuclear fusion releases enormous energy from small amounts of matter. Hydrogen has three isotopes of mass numbers 1, 2, 3. Hydrogen is mass number 1, deuterium (D) has mass number 2, and tritium (T) is number 3. At present, the deuterium-tritium (D-T) reaction is under intense development in the fusion-reaction field, because one cubic meter of water contains about 10^{25} atoms of deuterium with a mass of 34.4 g and a potential fusion energy of 7.94×10^{12} J. This is equivalent to the heat of combustion of 300 metric tons of coal, or 1500 bbl of crude oil.

It is for this reason that a race is on in the world to harness fusion power. Figure 6.8 is a sketch of the University of Princeton's tokamak-type fusion reactor. It is rated at 30 MW and will be used for engineering research and development. A D-T fuel mixture is to be employed.

CHAPTER **SEVEN**

Materials, Welding, and Nondestructive Testing

Boilers and pressure vessels require material that meets specifications to assure a safe plant. The ASME boiler code is recognized as the chief source for specifications on permissible material, and these are made legal requirements by state jurisdictions when they adopt the code as a jurisdictional requirement also. Steel boilers are assembled by welding, and the welding process and the welders themselves must meet code requirements. Nondestructive testing is an important quality-control tool in checking material and fabrication for possible rejectable defects. It is also used in checking weld repairs. This chapter will attempt to review some of the above.

MATERIALS

Q Briefly describe cast-iron types and properties of the commonly used cast irons.

A Compared with steel, cast iron is decidedly inferior in malleability, strength, toughness, and ductility. The most important types of cast iron are the white and the gray cast irons.

White Cast Iron White cast iron is so known because of the silvery luster of its fracture. In this alloy, the carbon is present in combined form as iron carbide (Fe_3C), known metallographically as cementite.

Gray Cast Iron Gray iron is the most widely used of cast metals. In this iron, the carbon is in the form of graphite flakes that form a multitude of notches and discontinuities in the iron matrix. The appearance of the fracture of this iron is gray because the graphite flakes are exposed. The strength of the iron increases as the graphite-crystal size decreases and the amount of cementite increases. Gray cast iron is easily machinable because the graphite carbon acts as a lubricant for the cutting tool and also provides discontinuities which break the chips as they are cast. Gray iron, having a wide range of tensile strength, from 20,000 to 90,000 psi, can be made by alloying with nickel, chromium, molybdenum, vanadium, and copper. Permissible stresses per the ASME code, Section IV, are as follows:

Class	Ultimate tensile strength, kips/in^2	Allowable stress in tensile strength, kips/in^2
20	20.0	4.0
25	25.0	5.0
30	30.0	6.0
35	35.0	7.0
40	40.0	8.0

Q What are some physical defects found in cast iron?
A Some of the most common physical defects which may be present in cast iron and which will weaken it are blowholes, cracks, segregation of the impurities, and coarse-grain structure.

Sulfur, which combines with the manganese or the iron to form a sulfide, makes the cast iron brittle (hot-short) at high temperatures. It also increases shrinkage. Hence, its amount is usually limited to less than 0.1 percent in specifications for cast iron.

Phosphorus, in amounts of more than 2 percent, makes the iron brittle and weakens it. However, it also has the effect of increasing the fluidity and decreasing the shrinkage. So it is desirable in making sharp castings for ornamental parts where strength is unimportant.

Q What impurities may affect the physical properties of steel?
A The strength, ductility, and related properties of the iron-carbon alloys can be affected by the presence of harmful elements such as sulfur, phosphorus, oxygen, hydrogen, and nitrogen.

Sulfur Sulfur has the same effect on steel as on cast iron, making the metal hot-short, or brittle at high temperatures. As a result, it may be harmful in steel which is to be used at elevated temperatures or, more particularly, may cause difficulty during hot rolling or other shaping operations. Most specifications for steel limit its amount to less than 0.05 percent.

Phosphorus Phosphorus makes steel cold-short, or brittle at low temperatures; so it is undesirable in parts which are subjected to impact loading when cold. However, it has the beneficial effect of increasing both fluidity, which tends to make hot rolling easier, and the sharpness of castings. Since cast iron is brittle anyway, phosphorus is sometimes added to make the castings clean-cut. Most specifications for structural steel limit the phosphorus content to less than 0.05 percent.

Oxygen When iron is in the liquid state, any free oxygen combines readily with iron to form iron oxide. In the finished steel or iron, the iron oxide usually appears in the form of tiny inclusions distributed throughout the metal. These inclusions introduce points of weakness and increased brittleness, concentration from voids, and stress, which are undesirable, for they promote the formation of cracks that may result in progressive fracture.

Hydrogen Hydrogen, such as is produced when steel is immersed in sulfuric acid to remove mill scale before cold drawing, makes steel brittle. The hydrogen may be removed by heating the steel for a few hours, or hydrogen will gradually work out of the steel at ordinary temperatures. When present, hydrogen increases the hardenability of the steel.

Nitrogen Nitrogen has a hardening and embrittling effect on steel. This may be objectionable, or it may be desirable in producing a hard surface on the steel, under controlled manufacturing conditions. For example, in the nitriding process, the steel is exposed to ammonia gas at about 600°C (1112°F) to produce a hard, wear-resisting surface without lowering the ductility of the center of the piece.

Nonmetallic Inclusions Nonmetallic inclusions occur during the plate-rolling operation in steel making. Quite often, these inclusions cause laminations, or planes in the plate where metal separation or voids exist.

The quality of steel plate can be significantly affected by nonmetallic inclusions. The presence of inclusions, such as sulfides and oxides, primarily affects the ductile behavior of the steel. The quality of a particular grade of steel can be improved by eliminating or minimizing inclusions. Nondestructive testing is used extensively for critical-pressure applications, such as nuclear reactors, in order to find inclusions.

Q Briefly outline the alloys and the reason they are added to carbon steel.

A The purpose of adding alloying elements to carbon steel is to impart to the finished product desirable physical or chemical properties which are not available in carbon-steel parts fabricated by standard procedures. These properties could involve desirable electrical, magnetic, or thermal characteristics, as well as engineering considerations such as (1) high tensile strength or hardness without brittleness; (2) resistance to corrosion; (3) high tensile strength or high creep limit at elevated or subzero temperatures; (4) other desirable physical features required to resist special loading.

Carbon Up to 1.2 percent carbon in iron increases the strength and ductility of steel. When the carbon content is above 2 percent, graphite formation is promoted, which lowers both the strength and the ductility of steel. Carbon content above 5 or 6 percent causes the metal to be brittle with very low strength for any load-resistance application.

Manganese By combining with sulfur, this element prevents the formation of iron sulfide at the grain boundaries. This minimizes surface ruptures at steel-rolling temperatures (red shortness), which results in significant improvement in surface quality after rolling.

Nickel This element increases toughness or resistance to impact. In this respect it is the most effective of all the common alloying elements in improving low-temperature toughness.

Chromium This element contributes to corrosion resistance and heat resistance in alloy steels. A strong carbide-former, chromium is frequently used in carburizing grades and in high-carbon-bearing steels for superior wear resistance. An 18 percent chromium and 8 percent nickel (18-8 nickel-chrome) alloy is widely used as a high-strength stainless steel in pressure equipment requiring strength and corrosion resistance.

Molybdenum Like nickel, molybdenum does not oxidize in the steel-making process, a feature which facilitates precise control of hardenability. It markedly improves high-temperature tensile and creep strength and reduces a steel's susceptibility to temper brittleness.

Boron Since boron does not form a carbide or strengthen ferrite, a particular level of hardenability can be achieved without an adverse effect on machinability and cold formability that may occur with the other common alloying elements.

Aluminum In amounts of 0.95 to 1.30 percent, aluminum is used in nitriding steels because of its strong tendency to form aluminum nitride, which contributes to high surface hardness and superior wear resistance.

Silicon Silicon combines with carbon to form hard carbides which, when properly distributed throughout the alloy, have the effect of increasing the elastic strength without loss of ductility.

Tungsten Tungsten forms hard stable carbides when it is added to steel. It raises the critical temperature, thus increasing the strength of the alloy at high temperatures.

Vanadium This element acts as a deoxidizing agent as aluminum does on molten steel. It forms very hard carbides, thus increasing the elastic and tensile strength of low-carbon and medium-carbon steels.

Other elements may be added to improve the rolled strength and toughness of steels in the high-strength low-alloy category.

Q What is the effect of heat treatment on steels?

A When steel is heated to certain temperatures and then rapidly or slowly cooled, its physical properties such as elastic limit, ultimate strength, and hardness can be changed.

Heat treatments fall into two general categories: those which increase strength, hardness, and toughness by quenching and tempering and those which decrease hardness and promote uniformity by slow cooling from above the transformation range or by prolonged heating within or below the transformation range, followed by slow cooling.

Annealing consists of heating steel to a certain temperature and cooling it by a relatively slow process. Annealing may be used to remove stresses such as are produced in forgings and castings, to refine the crystalline structure of steel, or to alter the ductility or toughness of steel.

Stress-relief annealing involves heating to a temperature approaching the transformation range, holding for a sufficient time to achieve temperature uniformity throughout the part, and then cooling to atmospheric temperature. The purpose of this treatment is to relieve residual stresses induced by normalizing, machining, straightening, or cold deformation of any kind. Some softening and improvement of ductility may be experienced, depending on the temperature and time involved.

Normalizing involves heating to a uniform temperature about 100 to 150°F above the transformation range, followed by cooling in still air.

Hardening consists of heating steel to above its transformation temperature range and cooling it suddenly by quenching in water, oil, or some other cooling medium that absorbs heat rapidly.

Quenching is defined as a process of rapid cooling from an elevated temperature, by contact with liquids, gases, or solids. Quenching increases the hardness of steel if its carbon content is 0.20 percent or higher. It also raises the elastic limit and ultimate strength and reduces the ductility; however, it induces internal stresses, and the metal is apt to become brittle.

STEEL TYPES

Q What is "plain" carbon steel?

A This is a broad range of steels representing about 90 percent of all steel produced. The term "plain" carbon steel denotes a steel whose properties depend primarily on carbon content, whereas alloy steels owe their properties primarily to the alloys added. Carbon content in plain carbon steels ranges from 0.10 to 0.60 percent for some grades of carbon-steel pipe.

Q What is rimmed steel?

A Rimmed steel is a carbon steel with less than 0.20 percent carbon. The name "rimmed" was derived from the soft rim which forms on the steel ingot during steel manufacture. When rimmed steel is heated to its melting point, such as in heli-arc welding, the iron oxide in solution reacts with the carbon in solution and forms carbon monoxide and carbon dioxide gas bubbles. Thus rimmed steel that is welded may have porosity in the weld from these escaping bubbles. To prevent this, oxyacetylene welding may be used, because this welding process permits the bubbles to escape and not cause voids, or gas pockets in the metal. Covered electrodes with silicon or other deoxidizers also help to reduce the bubbling by combining with the oxygen in the rimmed steel.

Q What is a "killed" steel?
A Killed steels contain enough deoxidizers to remove all the dissolved oxygen in the steel and thus prevent gases from forming. The sulfur and phosphorus content are usually 0.05 and 0.04 percent, respectively, with the silicon content being a minimum of 0.1 percent. Killed steel is a fine-grained steel that has a high yield strength and is used in pipes and tubes. Grades A106 and A192 are used for boiler tubes.

Q What is a semikilled steel?
A Semikilled steels are partially deoxidized and are intermediate between rimmed and killed steels, with less than 0.25 percent carbon and less than 0.90 percent manganese with low sulfur and phosphorus content. Porosity is not likely to be encountered in welding, since this steel has a silicon range from 0.05 to 0.09 percent.

Q How are low-alloy steels defined?
A Low-alloy steels are always killed steels with suitable alloys added. For example, the chromium-molybdenum grades A200 and A235 of ASTM specification are frequently used in high-temperature steam and oil-refinery-plant applications. These steels are usually preheated before welding, and some form of postweld heat treatment is also required to remove residual stresses from any welding.

Q Describe briefly stainless steels.
A Stainless steels are high-alloy killed steels which were developed to resist atmospheric and high-temperature oxidation and also to resist the corrosive effects of many salts and acids. The corrosion-resistance property is derived primarily from the principal alloys used—chromium and nickel. Thus, an 18.8 stainless steel has 18 percent chromium and 8 percent nickel.

Q What is meant by carbide precipitation in welding some stainless steels?
A When unstabilized stainless steels are heated to 800 to 1500°F, the chromium in the steel combines with the carbon to form chromium carbides, and this tends to form along the grain boundaries of the metal,

thus the term carbide precipitation. This lowers the dissolved chromium in the grain boundary and makes this area much less resistant to corrosion. The corrosion that does take place is intergranular. Another term for carbide precipitation is sensitization—the migration of chromium in solution to form chromium carbides and thus leave the chromium-depleted region "sensitive" to local corrosion where the depletion has taken place.

Q How can carbide precipitation be minimized in stainless steels?
A There are three methods:

1. Postweld heat treatment, but this requires temperatures of 1850 to 2050°F, which is difficult to attain in the field.

2. By keeping the carbon content of the stainless steel low so that very little is available to combine with the chromium.

3. By substituting or adding titanium and columbium so they will combine with the carbon and thus leave the chromium in solution to resist corrosion. The stainless steel is then classified as being "stabilized" under these conditions. AISI 347 and 321 are considered stabilized stainless steels. AISI 316 and 317 also contain molybdenum, and these steels are used for corrosive service and for steam-power-plant applications for temperatures above 1100°F.

Q What are the main classifications of stainless steels?
A Several hundred stainless steels are produced, but they can be classified into the following types:

Austenitic These steels can be hardened by cold working but not by heat treatment. When annealed, all are nonmagnetic. They have excellent corrosion-resistance properties, and most are readily weldable. Type 304 is a typical grade and is an 18-8 stainless steel. Type 316 is also an 18-8 stainless steel but has 2 percent molybdenum added for improved pitting and crevice corrosion resistance.

Ferritic These are straight chromium-type stainless steels with 11 to 30 percent chromium content. They are good resistors of chloride stress-corrosion cracking. Ferritic stainless steels cannot be hardened by heat treatment but can be moderately hardened by cold working. They are magnetic steels with good ductility. However, ferritic steels may lose their ductility at long exposures between 750 and 1000°F. The code limits these steels to 700°F maximum temperature as a result.

Martensitic Martensitic stainless steels are austenitic at elevated temperatures that can be hardened (hence the term martensitic) by suitable cooling to room temperature. They usually have 11 to 18 percent chromium, with type 410 stainless steel being a typical martensitic steel. These stainless steels do not have the corrosion resistance of the ferritic or austenitic types. The boiler code limits these steels to 700°F, since they have a tendency to oxidize at high temperatures.

Duplex These stainless steels are a mixture of austenitic and ferritic. They are resistant to chloride stress-corrosion cracking and are used in tube sheets of condensers and heat exchangers.

Precipitation-Hardening These are chromium-nickel stainless steels containing other alloying elements such as copper or aluminum. The addition of these alloys results in the formation of precipitates during processing of the steel. They are used primarily in gears and aircraft parts.

Q What steels are used for nuclear reactors?

A Pressurized-water reactors with pressures of 2000 psi and temperatures of 550 to 600°F have been fabricated from alloy steels of the manganese-molybdenum-nickel composition, generally conforming to ASTM A302 and A533. The thickness required ranges from 4 to 12 in, with reactor diameters of 11 to 15 ft.

One item that must be considered in nuclear-material selection, which is not required in fossil-fuel-fired plants, is the long-term irradiation effect on metals, especially as it may affect embrittlement.

Boiling-water reactors are 18 to 21 ft in diameter, operate in the 1250-psi range, and are about 3 to 8 in thick in the shell. Manganese-nickel-molybdenum steels conforming to ASTM A533 Grade B and A302 Grade B are commonly used.

Q What does the term *vacuum degassing* mean in the manufacture of steel?

A Vacuum is used over the furnaces in the steel-making process in order to withdraw harmful gases formed in steel making. Oxygen and hydrogen are pulled off, and this helps to produce steels that are free of voids, pockets of impurities, and porosity. It also permits adjusting the final chemical analysis of the steel to within close tolerances.

PROPERTIES OF MATERIALS

Q What are the physical and mechanical properties that are determined and considered when building power equipment?

A Properties of materials for engineering structures are generally described under the following headings:

1. *Physical.* These would describe the material's composition, structure, homogeneity, specific weight, thermal conductivity and ability to expand and contract, and resistance to corrosion.

2. *Formability.* These would relate to the manufacture of the material, such as fusibility (welding), forgeability, malleability, and ability to shape the material by bending or machining.

3. *Mechanical.* Mechanical properties describe the ability of a material to resist applied loads and are usually obtained from tests.

The basic mechanical properties are elastic limits, moduli of elasticity, ultimate strengths, endurance limits, and hardness. Secondary mechanical characteristics determined from the basic ones or simultaneously with them are resilience, toughness, ductility, and brittleness.

Strength depends on the type and nature of loading. The static strength of a material is expressed by the corresponding elastic limit stress. The impact strength is measured by the corresponding modulus of resilience. The endurance strength is expressed by the corresponding endurance limit. See later chapters on stress, pressures, and forces.

Q What qualities other than strength of material are important for boiler-component applications?

A Qualities other than strength are also important.

Hardness is a relative characteristic. There are several methods of measuring it, all of an arbitrary nature. The Brinell hardness number (Bhn) is obtained as follows: A hardened steel ball 10 millimeters (mm) in diameter is pressed under a certain load, F kilograms (kg), into the smooth surface of the material to be tested; the diameter D of the indentation is measured in millimeters, and the depth h is calculated from it. The hardness number, Bhn, is then expressed as

$$\text{Bhn} = \frac{F}{10\pi h}$$

In using the Shore scleroscope, a small cylinder of steel with a hardened point is allowed to fall on the smooth surface of the material, and the height of the rebound of the cylinder is taken as the measure of hardness.

The hardness number obtained with the Rockwell instrument is based on the additional depth to which a test point is driven by a heavy load beyond the depth to which the same penetrator has been driven by a definite, lighter load.

A material is *ductile* if it is capable of undergoing a large, permanent deformation and yet offers great resistance to rupture. The measure of ductility is the percentage of elongation or the percentage of reduction of area during a tensile test carried to rupture, and it is used as a relative measure. Ductility helps to relieve localized stress concentration through local yielding. It is a necessary characteristic of a material used to take live loads, especially where concentrated stresses may occur.

Brittleness is a characteristic opposite to ductility and toughness. A material may be considered brittle if its elongation at rupture through tension is less than 5 percent in a specimen 2 in long.

Toughness is a term used to denote the capacity of a material to resist failure under dynamic loading. The *modulus of toughness* is defined as the amount of energy per unit volume that a material can withstand or absorb without fracture occurring. The modulus of toughness is useful as an index for comparing the resistance of materials to dynamic loads, and it is especially applicable in the design of moving parts of machinery.

Q What procedures must a fabricator or repairer follow on boiler code material requirements?

A Certain procedures must be followed by a fabricator or repairer in order to make sure only code-specified material is used in boiler construction. It also is part of the responsibility of the authorized inspector to help implement a quality-control procedure in order to make sure code material is used. These controls may include the following:

1. The material to be used for a boiler or pressure vessel must be specified in the section of the code under which the boiler or pressure vessel is built. For example, if a high-pressure boiler is involved, it must be listed as a permissible material in Section I (Power Boilers), or data must be presented to show that it has the same chemical and physical characteristics as a code-listed material.

2. The fabricator of the boiler or pressure vessel generally orders permissible code material from the steel mills. The steel mill is responsible for making the necessary tests to the specifications given in Section II of the ASME Boiler and Pressure Vessel Code.

3. Section II test requirements that the steel manufacturer may have to perform include the following:

a. Chemical analysis of the steel to determine if it is within code limits for the specification.

b. Tests to determine if the metallurgical grain structure is within code-specified limits.

c. Inspection of plate or tube to note if defects such as blowholes, slag, laminations, and any other imperfections may be present and whether these are within code-permissible tolerance.

d. Tension and bend tests as stipulated in the code to note if these are within code specifications.

e. Notch toughness tests to check on fatigue-failure strength.

f. Mill test report showing that the material complies to code specifications; this must be certified by a responsible person of the material testing laboratory of the steel manufacturer.

Q What are some code material requirements in regard to high-pressure boilers?

A Plate steel for any part of a boiler subject to pressure and exposed to the fire or to products of combustion must be of firebox quality. If not exposed to fire or the products of combustion, the plate can be of flange quality. Some firebox-quality steels are specification SA-201 carbon-silicon steel, specification SA-202 chromium-manganese-silicon steel, and specification SA-204 molybdenum steel. Check the code for other firebox-quality steels and refer to the ASME Material Specification, Section II, for their physical and chemical characteristics. Seamless steel drum forgings made in accordance with specifications SA-266 and SA-336 for alloy steel can be used for any part of a boiler for which either firebox or flange quality is permitted.

Pipes and tubes may be made of open-hearth, electric-furnace, basic-oxygen, or acid deoxidized bessemer steel pipe or tubing, according to code specifications.

Superheater pressure parts, whether the integral type or separately fired, must be of wrought steel, puddled or knobbled charcoal wrought iron, or carbon or alloy steel, according to code specifications.

Rivets must be of steel or iron of the quality designated under specification SA-31 or SA-84 for wrought iron.

Stays and stay bolts fabricated by forge welding must be of SA-84 stay-bolt wrought iron. Threaded stay bolts must be of wrought iron of SA-84 grade, of steel complying with specification SA-31, or of annealed nickel-copper alloy of specification SB-164.

If the pressure does not exceed 250 psi and the temperature does not exceed 450°F, specification SA-278 gray-iron castings may be used for power boiler parts such as pipe fittings, water columns, and valves and their bonnets. Cast iron cannot be used for nozzles or flanges for any pressure or temperature. But this does not apply to low-pressure boilers. The same pressure parts as enumerated for cast iron can be made of malleable iron, except that the pressure is limited to a maximum of 350 psi and the temperature to 450°F.

The minimum thickness is ¼ in for plate subjected to pressure. An exception is for miniature boilers of seamless construction, where the minimum plate thickness may be $^3/_{16}$ in. The minimum thickness of tube sheets is ⅜ in, except on miniature boilers, where it is $^5/_{16}$ in. The plate material must be not more than 0.01 in thinner than that required for the plate by the formula used to calculate its strength, provided the tolerance in fabrication (or when the plate is ordered) also has this tolerance of not less than 0.01 in.

Q Does the code allow cast iron to be used for nozzles or flanges attached directly to a pressure part of a high-pressure boiler?

A No. Cast iron cannot be used for nozzles or flanges for any pressure or temperature. But this does not apply to low-pressure boilers.

Q For what part of an HP boiler may cast iron be used?
A If the pressure does not exceed 250 psi and temperature does not exceed 450°F, specification SA-278 gray-iron castings may be used for power boiler parts such as pipe fittings, water columns, and valves and their bonnets.

Q Do the above restrictions apply to malleable iron?
A The same pressure parts as enumerated for cast iron can be made of malleable iron, except that the pressure is limited to a maximum of 350 psi and the temperature to 450°F.

Q What is a mill test report?
A This is a report by the steel mill attesting to the chemistry and physical properties of the material. In the case of plate steel, it shows the heat, or slab number, from which the plate was made, which is stamped on the plate. It also gives the specification and thickness.

Q How are the ends of shell plates and butt straps formed for the longitudinal seam?
A Forming of these ends is by rolling or pressing and not by blows as with a hammer, which could weaken them.

Q Who is responsible for stamping or certifying the steel plate or material to be used on a boiler?
A The manufacturer of the plate or materials.

Q What are the three main methods of joining boiler elements?
A Welding, forge welding, and riveting. Forge welding of joints is limited by the power boiler code to an ultimate strength of 35,000 psi, with steel plates manufactured in accordance with SA-285 Grades A and B steel. Welding is the predominant method of joining boiler pressure parts. But riveting on numerous existing boilers will continue to require repair and alterations. Refer to the ASME Boiler and Pressure Vessel Code, Power Boilers, Section I, for detailed requirements on old riveted boiler joints.

Q Are threaded connections allowed on high-pressure boilers?
A The code states that threaded connections larger than 3-in pipe size shall *not* be used when the maximum allowable pressure exceeds 100 psi. But this 3-in pipe size restriction does not apply to (1) plug closures used for inspection openings and (2) end closures used for similar purposes. The number of threads that must be engaged and the minimum plate thickness required are shown in Fig. 7.1.

Q Are expanded connections allowed on high-pressure boilers?
A Yes, provided the pipe, tube, or forging does not exceed a 6-in OD,

Fig. 7.1 Minimum number of threads required for pipe connections on high-pressure boilers.

Pressures up to and including 300 psi:								
Size of pipe connection, in	1, 1¼	1½, 2	2½–4	4½–6	7, 8	9,10		12
Threads engaged	4	5	7	8	10	12		13
Min plate thickness required, in	0.348	0.435	0.875	1.0	1.25	1.5		1.625
Pressures above 300 psi:								
Size of pipe connection, in	½, ¼	1–1½	2	2½, 3	4–6	8	10	12
Threads engaged	6	7	8	8	10	12	13	14
Min plate thickness required, in	0.43	0.61	0.70	1.0	1.25	1.5	1.62	1.75

so that the opening meets all reinforcement requirements. Also, the expanded connections must meet all requirements on expanded tubes for fire-tube and water-tube boilers.

WELDING

Q How is welding defined?
A Welding is a localized coalescence (fusing together) or consolidation of metal where joining is produced by heating to fusion temperatures, with or without the application of pressure, and with or without the use of a filler metal. The filler metal (when used) has a melting point of approximately that of the pieces (base metal) joined together. The weld is that portion which has been melted during welding. And the welded joint is the union of two or more members produced by the welding process.

Q What is the most common method of welding?
A The most common method of welding pressure parts is by *fusion* (melting) of the metal, the heat being supplied in one of several different ways. In fusion welding, no pressure is applied between the pieces being welded. Arc welding, gas welding, and Thermit welding are classified as fusion welding, but arc welding is the most common.

Q Define arc welding.
A Arc welding is a localized progressive melting and flowing together of adjacent edges of the base-metal parts, caused by heat produced by an electric arc between a metal electrode, or rod, and the base metal. Both the welding material (welding rod or electrode) and the adjacent base metal are melted. On cooling they solidify, thus joining the two pieces with continuous material.

Q Describe the most common fusion welding methods.
A See Fig. 7.2.

a. The *oxy-fuel gas* method offers many advantages in depositing smooth, precise, and extremely high-quality welds. Low base-metal dilution is particularly important where filler and base metals differ considerably, as with cobalt-base filler metals applied to steels.

b. The *shielded-metal-arc* welding method uses a covered electrode whose coating during welding decomposes to form a slag and shielding gas.

c. The *gas-metal-arc welding* (MIG) process produces weld metal by fusion in an arc between the ends of a continuously fed bare electrode and the work; this progressively melts the electrode and the work. The welding current is carried to the electrode through a gun. The

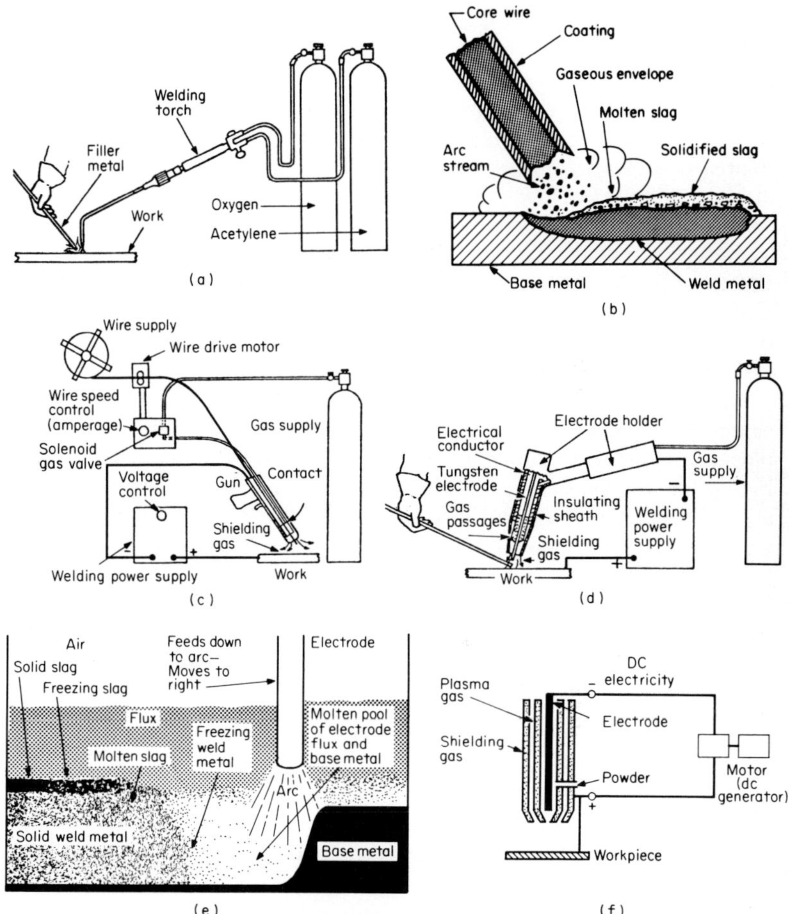

Fig. 7.2 Some types of weld fusion processes. (*a*) Oxyacetylene welding; (*b*) shielded-metal-arc welding; (*c*) gas-metal-arc welding (MIG); (*d*) gas-tungesten-arc welding (TIG); (*e*) submerged-arc welding; (*f*) plasma-arc welding. (*Courtesy of Arcos Corp.*)

arc is protected by an externally supplied shielding gas, which is usually argon or helium. Some flux-cored wires do not require separate shielding. This process permits fully automatic welding, with the gun traveling on a mechanized carriage.

d. In the *gas-tungsten-arc* (TIG) process, the electrode is not consumed to form the weld. It carries the current between the tungsten electrode and the work with virtually no change in the tip of the tungsten. The filler metal is separately fed to the arc, which is shielded by argon, helium, or a mixture of the two. This process can also be fully automatic.

e. The *submerged-arc* process is particularly suitable for automatic welding because it permits continuous feeding of the filler wire and at relatively high input currents. The arc is completely shielded by a layer of loose granules of flux.

If composite wires are used, this process permits the application of high-alloyed metals. The metal-cored wires have the main alloy in the sheath with all other alloys in the core. A special flux shields the arc from contamination from the air during deposition and also provides a protective slag blanket during solidification. The flux can also provide alloying ingredients to produce the desired weld composition.

f. The *plasma-arc weld* is a true welding process (not a metal spray or coating process). To accomplish this, one employs the transferred plasma-arc process shown in Fig. 7.2. It is particularly suitable for highly alloyed hard surfacing. The deposit is formed from powdered alloys which are conveyed into the arc by the gas stream. Plasma-arc weld surfacing fits especially well into mechanized production applications requiring thin weld overlays, but heavy deposits can also be made at relatively fast speeds. Its high deposition rates, along with smooth deposits that reduce material usage and require less finishing, lower costs significantly over gas-tungsten-arc and oxy-fuel-gas methods.

Q Define heli-arc welding.

A In heli-arc welding, an electric arc is struck between a tungsten electrode within the heli-arc torch and the workpiece. The arc provides the heat to melt metal, while simultaneously a stream of shielding gas covers the electrode and the weld zone, protecting both from contaminants in the atmosphere. An inert gas such as helium is used—hence the term heli-arc. The inert gas protects the weld puddle by preventing oxygen and nitrogen from the air from entering the puddle and thus avoids weak and possibly porous welds.

BOILER WELDING CODE

Q What is meant by a qualified welder?

A Before welders are permitted to work on a job covered by a welding code or specification, they must become certified under the code that applies. Many different codes are in use today; so the specific code must be referred to when qualification tests are taken. In general, these types of work are covered by codes: (1) boiler and pressure vessels and pressure piping, (2) highway and railway bridges, (3) public buildings, (4) tanks and containers for flammable or explosive materials, (5) cross-country pipelines, and (6) aircraft ordnance.

Certification differs under the various codes; thus one code may not qualify a welder to weld under a different code. And in most cases, certification for one employer will not qualify the welder to work for another employer. Also, if the welder uses a different process, or if the procedure is altered drastically, recertification is required for that procedure in most codes. But if the welder is continually employed as a welder, recertification is not required, providing the work performed meets the quality requirements. An exception is the military aircraft code, which requires requalification every 6 months.

Qualification tests may be given by responsible manufacturers or contractors. On pressure-vessel work, the welding procedure of the fabricator or contractor must also be qualified *before* the welder can be qualified. Under other codes this is not necessary. To become qualified, the welder must make specified welds using the required welding process, type of metal, thickness, electrode type, position, and joint design. Test specimens must be made to standard sizes and under the observation of a qualified person. In most government specifications, a government inspector must witness the making of welding specimens. Specimens must also be properly identified and prepared for testing. The common test is the guided-bend test. However, x-ray examinations, fracture tests, or other tests are also used. Satisfactory completion of test specimens (provided that they meet acceptability standards) will qualify the welder for specific types of welding. Code certification, in general, is based on the range of thicknesses to be welded, the positions to be used, and the materials to be welded.

> NOTE: Qualification of welders is too technical to be fully covered here. See Section IX of the ASME Boiler and Pressure Vessel Code.

Q For what five welding positions can a welder be qualified?
A Figure 7.3 shows flat, horizontal, vertical, overhead, and (for groove welds only) horizontal fixed positions. A welder who qualifies in the horizontal, vertical, or overhead position automatically qualifies for the flat position. For groove welds, qualifying in the fixed horizontal position automatically qualifies the welder for the flat, vertical, and overhead positions. Qualifying in the horizontal, vertical, and overhead positions automatically qualifies the welder for *all* positions. Angular deviations from these positions are plus or minus 15°.

Q What is meant by welding-procedure qualification?
A Manufacturers or contractors who are to do boiler code welding are

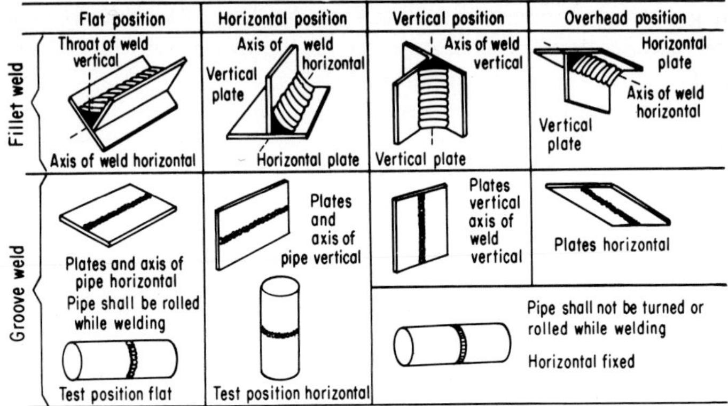

Fig. 7.3 Welding positions to perform for welder qualification tests.

required to record in detail the procedure they are to use. Each procedure requires testing of the welds to be made by reduced-section test specimens and guided-bend specimens. The variables requiring a new procedure and new test plates are very numerous. Among these are changes in base materials, grouped in ASME Welding Qualifications (Section IX) into P numbers. For example, P-1 includes mostly carbon steels, P-2 consists of wrought iron, P-3 consists of chrome-moly steels with chromium content below ¾ percent, and with a total alloy content not exceeding 2 percent. The P numbers range to P-10; so the base material variable on procedure qualification is large.

The next variable is the electrode and welding-rod selection, which ranges from F-1 to F-7. Any change in electrode or welding-rod selection requires a new set of test plates or procedure qualification. Weld metal is classified by weld metal analysis numbers A-1 to A-8. These are related to equivalent P numbers of the base material. Again, changes of weld metal from equivalent base metal classification require a new set of test plates or procedure qualification.

The thickness of the plate, or pipe, to be welded is another variable. Classification is from $\frac{1}{16}$ to ⅜ in, from ⅜ to ¾ in, and over ¾ in. Each classification requires new test plates as shown in the code. The ASME Welding Code specifies other variables to consider in requiring a new procedure qualification test, and these should be consulted for specific variables.

REMEMBER: Welding boiler parts is under close control of the many variables involved in welding; so check the ASME Welding Code.

Q What test plates for each position must a welder prepare and pass? And what test plates are required for procedure qualification?

A On groove welds, one face-bend test and one root-bend test are required for operators' qualification for each position they are to weld. For fillet welds, a test plate is required according to the code, but passing the groove-weld test will also qualify a welder for fillet welding. Procedure qualification requires two face-bend tests, two root-bend tests, and two reduced-section tension tests, as illustrated in the Welding Qualification Code of the ASME.

Q What is meant by reduced-section tension test, free-bend test, root-bend test, face-bend test, and side-bend test?

A See ASME code, Section IX, for typical weld test specimens. The reduced-section tension test is used for qualifying the procedure that the shop, or contractor, is to use in welding. When broken in tension, it must have an ultimate tensile strength at least that of the minimum range of the plate which is welded (base material), and the elongation of stretch must be a minimum of 20 percent.

The side-bend test is used for qualifying operators. The specimen is subjected to bending against the side of the weld. In the face-bend test, the specimen is subjected to bending against the surface, or face, of the weld. In the root-bend test, the specimen is subjected to bending against the bottom, or root, of the weld. The free-bend test is a shop- or contractor-qualifying procedure test. The test consists of bending the specimen cold, and the outside fibers of the weld must elongate at least 30 percent before failure occurs.

In order to pass each test, guided-bend specimens must have no cracks or other open defects exceeding ⅛ in measured in any direction on the convex surface of the specimen *after bending*, except that cracks occurring on the corners of the specimen during testing are not considered unless these occur from slag inclusions or other welding-technique defects.

Q 1. Who is responsible for conducting tests of welding procedures and for qualifying welding operators?

2. How long does a qualified welder's approval test remain in effect?

3. Who keeps the records for procedure and operator's approval tests?

A 1. Manufacturers or contractors who are to do code welding on boilers and pressure vessels (or nuclear vessels) are responsible for conducting procedure qualification tests, and welding operator's qualification tests for work to be done by their organization.

2. Operators' qualifications remain in effect as long as they are employed by the same manufacturer or organization and do welding on a continuous basis. But if they change employment, they are no longer considered qualified and thus must take the test again. If they have not

done any welding for a period of over 6 months in the position, material, etc., for which they are qualified, they must be requalified.

3. The manufacturer or contractor is responsible for keeping all records of procedure and operator's qualification tests. These are needed as evidence of the shop's, or welder's, ability to do acceptable code work. Inspectors have the right, however, to ask for retests if they believe the welding is not acceptable by code requirements. Figure 7.4 is a typical recommended ASME Welding Procedure Qualification Test recording form. Figure 7.5 is a typical recommended ASME Welder Performance Qualification Test recording form.

Q Can an unqualified welder do any welding on a boiler?
A Only if the strength of the boiler does not depend upon the weld in any way, and the effect of heating caused by welding will in no way affect the boiler component's strength or set up stresses, caused by welding, on adjoining parts.

Q What welding process is used on code boilers?
A Code work for arc or gas welding is restricted to shielded carbon arc, shielded metal arc, inert-gas metal arc, atomic-hydrogen metal arc, oxyhydrogen, and oxyacetylene processes. Pressure welding is restricted to flash, induction, resistance, pressure Thermit, and oxyacetylene processes.

Q Is it permissible to weld any kind of steel for use on boiler pressure parts?
A No. Only carbon or alloy steel having a carbon content of not more than 0.35 percent can be used in welded construction or be shaped by oxygen cutting or other thermal-cutting process.

Q Is preheating required in boiler code welding?
A Yes, and the temperature of preheating depends on the P number of the base material to be welded. Preheating varies from 175°F in certain P-1 (low-carbon steel) materials to 450°F maximum for P-10 material (stainless-steel tubing and nickel-steel plate). Always consult the ASME Welding Code.

Q What is the maximum permissible offset of plates to be welded in the longitudinal and circumferential direction on a welded boiler drum?
A The maximum permissible offset of two adjoining plates in the longitudinal direction is ⅛ in; in the circumferential direction it is ¼ in.

Q What is the maximum distortion permissible on a welded drum of a power boiler?
A The drum must be circular at any section within a limit of 1 percent of the mean diameter. If necessary to meet this requirement, plates may

Specification No. .... Date ....
Welding Process .... Manual or Machine ....
Material—Specification .... to .... of P-No. .... to P-No. ....
Thickness (if pipe, diameter and wall thickness) ....
Thickness Range this test qualifies ....
Filler Metal Group No. F- ....
Weld Metal Analysis No. A- ....
Describe Filler Metal ....
....
For oxyacetylene welding—State if Filler Metal is silicon or aluminum killed.

FLUX OR ATMOSPHERE

Flux Trade Name or Composition ....
Inert Gas Composition ....
Trade Name .... Flow Rate ....
Is Backing Strip used? ....
Preheat Temperature Range ....

WELDING PROCEDURE

Single or Multiple Pass ....
Single or Multiple Arc ....
Position of Groove ....
(Flat, horizontal, vertical, or overhead; if vertical, state whether upward or downward)

Postheat Treatment ....

For Information Only

Filler Wire—Diameter ....
Trade Name ....
Type of Backing ....
Forehand or Backhand ....

WELDING TECHNIQUES

Joint Dimensions Accord with ....
amps .... volts .... inches per min. ....

REDUCED SECTION TENSILE TEST

Specimen No.	Dimensions		Area	Ultimate Total Load, lb.	Ultimate Unit Stress, psi	Character of Failure and Location
	Width	Thickness				

GUIDED BEND TESTS

Type and Figure No.	Result	Type and Figure No.	Result

Welder's Name .... Clock No. .... Stamp No. ....
Who by virtue of these tests meets welder performance requirements.

Test Conducted by .... Laboratory—Test No. ....
per ....

We certify that the statements in this record are correct and that the test welds were prepared, welded and tested in accordance with the requirements of Section IX of the ASME Code.

Signed ....
(Manufacturer)

Date .... By ....

(Detail of record of tests are illustrative only and may be modified to conform to the type and number of tests required by the Code.

NOTE: Any essential variables in addition to those above shall be recorded.

Fig. 7.4 Manufacturer's record of welding-procedure qualification test.

be reheated, rerolled, or reformed. Furnaces must be rolled (scotch marine boilers), with a maximum permissible deviation from the true circle of not more than ¼ in.

Q Is it permissible to weld plates of unequal thickness?

Welder Name..Clock No...........................Stamp No............

Welding Process.......................................

Position (If vertical state whether upward or downward)....................
(Flat, horizontal, vertical, or overhead)

In accordance with Procedure Specification No...

Material—Specification................to................of P-No..................to P-No.........................

Diameter and Wall Thickness (if pipe) otherwise Joint Thickness...

Thickness Range this qualifies............................

FILLER METAL

Specification No...............................Group No. F.................

Describe Filler Metal ....

....

Is Backing Strip used?..................

For Information Only

Filler Metal Diameter and Trade Name.................... Flux for Submerged Arc or Gas for Inert Gas Shielded Arc
.... Welding...

GUIDED BEND TEST RESULTS

Type and Figure No.	Result	Type and Figure No.	Result

Test Conducted by..................................... Laboratory—Test No....................................

per.....................................

We certify that the statements in this record are correct and that the test welds were prepared, welded and tested in accordance with the requirements of Section IX of the ASME Code.

Signed..
(Manufacturer)

Date.................. By..

(Detail of record of tests are illustrative only and may be modified to conform to the type and number of tests required by the Code.
NOTE: Any essential variables in addition to those above shall be recorded.

Fig. 7.5 Manufacturer's record of welder-performance qualification tests on groove welds.

A Yes, provided a tapered transition section, having a length not less than three times the offset between the adjacent plate surfaces (Fig. 7.6), is provided at joints between plates that differ in thickness by more than one-fourth of the thickness of the thinner plate, or by more than ⅛ in.

Q Can stay bolts be welded instead of being threaded or riveted?
A Yes, provided they are inserted into countersunk holes through the sheets and attached by full-penetration welds according to the code. But the face of the weld cannot be below the outside surface of the weld, and the ends of stays inserted through the plate cannot project more than ⅜ in beyond the surfaces exposed to the products of combustion. Diagonal stays may be attached to the inner surfaces of plates (but not the head) by fillet welds as provided in the code. Stress relieving is required of welded-in stay bolts, but not radiographing.

Q The intent of the welding requirements is generally to obtain good penetration and thus a good joint. Illustrate some typical nozzle and

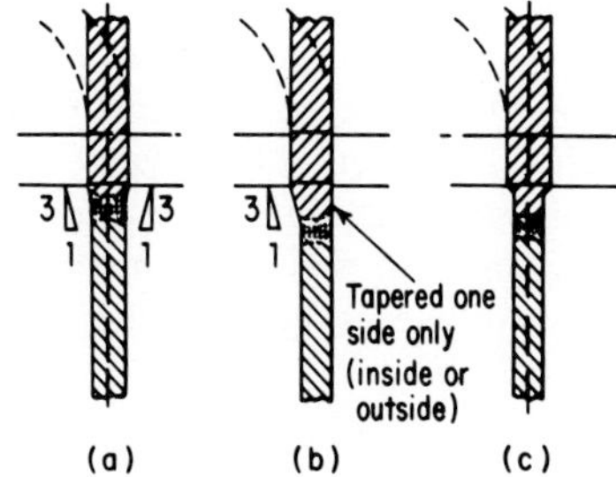

Fig. 7.6 Butt welding of plates of unequal thickness requires a tapered transition. (*a*) Preferred method (centerlines coincide); (*b*) permissible (circumferential joints only); (*c*) not permissible.

attachment welding details required by the boiler code to provide this full penetration.

A Figure 7.7 shows examples of acceptable types of welded nozzles and other connections to shells, drums, and headers. See ASME Power Boilers, Section I, for specifics.

Q Should weld reinforcements be removed substantially flush with the plate?

A On longitudinal joints of welded power boilers, weld reinforcements should be removed substantially flush with the surface of the plate. If they are not, the code provides a penalty on the permissible longitudinal joint efficiency. But removal of weld reinforcements is not mandatory on other joints.

Q What is the efficiency of a welded joint, and what value does the code allow?

A The efficiency of a welded joint is defined the same way it is on a riveted joint, namely, the strength of the joint divided by the strength of the solid plate. However, since a welded joint cannot be calculated the same way as a riveted joint, the code allows the following if all code

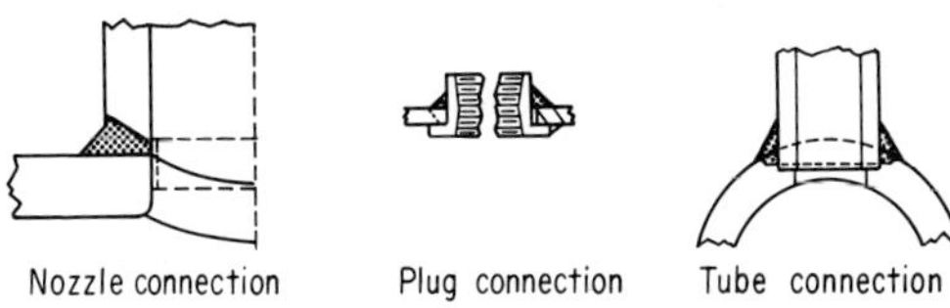

Fig. 7.7 Code-acceptable welded nozzle and other connections to shells, drums, and headers generally require full-penetration welds.

requirements are met:

1. Ninety percent longitudinal joint efficiency if the joint is stress-relieved and radiographed but the weld reinforcement is not removed substantially flush with the surface of the plate.

2. One hundred percent longitudinal joint efficiency if it is stress-relieved, radiographed, and the weld reinforcement is removed substantially flush with the surface of the plate.

OLD RIVETED JOINTS

Q What forms of riveted joints were used in boiler construction?
A Lap joints and butt joints. In the lap joint, edges of plates overlap. Because in the butt joint the plate edges meet or butt together, cover straps (called straps) are used.

Q Describe and sketch a single-riveted lap joint.
A In this joint, plate edges were lapped over and secured by one row of rivets. Figure 7.8*a* shows a side view and end-section view of a single-riveted lap joint.

Q Describe and sketch a double-riveted lap joint.
A Figure 7.8*b* shows how plate edges of this joint are lapped and secured by two rows of rivets.

Q Describe and sketch a double-riveted butt joint with double straps.

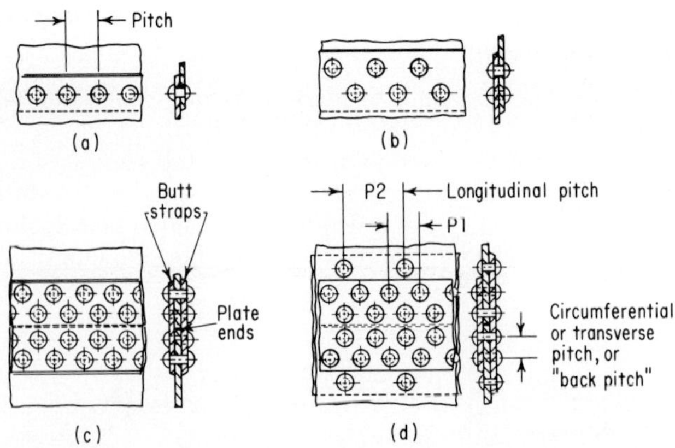

Fig. 7.8 Riveted joints used in the past. (*a*) Single-riveted lap joint; (*b*) double-riveted lap joint; (*c*) double-riveted butt joint with double butt straps; (*d*) triple-riveted butt joint with double butt straps.

A In this joint, plate edges are butted together and butt straps are placed inside and outside. Figure 7.8*c* shows a double-riveted double-strap joint with butt straps of equal width.

Q Describe and sketch a triple-riveted butt joint with double butt straps of unequal width.
A Figure 7.8*d* shows that this joint's plate edges butt together and butt straps are placed inside and outside. However, the outside butt strap is narrower than the one inside. Notice that the outer rows of rivets pass through the inner butt strap and shell plate, but *not* through the outer butt strap. Also, alternate rivets are omitted in these outer rows.

Q What are the maximum size and pressure for construction of an HP boiler with a longitudinal lap joint?
A Lap joints on longitudinal joints cannot be used on HP boilers over 36 in in diameter. And the pressure is limited to a maximum of 100 psi. If the boiler is over 36 in in diameter, or over 100 psi, a butt double-strap joint is required.

Q Why is a butt joint preferable to a lap joint?
A In the butt joint, shell plates form a true circle. Thus there is no tendency for pressure to distort the plates. In the lap joint the plates do not form a true circle at the lap; so internal pressure tends to pull them out of position. Also, in time this bending action creates a stress concentration at the edge of the plate. This in turn may cause grooving or cracking of the plate along the calked edge and thus create a dangerous condition (see Fig. 7.9). Properly designed butt joints have much higher efficiencies than lap joints with the same number of effective rows of rivets because in the butt joint all or most of the rivets pass through three plate thicknesses and so are in double shear. By contrast the rivets in lap joints are all in single shear. Rivets in double shear have twice the strength of rivets in single shear.

NONDESTRUCTIVE TESTING

Q In what manner is the quality of a weld checked?
A Depending on the type of vessel, various nondestructive techniques are used. In power boiler work, besides visual examination of the weld, x-rays and gamma rays are used and required. In nuclear-vessel work, these are supplemented by magnetic-particle tests, dye checks, and ultrasonic testing. Not to be forgotten is the standard hydrostatic test of 1½ times the design pressure. The field of nondestructive testing is expanding. New methods are being perfected and are finding increasing use in boiler and pressure-vessel work, during construction and during field testing after the equipment is in use.

Fig. 7.9 Old riveted boiler with a lap joint that developed a longitudinal crack from cyclic stressing. (*Courtesy of Hartford Steam Boiler Inspection and Insurance Co.*)

Q What is meant by a nondestructive test?
A This is a test to check for the soundness or quality of a material or a joint of materials without affecting it physically or chemically. It includes radiography (x-rays), dye checks, and ultrasonic and hydrostatic testing.

Q Are destructive tests also used to determine material soundness?
A Besides the well-known tensile test to determine the strength of different metals in tension or compression, destructive tests can also be grouped into chemical, metallographic, and mechanical. In these tests the material under test is damaged; therefore, they are considered destructive.

Chemical tests are used to determine the amount of carbon, phosphorus, and similar constituents of a metal or weld, especially if there is some question of the component integrity. These tests are also used to determine the corrosion resistance of a material or welds.

Metallographic tests of metals and welds require the specimen to be etched and ground for further visual examination, sometimes by magnification. The purpose of these tests is to determine the presence of cracks, porosity, and similar defects. These may be in the crystalline structure of the metal, in the heat-affected zone of a weld or base metal. Analysis can assist in determining the reasons for the defect, such as improper fit-up in fabrication, poor welding technique, incorrect weld-rod selection, or service conditions such as corrosion.

Hardness tests are a form of destructive test, since an impression is left in the specimen. This test is also used to check on the welding process as respects heat-affected zone hardening, the need for prewelding heat and postweld heat treatment, the possible effect of cold working, and similar metallurgical concerns. Figure 7.10 shows the relationship between some hardness numbers (columns 1, 2, 3) and the ultimate tensile strength (column 4) of the material under test. The Vickers hardness test is used extensively in Great Britain.

Mechanical tests are considered destructive and include tensile tests, yield-stress determination, endurance, and similar stress tests. See the next chapter.

Q Briefly describe the radiographic NDT method.

A Radiographic inspection includes x-ray and gamma rays obtained from isotopes, such as cobalt 60 and iridium 192, where the resultant radiant energy can be safely controlled. The radiographic method of testing basically involves passing rays through materials to be tested. The rays impinge on a film or screen, and by noting the contrast of the film, it is possible for an experienced radiographer to detect and detail the internal structure of the object under test. The focal spot is a small area in the x-ray tube from which radiation emanates. In gamma radiography, an isotope like cobalt 60 is the radiation source. When radioactive isotopes are used, strength in curies is important, as is the physical size of the source The smaller the radiation source, the closer to the material it can be placed; at the same time, the smaller the size, the weaker the source in curies and the longer the exposure time needed.

Q How are interpretations of films made in radiographic inspections?

A 1. The first requirement is visual inspection of the weld. Joints must have complete joint penetration and be free of undercutting, overlaps, or abrupt ridges and valleys. Weld reinforcement is specified not to exceed the following:

Plate thickness, in	Maximum thickness of reinforcement, in
Up to ½-in inclusive	1/16
Over ½- to 1-in inclusive	3/32
Over 1- to 2-in inclusive	1/8
Over 2-in inclusive	5/16

2. Welded joints to be radiographed must be free of ripples or weld surface irregularities to such a degree that the resulting radiographic contrast due to any irregularities cannot mask or be confused with the image of any objectionable defect.

3. Gages called *penetrameters* must be used for every exposure of a film. The penetrameter serves as a comparison gage on the film to

Fig. 7.10 Hardness number determination also permits approximate tensile-strength determination of a material as shown in the tensile-strength columns. (*Courtesy of American Welding Society.*)

Rockwell hardness		Vickers hardness,*	Brinell hardness†			Rockwell hardness		Vickers hardness,*	Brinell hardness†		
C, 150-kg load, diamond	B, 100-kg load, $^1/_{18}$ ball	diamond pyramid	Tungsten carbide ball	Steel ball	Tensile strength, 1000 psi	C, 150-kg load, diamond	B, 100-kg load, $^1/_{18}$ ball	diamond pyramid	Tungsten carbide ball	Steel ball	Tensile strength, 1000 psi
67	. . .	918	820	. . .	. . .	20	98.9	240	231	225	107
66	. . .	884	796	. . .	. . .	19†	98.1	235	226	220	106
65	. . .	852	774	. . .	. . .	18	97.5	231	222	215	103
64	. . .	822	753	. . .	. . .	17	96.9	227	218	210	102
63	. . .	793	732	. . .	. . .	16	96.2	223	214	206	100
62	. . .	765	711	. . .	. . .	15	95.5	219	210	201	99
61	. . .	740	693	. . .	. . .	14	94.9	215	206	197	97
60	. . .	717	675	. . .	. . .	13	94.1	211	202	193	95
59	. . .	694	657	. . .	. . .	12	93.4	207	199	190	93
58	. . .	672	639	. . .	. . .	11	92.6	203	195	186	91
57	. . .	650	621	. . .	. . .	10	91.8	199	191	183	90
56	. . .	630	604	. . .	. . .	9	91.2	196	187	180	89
55	. . .	611	588	. . .	. . .	8	90.3	192	184	177	88
54	. . .	592	571	. . .	. . .	7	89.7	189	180	174	87
53	. . .	573	554	. . .	283	6	89	186	177	171	85
52	. . .	556	538	. . .	273	5	88.3	183	174	168	84
51	. . .	539	523	500	264	4	87.5	179	171	165	83
50	. . .	523	508	488	256	3	87	177	169	162	82
49	. . .	508	494	476	246	2	86	173	165	160	81
48	. . .	493	479	464	237	1	85.5	171	163	158	80
47	. . .	479	465	453	231	0	84.5	167	159	154	78
46	. . .	465	452	442	221	. . .	83.2	162	153	150	76
45	. . .	452	440	430	215	. . .	82	157	148	145	74

44	. . .	440	427	419	208	. . .	80.5	153	144	140	72
43	. . .	428	415	408	201	. . .	79	149	140	136	70
42	. . .	417	405	398	194	. . .	77.5	143	134	131	68
41	. . .	406	394	387	188	. . .	76	139	130	127	66
40	. . .	396	385	377	181	. . .	74	135	126	122	64
39	. . .	386	375	367	176	. . .	72	129	120	117	62
38	. . .	376	365	357	170	. . .	70	125	116	113	60
37	. . .	367	356	347	165	. . .	68	120	111	108	58
36	. . .	357	346	337	160	. . .	66	116	107	104	56
35	. . .	348	337	327	155	. . .	64	112	104	100	54
34	. . .	339	329	318	149	. . .	61	108	100	96	52
33	. . .	330	319	309	147	. . .	58	104	95	92	50
32	. . .	321	310	301	142	. . .	55	99	91	87	48
31	. . .	312	302	294	139	. . .	51	95	86	83	46
30	. . .	304	293	286	136	. . .	47	91	83	79	44
29	. . .	296	286	279	132	. . .	44	88	80	76	42
28	. . .	288	278	272	129	. . .	39	84	76	72	40
27	. . .	281	271	265	126	. . .	35	80	72	68	38
26	. . .	274	264	259	123	. . .	30	76	67	64	36
25	. . .	267	258	253	120	. . .	24	72	64	60	34
24	. . .	261	252	247	118	. . .	20	69	61	57	32
23	. . .	255	246	241	115	. . .	11	65	57	53	30
22	100.2	250	241	235	112	. . .	0	62	54	50	28
21	99.5	245	236	230	110	. . .					

* *Vickers load*: 50 kg above 171 value, 30 kg 95 to 170 and 10 kg 62 to 94.

† *Brinell load*: 3000 kg except tungsten carbide values 159 to 86 at 1500 kg and 85 to 54 at 500-kg load.

compare faults in the weld. This is done by making a strip of metal for each exposure and drilling holes in the strip prior to exposure; these strips serve as guides to detect flaws within 2 percent of the plate thickness being welded. These holes are usually drilled with a minimum hole diameter of $\frac{1}{16}$ in specified. The proper method of penetrameter location is shown in Fig. 7.11.

When an x-ray picture is taken, the penetrameter will be included, and the holes will serve as a guide for detecting faults right on the film. For comparing the holes on a penetrameter with the holes noted on a weld when the film is developed, an immediate comparison gage is included on each film. The code stipulates tolerances for weld rejects based on penetrameter data.

The code also provides maximum porosity indications and porosity charts which serve as guides for the permissible number of pores (voids) and sizes of voids permitted in each 6-in length of weld. These should be referred to in determining unacceptable voids.

Defects such as cracks, slag inclusions, lack of penetration, voids, and others appear as darkened areas on the film because they have a lower density than solid metal. Typical faults are noted as follows. *Cracks* appear as dark irregular lines. *Slag inclusion* show up as small, dark spots with irregular outlines. *Gas pockets* appear as small, dark spots with smooth outlines with occasional teardrop tails. *Lack of penetration* is evidenced by a smooth, dark line most often located in the middle of a weld.

Q Briefly describe *magnetic-particle* testing.

A Magnetic-particle testing is used to detect surface faults by means of setting up a magnetic field or magnetic lines of force between two electrodes. Powdered magnetic material is sprinkled over the work to be tested. The magnetic field will affect the magnetic powder, and these particles will align themselves in a fault, as shown in Fig. 7.12. But the correct interpretation of the gatherings of the magnetic powder requires experience and practice. Magnetic-particle inspection is a practical means for spotting close-lipped discontinuities at or near the surface of a part. Both the wet and dry methods are presently available.

Discontinuities in a magnetized material give rise to localized leakage fields. And these fields attract finely divided magnetic particles. The latter point a finger at the defect and mark its extent on the surface of the part under inspection.

Both direct and alternating currents are used for magnetizing. Direct current (dc) is useful in finding subsurface discontinuities and is commonly used for inspecting welds and castings. Alternating current (ac) is usually employed when highly finished machined parts are checked.

Generally *dc magnetization* is considered where subsurface and

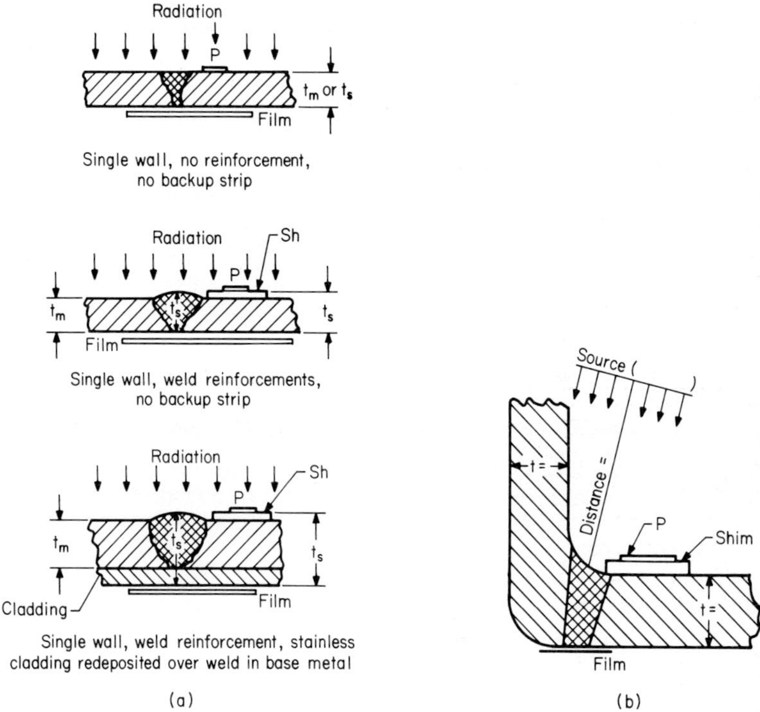

Fig. 7.11 Radiography requires penetrameters to be placed in line with film. (*a*) For butt welds; (*b*) for nozzle welds. *P* = penetrameter; *Sh* = shim; t_m = design material thickness upon which the penetrameter is based; t_s = specimen thickness. (*Courtesy of Spring and Kohan, Boiler Operator's Guide, McGraw-Hill.*)

surface cracklike defects must be found. With dc, the magnetic field extends within the part itself, and magnetic leakage fields are produced on the surface by interruption in the magnetic path below the surface.

The principal limitation of the magnetic-particle method is that it applies to magnetic materials only and is not suited for very small, deep-seated defects. The deeper the defect is below the surface, the larger it must be to show up. Subsurface defects are easier to find when they have a cracklike shape, such as lack of fusion in the weld. On large, heavy objects, when extremely sensitive inspection is desired, the operation takes more time. This holds true on medium-sized critical parts such as aircraft propellers. With magnetic-particle testing, the surface to be inspected must be available to the operator. This means shafts or other equipment cannot be inspected without removal of pressed wheels, pulleys, or bearing housing.

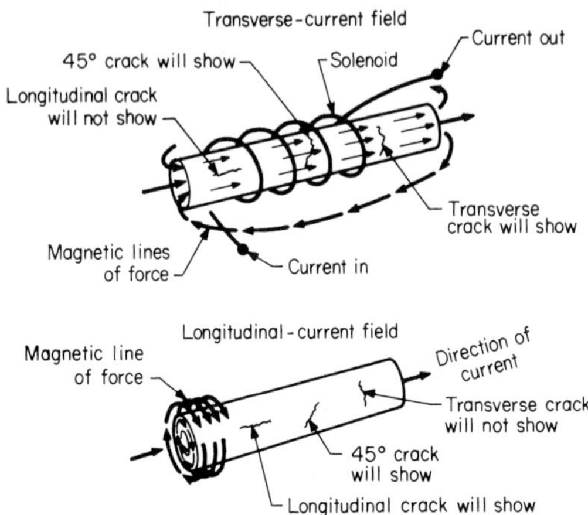

Fig. 7.12 Magnetic-particle inspection requires a magnetic field at right angle to defect to be detected by the magnetic particles. (*Courtesy of Power magazine.*)

Q Describe the *liquid (dye)-penetrant* inspection technique.

A Liquid-penetrant (dye) inspection testing is used somewhat like magnetic-particle testing, except that it is used primarily on nonmagnetic material. But it can be used on magnetic material. The dye penetrant contains a visible dye, usually red. Indications of defects appear as red lines or dots against the white developer background. It is primarily a surface-defect indicator, and it is applied as follows: A dye penetrant is applied to the part by dip, brush, or spray and is allowed to sit for a specified time. After suitable penetration time, the excess penetrant is removed from the surface, and a developer is applied. The penetrant becomes entrapped in a defect and is brought to the surface by the action of a developer. Cracks are detected by noting the contrast between the white color of the developer and the red penetrant.

Another penetrant method used is the fluorescent penetrant method, which contains a material that fluoresces brilliantly under black lights. Indications of defects appear as fluorescent lines or dots against a nonfluorescent background.

The advantages of the dye-penetrant method are as follows: It provides fast, on-the-spot inspection during overhaul or shutdown periods; the initial cost of the test is relatively low. A perfectly white or blank surface indicates freedom from cracks or other defects that are open to the surface. The disadvantages are that it is not practical on very rough surfaces and color contrast is limited on some surfaces. Also it detects only defects open to the surface.

Q What are the basic principles in *ultrasonic testing*?

A Ultrasonic testing makes use of high-frequency sound waves in the range of 0.5 to 10.0 megahertz (MHz) for the inspection of material for flaws and also for thickness testing. The basic principle used in an ultrasonic system is the transformation of an electric impulse into mechanical vibrations, and then the transformation of the mechanical vibrations into electric pulses that can be measured or displayed on a screen called cathode-ray-tube (CRT) screen. The transfer of mechanical energy to electric energy is performed by means of a transducer, which is capable of transforming one form of energy into another.

Ultrasonic tests are grouped into three basic categories: pulse-echo testing, through transmission, and resonance testing. The pulse-echo method involves transmitting a short burst of high-frequency sound through the piece being tested and then detecting the echoes that are received from either a construction detail, such as a shoulder or hole, or a defect in the material with a separation or void in the material sufficiently large that sound cannot be transferred across the interface. In operation, a pulse-echo unit will produce, through an electronic pulser, a short burst of high-frequency electric signal. This is transmitted to the transducer, which is forced to vibrate, usually at its resonant frequency. The probe must be coupled to the test piece with oil, water, or some other liquid or grease. The sound wave train then travels through the test piece until some form of discontinuity or boundary is encountered. This interruption in the medium then causes the sound wave to be reflected to the receiver transducer. The vibrational energy of the sound wave sets the transducer in motion to produce an electric impulse which is fed into an amplifier. The output of the amplifier is displayed on a cathode-ray-tube which shows the signals on a linear time baseline. If a linear amplifier is used, the amplitude of the returned echoes can be used as a measure of the area producing the reflected signal.

Resonance testing makes use of a tunable, continuous wave system. This method is usually employed for measuring small or thin walls to 2 or 3 in. The resonance of the crystal is tuned to the piece under test. In practice, a loud pip is heard and can be also seen because the electronic circuit is also in resonance electrically. By proper calibration, direct thickness readings can be made.

Q What is *eddy-current* testing?

A The underlying principle of eddy-current testing is the measurement of impedance of electron flow in the part being tested. Weak electric currents or eddy currents are induced by a probe containing inducing and sensing coils. Any changes in the geometry of the part such as a pit, crack, or thinning will affect the flow of the eddy currents. This change in the flow of the eddy currents will be detected by the sensing coil and displayed on a strip chart, a CRT display, or both. The accuracy of characterization of the detected flaw depends on the quality of the standard used in calibrating the eddy-current instrument. A piece of tubing of material identical to the tubes being inspected should be used for calibration. The standard should have a range of defects similar to those expected to be found in the tubes being tested.

Q Briefly describe the term *acoustic emission.*
A This method of nondestructive testing is being developed to monitor large pressure vessels such as digesters and nuclear reactors for crack growth. Acoustic emission is based on the principle that a growing defect releases bursts of energy or stress waves that can be detected by sensitive and suitably designed transducers. If a transducer is strategically installed in known highly stressed areas of the pressure vessel, it can convert the minute sound emitted when a material "gives" into electric signals. The signals can be recorded on a computer for immediate analysis or future use. When transducers are installed in a triangular mode, the source of any abnormal sound can be determined by quick trigonometric calculations, and thus the defect can be located for analysis and repair.

Q Explain the interrelationship between a code-qualified inspector and a qualified NDT examiner.
A Nondestructive testing examination and inspection require extensive experience in the method of NDT to be applied for flaw detection. This specialized experience is detailed in SNT-TC-1A, *Recommended Practice for Nondestructive Testing Personnel Qualification and Certification*, published by the American Society for Nondestructive Testing (ASNT). Three levels or grades of qualifications are possible in each of the NDT methods previously described. The Level 3 person is the most qualified; that level generally requires not only knowledge of operating an instrument or applying a method but also theoretical knowledge of the NDT methods, their advantages and shortcomings, and the interpretation of test results. The ASME Boiler and Pressure Vessel Codes refer to the ASNT society as the source for details on qualifying and certifying NDT "examiners." The ASME code inspector must still make sure that the written procedures can detect the discontinuities by the NDT method to be used which are not acceptable in the section of the code to which a boiler or pressure vessel is being built. For example, there are radiographic standards in Section I that may differ somewhat from those involving a nuclear pressure vessel. The "authorized" inspector making

the code inspections will review the NDT results in order to make sure discontinuities found by NDT are within code-permissible limits. Close coordination is thus essential between the NDT examiner and the authorized inspector.

CHAPTER **EIGHT**

Pressure, Stresses, and Boiler-Strength Calculations

Boilers and pressure vessels are required to resist the forces imposed on them by the pressure of fluids and gases in a closed container. In addition, the materials used and the design must also consider expansion and contraction effects as equipment is brought up to high operating temperatures. Environmental effects on material must also be considered, such as flue-gas attacks on boiler components. It is necessary to review the methods used in calculating some basic strength requirements on boiler components, and thus to be able to compare a questionable condition found in service.

The ASME boiler code is considered a minimum-requirement code; thus good engineering practice must be used where the code may not cover a particular problem or condition. Generally, boiler manufacturers and consulting engineering firms specializing in power-plant design can offer assistance for abnormal cases. The ASME code is periodically changed to conform with advancement in knowledge of materials or the development of new materials and fabricating practices as well as other technological changes that are continually occurring. Therefore, the latest code should be referred to, especially in regard to new designs.

PRESSURE, STRESSES, AND FORCE

Q What is meant by *pressure*?

A As stated in Chap. 1, pressure is that force in pounds per square inch (psi) caused by a fluid (liquids and gases are fluids) that is confined in a pressure vessel or piping.

Gage pressure, as also stated in Chap. 1, is the pressure noted on a pressure gage when it is installed at any opening into the pressure part of the pressure vessel. It indicates the pressure inside the vessel. But on the outside of the vessel, the atmosphere is also exerting a pressure on the outer surface of the shell. *Absolute pressure* is thus the total pressure of gage pressure *plus* atmospheric pressure.

The relationship between force and pressure is indicated by the following example:

EXAMPLE: If a plate of 20 × 20 in area has 200 psi acting on it, the force on the plate is found by the following equation:

$$F = P \times A$$

where F = force
P = pressure
A = area (upon which pressure acts)

For the plate in this problem, force = 200 × (20 × 20) = 80,000 lb

Q What is meant by *stress*?

A Any body or material subject to external forces on it will resist these external forces. This resistance of the material comes from *within* the material. The internal structure of the material is subject to intercrystalline loading when an external force is applied. Thus, *stress* is defined as the internal force per unit area on the material which resists external forces on the material. It is expressed in pounds per square inch, but the notation psi is *not* used for stress as in pressure notation. Stress is always expressed by engineers with the notation lb/in^2 to differentiate it from the psi designation of pressure, which is an *external force* per unit area on the material.

Q Name some stresses.

A There are several general classifications of stresses that affect materials. A *normal stress* is a *stress* on an area of a material produced by a force at a right angle to the area acted upon. Normal stresses are further classified as *tension stresses* or *compression stresses*. In Fig. 8.1, a 1-in-diameter bar is pulled by a force F. This force produces a normal stress at a right angle to the cross-sectional area of the bar. Since the force tends to pull (stretch) the bar apart, it is called a *tensile* stress. Within the material, the intercrystalline structure (assuming it is steel) is also being stressed. The tensile stress of the bar is found by the following equation, based on the definition of stress as pounds per square inch, or internal force per unit area within the material.

$$F = aS_T$$

where F = external force
a = cross-sectional area of material resisting F
S_T = tensile (internal) stress on the material

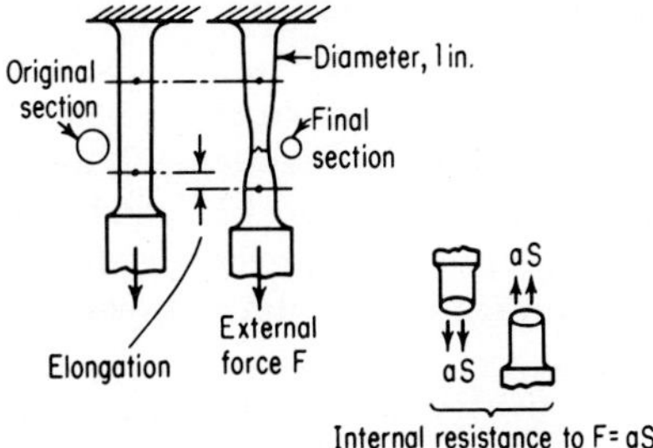

Fig. 8.1 External force acting on bar creates internal stress. a = cross-sectional area; S = tensile internal stress on bar.

From this, the equation for stress S_T is

$$S_T = \frac{F}{a}$$

This equation shows that the tensile stress S_T is found by dividing the external force F, acting normal to the cross-sectional area, by the cross-sectional area of the material resisting the force.

If the force were acting in the opposite direction in Fig. 8.1, a compressive stress (known also as a bearing stress) would be imposed on the material. But the compressive stress would be found by the same equation:

$$F = aS_C$$

where S_C = compressive stress.

Assume that in Fig. 8.1 the tensile force is 15,000 lb and the rod is of 1 in diameter. What would be the tensile stress on the bar?

$$S_T = \frac{F}{a}$$

$$F = 15{,}000 \text{ lb}$$

$$a = \frac{\pi(1)^2}{4} = 0.7854 \text{ in}^2, \text{ where } \pi = 3.1416$$

Substituting,

$$S_T = \frac{15{,}000}{0.7854}$$

$$= 19{,}099 \text{ lb/in}^2$$

Q What is a *shear stress*?

A If a force acts tangent (sideways) to the area of a material, a shear stress is produced. This is illustrated in Fig. 8.2 in which a force F acts on the rivet area tangent to its cross-sectional area, thus producing a shear stress on the rivet. The shear stress is found by the equation $F = a \times s$, but a is the cross-sectional area resisting shear. In Fig. 8.2, since only one area of the rivet is resisting the external force F, it is in *single shear.*

In Fig. 8.3, a butt-riveted joint is shown with butt straps on each side of the butting plates. The rivets in this illustration are in double shear, as *two* areas of the rivet are resisting the load F. An example will illustrate the significance of single shear and double shear.

Assume that in Figs. 8.2 and 8.3 only one rivet is being considered for analysis, but one is in single shear, the other in double shear. Assume the rivet to be 1 in in each case and the load F to be 15,000 lb on each rivet. What is the shear stress for each rivet?

Single-shear rivet:

$$S_S = \frac{F}{a}$$

$$F = 15{,}000 \text{ lb}$$

$$a = \frac{\pi D^2}{4}$$

$$= \frac{\pi(1)^2}{4}$$

$$= 0.7854 \text{ in}^2$$

$$S_S = \frac{15{,}000}{0.7854}$$

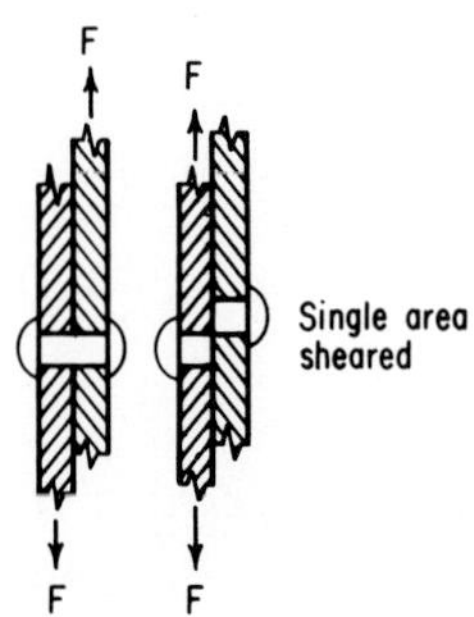

Fig. 8.2 Rivets in single shear.

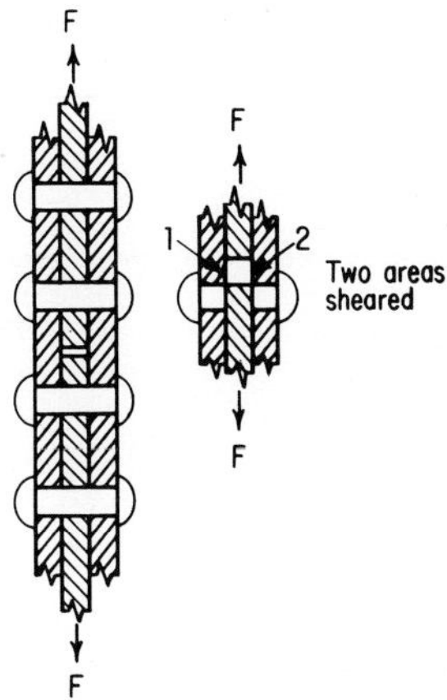

Fig. 8.3 Rivets in double shear.

$$= 19{,}099 \text{ lb/in}^2$$

Double-shear rivet:

$$S_S = \frac{F}{a}$$

$$F = 15{,}000 \text{ lb}$$

$$a = 2\left(\frac{\pi D^2}{4}\right)$$

$$= 2\left[\frac{\pi(1)^2}{4}\right]$$

$$= 1.5708 \text{ in}^2$$

$$S_S = \frac{15{,}000}{1.5708}$$

$$= 9549.5 \text{ lb/in}^2$$

This shows that a rivet in double shear is *twice* as strong as a rivet in single shear.

Q What is a *bending stress*?

A Another stress is that due to bending. A beam supported on each end and loaded in the middle will develop a bending stress. The beam, when bending, will actually be under a tension stress on one side and a compression stress on the opposite side. This is illustrated in Fig. 8.4.

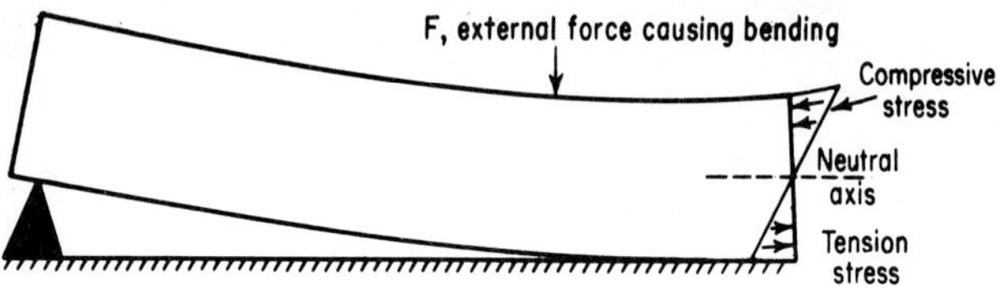

Fig. 8.4 Bending stresses consist of compression and tension stresses on a loaded beam.

Flat plates and stayed surfaces in boilers are some elements that are subjected to bending stresses.

> NOTE: Stress due to torsion, such as in an axle being rotated and transmitting power, is another stress considered in an analysis of resistance of materials. Stress analysis is beyond the scope of this book. However, a knowledge of *tension, compression, shear,* and *bending* stresses is essential in understanding how pressure is contained in a pressure vessel, pipe, or any other apparatus made of material designed to confine that pressure within safe limits.

Q What is *strain*?

A When a body or material is subjected to external forces, internal stresses resist these forces, but there is always some deformation with load. For example, a steel rod will stretch when pulled upon by an external force. The *total stretch* is expressed in a length measurement such as inches or centimeters. Strain is defined as the stretch *per unit length,* or deformation of a body per unit length, and is always expressed as inches per inch, centimeters per centimeter, etc. For example, assume that a steel rod 10 in long stretches 0.010 in with load; then the unit strain will be 0.010/10 = 0.001 in/in.

Q Is there any relationship between *stress* and *strain* as they apply to boiler application?

A Stress and strain are very important in all structural materials designed to resist dangerous forces. In the design, inspection, and operation of boilers, pressure and temperature are the key elements that create dangerous forces. There are others that will be mentioned later. These forces always create stresses and strains. But the chief relationship between stress and strain can be best understood by looking at a material that is to be used for any structural support, not only boilers.

Some of the fundamental properties of all structural materials that must be determined are found by means of *stress-strain diagrams.* From the stress-strain diagram, very important structural properties are determined, such as (1) proportional limit, (2) yield point, (3) ultimate strength, and (4) modulus of elasticity.

Q How are stress-strain diagrams determined?

A Modern engineering practice requires testing of materials so as to specify and identify their physical properties. This is particularly true for materials intended to be used in boilers, pressure vessels, and nuclear reactors. Steel manufacturers' laboratories run tests, and the boiler code shows sketches and requirements for preparing samples to run tensile tests and bending tests on boiler materials.

A test specimen of a specified grade of steel is cut out from a rolled stock, then machined to fit a test machine. A test specimen for a tensile test is shown in Fig. 8.5. The square ends are clamped in jaws of a tension-test machine. The ½-in round center piece is marked off in a 2-in gage length. An extensometer is attached to the 2-in gage length. This tension-test machine has a dial gage indicating force F applied to pull the rod apart. Incremental loading is applied. At each increment, the force F is recorded and the total amount of length l is taken at that incremental loading. This procedure is followed until rupture occurs. The data are tabulated, and then by the relationship F/a where a is the original ½-in-diameter area, stresses are found for each increment of load.

The strain, or stretch, is found by calculating the amount of stretch from the original 2-in length to obtain the unit strain ε. These values are then plotted as in Fig. 8.6, showing a stress-strain diagram for ductile steel. The table in Fig. 8.7 shows a typical test-data recording during test. Note where a permanent set developed during this test.

Q How is the *proportional limit* determined from this chart, and what is the proportional limit?

A The stress-strain diagram shows a sloping straight line extending from zero upward to the point marked *PL*. The reason for this is that as the load is increased, the imposed stress S increases, as does the strain ε. The increase for both is in the same ratio, meaning that if the load is doubled, the stress is doubled, and so is the strain. This is why a straight line is drawn, not a curve. The *proportional limit* of a material is thus the maximum unit stress that can be developed in the material without causing a deviation from the law of proportionality of unit stress to unit

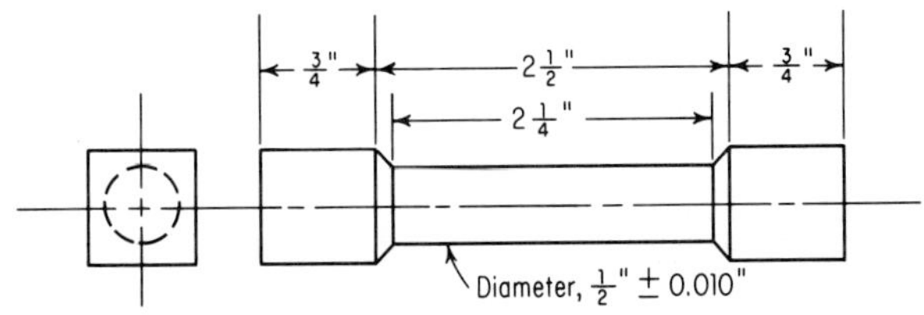

Fig. 8.5 Specimen prepared for a tension test.

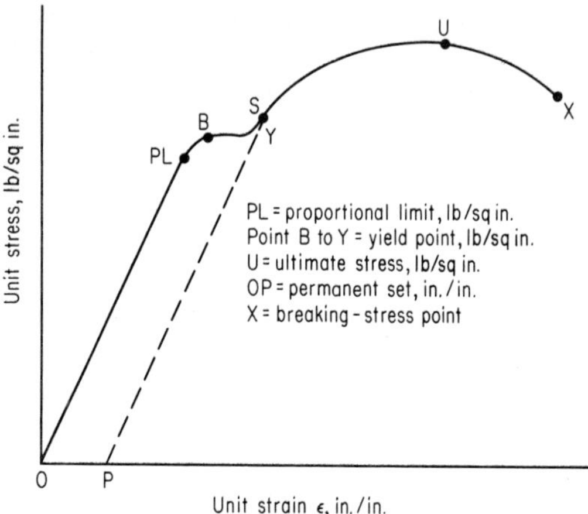

Fig. 8.6 Stress-strain diagram for a ductile material.

strain. Of even more significance, it implies that if the load is decreased, the material will return to its original length without having a permanent *set* as a result of the loading. The material will not be permanently deformed but will return to its original shape as long as the proportional limit is not exceeded.

Fig. 8.7 Typical tension test data on 2-in specimen of 0.5-in diameter.

Load, lb	Extensometer reading, in.
500	0.0
1,500	0.0004
500	0.0
3,000	0.0008
500	0.0
4,400	0.0014
500	0.0
5,980	0.0018
500	0.0
7,510	0.0024
500	0.0
8,630	0.0029
500	0.0002*
9,500	0.0075
500	0.004
9,600	0.0130

* Permanent set.

Q What is meant by *yield point*, and how is this found?

A As the load on our test specimen is increased further, causing a stress greater than the proportional limit, a unit stress is reached, at which point the material continues to stretch *without* an increase in load, assuming it is a ductile steel. The unit stress at which this stretch-without-load occurs is called the *yield point* and is represented by the short horizontal line *B* to *Y* on the stress-strain diagram. The yield point of a material is defined as the *minimum unit stress in the material at which the material deforms or stretches appreciably without an increase of load.*

If a material is stretched or loaded slightly beyond the yield point, a permanent set or deformation occurs in the material. For example, in Fig. 8.6, if the load is reduced to zero after just passing the yield point, the extensometer will show a permanent stretch or deformation. This is found by drawing a line parallel to the proportional limit line, and the set will be length 0 to *P* per inch of test specimen.

Q What is the *ultimate strength*, and how is this determined?

A If the loading on our test specimen is increased, as indicated by the curve *S* to *U*, a point of maximum unit stress is reached. Then the unit stress declines with slight additional loading and stretches until it breaks. This is particularly true of ductile material, which *necks down* very rapidly after reaching its maximum unit stress because of reduced area in the neck section, which requires less load to cause rapid stretching to complete breakage. See Fig. 8.8.

The *ultimate strength* of a material is defined as the maximum unit stress that can be developed in the material as determined from the *original* cross section of the material. It is point *U* in the stress-strain diagram. The curve from *U* to *X* is a rapid, unstable testing condition, with point *X* called the *breaking-load point*. The maximum unit stress is at point *U*, and this is the ultimate stress designated for the material.

Q A common problem involving testing is as follows. Data supplied

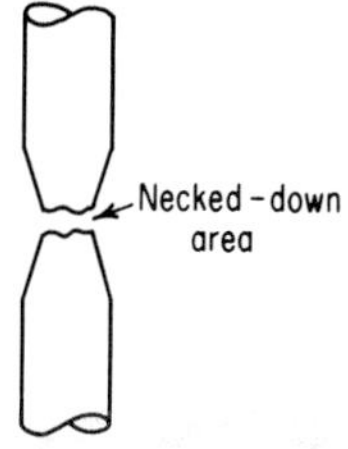

Fig. 8.8 Tension test specimen shows ductile failure in necked-down area.

on a tested specimen: Total length, 19 in; sections at each end of a specimen are 1 × 1 × 6 in long; center section $\frac{7}{16}$-in diameter by 7 in long with center section concentric with the square sections at each end; center 7-in section has a 2-in gage length marked.

The specimen here being tested broke when the load reached a maximum of 11,274.75 lb, and the break was through the original $\frac{7}{16}$-in diameter. But the diameter was now ¼ in. The cross-sectional area of a $\frac{7}{16}$-in-diameter bar is 0.15033 in^2. The length of the 2-in gage section had stretched to 2.55 in.

1. Find the ultimate strength of the material.

2. What is the ultimate strength in pounds per square inch of the 1 × 1 in section at each end?

3. What is the percentage elongation of the 2-in gage section?

A 1. See Fig. 8.5. In this problem the break was the $\frac{7}{16}$-in-diameter section; so the stress S is

$$S = \frac{F}{a}$$

where a = original cross-sectional area.

$$S = \frac{11{,}274.75}{0.15033}$$

$$= 75{,}000 \text{ lb/in}^2 \text{ ultimate stress}$$

2. The ultimate stress at the 1 × 1 in section is the same (75,000 lb/in^2) because it is still the same material. It did not break at this section, because the cross-sectional area is larger than at the $\frac{7}{16}$-in-diameter section.

3. The percentage elongation is found as follows:

$$\frac{(\text{Final length} - \text{original length}) \times 100\%}{\text{Original length}} = \%\text{ elongation}$$

$$\frac{(2.55 - 2) \times 100\%}{2} = \frac{55\%}{2} = 27.5\%$$

Q What is meant by *modulus of elasticity*, and how is this determined?

A This is also known as Hooke's law, which states that the unit stress in a material is proportional to the accompanying unit strain, provided that the unit stress does not exceed the proportional limit. In different words, it states that the ratio of stress to strain for a certain material is always a constant, called E, the modulus of elasticity, or in equation form

$$E = \frac{\text{stress}}{\text{strain}} = \frac{S}{\varepsilon} = \text{constant}$$

For steel, the modulus of elasticity is usually taken as 30,000,000 and is written 30×10^6 lb/in^2. This is the modulus of elasticity for normal or axial loads. There is also a shear modulus of elasticity. For steel it is 12,000,000 lb/in^2.

Q How is the modulus of elasticity applied to boiler work?

A Designers and engineers require this for many calculations. However, one use in field and test applications involves the use of *strain gages*. Strain gages are used to determine stresses at critical areas of boilers, nuclear reactors, and pressure vessels for which exact calculations cannot be made. With the following relationship between stress, strain, and the modulus of elasticity of the material, the stress can be calculated. It is much easier to measure strain, or the deformation of a material under load, than to measure stress.

We explained that stress for normal loads is F/a, where F = imposed load and a = original area of material resisting the load.

We also explained that unit strain ε is e/l, where ε = strain in inches per inch, e = amount of strain from original length l, and l = original length.

Now

$$E = \frac{\text{stress}}{\text{strain}}$$

Substituting the above values

$$E = \frac{F/a}{e/l}$$

Rewriting this in terms of stress S,

$$E = \frac{S}{e/l}$$

as $$S = \frac{F}{a}$$

or $$S = \frac{Ee}{l} = E\varepsilon$$

as $$\frac{e}{l} = \varepsilon$$

It can be seen that if strain is measured, stress can be calculated by knowing the modulus of elasticity of the material, which is usually a constant for the class of material being considered.

Q What does the modulus of elasticity reveal in a physical sense?
A The modulus of elasticity is a measure of the *stiffness* of a material. For example, if one material has a modulus of elasticity twice as large as another material, the elastic unit strain in the one material for a given unit stress is one-half as large as that in the other material. Thus, one material is considered twice as stiff as the other. Some common E values are steel, 30,000,000; cast iron, 15,000,000; aluminum, 12,000,000; concrete, 3,000,000.

Q What is meant by the *elastic limit* of a material?
A The *elastic limit* is the maximum unit stress that can be developed in the material without causing a permanent set. Test results show that for most structural metals the elastic limit of the material has about the same value as the proportional limit, and in most technical literature the elastic and proportional limits are considered identical. A small difference is apparent in testing work, but for practical purposes they can be treated as identical quantities.

LONGITUDINAL AND CIRCUMFERENTIAL STRESS

Q The terms *longitudinal stress* and *circumferential stress* are often used with cylinders and drums. How are these determined, and what is the relationship between the two stresses?
A Thin-walled cylinders, meaning those where the thickness of the shell does not exceed one-half the inside radius, have two stresses, called (1) *longitudinal* and (2) *circumferential*. The latter sometimes is called the *transverse stress*. Thick-walled cylinders have these stresses also, but they are determined differently. Both stresses are known by these names because of the loading they resist in a cylinder. Both are fundamentally tensile stresses.

Figure 8.9*a* shows a seamless cylinder with an inside diameter D, shell thickness at t, length L, and with a uniform pressure P acting inside the cylinder. Pressure acts on the cylinder walls; so the resultant force created tends to split the cylinder along its long axis. Thus the first stress to be considered is the longitudinal stress resisting this force tending to split the cylinder along this axis. The pressure acts in all directions. But if we cut the cylinder as in Fig. 8.9*b*, which shows the external force on one side and also the internal material stress resisting this external force, the following is developed for a condition of equilibrium to exist.

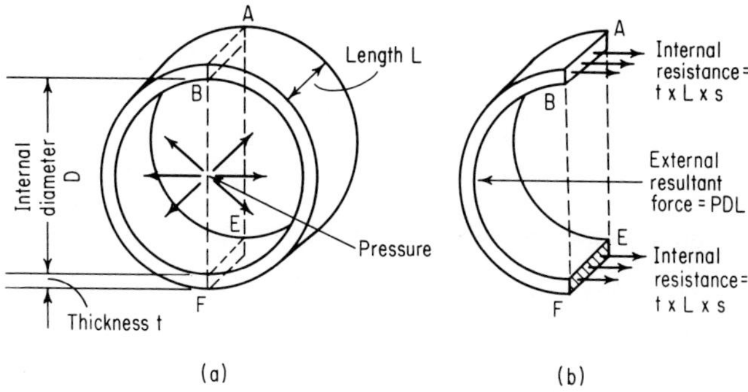

Fig. 8.9 Longitudinal force and stress on thin-walled cylinder.

The force tending to split the cylinder is area × pressure; that is,

$D \times L \times P =$ force acting on one side

where $D \times L =$ projected effective area. The internal force of the material resisting this force is stress × material area; that is,

$S_L \times t \times L \times 2 =$ resisting force

where $t \times L =$ one area of the material. But since there are two material areas resisting the force, it is multiplied by 2. Equating the two forces,

$$D \times L \times P = S_L \times t \times L \times 2$$

From this,

$$\text{Longitudinal stress } S_L = \frac{DP}{2t}$$

The force tending to split the cylinder endwise, or around its circumference, is shown in Fig. 8.10. Pressure acting on each end creates a force which is equal to the end area (circle) times the pressure, or

$$\frac{\pi D^2}{4} \times \text{pressure} = \text{end force}$$

The material resists this by a force equal to the end area of the material times the stress, or

$\pi D t S_c =$ resisting force

where $S_c =$ circumferential stress. Equating the two forces for equili-

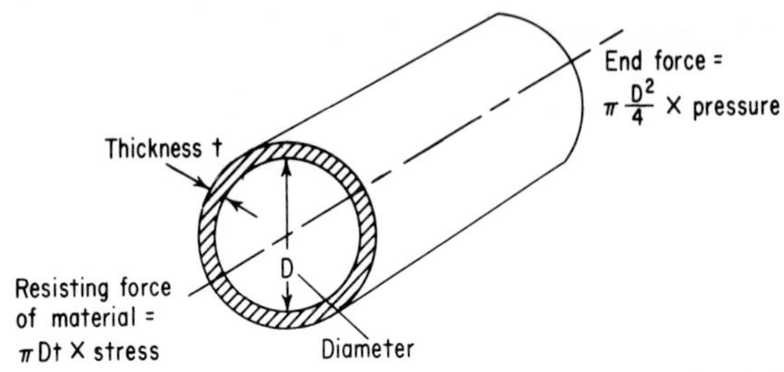

Fig. 8.10 End force on cylinder is resisted by circumferential stress in material.

brium gives

$$\frac{\pi D^2}{4} P = \pi D t S_c$$

by elimination, and solving for S_c, circumferential stress, gives

$$S_c = \frac{DP}{4t}$$

If we compare this with the longitudinal stress, we find the circumferential stress is one-half the longitudinal stress.

The two equations for longitudinal and circumferential stresses are fundamental strength-of-material equations. They are modified somewhat by the ASME Boiler and Pressure Vessel Code to take into account manfuacturing and experience factors.

The equations developed are for seamless construction, meaning that no welded or riveted joint is present. Later calculations will show how the joint efficiency has to be considered to modify these equations. Note that equations for both longitudinal and circumferential stresses (due to pressure) are independent of the length of the vessel. But if a vessel is very long, the bending stress will have to be added to the stress due to pressure. This is especially true of a vessel filled with a substance of considerable weight.

The significance of the circumferential stress being one-half the longitudinal stress in a cylinder enters many problems in boiler design and calculation. For example, riveted circumferential joints do not have to be as strong in this direction as they do longitudinally. But in many calculations it is extremely important to check a cylinder both longitudinally and circumferentially, so as to make sure that the strength circumferentially is at least one-half the strength longitudinally. This is brought out in problems in this chapter.

TEMPERATURE EFFECTS

Q We know that pressure in a boiler causes the material confining that pressure to be stressed. What is the effect of temperature?

A Temperature above designed limits has the immediate effect of lowering the permissible stress on a material. For example, SA-30 Grade A quality carbon plate firebox steel has an allowable stress of 12,000 lb/in^2 for temperatures from −20 to 400°F. At 900°F the allowable stress is only 5000 lb/in^2. Assuming the same pressure at both temperatures, it can be seen that a boiler designed for 12,000 lb/in^2 normal stress will be weakened to 5000/12,000 or 41.7 percent of its original strength with a temperature increase to 900°F.

Certain parts of boilers, particularly tubes, tube sheets, furnaces in scotch marine boilers, and cast parts in CI boilers are very susceptible to temperature or overheating damages. A large temperature increase in a material, with accompanying *lower permissible stress levels*, is one of the most common causes of boiler damages. Low water, poor circulation, and scale are some causes of overheating of the material beyond safe stress levels. Let us not forget that the firing side of boilers is hot enough to melt steel. And with existing pressure on the water or steam side, it does not require much overheating to cause ruptures, bulges, and other deformation. Thus if the material is stressed well beyond the yield stress at high temperatures, permanent deformation will take place. In severe cases, the ultimate stress of the material is reached leading to complete rupture of the affected parts of the boiler.

Q What effect does temperature have on boiler stresses caused by expansion and contraction resulting from temperature changes?

A If the boiler part is free to expand or contract, no increase in stress occurs, unless the stress is influenced by too high a temperature rise, thus lowering the permissible stress due to physical changes caused by temperature. But expansion or contraction, even if the part is free to move, can be considerable. This can be calculated by the following equation, in which all units must be the same:

$$e = nl(T_2 - T_1)$$

where e = change in length
l = original length
T_1 = original temperature, °F
T_2 = final temperature, °F
n = coefficient of expansion (change in length per unit of length per degree change in temperature)

EXAMPLE: Steel has a coefficient of thermal expansion of 0.0000065 in/(in · °F). To show the possible rate of expansion to be considered, assume that a stay in an HRT boiler running from tube sheet to tube sheet is 30 ft long. How much will this rod expand with a temperature change from 70 to 300°F, assuming free expansion?

Substituting in the equation,

$$e = 0.0000065\,(30)(12)(300 - 70)$$
$$= 0.538 \text{ in stretch, which is over } \tfrac{1}{2} \text{ in}$$

Q If we assume that the stay rod was fixed at each end and that the tube sheets would *not give*, what compressive stress would be imposed on the rod, neglecting the column effect of a long rod?

A This is calculated from the modulus of elasticity equation

$$S = E\varepsilon$$

where $\varepsilon = \dfrac{\text{stretch in inches}}{\text{inch}}$

$$S = 30{,}000{,}000\,\frac{0.538}{30 \times 12}$$
$$= 30{,}000{,}000\,(0.001494)$$
$$= 44{,}820 \text{ lb/in}^2$$

This example illustrates the importance of considering temperature effects in boiler design and the rapid stress buildup when a part becomes accidentally overheated above design conditions. Remember that the stress developed is not calculated as simply as shown by the illustration. For example, we assumed that the shell and tube sheets would *not* expand because of temperature. This is obviously *not* true. If the shell is fixed or anchored, some relief will still be obtained from the expansion of the tube sheet. It does illustrate, however, the high stresses possible on stays and tubes. This is one of the chief reasons tubes start to leak around rolled joints or become bowed when a low-water condition develops in a boiler.

Also, on the long stay rod we ignored the column effect. But long, thin structures have to be treated as columns, involving the ratio of the length to the radius of gyration. Stay rods that bow from temperature effects are also influenced by the strength-of-material equations involving columns.

If a tube leaks at the rolled joint but does not become bowed, it is an indication that the expansion force is greater than the rolled joint's holding power. Rolled joints are equivalent to *press fits*, depending on the friction of contact areas to hold the tubes tight in a tube sheet. The exception, of course, is welded-in tubes, where a shear stress is imposed by expansion.

Q What is meant by *stress concentration*?
A If a structural material has an abrupt change in a section, for example, a flat plate containing an opening or sharp corner as shown in the rod in Fig. 8.11, the stress distribution is not uniform over the cross-sectional area of the material. Near the abrupt change the stress is much higher than calculated. The affected section is said to have a *stress-concentration* section, or area, and the ratio by which the normal stress has to be multiplied, *K* in Fig. 8.11, is called the *stress-concentration factor.*

Stress concentration plays an important part in structural members subject to repeated type of loadings, as the stress concentration can lead to cracks and fatigue failures. If the stress concentration is severe enough (even in normal loading), stresses may be induced far above the normal expected stress. Sharp corners in welded joints and other sharply formed shapes must be avoided. Thus openings cut into plates must be reinforced to strengthen the edges around the opening against stress concentration.

Q How are stress-concentration factors determined?
A The boiler code specifies permissible joint connections to avoid stress concentrations. Fillet radii are specified on formed shapes. Openings must be calculated by code rules. In analytical and design work, stress concentrations are determined by the *photoelastic method, stress-coat method*, and *strain-gage method* using the electrical resistant wire gage. In nuclear vessels one must carefully design the elements by the endurance limit and other stress-analysis methods.

Q What is meant by the *endurance limit* of a material?
A The endurance limit (also known as *fatigue limit*) is the maximum unit stress that can be imposed and repeated on a material through a

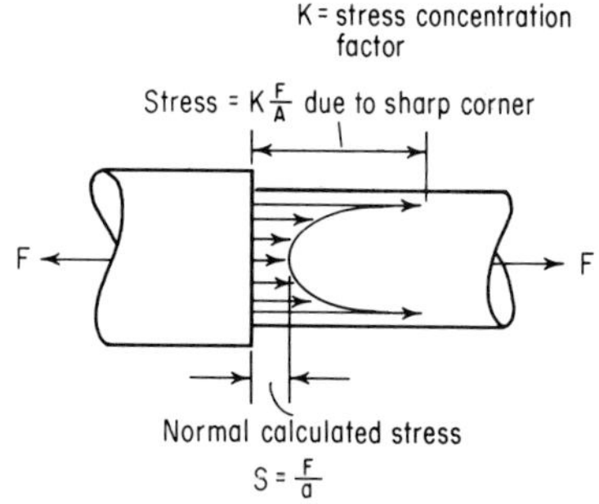

Fig. 8.11 Sharp corners magnify normal stresses by a stress-concentration factor *K*.

definite cycle, or range of stress, for an indefinitely large number of times without causing the material to rupture.

Q How is the endurance limit determined?

A By testing a material through a complete reversal of stresses; when stressed nearly to its ultimate strength, the specimen will rupture after a few cycles. If a second sample of the same material is again tested, but stressed slightly less than before, a large number of reversals, or cycles, can be imposed. This is continued until a stress value, known as the *endurance limit*, is reached where an almost indefinite cycle of stress can be imposed without causing rupture.

Figure 8.12 shows an *S-N* diagram, where stress-to-rupture is plotted on one side and number-of-cycles-to-failure on the other. The horizontal line obtained is the endurance stress for the material. In Fig. 8.12 this is 22,500 lb/in^2. Endurance limits are widely used in machine design work, and with the adoption of the Nuclear Vessel Code, Section III, it will receive increasing attention in the code.

ALLOWABLE STRESS

Q What is meant by an *allowable stress?*

A Sometimes called the *allowable working stress*, it is the maximum stress that is considered to be safe when the material is subjected to resisting loads that are assumed to be applied in service. In boiler applications, the term *allowable pressure* is often used. Actually, the allowable pressure is determined by applying the forces acting on a material, and then calculating the allowable pressure from the allowable stress on the material.

Q How is the allowable stress determined?

A It is determined from the ultimate strength of the material, which is divided by a safety factor to obtain the allowable stress. The safety factor used in modern boilers is 4. However, certain elements of a boiler, such as rivets, have to be designed with a safety factor of 5. Other parts have to be designed with a safety factor as high as 12.5, such as the rivets holding lugs on brackets on an HRT boiler to be suspended from a beam.

That is why the code must always be checked regarding the part of the boiler being considered to obtain the allowable stress. Data in Fig. 8.13 are taken from ASME, Power Boiler Code, Section I, and indicate permissible allowable stress values for various temperatures. The temperature indicated is for the maximum expected steam or water temperature.

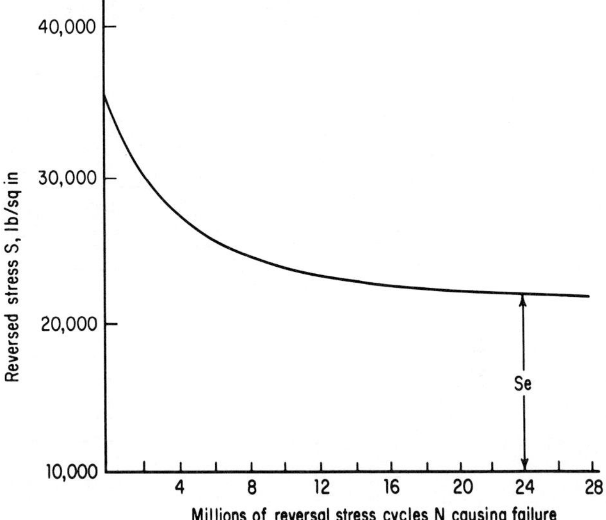

Fig. 8.12 *S-N* diagrams help determine endurance limits of a specimen.

NOTE: The *specified minimum tensile strength* is the ultimate tensile strength of the material. The allowable stresses are shown under the expected temperature, and this is equal to the ultimate stress divided by the safety factor.

SAFETY FACTOR

Q What is meant by *safety factor*?

A In boiler usage, the factor of safety is the ultimate strength divided by the allowable loadings, or the ultimate stress (S_u) divided by the allowable stress (S_a). In equation form

$$\text{Safety factor} = \frac{S_u}{S_a}$$

Another method of expressing the safety factor is by dividing the bursting pressure by the allowable pressure. This method is used on state inspection reports and on ASME data reports. In equation form, it is

$$\text{Safety factor} = \frac{\text{bursting pressure}}{\text{allowable pressure}}$$

Fig. 8.13 Some maximum allowable stress values for plate of ferrous material in pounds per square inch for different temperatures as stipulated by ASME code.

Spec number and grade		Nominal composition	Spec min tensile	Allowable stress for metal temperature, max, °F								
				−20 to +400	500	600	650	700	750	800	850	900
Plate steels:												
Carbon steel:												
SA-30	Flange	....	55,000	13,750	13,750	13,750	13,750	13,250	12,050	10,200	7,800	
SA-30	Firebox A	....	55,000	13,750	13,750	13,750	13,750	13,250	12,050	10,200	7,800	5,000
SA-30	Firebox B	....	48,000	12,000	12,000	12,000	12,000	11,650	10,700	9,000	7,100	5,000
Low-alloy steel:												
SA-202	A	Cr-Mn-Si	75,000	18,750	18,750	18,750	18,750	17,700	15,650	12,000	7,800	5,000
SA-202	B	Cr-Mn-Si	85,000	21,250	21,250	21,250	21,250	19,800	17,700	12,000	7,800	5,000
High-alloy steel:												
SA-240	304	18 Cr-8 Ni	75,000	......	15,100	14,900	14,850	14,800	14,700	14,550	14,300	14,000
SA-240	304	18 Cr-8 Ni	75,000	......	12,500	11,600	11,200	10,800	10,400	10,000	9,700	9,400
SA-240	316	18 Cr-10 Ni-2 Mo	75,000	......	17,200	17,100	17,050	17,000	16,900	16,750	16,500	16,000
SA-240	321	18 Cr-8 Ni-Ti	75,000	......	15,200	14,900	14,850	14,800	14,700	14,550	14,300	14,100
SA-240	347	18 Cr-8 Ni-Cb	75,000	......	15,200	14,900	14,850	14,800	14,700	14,550	14,300	14,100

Spec number and grade	Allowable stress for metal temperature, max, °F											
	950	1000	1050	1100	1150	1200	1250	1300	1350	1400	1450	1500
Plate steels:												
Carbon steel:												
SA-30 Flange												
SA-30 Firebox A												
SA-30 Firebox B												
Low-alloy steel:												
SA-202 A	*3,000*	*1,500*										
SA-202 B	*3,000*	*1,500*										
High-alloy steel:												
SA-240 304	13,400	12,500	10,000	7,500	5,750	4,500	3,250	2,450	1,800	1,400	1,000	750
SA-240 304	9,100	8,800	8,500	7,500	5,750	4,500	3,250	2,450	1,800	1,400	1,000	750
SA-240 316	15,100	14,000	12,200	10,400	8,500	6,800	5,300	4,000	2,700	2,000	1,500	1,000
SA-240 321	13,850	13,500	13,100	10,300	7,600	5,000	3,300	2,200	1,500	1,200	900	750
SA-240 347	13,850	13,500	13,100	10,300	7,600	5,000	3,300	2,200	1,500	1,200	900	750

Q Why is it necessary to have a safety factor? Why cannot the stresses be calculated accurately, and the allowable load based on the yield *stress be used*?

A In boiler design and usage and other critical structures where life and property may be at stake, there is a definite need for selecting working stresses and loadings considerably less than the ultimate, or the yield stress, for these reasons:

1. There is always some uncertainty in materials being used, how they were made, how they were assembled, and how they were joined or fabricated with other materials.

2. There is always some uncertainty as to the exact loading that a structure, or part of it, may have to resist and how it is abused in operation.

3. Calculations of all stresses possible in a fabricated structure are never that exact, not when one considers the variables to be encountered in service through the years.

Let us remember that the ASME code was drawn up when yield stresses for many materials were not available. Then again, certain brittle materials like cast iron do not have a *definite* yield point. Thus it was much easier to work from ultimate stress and apply a safety factor to this so as to obtain an allowable stress. In some countries of Europe, the *yield point* is the basis of design. The Nuclear Vessel Code uses the yield stress or endurance stress or both as a basis of design. With increasing technological changes, the present boiler code may also be changed on this matter in time.

Q Can a boiler that was originally designed with a safety factor of 5 be operated with a safety factor of 4, since the latest Power Boiler Code allows a safety factor of 4?

A Usually, the original safety factor of 5 remains. The safety factor of 4 was drawn up principally for seamless steel or welded boilers that met stiff quality-control and inspection requirements on welding. Older boilers may not meet this requirement; thus the original safety factor of 5 should govern the allowable pressure.

Q If a boiler is stamped for an allowable pressure of 275 psi and the SV is set at 150 psi, what is the safety factor? Assume a code-welded boiler meeting latest code requirements and an original design safety factor of 4.

A The bursting pressure of this boiler would be $4 \times 275 = 1100$ psi.

$$\text{Safety factor} = \frac{\text{bursting pressure}}{\text{allowable pressure}} = \frac{1100}{150} = 7.33$$

This question brings up an important consideration as to which pressure to use in calculating safety factors on existing boiler installations. For example, in this question, should the SV setting be used or the stamped allowable pressure? If the SV setting is *below* the stamped allowable working pressure, use the SV *setting*. If by chance the SV or SVs are set *higher* than the allowable pressure (above code limits), then a dangerous condition exists because SVs must always be set at or *below* the allowable, or working, pressure stamped on the boiler.

The reason the SV setting is used where the stamped allowable pressure is above the SV setting in calculating the safety factor is that the allowable pressure on the boiler *is the* SV *setting*. The boiler is *not* supposed to operate above this SV setting. If an increase in pressure is needed, other items will have to be checked out first before the pressure can be raised to the maximum allowable pressure stamped for the boiler. Then, a new SV will be required. Also, code specifications on valve ratings and water-column connections will have to be checked to see if they meet the code requirements for the new pressure.

Q Does the *original* safety factor continue for the *life of the boiler*?
A It depends on the type of boiler, the condition of the boiler, and even the state in which it is located. Assuming that internal and external inspections are satisfactory, the NB regulations state:

1. Lap-riveted longitudinal-joint boilers operating over 50 psi can be operated at this pressure for 20 years. After that, 50 psi or less is permissible, but if the boiler is relocated, only low-pressure service is permitted.

2. For boilers of butt construction, at the end of 25 years, and every 5 years thereafter, the safety factor must be increased by 0.5 unless a hydrostatic test of 1½ times the allowable pressure is imposed; if this test is satisfactory, no increase in the safety factor is necessary.

The best rule to follow on any safety factor changes is to check with an insurance company boiler inspector or legal jurisdiction inspector. *Never* assume that the boiler can be operated at a higher pressure by just changing SVs, even if the boiler is stamped for the higher pressure. There are other requirements to be met on feedwater, blowdown, water-column connections, low-water fuel cutoff, and service valve ratings that must be considered.

Q Name the parts of boilers that must be considered for *stress calculations*.
A Any pressure vessel and parts confining pressure must be analyzed per component by carefully considering the strength of the material being used, its physical characteristics as to type and grade, allowable stress, thickness, etc. The forces acting on this material must then be analyzed. This force is usually created by pressure but may also include

temperature, the weight it is supporting, and stress concentration such as around an opening. The problem then evolves to comparing the forces acting on the material and determining whether the material is being stressed beyond the allowable stresses governed by the boiler code rules. Elements to be considered depend on the type of boiler but will generally include shells or drums, tubes, tube sheets, heads, flat surfaces, stays, stay bolts, openings, furnaces, rivets, welded joints, structural supports, and connected piping and valves. Each of these is governed by boiler code rules as to allowable material, allowable stresses, and method of calculating forces to obtain the allowable pressure.

> NOTE: In boiler and pressure vessel application, the weakest element producing the lowest pressure then determines the *allowable pressure* for the boiler.

STRESS CORROSION

Q The term *stress corrosion* has often been applied to boiler and pressure-vessel elements that fail unexpectedly, even though *normal stresses* are within prescribed limits. What is stress corrosion?

A The endurance limit of a metal, when it is repeatedly stressed in the presence of a corroding agent such as water, may be greatly reduced. The result is often an unexpected failure, which is a form of fatigue failure due to corrosion. The process begins with corrosion pitting on a stressed part. As the pitting progresses, the stress in the affected part rises. This is caused by the pits serving as stress-concentration points. As corrosion continues, stress continues to rise, especially if repeated, until the endurance limit of the metal is reached. Fatigue cracks start at the bottom of corrosion pits and proceed until failure occurs. The damage is primarily due to sharp pits rather than general corrosion.

Pits formed under simultaneous stress and corrosion are always sharper and deeper than pits formed in the same time under stressless conditions. The more repetitive the stress, the faster will be the pitting action. At low-cycle repetitive stress, pitting will proceed entirely by normal corrosion, and the section will fail by the normal tension failure or when the resisting area is thinned down so the stress rises proportionately, assuming a constant load.

Stress corrosion then is primarily caused by repetitive stresses of high frequency in a corroding medium, leading to pitting caused by stress and corrosion. The pitting then develops high-stress concentration points which lead to fatigue failure.

Figure 8.14 shows the typical pattern of stress-corrosion cracking in

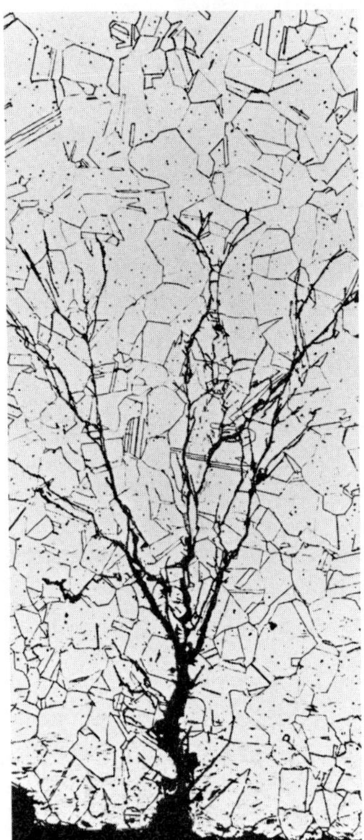

Fig. 8.14 Stress-corrosion cracking in a type 304 stainless steel which started in a pit. (*Courtesy of Babcock & Wilcox Co.*)

a type 304 stainless steel. Cracking started in a pit on the surface and progressed along the grain boundaries. Note the branching of the cracking from its original starting point.

Q Can stress corrosion be calculated for in design?

A No experimental data are available for determining the extent to which the endurance limit is reduced for most materials in combination with corroding solutions or media. But for boilers and pressure vessels the obvious precaution to take against stress corrosion is to make sure the water or medium being confined is free of any corrosive tendencies. This is determined by analysis of the water at regular intervals by personnel experienced in water analysis. This also points out the value of periodic internal inspections of pressure-containing parts and examination of surfaces for evidence of pitting and corrosion.

BOILER STRENGTH CALCULATIONS

Some typical examples of boiler code calculations are presented in the following questions and answers. All allowable stresses shown and the formulas used are from the code. But a word of caution. The code does change the allowable stresses and the formulas from time to time. So always refer to the latest code for the exact formula and stress allowed. The methods presented here show how each component of a boiler or other pressure vessel has to be considered to evaluate its strength. Refer to a good stress-analysis book to ascertain the derivation of the formulas used. See Appendix B for a review of how to transpose equations and how far to carry decimals.

Q What are the usual components of a boiler that require code calculations?

A These can be grouped into the following components: (1) tubes 5 in and under; (2) shells, pipes, and drums; (3) heads; (4) combustion chambers, furnaces, and flues for fire-tube boilers; (5) stayed surfaces; (6) miscellaneous—this would include safety valve size calculations, heating surfaces, areas to be stayed, lugs or brackets to support a boiler, and similar items.

It should be obvious that the code has minimum requirements. Many structural and wear factors are left to the discretion of good "engineering" practice, including firing rates vs. heating surfaces. The ASME boiler code is primarily a safety code. For this reason it is good practice to refer to the code whenever a weakened or deteriorated part is found in order to calculate the allowable working pressure with a code equation.

Q Figure 8.13 provides some allowable stresses for plate steel. Are there similar allowable stresses for tubes?

A See Fig. 8.15. For selecting the allowable stress for tubes, the operating temperature of the metal should be not less than the maximum expected mean wall temperature (the sum of the outside and inside surface temperatures divided by 2) of the tube. This in no case should be taken as less than 700°F for tubes absorbing heat. For tubes which do not absorb heat, the wall temperature may be taken as the temperature of the fluid within the tube, but not less than the saturation temperature.

TUBES

Q What are the code equations to be used on tube problems?

A There are four equations (ASME Power Boilers, Section I).

For WT boilers having ferrous metal economizer, superheater, and

Fig. 8.15 Steel tube material, allowable stresses per Section I of ASME code, in thousands of pounds per square inch.

Spec no.	Composition	Form	Ultimate	Allowable stress at temperature of				
				−20 to 650°F	700°F	800°F		
		A. Carbon steel—tubes						
SA 192	C-Si	Seamless	47.0	18.8	11.5	9.0		
SA 178A	C	Welded	47.0	11.8	11.5	7.7		
SA 226	C-Si	Welded	47.0	11.8	11.5	7.7		
SA 210 A-1	C	Seamless	60.0	15.0	14.4	10.8		
SA 178 C	C	Welded	60.0	15.0	14.4	9.2		
SA 210 C	C-Mn	Seamless	70.0	17.5	16.6	12.0		
		B. Low-alloy steel—tubes						
SA 209 T1b	C-½ Mo	Seamless	53.0	13.3	13.2	13.1		
SA 250 T1b	C-½ Mo	Welded	53.0	11.3	11.2	11.1		
SA 250 T1	C-½ Mo	Welded	55.0	11.7	11.7	11.7		
SA 209 T1	C-½ Mo	Seamless	55.0	13.8	13.8	13.7		
SA 213 T2	½ Cr-½ Mo	Seamless	60.0	15.0	15.0	14.4		
SA 423-1	¾ Cr-½ Ni-Cu	Seamless	60.0	15.0	15.0			
SA-213-T12	1 Cr-Mo	Seamless	60.0	15.0	15.0	14.8		
SA-213-T11	1¼ Cr-½ Mo-Si	Seamless	60.0	15.0	15.0	15.0		
SA-213-T3b	2 Cr-½ Mo	Seamless	60.0	15.0	15.0	14.7		
SA-213-T22	2¼ Cr-1 Mo	Seamless	60.0	15.0	15.0	15.0		
				−20 to 100°F	300°F	500°F	700°F	800°F
SA 213-T21	3 Cr-1 Mo	Seamless	60.0	15.0	15.0	15.0	14.8	14.5
SA 213-T5	5 Cr-½ Mo	Seamless	60.0	15.0	15.0	15.0	13.4	12.8
SA 213-T7	7 Cr-½ Mo	Seamless	60.0	15.0	15.0	14.5	13.4	12.5
SA 213-T9	9 Cr-Mo	Seamless	60.0	15.0	15.0	14.5	13.4	12.8
		C. High-alloy steel—tubes						
SA 268-TP405	12 Cr-1A1	Seamless	60.0	15.0	13.3	12.9	12.1	
SA 268-TP446	27 Cr	Seamless	70.0	17.5	15.6	14.5	14.1	
SA 213-TP304	18 Cr-8 Ni	Seamless	75.0	18.8	16.6	15.9	15.9	15.2
SA 213-TP316	16 Cr-12 Ni and 2 Mo	Seamless	75.0	18.8	18.4	18.0	16.3	15.9
SA 213-TP321	18 Cr-10 Ni and Ti	Seamless	75.0	18.8	17.3	17.1	15.8	15.5
SA 213-TP347	18 Cr-10 Ni and Cb	Seamless	75.0	18.8	15.5	14.9	14.7	14.7

generator tubes up to 5 in OD, use

$$P = S\frac{2t - 0.01D - 2e}{D - (t - 0.005D - e)}$$

or $$t = \frac{PD}{2S + P} + 0.005D + e$$

where P = maximum allowable pressure, psi
D = outside diameter of tubes, in
t = minimum required thickness, in
S = maximum allowable stress, lb/in^2
e = thickness factor for expanded tube ends, defined as follows:

$e = 0.04$ over a length at least equal to the length of the seat, plus 1 in for tubes expanded into tube seats. However, $e = 0$ for tubes expanded into tube seats, privided that the thickness of the tube ends over a length of the seat plus 1 in is *not* less than the following:

e = 0.095 in for tubes 1¼ in OD and smaller
e = 0.105 in for tubes above 1¼ in OD and up to 2 in OD
e = 0.120 in for tubes above 2 in OD and up to 3 in OD
e = 0.135 in for tubes above 3 in OD and up to 4 in OD
e = 0.150 in for tubes above 4 in OD and up to 5 in OD
e = 0 for tubes strength-welded to headers and drums.

For FT boilers using SA-83 and SA-178 material,

$$P = 14{,}000\frac{(t - 0.065)}{D}$$

where P = maximum allowable pressure, psi
t = minimum required thickness, in
D = outside diameter of tube, in

For FT boilers using copper tubes of SB-75 specification,

$$P = 12{,}000\frac{(t - 0.039)}{D} - 250 \text{ (the symbols are the same as above)}$$

Q Give an example of a code tube problem.
A The following specifications are for a tube strength-welded in a WT boiler: minimum thickness, 0.158 in: OD 2.5 in; material, SA-213-T11; temperature, 700°F. What is the maximum allowable working pressure?

Using the formula

$$P = S \frac{2t - 0.01D - 2e}{D - (t - 0.005D - e)}$$

where $S = 15{,}000$ (from Fig. 8.15)
$t = 0.158$ in
$D = 2.5$ in
$e = 0$ (strength-welded)

$$P = 15{,}000 \frac{2\,(0.158) - 0.01\,(2.5)}{2.5 - [0.158 - 0.005\,(2.5)]}$$

$$P = 15{,}000 \frac{0.291}{2.3545}$$

$$P = 15{,}000(0.1236) = 1854 \text{ psi} \qquad \textit{Ans.}$$

Q What thickness of tube is needed for an FT and a WT boiler, each having 3.5-in-diameter tubes, properly flared and/or beaded by code requirements, both to be designed for 250 psi, and both using SA-178A tube material with expected mean temperature below 650°F?

A Two equations must be used, one for FT and the other for WT boilers.

1. For WT boilers,

$$t = \frac{PD}{2S + P} + 0.005D + e$$

From Fig. 8.15, the allowable stress for SA-178A material is 11,800 lb/in^2

$D = 3.5$ in, $P = 250$, $e = 0.04$ (assume this value initially)

$$t = \frac{250(3.5)}{2(11{,}800) + 250} + 0.005(3.5) + 0.04$$

$$t = 0.0942 \text{ in required}$$

The assumption of $e = 0.04$ was correct, because for e to be zero, t would have to be at least 0.135 in.

2. For FT boilers,

$$P = \frac{14{,}000(t - 0.065)}{D}$$

Solving for t,

$$t = \frac{250(3.5)}{14{,}000} + 0.065$$

$$t = 0.1275 \text{ in}$$

Note that the fire tube has to be thicker than the WT boiler tube in order to resist the pressure on the external side of the tube.

Q Name the methods used to fabricate tubes.
A Three common methods of boiler-tube fabrication are used. (1) The seamless tube is pierced hot and drawn to size. (2) The lap-welded (forge-welded) tube consists of metal strip ("skelp") curved to tubular shape with the longitudinal edges overlapping. Heat is applied and the joint is forge-welded. (3) The electric-resistance butt-welded tube is formed like the second type, but as its name implies, the joint is butt-welded.

It is considered good practice by some to place the weld on welded tubes away from the radiant heat of the fire. Tubes for bent-tube-type boilers are bent usually by machine.

The diameter of boiler tubes always refers to nominal outside diameter, while pipe diameter refers to nominal inside diameter.

Q Explain the flaring and beading tube-expanding methods.
A Practically all boiler tubes have the ends expanded into the tube hole of the shell or drum. This is to make the tube tight against leakage and to give it a firm grip on the tube hole so that the tube may have a definite holding or staying effect.

The edges of the tube holes are chamfered about $\frac{1}{16}$ in after the holes are drilled so that there will be no sharp edges to cut into the tube when it is expanded.

Tube holes are finished $\frac{1}{32}$ in larger in diameter than the outside diameter of the boiler tube, except in the tube sheet of fire-tube boilers. Through this, tubes must be drawn during retubing, and therefore its holes are finished $\frac{1}{16}$ in larger in diameter so as to permit a tube that is coated with scale to be removed without damage to the tube sheet.

Thick drums may be counterbored in order to have a reasonably narrow circumferential strip of tube to expand. The diameter of the counterbore should be sufficient to allow for flaring the tube end according to requirements.

The counterbore may be from either the inside or the outside. When a drum of a water-tube boiler has tubes expanded in its upper side, it is best practice not to use the outside counterbore, for pockets for soot would thus be formed.

For water-tube boilers the tubes and nipples should extend through the tube hole $\frac{1}{4}$ to $\frac{3}{4}$ in and be flared to at least $\frac{1}{8}$ in larger than the tube-hold diameter.

Fire-tube boilers have the tube ends exposed to heat and products of combustion, and therefore the tube ends might soon be burned off if they were flared. In these boilers the tube ends are driven back into a bead after expanding the tubes, in order to protect them against overheating, although the bead does not increase the holding power of the tube appreciably.

SHELLS AND DRUMS

Q What are the two basic equations used for shells, piping, and drums?

A The two equations used have as their difference the question of whether inside backing strips are removed and the weld surface prepared for radiographic examination as required by the code. If the backing strip is not removed, and for old riveted boilers, use the following equation:

$$P = \frac{0.8\,SEt}{R + 0.6\,t} \tag{1}$$

or $$t = \frac{PR}{0.8SE - 0.6P}$$

where P = maximum allowable pressure, psi
S = maximum allowable stress, lb/in^2, for operating temperature of metal
t = minimum required thickness, in
R = inside radius of cylinder, in
E = efficiency of joint

E = efficiency of longitudinal welded joints or of ligaments between openings, whichever is lower. E = 1.00 for seamless cylinders, and E = 1.00 for welded joints, provided all weld reinforcement on the longitudinal joints is removed substantially flush with the surface of the plate. E = 0.90 for welded joints with the reinforcement on the longitudinal joints left in place. E = efficiency for riveted joints or ligament joints as calculated.

Where the backing strip is removed and the weld is properly prepared for radiographic examination, the following equation applies:

$$P = \frac{SE\,(t - C)}{R + (1 - y)(t - C)} \tag{2}$$

or $$t = \frac{PR}{SE - (1 - y)P} + C$$

P, S, t, R, and E are defined as above; y = temperature coefficient as follows:

Temperature, °F	Ferritic steel, y	Austenitic steel, y
900 and below	0.4	0.4
1000	0.5	0.4
1050	0.7	0.4
1100	0.7	0.5
1150 and over	0.7	0.7

C = factor for threading of pipe, or structural stability, as follows:

	C value
1. For threaded steel or threaded nonferrous pipe, ¾ in and smaller	0.065
2. For items in 1 over ¾ in size	Depth of thread
3. For plain steel and nonferrous pipe 3¾ in and smaller	0.065
4. For items in 3 over 3¾ in	0.0

Q How is the efficiency of the ligaments on the steam drums calculated? See Figs. 8.16 and 8.17.

A From the equation

$$E = \frac{p - nd}{p}$$

where E = efficiency
p = pitch of ligament, in
n = number of tube holes in ligament pitch
d = tube hole diameter, in

SHELL AND DRUM CALCULATION EXAMPLES

Q Give examples of shell and drum calculations according to the code.

A EXAMPLE: A boiler is 66 in in diameter and 16 ft long. The shell plate is $\frac{7}{16}$ in thick, the allowable stress is 13,750 lb/in^2, and the allowable pressure is 125 psi. What is the least permissible circumferential efficiency of the girth joint?

Using the formula, with S = 13,750, P = 125, t = 0.4375,

$$P = \frac{0.8\ SET}{R + 0.6t}$$

Fig. 8.16 Ligaments in drum of WT boiler weaken the drum and must be considered in design. (*Courtesy of Foster Wheeler Corp.*)

$$125 = \frac{0.8\ (13{,}750)\ (E)\ (0.4375)}{33 + 0.6\ (0.4375)}$$

$$E = \frac{125\ (33.26)}{11{,}000\ (0.4375)} = 86.6\%$$

$$\text{Circumferential } E = \frac{86.6}{2} = 43.3\% \qquad \textbf{\textit{Ans.}}$$

This is a riveted boiler. The circumferential strength of the shell must be always at least ½ that of the longitudinal joint.

EXAMPLE: What is the maximum OD allowed for a drum built to the following specifications: pressure, 300 psi; SA-30 flange steel; joint efficiency, 100 percent; thickness of drum plate, ¾ in; temperature, under 650°F.

Using equation 2, with $P = 300$, $C = 0$, $S = 13{,}750$, $E = 100$

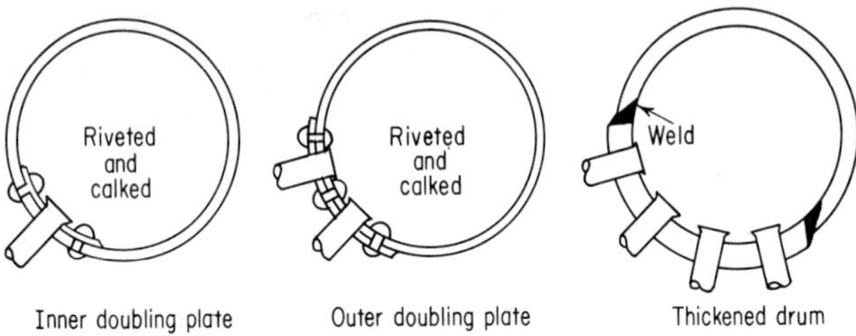

Fig. 8.17 Three methods of reinforcing drum ligaments.

percent, $y = 0.4$,

$$t = \frac{PR}{SE - (1 - y)P} + C$$

$$0.75 = \frac{300R}{13{,}750(1) - (1 - 0.4)300}$$

$$0.75 = \frac{300R}{13{,}570}$$

$$R = 33.925 \text{ in}$$
$$\text{OD} = 2(33.924) + 2(0.75)$$
$$\text{OD} = 69.35 \text{ in} \qquad \textit{Ans.}$$

EXAMPLE: The mud drum of a WT boiler has a 42-in ID. The tube sheet is ⅝ in thick and contains $3\frac{9}{32}$-in-diameter tube holes pitched horizontally $5\frac{5}{16}$ in in banks of three (Fig. 8.18) and two tubes with 6⅞ in between banks. The shell plate is ½ in thick. The joint efficiency between tube sheet and shell is 67 percent. What is the allowable working pressure for this drum if the material is SA-285C and the maximum temperature is 650°F?

NOTE: The tube sheet is the portion of the drum where the tubes are located (Figs. 8.16, 8.17, and 8.18). The holes weaken the solid plate by a factor called ligament efficiency.

Ligament efficiency must be calculated to see if it is lower than longitudinal joint efficiency.

$$E = \frac{p - nd}{p}$$

$$= \frac{29.8125 - 5(3.28125)}{29.8125}$$

$$= 0.448 \quad \text{(This is lowest joint efficiency.)}$$

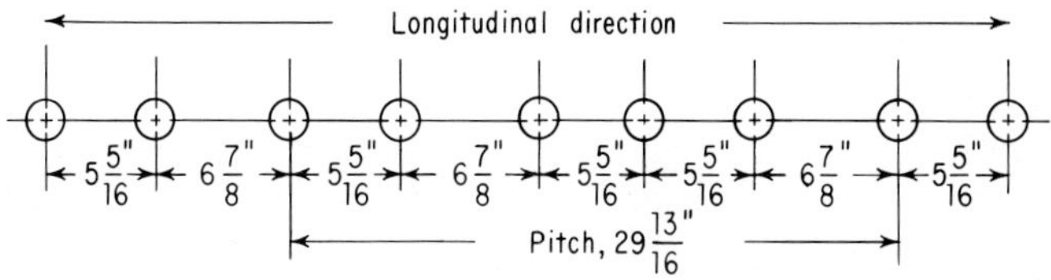

Fig. 8.18 Tube holes in shell or drum require ligament efficiency calculations.

Using the equation in the first example (½-in shell plate, riveted construction as efficiency is 67 percent) and tube-sheet data,

$$P = \frac{0.8SEt}{R + 0.6t}$$

$$= \frac{0.8(13{,}750)(0.448)(0.625)}{21 + 0.6(0.625)}$$

$$= 145 \text{ psi} \qquad \textit{Ans.}$$

EXAMPLE: What hydrostatic pressure should be made on a WT boiler, all-welded construction, of 36-in ID, shell plate ⅝ in, tube sheet ¾ in thick? Ligament efficiency of tube holes longitudinally is 43.5 percent, girthwise it is 20 percent; longitudinal weld efficiency is 90 percent; allowable stress is 13,750 lb/in² under 900°F.

Using equation 2, with $R = 18$, $t = 0.625$ in, $E_w = 0.90$, $E_{\text{lig}} = 0.40$ (or twice girth), $y = 0.4$, $C = 0$,

$$P = \frac{SE(t - C)}{R(1 - y)(t - C)}$$

Two calculations must be made as follows:

1. Based on weld efficiency,

$$P = \frac{13{,}750(0.9)(0.625)}{18 + (1 - 0.4)(0.625)}$$

$$P = 453.6 \text{ psi}$$

2. Based on ligament efficiency and tube sheet t,

$$P = \frac{13{,}750(0.4)(0.75)}{18 + (1 - 0.4)(0.75)}$$

$$P = 223.6 \text{ psi, which is the allowable pressure}$$

Hydrostatic pressure required = 223.6 × 1.5 = 335.4 psi *Ans.*

UNSTAYED HEADS OR DISHED HEADS

Q Name the blank unstayed heads that are allowed on power boilers.
A There are four: (1) segment of a sphere, (2) semiellipsoidal, (3) hemispherical, and (4) flatheads. The first three are bumped heads (Fig. 8.19). Some code flatheads and methods of attachment are shown in Fig. 8.20. Bumped heads are flanged, with a corner radius on the concave side of the head of not less than 3 times the head thickness and in no case less than 6 percent of the diameter of the shell for which the heads are to be attached.

Flanged-in access-hole openings in dished, or bumped, heads must be flanged to a depth of not less than 3 times the required thickness of the head for plate of up to 1½ in thickness. If thicker, the depth of the flange must be the thickness of the plate plus 3 in. The minimum width of the bearing surface for a gasket on an access hole must be $^{11}/_{16}$ in, and the gasket thickness when compressed must be less than ¼ in.

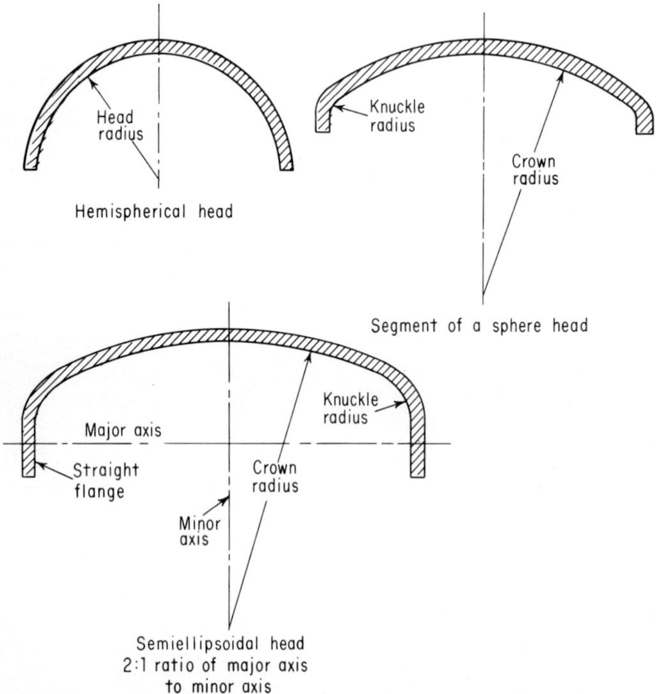

Fig. 8.19 Dished heads may be hemispherical, semiellipsoidal, or segment of a sphere.

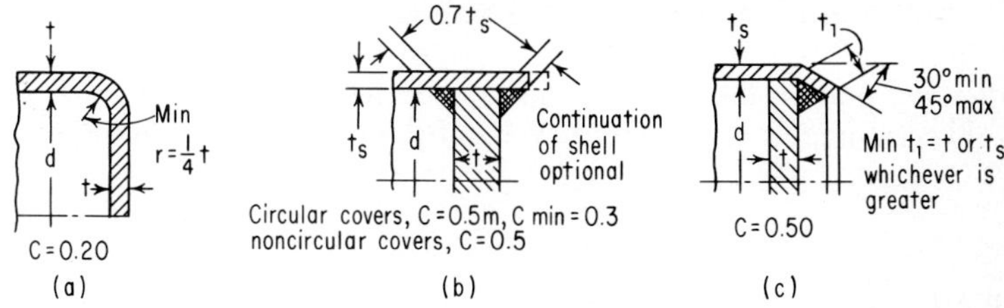

Fig. 8.20 Some code-approved flatheads and covers with *C* factors to be applied in calculations.

Q If a dished head is not designed according to code rules and is found to be too thin, must the head be replaced?
A Not if it is stayed for additional strength, provided other requirements of the code are met.

Q What formulas are used for the various heads?
A For *segments of a sphere head* without an access hole use the formula

$$t = \frac{5\,PL}{4.8\,SE}$$

where S = maximum allowable stress, lb/in^2
t = thickness of head, in
P = maximum allowed pressure, psi
L = radius on concave side to which head is dished, in
E = efficiency of weakest joint, but *not* head-to-shell joint

For *semiellipsoidal heads* with pressure on the concave side (without an access hole), use the shell formula for cylinders, but assume that the shell is seamless (no joint efficiency). For *hemispherical heads* with no access hole and pressure on the concave side, use the following formula:

$$t = \frac{PL}{1.6\,SE} \quad (1)$$

or $$t = \frac{PL}{2SE - 0.2P} \quad (2)$$

where t = required thickness, in
P = maximum allowable pressure, psi
S = maximum allowable stress, lb/in^2
L = radius to which head was formed, in
E = efficiency of weakest joint, including head-to-shell joint

Equation 1 is for heads up to ½ in thick. Equation 2 is for heads over ½ in integrally formed with the shell, or welded head-to-shell joints, provided that all the welding meets code requirements, including the weld reinforcement being removed substantially flush with the plate.

HEADS WITH ACCESS HOLES

Q Are the same equations used for dished heads with flanged-in access holes?

A When any of the heads, either segment of a sphere, semiellipsoidal, or hemispherical, has a flanged-in access hole or an access opening that exceeds 6 in in any dimension, it is computed on the following basis:

1. By the formula for a segment of a sphere head.
2. The thickness of the head must be increased by 15 percent but in no case less than ⅛ in after the thickness is obtained by the formula.
3. If the radius to which a head is dished is less than 80 percent of the diameter of the shell, the thickness of the head with a flanged-in access-hole opening must be found (or calculated) by making the dish radius equal to 80 percent of the diameter of the shell.

Q To what percentage may the knuckle of a dished head be thinned in forming?

A Not over 10 percent.

Q What formula is used for calculating *flatheads*?

A Consult the code, because there are several variations, depending on whether the flathead is round, square, rectangular, etc. The typical round flathead is calculated by this formula:

$$t = d\sqrt{\frac{CP}{S}}$$

where t = minimum required thickness, in
d = diameter, measured as indicated in code
C = a factor, depending on method of attachment (Fig. 8.20)
S = maximum allowable stress value, lb/in^2
P = maximum allowable pressure, psi

HEAD CALCULATION EXAMPLES

Q Illustrate some typical head calculations.

A EXAMPLE: A blank, unstayed dished head (segment of a sphere) with a flanged-in access hole and with pressure on the concave side is ⅞ in thick. The radius of the dished head is 36 in, the material is SA-285C, the temperature is under 650°F. What is the allowable pressure on this head?

Using the following formula, with $S = 13{,}750$, $L = 36$, $E = 1$, $t = 0.875/1.15 = 0.761$ (thickness must be reduced because of access hole),

$$t = \frac{5PL}{4.8SE}$$

$$0.761 = \frac{5(P)(36)}{4.8(13{,}750)(1)}$$

$$P = \frac{(0.761)(4.8)(13{,}750)}{(5)(36)}$$

$$= 279 \text{ psi} \qquad \textit{Ans.}$$

EXAMPLE: A 50-in-diameter boiler drum has dished heads with pressure on the concave side, butt-welded to the shell with a head joint efficiency of 90 percent. Working pressure is 420 psi; material used has an allowable stress of 17,500 lb/in^2 with working temperatures below 650°F. Calculate allowable minimum thickness of (*a*) a dished head of the semiellipsoidal type and (*b*) a dished head of the full hemispherical type with a dish radius of 25 in.

(*a*) Semiellipsoidal type requires using shell equation 2 with seamless construction, $C = 0$, $P = 420$, $S = 17{,}500$, $y = 0.4$, $R = 25$. Substituting in shell equation 2,

$$t = \frac{420(25)}{17{,}500(1) - (1 - 0.4)420}$$

$$t = 0.609 \text{ in} \qquad \textit{Ans.}$$

(*b*) Full hemispherical type requires using the first equation shown for this head since the 90 percent efficiency indicates weld reinforcement was not removed substantially flush. Substituting,

$$t = \frac{420(25)}{1.6(17{,}500)(0.9)}$$

$$t = 0.417 \text{ in} \qquad \textit{Ans.}$$

Q How is the dish radius calculated if the chord length and the bump are known? Assume chord length of 38 in and bump of 4 in.

A Refer to Fig. 8.21 and triangle *ABO*. By trigonometric rules, the

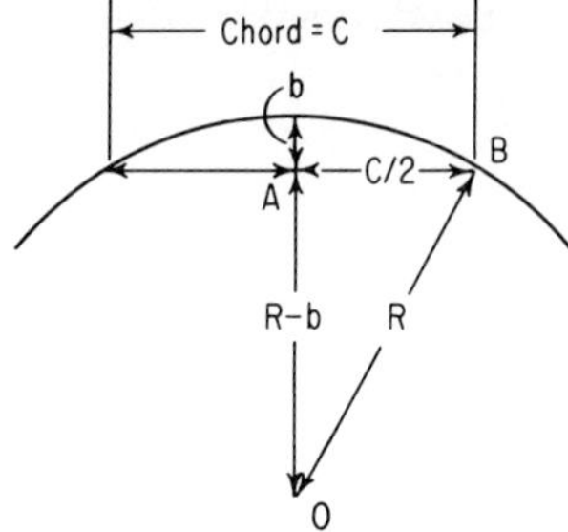

Fig. 8.21 Dish radius of head can be calculated if chord length and depth of bump are known. C = chord length; b = bump of head; R = dished radius.

following calculations are made:

$$R = \text{hypotenuse of right triangle}$$

$$R^2 = \left(\frac{C}{2}\right)^2 + (R - b)^2$$

$$2Rb = \frac{C^2}{4} + b^2$$

$$R = \frac{C^2}{8b} + \frac{b}{2}$$

Substituting,

$$R = \frac{(38)^2}{8(4)} + \frac{4}{2}$$

$$= 45.1 + 2$$

$$= 47.1 \text{ in} \qquad \textit{Ans.}$$

FLATHEAD CALCULATION

EXAMPLE: An unstayed flathead is attached as in Fig. 8.20*b*. All welding meets code requirements. The head is circular with a 16-in diameter and thickness of 1½ in, the material is SA-285-C, and the temperature is under 650°F. The shell to which the head is attached is ⅜ in, and the required shell thickness is $^5/_{16}$ in. What is the allowable pressure on this flathead?

Using the formula below and with $S = 13{,}750$, $d = 16$, $t = 1.5$, $C = 0.5m$, $m = 0.3125/0.375 = 0.833$, $C = 0.5(0.833) = 0.4165$ or 0.417,

$$t = d\sqrt{\frac{CP}{S}}$$

$$1.5 = 16\sqrt{\frac{0.417P}{13{,}750}}$$

$$\left(\frac{1.5}{16}\right)^2 = \frac{0.417P}{13{,}750}$$

$$P = \left(\frac{1.5}{16}\right)^2 \frac{13.750}{0.417}$$

$$= 289 \text{ psi} \qquad \textit{Ans.}$$

STAYED SURFACES

Q Explain the method of calculating stayed surfaces on code boilers.
A The code has various rules on pitch, thickness of plate for the pitch, and factors to be used for various constructions. Two fundamentals should be understood. Figure 8.22 shows pressure acting on a square-pitch flat plate, with a stay bolt in the middle to brace the plate. The force acting on the plate from pressure is the pitch squared minus the stay bolt area times the pressure.

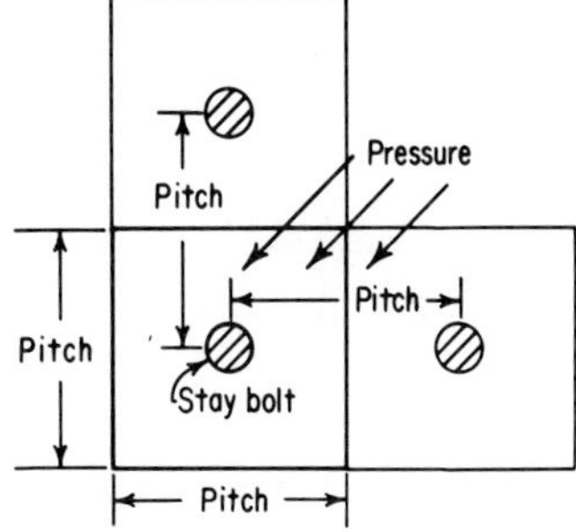

Fig. 8.22 Pressure on net area to be stayed creates a force that must be resisted by stay bolt.

In equation form, we have

$$(p^2 - a)P = F$$

where p = pitch, in
P = pressure, psi
a = stay bolt area, in^2
F = force on plate, lb

This force is resisted by the strength of the stay bolt. Assume that the stay bolt is threaded with a telltale hole of $\frac{3}{16}$-in diameter (the area of a $\frac{3}{16}$-in-diameter hole is 0.027 in^2) as shown in Fig. 8.23*b*. The symbol a is the net area of the stay bolt at the bottom of the threads. The strength of the stay bolt is then the net sectional area of the stay bolt times the allowable stress, or

$$(a - 0.027)S$$

where S equals the allowable stress on the stay bolt.

For equilibrium to exist, obviously the two forces must be equal, or

$$(p^2 - a)P = \frac{(a - 0.027)}{1.1} S$$

The 1.1 factor is required by the code. The net area obtained by dividing the load on a stay bolt by the allowable stress must be multiplied by 1.1.

In addition to the strength of the stay bolt, the strength of the plate between the stay bolts must be adequate, or the plate might buckle between the stay bolts. The code requires this to be checked by one of the following equations:

$$t = p\sqrt{\frac{P}{CS}} \qquad \text{or} \qquad P = \frac{t^2CS}{p^2}$$

where t = required thickness of plate, in
p = maximum pitch, in

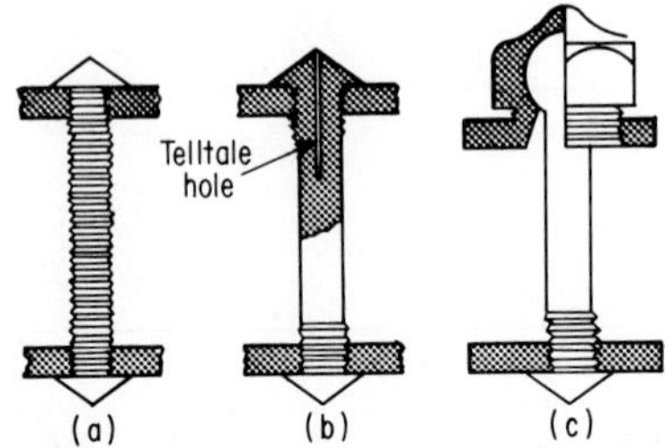

Fig. 8.23 Types of stay bolts—some with telltale hole.

P = maximum allowable pressure, psi
C = factor, depending on construction (2.1 for stays screwed through plates of not over $\frac{7}{16}$ in thickness; 2.2 for stays screwed through plates of over $\frac{7}{16}$ in thickness)
S = maximum allowable stress for plate

NOTE: The code shows other values of C, based on different stay construction.

Q Why are stays necessary in boiler construction?
A Since flat surfaces exposed to pressure tend to bulge outward, they must be supported by stays, as otherwise the flat plate required would be very thick. Because cylindrical or spherical surfaces do not tend to change their shape under pressure, they do not require staying.

STAYED SURFACE CALCULATION EXAMPLES

Q Illustrate some typical code stay bolt calculations.
A EXAMPLE: How many and what size of threaded stay bolts (12 V threads per in) are required to adequately support 374 in^2 of a stayed surface in a code boiler, if the stay bolts are pitched $5\frac{1}{4} \times 5\frac{1}{2}$ in, drilled with telltale holes of $\frac{3}{16}$ in (area of hole 0.027 in^2), and the working pressure is 150 psi?

$$\text{Number of stay bolts required} = \frac{374}{5.25 \times 5.5} = 13 \qquad \textit{Ans.}$$

Let a = net area of stay bolt; then

$$(5.25 \times 5.5) - a = \text{net area pressure is acting on}$$

or $$28.875 - a = \text{net area pressure is acting on}$$

$$\text{Resisting force of stay bolt} = \frac{a - 0.27}{1.1}(7500)$$

where 7500 = allowable stress on stay bolt.

Equating forces, we have

$$(28.875 - a)150 = \frac{7500(a - 0.027)}{1.1}$$

$$4331.25 - 150a = 6818.2a - 184.1$$
$$4515.35 = 6968.2a$$
$$a = 0.648 \text{ in}^2 \qquad \textit{Ans.}$$

Nearest standard stay bolt is a $1\frac{1}{16}$-in-diameter stay bolt with root area of 0.662 in^2.

EXAMPLE: The stay bolts in a firebox of a locomotive-type boiler are spaced 7 in horizontally and 6½ in vertically. The diameter of the 12 V threaded stay bolt is 1½ in, while the plate thickness is ½ in. If the telltale hole area is 0.027 in^2, what is the allowable pressure on this stayed area? Allowable stress on the plate is 15,000 lb/in^2 and on the stay bolt 7500 lb/in^2. Two calculations are required, one to check on the strength of the plate to resist buckling and the other to check on the strength of the stay bolt to resist the force on the plate caused by pressure. The check on the plate is really a check on whether the pitch of the stay bolts is not too far apart.

1. Plate calculations, use

$$P = \frac{t^2CS}{P^2}$$

$$P = \frac{(0.5)^2(2.2)15{,}000}{7 \times 6\ 1/2}$$

$$P = 181.3 \text{ psi}$$

where $t = 0.5$ in, $p^2 = 7 \times 6½ = 45.5$ in^2, and $C = 2.2$ (by code).

2. Stay bolt calculations with root area of threaded stay bolt being 0.960 in^2,

$$(45.5 - 0.960)P = \frac{(0.960 - 0.027)7500}{1.1}$$

$$1.1(44.5)P = 6997.5$$

$$P = 142.3 \text{ psi} \qquad \textit{Ans.}$$

Thus, the allowable pressure on this stayed surface is 142.3 psi *Ans.*

DIAGONAL STAYS

Q Sketch and describe a diagonal stay and explain where it is used.
A To stay the flat portions of heads that are not supported by tubes, diagonal stays are used above the tubes. This stay is not as direct as the through stay, and it throws stress on the shell plates as well. But the diagonal stay leaves more room above the tubes for inspection, repair, and cleaning. A common form of diagonal stay is shown in Fig. 8.24. Modern units are welded.

Q Explain how diagonal stays are installed.
A When the crowfoot is against the head, the holes in the shell should be so placed that the holes in the palm of the stay are about $\frac{1}{32}$ in shy of

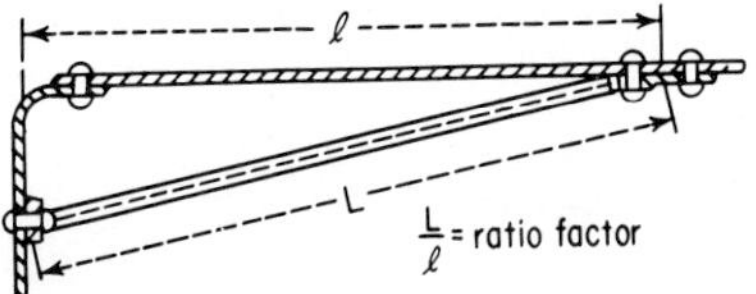

Fig. 8.24 If the *L/l* ratio is 1.15 or less, the holding power of a diagonal stay is calculated as a straight stay. Welding is used instead of rivets in modern boilers.

lining up. Often, the crowfoot is bolted to the head, and the shell or palm holes are marked off and drilled to meet this requirement. The crowfoot is then riveted to the head. The stay is elongated so that the shell and palm holes line up. This effect may be accomplished by heating, but a driftpin is often sledged in one hole instead. After the palm is riveted in position, the stay becomes in tension to support the head. Diagonal stays are installed before the tube is in place. The tube sheet must be held by a strong back to prevent buckling, until the tubes are inserted. Figures 8.25 and 8.26 illustrate diagonal- and through-stay details on an HRT boiler.

Q How are diagonal stays calculated for holding power to resist the force caused by pressure?

A Three important factors to consider on diagonal stays are:

1. What is the slant of the diagonal or its angle to the flat surface being supported?

2. Is it welded or riveted to the shell and head?

3. What is the construction on the ends of the stay where it is fastened to the shell or head: riveted, pins, split palms, or blades (crowfoot type)? See Fig. 8.27.

The code permits most diagonal stays to be calculated as straight stays similar to the stay-bolt method. This method calls for multiplying pressure times area on one side, with the holding power of the stay on

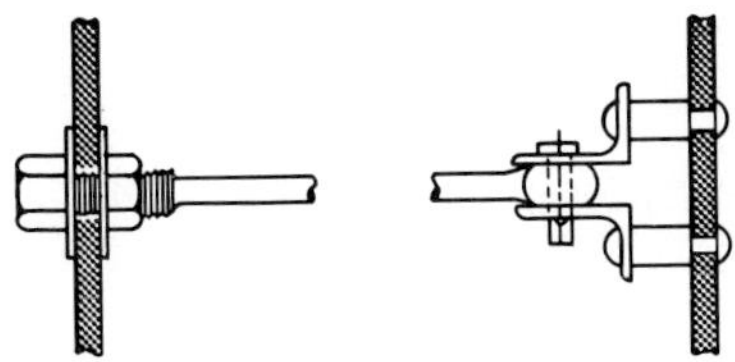

Fig. 8.25 Through stay end-attachment details.

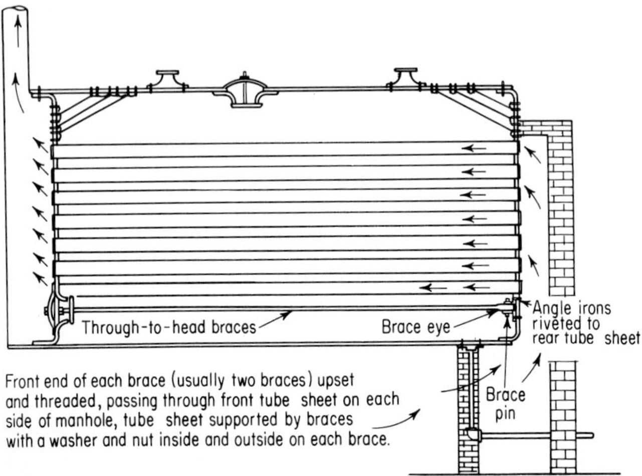

Fig. 8.26 Diagonal and through stays are used to strengthen the flat tube-sheet area without tubes in an HRT boiler.

the other side of the equation. For example, in Fig. 8.24, if the ratio of L/l is 1.15 or less (on an HRT boiler), the body of the stay is calculated as a straight stay. But the allowable stress to be used is 90 percent of that allowed for a straight stay. If the ratio of L/l is over 1.15, the body of the stay is calculated by increasing the area required on the body of the stay by the ratio of L/l. In equation form, this is expressed as follows:

$$A = \frac{aL}{l}$$

where a = cross-sectional area of direct stay body
A = cross-sectional area of diagonal stay body
l = length of right angles to area to be supported (see Fig. 8.24)
L = diagonal length of stay

The code rules on palms that are riveted on diagonal stays require the cross-sectional area of this part of the stay to be at least 25 percent greater than the body of the stay.

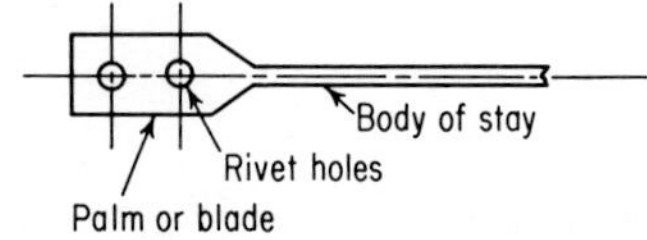

Fig. 8.27 Elements of a riveted diagonal stay. Modern units are welded.

Q Illustrate some typical code calculations on diagonal stays.

A EXAMPLE: The area of a segment to be stayed is 504 in^2. This area is supported by seven 1¼-in-diameter round diagonal braces (the area of the brace is 1.227 in^2). The length of the brace is less than 1.15 times the length of a direct stay. What pressure is allowed on the segment if the stays are welded and allowable stress is 6000 lb/in^2?

Deducting area of stays method, with allowable stress on welded stys being 6000, the area stayed by one stay = 504/7 = 72 in^2.

$$(72 - 1.227) \times P = 1.227 \times 6000 \times 0.9$$
$$70.773 \times P = 6625.3$$
$$P = 92.2 \text{ psi} \qquad \textit{Ans.}$$

EXAMPLE: The area to be stayed on the front tube sheet of an LFT boiler is 136 in^2. It is braced with two diagonal stays, weldless type, where L = 29¼ in, l = 28⅝ in. What diameter of brace is required to carry 165 psi? (Do not deduct the area of the stays.)

Allowable stress is 9500 lb/in^2.

$$136 \times P = \text{area of stays } (2a) \times \text{allowable stress}$$
$$136 \times 165 = 0.9 \times 9500 \times 2a$$

$$a = \frac{22{,}440}{17{,}100} = 1.312 \text{ in}^2 \qquad \text{(use } 1\tfrac{5}{16}\text{-in-diameter round stays)}$$

NOTE: L/l is less than 1.15.

Q How is the area to be stayed calculated on an HRT boiler for the part of the tube sheet above the tubes?

A This area is illustrated in Fig. 8.28*a* and *b*. Two equations can be used. Let A_s = area to be stayed.

1. For flanged heads:

$$A_s, \text{in}^2 = \frac{4(H - d - 2)^2}{3} = \sqrt{\frac{2(R - d)}{(H - d - 2)} - 0.608}$$

2. For unflanged heads:

$$A_s, \text{in}^2 = \frac{4(H - 2)^2}{3} = \sqrt{\frac{2R}{(H - 2)} - 0.608}$$

where H = distance from tubes to shell, in

d = outer radius of flange, not exceeding 8 times thickness of head, in

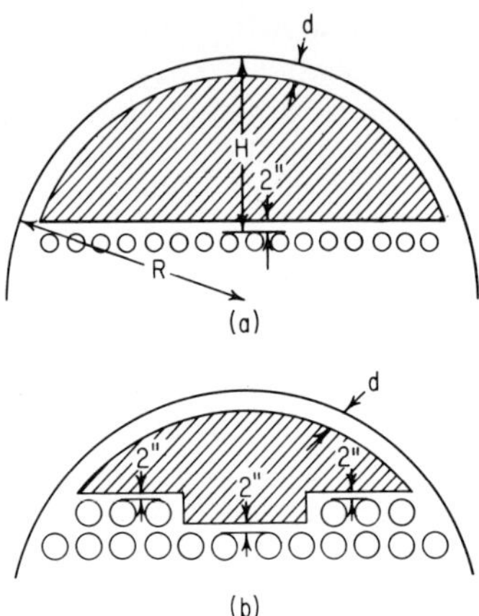

Fig. 8.28 Shaded area shows area of head that requires staying. (*a*) Regular head or tube sheet; (*b*) irregular head.

$$\text{or} \quad d = \frac{80t}{\sqrt{P}} \qquad \text{(use largest value of } d\text{)}$$

or $d = 0$ for unflanged heads
t = thickness of head, in
P = maximum allowable pressure, psi
R = radius of boiler head, in

NOTE: If $d = 3$ in, the code has a table for the area to be stayed for various lengths of H.

EXAMPLE: (a) A 66-in HRT boiler is to be built for 140-psi working pressure. The flanged heads are $\frac{9}{16}$ in thick. The distance from the upper tubes to the shell is 24 in, and $d = 3$ in (Fig. 8.28*a*). What is the area to be stayed?

(b) If this head is to be stayed by $1\frac{1}{4}$-in-diameter diagonal braces (weldless), how many braces will be required when L does not exceed l more than 1.15 times and 9500 lb/in^2 at cross-sectional area is allowed for a straight brace?

(*a*) The flanged-head equation applies:

$$A_s = \frac{4(H - d - 2)^2}{3}\sqrt{\frac{2(R - d)}{H - d - 2} - 0.608}$$

where $H = 24$, $d = 3$, $R = 33$.

Substituting,

$$A_s = \frac{4(24 - 3 - 2)^2}{3} \sqrt{\frac{2(33 - 3)}{24 - 3 - 2} - 0.608}$$

$$A_s = 481.3 \sqrt{2.550}$$
$$A_s = 768 \text{ in}^2 \qquad \textit{Ans.}$$

(*b*) Use the stay equation and substitute as follows:

$$768(140) = n(1.2272)(9500)(0.9)$$
$$n = 10.2$$

where n = number of stays.
Use 11 braces. ***Ans.***

Q Sketch and describe a girder stay.
A The girder stay (Fig. 8.29) was formerly used very extensively to support flat crown sheets in LFB units. But it has been largely superseded by the radial stay (Fig. 8.30) for this purpose. It is still used to support the tops of combustion chambers in boilers of the scotch marine (SM) type. The girder stay consists of a cast-steel or built-up girder with its ends resting on the side, or end sheets, of the firebox or combustion chamber. It supports the flat crown sheet (the top of the combustion chamber) by means of bolts.

Q If the girder stay (Fig. 8.29) has a span of 40 in from tube sheet to back connection plate, the bolts in each girder are pitched 5 in, and the center distance from girder to girder is 7½ in, what is the allowable pressure if the depth of the girder is 7½ in and the thickness of the girder is 2 in?
A Use the following equation:

$$P = \frac{Cd^2t}{(W - p)D_1W}$$

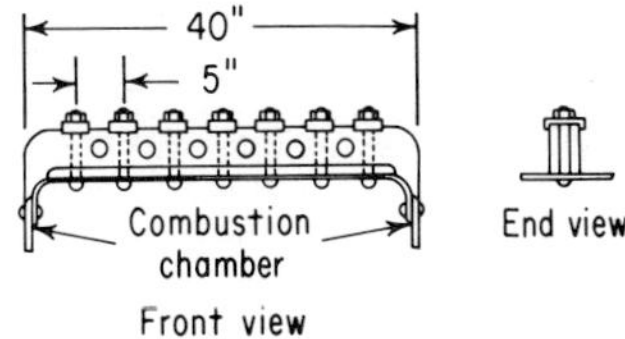

Fig. 8.29 Girder stays were used to stay crown sheets of locomotive and scotch marine boilers.

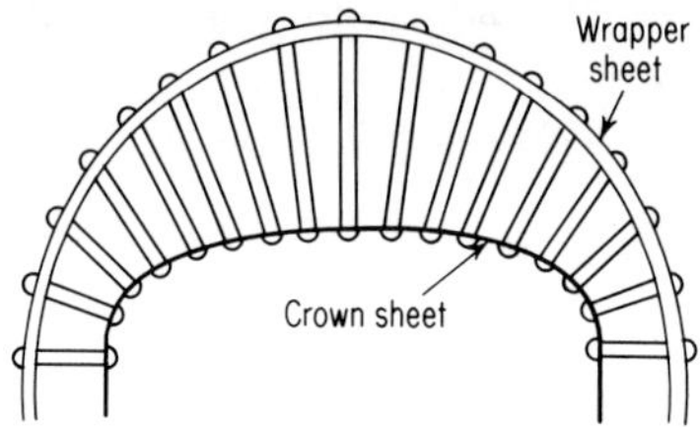

Fig. 8.30 Radial stays have replaced girder stays in bracing crown sheets.

where, with seven supporting bolts, as shown in Fig. 8.29,

C = 11,500 (see code)
d = depth of girder, 7½ in
t = thickness of girder, 2 in
W = distance from tube-sheet support to back plate, 40 in
D_1 = distance from center of girders, 7½ in
p = pitch of supporting bolts, 5 in

Then

$$P = \frac{11{,}500(7.5)^2(2)}{(40 - 5)(7.5)(40)}$$

$$= 123 \text{ psi} \qquad \textit{Ans.}$$

Q Does not the girder stay resting on the tube sheet affect the tube sheet's strength?

A Yes. This must be calculated according to the code, using the equation

$$P = \frac{27{,}000t(D - d)}{WD}$$

where P = maximum allowable working pressure, psi
t = thickness of tube plate, in
d = inside diameter of tube, in
W = distance from tube sheet to opposite combustion chamber sheet, in
D = least horizontal distance between tube centers on horizontal row, in

This equation applies only to tube sheets where the crown sheet is not supported from the shell of the boiler, which is the case with girder stays.

Q Give an example by using the above equation.

A What pressure is allowed on the rear tube sheet of an SM boiler where the crown is not stayed to the shell, the tubes are 6 in center-to-center horizontally, 4 in OD, 0.16 in thick, the tube sheet is $\frac{7}{16}$ in thick, and the depth of the combustion chamber is 30 in?

Substituting, with $d = 4 - 0.32, = 3.68$, $D = 6$, $W = 30$, $t = 0.4375$,

$$P = \frac{27{,}000(0.4375)(6 - 3.68)}{30(6)}$$

$$= 165 \text{ psi} \qquad \textit{Ans.}$$

Q Sketch and describe a gusset stay.

A This is a form of diagonal stay (Fig. 8.31) in which a plate is used instead of a bar. It consists of a heavy plate fastened by welding (or rivets) and angle bars to the head and shell. It is more rigid than the diagonal stay, takes up more room, and interferes to a greater extent with water circulation. Gusset stays are used very little in modern boiler construction. The gusset plate requires a cross-sectional area 10 percent greater than a typical diagonal stay.

Q How are curved locomotive-boiler crown sheets usually supported?

A By long threaded rods called *radial stays* (Fig. 8.30). These rods are screwed through both crown sheet and wrapper sheet and the ends are riveted over.

Q What are the advantages and disadvantages of radial stays and girder stays?

A The radial stays are more flexible and tend to hold less scale from circulation than do girders. About the only advantage of the girder stays is that they pass straight through the sheet rather than at an angle.

COMBUSTION CHAMBERS

Q Why must furnace and combustion chambers be calculated for strength against pressure since only the products of combustion flow against them?

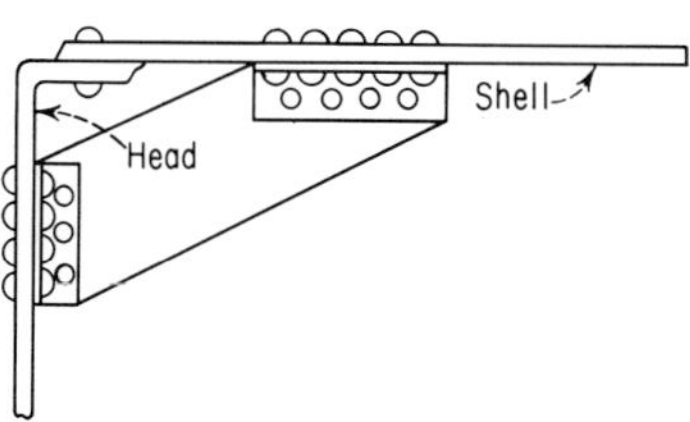

Fig. 8.31 Old gusset stay.

A There is pressure acting on the furnace from the waterside or steam side which might collapse the furnace inward.

Q Name four ways of supporting a circular furnace subjected to external pressure.
A The circular furnace may be (1) self-supporting, (2) stay-bolted, (3) corrugated, or (4) equipped with Adamson rings.

Q Give some typical code calculations for combustion chambers.
A EXAMPLE: What pressure is allowed on a seamless flue 16 in ID, thickness $^{13}/_{32}$ in?

Where the thickness of the wall is greater than 0.023 times the diameter, use the following (see ASME code):

$$P = \frac{17{,}300t}{D} - 275$$

where P = maximum allowable working pressure, psi
D = outside diameter of flue, in
t = thickness of wall of flue, in

$$\text{OD} = 16 + {}^{13}/_{16} = 16.8125$$
$$16.8125(0.023) = 0.387$$
$^{13}/_{32} = 0.40625$; so the formula applies

$$P = \frac{17{,}300(0.40625)}{16.8125} - 275$$
$$= 418 - 275 = 143 \text{ psi} \qquad \textit{Ans.}$$

EXAMPLE: An SM dry-back boiler has a cylindrical furnace which is 10 ft between rivet seams and has an Adamson ring midway between its length; OD is 42 in, pressure is 125 psi. What thickness of furnace is required?

Where the length does not exceed 120 times the thickness of the plate, use the following (see ASME code):

$$P = \frac{57.6(300t - 1.03L)}{D}$$

where P = maximum allowable working pressure, psi
D = OD of furnace, in
L = length of furnace section, in
t = thickness of furnace wall, in

$$125 = \frac{57.6(300t - 1.03 \times 60)}{42}$$

$$\frac{125(42)}{57.6} = 300t - 61.8$$

$$\frac{5250}{57.6} + 61.8 = 300t$$

$$\frac{152.946}{300} = t$$

$$t = 0.51 \text{ in} \qquad \textit{Ans.}$$

OPENINGS

Q Are openings in boiler shells of importance in the inspection of, and calculations on, boilers?

A Yes, because many failures start with cracks around openings. In cutting an access hole in the shell of boiler, it is necessary to compensate for the metal removed. This is done by installing an access-hole frame if needed.

The minimum-size elliptical access hole permitted by the ASME code is 11 × 15 in. In cutting the shell for a frame having an opening of this size, the shorter dimension is placed along the longitudinal axis of the boiler so that less frame material will be required for replacement in this weaker directional axis.

In considering a cross-sectional plane of the boiler shell plate in the vicinity of an access-hole "cutout," it is necessary to find the total area of metal removed, including rivet holes, and to provide an access-hole frame having an equal cross-sectional area in the same plane if the shell does not have excess thickness.

Q How does the code permit calculating the weakening effect of cutting a hole in a shell or header?

A It is possible to use stress analysis around an opening and determine the stress-concentration factor to be applied to the normal calculated stresses, and this is done by designers on critical components. The code has developed a compensating method around openings, which so far has proved satisfactory in service. Essentially, it requires checking if the cross-sectional area removed by a finished opening has been replaced by (1) extra thickness in shell or nozzle inserted into the opening or (2) the addition of a reinforcing plate around the opening.

For the 1 condition, the code permits any extra thickness along the shell wall for a distance of $2d$ or $d + 2x$, the combined thickness of the shell and nozzle, whichever is greater. This length is multiplied by the extra thickness to see how much area is available from the shell wall. A similar procedure is followed for the nozzle, except that the length along the nozzle wall is the smaller of 2½ times the shell or nozzle thickness.

The Power Boiler Code has detailed requirements on openings cut into shells or headers and how to calculate if reinforcement around the opening is necessary. The following procedure generally applies. See Fig. 8.32. The area required to be restored by the finished opening d is

$$A = d \times t_r \times F$$

where d = diameter of finished opening in given plane, in
t_r = required thickness of seamless shell for the pressure
F = factor that considers axis of the nozzle, usually 1.00

To determine the metal available from the *shell*,

$$A_1 = (E_1 t_s - F t_r)d \qquad \text{or} \qquad A_1 = 2(E_1 t_s - F t_r)(t_s + t_n)$$

The larger value is used.

To determine the metal available from the *nozzle*,

$$A_2 = (t_n - t_{rn})5t_s \qquad \text{or} \qquad A_2 = (t_n - t_{rn})(5t_n + t_e)$$

where t_s = actual thickness of shell
t_r = required thickness of a seamless shell
t_n = actual thickness of nozzle attached to shell
t_{rn} = required thickness of seamless nozzle

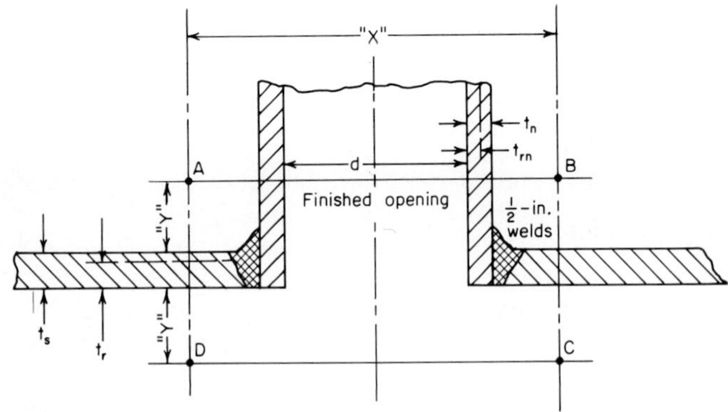

Fig. 8.32 Area to be considered within reinforcement credit must be within rectangle *ABCD*. Along shell wall, this extends out to larger of $2d$ or $d + 2(t_s + t_n)$ shown by *X*. Limit along nozzle wall, or *Y* distances shown, is 2 1/2 t_s or 2 1/2 t_n, whichever is smaller. Weld and reinforcement pad in this area can be calculated as reinforcement. t_s = actual shell thickness; t_r = required shell thickness; t_n = actual nozzle thickness; t_{rn} = required nozzle thickness.

E_1 = longitudinal joint efficiency of weld if opening is through the longitudinal welded joint; otherwise E_1 is 1

Use the smaller value. Note the equations given are for the excess area available along the shell and nozzle walls as previously described.

The t_e factor is for the thickness of a reinforcing pad that may be added as opening reinforcement.

An example will illustrate a typical reinforcement calculation. A 5-in extra heavy pipe nozzle SA-53B is welded to a shell similar to that shown in Fig. 8.32 (½-in welds). The shell has an inside diameter of 30 in, a thickness of $^7/_{16}$ in, and a working pressure of 200 psi. The material is SA-285C. Assume that all welds are in accordance with the code and that the welds are strong enough for this installation. The outside diameter of the 5-in pipe is 5.563 in, and the thickness is 0.375 in. With an allowable stress of 13,750 for the shell and nozzle material, does the opening shown in Fig. 8.32 meet code requirements? The required shell thickness is

$$t_r = \frac{200(15.0)}{13{,}750(1.0) - 0.6(200)}$$

$$t_r = 0.220 \text{ in}$$

The nozzle required thickness is

$$t_{rn} = \frac{200(2.407)}{13{,}750(1.0) - 0.6(200)}$$

$$t_{rn} = 0.035 \text{ in}$$

And so the area of reinforcement required is

$$\begin{aligned} A &= d \times t_{rs} \times F \\ &= 4.813(0.22)(1.0) = 1.059 \text{ in}^2 \end{aligned}$$

The area of reinforcement provided, for $E_1 = 1.00$, is

$$\begin{aligned} &A_1 \text{ (shell)} = [1.0(0.438) - 1.0(0.220)]4.813 = 1.049 \text{ in}^2 \\ &A_2 \text{ (nozzle)} = 2.(2.5)(0.375)(0.375 - 0.035) = 0.638 \text{ in}^2 \\ &A_3 \text{ (welds)} = 1(0.50)^2 = 0.250 \text{ in}^2 \\ &\qquad \text{Total} \quad 1.937 \text{ in}^2 \end{aligned}$$

Construction meets code requirements because the area of reinforcement provided exceeds the area removed.

These examples of minimum code strength calculations should make it apparent that any repairs to boilers and pressure vessels must include a review of the code so that the boiler or pressure vessel retains its original strength for safe and continuous service. An authorized inspector should therefore be consulted before any major repairs involving the structural integrity of the boiler are made.

Figure 8.33 shows a code-authorized inspector reviewing a boiler being erected. Note round access hole in head was flanged.

Q Where are stampings usually located to indicate a boiler was built to code requirements?

A An identifying stamp on a power boiler is required by states and municipalities. The National Board standard form of stamping is sufficient for the boiler to pass rules and regulations for construction in practically all sections of the United States. This stamp consists of the ASME symbol above the manufacturer's serial number, the manufacturer's name or approved abbreviation, the maximum pressure for which the boiler was built, the water heating surface in square feet, and the year built.

"National Board" followed by a serial number stands for the National Board of Boiler and Pressure Vessel Inspectors. It is an enforcement body for the ASME Power Boiler Code, and a copy of the manufacturer's constructional data is filed with this board under the serial number. The National Board stamp indicates that the boiler is of ASME standard construction and that its construction was followed in the shop by a qualified inspector.

Fig. 8.33 Drum with round flanged manhole in head. *(Courtesy of E. Keeler Co.)*

On horizontal fire-tube boilers (externally fired), the stamping should be in the middle of the front tube sheet, above the top row of tubes. On horizontal fire-tube boilers of the firebox type, the stamping should be located above the center or right-hand furnace or above a handhole at the furnace end. On vertical fire-tube boilers, the stamping should be located over the furnace door. Water-tube boilers have the stamping on the drum heads above the access hole flange.

It is very important when purchase or relocation of a boiler is contemplated to ascertain the boiler laws of the state. Many states require filing a special form and following a definite procedure before relocating a boiler.

CHAPTER **NINE**

Boiler Safety Appurtenances, Connections, and Auxiliaries

Safety appurtenances on boilers include safety valves, low-water fuel cutoffs, water gages, and similar devices stipulated by code requirements for safe operation of the equipment.

Boiler auxiliary systems include feed pumps, forced-draft and induced-draft fans, precipitators, and similar equipment that is necessary to provide feedwater to the boiler, combustion air for the burners, and removal equipment to dispose of the flue gases developed in the combustion process.

Fuels, burners, and combustion controls merit separate chapters. This chapter will concentrate on the water and steam safety appliances and auxiliary equipment involved with this loop to the boiler.

Q What are the minimum appliances or appurtenances necessary for safe boiler operation?
A The pressure gage and test connection, safety valve, blowdown valve, gage glass, gage cocks, stop valve in the steam line, and the stop and check valves in the feed line.

Q What are boiler fittings?
A They are valves, gages, and other connections or devices attached directly to the boiler which are necessary for safe and efficient operation.

Q Name the most important boiler appliance and its function.
A The safety valve (SV). It prevents the boiler pressure from going above a safe predetermined pressure by opening to allow the excess steam to escape into the atmosphere when the set point is reached. This guards against a possible explosion from excessive pressure. Each boiler must have at least one SV, the minimum size being a ½-in valve for miniature boilers, a ¾-in valve for others. A boiler with more than 500 ft^2 of heating surface and an electric boiler with a power input over 500

kW should have two or more SVs. In any case, SV capacity must be such as to discharge all the steam the boiler can generate without allowing pressure to rise more than 6 percent above the highest pressure at which any valve is set, and in no case more than 6 percent above the maximum allowable working pressure. All SVs used on steam boilers must be of the direct spring-loaded pop type.

Q Name some causes of overpressure on automatically fired boilers.
A Many causes of control malfunction can lead to overpressure by a runaway firing condition. Often the safety relief valve is the last means of preventing a dangerous overpressure. Some causes are:

1. Fused contact points on electrically operated pressure cutout switches. Then the burner keeps operating because the current to the burner motor cannot be interrupted.
2. Shorted wires to the pressure switch, thus bypassing the on-off feature of the pressure switch, permitting the burner to keep operating. A similar cause is a loose terminal connection on the electrically operated pressure switches, which might complete a circuit by shifting to the other terminal of the connection, again bypassing the on-off switch.
3. Leaking solenoid gas valves or diaphragm-operated valves on the fuel line to the burner. Thus the burner keeps operating with a resultant pressure increase, if the connected load cannot dissipate the energy input.
4. In many installations, the pressure switch (or thermostat) is electrically connected to a motor controller of the oil-burner motor. Contacts on the controller can fuse, again not permitting the oil-burner motor to stop the oil burner at the set pressure cutout setting.
5. On gas burners, manual bypass lines have been found, permitting the solenoid or diaphragm-operated gas shutoff valve to be bypassed so as to permit firing the boiler under manual control. The unattended boiler (in this condition) has no high-pressure cutout except the SV.
6. Obstructed tubing to pressure sensing switches can block the signaling of pressure to the pressure cutout switch, again permitting the burner to operate with the SV the last means of preventing overpressure.

> EXAMPLE: Consider a cock between the boiler and pressure switch (usually where the pressure switch is not mounted directly on the boiler but in a control cabinet). If the cock is inadvertently closed, the controls will sense no pressure buildup. Again the SV is the last means of preventing overpressure.

7. On closed hot-water-heating systems with only a temperature cutout switch, a water-clogged expansion tank can lead to high pressure

on the boiler and piping. Again the SV is the only protection against overpressure.

8. Low water in a boiler due to faulty feed devices or improper interlocking between feed pumps and burner can lead to overpressure if it is combined with a defective pressure cutout switch.

These examples show why constant checking is needed for proper control functioning on automatically fired boilers.

> REMEMBER: Never assume that the automatic features cover all possible means of failure.

SAFETY VALVES AND RELIEF VALVES

Q What type of SV should be installed on a boiler?
A An ASME- or NB-approved and registered direct spring-loaded pop type, properly marked as to pressure and capacity and equipped with a testing lever. The pressure setting must match either the maximum allowable pressure for which the boiler is designed or, on older boilers, the maximum pressure allowed by state or city law. The capacity of the SV should be at least equal to the maximum steam that can be generated by the boiler. Figure 9.1 gives the marking required on the SV of an HP steam boiler. Similar rules govern high-temperature hot-water (HTHW) boilers and low-pressure steam and HW heating boilers.

Q How can one tell an ASME-approved SV, and how does one secure permission from the ASME to make approved SVs?
A Figure 9.2 shows the official ASME stamp of approval. Permission

Fig. 9.1 ASME markings required on high-pressure steam safety valves.

1. The name or identifying trademark of the manufacturer
2. Manufacturer's design or type number
3. Size, in. Seat diameter, in.
   (The pipe size of the valve inlet)
4. Pressure, lb
   (The steam pressure at which it is to blow)
5. B.D., lb
   (Blowdown—difference between the opening and closing pressure)
6. Capacity, lb per hr
   (In accordance with Pars. in code, and with the valve adjusted for the blowdown given in the preceding item)
7. Capacity lift, in.
   (Capacity lift—distance the valve disk rises under the action of the steam when the valve is blowing under a pressure of 3 percent above the set pressure)

Fig. 9.2 Official ASME symbol to show safety valve conforms to code standard.

to use the symbol is granted by the ASME to any manufacturer complying with the provisions of the code. The manufacturer must agree (upon forms issued by the society) that any SV to which the symbol is applied will be constructed in accordance with the code and that it has the capacity stamped upon the valve under the stated conditions. Manufacturers must also agree that they will not misuse or allow others to use the stamp by which the symbol is applied.

Q Explain the difference between a relief valve and an SV. What is a safety relief valve (SRV)?
A A relief valve is used primarily for liquid service and is an automatic relieving device actuated by the static pressure upstream of the valve, which opens farther with an increase in the pressure over the opening pressure (no pop action).

An SV is used for gas or vapor service and is an automatic pressure-relieving device, actuated by the static pressure upstream of the valve, and which opens with a full pop action once the upstream pressure activates the valve.

An SRV is an automatic pressure-relieving device actuated by the pressure upstream of the valve, and which opens by pop action with further increase in lift of the valve when pressure increases over the popping pressure. It thus combines the feature of pop action and further lift with pressure increase.

Q What is the difference between a spring-loaded SV and a pop SV?
A Both valves are spring-loaded, but the pop type has a lip or slight extension on the disk of the valve which extends beyond the seat surface and provides a huddling chamber (Fig. 9.3). As the valve opens, this huddling chamber is filled with steam, thus building a static pressure on the lip because of the increased disk area exposed. This extra force upward suddenly lifts the disk against the compression spring and causes it to open wide almost instantaneously with a popping noise.

Spring-loaded valves without this pop feature open slowly, and lift is more dependent on pressure increase than sudden complete pop release. This can cause wire drawing of the disk and seat, resulting in leakage. The valve disk is held firmly on its seat by a heavy coil spring.

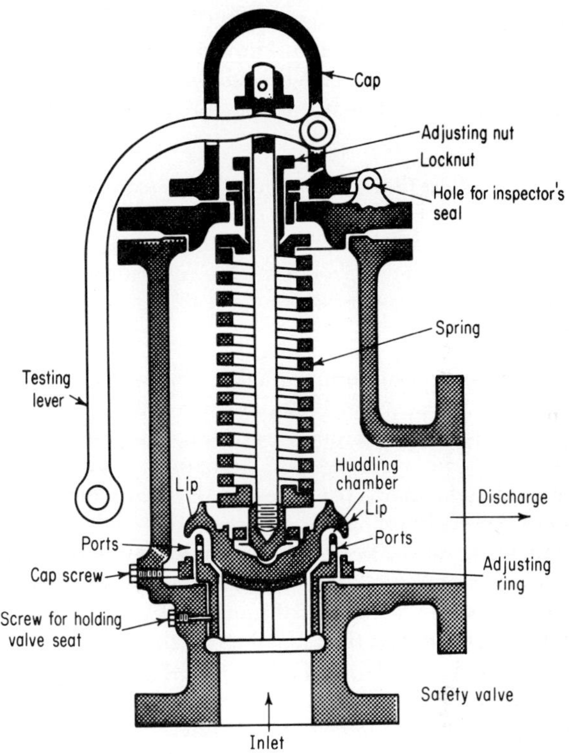

Fig. 9.3 Pop-type safety valve has extended lip over seating surface to provide lifting huddling chamber.

The point at which the valve lifts and relieves the pressure is set by screwing the adjusting nut up or down and so decreasing or increasing the compression of the spring. A maximum adjustment of 10 percent from the stamped pressure is permitted. Once it is set, a lock nut keeps the adjusting nut from moving. The cap on top of the valve may be sealed, if desired, to prevent access to the adjusting nuts. A hand lever lifts the valve from its seat for testing purposes so as to make sure the valve is free. And a blowdown-adjusting arrangement regulates the number of pounds that the valve blows down before it closes.

Q How do the requirements of superheater SVs discharging steam at over 450°F differ from the requirements of those on the boiler drum?

A Superheater SVs should have a flanged connection or a welding-end inlet connection for all sizes. They should be constructed of steel or alloy steel suitable for the maximum temperature. The spring should be exposed outside the valve casing so that it will not come in contact with high-temperature steam (Fig. 9.4). The springs are made of high-speed steel, the disk of Inconel-X, and the seating surface of the bushing of Stellite. For better adjustment, a ball thrust bearing is used on the compression screw of some valves.

Q Do SVs use other than the increased area exposed by the huddling chamber for the major lift?
A Yes, it is known as the reaction principle. The disk of the valve is curved so that when the valve opens, the reversed flow off the curved surfaces provides a reactive force upward, aiding the valve to lift against the spring.

Q What is meant by the term *blowdown* of an SV, and what are the ASME code requirements for blowdown settings?
A Blowdown is the difference between the opening and closing pressure of the SV. For example, an SV that pops at 200 psi and reseats at 195 psi has a blowdown of 5 psi. The ASME rule on blowdown of SVs is as follows: Safety valves shall operate without chattering and shall be set and adjusted to close after blowing down not more than 4 percent of the set pressure, but not less than 2 psi in any case. For a spring-loaded pop SV for pressures between 100 and 300 psi, both inclusive, the blowdown shall be not less than 2 percent of the set pressure. To ensure guaranteed capacity and satisfactory operation, the blowdown as marked upon the valve shall not be reduced.

Safety valves used on forced-circulation boilers of the once-through type may be set and adjusted to close after blowing down not more than 10 percent of the set pressure. The valve for this special use must be so adjusted and marked, and the blowdown adjustment shall be made and sealed by the manufacturer.

Q What popping tolerance is required to be met on SVs?
A The popping point tolerance that a valve must meet is the following on a plus or minus basis:

1. Two psi for pressures up to and including 70 psi
2. Three percent for pressures from 71 to 300 psi
3. Ten psi for pressures over 301 to 1000 psi
4. One percent for pressures over 1000 psi

Q What are the rules on adjusting or replacing the springs which will vary the pressure setting of the SV?
A A spring in an SV or SRV in service for pressures up to and including 250 psi cannot be used for any pressure more than 10 percent below the pressure for which the SV is marked on its nameplate. For pressures above 250 psi, the rule in spring adjustment or changing is 5

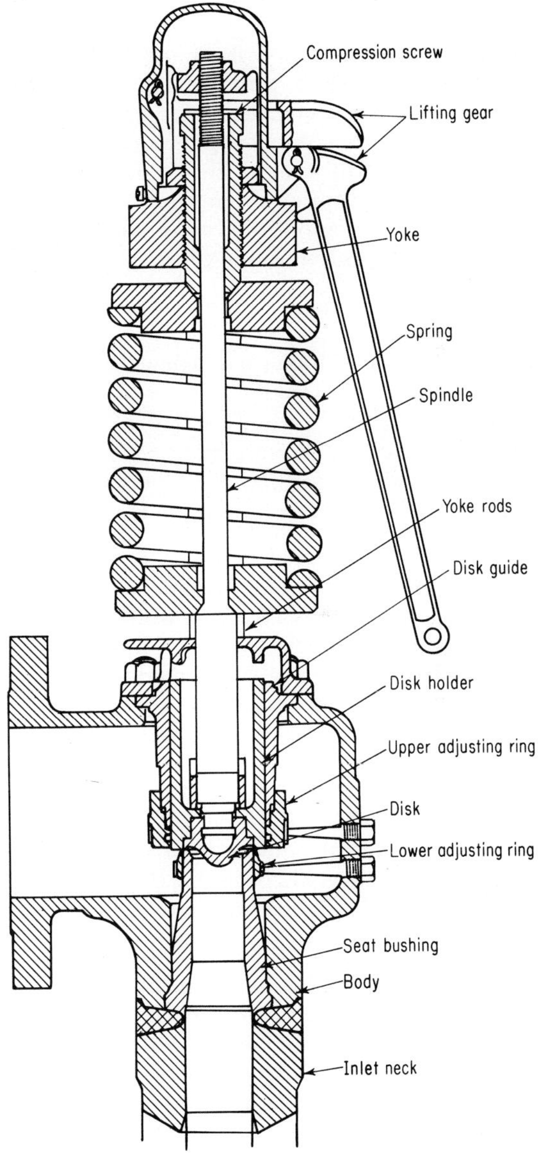

Fig. 9.4 High-pressure, high-temperature safety valve has springs outside the valve casing to avoid contacting steam.

percent above or below the pressure for which the SV is marked; if this limit is exceeded, it cannot be used.

New springs must be installed if any of the above conditions is exceeded, and the valve must be adjusted by the manufacturer of the valve or an authorized representative. Then a new nameplate has to be installed on the SV indicating the new pressure setting.

The main purpose of this rule, so often violated by some operators, is to prevent the coils of the springs from being compressed to the point where the spring cannot act and lift the valve when needed, or at best, where the lift is reduced and, thus, the capacity of the SV.

Q Give the code rules for range of pressure settings for two or more SVs on an HP steam boiler.

A 1. One or more SVs must be set at or below the maximum allowable pressure.

2. The highest pressure setting of any SV cannot exceed the maximum allowable working pressure by more than 3 percent.

3. The range of pressure settings of all the saturated steam SVs on the boiler cannot exceed 10 percent of the highest pressure setting to which any valve is set.

Q Illustrate the above rule with an example.

A A WT boiler has a maximum allowable working pressure of 200 psi and is equipped with three SVs. What is the highest permissible pressure setting? Also, at what pressure would the other valves be set according to code rules? First, the highest setting would be 200 + 0.03(200) = 200 + 6 = 206 psi. One valve would have to be set at the maximum allowable pressure or below the maximum allowable pressure. Let one valve be set at 200 psi. The lowest setting then for the third valve would be 206 − 10 percent of 206 = 206 − 20.6 = 185.4 psi. If we assume a range of 10 percent from the maximum allowable pressure, the lowest setting would be 200 − 10 percent (200) = 180 psi. The other two valves under this condition could be set between 180 and 200 psi.

Q When is more than one SV required on a boiler?

A Each boiler requires at least one SV, but if the heating surface exceeds 500 ft^2, or the boiler is electric with a power input over 500 kW, the boiler must have two or more SRVs. When not more than two valves of different sizes are mounted singly on the boiler, the smaller valve must be not less than 50 percent in relieving capacity of the larger valve.

Q Can SVs on HP boilers be attached to drums or headers by welding?

A Yes, provided the welding is done according to code welding requirements, including those covering such factors as the use of qualified welders, post-weld heat treating, and radiographing.

Q What is the code rule for determining SV relieving capacity?

A The following rules on HP boilers must be followed:

1. The SV capacity on a boiler must be such that the SV (or valves) will discharge all the steam that can be generated by the boiler (this is assumed to be the maximum firing rate) without allowing the pressure to rise more than 6 percent above the highest pressure at which any valve is set, and in no case more than 6 percent above the maximum allowable pressure.

2. The minimum SV relieving capacity, for other than electric boilers, must be determined on the basis of pounds of steam generated per hour per square foot of boiler heating surface and waterwell heating surface, as given in Fig. 9.5. For electric boilers, the relieving capacity is determined by multiplying the kilowatt input by 3½ to obtain the pounds per hour of steam-relieving capacity.

3. For HTHW boilers, the required steam-relieving capacity in pounds per hour is determined by dividing the maximum Btu output (for the fuel being fired) of the boiler by 1000.

Relieving capacity rules for LP steam boilers are:

1. Low-pressure steam boilers require an SV capacity such that with the fuel-burning equipment installed, the pressure cannot rise more than 5 psi above the maximum allowable working pressure of the steam boiler.

2. The minimum relieving capacity is determined by multiplying the heating surface by the values shown in Fig. 9.5, or by dividing by 1000 the maximum Btu (for the fuel being fired) output of the boiler.

Fig. 9.5 Minimum relieving capacity (pounds of steam per hour) per square foot of heating surface for different fuels for FT and WT boilers.

Surface	Fire-tube boilers	Water-tube boilers
Boiler heating surface:		
Hand-fired	5	6
Stoker-fired	7	8
Oil-, gas-, or pulverized-fuel-fired	8	10
Waterwall heating surface:		
Hand-fired	8	8
Stoker-fired	10	12
Oil-, gas-, and pulverized-fuel-fired	14	16

NOTE: When a boiler is fired only by a gas having a heat value not in excess of 200 Btu per cu ft, the minimum safety-valve relieving capacity may be based on the values given for hand-fired boilers above.

Relieving capacity rules for LP hot-water boilers are:

1. The same rule of minimum capacity of relief valves applies to hot-water boilers: The capacity is determined by taking the values shown in Fig. 9.5 times the heating surface or by dividing by 1000 the rated Btu output of the boiler for the fuel being fired.

2. The capacity must be such that the pressure cannot rise more than 20 percent above the highest maximum allowable working pressure up to and including 30 psi, and a 10 percent rise is permitted for pressures over 30 psi.

HEATING SURFACES

Q What area of the boiler should be computed as heating surface?

A That side of the boiler surface exposed to the products of combustion, exclusive of superheating surface. The areas to be considered for this purpose are tubes, fireboxes, shells, tube sheets, and the projected area of headers. For vertical FT steam boilers, compute only the portion of the tube surface up to the middle gage cock.

Q Calculate the heating surface and relieving capacity of the SVs required for an oil-fired, FT boiler of 100 tubes, each 2½ in in diameter, No. 20 gage in thickness, each 15 ft long. The remaining heating surface of fire sheet and tube sheet totals 130 ft^2 at a working pressure of 125 psi. How many and what size valves should be installed?

A Use the ID of the tubes as heating surfaces. Heat transfer is from the inside of the tube through the tube thickness to the waterside. No. 12 gage tube has a wall thickness of 0.105 in, thus the ID of the tube equals $2.5 - (2 \times 0.105) = 2.29$ in.

The area in square feet of all tube equals the circumference times the length times the number of tubes $= \pi(2.29)/(12) \times 15 \times 100 = 897$ ft^2.

$$\text{Total heating surface} = 897 + 130 = 1027 \text{ ft}^2$$

From the table in Fig. 9.5, we see that 8 lb of steam per hour per square foot of heating surface is the minimum pounds per hour relieving capacity required for the SV. Thus, the minimum relieving capacity $= 8 \times 1027 = 8216$ lb/h. Because the boiler has over 500 ft^2 of heating surface, two or more SVs are required. Refer to the ASME power code (also steam tables). It shows that for 125 psi, $v = 3.220$.

$$\text{Use } A = \frac{Hv}{420}$$

where A = area of opening required
H = boiler heating surface, ft^2
v = specific volume of steam, ft^3/lb at allowable pressure

$$A = \frac{1027 \times 3.220}{420} = 7.874 \text{ in}^2$$

Referring to the ASME Power Boiler Code, a 2-in- and a 2½-in-diameter SV are needed to give the required area of opening.

NOTE: This value must also be used to obtain the minimum size of SV connections.

Q If a boiler has no stamping as to capacity and there is a question as to what relieving capacity is needed on an SV, how is the proper relieving capacity for the SV determined?

A 1. If the boiler has SVs, the boiler can be fired at its maximum rate. Isolate the boiler from connected load; then note if the pressure rises more than 6 percent above the highest pressure to which any valve is set. If the pressure rises above 6 percent (with SVs discharging), the SVs are too small, and greater relieving capacity is needed. This is an accumulation test.

WARNING: Because of the dangers of damaging the boiler from lack of circulation, this test is not permitted on boilers with superheaters or reheaters, or on HT water boilers.

2. Capacity can also be determined by measuring the maximum amount of fuel that can be burned and then working back to calculate the output of the boiler. The heating value of the fuel must be known. For example, assume that crude oil is burned with a heating value of 18,500 Btu/lb. Assume that the boiler uses 1000 lb/h at the maximum firing rate. Use the code equation:

$$W = \frac{C \times H \times 0.75}{1100}$$

where W = weight of steam generated per hour, lb
= minimum SV capacity
C = maximum weight of fuel burned per hour, lb
H = heating value of fuel, Btu/lb

then

$$W = \frac{1000 \times 18{,}500 \times 0.75}{1100}$$

$$= 12{,}620 \text{ lb/h minimum} \qquad \textit{Ans.}$$

3. The third method for determining the relieving capacity of the SVs is based on the fact that input equals output, with respect to water and steam flow. Measure the water fed into the boiler in pounds per hour at the maximum firing rate. The maximum outlet of steam is assumed to be equal to this. The relieving capacity of the SVs must at least equal this flow in pounds per hour. But do not use this method on hot-water boilers as no evaporation takes place. Also, the pump used to circulate the water through the boiler is not truly a feedwater pump.

Q What are the minimum diameters of SVs permitted on a miniature steam boiler, a power boiler, an LP steam boiler, an HW-heating boiler, and an HW-supply boiler?
A The ASME code requires the following minimum-diameter SRVs: miniature steam boiler, ½ in; power boiler, ¾ in; LP steam boiler, ¾ in; HW-heating boiler, ¾ in; HW-supply boiler, ¾ in.

Q When are SVs required on superheaters, and can the capacity of the superheater SVs be included in the total SV capacity required on a boiler?
A Every superheater attached to a boiler with no intervening valves between the superheater and boiler requires one or more SVs on the superheater outlet header. With no intervening stop valves between the superheater and the boiler, the capacity of the SVs on the superheater may be included in the total required for the boiler, provided the SV capacity in the boiler is at least 75 percent of the aggregate SV capacity required for the boiler.

Q Illustrate the previous answer with an example.
A Assume a boiler needs 50,000 lb/h relieving capacity. A minimum of two valves is required on the boiler, with a total relieving capacity of 75 percent × 50,000 = 37,500 lb/h. The superheater would then require an SV with a capacity of 50,000 − 37,500 = 12,500 lb/h.

Q If there are stop valves between the boiler and the superheater, what is the code rule for SVs on the superheater?
A The superheater now may become an independently fired superheater or fired pressure vessel (isolated). The SVs required on the superheater must be determined on the basis of 6 lb of steam per square foot of heating surface on the superheater surface exposed to the hot gases.

Q Which SVs should blow first, the superheater or drum SVs? Why?
A The superheater SVs should always be set at a lower pressure than the drum SVs so as to ensure steam flow through the superheater at all times. If the drum SVs blow first, the superheater could be starved of cooling steam, leading to possible superheater tube overheating and rupture.

Q Do reheater SVs requirements follow the rule on superheater SV requirements?
A No. The capacity of the reheater SVs cannot be included in the total SV capacity required for the boiler and superheater. The relieving capacity of the reheater outlet SV must be not less that 15 percent of the required total on the reheater. And the total capacity on the reheater must be at least equal to the maximum steam flow for which the reheater is designed. One SV must be in the reheater outlet.

Q Name some points to consider when attaching an SV to a steam boiler.
A See the latest ASME boiler code, Sections I and IV. Pertinent excerpts are:

1. The SV or SVs shall be connected to the boiler independent of any other steam connection and attached as close as possible to the boiler, without any unnecessary intervening pipe or fitting.
2. Every SV shall be connected so as to stand in an upright position, with spindle vertical.
3. The opening or connection between the boiler and the SV shall have at least the area of the valve inlet. No valve of any description shall be placed between the SV or SVs and the boiler, or on the discharge pipe between the SV and the atmosphere. When a discharge pipe is used, the cross-sectional area of the outlet pipe shall be not less than the full area of the SV outlet, or of the total of the areas of valve outlets discharging thereinto, and shall be as short and straight as possible, and so arranged as to avoid undue stresses on the valve or valves.
4. All SV discharges shall be located on pipes so as to be carried clear from running boards or platforms. Ample provision for gravity drain shall be made in the discharge pipe at or near each SV, and where water of condensation may collect. Each valve shall have an open gravity drain through the casing below the level of the valve seat. For iron and steel-bodied valves exceeding 2-in size, the drain hole shall be tapped not less than ⅜-in pipe size.
5. If a muffler is used on an SV, it shall have sufficient outlet area to prevent backpressure from interfering with the proper operation and discharge capacity of the valve. The muffler, or plates, or other devices shall be so constructed as to avoid any possibility of restriction of the steam passages due to deposit.

Q What causes an SV to stick to its seat?
A Mostly corrosion and deposits on valve and valve seat due to the SV not having lifted for a long period. To avoid this most dangerous condition on automatic fired (especially LP) boilers, the SV should be periodically raised by using the hand lever, or preferably by raising the steam pressure to the popping point. The latter practice should be done only with constant attendance at the boiler, and then only under the

supervision of trained personnel who will carefully watch boiler pressure and immediately shut the boiler down if the pressure starts exceeding the maximum allowable. The lever testing of SVs should be done with at least 75 percent boiler pressure on the SV.

Q How often should SVs be tested?

A 1. On LP automatic nonattendant-fired boilers, at least once a month, with a yearly test by raising the steam pressure to the popping point under the supervision of trained personnel. The same rule should be followed for the smaller package-type HP boilers that have infrequent operator attendance.

2. On large HP boilers in industrial plants with integrated and interlocked controls, once every 6 months, with an actual popping-point test under pressure prior to shutting the unit down for the yearly internal inspection.

3. On utility boilers with good interlocked controls and constant supervision by trained personnel, testing need only be done yearly. Here the popping test is under full pressure prior to shutting the unit down for the yearly internal inspection. Thus we see that SV testing is largely governed by surroundings and maintenance practices carried out in a plant.

Surroundings can affect the SV by dirt accumulation around the valve, chemical fumes, and sludge, any of which may cause a valve to freeze closed. Water treatment, and in many cases, lack of treatment of boiler water, also contribute to valves sticking closed. Scale can easily bind the valves by adhering to the water and steam sides of SVs, making them inoperative when needed. Uncorrected leaks on the SVs or on overhead lines can affect the valve very quickly if uncorrected. Where conditions are poor, daily checking of SVs is very important.

Q If it is true that all boilers furnished today are provided with ASME-approved SVs, why should they not always work?

A ASME certification is based on new valves meeting pressure-setting, capacity, and blowdown requirements. But valves in service are subject to material deterioration. And while the ASME code stipulates using materials suitable for intended service, no design can cope with the many factors that can lead to SV malfunction. These are due to surroundings, maintenance practices, and untrained personnel. Thus periodic testing of SVs is still the best assurance of their working when needed.

Q Is it true that certain SV designs are more prone to failure than others?

A Some older ASME- and NB-approved valves of the top outlet type have caused trouble. They had a tendency to leak, thus rusting and scaling the cup, and if not equipped with a drain, or if the drain had become plugged, the valve would freeze. Many states have adopted laws requiring side-outlet-type SRVs. Diaphragm-operated relief valves (Fig. 9.6) for liquid service have caused trouble because of diaphragms cracking or leaking. This causes equal pressure on each side of the diaphragm, reducing the lift capacity of the valve.

Q Is there a foolproof, nonmaintenance-type SV suitable for automatic nonattended boilers of the low-pressure type?

A Figure 9.7 shows a valve that has poppet action and a fusible eutectic element that ruptures at a given temperature. It is for LP heating service. But so far there is no valve that does not require periodic testing. The valve shown has a nonleak-type valve face and a Monel-400 corrosion-proof seat. The poppet disk is suspended by a dual-pivot system to ensure alignment and free movement. The bronze adjusting cap is set at the factory for 15 psi, then pinned and sealed. If the unit is popped by allowing steam pressure to rise to 15 psi, the valve will reseat itself before pressure has dropped to 12 psi.

When the eutectic relief blows, it generally indicates that the function of the poppet valve is impaired. An audible-visible alarm signals when the eutectic element blows out. When the special alloy melts out of the eutectic relief at a predetermined temperature, it can be replaced only by a new one from the manufacturer. In normal installations, the valve outlet is piped to a location which allows free discharge with safety. Should this line become accidentally plugged or icebound, another feature comes into play—the removable cover plate has a spring-loaded fastening. As pressure behind the plate builds up because of a blocked exhaust line, the cover plate moves away from the valve body, permitting steam to exhaust to the boiler room.

Q List the most common SV installation mistakes.

A The following mistakes are repeatedly made by contractors and operators:

1. Installing noncode SRVs with no pressure or capacity stamping

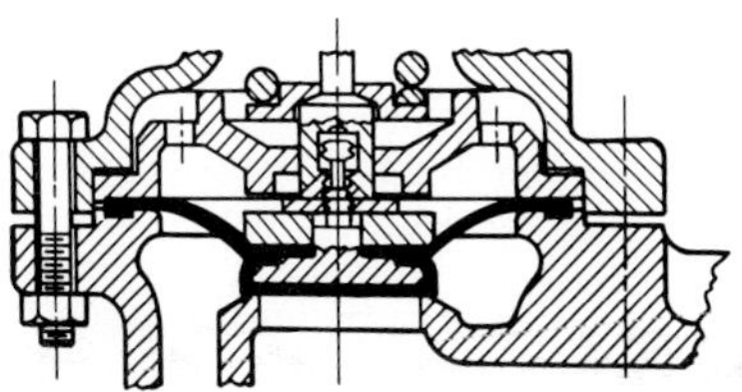

Fig. 9.6 Details of diaphragm-type safety valve.

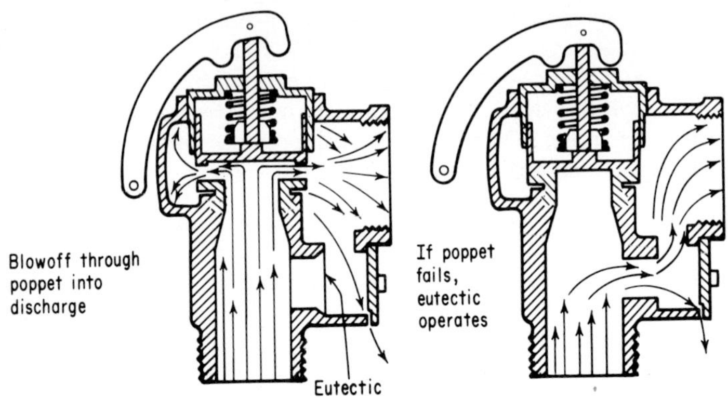

Fig. 9.7 Combination pop- and eutectic-type safety valve for low-pressure service.

on the valve. Thus it is impossible to know if the SV is suitable for overpressure protection for the boiler. This is especially true in states having no boiler codes.

2. The SVs do not match pressure rating and capacity with that of the boiler. This is dangerous if too small a valve is installed.

3. The relief valve is installed in the wrong place on feed piping in HW systems, for example. It should be installed on the boiler proper. Many contractors and plumbers believe the function of the relief valve on a hot-water-heating boiler is to protect the boiler against overpressure from the city water supply. Its main function is to protect the boiler against overpressure due to a runaway firing condition.

4. Hot-water-heating and hot-water-supply boilers require ASME pressure-rated, Btu-rated valves. But Btu rating should match the Btu output of the boiler for the fuel being fired. Many relief valves are installed with the Btu capacity not given, just the pressure setting and diameter of the valve.

5. Undersize inlet and outlet nipple connections of the SV restricting flow. This is usually found where the outlet from the SV is piped to a safe point of discharge. And smaller-diameter pipe or tubing is also often used.

> WARNING: Never use smaller inlet and outlet connections than the minimum diameter of the SV.

6. Outlet piping of SVs not sufficiently braced against the reaction forces when an SV discharges (Fig. 9.8). This often leads to flanged connections breaking.

Excerpt from ASME Power Boiler Code states: When a discharge

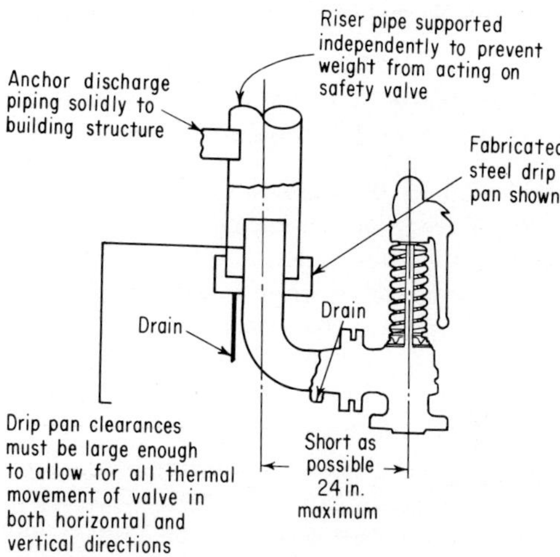

Fig. 9.8 Safety-valve outlets to roofs need bracing to prevent discharge-pipe failures.

pipe is used, the cross-sectional area shall be not less than the full area of the valve outlet, or of the total of the areas of the valve outlets discharging therein. And it shall be as short and straight as possible and so arranged as to avoid undue stresses on the valve or valves.

The usual method of handling the valve effluent, where it can be discharged directly to the atmosphere, is shown for valves *A* and *B* in Fig. 9.9. But to use less reinforcement, the tee discharge shown for valve *C* is the answer. The tee produces no reaction on the valve or valve

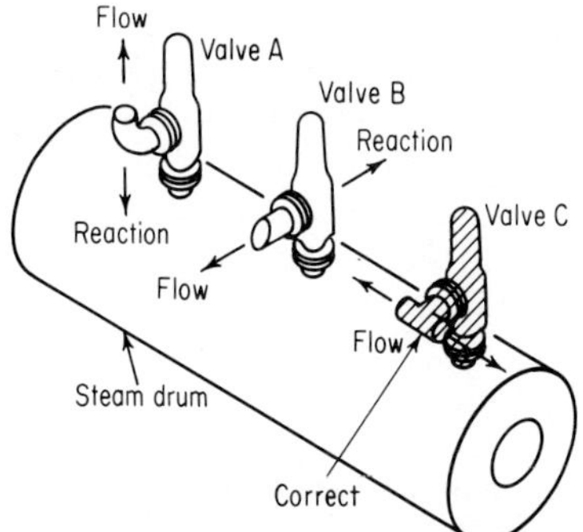

Fig. 9.9 Double exit from safety-valve outlet avoids reaction force on valve.

nozzle because it discharges in opposite directions, thus equalizing the thrust.

> **REMEMBER:** Discharge of an SV should be located so there is no danger of scalding a person standing nearby.

Escape pipes should be used if the discharge is located where workers might be scalded. A proper escape pipe is as essential to the safety of plant personnel as the safety valve is to the boiler. Too often a worker has been opening a stop valve when a safety valve, having no escape pipe and pointing directly at the person, pops. To be standing in the path of a high-pressure 3- or 4-in jet of steam is usually fatal.

Every escape pipe should be at least 6 ft high. If headroom makes it impossible to terminate the escape pipe within a reasonable distance from the ceiling, it should extend out through the building wall or roof. If it is a flat roof where workers may be, the escape pipe should extend at least 6 ft above it. If a horizontal escape pipe is more practical, it should discharge at a safe location.

It is essential that the escape pipe diameter be at least equal to the size of the safety valve. If a length of over 12 ft is necessary, it is better to use a diameter ½ in larger for each 12 ft in length. A long line with no increase in diameter will cause a backpressure because of flow friction and may cause serious chattering of the safety valve. All 90° bends should be avoided if possible.

The escape pipe should be supported independently of the safety valve. Serious stresses may be set up in the safety-valve body, connection, or boiler nozzle by the weight of a heavy, unsupported escape pipe.

After a safety valve has blown many times, it is not uncommon for slight leakage to develop. Condensation of this leakage may gradually fill an undrained escape pipe with water. This condition alone prevents the safety valve from blowing at its set pressure. The popping point will be increased 1 lb for every 2.3-ft elevation of water in the escape pipe. Also, in an outdoor escape pipe exposed to severe winters, ice may form and seriously interfere with proper safety-valve operation. Every escape pipe should have a ⅜- or ½-in open drain at its lowest point. This drain should be conducted off the boiler top in order to prevent external corrosion induced by dampness. Figure 9.10 shows a correctly installed safety valve.

HOT-WATER-HEATING BOILER EXPLOSIONS

Q We know that a steam boiler is potentially dangerous as a source of explosion, but how can a hot-water boiler *explode*, since it is filled with water at all times?

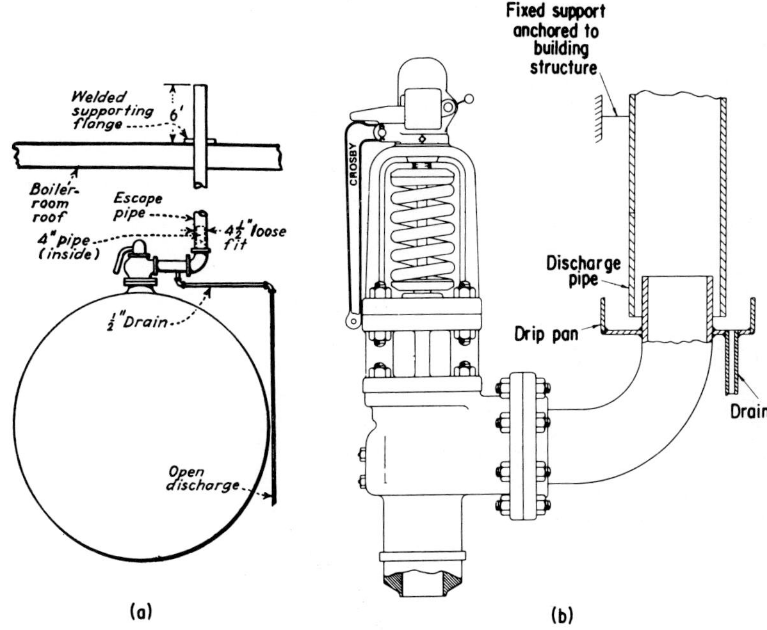

Fig. 9.10 Escape pipe from the safety valve should have a drain to remove condensate. (*a*) Low-pressure boiler; (*b*) high-pressure boiler.

A Figure 9.11 illustrates a typical hot-water-heating boiler setup. Explosions occur in hot-water boilers from the following two very basic elementary sources (assuming improper safety-valve protection).

1. In a runaway firing condition (defective temperature cutout switch) the water turns into steam. Then the boiler becomes a high-pressure steam boiler for which it was not designed. Thus the pressure buildup can lead only to an explosion, with the compressed gas (steam) causing destruction until reduced to atmospheric pressure. For example, take a basement wall measuring 7 × 20 ft (140 ft²). If 100 psi acts on this wall, a force of 2,016,000 lb will push against the wall (1013 tons), which it was not built to withstand.

2. Even if water does not reach the steam state, in a runaway firing condition, the water will get hotter and hotter, and the pressure will build up as more heat is applied. If the boiler ruptures while full of water above 212°F, it will flash into steam when relieved at atmospheric pressure. The flashing will be so spontaneous that pressures corresponding to saturation temperature will build up in the room where the boiler is located. This pressure buildup will depend on the amount of water in the boiler, the size of the room, the temperature of the water, leakage from the room, etc.

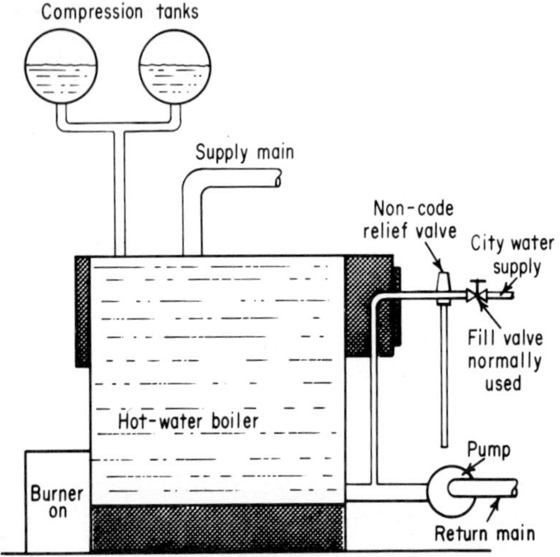

Fig. 9.11 Hot-water-heating boiler has relief valve on feed instead of on boiler proper.

Q What can cause an overfiring condition of a *hot-water boiler*?

A Overfiring may be caused in numerous ways, such as:

1. Failure of a limit control to stop the burner because of a relay or mechanical defect.

2. Mechanical failure of a fuel valve or dirt lodged in a valve so it cannot close.

3. Burner on manual operation with no one watching the temperature.

4. Residual heat with coal firing, with no one watching the temperature.

5. Burner considerably oversized in relation to the boiler and the system. Also if demand is mild on a day of use with pump not operating.

6. Wiring short, causing controls to be bypassed.

7. Fusing of contacts on a *stop-go* switch into the *go* position.

8. Solenoid- or air-operated valves isolating the boiler from the load because of mechanical or electrical defect of the controls on the solenoid or on the air *stop-go* device.

Q If a *liquid relief valve* is installed on a hot-water boiler, would not this release the increased volume of steam?

A No, because such a valve may be sized to handle liquids only. In a runaway firing condition, the water will flash into steam with approximately a thousandfold increase in volume. Obviously, the lifting of a typical liquid-type relief valve is not adequate to handle this volume of steam. Thus the pressure would still build up, causing rupture and property destruction.

Q Is any device for overpressure protection required on a hot-water boiler or on a hot-water heater?
A Yes. A safety relief valve of ASME-approved design is required, which must be set for the highest pressure allowed on the boiler. And it must be stamped to show Btu capacity (in Btu per hour) to match the Btu output (in Btu per hour) of the boiler or heater.

Q If a hot-water storage tank is heated by a *steam* coil, is a Btu-type safety relief valve required on the tank?
A Yes. If the steam coil uses steam from a low-pressure boiler and the boiler has a proper 15-psi safety valve, a safety relief valve (minimum 1-in diameter) must be installed on the tank. But the relief valve on the tank must be set *at* or *below* the allowable working pressure of the tank. If the tank is supplied by steam from a steam line, the following precautions are required:

1. The pressure of the steam used in the coil cannot exceed the safe working pressure of the hot-water tank.
2. A combination pressure-Btu (ASME type) relief valve is required on the tank, set at or below the maximum pressure allowed on the tank. The relief valve must have a relieving capacity based on the steam coil Btu heat-transfer rating.

Q Can *low water* occur in a hot-water-heating-type boiler?
A Yes. There are numerous reasons, such as the following: (1) loss of water due to carelessness in (*a*) draining the boiler for repair or summer lay-up without eliminating the possibility of firing, (*b*) drawing hot water from the boiler; (2) loss of water in the distribution system because of (*a*) leaks in the piping caused by expansion breakage or corrosion, (*b*) leaks in the boiler, (*c*) leaks through the pump or other operating equipment; (3) relief valve discharge caused by overfiring; (4) closed or stuck city makeup line.

> NOTE: Many boilers are damaged because of a common misconception that a pressure-reducing valve, used to fill a hot-water system initially, will keep the boiler and system full under all circumstances. With a 30-psi relief valve and a pressure-reducing valve which opens at 12 psi and closes at 16 psi, it becomes obvious that the pressure-reducing valve cannot supply water during the time the relief valve is functioning.

> CAUTION: If a hand-fill valve is used, then any leak in the system can quickly cause a low-water condition. Also don't forget that the

makeup water may not match the burner's capability for generating steam, as on larger heating boilers.

Q In addition to the ASME pressure Btu safety relief valve, what other *safety relief device* should be installed on a hot-water-heating boiler?
A A low-water fuel cutoff for an automatic-fired boiler, hooked up as shown in Fig. 9.12. Note that the safety relief valve is connected to the boiler proper. A low-water fuel cutoff is also included to guard against low-water failure.

WATER GAGES AND WATER COLUMNS

Q What is the purpose of a water gage, and how should it be attached to a steam boiler?
A The water gage shows the proper water level that must be maintained on each boiler to avoid overheating damage. For boilers like the locomotive or VT type, it is usual to attach water gage glass fittings directly on the boiler head or shell. But in HRT and dry-back marine boilers, the setting or smokebox prevents direct attachment; so a water column is used. Water columns are also used on WT boilers where direct attachment to the drums is not convenient. The current practice is to combine water gage and water column so the water column acts as a

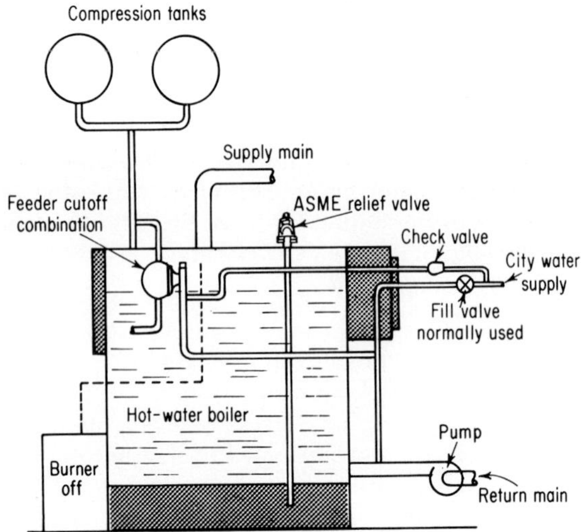

Fig. 9.12 Relief valve is installed on boiler proper on this hot-water-heating boiler system.

stabilizer. That prevents water from fluctuating severely in a gage glass connected directly to a drum.

Q What is the first appliance to observe on a steam boiler when checking operation?
A The gage glass and the level of water in the gage glass. This is extremely important, as most boiler damage is due to low water.

Q Where should the lowest visible part of the water glass be located?
A For HRT boilers, at least 3 in above the highest point of the tubes, flues, or crown sheets (highest heating surface). For locomotive boilers of over 36-in diameter, 2 in above the highest heating surface. On all other boilers, at least 2 in above the lowest permissible water level which can prevent overheating any part of the boiler.

Q What types of boilers require no gage glass or gage cocks?
A Forced-flow steam generators with no fixed steam line and waterline (once-through boilers) and high-temperature water boilers of the forced-circulation type. The same applies to once-through hot-water-heating and hot-water-supply boilers having no fixed steam line and waterline.

Q Where is more than one water-gage glass required on a boiler?
A The ASME rule on gage-glass connections for HP boilers indicates: Each boiler shall have at least one water-gage glass, except boilers operated over 400 psi shall have two water-gage glasses, connected to a single water column, or directly to the drum. For power boilers with all drum SVs set at or above 900 psi, two independent remote level indicators may be used instead of one of the two gage glasses for boiler drum water-level indication. When both remote level indicators are in reliable operation, the gage glass may be shut off but shall be maintained in serviceable condition. When the direct reading of the gage glass water level is not readily visible to the operator in the working area, two dependable indirect indications shall be provided, either by transmission of the gage glass or by remote level indicators.

Figure 9.13 shows a typical HP remote level indicator. It is mounted on the boiler-room instrument panel, or installed at any other eye-level location convenient for the operator. The indicator is connected to fittings on the boiler drum by two small tubes. Changes in the boiler-water level cause corresponding change in static head in one of these tubes; static head in the other tube remains constant. Variations in the differential pressure at the indicator cause movement of the pointer, which accurately indicates water level. The indicator is operated by the boiler water itself, using the pressure differential between a constant head of water and the varying head of water in the boiler drum. By

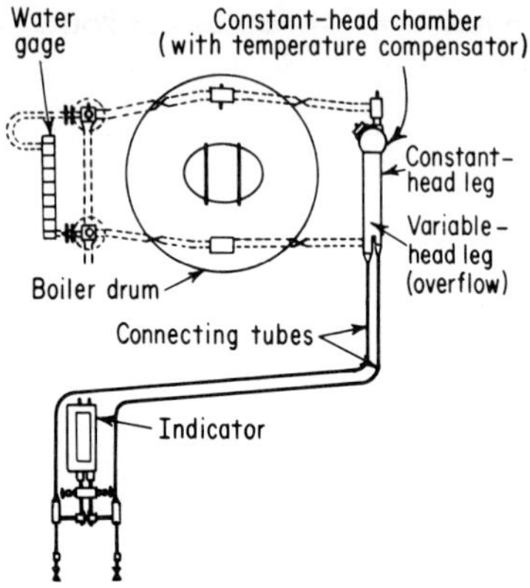

Fig. 9.13 Remote water-gage-level indicator for HP boiler.

means of a diaphragm-operated mechanism, with the diaphragm sides connected to high and low levels by tubing connection, the water level is shown by a graduated scale on the instrument.

Q What is a water column, and how should it be connected to a boiler?
A A water column is a hollow casting, or forging, connected by pipes at top and bottom to the boiler's steam and water spaces (Fig. 9.14). The steam-pipe connection to the top of the water column must not be lower than the top of the glass, and the water-pipe connection to the column must not be higher than the bottom of the glass. The minimum size of these connecting pipes must not be less than 1 in. Use plugged tees or crosses at right-angle turns, so that all piping may be easily examined and cleaned by removing the plugs. Valves, if used on steam and water connections to the water column, must be outside screw-and-yoke, lever-lifting gate valves or stopcocks with a level handle. Or they must be other valve types that offer a straightway passage and show by position of the operating mechanism whether they are open or closed. Always lock these valves or cocks open, or seal them open. If this is not done, the whole purpose of the water column and gage-glass connection will be destroyed, and the true level of water in the boiler cannot be determined.

The water-gage glass with its steam, water, and drain valves is placed on the water column as shown, and also the required number of gage cocks. Damper regulators, feedwater regulators, steam gages, and other pieces of apparatus that do not require or permit escape of an appreciable amount of steam or water may be connected to the pipes leading from the water column to the boiler. Cast-iron water columns may be used for pressures not exceeding 250 psi, and malleable-iron columns for pressures not exceeding 350 psi. Above that pressure, steel columns are used.

Q Why is a drain required on a water column, and what should be its size?
A A drain is needed to remove sediment which might block the lower connection and thus cause a false water-level indication. The drain should be of at least ¾-in diameter so that it does not easily become obstructed with sediment.

Q Name the attachments that are permitted to pipe connections of a water column. Why limit the number of these attachments?
A These are pressure gage, damper regulator, feedwater regulator, drains, level indicators, and any other connections that need only a slight flow of water to operate. The reason is that a heavy flow would cause a false water-level indication in the gage glass.

Q Is there a desirable location for the gage glass?
A Yes. The gage glass should be easily seen from the operating floor, with its lowest visible point at least 2 in above the lowest safe water level in the boiler.

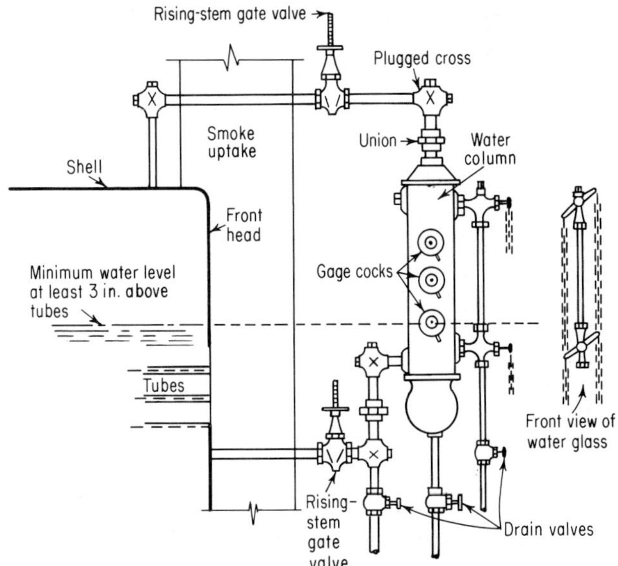

Fig. 9.14 Water-column and gage-glass connection to HRT boiler.

Q Why is a globe valve not desirable for water-column drain control?
A Because the dam or pocket in this type of valve forms a natural trap for sediment and scale.

Q How do you test the water column and water-gage glass to prove that all passages are clear, while the boiler is in operation?
A The four basic steps are:

1. Close the top valve on the column and the top valve on the glass. Then open the drain valve on the glass. If water blows freely from the drain, the water passages from the boiler to the column and from the column to the glass are clear.

2. Close the bottom valves on the column and glass; then open the top valves. If the steam blows freely from the drain valve at the bottom of the glass, the steam passages from the boiler to the column and from the column to the glass are clear.

3. Close the drain valve on the glass and open the drain valve on the column. If the steam blows freely from the column drain, the column itself is clear.

4. Close the column drain valve and open the bottom valves to the column and glass. Note whether the water rises quickly to correct the level. If the action is sluggish, there may be some obstruction in the pipes or valves. Make sure all drain valves are tightly closed and all other valves are wide open. Also make sure that the seal valves on the water-column pipe connections are in the open position.

Q What are gage cocks, and where are they used?
A Gage cocks are small globe valves with side outlets. They are a check on the water gage or a temporary means of finding the water level when a gage glass breaks or if it is plugged. Because of water flashing into steam when a gage cock is open, it is very difficult to tell whether water or steam blows out when a cock is open.

The ASME code stipulates that each boiler must have three or more gage cocks located within the visible length of the water glass, except when a boiler has two water glasses independently connected to the boiler at least 2 ft apart. Locomotive boilers of not over 36-in diameter or other firebox or water-leg boilers with not more than 50 ft^2 of heating surface need have only two gage cocks. The bottom gage cock is placed level with the visible bottom of the water glass, while the others are spaced vertically at suitable distances.

Q What design of shutoff valves is permitted in connecting pipes between the boiler and water column?
A The rising-stem outside-screw-and-yoke type, or the straightway type, or cocks that are marked plainly for their open and closed positions. If used, they must be locked or sealed open so as to always provide a means of establishing the level of water in a boiler.

VALVES AND PIPING

Q Where are valves generally used on boilers?
A Valves on boilers include steam valves on the main headers; feed valves on the water feed to a boiler; drain valves on water columns, gage glass and drain connections; blowdown valves for both surface blowoff and bottom sediment blowoff; check valves on feed lines; and nonreturn valves on steam mains.

Q What do boiler codes stipulate for stop valves used on steam boilers and steam mains?
A Each main or auxiliary discharge steam outlet, except the SV and superheater connections, must have a stop valve placed as close to the boiler as possible. When the outlet size is over 2 in (pipe size), the valves must be the OS&Y type to indicate by the position of the spindle whether the valve is open or closed. When two or more boilers are connected to a common steam main, the steam connection from each boiler having an access hole must have two stop valves in series, with an ample free-blowing drain between them. The discharge of the drain must be in full view of an operator who is opening or closing the valves. Both valves may be of the OS&Y type, but one should be an automatic nonreturn valve. This should be placed next to the boiler so that it can be examined and adjusted or repaired when the boiler is off the line. Steam mains going into a plant from the boiler should be adequately supported.

Q Explain the operation of an automatic nonreturn valve.
A In addition to the OS&Y stop valve, automatic nonreturn valves are usually placed on the main stream outlets of boilers installed in battery with others. This valve (Fig. 9.15) is closed by screwing down the outside stem but can be opened only by boiler-steam pressure, because the outside stem is not attached to the valve. The dashpot on top of the valve spindle cushions the valve movement and prevents chattering. When a boiler is about ready to cut in, the OS&Y stop valve is opened. As soon as the boiler pressure rises a little above the pressure in the steam main, it raises the nonreturn valve disk and automatically puts the boiler on the line.

During operation, if for any reason the boiler pressure falls below the main header pressure, the nonreturn valve closes and cuts the boiler out. This valve can also be used in this way to cut out the boiler when it is being taken off the line for cleaning or repair. It really acts as a check valve, allowing steam flow from the boiler to the main, but preventing steam flow from the main to the boiler.

Q Does piping connected to a boiler by welding have to be radiographed? What other requirements have to be met?

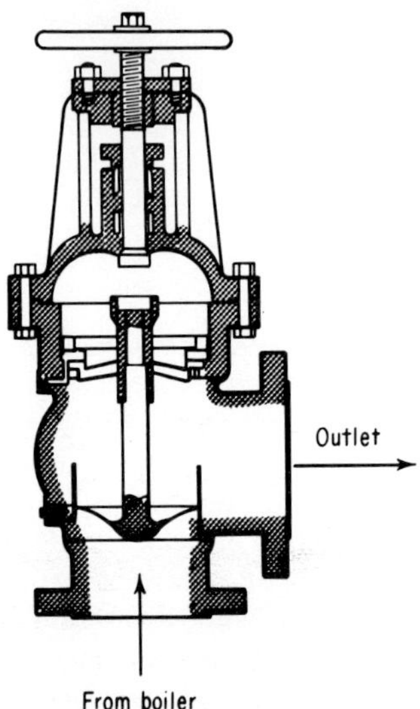

Fig. 9.15 Automatic nonreturn valve is used for boilers in battery.

A The welding of circumferential joints on pipes or headers is often ignored by boiler installers. The boiler code limits on steam pipe extends to the valves as shown in Fig. 9.16. All pipe welded within this limit must be in accord with the boiler code. Beyond this, it falls into the piping code. The contractor doing this installation of steam piping

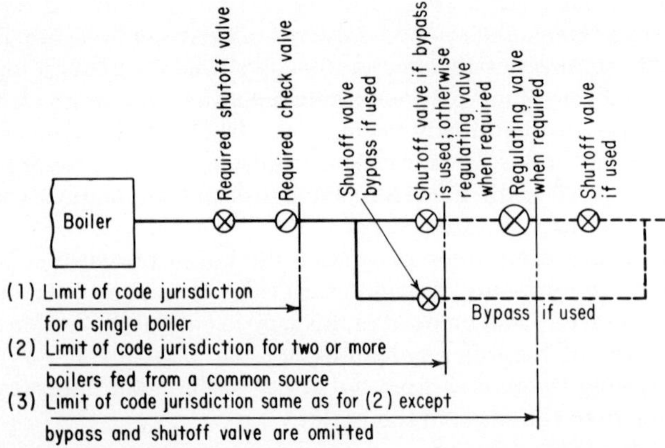

Fig. 9.16 The boiler code extends to stop valves shown. The piping code of ANSI has requirements beyond the boiler code.

within this limit must have a PP, and A, or an S stamp for HP boilers. The welders must be qualified for the position, material, and welding rod to be used.

If the steam piping exceeds 16 in in nominal diameter or 1⅝ in in wall thickness, the circumferential weld must be radiographed for the entire length. But this weld cannot be in contact with furnace gases. If they are in contact, pipe over 6 in in diameter or ¾ in in thickness needs radiographing. If a pipe is in contact or subjected to radiation from the furnace, radiographing is required whenever the pipe is over 4 in in nominal diameter, or over ½ in in wall thickness.

> NOTE: Water piping need not be radiographed if it does not exceed 10 in in nominal pipe size or 1⅛ in in wall thickness.

Q What can cause water hammer in steam and condensate return lines?

A If the steam main is pitched incorrectly when the line is not dripped, water hammer may occur as shown in Fig. 9.17. To prevent water hammer, (1) pitch pipes properly, (2) avoid undrained pockets, and (3) choose a pipe size that prevents high steam velocity when condensate flows opposite to the steam, or where condensate has a chance to collect during ideal periods.

Q What factors must be considered in piping and valve installations for power-plant applications?

A Service conditions, such as the following:

1. Expected service conditions with respect to pressure, temperature, and expected fluctuations of these in operation.

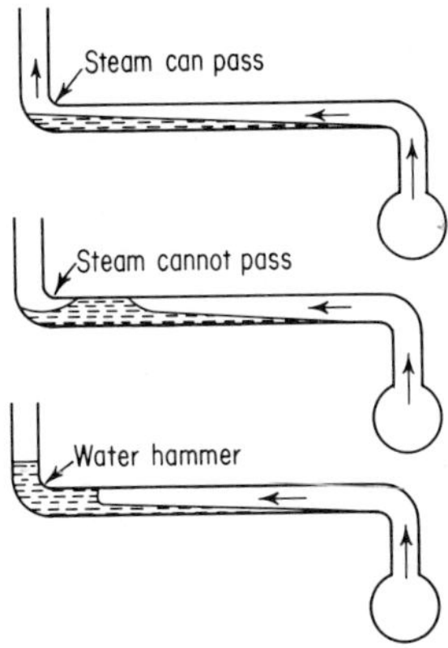

Fig. 9.17 Pockets in piping can cause destructive water hammer.

2. Flow requirements and corresponding controls for these flows.

3. Pressure drops in piping due to lengths, elbows, and similar restrictions.

4. Necessary provisions for blowdown and drainage on start-up. This is especially important where steam-utilizing machinery is connected to the steam piping. Many turbines have been wrecked from water slugging.

5. Seismic disturbances. These are important in some parts of the United States and are a mandatory design consideration on nuclear plants.

6. Any machinery connected to steam piping. It may transmit vibration to the connected piping. Expansion of machinery with temperature change can also impose strain on piping, which must be considered in the design. Figure 9.18 shows some methods of supporting power-plant piping.

Figure 9.19 shows some common types of valves found around boiler systems.

Q Why are steam traps installed on steam lines?

A Steam traps should be installed in lines wherever condensate must be drained as rapidly as it accumulates and wherever condensate must be recovered from heating, for hot-water needs, or for return to boilers. They are a must for steam piping, separators, and all steam-heated or steam-operated equipment.

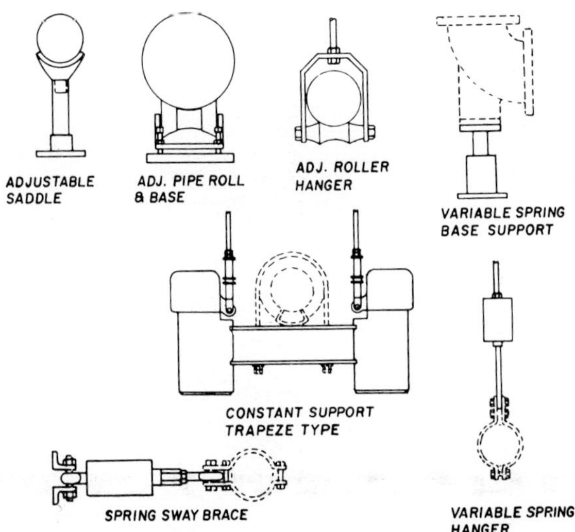

Fig. 9.18 Some methods of supporting power-plant piping.

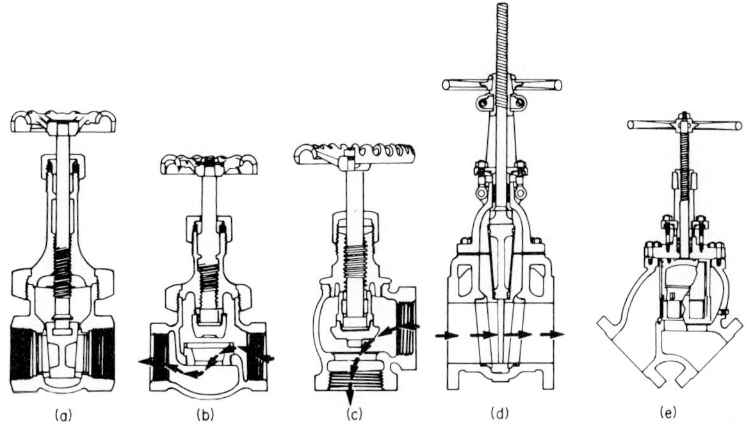

Fig. 9.19 Types of valves. (*a*) Non-rising-stem type of gate valve; (*b*) globe valves, used to regulate flow; (*c*) angle valve; (*d*) outside screw-and-yoke type of gate valve; (*e*) nonreturn valve for steam line. (*Courtesy of Crane Co.*)

Inverted open-float steam traps can be used to return condensate to regions where the pressure is less than that in the trap. They can lift condensates against a head not exceeding the pressure in the trap. Steam traps should be selected on the basis of their discharge capacity—and not by pipe size. Capacity is determined by the design and construction of the trap, the size of the orifice, and the effective seat pressure.

Nonpumping traps are used to drain condensate from steam lines, separators, and steam chambers of a wide variety of machinery. An efficient trap should expel condensate but prevent the wasteful blowing through of steam. Many types are available. See Fig. 9.20*c*, *d*, and *e*.

Q What is the purpose of check valves?

A *Check valves* are used where unidirectional flow is essential, as when feedwater flows into a boiler. The swinging disk in the valve (Fig. 9.20*a*) closes against its seat if the flow tends to reverse. By bleeding pressure from the piping and removing the bonnet and side plug, the valve and seat may be ground to a new face when worn. Also, all these parts are renewable. Being nonreturn valves, check valves are used to prevent backflow in lines. In operating principle, all check valves conform to one of two basic patterns. Shown is the swing check type (Fig. 9.20*a*). Flow moves through these valves in approximately a straight line comparable with that in gate valves. In lift check valves (Fig. 9.20*b*), flow moves through the body in a changing course, as in globe and angle valves. In both swing and lift types, flow keeps the

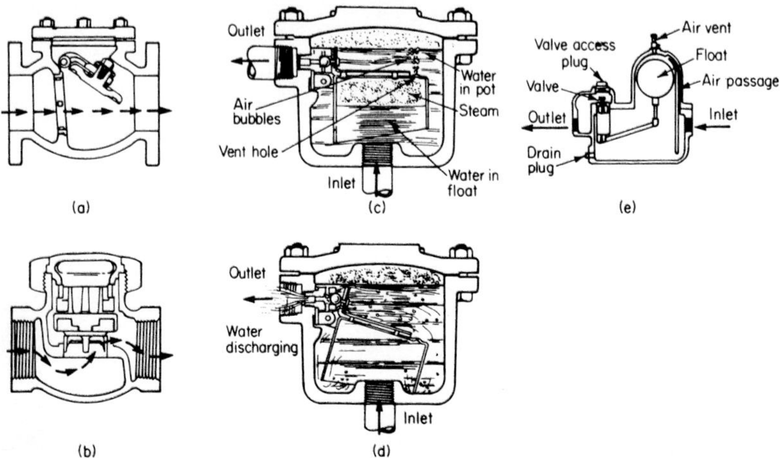

Fig. 9.20 Check valves and traps used on steam lines. (*a*) Swing check valve; (*b*) lift check valve; (*c*) trap closed, incoming steam under the float buoys the float up, closing the valve; (*d*) trap open, the incoming condensate fills the float, sinks it, and opens valve; (*e*) ball float trap. (*Courtesy of Crane Co.*)

valve wide open while gravity and reversal of flow closes it automatically.

BLOWDOWN VALVES

Q How many and what kinds of valves are required on a boiler blowdown line, and where should the blowdown valves be located?
A On all boilers, except those used for traction or portable purposes or both, when the allowable working pressure exceeds 100 psi, the bottom blowoff pipe must have two slow-opening valves, or one slow-opening valve and one quick-opening valve, or a cock complying with the code. The blowdown connection to the boiler must be at the lowest point of the boiler so as to drain it properly.

Q What is a slow-opening type of valve?
A A slow-opening type of valve is one requiring at least five 360° turns of the operating mechanism to change from full-closed to full-opening and vice versa.

Q Is an ordinary type of globe valve permissible for blowoff service?
A No. A globe valve has a dam or pocket where sediment can collect and which might interfere with the proper closing of the valve. A

straightway Y valve or angle valve may be used for blowdown service, provided it meets the proportions shown in Fig. 9.21.

Q Explain why blowdowns are necessary on boilers.

A Blowing down does three jobs:

1. Rapidly lowers the boiler-water level in case it accidentally rises to high. This action reduces the possibility of slugs of water passing on with the steam to wreck machinery.

2. Permits removal of precipitated sediment or sludge while the boiler is in service. Otherwise it might be necessary to take the boilers off the line frequently to wash out sludge accumulations.

3. Controls the concentration of suspended solids in the boiler. The solids would settle on metal parts, reducing heat transfer and causing metal overheating where the scale is located. Rupturing of tubes, shells, and tube sheets may then occur.

Q Without periodic blowdown, what conditions could develop?

A Assume that the feedwater averages 100 parts per million (ppm) of suspended solids. And suppose that no solids are carried over with steam and there is no provision for deconcentration. If all the water in the boiler is evaporated and replaced once each hour, the concentration at the end of 24 h will be $24 \times 100 = 2400$ ppm. And after 100 days it will be 240,000 ppm. Thus, 25 percent of the boiler contents would be sludge, scale, or suspended solids. So blowing down is necessary to remove enough of the suspended solids to maintain a safe limit.

Q What is a typical blowoff-valve arrangement?

A Figure 9.22 shows an angle valve (sliding cylinder with ports) in series with a quick-opening valve. Plug cocks are also used. But any plug cock used for blowoff must have a guard, or gland, to hold the plug in place. And the end of the plug must be marked in line with the

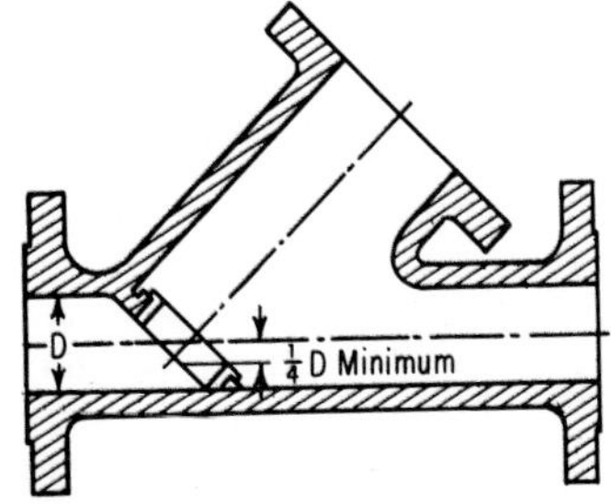

Fig. 9.21 Angle valve needs minimum distance from centerline shown to avoid dams or pockets for sediment collection.

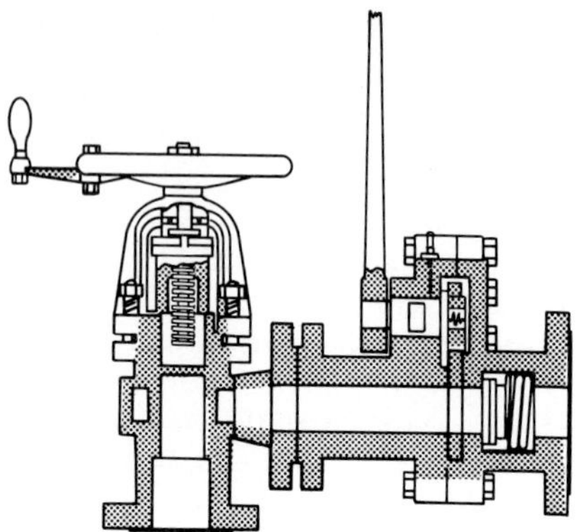

Fig. 9.22 Angle and quick-opening straightway blowoff valve.

passage through the plug to indicate whether the valve is open or closed.

Q What must the pressure rating be of blowdown valves?
A The pressure rating must exceed the maximum allowable pressure on the boiler by 25 percent. If the pressure exceeds 100 psi, the fittings between the boiler and the blowdown valves must be of steel. For pressures over 200 psi, the valves must be of steel.

Q What precautions should be taken with blowoff valves?
A Avoid accidental opening of a blowoff valve, especially of the quick-opening type. Remove the handle or lock it in a closed position when the valve is not in use. Open and close the valve slowly to prevent water hammer and possible rupture of pipes, valves, or fittings. If a blowoff valve appears to leak when closed, open it again so that boiler pressure will remove whatever is holding it open. Forcing only damages the valve. When boilers are in battery, and one boiler is open for cleaning, always "break" the connection between the idle boiler and the blowdown line (if installed). Otherwise blowing down the operating boiler will blow back into the idle boiler, scalding personnel inside.

Q Why and when is a *blowoff tank* necessary?
A A blowoff tank is necessary when there is no open space available into which blowoff from the boilers can discharge without danger of accident or damage to property. For example, discharging to a sewer would probably damage the sewer by blowing hot water under high pressure directly into it. A good blowoff-tank installation is always nearly full of water (Fig. 9.23).

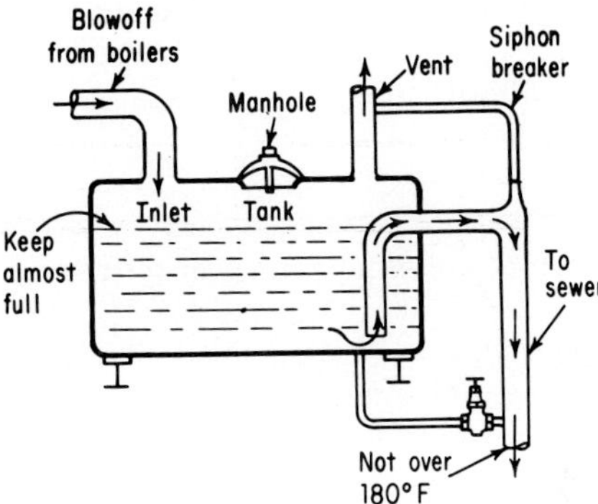

Fig. 9.23 Blowoff-tank details. Use of blowoff tanks will prevent possible scalding accidents.

Q When are two means of feeding water into a power boiler required?
A Boilers having more than 500 ft^2 of heating surface require two means of feed. But the code has been changed recently to allow one means of feed under these conditions: Boilers fired by gaseous, liquid, or solid fuel in suspension may be equipped with a single feedwater system, provided means are furnished for the immediate shutoff of heat input if the water feed is interrupted. If the boiler furnace and fuel systems retain sufficient stored heat to cause damage to the boiler if the feedwater supply is interrupted, two means of feed are still needed.

For boilers firing solid fuel not in suspension, one means of feed must be steam-operated. The source of feed must be such as to supply water to the boiler at a pressure at least 6 percent higher than any SV setting.

Q What is an *evaporator*, and why is it used for boiler operation?
A Evaporators are used to distill makeup water required in boilers as a result of leakage, process, or other unavoidable losses. The use of distilled water almost eliminates the formation of scale and other feedwater difficulties associated with raw water being pumped into a boiler. Evaporators are classified by the method of vaporization used as:

1. Flash type. Hot water is pumped or injected into a chamber under vacuum, where the water flashes into steam.
2. Film type. Water in a thin film is passed over steam-filled tubes.
3. Submerged type. Steam-filled tubes are submerged in the water to be evaporated.

Deaerators are also used. Air, oxygen, carbon dioxide, or other such entrained gases are carried by water into a boiler. These may come from raw water, from leakages within a system, or by chemical reactions of water and metals in a boiler loop system. The deaerator's main function is to remove these gases from the boiler water so as to prevent corrosion of metal parts in the boiler loop.

OIL HEATERS

Q What is a constant potential source of oil getting into an oil-fired boiler?

A Oil leakage does occur in steam plants, especially during start-up. A tube failure inside the oil heater can cause a boiler to be contaminated with oil. If the oil pressure is higher than the steam pressure, oil will be forced into the steam side of the oil heater and then travel to the waterside of the boiler. To avoid this, place a check valve on the steam line to the oil heater so reverse flow cannot take place. Where condensation is returned to the boiler, it should be piped to a condensate receiver equipped with a gage glass. Then oil, if any, can be seen in the glass. Double-shell (or double-tube)-type oil heaters are often used for extra safety. See Fig. 9.24.

The shell of the heater is of steel plate, with steel flanges; all connections are arc-welded. The steam chamber is of cast iron; tube sheets are of rolled steel; oil-heating tubes are of steel, expanded into the steel tube sheet, with tube ends swaged to shape and sealed. The steam tubes fitted inside the oil-heating tubes are of copper, with free ends supported by the swaged ends of the oil-heating tubes. The construction of this heater is such that the tube bundle can be easily removed for inspection or cleaning, by disconnecting the steam and condensate return lines and removing the steam-chamber flange bolts.

FEEDWATER SYSTEMS

Q What is the minimum pipe size of feedwater per code requirements?

A Boilers with not more than 100 ft^2 of heating surface require the

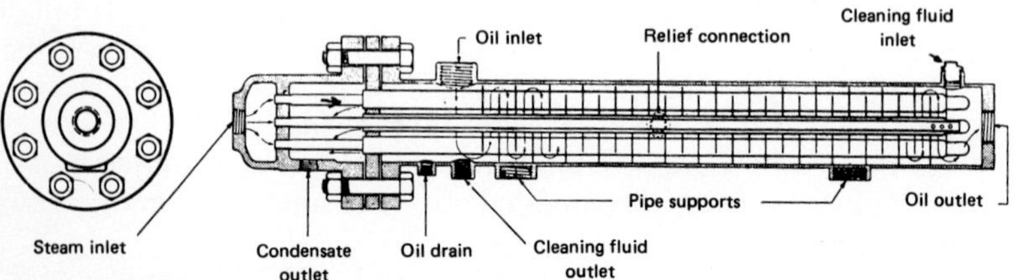

Fig. 9.24 Double-tube-type oil heater. Steam tubes are rolled into one tube sheet (or welded), condensate return in another. (*Courtesy of Cleaver Brooks Co.*)

feed connection to be not smaller than ½ in. For boilers over 100 ft^2, the feed-pipe size must be at least ¾ in.

Q What is the purpose of feedwater heaters?
A Feedwater heaters are used to bring feedwater nearer to the temperature of the boiler water. Each 10°F rise in feedwater temperature increases the overall boiler efficiency about 1 percent, owing to savings in fuel that would have been required to heat the boiler water an equal amount. An added advantage is that temperature stresses in the boiler may be avoided by feeding water at higher temperatures.

Q What is the difference between an open and closed feedwater heater?
A Two general classes of feedwater heater are used: open and closed. The open heater is sometimes classed as a *direct-contact* heater in that the water and steam mix, and the closed heater is sometimes termed an *indirect* heater because the steam and water are separated by tubes and the water is heated by conduction.

Under these classifications, the direct-contact heater has two definite subdivisions, namely, the standard open heater and the deaerating heater. The open heater was originally designed to utilize exhaust steam for feedwater heating and is essentially a low-pressure heater. It is always located on the suction side of the feed pump, and the heater must be at a sufficient elevation above the pump suction to prevent steam binding. (When hot water is subjected to vacuum, it flashes into steam. Thus, a pump handling hot water must have its suction fed under positive pressure, or no water will flow to the pump. A steam-bound pump will race and so may be damaged.) The required elevation depends on the maximum water temperature.

The principle of the open heater is to pass cold makeup water from the top down over a series of metal trays. Low-pressure steam enters between these trays, condensing and mixing with the water.

Q Besides raising the temperature of the feedwater, what other functions does an open heater perform?
A Important functions performed by the open heater in addition to raising the water temperature are:

1. Depositing solids causing "temporary" hardness in the water.
2. Removing a considerable proportion of free oxygen by bringing the water to the boiling point and venting the gases to the atmosphere.

Step 1 may reduce scale formation in the boiler; step 2 helps to reduce corrosion and pitting, which are accelerated by free oxygen.

The steam supply to open heaters is often exhaust from reciprocating engines or pumps. The pressure is seldom over 3 to 5 psi, and the heater shell is usually vented to the atmosphere through a small line.

Thus, the maximum temperature attainable is slightly over 212°F. The shell should be protected against excessive pressure by an atmospheric relief valve (large-diameter safety valve) set at not over the maximum pressure for which the heater was constructed. This is often 15 psi.

The *deaerating heater* (Fig. 9.25*a*) is a development of the open heater and increases its oxygen-removal function by operating at temperatures corresponding to pressures above atmospheric. Although for this reason it is no longer an "open" heater, it is nevertheless still a direct-contact heater. It is used with excellent results in moderate- to large-sized plants where a sufficient volume of low-pressure steam (5 to 50 psi) is available for the heating process.

Oxygen and noncondensable gases are vented with steam through a

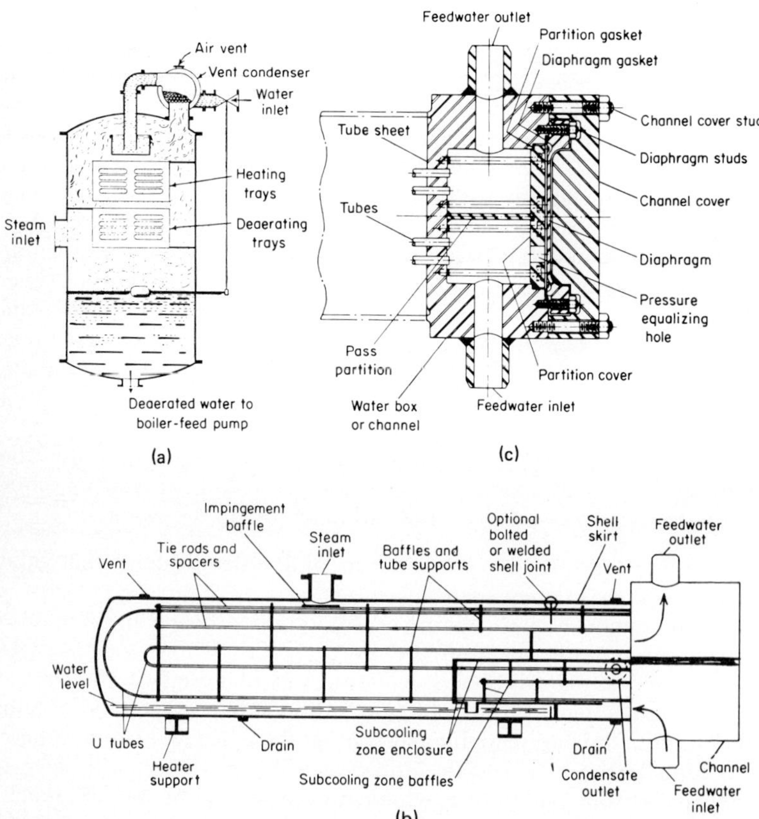

Fig. 9.25 (*a*) Deaerating heater; (*b*) U-tube closed feedwater heater; (*c*) channel-cover details of closed feedwater heater. (*Courtesy of Foster Wheeler Corp.*)

vent condenser on top of the heater. Here the steam condenses and the condensate returns to the system, with the oxygen and other noncondensable gases being vented through a vacuum pump to the atmosphere.

Q Why were closed feedwater heaters developed?

A The closed feedwater heater (Fig. 9.25*b*) was developed originally to operate where, because of oil contamination, the steam conditions were not satisfactory for mixing with feedwater. Since it is an indirect heater, the condensed steam was usually wasted, a condition seldom found in modern plants using this type of heater.

In large steam plants, closed feedwater heaters are frequently operated in series or cascade. In this manner, the ultimate temperature is limited only by the temperature of available steam and the efficiency of heat transfer. Needless to say, clean steam is used in these installations, when extracted from a turbine. All condensate is reclaimed by trapping it back into the system. This extraction practice is common in high-pressure plants, for it permits the use of a smaller condenser for the turbine. Closed feedwater heaters used in this manner are often referred to as *bleeder* or *extraction* heaters.

Q How are feedwater pumps classified?

A *Feedwater pumps* in general use may be divided into general classes, *reciprocating*, *rotary*, and *centrifugal* types. The reciprocating type makes use of a water cylinder and a plunger directly mounted on a common rod from a direct-connected steam cylinder. One or two water (and steam) cylinders in parallel, known as *simplex* and *duplex* pumps, respectively, are the most common types of reciprocating feed pumps (Fig. 9.26*a*). Triplex and quadruplex feed pumps often have each plunger rod connected by cranks to a mechanically driven crankshaft.

All centrifugal pumps are designed to operate on liquids. Whenever they are used on mixtures of liquid and vapor or air, shortened rotating-element life can be expected. If the liquid is high-temperature or boiler feedwater with vapor (steam) present, rapid destruction of the casing can also occur. This casing damage is commonly called *wiredrawing* and is identified by wormlike holes in the casing at the parting which allow liquid to bypass behind the diaphragms or casing wearing rings.

Whenever wiredrawing is detected, an immediate check of the entire suction system must be made to eliminate the source of vapor. Vapor may be present in high-temperature water for several reasons. The net positive suction head (NPSH) available may be inadequate, resulting in partial or serious cavitation at the first-stage impeller and formation of some free vapor. The pump may be required to operate with no flow, resulting in a rapid temperature rise within the pump above the flash point of the liquid, unless a proper bypass line with

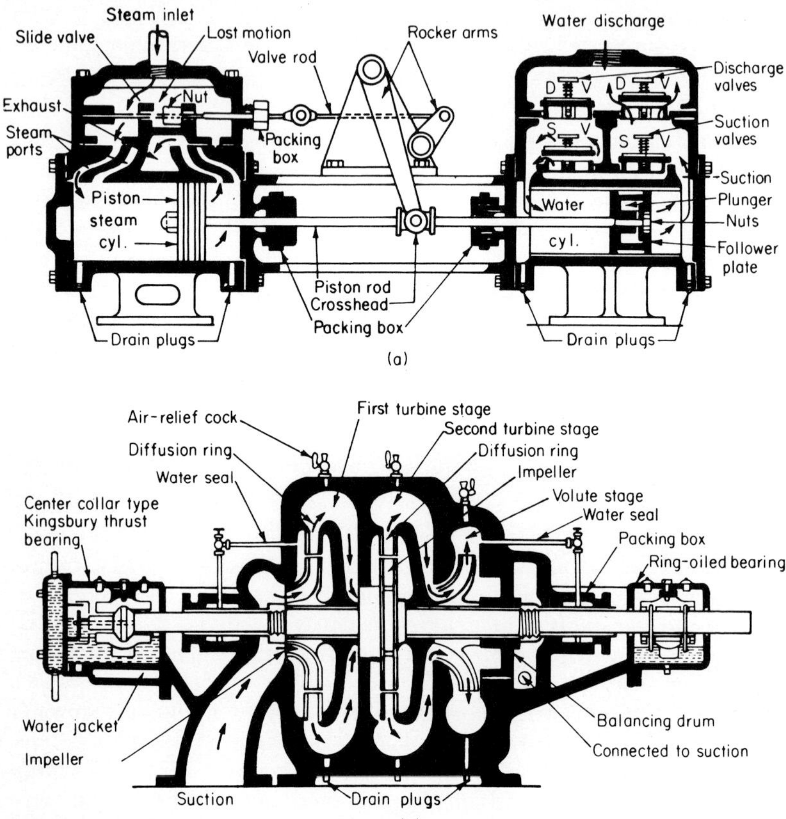

Fig. 9.26 Boiler-feed-pump types most used. (*a*) Duplex direct-acting steam type; (*b*) multistage centrifugal. (*Courtesy of Spring and Kohan, Boiler Operator's Guide, McGraw-Hill.*)

orifice is connected and is open. (This can also cause seizure of the rotating element.) The submergence over the entrance into the suction line may be inadequate, resulting in vortex formation and entrainment of vapor or air.

When a pump becomes vapor-bound or loses its prime, a multistage pump becomes unbalanced and exerts a maximum thrust load on the thrust bearing. This frequently results in bearing failure; if it is not detected immediately, it may ruin the entire rotating element because of the metal-to-metal contact when the rotor shifts and probable seizure in at least one place of the pump.

Q How are *injectors*, or inspirators, used in feeding water to a boiler?
A *Injectors, or inspirators*, are used commonly for feeding water to small boilers or as an auxiliary means of mechanical feed for medium-sized boilers. They were quite common in railroad locomotive practice. They make use of an elongated nozzle, or *venturi tube*, so that steam may feed water back against its own pressure.

Steam enters one end of the venturi tube in a jet. The vacuum produced around this entering jet draws the feedwater fed to the jet chamber into the steam flow. As the steam-and-water mixture passes through the reduced area of the throat of the tube, a very high velocity of flow is produced. The weight of the water content in this steam-and-water mixture attains sufficient momentum to open the feed-pipe check valve against boiler pressure, with water being thus fed to the boiler.

Q What is *cavitation* on a pump?
A This refers to a phenomenon within a pump when the fluid being handled drops in pressure in a part of the pump *below* the vapor pressure of the fluid being handled. This results in local vapor bubbles forming. As these bubbles reach the higher regions of pump pressure, they start to collapse. The fluid fills the void with high impact force, similar to water hammer. The result is excessive wear on the pump with severe end thrusts developed that can wipe out seals and pump blading. On boiler feed pumps, it is necessary to prevent the hot feedwater from flashing into steam. This can occur when the pump's net positive suction head is exceeded.

Q How is low net positive suction head avoided on high-pressure feed pumps?
A Recirculation from the pump discharge to the suction side of the pump can be programmed to operate automatically whenever the suction pressure approaches the net positive suction head of the pump. A recirculation control valve is used that senses boiler-feedwater temperature and then adjusts recirculation to a corresponding safe suction pressure to avoid flashing.

Another method is to use a booster pump that raises the condensate to a pressure at the suction side of the boiler feed pump so that flashing is avoided. Figure 9.27 shows a booster, heater, and boiler-feed-pump arrangement for high-pressure power-plant application.

Q Describe the function of a desuperheater in a power plant.
A Large power-generating units are designed to operate more efficiently with a high degree of superheat. But small steam auxiliary units are often designed to operate with saturated-steam temperatures, for the use of superheated steam necessitates higher costs of construction with

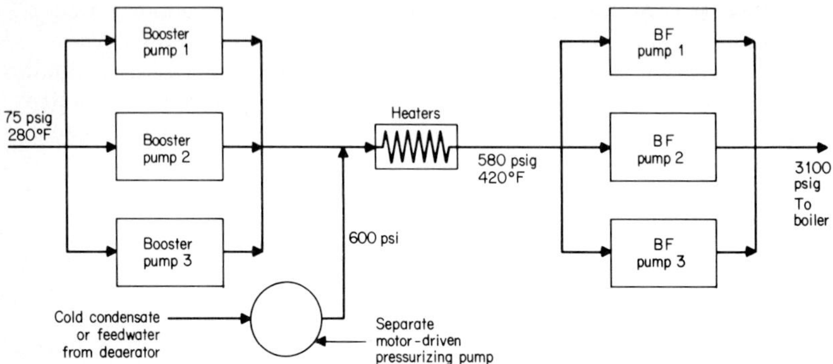

Fig. 9.27 Booster pumps are used to supply adequate suction pressure to boiler feed pumps to avoid flashing in high-pressure plants. (*Courtesy of Power magazine.*)

respect to close clearance and rotating expansion control than would be warranted, even when compared with the possible operating-expense reduction.

Rather than run a separate steam line from the boiler, independent of the superheater, it is often more practical (especially for temperature control) to pass all steam through the superheater and tap off a small line, from the superheater steam header, for auxiliary use. This small line passes steam through the desuperheater (Fig. 9.28*b*) which sprays a carefully proportioned amount of water into the flow. This proportion is regulated so that the amount of superheat to be removed will equal (or not quite equal) the amount of heat necessary to evaporate all water added to saturated steam.

It is important to use condensate or treated water for the water spray in order to avoid steam chemistry problems on turbines. On-off spraying may cause thermal shocking of associated piping connected to the desuperheater and as a result may cause cracks in superheater headers and piping.

Q Describe the difference between forced-draft and induced-draft fans.

A Induced-draft fans and forced-draft fans for steam generators require reliable service and an availability at all times in order to keep steam generators operating. The forced-draft fan supplies air for combustion of fuel as well as draft, while the induced-draft fan pulls the flue gas out of the boiler and into a stack. Centrifugal and axial fans are used for forced-draft, induced-draft, gas-recirculation, and primary air fans. Interlocks are needed between fans and the combustion or burner equipment in order to avoid boiler combustion problems.

Q What wear problem may exist on induced-draft fans employed in coal-burning plants?

A High-velocity particle erosion from the fly ash may cause abnormal wear on internal components of the fan. Blades are especially vulnerable. Hard steel material specially made for abrasive service can be specified when a fan is purchased. Also employed are the following to arrest the erosion:

1. Flame-sprayed metallurgical coatings on internal fan parts. Usually nickel aluminate is used.

2. Plasma-sprayed metallizing has been used with heat-resistant carbides.

3. Furnace-brazed coatings have been applied by using chrome-nickel-boron-silicon alloy combinations in paste form on the internal surfaces of the fan before fan assembly. The covered paste parts are heated in a 2000°F inert-atmosphere furnace, and the paste brazes itself to the part to produce a hard, thin overlaid surface.

4. Welding chrome carbide over the tough and ductile steel has also been used to provide erosion resistance to induced-draft fans.

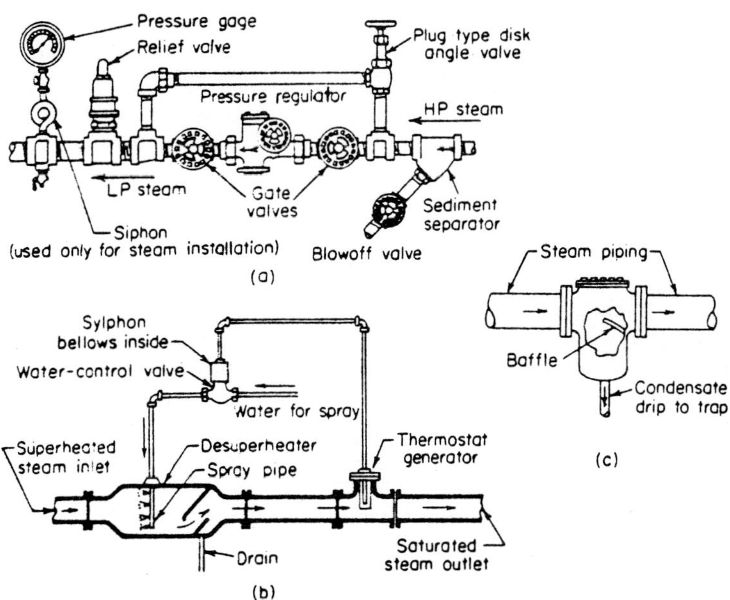

Fig. 9.28 Steam piping auxiliaries. (*a*) Pressure-reducing valve arrangement; (*b*) desuperheating steam with water spray; (*c*) steam separator to eliminate slugs of water in steam line.

Vibration monitoring is important on induced-draft fans in order to detect unbalance from wear or deposits.

Q What is used to remove fly ash and other particulate matter from flue-gas exhausts?

A Coal-burning and other solid-fuel-burning boilers require auxiliary equipment to remove fly ash and other particulates being emitted to the surrounding atmosphere. The equipment commonly used includes the following:

1. Baghouse employing fabric filters now usually made of fiberglass that can withstand flue-gas temperatures of 275 to 550°F
2. Scrubbers that wash particulate emissions out of the flue gas and form a sludge that is disposed of in landfills
3. Electrostatic precipitators

These precipitators produce an electric charge between two electrodes through which flue gas is passed. The particles in the flue gas become charged and are attracted to the positively charged and grounded collecting electrode. The particles so collected are discharged into hoppers by rapping the collecting electrode.

Q What is meant by the term *implosion*?

A Large induced-draft fans on large water-tube boilers have caused high negative furnace pressures, especially when the forced-draft fan was lost and when the interlock systems failed. The large negative pressures have caused *implosion* to occur in furnaces that included the cave-in of waterwall tubes.

Q What boiler fittings are usually supplied with a packaged FT boiler of moderate size of the low-pressure type?

A A prominent manufacturer includes the following with the boiler:

1. **Water column**—complete with gage, three trycocks, blowdown valve, test valve, and vacuum breaker.
2. **Low-water cutoffs**—mounted on or cross-connected to water column. Auxiliary low-water cutoff mounted on opposite side of boiler from water column. Both controls wired in burner control circuit prevent burner operation with boiler water below safe level.
3. **Boiler feed control**—included as an integral part of primary low-water cutoff control mounted on or cross-connected to water column. Automatically actuates motor-driven boiler feed pump to maintain boiler water level within normal high and low limits. Pump control circuit has a momentary contact switch for manual operation of boiler feed pump. (Boiler feed pump is not included with boiler unless specifically mentioned in the proposal.) On LP units a water feeder can replace the pump control at customer's option.

4. **Steam pressure gage**—located at the front of the boiler shell, this includes siphon and cock.

5. **Pop safety valves**—one or more valves of a type and size to comply with code specifications.

6. **Injector**—mounted on side of boiler within full view of water column. All necessary valves are provided (not included on LP units).

7. **Stack thermometer**—dial type, mounted in exhaust-gas stream on all high-pressure units.

CHAPTER **TEN**

Boiler Instruments, Control Theory, and Applications

Boiler operation and control involve instruments, sensors, and actuators integrated in a logical manner to regulate fuel, air, and water flow in order to obtain the desired output under variable load demands. Automatic operation of boilers has grown with the development of controls and now minicomputers and processors. Pollution control has expanded the needs for instruments, sensors, and actuators in order not to exceed governmental pollution standards. Upgrading of existing plants is an economic way of improving performance on older equipment, and this can be achieved by installing modern monitoring and diagnostic instrumentation and controls. This chapter will review the instruments required and employed and the corresponding control hardware.

Q Why are instruments important for properly operating a steam generator or boiler?
A Use of instruments and a knowledge of such factors as normal operating pressures, temperatures, flows, draft, CO_2 content, Btu input, and permissible variations in the readings observed are the most important responsibilities of any operator or owner of a boiler. Instruments indicate conditions in the boiler. Thus any variations from design or stipulated conditions serve as warnings of impending danger or inefficient operation.

Q Name the two general types of instrumentation found on boilers.
A (1) Recording and (2) indicating. Recording instruments provide a permanent record of readings. Indicating instruments provide only visual observation of readings. Boiler recording instruments cover steam pressure, steam flow, airflow, flue-gas temperature, feedwater temperature, fuel flow, and fuel temperature. In many cases three items are recorded on a single chart.

Indicating instruments are pressure gages, draft gages, fuel-oil meters, and thermometers for fuel, feedwater, and flue-gas temperatures, etc. Any installation will be improved by the use of instruments when trained operating personnel are in attendance and make intelligent use of the data provided. Instrumentation of larger packaged boilers should include an Orsat apparatus for obtaining flue-gas analysis and determining combustion efficiency.

Q Name some basic instruments needed for large power boilers.
A As a minimum, the ASME boiler code recommends the following: (1) steam pressure gage, (2) feedwater pressure gage, (3) furnace draft gage, (4) an outlet pressure gage on the forced-draft fan and an inlet pressure gage on the induced-draft fan, (5) steam flow recorder for checking boiler output, (6) CO_2 recorder to check on combustion, (7) superheater inlet and outlet temperature recorder, (8) inlet and outlet temperature recorders for air heaters, (9) thermometers indicating inlet and outlet steam temperatures for boiler reheaters, (10) feedwater temperature recorders for checking degree of deaeration and economizer operation, (11) pressure gages on pulverizers to check differential pressure for fuel-air mixtures to burners, (12) pressure gages for oil-fired boilers on oil lines to burners and temperature gages before and after any oil preheaters, (13) pressure gages for gas-fired boilers on the main gas line to burners and on individual burners.

Q How do pressure gages function?
A The two main types of pressure gages are the Bourdon tube and the diaphragm type. Figure 10.1 shows the interior mechanism of the single-tube Bourdon gage with the dial removed. The bent tube of oval cross section is closed at one end and connected at the other to boiler pressure. The closed end is attached by links and pins to a toothed quadrant, which in turn meshes with a small pinion on the central spindle. As pressure builds up inside the oval tube, it attempts to assume a circular cross section, thus tending to straighten out lengthwise. This action turns the spindle by the links and gearing, causing the needle to move and register the pressure on a graduated dial.

Q Where and why is a siphon required in steam-pressure-gage lines?
A The siphon is simply a pigtail or drop leg in the tubing to the gage for condensing steam, thus protecting the spring and other delicate parts from high temperatures. Three forms are shown in Fig. 10.2. If there is danger of freezing during long periods of shutdown, the siphon should be removed or drained.

Q How are steam gages tested?
A In these three ways:

1. By comparison on the boiler with a good factory-tested gage.

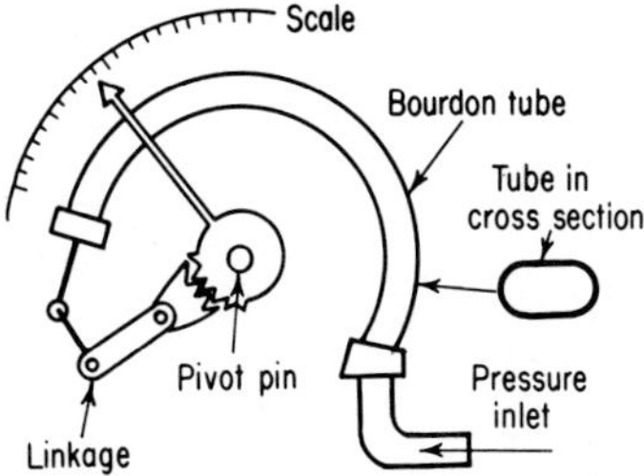

Fig. 10.1 Bourdon-tube steam-pressure gage.

Here the test gage is attached to the boiler and the two gages are compared as boiler pressure rises or falls.

2. With a screw plunger pump (Fig. 10.3*a*). The test gage and gage to be tested are attached to a screw plunger tester. To operate, unscrew the top cover, fill the cylinder with water or light oil, screw on the top cover, and force the plunger downward by turning the handle on the threaded rod. This puts equal pressure on both gages.

3. With a deadweight plunger pump (Fig. 10.3*b*). The plunger, working in a cylinder, floats on oil and is loaded by weights.

Q List the code requirements for pressure gages on a steam boiler.

A The boiler must have at least one pressure gage so located and of such size that it is easily readable and which at all times indicates the boiler pressure. A valve or cock must be placed in the gage connection

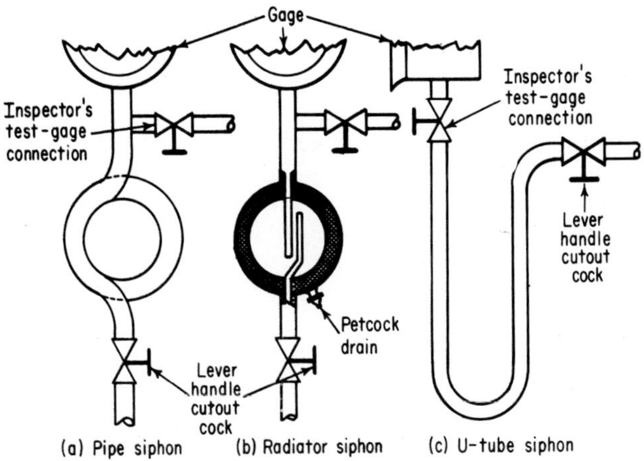

Fig. 10.2 Types of siphons used on steam-pressure gages.

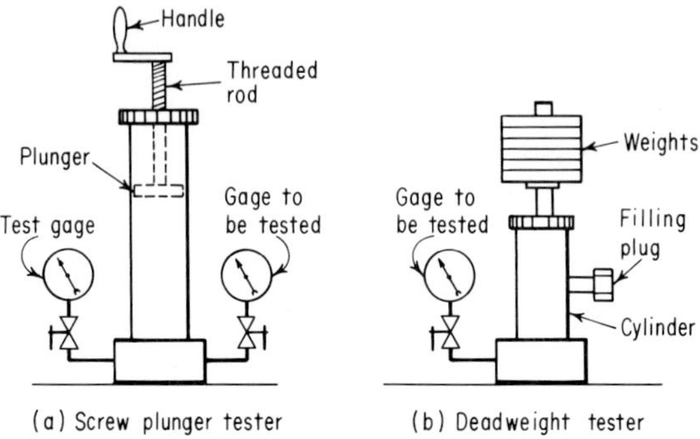

Fig. 10.3 Two types of pressure-gage testers.

(Fig. 10.2) adjacent to the gage so it can be removed for repairs. The gage must be connected to the steam space or to the water column or its steam connection. For a steam boiler the gage or its connection must have a siphon for maintaining a water seal to prevent steam from entering the gage tube. The connection of a pressure gage must be a minimum of ¼ in ID.

For temperatures over 406°F, no brass or copper tubing should be used. The pressure-gage dial should be graduated to twice the SV setting, but in no case less than 1½ times this setting. A valve connection of at least ¼-in pipe size must be installed on the boiler for the exclusive purpose of attaching a test gage, when the boiler is operating, for checking the accuracy of the boiler pressure gage. This connection is known as the inspector's connection.

The pressure gage must be illuminated, free from objectionable glare or reflection that can in any way obstruct an operator's view while noting the setting on the gage. The pointer on the gage must be in a near-vertical position when indicating the normal operating pressure. This is also true of other pressure gages in the boiler room that are used on auxiliaries. Pressure gages must not be tilted forward more than 30° from vertical, and then only when it is necessary for proper viewing of the dial graduations.

Q What are the common or primary measurements used in controlling boilers?

A Although manufacturers differ in approach, the following factors must be considered in any control used on a boiler: (1) steam pressure and flow, (2) furnace pressure and draft, (3) air pressure and flow, (4) feedwater pressure and flow (including low water), (5) flue-gas flow and composition, and (6) proper ignition and burner-flame control.

Q What are the common components in an automatic control loop used on boilers?

A A control loop, whether pneumatic or electric, is usually made up of the following basic components:

1. Primary-measurement elements as enumerated in an earlier question.
2. An error detector or comparator which compares the measurement with prefixed set points and makes adjustment signals to get the measured value to the preset points (or efficiency point in comparator application). This is actually the controller.
3. An actuator which responds to the controller signals showing deviation from set points.
4. The final control element such as a valve, damper, or variable-speed motor that makes adjustments as signaled from the controller and actuator so as to bring the variable within preset limits.

Q Name some safety controls on a boiler.

A Safety controls generally are those that limit energy input and thus shut down the equipment when unsafe conditions develop. They are (1) pressure-limit or temperature-limit switches, (2) low-water fuel cutoffs, (3) flame failure safeguard systems, (4) automatic ignition controls, (5) oil and gas fuel shutoff valve controls, (6) air and fuel pressure interlock controls, and (7) feedwater regulating controls.

The SV (or relief valve) is the most important safety device. While not considered a control in the usual sense, it is the last measure against a serious explosion.

Q What do safety controls guard against?

A 1. Overpressure leading to explosions from the waterside or steam side.

2. Overheating of metal parts, possibly also leading to explosion in a fired boiler (mainly due to low water or poor circulation).

3. Fireside explosions (furnace explosions) due to uncontrolled combustible mixtures on the firing side.

These types of accidents are considered major and may lead to loss of life and serious property damage. Other potential sources of accidents are cracking, bulging from local overheating because of scale, deforming such as tubes bowing, thinning of vital pressure parts, which can lead to cracking or localized rupture, and expansion an contraction failures causing cracking or rupturing of metal parts.

Manufacturers and state laws are trying to prevent, with safety control equipment, the three major types of accidents of overpressure, dry firing, and furnace explosions. While the other types of failures are controlled somewhat by automatic controls, prevention is mostly by legal inspection requirements and by proper operation and maintenance practices expected from the owner-user of a boiler. Included are good feedwater treatment and testing of controls at periodic intervals, including safety relief valves.

Q What are the differences between operating controls and safet controls?

A See Fig. 10.4. Operating controls regulate the boiler so it operates under certain set conditions involving load, feedwater, and combustion. The aim is to obtain the most efficient output from the boiler for each pound of fuel burned. Operating controls can be viewed as governing controls that make adjustments of feedwater, fuel, air, flue gas, and steam flow as demanded by the load. The controls are often interlocked with such auxiliaries to the boiler as feedwater pump, draft fans (depending on the size of the boiler), and the programming controls set up for the unit.

Safety controls provide the upper or lower limits set for the safe operation of the boiler. If the variable conditions involving the boiler go

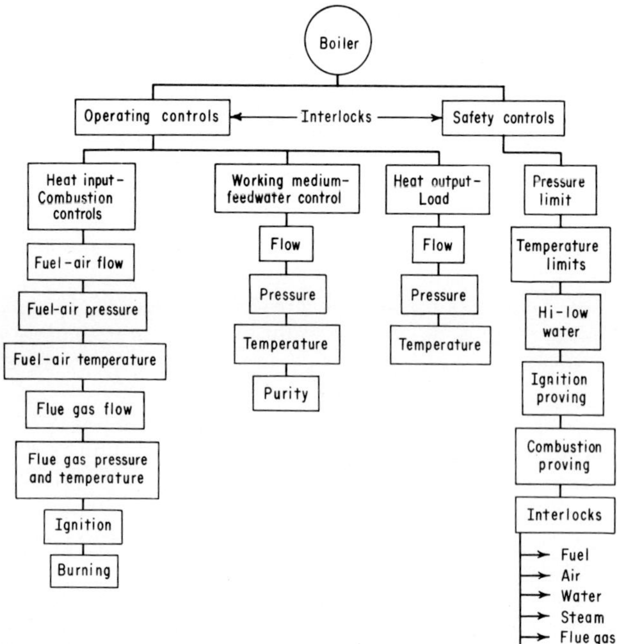

Fig. 10.4 Both operating and safety controls are important to safe boiler operation, and require interlock systems.

beyond a certain safe limit, an alarm warns that corrective actions must be taken or the boiler is shut down. The larger and more complex the boiler, the more safety controls may be needed.

Q When should interlock controls trip a boiler off?
A Critical interlock control trips are needed where there is a danger of serious damage to the boiler, and where personnel safety may be jeopardized. Generally, trips would be initiated for the loss of power, water, air for combustion, and fuel.

Loss of combustion air can result if the forced-draft fan fails; and if the fuel supply is not tripped, the furnace may fill with fuel, which could result in a serious furnace explosion from delayed ignition.

Lack of feedwater should also trip the fuel supply to avoid dry firing the boiler.

On balanced-draft boilers, an induced-draft fan failure should also trip the fuel supply to avoid a possible furnace or implosion incident, and the forced-draft fan should also be tripped to prevent the products of combustion from being forced out of the boiler settings owing to pressurized furnace operation.

CONTROL THEORY

Q What is meant by mode of control?
A Mode of control means the manner in which the automatic controller acts and reacts to restore a variable quantity on a boiler, such as pressure, flow, or temperature, to a designed control or desired value.

Q Name the three controller systems used to control a boiler.
A (1) Pneumatic, (2) electric, and (3) electronic computer.

Q Name some common modes of control.
A Assume that human operators must regulate the turbine stop valve as load demanded increases. They can instantly open the valve wide, open it slowly at constant speed as demand increases, open it more when demand rises rapidly, or open it a constant amount for each unit of demand change. Similarly, a pneumatic or electric controller can follow any of the following five principal modes, either singly or in combination:

1. Two-position, or on-off.
2. Fixed-speed floating. This drives the final control element at a fixed speed between its limit-switch contacts.
3. Proportional position. Here the controller changes the final control element's position in proportion to the measured variable's deviation, shown diagrammatically in Fig. 10.5. But since proportional

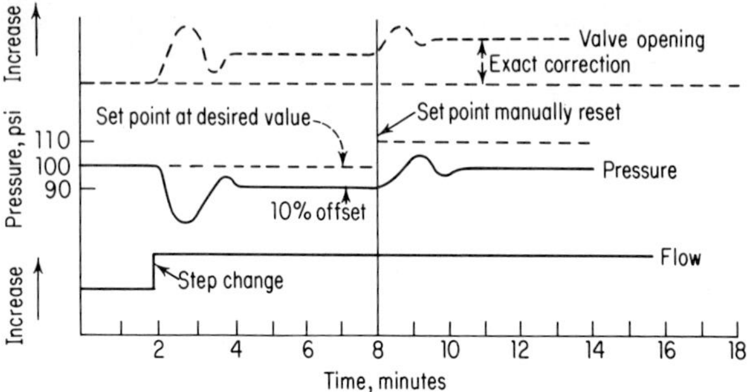

Fig. 10.5 Proportional-control mode has some offsets or droop in control.

control must sense a change in deviation (in this case, pressure drop) so as to produce a new valve position, it provides exact correction for only one load condition; at all other loads some deviation must remain. This error is called offset, or droop. Thus in Fig. 10.5 a 10 percent pressure error remains even after the valve has come to rest in its new position. The only correction is to manually reset the controller's set point, thus compensating for its inherent offset characteristic.

4. Variable-speed floating mode. Here the position of the final control element is changed at a rate which is proportional to the measured variable's deviation. The greater the deviation, the faster the valve moves. But as long as deviation continues, so does the correction to valve position. Thus the controller continues to operate until an exact correction has been made for any load change. And since valve position varies as an integral of deviation and time, variable-speed floating is often referred to as integral mode. It is usually applied in combination with proportional control to produce the proportional-plus-integral mode shown in Fig. 10.6.

5. Derivative, or rate control (Fig. 10.7). This positions the valve in proportion to the rate of change of deviation. It is sensitive to direction. Thus if pressure is rising, the controller tends to close the valve and has a great stabilizing effect. But since it cannot sense a datum, it must be combined with some other control mode, as shown in Fig. 10.8, into a three-element control mode.

Q Where are on-off combustion controls used, and how do they operate?

A On-off controls are limited to FT and small WT boilers. As the name implies, a drop in pressure actuates a pressurestat or mercury switch to start the stoker or burner and open the air damper, or to reverse the process when pressure rises again. Since control is limited to varying the lengths of on and off periods, combustion efficiency is low. See Fig. 10.9. Figure 10.10 shows a typical on-off combustion-control circuit for a low-pressure steam-heating boiler. Note the low-water cutoff control and pressure control switches hooked up in electric series. Thus if either control opens, the current to the burner motor is interrupted. The primary control consists of an electromagnetic relay that is energized by the thermostat.

On demand for heat in a steam system, the thermostat will actuate (by means of low voltage) the clapper in the relay. As the clapper is pulled in, it transfers current from the No. 1 to the No. 3 terminal on the primary control (see the sketch in Fig. 10.10). The burner motor and ignition will then come on, operating the unit. Either one of two controls determines the sequence of normal operation of the system. These are the thermostat and pressure control. If the thermostat is satisfied by a rise in room temperature, it will break its contact and deenergize the relay, thus interrupting the flow of current from the No. 1 terminal to the No. 3, and thereby stopping the burner. But if a longer period of time is required to bring up enough heat to satisfy the thermostat, the pressure (limit) control setting may be reached. This control will then shut off the current to the No. 1 terminal on the primary control. This will deenergize the primary of the step-down transformer.

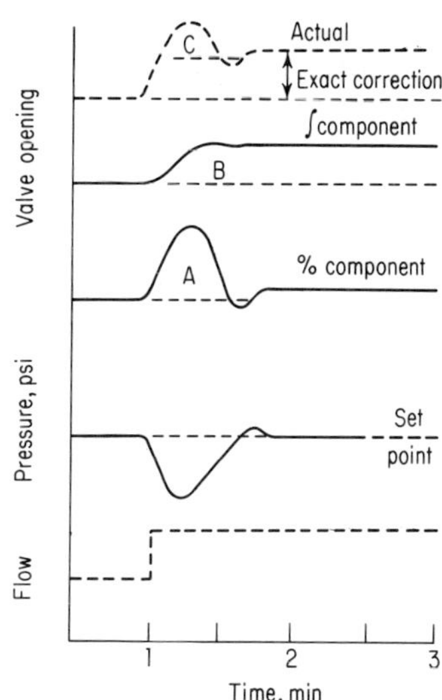

Fig. 10.6 Proportional plus integral mode control has less droop in controlling variables.

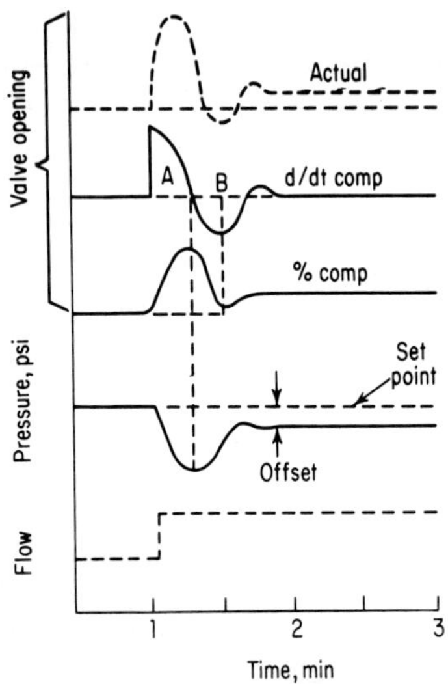

Fig. 10.7 Proportional plus derivative mode control positions actuator in proportion to rate of change of the controlled variable.

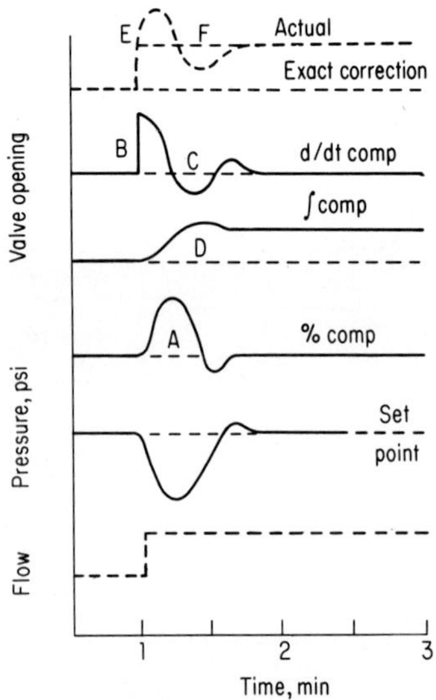

Fig. 10.8 Three-element control mode combines the control of three related variables for finer control under different loads.

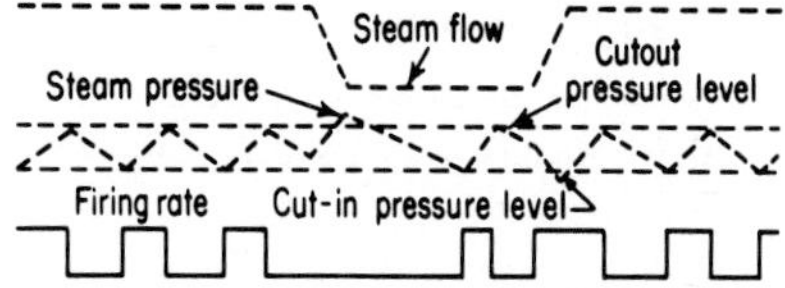

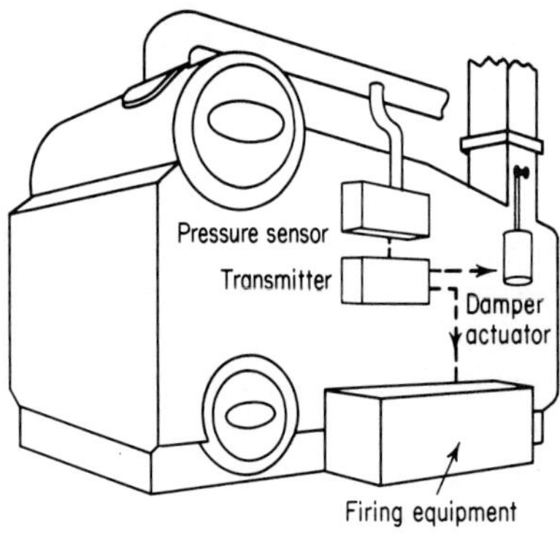

Fig. 10.9 On-off control gives "sawtooth" control of pressure, firing rate, and steam flow. Chief advantages are simplicity and low cost.

Then the thermostat will no longer be able to hold the clapper in, even though it is making contact. Thus the primary control is deprived of all current, and the burner will stop.

When the steam pressure drops, the circuit will be restored, and if

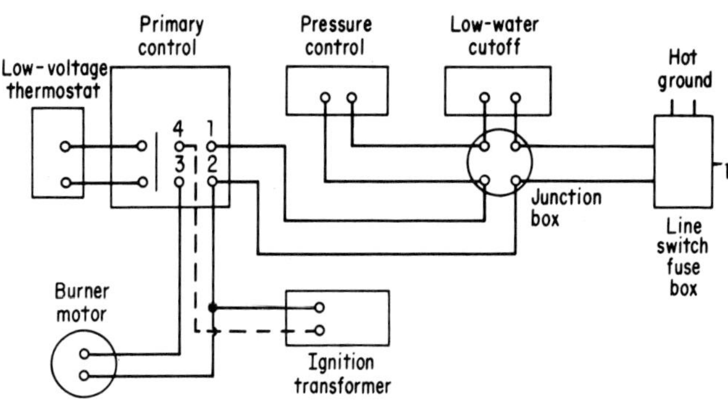

Fig. 10.10 On-off electric circuit for combustion control of heating boiler.

the thermostat is still calling for heat (making contact), the unit will resume operation. In case of abnormal operation (no fire or insufficient fire appearing), the safety function of the primary control takes over and the unit shuts down, going into *safety*.

Q What combustion control is generally used on packaged boilers?

A A positioning control system, because it is more flexible. Steam pressure is the measured variable, and a master pressure controller responds to changes in header pressure and (by means of power units or actuators) positions the forced-draft damper to control airflow and the fuel valve to regulate fuel supply. An independent controller, positioning the uptake damper, maintains furnace draft within the desired limits.

Although positioning-type control systems (Fig. 10.11) are an improvement over the on-off type, airflow and fuel supply are at their theoretically correct ratio at only one setting. This is usually the point at which they were calibrated on installation. Positioning control also assumes that a given output signal from the master controller always produces the same change in the flow of combustion air, in stoker speed, or in fuel-valve setting. But stoker speed might be affected by line

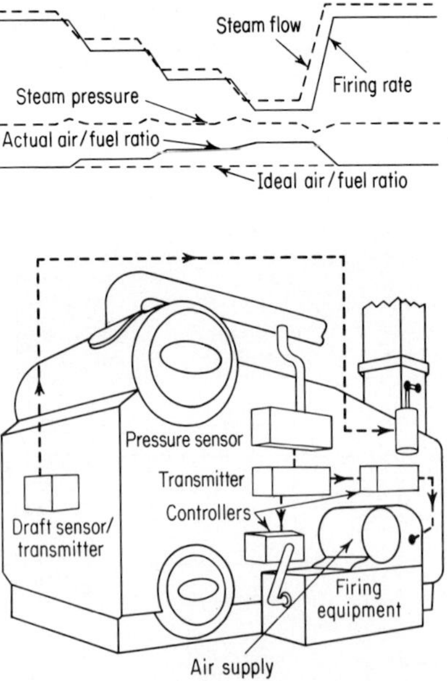

Fig. 10.11 Positioning control holds pressure in narrow range but not the air/fuel ratio.

voltage variations, and airflow by boiler slagging or barometric conditions. Thus manual adjustment is still necessary, not only on load changes but to counteract these longer-term effects.

Q Is there a combustion control considered superior to a positioning control?

A Yes. A metering control (Fig. 10.12) measures the fuel flow and airflow, then modifies the valve and damper positions to maintain these measured flows rather than implied ones. Thus it holds an optimum air/fuel ratio over a wide load range without manual intervention. Especially valuable is its inherent compensation for such variables as boiler cleanliness, voltage swings in electric actuators, lost motion in mechanical or pneumatic devices, and changes in fuel quality.

For simplicity, let us consider one system having a pressure-responsive primary element controlling fuel feed and airflow, plus an independent furnace-pressure controller. The output signal from this primary element, or master pressure controller, is frequently modified by elements sensitive to steam flow, airflow, fuel feed rate, flue-gas

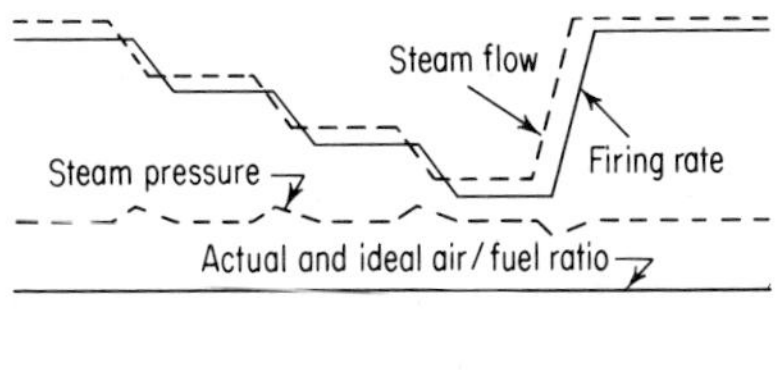

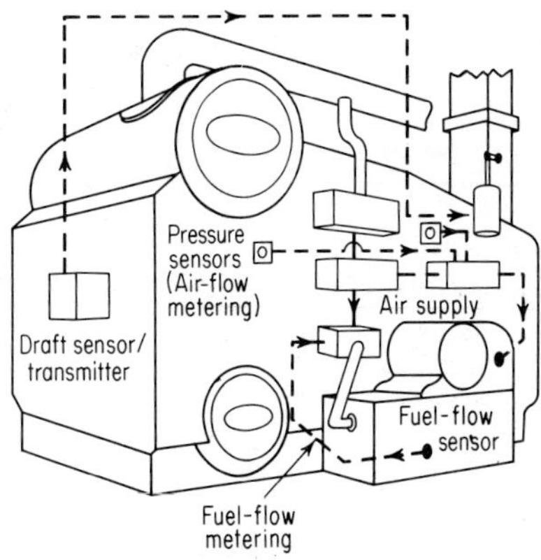

Fig. 10.12 Metering control measures fuel flow and air/fuel ratio to maintain optimum air/fuel ratio over broad load range.

analysis, or other variables. And these elements may be combined in feed-forward control in various ways.

Q Why are combustion controls geared to pressure variation in a boiler?

A To begin, pressure variation is caused by:

1. Load on the boiler. An increase in load without additional fuel input causes a pressure drop. A decrease in load without an accompanying decrease in fuel input causes a pressure rise.

2. Fuel input to the boiler. Too high an input will cause a pressure rise, while too low an input will cause a pressure drop.

Thus pressure regulation and fuel regulation, or combustion controls, are directly related. And for this reason combustion controls are geared for modulation by pressure variations within close limits. While airflow and exhaust flow usually follow fuel flow, the latter is determined by the pressure-set limits within which a boiler is to operate. Figure 10.13 shows combustion controls geared to pressure variation on a boiler.

Q How is furnace draft measured and controlled?

A Draft-measuring elements may be combined with a controller (Fig. 10.14*b*). The spiral Bourdon tube (Fig. 10.14*a*) handles draft measurements with greater sensitivity than the fairly rigid element used for

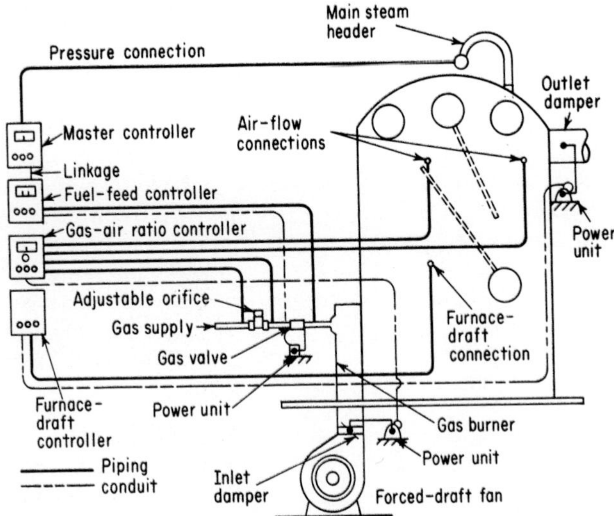

Fig. 10.13 Combustion controls in this boiler control fuel and air flow with variations in boiler pressure.

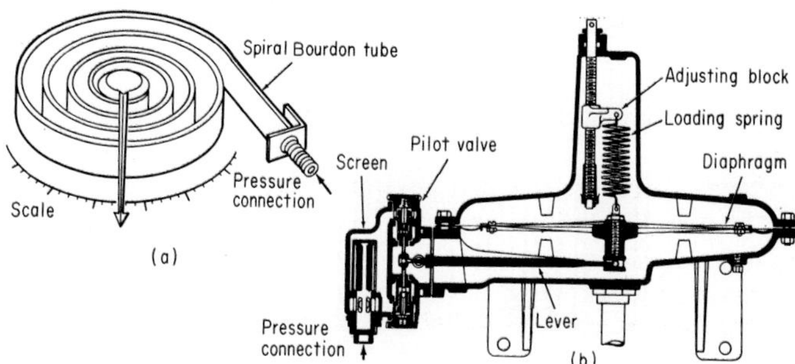

Fig. 10.14 Draft-measuring element is combined with controller by (*a*) spiral Bourdon tube and (*b*) diaphragm mechanism.

high-pressure applications. The diaphragm mechanism (Fig. 10.14*b*) converts, through a pilot valve, furnace-draft variations into proportional changes in the output signal.

Q Explain how steam flow is measured and controlled.
A The principle underlying flow measurement is shown in Fig. 10.15*a*. A pressure drop across an orifice, in this case in the steam line, can be measured by tapping the pipe at each side of the restriction. The resulting pressure differential is proportional to the square of the fluid velocity. But correct location of the tapping points is important. By using nozzle restriction (Fig. 10.15*b*), the best result (largest pressure differential for a given flow) is obtained when connections are located about one pipe diameter upstream and one-half diameter downstream from the nozzle's inlet face.

Converting differential pressure to a usable output signal may be done in various ways. Two widely used secondary elements are a mercury-float manometer (Fig. 10.15*c*) or a Ledoux bell mechanism (Fig. 10.15*d*). In Fig. 10.15*c* the two pressure tappings from the primary element are connected to two mercury chambers, joined by a U tube. Pressure variations raise and lower the float. A pressure-tight shaft conveys this movement to mechanical linkage within the controller. Check valves in the two legs of the U tube prevent damage to the mechanism resulting from sudden changes or reversals in pressure differential.

Q Explain a pneumatic differential pressure transmitter.
A Figure 10.16 illustrates a pneumatic system used for differential pressure transmission which is converted to steam flow in an indicating, recording, or control device. Diaphragm motion moves the force beam and vane. Movement of the force beam changes the vane-nozzle relationship, thus changing nozzle backpressure (NBP) in chamber B of

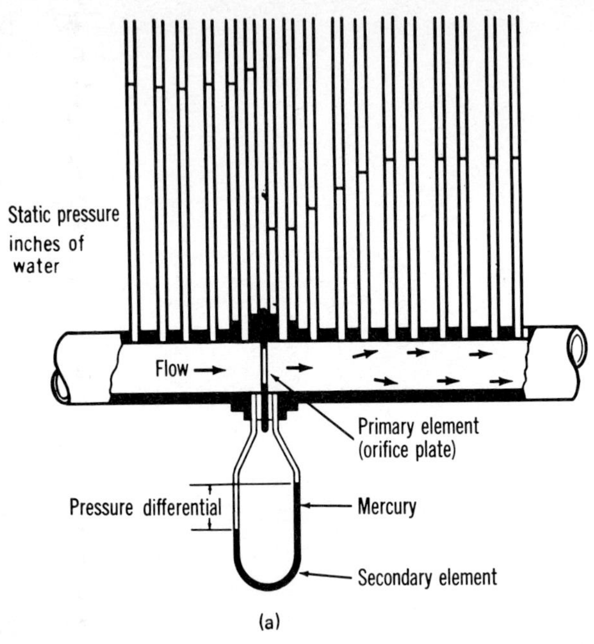

Static pressure
inches of
water
Flow
Primary element
(orifice plate)
Pressure differential
Mercury
Secondary element

(a)

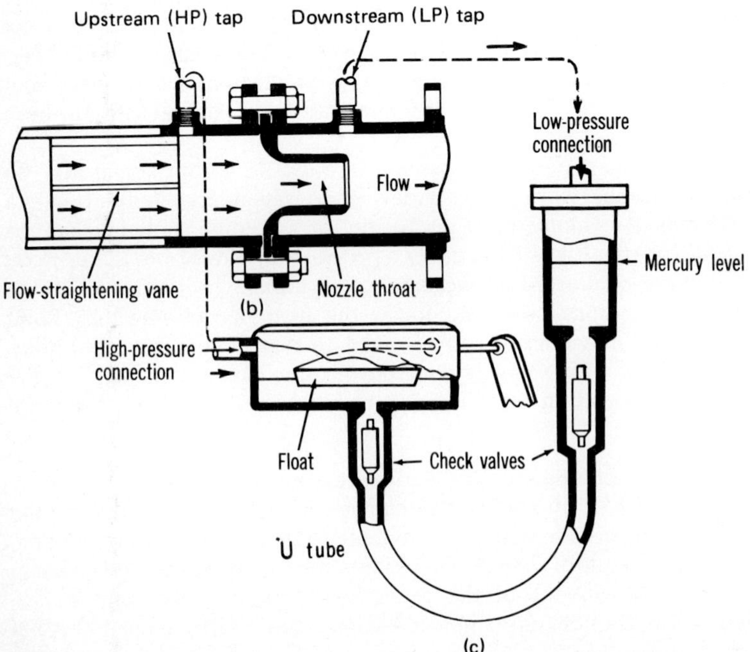

Upstream (HP) tap
Downstream (LP) tap
Low-pressure
connection
Flow
Mercury level
Flow-straightening vane
Nozzle throat
(b)
High-pressure
connection
Float
Check valves
U tube

(c)

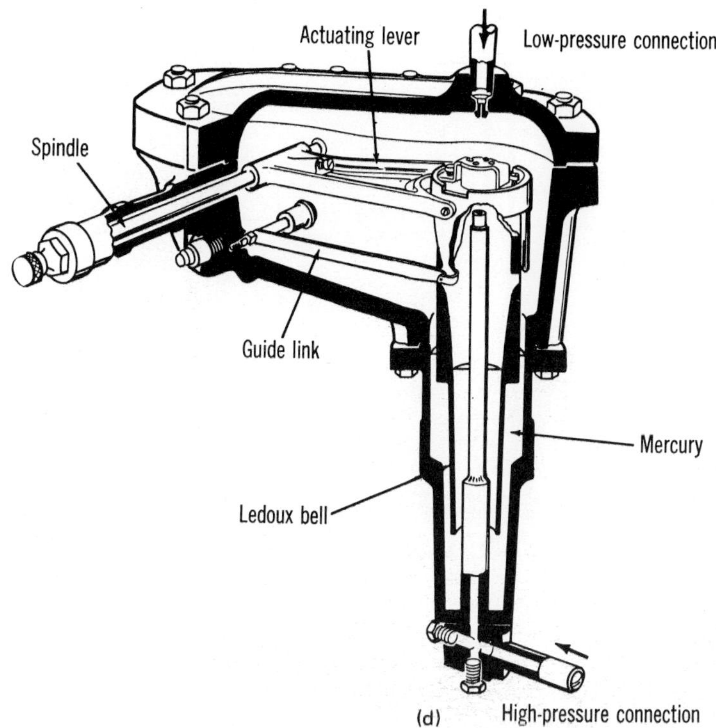

Fig. 10.15 Methods of converting differential pressure to usable signal. (*a*) and (*b*) Orifice plate and nozzle measure differential pressure; (*c*) and (*d*) mercury-float manometer and Ledoux bell mechanism.

the reset booster relay. This opens either the inlet valve (increase in NBP) or the exhaust valve (decrease in NBP), producing a change in booster output pressure. Booster output pressure goes to the restoring bellows and repositions the restoring beam. The restoring beam, in contact with the force beam through range adjustment, moves the measuring diaphragm back to the normal centered position and restores vane-nozzle *at balance* distance. The booster output pressure, which is proportional to the measured level, is also transmitted to indicating, recording, and/or controlling equipment.

Q Describe an electric pressure transmitter.

A A true force-balance mechanism characterizes the dc transmitter shown in Fig. 10.17. Deflection of the Bourdon tube exerts a force on the beam balance system; as the beam starts to move, the air gap, and thus the inductance of the position detector, changes proportionally and in turn varies the oscillator-amplifier output signal. The null-balance

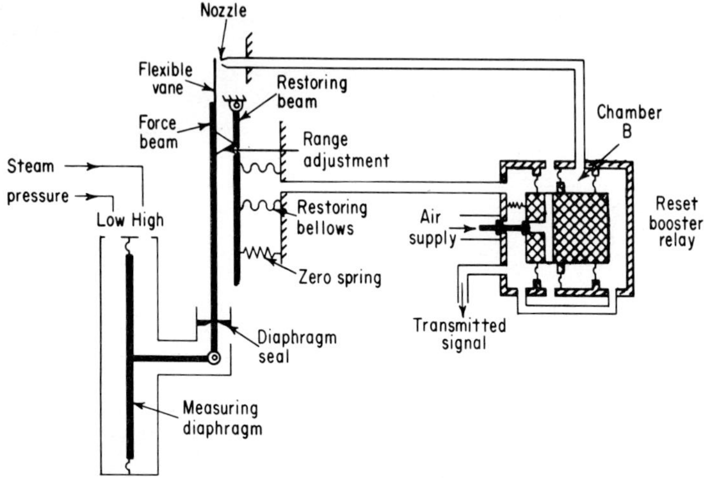

Fig. 10.16 Force-balance system used for differential pressure transmission which is converted to steam flow on recording device.

force motor, a wire coil maintained in an air gap between permanent-magnet poles, develops a feedback force balanced against the input force from the Bourdon tube, thus restoring null balance.

No matter how the output signal is generated or what measured variable it represents, whether it be alternating or direct current, it forms the input to an electric or electronic controller whose circuits perform the same functions as the bellows, baffles, and nozzles of the pneumatic controller. The unit illustrated handles pressures up to 10,000 psig. As the Bourdon-tube deflection reacts on the balance beam,

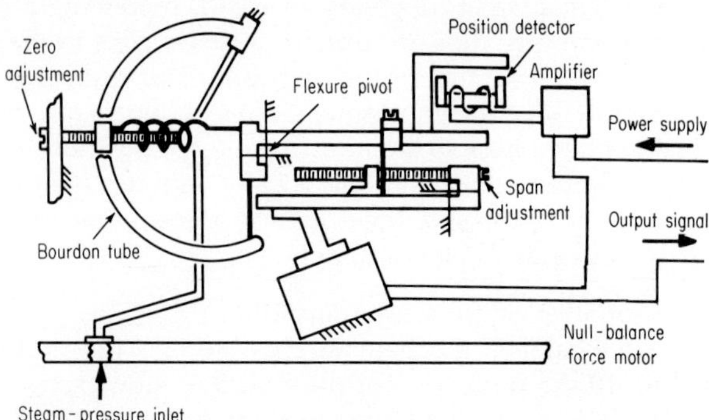

Fig. 10.17 Explosionproof electric dc pressure transmitter.

its movement is sensed by the position detector and converted into a proportional change in the 10 to 50 mA output current. The force motor, in series across this output, acts as feedback in opposing beam movement.

Q How may a pressure measurement be transmitted electrically to indicating, recording, and control equipment at a remote point and then show measured pressure at that point?

A In one device a Bourdon tube positions a movable core in a transformer. The core is adjusted to travel the same distance for all pressure ranges, thus making it possible to transmit to standard electronic receivers. The core position (Fig. 10.18) determines the magnetic flux linkage between the primary and the secondary windings.

The voltage induced in each secondary winding is proportional to the displacement of the core from its center position. Thus, the core

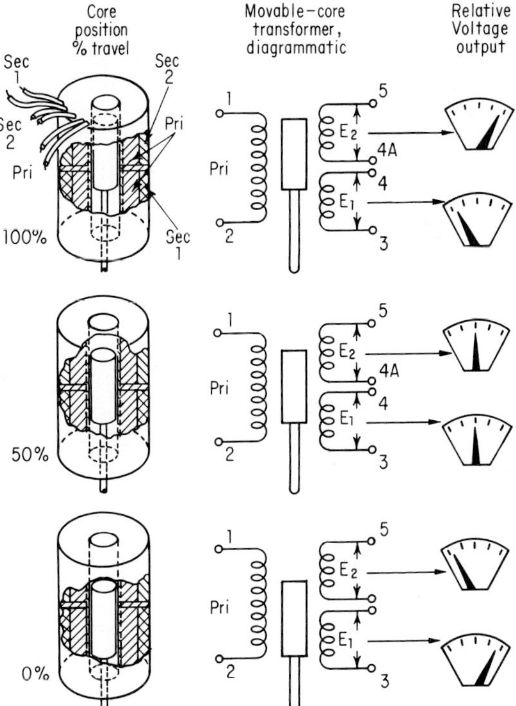

Fig. 10.18 Method of transmitting pressure or other parameter by electric signals. E_2, output voltage of secondary 2; E_1, output voltage of secondary 1.

position determines the signal voltage output. At 100 percent travel, voltage E_2 is larger than E_1 since the core is near the top of the transformer. Downward motion of the core causes E_2 to decrease linearly and E_1 to increase linearly until at 0 percent travel, the voltage magnitudes have reversed. At 50 percent travel, voltages E_2 and E_1 are equal since the core is centered between the two secondary windings. Within the operating range, a definite linear relationship exists between each of the two secondary voltages and the core position.

Utilizing a movable-core transformer with a duplicate set of secondary windings, changing the external connections, or doing both permits a wide range of applications using the same basic transmitter. The output voltage on the secondary windings can be calibrated for a proportional pressure reading on the receiving end.

Q What is the trend in feedwater control in modern boilers?
A Most boilers today use two-element (drum level and steam flow) or three-element controls in which steam flow and water flow form the primary measured variables, with water level acting as a third input. But the single-element (drum level) regulator (Fig. 10.19) is still popular. This is a thermohydraulic device based on the principle that the volume of a given weight of low-pressure steam is far greater than that of the water from which it is generated.

The regulator forms a closed hydraulic system, including the annular space between inner and outer tubes of the steam generator, the

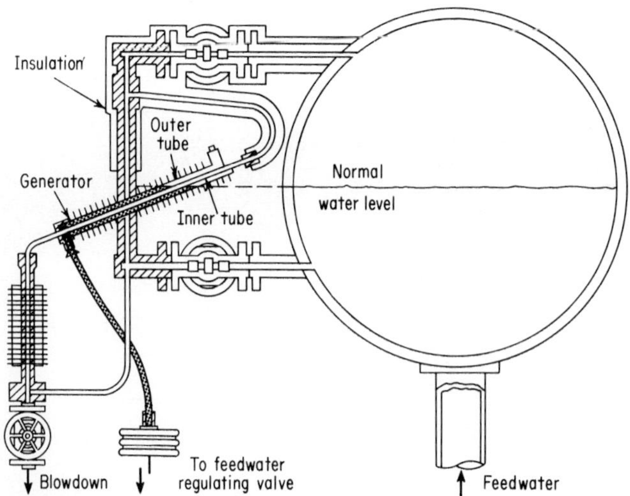

Fig. 10.19 Single-element thermohydraulic device for feedwater flow regulation.

connecting tubing, and the metal bellows of the regulating valve. Heat from steam in the upper portion of the inner tube causes the surrounding water (in the space between the two tubes) to flash into steam. The remaining water is thus forced out of this space until the water levels in the two tubes are equal. The displaced water passes into the actuator bellows, thus partially opening the regulating valve.

As steam demand increases, the water level in the drum and generator will fall; more water in the annular space between the tubes flashes into steam, and the regulating valve opens still farther. Thus if the drum level rises, the water in the generator rises also; cooled by the radiating fins, this water condenses part of the steam in the annular space and in turn permits the regulating valve to partially close. Since this is a proportional device, it is suitable only for small boilers with relatively stable steaming rates.

In larger units, drum level measurement forms a trimming signal, with steam and water flow as the primary variables. Displacer-type units are also replacing the self-acting, thermohydraulic device. In these, a cylindrical float or displacer is suspended in the measuring chamber; as the water level rises, the displacer is lightened by the weight of the liquid it displaces. This change in weight is detected by either a torque-tube or a force-balance system (Fig. 10.20) and converted into a pneumatic or electric output signal to a controller.

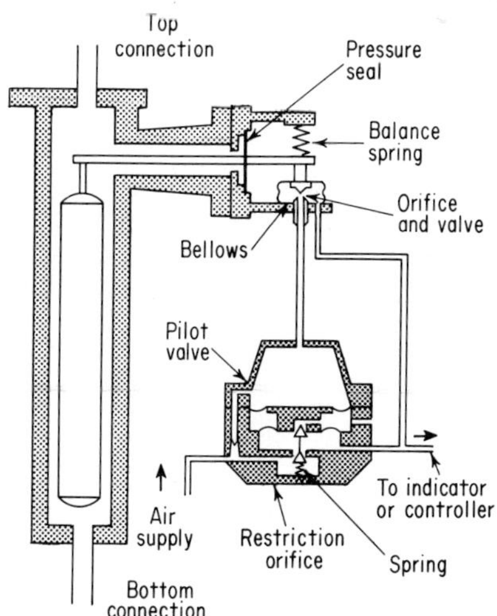

Fig. 10.20 Force-balance system converts level change to pneumatic or electric output signal.

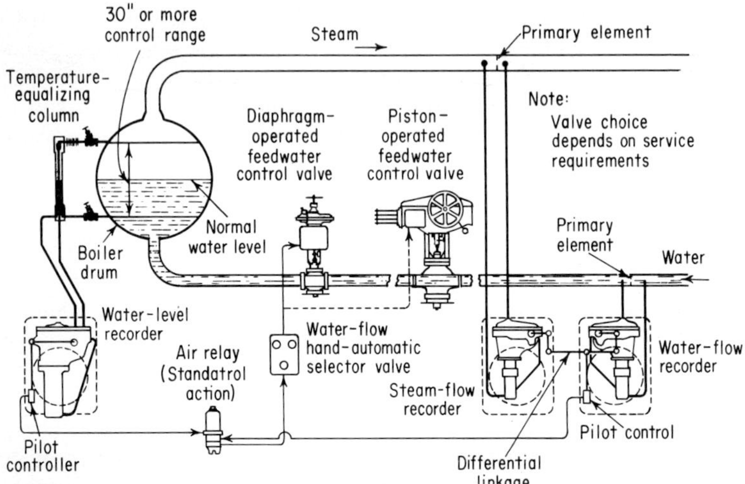

Fig. 10.21 Three-element feedwater control system measures steam and feedwater flow as well as water level.

Q How does the three-element feedwater control system work?

A In this system (Fig. 10.21) steam flow, feedwater flow, and water level are measured and recorded by mechanically operated meters. Measurements of steam flow and water flow are balanced against each other with differential linkage. A pilot control is connected to the linkage so that any difference between the amounts of steam flow and of water flow causes a change in the pneumatic output signal. This signal is transmitted to an air relay where it is combined with the pneumatic signal from the water-level recorder.

A change in boiler load unbalances the differential linkage, thus producing a change in the output of the pilot control. That in turn changes the output of the air relay. This new signal repositions the feedwater control valve, admitting required water into the boiler equal to the steam flow out of the boiler. The resulting change in feedwater flow rebalances the differential linkage and brings the pilot-control signal back to its neutral point. As a final check, and to ensure having the proper drum level, the signal from the pilot control in the water-level recorder readjusts the feedwater control valve, if required. The selector valve in the system provides automatic or remote manual control. Under normal operating conditions the control pressure gage on the selector valve is an indication of valve position.

Q How is the water level usually controlled on boilers of the heating, small commercial, and industrial types?

A On a closed-loop heating boiler, it is common to find no automatic water feed to the boiler. The attendant checks the gage glass periodically and maintains the water level manually if required. The owner relies on the automatic low-water fuel cutoff to save the boiler in case of a low-water condition.

Water level is usually controlled by:

1. Manual feed (Fig. 10.22*a*).
2. Automatic feeder, either a mechanical or an electrical solenoid-operated valve, actuated by a float or electrode in the water space of the boiler (Fig. 10.22*b*).
3. Combination low-water fuel cutoff and feeder, based on the water level in the boiler (Fig. 10.22*c*).

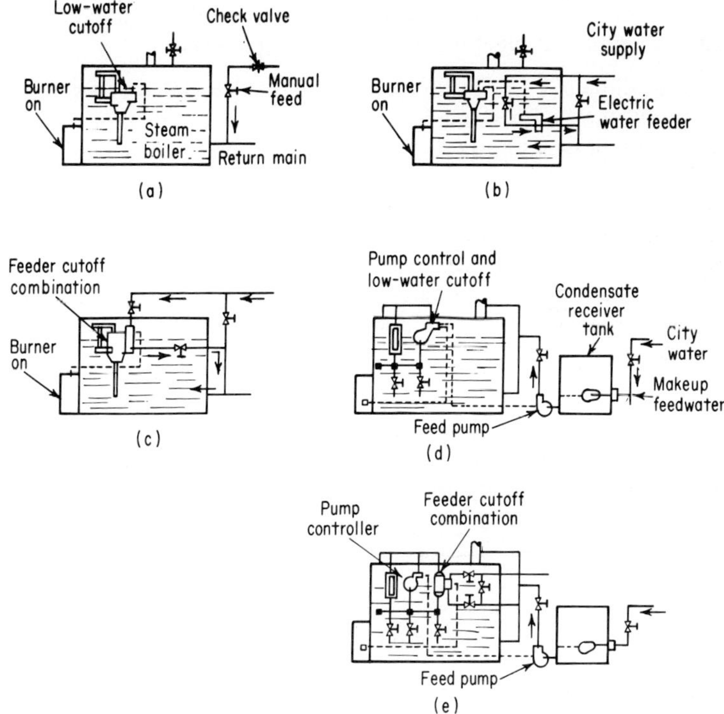

Fig. 10.22 Small boiler feed systems. (*a*) Manual makeup feed system; (*b*) solenoid-operated makeup water feeder; solenoid, when actuated from switch, contacts a float-operated low-water cutoff; (*c*) combination water feeder and low-water cutoff; (*d*) combination low-water cutoff and pump-control feed; (*e*) combination low-water cutoff and feeder and pump controller with makeup feeder on condensate tank.

4. Combination low-water cutoff and pump control (Fig. 10.22*d*). This system is installed at the normal boiler waterline. The water level is maintained by connecting the pump control directly to the condensate pump. For the additional makeup water required, a receiver tank makeup feeder is installed on the condensate receiver tank. In addition to functioning as a pump control, a second set of electrical contacts installed in the controller acts as a low-water cutoff. This stops the burner if a low-water condition occurs. This type of water feed is used also on packaged boilers of up to about 10,000 lb/h and 200 psi pressure.

> NOTE: Feed is actuated electrically by a float-switch combination turning on and off the pump. If the valve from the receiver tank to the boiler is closed, this circuit will not sense the need for more water, because the pump could still operate but obviously deliver no water. Then the boiler would have to be shut down by the low-water cutoff to protect it from overheating damage.

5. Combination low-water cutoff and feeder actuated by direct contact with the boiler water level and pump controller (Fig. 10.22*e*). This is an improvement upon the system described in No. 4 above.

A state code for low-pressure steam boilers requires adding a mechanical water feeder to an electrical pump system in these words: "Where water of condensation of all the steam generated by the boiler cannot be returned by gravity to the boiler, and pumps, traps, or other devices are used for this purpose, or steam is used for processes where there would be no water of condensation to return to the boiler, an automatic water feeding device should be installed."

Q Describe three types of low-water cutoffs used in packaged boilers for the prevention of low water.

A The low-water cutoff (Fig. 10.23), separate from the programming-sequence control, immediately shuts down the boiler if the water drops

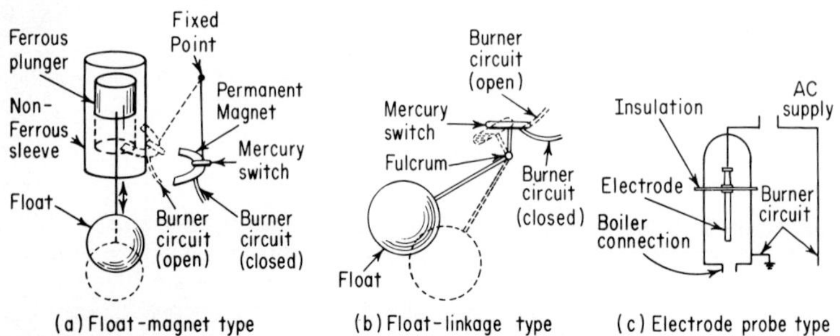

Fig. 10.23 Three popular types of low-water fuel cutoffs.

to a dangerously low level. The three types are as follows:

1. The float-magnet type (Fig. 10.23*a*) has a ferrous plunger on one end of a float rod. The plunger slides within a nonferrous sleeve. A permanent magnet, with a mercury switch affixed, is supported by a pivot adjacent to the nonferrous sleeve. Under normal water conditions, the ferrous plunger is above and out of reach of the magnetic field. In this position the mercury switch is in a horizontal plane, keeping the burner circuit closed. But if the boiler-water level drops, the float also drops, bringing the ferrous plunger within the magnetic field. Then the magnet swings through a small arc toward the plunger; the mercury switch tilts, opening the burner circuit.

2. The float-linkage type (Fig. 10.23*b*) has a float connected through linkage to a plate supporting a mercury switch. Because the plate is horizontal in the normal water-level position, the switch holds the burner circuit closed. If the water level drops, the float drops, tilting the plate so the switch opens the circuit.

3. The submerged-electrode type (Fig. 10.23*c*) uses boiler water to complete the burner circuit. If the water level drops below the electrode tip, current flow is interrupted, shutting down the burner. On FT boilers, the low-water cutoff generally includes an intermediate switch that controls the feed pump.

Q What is the correct way to hook up two low-water cutoffs, one of the electrode type, the other of the float type?

A For maximum protection, the second additional low-water cutoff should be mounted on the opposite drumhead, with independent connections and blowdown piping and valving. And both units should be wired and properly interlocked.

> **REMEMBER:** Dual low-water cutoffs provide dual protection only when independent drum connections are provided.

A common mistake on heating boiler installations utilizing a combination feed and low-water cutoff is to connect the makeup-water line to the water-column cross-pipe connection as shown in Fig. 10.24*a*. This horizontal pipe connection can become obstructed, and both the gage glass and the combination feed and low-water cutoff will not sense the true level of water in the boiler. Figure 10.24*b* shows the proper connection to avoid this pipe-obstruction possibility, with the cross-tee connections having accessible plugs for cleaning out the horizontal pipe connections to the boiler.

Q How may scale and sediment impair the operation of a float and electrode-probe-type low-water fuel cutoff?

A See Fig. 10.25. The float-type low-water fuel cutoff (LWCO) pre-

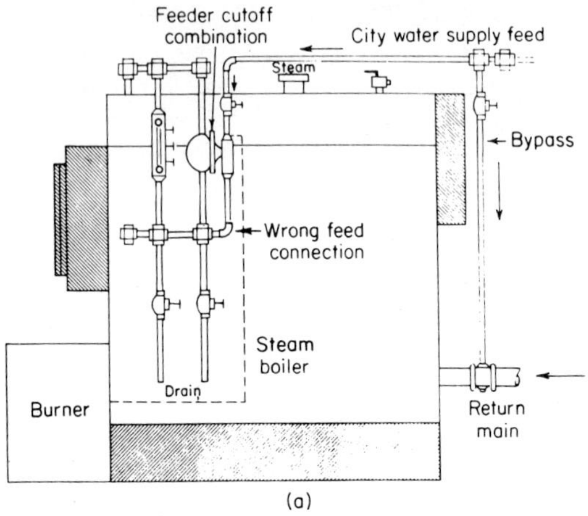

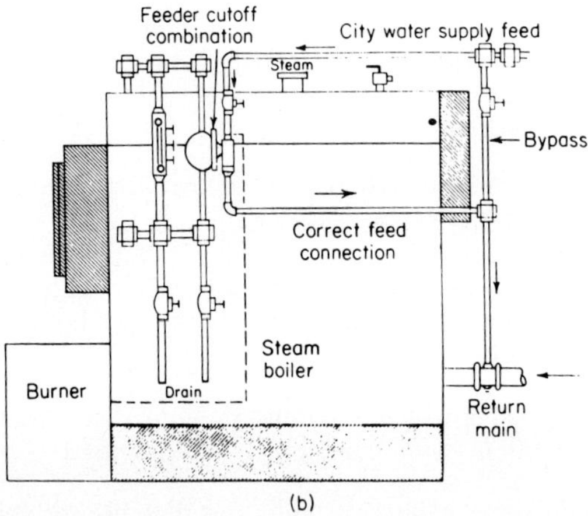

Fig. 10.24 Feeder–low-water fuel cutoff combinations can be connected incorrectly. (*a*) Makeup is at the wrong connection; (*b*) the preferred method of feed connection is through return lines on heating boilers.

sents two problems with sediment deposits in the bowl. The first is that the float itself may become inoperative so that it sticks in the normal water level. If a low-water situation were to occur, the float would not drop and the limit would remain closed. The second problem is that the sediment may plug up the water return and hold the water level in the float chamber at a normal level no matter what the level in the boiler is.

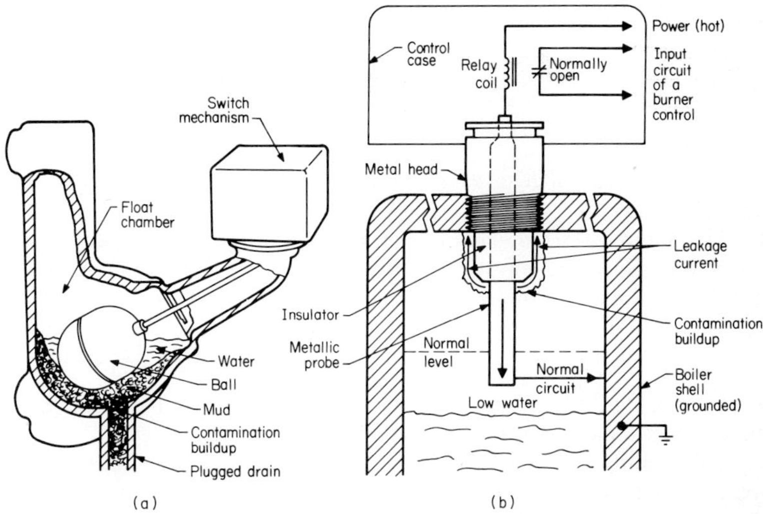

Fig. 10.25 Scale and sediment may affect proper operation of low-water fuel cutoffs when needed. (*a*) Float-type LWCO stuck in sediment; (*b*) probe-type LWCO has leakage current through scale to complete circuit, thus providing no low-water protection.

In this case, the float limit is operating correctly but the overall effect is unsafe. There is a simple solution to these problems: float-type low-water cutoffs require frequent and regular maintenance. When the boiler pressure is used, the float chamber must be blown down, sometimes as often as once every shift. In practice, this maintenance is not performed correctly, often enough, or at all.

This has resulted in many dry-fired incidents each year that were unnecessary. Floats that control both feed to the boiler and the low-water fuel cutoff switch are especially vulnerable to causing major dry-firing damage to a boiler, because one float controls both the feed (an operating control) and the low-water fuel cutoff (a safety control). This is why an independent low-water fuel cutoff is needed; however, both feed and low-water controls must still be maintained free of any scale or sediment buildup in or on their mechanisms.

The probe-type LWCO shown in Fig. 10.25*b* also presents problems from sediment or scale buildup. Contamination buildup on the surface of the insulator will allow leakage current to flow across it from the metallic probe to the metal head and boiler shell. The leakage current is very small at first but increases as the layer of contamination becomes thicker. Eventually, the leakage current becomes large enough to hold in the relay. When this happens, water is no longer necessary to complete the circuit. The water level can then drop to a dangerous level (below the tip of the probe), and the relay will stay pulled in. The normally open contact will stay closed and the burner wil be permitted to run, creating an unsafe condition. Obviously the probe-electrode-type LWCO must also be maintained free of deposits.

Q Why are manual reset controls useful on pressure or temperature high-limit controls?

A The manual reset mechanism on the high-limit control calls attention if the operating control (not high-limit) has malfunctioned, thus prohibiting further boiler operation until corrected. At times, this malfunction may be due to fused contacts, a leaking gas valve, a shorted wire, etc. Thus the boiler should not be operated until this is corrected.

Q How is the fuel/air ratio controlled in small industrial and commercial automatic packaged-type boilers?

A Valve control of the fuel/air ratio is achieved by use of constant-pressure variable areas. A simple mechanism can be used to cause the opening area of two valves to vary in proportion to one another. If the valve characteristics are not the same, the fuel and air flows will match at only two points throughout the range. If the movement is not directly proportional, the mixture will be lean at some firing rates and rich at others.

Figure 10.26*a* shows two rotary-type valves on a common shaft. Figure 10.26*b* shows two rotary-type valves driven by a parallel-arm linkage. One or preferably both of the valves should embody manual adjustment of the valve opening (in addition to the handle adjustment) so as to permit adjustment of the fuel/air ratio. The valve control system

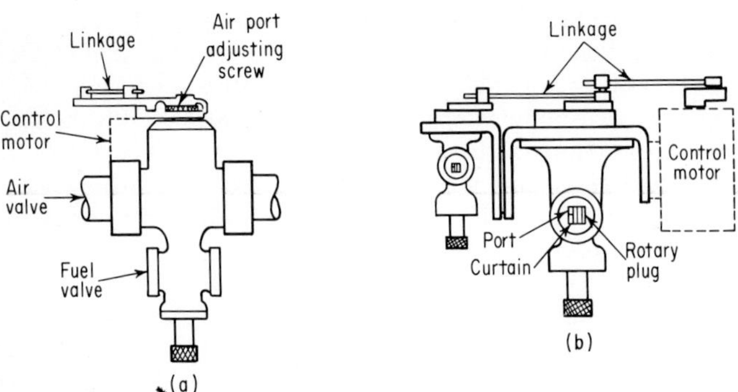

Fig. 10.26 Fuel/air ratio controllers used in industrial boilers. (*a*) Two rotary-type valves control air and fuel flow; (*b*) this mechanism has parallel arms linkage driving the two rotary valves.

requires an air blower with a constant-pressure characteristic and oil or gas pressure regulators ahead of the control valve. Thus the upstream pressures for both air and fuel must be constant at the valve because variations in the oil viscosity would affect the flow rate.

Q Describe a typical control system for pulverized-coal-fired boilers.
A The basic principle is to control the fuel and combustion air simultaneously with changes in steam pressure and readjusting airflow as needed to maintain the optimum air/fuel ratio. Instead of directly connected controllers, actuated by the process variable being measured, transmitters sense the variables directly and then transmit appropriate pneumatic or electric signals to the controllers. The unit shown in Fig. 10.27 is equipped with pressurized, ball-type pulverizers. Preheated combustion air is supplied by the forced-draft fan. In passing through the pulverizer, this air picks up the coal and carries it to the burners. Thus fuel supply is controlled by regulating the flow of primary air.

Combustion air from the forced-draft and primary-air fans is controlled from steam header pressure, positioning a damper at the inlet to each pulverizer. Sensing this airflow is a feeder control subloop (self-contained control system), which regulates the supply of raw coal leaving the feeder, relative to airflow, in order to ensure a constant coal level in the pulverizer. The air/fuel ratio is maintained by means of an air/steam-flow ratio controller, sensing steam header flow and, in turn, positioning the forced-draft fan outlet damper. Thus varying airflow through the pulverizers meets load changes.

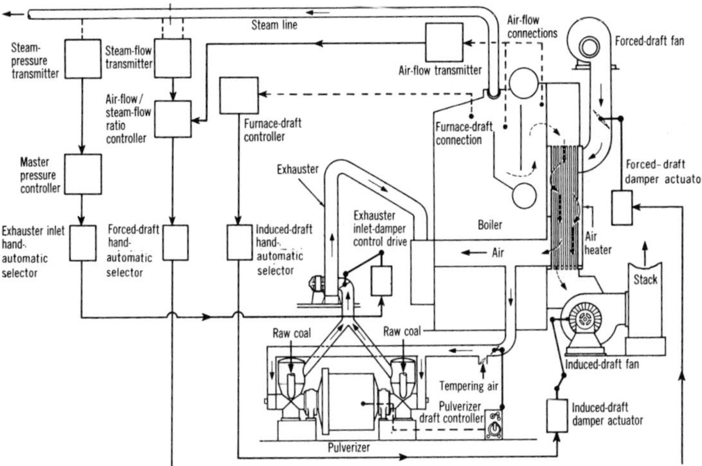

Fig. 10.27 Control-panel layout for industrial pulverized-coal-fired boiler.

A second subloop (coal-air temperature control) senses the pulverizer outlet temperature and regulates the proportion of tempering air combined with that from the air heater before it enters the primary air fan. As with the other control circuits, a hand auto-selector enables the operator to assume manual control when required. When two or more pulverizers are operated in parallel, either manual or automatic compensation is made for the number in service, so that the total load is shared equally by each.

Q How are superheat and reheat steam temperatures controlled?

A Today steam temperature of 1000°F is common and units are being installed for 1050 and 1100°F. Because these high temperatures are limited only by metallurgy, steam temperatures must be held to close limits for safety as well as for economy. Six basic methods are used for controlling the temperatures of steam leaving the boiler:

1. Bypass damper control with a single bypass damper or series-and-shunt damper arrangement for bypassing flue gas around the superheater as required.
2. Spray-type desuperheater control where water is sprayed directly into the steam with a spray-water control valve for temperature regulation.
3. Attemperator control where a controlled portion of the steam passes through a submerged tubular desuperheater and a control valve in the steam line to the desuperheater or attemperator is used.
4. Condenser control with desuperheating condenser-tube bundles located in the superheater inlet header and water-control valve or valves to regulate a portion of the feedwater flow through the condenser as required.
5. Tilting-burner control where the tilt angle of the burners is adjusted to change the furnace heat absorption and resultant steam temperature.
6. Flue-gas-recirculation control where a portion of the flue gas is recirculated into the furnace by means of an auxiliary fan with a damper control to change the mass flow through the superheater and the heat absorption in the furnace, as required to maintain steam temperature.

Q What basic controls are used on once-through boilers?

A Unlike the natural-circulation boiler, where steam-temperature control is best attained by varying the ratio of heat absorption between steam-generating surface and superheating surface, the once-through unit can maintain the desired steam temperature over a much wider load range. If the ratio of heat input to fluid flow is correct, heat absorption is self-adjusting. Spray attemperation then provides rapid temporary control of temperature on load swings.

Figure 10.28 shows a simplified integrated control system for normal operation. The firing rate is closely tied to feedwater flow control. And these are also being controlled to maintain both desired load and desired steam temperature. The turbine governor controls are also tied into the system to give quick response and stability. A small amount of spray desuperheating is used in the up and down deviation of load for better steam-temperature control during transient conditions.

Q What steps are being taken to help power-plant operators to understand better the functions of the computer in aiding them in operating a plant?

A The complexity of a modern power plant has led control engineers to search for better communication between the computer and the human operator. One result is the use of lighted, color-coded push buttons grouped adjacent to their respective edgewise indicators. Actuation of "increase" and "decrease" buttons replaces the operation of conventional control handles. Flashing alarm lamps, similarly located in the push buttons, draw instant attention to the pieces of equipment concerned, saving the time otherwise wasted in relating an alarm signal to its source.

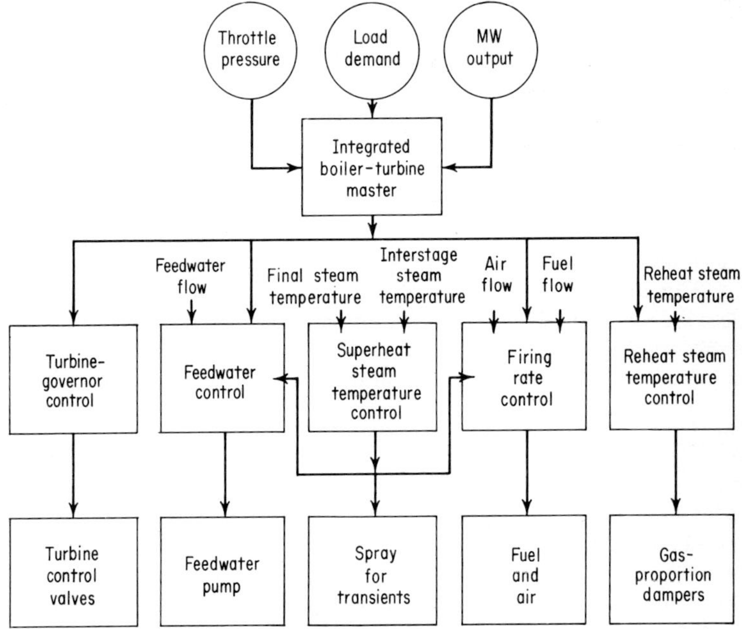

Fig. 10.28 Block diagram of integrated boiler-turbine control system.

Another solution to the same problem is to link the process computer with an audible alarm annunciator. One such power-plant installation has an audible annunciator coupled with a continuous loop 16-track tape-recording unit. When the computer recognizes an out-of-limit input signal, it simultaneously addresses both an alarm printer and the tape unit. While the printer types a statement of the alarm occurrence, the tape unit selects, from several hundred prerecorded vocal messages, the one message stating what alarm signal has been received and broadcasts it over the annunciator.

Should out-of-limit inputs occur faster than they can be announced, the computer is programmed to store the signals and announce them either in the order in which they occur or in order of their priority. Both alarm warnings and remedial instructions can be prerecorded on the tape. A microphone data-input facility lets the programmer or operator change this taped information at will.

Q What are some of the advantages of the new developing electronic controls?

A Because of the increased use, costs are beginning to be less than for mechanical and pneumatic controls. Their chief attractions are in the fine control that is possible provided sensors are available with similar accuracy, and the fact that solid-state electronic devices have a good reliability record. Their speed of response has improved efficiency of boiler fuel consumption.

The development of reliable sensors to monitor pressure, temperature, flows, and similar power-plant variables has permitted rapid transmission of these measurable quantities to electric actuators, such as motor-controlled or solenoid-actuated valves and dampers in order to quickly and accurately regulate pressure, temperature, flow, and similar variables within established set points.

Electronic signals from sensors can also be quickly displayed on data loggers and CRT screens for operator review, analysis, and adjustment (if needed) of the variables being controlled. This assists the operator in diagnosing troubles on equipment in the power cycle.

Q How are microprocessors being applied in power-plant operation?

A They are being applied on a segmented basis instead of on a total plant computer basis, for example, for fine air/fuel ratio control. Energy is lost in combustion when the proportions of fuel and air are not in the correct chemical or stoichiometric ratio (covered in the next chapter). Sensors have been developed, such as optical-opacity monitors, zirconium oxide fuel cells that sense oxygen, and infrared carbon monoxide analyzers.

These sensors can be employed with small or microprocessor computers which can quickly calculate the chemical equation for the proper air/fuel ratio and feed this information to appropriate actuators (valve, damper, speed control on fans, etc.) in order to obtain the desired air/fuel ratio. They are considered very efficient in load swings. Many of the developing microprocessors have built-in calibration, self-diagnostic circuits, and integral alarms.

The biggest advantage of microprocessor units is in their ability to control variables within narrow set points so that there is very little drift in the controlled parameters. This can result in fuel savings by avoiding fuel-consumption swings that occur under normal mechanical or pneumatic controllers.

Benefits cited in retrofitting existing systems with modern electronic controls are:

1. Existing controls may be obsolete, with parts difficult to obtain.
2. Some existing controls may not be up to present safety standards. This applies especially to flame-safeguard systems (see later chapters).
3. Additional maintenance may be required on old, worn controls.
4. Efficiency is increased in maintaining boilers on the line within closely defined fuel-control limits.

CHAPTER **ELEVEN**

Fuels, Firing, Combustion, and Pollution-Control Equipment

Steam generators are designed to burn all types of fuels under all types of demand within defined pressure and temperature limits. The burning process is a dynamic one, with fuel and air constantly being added or deleted depending on load conditions. Heat-transfer surfaces may affect the amount of fuel required to obtain the desired output. Recirculation and superheater control on large units also require operator attention. The products of combustion must be disposed of in a safe and legal manner.

This chapter will review the fuels available for burning in steam generators, the burners used, and the advent by federal law of pollution-control equipment. The next chapter will review combustion-control safeguard systems.

COMBUSTION

Q Explain how any material starts to burn.

A Some things burn more easily than others, depending on how easy it is to turn the substance into a gas, because nothing truly burns until it is a gas. We know it is easier to start wood burning than coal and easier to ignite a twig than a log. To make wax in a candle burn, we first turn it into a gas. With a wick, heat from a match flame lights it easily. But without a wick the wax won't light because the heat of the match is too small in comparison with the mass of wax; so no wax is vaporized. But the wick draws (by capillary action) an amount of melted wax so small in relation to the match flame that enough heat is available to raise the wax to the temperature needed for vaporization. Once wax is vaporized, or made like a gas, burning is relatively easy to start. Then the burning wax gives off enough heat to continue the process of melting, vaporizing, and igniting.

Q What is meant by the term *ignition temperature*?
A When we have a combustible gas and the exact amount of air to burn it completely, we still need another element—heat. If such a perfect mixture is heated gradually, the rate of chemical combination increases until a point is reached where the reaction no longer depends on heat from an outside source. The lowest temperature at which this happens is the ignition temperature, the temperature at which heat is generated by the reaction faster than it is lost to the surroundings, and combustion thus becomes self-propelling. Below this point, the gas-air mixture will not burn freely.

Q How is fuel converted for burning in a boiler?
A It is the furnace heat that does the conversion. The firing equipment puts the fuel into shape that makes the best use of this heat. For suspension firing of both oil and coal the answer is the same: fuel must be broken up into many small particles to expose as much surface as possible. With oil, atomization is obtained in a variety of ways by the burner. With coal, the pulverization is handled in a separate unit, the burner merely mixing the finely ground particles with air and injecting them into the furnace.

Q What three requirements are needed for the proper chemical reaction to take place in combustion?
A 1. Proper proportioning of fuel and oxygen (or air) with the fuel elements as shown by chemical equations.

2. Thorough mixing of fuel and oxygen (or air) so a uniform mixture is present in the combustion zone and so every fuel particle has air around it to support the combustion. Solid fuels will generally be converted to gas first by the heat and presence of air. Liquid fuels will vaporize into gases and then burn. Atomization of liquids increases the mixing with air and increases the vaporization into a gas. Pulverization of coal will have the same effect.

3. The ignition temperature must be established and monitored so that the fuel will continue to ignite itself without external heat when combustion starts.

Q What are the chief heat-producing elements in solid, liquid, or gaseous fuels?
A The chief heat-producing elements in fuels (except for atomic reaction and electricity) are carbon, hydrogen, and their compounds. Sulfur, when rapidly oxidized, is also a source of some heat energy, but its presence in a fuel has bad effects. The burning of coal, oil, or gas is a chemical reaction involving the fuel and oxygen from the air. Air is 23 percent oxygen by weight and 21 percent by volume. The remainder of air is mostly nitrogen, which takes no actual chemical part in combus-

tion but does affect the volume of air required. The table in Fig. 11.1 represents some typical combustion reactions for various fuel constituents.

> **NOTE:** It is always the carbon, hydrogen, or sulfur that produces the chemical reaction for heat by combining with oxygen.

Q How can the amount of air required for combustion be determined from combustion equations where only oxygen is being actually used?
A Since oxygen in the air is known to be 23.15 percent by weight and 21 percent by volume (from the combustion equation), the amount of air required can be calculated. For example, in the complete combustion of carbon, it has been determined that 2⅔ lb of oxygen is required to burn 1 lb of carbon. The amount of air required to burn 1 lb of carbon would then be

$$\frac{\text{Amount of oxygen}}{\text{\% oxygen in air by weight}} = \frac{2.67}{0.2315} = 11.52 \text{ lb}$$

The table in Fig. 11.2 shows some typical combining proportions for various fuel elements for perfect combustion.

Q How does nitrogen affect the combustion process?
A Because nitrogen does not burn although occupying 79 percent by volume of air used in burning fuels, it increases the amount of air required to support combustion. Oxygen occupies only 21 percent of the air's volume. Although nitrogen will not burn, it will absorb released heat. Thus nitrogen affects combustion as to the temperature and time needed to complete the burning of the fuel. With pure oxygen, combustion is more spontaneous and rapid. For every pound of oxygen, 4.32 lb of air must be supplied, and for every cubic foot of oxygen, 4.78 ft^3 of air is needed for combustion.

Q What is the effect of too little or too much air on combustion?
A If not enough oxygen or air is supplied, the mixture is rich in the fuel; thus the fire is reduced, with a resultant flame that tends to be longer and smoky. The combustion also is not complete, and the flue gas (products of combustion) will have unburned fuel such as carbon particles or carbon monoxide instead of carbon dioxide. Less heat will be given off by the combustion process. If too much oxygen or air is supplied, the mixture and burning are lean, resulting in a shorter flame and cleaner fire. The excess air takes some of the released heat away from the furnace and carries it up the stack.

Q Are most fuels burned in a boiler with lean mixtures of excess air or with rich mixtures of too much fuel?

Fig. 11.1 Chemical reactions of fuel with air produces combustion products and heat.

Combustible element	Symbol	Chemical reaction	Combustion product	Volumes	Weights/lb of combustible				Heating value, Btu/lb
					Oxygen, lb	Nitrogen, lb	Air, lb	Gaseous products, lb	
Carbon..........	C	$C + O_2 \rightarrow CO_2$	Carbon dioxide	1 vol C + 1 vol O_2 = 1 vol CO_2	2.67	8.85	11.52	12.52	14,600
Carbon..........	C	$2C + O_2 \rightarrow 2CO$	Carbon monoxide	2 vol C + 1 vol O_2 = 2 vol CO	1.33	4.43	5.76	6.76	4,440
Carbon monoxide.	CO	$2CO + O_2 \rightarrow 2CO_2$	Carbon dioxide	2 vol CO + 1 vol O_2 = 2 CO_2	0.57	1.90	2.47	3.47	10,160
Hydrogen........	H	$2H_2 + O_2 \rightarrow 2H_2O$	Water	2 vol H_2 + 1 vol O_2 = 2 vol H_2O	8	26.56	34.56	35.56	62,000
Methane.........	CH_4	$CH_4 + 2O_2 \rightarrow CO_2 + 2H_2O$	Carbon dioxide and water	1 vol CH_4 + 2 vol CO_2 = 1 vol CO_2 + 2 vol H_2O	4	13.28	17.28	18.28	23,850
Ethylene.........	C_2H_4	$C_2H_4 + 3O_2 \rightarrow 2CO_2 + 2H_2O$	Carbon dioxide and water	1 vol C_2H_4 + 3 vol O_2 = 2 vol CO_2 + 2 vol H_2O	3.43	11.38	14.81	15.81	21,600
Ethane..........	C_2H_8	$2C_2H_8 + 7O_2 \rightarrow 4CO_2 + 6H_2O$	Carbon dioxide and water	2 vol C_2H_8 + 7 vol O_2 = 4 vol CO_2 + 6 vol H_2O	3.73	12.40	16.13	17.13	22,230
Sulfur...........	S	$S + O_2 \rightarrow SO_2$	Sulfur dioxide	1 vol S + 1 vol O_2 = 1 vol SO_2	1	3.32	4.32	5.32	4,050

Fig. 11.2 Weights and volumes of air and O_2 per pound or cubic foot of fuel.

			Weights and volumes per lb or cu ft of fuel					
Combustible element	State	Symbol	Cu ft O_2/cu ft fuel	Cu ft air/cu ft fuel	Lb O_2/lb fuel	Lb air/lb fuel	Cu ft O_2/lb fuel	Cu ft air/lb fuel
Carbon	Solid	C	...	.....	2.67	11.51	31.65	151.3
Hydrogen	Gas	H_2	0.5	2.39	8.0	34.5	94.8	453
Carbon monoxide	Gas	CO	0.5	2.39	0.572	2.47	6.79	32.5
Sulfur	Solid	S.	...	.....	1.0	4.52	11.87	56.7
Methane	Gas	CH_4	2.0	9.56	4.0	17.28	47.4	226.5
Ethane	Gas	C_2H_6	5.5	16.72	3.735	16.12	44.5	212
Propane	Vapor	C_2H_8	5	23.9	3.655	15.68	43.1	206.5
Butane	Vapor	C_4H_{10}	6.5	31.1	3.585	14.48	42.6	205.5
Octane	Liquid	C_8H_{18}	...	.....	3.51	15.15	41.6	199

A Burning should always be with excess air to ensure that all the fuel is properly burned and thus to attain better efficiency in heat release. This also reduces smoke formation and soot deposits, which today, with stricter pollution laws, is important.

Q How can one determine whether the combustion in a boiler is with too little air or excess air?

A When flue gas comes out of a stack as black smoke, it is an indication of insufficient air. Too much air usually causes a dense, white smoke. A faint, light-brown haze coming from the stack is a sign of a reasonably good air/fuel ratio. Of course a more exact analysis is made with a flue-gas analyzer, such as an Orsat apparatus. From this analysis, the percentage of either excess or insufficient air can be determined.

Q What does a flue-gas analyzer measure?

A It measures the percentages of volume of carbon dioxide, carbon monoxide, and oxygen. Because air contains 21 percent oxygen and 79 percent nitrogen by volume and because nitrogen goes through a combustion process unchanged, the maximum percentage of CO_2 in flue gases that is possible (discounting nitrogen) is 21 percent. And since one volume of carbon combining with one volume of oxygen produces one volume of carbon dioxide, any unburned oxygen (excess) will reduce the percentage of carbon dioxide in the flue gas.

EXAMPLE: 1. If there is no excess air, the flue-gas analysis will be

Carbon dioxide	21%
Oxygen	0
Nitrogen	79
Total	100%

2. If there is excess air of 100 percent

Carbon dioxide	10.5%
Oxygen	10.5
Nitrogen	79.0
Total	100.0%

NOTE: The 21 percent maximum possible is now split evenly between oxygen and carbon dioxide at 10.5 percent each (100 percent excess air).

3. If there is excess air of 50 percent

Carbon dioxide	14%
Oxygen	7
Nitrogen	79
Total	100%

In this case, carbon dioxide and oxygen percentages still add up to 21 percent, but the oxygen is 50 percent of the carbon dioxide (50 percent excess air). It is also possible to calculate the amount of air used per pound of fuel from the flue-gas analysis.

Q Why is a high percentage of CO_2 in boiler flue gas efficient?
A A high percentage of CO_2 in boiler flue gas is a good thing, within limits, and not because of perfect combustion. While carbon, completely burned, turns to CO_2, this does not mean that a high percentage of CO_2 indicates complete combustion. Actually, combustion can be complete with 6 percent CO_2 and incomplete with 15 percent. So what is wrong with a low percentage of CO_2? A low percentage of CO_2 is proof that the flue gas is heavily diluted with excess air. Since this excess air goes to waste up the stack at a fairly high temperature, it is a great loss in efficiency (heating the sky).

As an illustration, let us start with 100 ft^3 of air. That is the amount theoretically needed to burn ⅔ lb of carbon, our fuel supply for this example. Of the 21 percent oxygen by volume in air, only the oxygen burns with the coal. The remaining 79 percent is nitrogen and goes along for a free ride. And when oxygen burns with carbon, the volume of CO_2 produced (figured back to room temperature) equals the volume of oxygen consumed.

Q What are the usual percentages of excess air in burning the various common fuels in boilers?
A For coal, usually 50 percent excess air is used. With oil, gas, or pulverized coal, excess air is 10 to 30 percent.

Q Illustrate and explain a typical Orsat flue-gas analyzer.
A The Orsat (Fig. 11.3) is a portable flue-gas analyzer. The leveling bottle and measuring burette contain pure water. The first pipette, A, is packed with steel wool, kept wet with caustic-potash solution from the container below. The steel wool in pipette B is wet with pyrogallic solution from the lower container. The third pipette, C, contains copper strips wet with cuprous (copper) chloride solution. Pipettes A, B, and C are for the absorption of CO_2, O_2, and CO. But other chemicals than those indicated are sometimes used for these absorptions.

The Orsat is connected to the sampling tube by rubber tubing in series with an aspirator rubber bulb to pump the gas. The gas line connects to the back side of the three-way cock that has three positions for these purposes: (1) to connect the sampling tube to the measuring burette, (2) to connect the burette to the atmosphere, and (3) to connect the sampling tube to the atmosphere.

Q Explain how a flue-gas analysis is made with an Orsat.

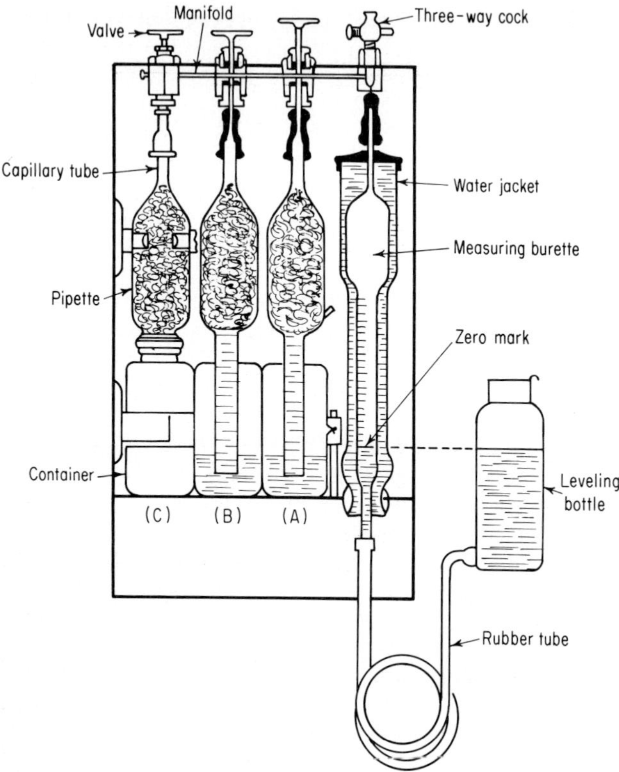

Fig. 11.3 Orsat flue-gas analyzers are still used to check flue-gas constituents in power plants.

A Always study the instructions given with your instrument. Here are the main steps to use for this one (Fig. 11.4).

Step 1. With the hand aspirator, pump gas through the measuring burette. Gas bubbles out of the measuring bottle as shown. Keep pumping to displace all air.

Step 2. Set three-way cock to connect the burette to the atmosphere as the leveling bottle is slowly raised until water is at the zero mark near the bottom of the burette. As this is level with the water in the bottle, the sample is measured at atmospheric pressure. (All gas measurements must be made with the water level the same in the burette as in the leveling bottle.)

Step 3. Lift the bottle high to absorb the CO_2 after closing the three-way cock. Open the valve to the CO_2 pipette, so that the gas displaces the liquid and contacts the caustic solution on the steel wool, which absorbs CO_2. Pinch the tube to check the flow when water rises to the mark in the capillary tube above the burette. Allow a few seconds for CO_2 absorption.

Step 4. With the tube pinched, lower the bottle. Then release the rubber tube slowly until the liquid rises exactly to the mark in the capillary tube on top of the pipette. After closing the pipette valve, raise or lower the bottle until the water level in the bottle is exactly the same as in the burette. Read the graduation at the water level in the burette. This is the CO_2 percentage. Repeat Steps 3 and 4 to make sure that all

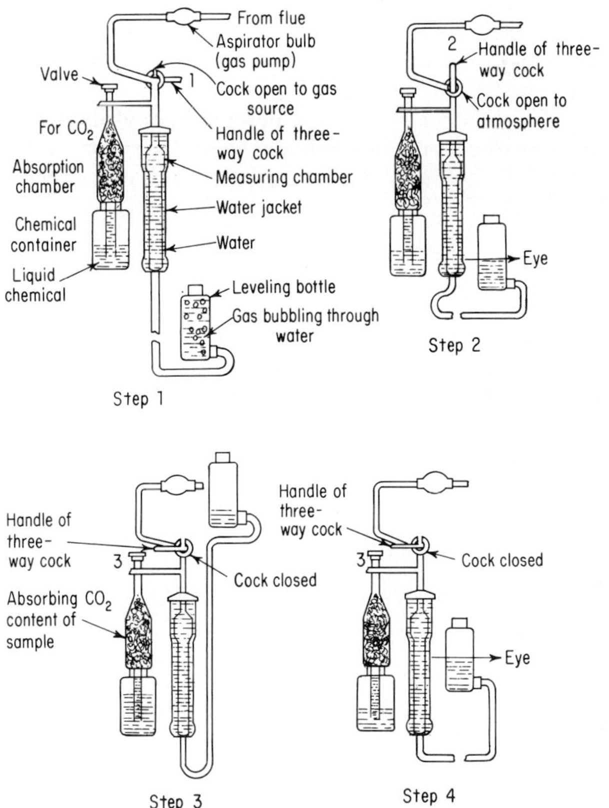

Fig. 11.4 Procedure in using Orsat apparatus: Step 1: Pumping sample of flue gas into measuring burette. Step 2: Bringing gas sample to zero mark at atmospheric pressure. Step 3: Transferring gas sample to first absorption pipette. Step 4: Returning gas sample to burette to measure percentage of CO_2.

the CO_2 is absorbed. To absorb the O_2, repeat Steps 3 and 4 with pipette B. The increase in the burette reading is the percentage of O_2. For pipette C to absorb the CO, repeat Steps 3 and 4. The further increase in burette reading is the percentage of CO.

Q Explain how the actual weight of air used per pound of fuel can be calculated from a flue-gas analysis.

A Use the equation

$$W_A = \frac{28N_2}{12\,(CO_2 + CO)(0.769)}\left(\frac{W_f C_f - W_r C_r}{W_f \times 100}\right)$$

where CO_2 = percent of carbon dioxide in flue gas by volume
CO = percent of carbon monoxide in flue gas by volume
N_2 = percent of nitrogen in flue gas by volume
W_f = weight of fuel fired, lb
C_f = carbon content of fuel, percent from ultimate analysis
C_r = carbon content of ash and refuse, percent
W_r = weight of ash and refuse from W_f pounds of fuel, lb
W_A = actual weight of air per pound of fuel burned, lb

Q Illustrate the use of this equation.

A Assume that 700 lb of coal is fired in a boiler. The carbon content of this coal is 68 percent. Ash and refuse after burning amounted to 60 lb, with the carbon in refuse being 7.8 percent.

Flue-gas analysis showed the following percentages by volume:

CO_2 (carbon dioxide) = 11.9%
N_2 (nitrogen) = 80.9%
CO (carbon monoxide) = 1%
O_2 (oxygen) = 7.3%

What is the actual weight of air used to burn this coal?

Substituting values, we have

$CO_2 = 11.9$
$CO = 1.0$
$N_2 = 80.9$
$W_f = 700$
$C_f = 68$
$C_r = 7.8$
$W_r = 60$

Substituting in the equation,

$$W_A = \frac{28\,(80.9)}{12\,(11.9 + 1.0)(0.769)}\left[\frac{700\,(68) - 60\,(7.8)}{700 \times 100}\right]$$

$$= 19.0(0.68)$$
$$= 12.9 \text{ lb of air per pound of fuel fired}$$

or 700 × 12.9 = 9030 lb of air for 700 lb of coal

Q Can the weight of *flue gas* W_{fg} be obtained per pound of fuel fired from a flue-gas analysis?

A Yes, use the equation

$$W_{fg} = \frac{4CO_2 + O_2 + 700}{3\ (CO_2 + CO)} \left(\frac{W_f C_f - W_r C_r}{W_f \times 100} \right)$$

Also use the same symbols as in the previous equation.

EXAMPLE: The same coal is being burned with the same analysis as shown in the previous problem.

Substituting

$$W_{fg} = \frac{4\ (11.9) + 7.3 + 700}{3\ (11.9 + 1.0)} \left[\frac{700\ (68) - 60\ (7.8)}{700 \times 100} \right]$$

$$= 20.06(0.68)$$
$$= 13.6 \text{ lb flue gas per pound of fuel fired}$$

or 13.6 × 700 = 9520 lb of flue gas for 700 lb of coal

Q Can the amount of heat lost by flue gases be calculated?

A Yes, by using the specific heat (Btu per pound per degree temperature rise for a gas), it is possible to calculate the heat lost up the stack. Use the equation

$$H_L = W\ C_p(T_2 - T_1)$$

where H_L = heat lost by flue gas
W = weight of flue gas going up the stack, usually lb/h
C_p = mean specific heat of flue gas; can be taken as approximately 0.25 Btu/(lb · °F)
T_2 = stack temperature, °F
T_1 = temperature of air entering furnace, °F

EXAMPLE: Assume that in the previous problem 9520 lb of flue gas per hour went up the stack. The stack temperature is 650°F and the air inlet temperature is 80°F. Then the heat lost up the stack is

$$H_L = 9520(0.25)(650 - 80)$$
$$= 1{,}356{,}600 \text{ Btu/h}$$

Since 700 lb of coal was fired, with an assumed average of 14,500

Btu/lb, the total energy input is

$$700 \times 14{,}500 = 10{,}150{,}000 \text{ Btu/h}$$

The percent of heat input going up the stack is then

$$\frac{1{,}356{,}600}{10{,}150{,}000} = 13.4\% \qquad \textit{Ans.}$$

This indicates a boiler efficiency of $100 - 13.4 = 86.6\%$

Q Explain the use of charts showing excess air, CO_2 percentage, and heat-loss percentges.

A Charts are only for broad answers and are approximate percentages. See Fig. 11.5 to solve the following: What is the percentage of stack loss in hot gases for a bituminous coal if the stack-gas temperature is 580°F, the room temperture is 80°F, and the CO_2 in the flue gas is 14.2 percent?

First, a given CO_2 analysis means there is one specific percentage of excess air when burning anthracite coal, less for bituminous coal, and still less for fuel oil. For 12 percent CO_2, Fig. 11.5*a* shows 61 percent excess air for anthracite, 53 percent for bituminous, and 24 percent for fuel oil. The actual stack loss in the hot gases depends on both the excess air and the stack temperature.

The scale on this chart for bituminous coal shows that the excess air is 30 percent for 14.2 percent CO_2. The temperature rise is the stack temperature minus the room temperature, or 580°F − 80°F = 500°F. Now to find the loss from the chart in Fig. 11.5*b*.

Run a straightedge from 500 on the left scale through 30 on the center scale (dashed line). It cuts the right scale at 12.2 percent. This means that 12.2 percent of the coal's heat is wasted in heating the sky.

Q How does the furnace provide for proper combustion?

A For fuel to be burned efficiently, the furnace must have adequate combustion space to ensure a thorough mixing of air and fuel. The furnace must also maintain a high enough temperature for complete combustion. The furance must be tight so that air cannot leak into, or out of, the casing and thus affect the air/fuel ratio. Too high a furnace temperature must also be avoided, as this can lead to rapid deterioration of linings (if installed) or possible overheating of vital pressure parts such as tubes or combustion furnaces in scotch marine boilers. Also, the reaction of combustion should be completed before the flue gases leave the combustion chamber so as to avoid flame impingement on tubes and possible overheating.

Q What causes soot and smoke?

A With either coal or oil, some carbon formerly associated in hydrocarbon compounds breaks away as free carbon. Thus we always have the

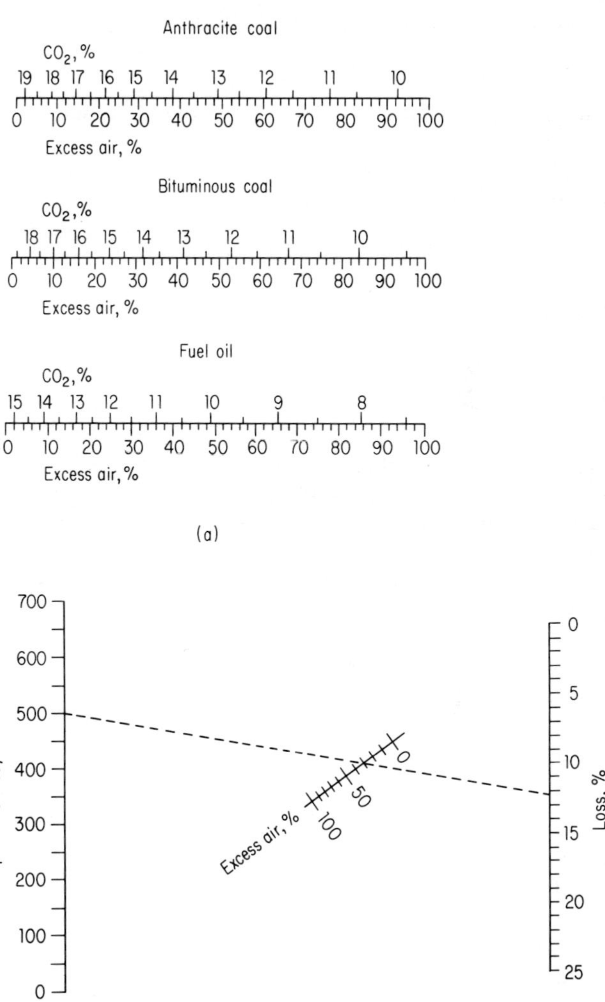

Fig. 11.5 (*a*) For any values of CO_2, these scales give corresponding values in percentage of excess air involved. (*b*) For hard and soft coal and for fuel oil, this chart gives percentages of fuel's heat wasted in hot flue gas.

problem of burning carbon particles, which proves difficult even when they have ample time to pass through the furnace. Incomplete combustion of carbon is the germ of smoke and soot. If the furnace is small and the relatively cold areas (boiler tubes and other heating surfaces) are badly located, carbon cannot possibly burn to completion. So all these steps, bringing air and fuel together, raising the mixture to ignition temperature, sweeping away successive layers of gas from fuel particles, and burning carbon as far as possible, occur while the fuel and air travel from burner to furnace outlet. But this is an extremely short time; thus complete combustion also depends on the distance traveled, the speed, and whether or not the flow is turbulent.

Q What are primary and secondary air in combustion of fuels in a boiler?
A See Fig. 11.6. Primary air is the air mixed with the fuel at or in the burner. Secondary air is air usually brought in around the burner or through the openings in a furnace wall or floor. The primary air ensures instant combustion as the fuel enters the furnace. The secondary air provides the oxygen to complete the combustion in the furnace.

Q What is the harmful effect of sulfur in a fuel?
A Sulfur burns to sulfur dioxide, which when mixed with water or water vapor, forms sulfurous acid which is corrosive to tubes, breechings, and economizer sections on larger boilers. The dew point (temperature at which water vapor condenses) of the flue gas has to be watched (especially with high-sulfur-content fuel) as the gas becomes cooler and cooler while going through a furnace, so as to prevent the combination of water vapor with sulfur dioxide. Recent air-pollution laws are tough on high-sulfur-content fuels, because sulfur dioxide is known to pollute the air. Percentages of permissible sulfur content in fuels are slowly being lowered from the previous 4 percent maximum to 1 percent in the near future.

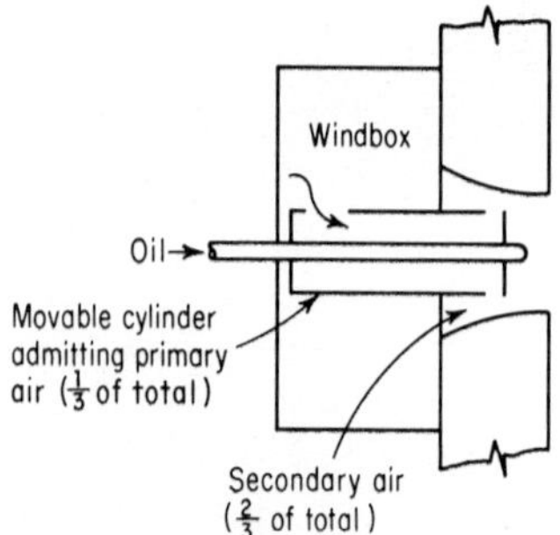

Fig. 11.6 Primary and secondary air for oil burner. Secondary air completes the combustion of the gas from the fuel inside the furnace.

Sulfur in oil-fired boilers leads to fireside corrosion of tubes, especially in those boilers cycling on and off. In cooling on the off cycle, the sulfur dioxide combines with water vapor and with water from leaks to attack the tubes by means of the resultant acid formed.

Q What are ash and slag? What problems do they cause?
A Ash and slag are impurities that do not burn to a gas and usually trouble coal-fired boilers. The solid particles at high velocity are carried through the boiler with gas in suspension. The general term *fly ash* is used for this slag and ash. Ash and slag can be very abrasive to tube sections if flow distribution is concentrated in the convection passages of a boiler.

Q What function does draft play in the combustion of fuel?
A Draft provides the differential pressure in a furnace to ensure the flow of gases. Without draft, stagnation in the burning process would result, and the fire or process of combustion would die from lack of air. Draft pushes or pulls air and the resultant flue gas through a boiler and up into the stack. The draft overcomes the resistance to flow of the tubes, furnace walls, baffles, dampers, and the chimney lining (also slag).

Natural draft is produced by a chimney into which the boiler exhausts. The cool air admitted to a furnace (by means of damper openings) rushes in to displace the lighter hot gases in the furnace. Thus the hot gases rise (chimney effect), causing a natural draft.

Mechanical draft is produced artificially by means of forced- or induced-draft fans. The chimney is still necessary on mechanical-draft installations for venting the products of combustion high enough not to be offensive to the surroundings. Most modern boilers, including the domestic type, use some form of mechanical draft. Domestic burners may have a fan built into the burner unit.

Q How does a forced-draft fan differ from an induced-draft fan?
A A forced-draft fan pushes, or forces, air into the furnace, usually at a pressure higher than atmospheric pressure, whereas an induced-draft fan draws the air out of the furnace by creating a partial vacuum on the suction side of the fan. The blades of induced-draft fans are prone to rapid wear because they have to handle hot, corrosive gases, possibly hot unburned cinders or fly ash. Thus they require periodic cleaning and dynamic balancing to prevent excessive vibration.

Q What is meant by balanced draft?
A A boiler using both forced-draft fans and induced-draft fans can be regulated and balanced in the amount of air and flue gas handled so that furnace pressure is almost atmospheric. This results in better control of

air leakage from the furnace and thus control of the fuel/air ratio in the furnace.

Q How is draft *measured*?
A Air and gas under flow conditions are measured in *inches of water* (balanced by air or gas pressure) usually with a U tube (manometer). One side of the tube is connected to the chimney or furnace. The other side is open to the atmosphere. Thus the difference in the water level in the two columns indicates the inches of water, which is a measure of the difference in pressure (Fig. 11.7). If the furnace pressure is greater than atmospheric, the water in the column connected to the furnace will be pushed down, and on the column connected to the atmosphere the water will rise the same amount. If the pressure is less than atmospheric, the reverse will take place.

Q Convert 3 in of draft into pounds per square inch or into ounces per square in.
A Since water at normal temperature weighs 62.4 lb/ft^3, dividing this by the number of cubic inches in a cubic foot gives

$$\frac{62.4}{1728} = 0.036 \text{ psi per in}^3 \text{ of water}$$

Thus for 3 in, pressure = 3 × 0.036 = 0.108 psi. And since there are 16 oz in a pound, the ounces per square inch = 0.108 × 16 = 1.73 oz pressure per square inch. *Ans.*

Q Explain how electronic sensors are being used to optimize air/fuel ratios by fine "tuning" of burner systems.
A Early combustion-optimization systems were manually oriented. Visual observation of smoke indicated presence of unburned carbon and need for the operator to open the damper and increase the proportion of air. Advances in sensing technologies now make it practical to measure stack-gas composition directly and use the signals to trim the air/fuel ratio to the point where the excess oxygen is just enough to ensure essentially complete combustion.

Oxygen concentration yields a direct indication of excess air that can be fed back to the control system. The chief advantage of using this variable is simplicity. Limitations include need to change the excess oxygen set point for different fuels or boiler loads, relatively low sensitivity close to the control point, and bias introduced by air infiltration into the system downstream of the combustion chamber.

Carbon dioxide measurement is closely correlated with oxygen concentration near the control point and can be employed to infer excess air. The measurement offers the same relative advantages and disadvantages as oxygen for control.

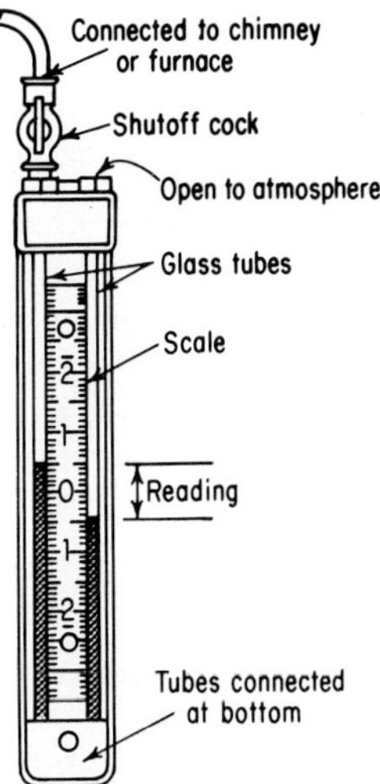

Fig. 11.7 U-tube draft gage measures pressure differential in inches of water between stack and atmosphere.

Carbon monoxide concentration is a sensitive indicator of the onset of incomplete combustion, even when excess air is above the set point. This approach is independent of fuel or load factors and is not subject to errors caused by infiltration of air into the ductwork downstream of the combustion chamber because the gas does not occur in the atmosphere. Use of carbon monoxide as the process variable has the added benefit that smoking is curtailed because strategies inherently eliminate unburned carbon. Limitations are chiefly in the narrow spans of concentrations over which control can be exerted.

Concentration measurements can be used in combination as well as singly for feedback control. For instance, oxygen or carbon dioxide concentration can be used for coarse adjustments with carbon monoxide for fine tuning. Alternatively, carbon monoxide can be utilized for control with a high-oxygen alarm to help identify dirty burners, faulty air registers, and incorrect fuel/air mixing. Variables such as exhaust gas flow, temperature, and opacity or particle density can also be factored into the strategies.

Q How are microprocesser-based analyzers being used in stack-gas monitoring systems?

A Microprocessors and microcomputers are being introduced into stack monitoring systems and may include digital circuits that accept measurement from standard analyzers and provide the logic necessary for sophisticated control or reporting. The microprocessors can also operate directly from the sensors, producing such outputs as converting engineering units, checking if alarm conditions may be developing, compensating for drift on actuators, and similar assigned functions.

Q How are stack analyzers calibrated?
A Units with microprocessors may have means for automatic calibration by computer logic. Other approaches include replaceable span-gas canisters and internal filters so that units can be calibrated in place. One company offers manual and automated units that regulate the flow of reference air supply to the probe for zero adjustment and also provide span gas to the sensor.

Q What determines whether continuous or periodic monitoring of stack gas should be employed to optimize combustion?
A The size of a boiler and the sophistication of its controls generally dictate whether continuous monitoring or periodic inspection would be more appropriate for energy-combustion optimization. Continuous instruments are frequently installed in environmentally regulated areas adjacent to the boiler or stack. Probes, inserted in the flue-gas passage, extract samples of the gas for transport to the instrument or processing unit. Extractive sampling has the advantage of safeguarding instrument performance, since the analyzer need not be exposed to the conditions in the stack. Another advantage is that a single analyzer can often be used with multiple samplers, to measure the flue-gas composition of different units.

Q What gases are measured for pollution control instead of for optimizing combustion efficiency?
A Actually efficient combustion will also assist in limiting pollution. However, pollution-control measurement involves checking for the presence of sulfur dioxide, hydrogen sulfide, unburned carbons or carbon monoxide, and nitric and nitrous oxides. Visible emissions such as smoke are usually monitored by opacity meters.

Q On boilers equipped with automatic combustion systems, why should an operator be familiar with both automatic and manual operation?
A There is always the problem that automatic devices will fail or malfunction or be in error in maintaining proper combustion. If operators recognize improper combustion when in the automatic mode, they can take corrective actions in the manual mode in order to correct the malfunction that may be present.

Q On heating-boiler installations, what rule of thumb is generally followed in judging whether sufficient fresh air is being supplied to a boiler?

A Insufficient air is often provided for heating-boiler rooms on the assumption that a "closed" boiler room will provide preheated air for the burner and thus will save fuel. A decidedly acrid odor is often noticeable in such boiler rooms. Lack of sufficient combustion air can result in soot deposits on heat-transfer surfaces that may block flue-gas passage. The formation of carbon monoxide from incomplete combustion can be dangerous if the setting leaks and this odorless gas starts to fill the boiler room. A free and clear fresh-air opening to the boiler room will help prevent such a condition from developing. Generally, 1 in^2 of outside air opening per 2000 Btu of fuel input is a good rule of thumb to follow.

EXAMPLE: An oil burner burns 5 gal/h. If this fuel has a heating value of 160,000 Btu/h, the required fresh-air intake would be

$$\frac{5 \times 160{,}000}{2000} = 400 \text{ in}^2 \text{ or a } 20 \times 20 \text{ in opening}$$

HEATING VALUES OF FUELS

Q Explain the term *heating value of fuel*.

A The heat liberated by the complete and rapid burning of a fuel per unit weight or volume of the fuel is the heating, or calorific, value of the fuel. For solid and liquid fuels, this is usually expressed in Btu per pound. For gaseous fuels, it is expressed in Btu per cubic foot at a standard temperature and pressure, usually atmospheric pressure at 68°F.

Q What is meant by the higher and lower heating values of a fuel?

A Fuels which contain hydrogen have two heating values, higher and lower. The reason for this is that the burning of hydrogen produces superheated water vapor, which escapes at the temperature of the chimney gases. The lower heating values is the net heat liberated per pound of fuel after the heat necessary to vaporize and superheat the steam formed from the hydrogen (and from the fuel) has been deducted. The higher heating value is the one indicated by a fuel calorimeter and is usually used in engineering work. See Fig. 11.8.

A fuel calorimeter is a meter (also called oxygen bomb) to determine the heating value of 1 lb of fuel by burning a sample of the fuel under controlled conditions.

Q Can the heating value of a fuel be determined by calculations?

A Yes, but the following must first be established. We know that the main combustible elements in a fuel are carbon, hydrogen, and sulfur, which combine with oxygen. When each is burned separately, the

Fig. 11.8 Higher and lower heating values of typical fuels.

Gas	Symbol	Heating value, Btu/cu ft	
		Higher	Lower
Carbon monoxide	CO	318	318
Hydrogen	H_2	320	269
Methane	CH_4	985	886
Ethylene	C_2H_4	1551	1451
Ethane	C_2H_6	1721	1571

following heating values are obtained for 1 lb of the element: carbon, 14,600 Btu; hydrogen, 62,000 Btu; sulfur, 4050 Btu.

The heating value of 1 lb of fuel can be calculated by Dulong's formula:

$$HV = 14{,}000C + 62{,}000\left(H - \frac{O}{8}\right) + 4050S$$

where HV = heating value, Btu/lb of fuel
C = weight of carbon per lb of fuel
H = weight of hydrogen per lb of fuel
O = weight of oxygen per lb of fuel
S = weight of sulfur per lb of fuel

EXAMPLE: An ultimate analysis of a coal shows the following percentages:

Carbon 82% = 0.82 lb/lb of coal
Hydrogen 4.5% = 0.045 lb/lb of coal
Oxygen 2.2% = 0.022 lb/lb of coal
Sulfur 1.8% = 0.018 lb/lb of coal
Ash 9.5% = 0.095 lb/lb of coal

Substituting,

$$HV = 14{,}600\,(0.82) + 62{,}000\left(0.045 - \frac{0.022}{8}\right) + 4050\,(0.018)$$

$$= 11{,}972 + 2635 + 729$$
$$= 15{,}336 \text{ Btu/lb}$$

Q Illustrate some common heating values of fuels.

A The ASME Power Boiler Code permits the values shown in Fig. 11.9 to be used in calculating SV capacities where the exact heating value of the fuel is not known. H is the heating value.

Fig. 11.9 Heating values of some fuels for safety-valve capacity checking by the ASME code.

	H = Btu/lb
Semibituminous coal	14,500
Anthracite	13,700
Screenings	12,500
Coke	13,500
Wood, hard or soft, kiln dried	7700
Wood, hard or soft, air dried	6200
Wood shavings	6400
Peat, air dried, 25% moisture	7500
Lignite	10,000
Kerosene	20,000
Petroleum, crude oil, Pennsylvania	20,700
Petroleum, crude oil, Texas	18,500
	H = Btu/cu ft
Natural gas	960
Blast-furnace gas	100
Producer gas	150
Water gas, uncarbureted	290

FUELS AND BURNERS

Q Name the principal forms of fuels used in boilers.

A The common fuels used for production of heat in commercial boilers are in either a solid, liquid, or gaseous state. The solid fuels include coal, lignite, coke, wood, wood wastes, and vegetable wastes such as bagasse. Fuels in the liquid form include petroleum oil, black liquor in kraft paper mills, and some other liquid chemical derivatives in process industries. Gas fuels include natural, blast-furnace, coke-oven, and waste-heat gases from process industries. Other energy sources such as atomic reactors and electricity make heat, but these do not involve combustion.

SOLID FUELS

Q What are the three main methods of burning solid fuels?

A Solid fuels are burned on a fuel bed such as in stoker firing, in suspension firing similar to pulverized-coal-firing systems, and on atmospheric-combustion fluidized-bed systems.

Q Describe some of the principal classes of coal.

A Anthracite coal is very hard, is noncoking, and has a high percentage of fixed carbon. It ignites slowly, unless the furnace temperature is high, and requires a strong draft. The heating value is around 14,000

Btu/lb. Bituminous coal is soft, has a high percentage of volatile matter, burns with a yellow, smoky flame, and has a heating value of 11,000 to 14,000 Btu/lb. Semibituminous coal is the highest grade of bituminous. It burns with little smoke, is softer than anthracite, and has a tendency to break into small pieces when handled. The heating value is 13,000 to 14,500 Btu/lb. Subbituminous (black lignite) is a low grade of bituminous coal with a heating value between 9000 and 11,000 Btu/lb. Lignite is between peat and subbituminous coals, with a wood structure and claylike appearance. The heating value is 7000 to 11,000 Btu/lb.

Q What are caking coal and free-burning coal?
A A caking coal is one which fuses at the surface when burning to form a more or less heavy crust The term *coking* is also used. A free-burning coal does not form a crust and is friable (easily crumbles) throughout the combustion process.

Q Name the two methods of analyzing coal.
A (1) Ultimate analysis and (2) proximate analysis. Ultimate analysis gives the percentages of the various chemical elements of which the coal is composed. Proximate analysis determines the percentage of moisture, volatile matter, fixed carbon, and ash with a fair degree of accuracy.

Q How is the ultimate analysis made?
A This analysis requires a laboratory and a skilled chemist. If a sample of coal is separated into its elements, certain proportions of oxygen, hydrogen, carbon, etc., will be found. These proportions are generally expressed as percentages of the weight of the original sample, the unit weight being 100 percent. The heating value of coal is estimated from the ultimate analysis by getting the percentages of carbon, oxygen, hydrogen, and sulfur in the coal, and by measuring the heat of combustion available in one pound of coal.

Other conditions reported in a coal analysis are (1) as-received, (2) air-dried, (3) moisture-free, (4) moisture- and ash-free, and (5) moisture- and mineral-free.

Q What are the mineral impurities of coal?
A One is ash, which is the incombustible mineral matter left behind when coal burns completely. The amount and character of the ash constitute the biggest single factor in fuel-bed and furnace problems like clinkering and slagging. An increase in ash content usually means an increase in carbon carried to waste or imperfect combustion. Next are the incombustible gases like carbon dioxide and nitrogen. When the volatile matter distills off, a solid fuel is left, consisting mainly of carbon, but containing some hydrogen, oxygen, sulfur, and nitrogen that are not

driven off with the gases. Sulfur in coal burns but is undesirable. Besides causing clinkering and slagging, it corrodes air heaters, economizers, breachings, and stacks. It also causes spontaneous combustion in stored coal.

Q How is the heating value of coal determined?
A By burning a coal sample in a calorimeter (bomb) filled with oxygen under pressure.

Q How is anthracite coal sized?
A Standard sizes are broken, 4⅜ to 3¼ in; egg, 3¼ to 2$\frac{7}{16}$ in; stove, 2$\frac{7}{16}$ to 1⅝ in; chestnut, 1⅝ to $\frac{13}{16}$ in; pea, $\frac{13}{16}$ to $\frac{9}{16}$ in; No. 1 buckwheat, $\frac{9}{16}$ to $\frac{5}{16}$ in; No. 2 buckwheat rice, $\frac{5}{16}$ to $\frac{3}{16}$ in; No. 3 buckwheat (barley), $\frac{3}{16}$ to $\frac{3}{32}$ in; culm or river coal, refuse from screening anthracite into prepared sizes.

Q How is bituminous coal sized?
A There is little standardization of either screen opening or names given to sizes. Run-of-mine is unscreened coal as it comes from the mine. A 2-in nut-and-slack normally means that all coal passes a 2-in screen, but the amount of different sizes present may vary widely. The so-called between-screen sizes include everything passing one screen and retained by another. This gives a closer idea unless the spread between the screens is large.

> REMEMBER: Coal size affects the nature of the fuel bed, the draft required, the density of coal formed, and the amount of unburned-carbon loss.

Q What do the specimen analyses of typical coals indicate?
A The average, or specimen, analyses (Fig. 11.10) supplied by the Bureau of Mines can only be broadly representative and hence must be used with caution. These analysis figures are condensed and cover each bed, or seam, of coal in each county, state, and district. But remember that preparation changes the coal's characteristics. Bureau of Mines figures are from run-of-mine samples. Screenings would probably show slightly more ash. Prepared coals would show slightly less ash and sulfur, higher heating values, and probably higher fusion temperatures. Grindability data, Hardgrove (HG), does not relate directly to the analyses but serves to give a general picture.

Q Name three methods of feeding coal for combustion into the furnace of a boiler.
A By hand shoveling, with stokers, and with pulverizers. Hand firing is inefficient and is slowly disappearing.

Fig. 11.10 Constituents of some typical coals from different parts of the United States.

Specimen Analyses of Some Typical Coals

Bed or seam	State	County	% M*	% V*	% FC*	% Ash	% S*	Htg val, Btu/lb	Ash-fusion temp, °F	HG†
District 1:										
Barton	Md.	Allegany	2.5	15.8	68.3	13.4	2.1	13,020		
Brookville	Pa.	Somerset	4.3	18.0	69.0	8.7	1.2	13,340		
District 2:										
Brookville	Pa.	Butler	3.7	37.0	52.4	6.9	1.6	13,550		
L Freeport	Pa.	Butler	3.1	35.0	55.2	6.7	1.1	13,710		
District 3:										
Bakerstown	W. Va.	Preston	2.0	28.9	61.4	7.7	1.8	14,010	2200–2500	
L Kittanning	W. Va.	Randolph	2.5	29.5	57.4	10.6	1.3	13,410	2600–3000	51
District 4:										
Clarion	Ohio	Vinton	5.5	39.6	41.9	13.0	3.7	11,690		
L Freeport	Ohio	Jefferson	3.8	37.6	49.8	8.8	2.8	13,050		
District 7:										
Beckley	W. Va.	Raleigh	2.7	17.1	74.8	5.4	0.7	14,390	2500–3000	107
Big Eagle	W. Va.	McDowell	3.0	29.2	62.0	5.8	0.8	14,150	....	69
Campbell Creek	W. Va.	Fayette	2.6	32.6	59.4	5.4	1.2	14,070	2400–2800	
District 13:										
Clark	Ala.	Bibb	2.9	34.8	56.1	6.2	0.8	13,830	2100–2400	59
Jagger	Ala.	Fayette	3.1	33.9	52.5	10.5	0.7	12,790		
Black Creek	Ala.	Jefferson	2.9	31.8	61.5	3.8	0.7	14,410	2500–2800	64–70

* M = moisture, V = volatile matter, FC = fixed carbon, S = sulfur.

† Grindability data (Hargrove).

Q How are stokers classified?

A See Fig. 11.11. Two broad classes are (1) overfeed, in which the fuel is carried into the furnace above the stoker, and (2) underfeed, where the fuel is carried by the stoker underneath. Overfeed stokers are further classified into spreader and chain-grate stokers.

Q How does the underfeed stoker work?

A Raw coal is pushed by a ram (Figs. 11.11*c* and 11.12) into the furnace along a feed trough. The fresh coal pushed in causes the coal in the furnace to rise, exposing more coal to the air coming from the

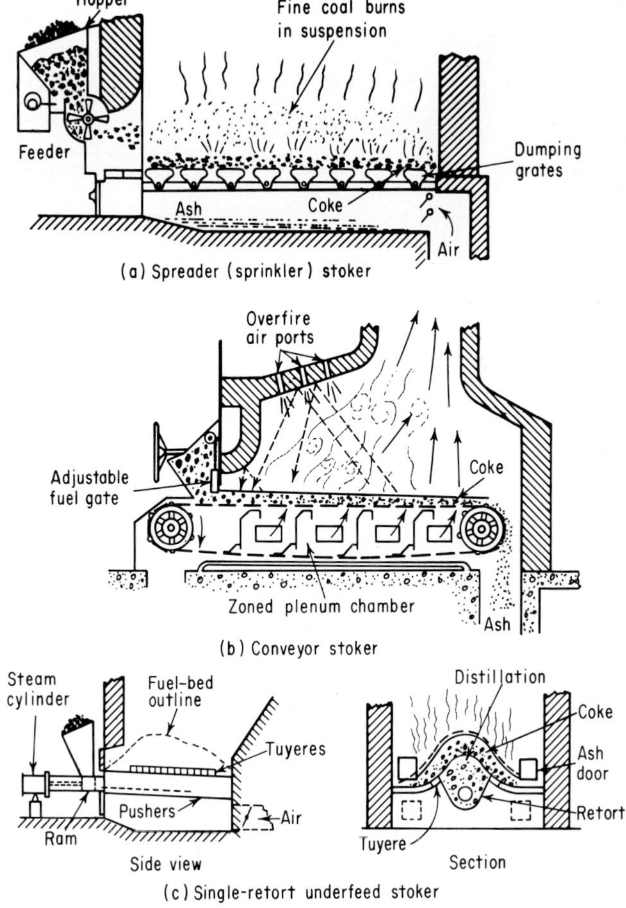

Fig. 11.11 Types of stokers.

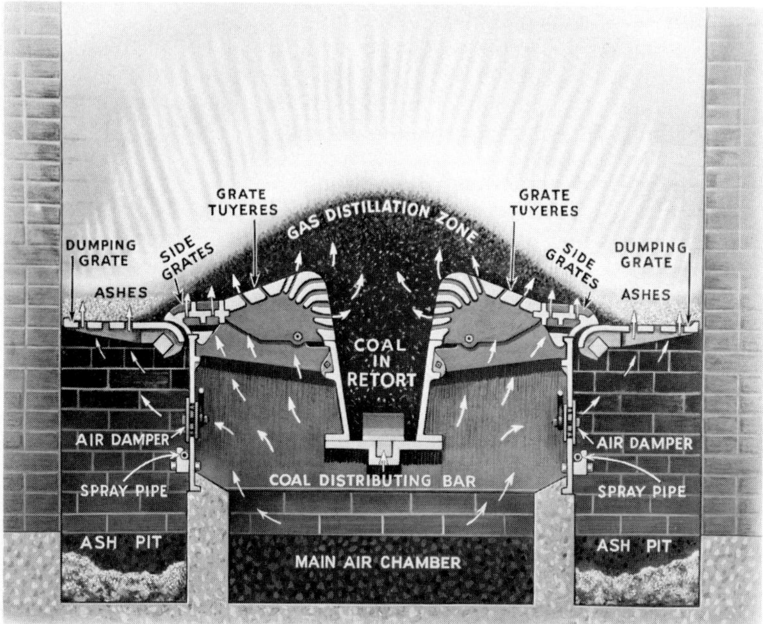

Fig. 11.12 Front sectional view of single-retort underfeed stoker. (*Courtesy of Detroit Stoker Co.*)

tuyeres (openings) in the grate section. The raw coal is heated by the furnace heat and the incoming air, and thus ignites and burns as it moves up toward the fuel-bed outline. The burning coke then moves slowly to the ash-discharge end, pushed that way by either the pressure of the incoming coal or the motion of the grate.

Q Name some underfeed stoker types.

A The name of each type is determined by the mechanism used to move the coal, such as single retort, multiple retort, screw feed, or ram feed. Single-retort units handle up to 50,000 lb/h; multiple-retort designs handle up to 500,000 lb/h.

Figure 11.13 shows a side sectional view of a ram-type underfeed stoker that provides an undulating or wavelike motion to the grate. This type of stoker is used with high-coking coals that have a tendency to cause the fuel bed to arch and rise off the grate. This can result in inadequate distribution of air which produces blowholes and grate deterioration. The undulating grate action breaks up such arch formations, thus keeping the fuel bed porous and free-burning without manual poking. This stoker may have 31 or more fuel feed rates between maximum and zero feed, thus making it possible to link coal feed and combustion air to an automatic combustion-control system. See Fig. 11.14, showing coal feed to a coal hopper in an automatic system.

Q Can the amount of coal, in tons per hour, be determined for a coal-firing boiler that produces an average of 885,000 lb/h of steam at a 4-h test period at a pressure of 400 psi and 700°F, and with feedwater temperature being an average of 280°F? The coal has a heating value of 13,850 Btu/lb. Boiler efficiency is a consistent 82.5 percent.

A It is necessary to match output with input and solve for the tonnage. Average steam produced per hour is 885,000/4 = 221,250 lb/h. From steam tables, at 400 psi and 700°F, enthalpy is 1362.1 Btu/lb. At 280°F, saturated liquid is 249.1 Btu/lb.

$$\text{Btu output} = 221{,}250(1362.1 - 249.1) = 246{,}000{,}000 \text{ Btu/h}$$

$$\text{Boiler input} = \frac{\text{Btu output}}{\text{boiler efficiency}} = \frac{246{,}000{,}000}{0.825} = 298{,}000{,}000$$

$$\text{Coal required} = \frac{298{,}000{,}000}{13{,}850(2000)} = 10.75 \text{ tons/h} \qquad \textit{Ans.}$$

Q What type of fuel is best adapted to the chain- or traveling-grate stoker and to underfeed stokers?

A For chain- or traveling-grate stokers, noncaking or free-burning coal. Typical Illinois bituminous and Pennsylvania anthracite are good examples. In underfeed stokers, coking coals such as West Virginia and Pennsylvania bituminous and many other eastern coals are used.

Q What do we mean by fuel-bed firing on stoker-fired boilers?

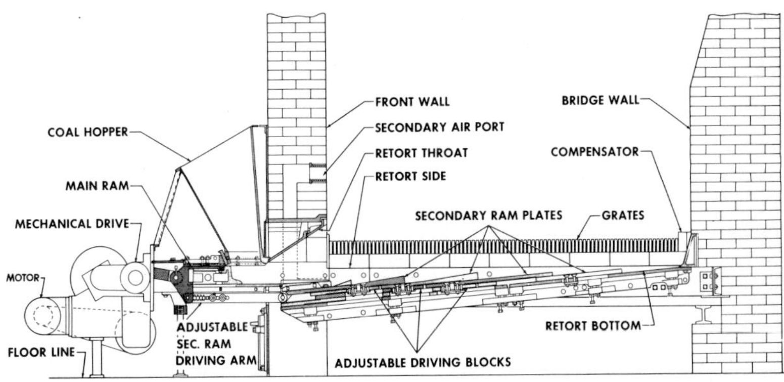

Fig. 11.13 Ram-type single-retort undulating underfeed stoker. (*Courtesy of Detroit Stoker Co.*)

Fig. 11.14 Coal feed is automated with combustion control on this stoker-fired boiler.

A In fuel-bed firing, fuel is pushed, dropped, or thrown on a grate in a high-temperature region within the furnace. Air flows upward through the grate and through the fuel bed that forms on it. The green coal is heated, volatile matter distills off, and coke is left on the grate. As the coke burns to a mixture of carbon dioxide and carbon monoxide, ash remains. The volatile matter of the coal and the carbon monoxide from the coke burn over the fuel bed with air that has come up through it. Usually, secondary air is admitted to the furnace over the grate. Thus anywhere from 40 to 60 percent of the coal's heat is liberated over the fuel bed. So burning of gases is a big part of the job even with fuel-bed firing.

To understand what happens in the fuel bed, divide it into four zones as in Fig. 11.15, one above the other. In actual beds, the location and shape of these zones vary, depending on grate construction and how the fuel reaches it. For simplicity, we can visualize the beds as more or less uniform layers extending across the grate. From the top down, these zones are distillation, reduction, oxidation, and ash.

Q Explain what takes place in the fuel-bed zone on stoker-fired boilers.

A At the top of the fuel-bed zone, green coal receives furnace heat,

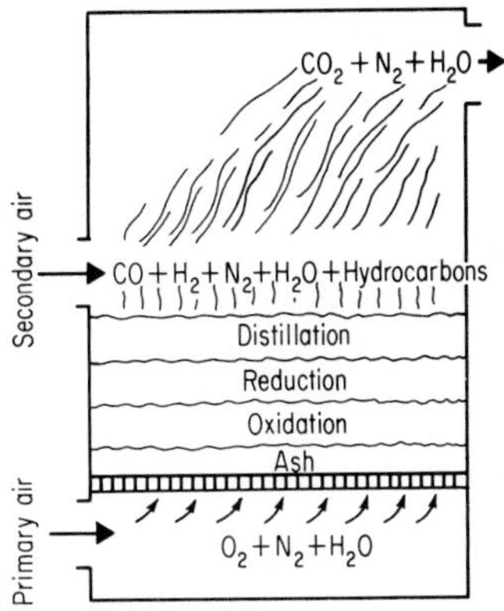

Fig. 11.15 Fuel-bed zones of a stoker-fired boiler can be broken up into oxidation, reduction, distillation, and volatile-gas burning.

volatile matter distills off, and the coke thus formed works down through the bed as the lower layers burn out. In the oxidation zone (Fig. 11.15), coke is burned to carbon dioxide with the primary air rising through the bed. This carbon dioxide travels up through the zone above and is partly reduced to carbon monoxide by contact with the hot coke. Ash at the bottom protects the grate from excessive heat. Oxygen in the primary air is mostly used up a few inches above the grate (oxidation zone).

Only a little carbon monoxide is formed until all oxygen is gone, but the rate of CO formation becomes rapid at the beginning of the reduction zone. As the distance from the grate increases, CO continues to form, but at a slower rate. Thus the thicker the fuel bed, the more CO is formed. Because reduction to CO_2 depends on the time of contact between gas and coke, higher air velocities mean less CO. The fuel-bed temperature depends largely on the rate of firing, being higher at high rates. Usually the temperature within the bed is somewhat greater than at the surface. If this temperature is above that at which ash fuses, clinkers may form.

PULVERIZED FIRING

Q Explain pulverized-coal firing.

A Coal is ground to almost the fineness of flour, and then flows by means of air and coal (in suspension) through ducts or pipes into the furnace to burn like gas or oil (Fig. 11.16). The grinding of the coal exposes the fuel elements in the coal to rapid oxidation (burning) as the ignition temperature is reached. More complete burning is thus possible than with fuel-bed burning.

As these fine particles enter the furnace and become exposed to radiant heat, the temperature rises, and the volatile matter of the coal is distilled off in the form of a gas. Enough primary air is introduced at the burner to intimately mix with the stream of coal particles, which thus support combustion. The volatile matter burns first and then heats the remaining carbon to incandescence. Secondary air is introduced around the burner, which supplies the oxygen to complete the combustion of carbon particles in flames several feet long.

Q Name three pulverizer burning arrangements.

A Figure 11.17 shows pulverized-coal furnaces arranged for (1) long-flame-system firing, (2) shelf-system firing, and (3) corner (tangential) system firing.

Q How are pulverizers classified?

A In general, pulverizers (sometimes called mills) may be classified as attrition or impact types. To these might be added the shearing type, which is a form of either the attrition type, the impact type, or both. The impact mills generally have some attrition action. And conversely, while attrition may be the primary action of a mill, impact is usually present as a secondary action. Thus we have (1) impact mills, including ball mills and hammer mills, and (2) attrition mills, including bowl mills and ball and race mills.

Q Illustrate and explain a typical bowl mill.

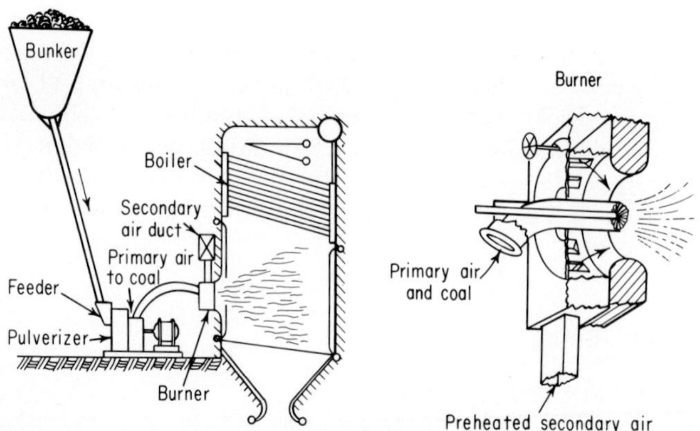

Fig. 11.16 Pulverized-coal firing systems use suspension fuel burning of coal and air instead of fuel-bed burning.

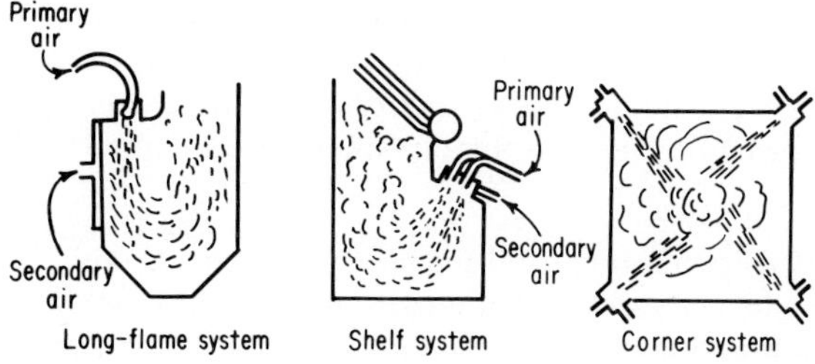

Fig. 11.17 Three methods of suspension firing of pulverized coal: long-flame system, shelf system, and corner system.

A A bowl mill (Fig. 11.18) has a grinding ring carried by a revolving bowl whose rim speed is about 1200 ft/min. Stationary rolls, preloaded by springs, are set so they do not contact the ring. When incoming coal reaches the spinning bowl, centrifugal force throws it against the grinding rim and into contact with the rolls. The ground coal is thrown

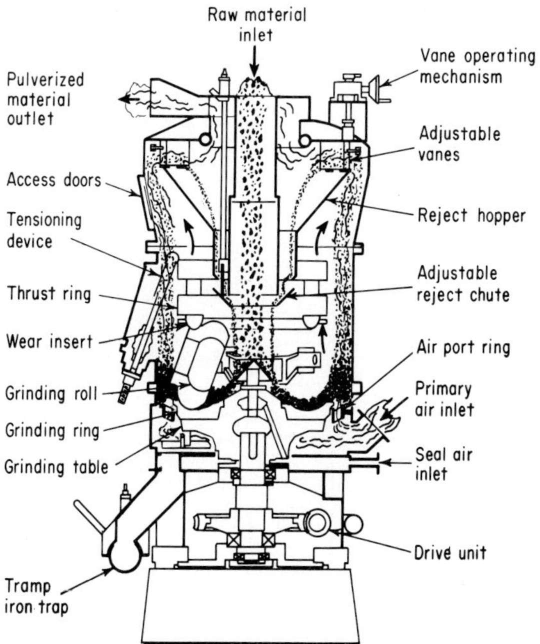

Fig. 11.18 Bowl-mill pulverizer has grinding stationary rolls that crush the coal against rotating grinding ring.

from the rim into annular passages, where heavy particles are rejected to the grinding ring by deflection plates. The remaining coal particles are carried along in the exhauster-induced airstream into the classifier section. Adjustable vanes on the classifier set the fineness of a pulverized coal.

Q Describe the burners used for pulverized-coal firing.
A Burners for pulverized coal must supply air and fuel to the furnace in a manner that permits (1) stable ignition, (2) effective control of flame shape and travel, and (3) thorough and complete mixing of fuel and air. The air used to transport the coal to the burner forms the primary air; secondary air may be introduced in the burner (turbulent burners) or around or near the burner (nozzle burners). The turbulent burner (Fig. 11.19) imparts a rotary motion to (1) the coal-air mixture in a central nozzle and (2) the secondary air issuing from a chamber around that nozzle, all within the burner. This gives some premixing for coal and air, and considerable turbulence. In some burners the coal-air mixture issues from a series of nozzles to mix within the furnace with secondary air admitted through separate openings.

Most burners fire into the furnace horizontally, usually from only one wall, and a burner block of refractory surrounds the burner opening. If the wall carrying the burners is water-cooled, the tubes bend around the burner opening.

Vertical burners are also used. Early designs had the primary-air

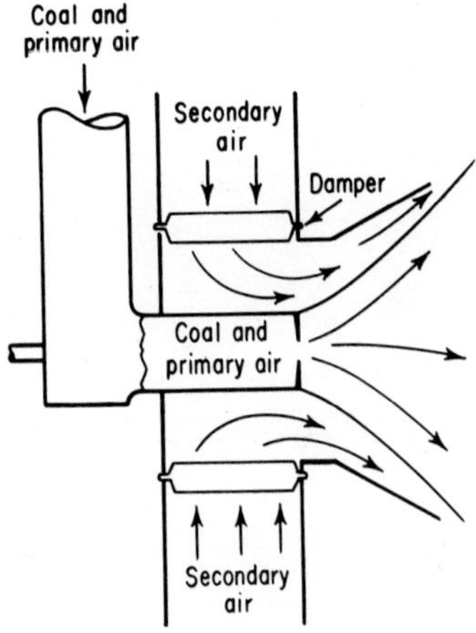

Fig. 11.19 Turbulent burner for pulverized-coal burning imparts a rotary motion to coal-air mixture prior to furnace entrance.

and coal nozzle at the top of the furnace chamber pointing downward. Secondary-air nozzles direct air into the furnace horizontally for flame control and turbulence. A typical vertical burner of today (Fig. 11.20) leads secondary air around the burner nozzles in baffled streams. Horizontal burners may be located in such relation to each other as to promote turbulence. In opposed firing, burners in opposite walls of the furnace play their flames against each other.

Tangential firing (Fig. 11.17) has burners in furnace corners that direct their flames tangentially to an imaginary circle in the furnace space. Turbulence can be set up in this fashion, and there is a tendency for unburned combustible material in the tail of one flame to be caught up in the second. A special form of tangential firing appears in Fig. 11.21. These burners are adjustable to shift the flame zone vertically and so regulate the temperature of the furnace exit gas according to load. This, in turn, controls superheat over a wide load range. Newer installations use fully automatic control of burner inclination.

A combination firing unit (not shown) handles more than one fuel. Here coal and oil, in which primary, secondary, and tertiary (third) air, under separate control, are used. Many other combinations are available. Because these burners are designed to give peak performance at, or near, capacities and fall off at low loads, the problem of lighting off is a major one. Usually, an auxiliary oil or gas burner, electrically ignited, is used for this purpose. One design has a mechanical-atomizing oil burner. Both burner and igniting electrodes are retractable, and a safety interlocking system prevents oil flow until the burner and electrodes are extended.

Q Describe the cyclone burner illustrated in Fig. 11.22.

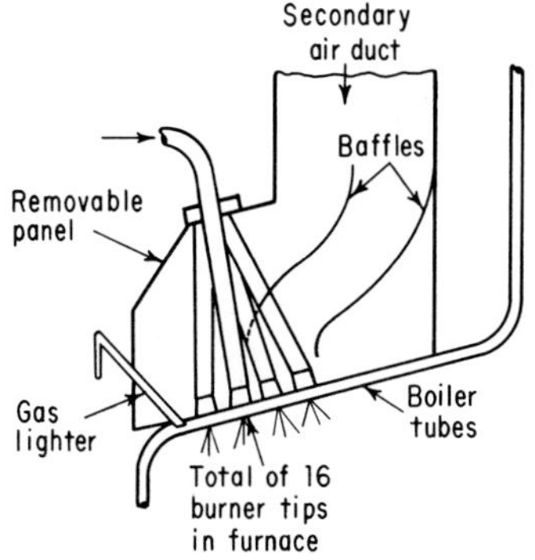

Fig. 11.20 Vertical burners for pulverized coal direct coal-air mixture toward furnace bottom.

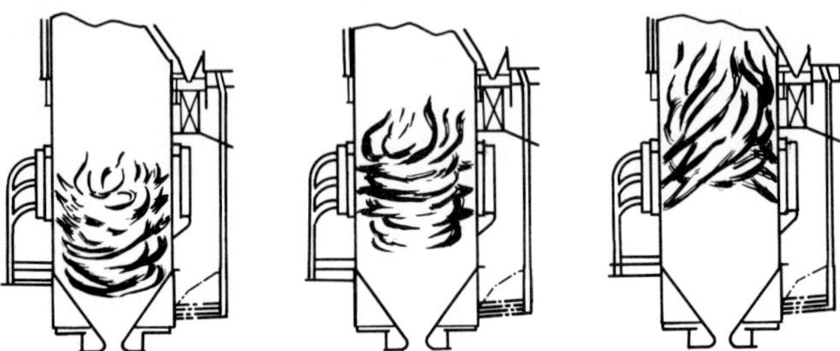

Fig. 11.21 Tangential firing from corners of the furnace with the burners having the capability to direct flame up and down for exit-flue-gas temperature control.

A In this furnace design, what might be termed deliberate slagging is used. The burner receives crushed (not pulverized) coal in a stream of high-velocity air tangent to the circular burner housing, which forms a primary water-cooled furnace. Coal thrown to the rim of the furnace by centrifugal force and held by a coating of molten ash is scrubbed by fast-moving air. Secondary air enters at high velocity also, and parallel to the path of the primary coal-air mixture. The coal in the sticky slag film burns as if it were in a fuel bed. Volatiles are distilled off, and carbon is burned out to leave ash. Combustion of volatile matter begins in the burner chamber and is completed in the secondary furnace into which the burner chamber discharges. Molten ash, under centrifugal force, clings to the burner-chamber walls and the slight inclination causes slag to discharge continuously. The nature of this burning tends to reduce greatly the amount of ash carried in suspension, and hence fly-ash emission is negligible.

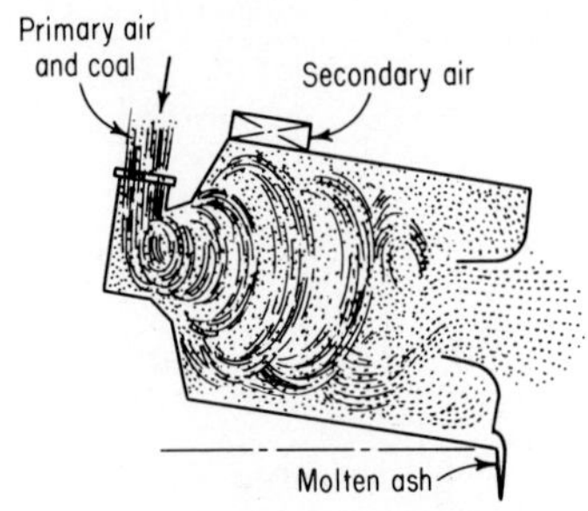

Fig. 11.22 Cyclone burner uses crushed coal in a stream of high-velocity air to obtain good turbulent burning in the furnace.

Q How is the degree of fineness of the pulverized coal checked?
A A sample drawn from the primary outlet is put through a stacking machine with different mesh screens stacked on top of one another. For a given weight of the sample, the amount remaining on each mesh can then be calculated as a percentage of the original weight. This can then be compared with the specification fineness required on the pulverizer and adjustments made to obtain this fineness.

Q Describe the fire-detection system used on coal pulverizers.
A In the past, thermocouples were used to detect a heat rise in the pulverizer that may be due to burning in the pulverizer. Though the basic design of pulverizers is still essentially the same, they are becoming more heavily instrumented as sensors are developed for power-plant applications. Monitoring of the level of carbon monoxide in pulverizers in order to detect fires and prevent explosions in the pulverizer is supplementing thermocouple detection systems. The CO monitoring system has been tested in several stations and is reported to give operators an hour or more of lead time upon signals of the presence of CO. This permits them to adjust operations, apply fire extinguishers if required, and take a mill out of service to prevent a coal-gas explosion. The CO monitoring system has a higher degree of sensitivity than does the thermocouple method.

Q Describe low NO_x burners on pulverized-coal systems.
A The refinement of existing burners and the development of new ones has concentrated in reducing NO_x emission without adversely affecting thermal efficiency or increasing furnace corrosion rates. Low NO_x burners control the mixing of fuel and air in a pattern that holds the flame temperature low and dissipates heat quickly. Conventional burners mix secondary air with the primary air/coal stream in the furnace, and this promotes high-intensity combustion and NO_x formation. The low NO_x burners establish distinctly separate primary and secondary combustion zones. This reduces peak flame temperatures and results in lower NO_x flue-gas emissions. There has also been an attempt to reduce the use of auxiliary fuels for either load purposes or ignition purposes by the development of high-energy ignition systems (electric igniters) that are capable of firing or igniting pulverized coal directly.

OTHER SOLID FUELS

Q Describe some other solid fuels.
A Wood from lumber and woodworking industries in the form of sawdust, slabs and shavings, and hog wood. Hog wood is wood refuse cut to uniform size before burning. Bark from pine, oak, and hemlock

trees is burned in special furnaces. The heating value of wood varies from 2500 to 3000 Btu/lb. Bagasse is the crushed stalks of sugarcane from which the sap has been extracted. The heating value is from 3500 to 4500 Btu/lb. Coke is the solid remains after the destructive distillation of either petroleum oils or certain bituminous coals. The heating value of petroleum coke is from 11,500 to 15,000 Btu/lb.

Q What are some of the problems in burning lignite?
A Lignite is a low-sulfur coal but produces large amounts of alkaline ash that must be disposed of in an environmentally safe manner. Ash fouling of tube passages is a problem, requiring extensive use of soot blowers, precipitators, and baghouses in order to eliminate this high-ash problem.

Q How do lignite- and peat-burning boilers compare in size with boilers that burn other fuels?
A See Fig. 11.23 for a comparison, assuming same capacities for the boilers. Lignite boilers need large boiler volumes in order to handle larger flue-gas volumes. This also requires larger flue-gas cleaning equipment and greater ash-handling capacities than a coal-burning unit. This can add to initial plant construction costs.

Most lignite boilers are installed near lignite-mine sites because of the large volume of fuel needed in order to avoid high fuel-transportation costs. Figure 11.24 illustrates a 550-MW lignite-mine-mouth boiler installed in the North Dakota lignite fields. The unit burns lignite with a heating-value range from 5780 to 7100 Btu/lb.

Q Explain the fuel-burning system for the lignite boiler illustrated in Fig. 11.24.
A It requires eight bowl-mill pulverizers with 110-in-diameter bowls to pulverize about 800,000 lb of lignite per hour. Tangential firing is used. The furnace heat-release rate is 47,500 Btu/(h · ft^2). It is a controlled-circulation boiler.

FLUIDIZED-BED BURNING

Q What are the advantages of burning some solid fuels in a fluidized bed?
A Coal-fired fluidized-bed boilers burn particles of coal in a bed of limestone through which air is passed. See Fig. 11.25. The velocity of the air is maintained through the nozzles so that the limestone becomes suspended, and the bed becomes or resembles a boiling fluid. The limestone captures any sulfur dioxide formed by the coal burning. This eliminates the need for flue-gas desulfurization systems. The tempera-

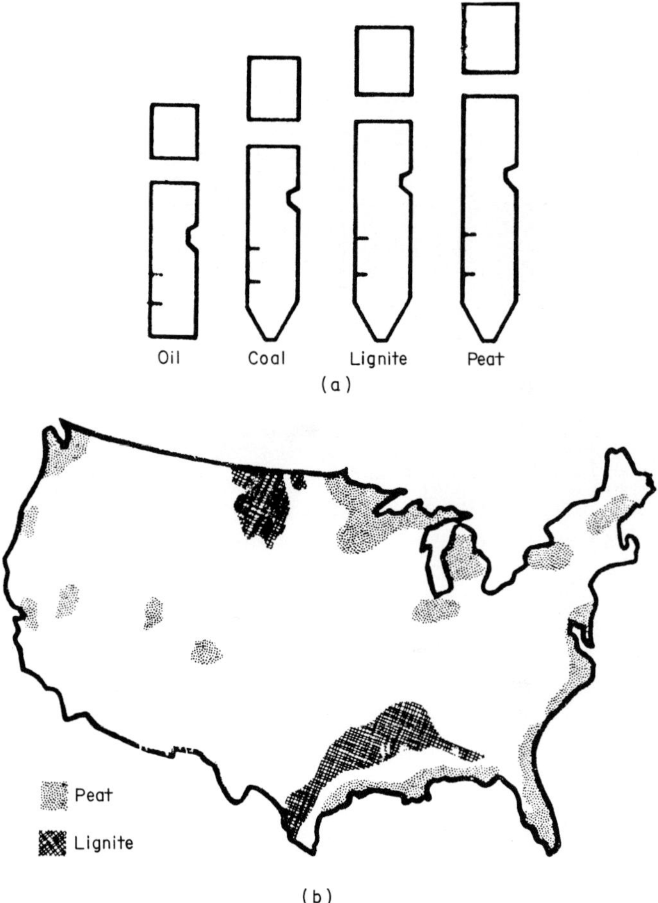

Fig. 11.23 (*a*) Comparative size of boilers with same output but firing different fuels, (*b*) U.S. peat and lignite deposit areas.

ture of the fluidized bed is low enough, thus minimizing the formation of nitrogen oxides. Slag is also reduced when coal is fired in a fluidized-bed boiler.

Q How does ash removal affect fluidized-bed burning?
A Both bottom ash and fly ash must be considered in fluidized-bed firing. Spent bed material must be removed at a controlled rate to maintain proper bed conditions in order to continue proper chemical combination of sulfur dioxide and lime; otherwise corrosion and other

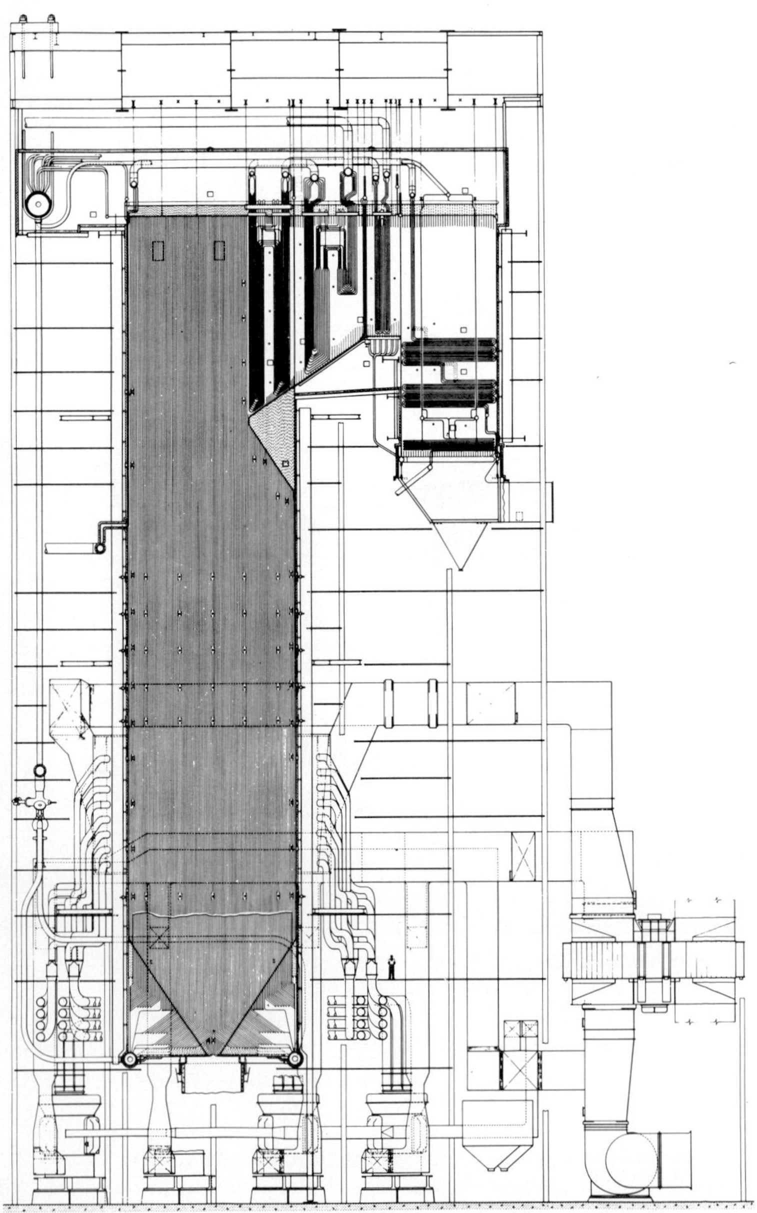

Fig. 11.24 550 MW, 2620 psi, 1005°F, 3,764,000 lb/h lignite-burning boiler. (*Courtesy of Combustion Engineering, Inc.*)

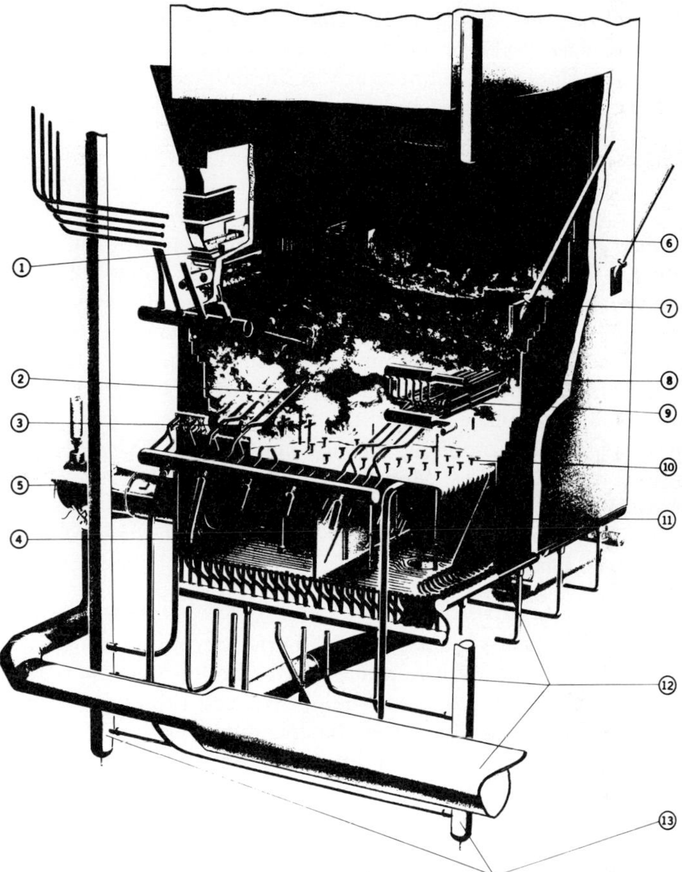

Fig. 11.25 Foster Wheeler Corp. design of a fluidized-bed furnace. (1) Rotor flipper coal feeder; (2) sloped in-bed boiler tubes; (3) and (4) cyclone ash reinjection nozzles; (5) in-duct burner; (6) limestone gravity-feed pipe; (7) partition wall between cells A and B; (8) finned-tube furnace panel; (9) primary superheater; (10) air nozzle; (11) plenum chamber; (12) air supply; (13) downcomers. (*Courtesy of Foster Wheeler Corp.*)

equipment problems may develop. Erosion of tubes in the fluidized-bed area also requires attention.

Q How are fluidized-bed burners started?

A An oil-fired ignition burner can be used to raise the temperature of a portion of the limestone bed to autoignition temperature of the coal, or solid fuel may be employed. This temperature is achieved before the fuel is introduced. The start-up zone in the fuel bed and then the remainder of the bed is brought into service once a self-sustaining fire is established. Operating temperature in the fluidized bed is maintained by adjusting coal feed rate and the fluidizing air velocity.

WOOD BURNING

Q What are some considerations in burning wood products in a boiler?
A Like all fuels, chemical and physical characteristics of the wood product to be burned must be known. Figure 11.26 shows some of these properties in comparison with coal. However, one of the most variable items in wood burning is moisture content of the wood when the wood to be burned comes directly from the forests. Because moisture content can be as high as 65 percent, making combustion almost unstable, and burning may take place beyond the furnace proper. This can result in flame-impingement problems on shell and tubes that were designed to handle convection gases. Moisture content of the wood can vary with the seasons of the year. Predrying of the wood external to the furnace with flue-gas heat will prevent many burning and boiler problems that use wet wood as a fuel.

Q What danger exists if a pressurized furnace develops a casing leak?
A Since the furnace is above atmospheric pressure, a casing leak may permit products of combustion to enter the boiler-room operating area. This could result in personnel being exposed to hot gases and also perhaps inhaling toxic flue gases. Balanced-draft boilers may also have this hazard from a casing leak if the induced-draft fan should shut down unexpectedly. Under these conditions, the balanced-draft boiler may become a pressurized boiler if the forced-draft fan continues to operate.

FUEL OILS

Q Briefly describe fuel oils.
A Fuel oils are derived from petroleum, with the chief combustible ingredients being carbon and hydrogen. Their specific gravities vary with temperature and origin. The range at 60°F is from 0.84 to 0.96. The table in Fig. 11.27 gives some properties at 60°F.

In this table, the API scale is the American Petroleum Institute scale for showing specific gravity. The API scales fixes a reading of 10°F as equal to a specific gravity of 1.00. Readings greater than 10°F indicate a specific gravity of less than 1.0, or an oil which is lighter. To obtain the actual specific gravity in relation to water from the API reading, use:

Fig. 11.26 Physical and chemical properties of wood vs. coal as a fuel. (*Courtesy of Power magazine.*)

	Chemical composition, % by wt (dry basis)										
	Bark				Wood			Coal			
Fuel characteristics	Pine	Oak	Spruce[a]	Redwood[a]	Redwood	Pine	Fir/Pine[b]	Lig[c]	Sub[d]	Bit[e]	Bit[f]
Proximate analysis:											
Volatile matter	72.9	76.0	69.6	72.6	82.5	79.4	75.1	44.1	39.7	35.4	16.0
Fixed carbon	24.2	18.7	26.6	27.0	17.3	20.1	24.5	44.9	53.6	56.2	79.1
Ash	2.9	5.3	3.8	0.4	0.2	0.5	0.4	11.0	6.7	8.4	4.9
Ultimate analysis:											
Hydrogen	5.6	5.4	5.7	5.1	5.9	6.3	6.3	4.6	5.2	4.8	4.8
Carbon	53.4	49.7	51.8	51.9	53.5	51.8	50.7	64.1	67.3	74.6	85.4
Sulfur	0.1	0.1	0.1	0.1	0	0	0	0.8	2.7	1.8	0.8
Nitrogen	0.1	0.2	0.2	0.1	0.1	0.1	2.4	1.2	1.9	1.5	1.5
Oxygen	37.9	39.3	38.4	42.4	40.3	41.3	40.2	18.3	16.2	8.9	2.6
Ash	2.9	5.3	3.8	0.4	0.2	0.5	0.4	11.0	6.7	8.4	4.9
Heating value, dry basis, Btu/lb	9030	8370	8740	8350	9220	9130	8795	11,084	12,096	13,388	15,000

[a]Logs stored in saltwater. [b]Sanderdust. [c]Texas lignite. [d]Wyoming subbituminous B. [e]Illinois bituminous (high-volatile A). [f]West Virginia bituminous (low-volatile).

Sources: Babcock & Wilcox Co. Combustion Engineering Inc., Coen Co.

Fig. 11.27 Some properties of fuel oil.

Gravity, deg API	Sp gr	Lb/gal	Btu/lb	Btu/gal	Lb/42-gal bbl	Lb/cu ft
3	1.0520	8.76	18,190	159,340	368.00	65.54
5	1.0366	8.63	18,290	157,840	362.62	64.59
7	1.0217	8.50	18,390	156,320	357.37	63.65
9	1.0071	8.39	18,490	155,130	352.46	62.78
11	0.9930	8.27	18,590	153,740	347.71	61.93
13	0.9792	8.16	18,690	152,510	342.88	61.07
15	0.9659	8.05	18,790	151,260	338.22	60.24
17	0.9529	7.94	18,890	149,980	333.64	59.42
19	0.9402	7.83	18,980	148,610	329.23	58.64
21	0.9279	7.73	19,060	147,330	324.91	57.87
23	0.9159	7.63	19,150	146,110	320.71	57.12
25	0.9042	7.53	19,230	144,800	316.59	56.39
27	0.8927	7.44	19,310	143,670	312.60	55.68
29	0.8816	7.35	19,380	142,440	308.70	54.98
31	0.8708	7:26	19,450	141,210	304.92	54.31
33	0.8602	7.17	19,520	139,960	301.18	53.64
35	0.8498	7.08	19,590	138,690	297.57	53.00
37	0.8398	7.00	19,650	137,550	294.04	52.37
39	0.8299	6.92	19,720	136,400	290.64	51.76
41	0.8203	6.83	19,780	135,090	287.23	51.16

Actual specific gravity = 141.5/(131.5 + API deg)

Q How are fuel oils sold?

A Fuel oils are sold in six standardized grades, under the numbers or grades of 1, 2, 3, 4, 5, and 6. Grades 1, 2, and 3 are light, medium, and heavy domestic fuel oils. These usually do not require heating prior to burning in a furnace. Grades 4, 5, and 6 correspond to federal specifications for Bunkers A, B, and C, respectively. These oils are heavy and viscous and thus require heating before being sprayed into a furnace.

Q Describe some physical properties of fuel oil.

A Since hydrogen has a much higher heating value and lower atomic weight than the other principal elements in fuel oil, the proportions of carbon and hydrogen affect both specific gravity and heating value. Because of this, specific gravity forms a reliable guide to an oil's heating value. Some typical physical properties are:

Specific gravity in degrees API is found by dividing specific gravity with respect to water (at 60°F) into 141.5 and subtracting 131.5 from the answer. Specific gravity *in degrees Baumé* (°Bé) is found in the same way except that the numbers are 140 and 130, respectively. For practical purposes, the two specific gravity scales may be considered the same.

Viscosity is the relative ease or difficulty with which an oil flows. It is measured by the time in seconds a standard amount of oil takes to flow through a standard orifice in a device called a *viscosimeter*. The usual standard in this country is the Saybolt Universal, or the Saybolt Furol, for oils of high viscosity. Since viscosity changes with temperature, tests must be made at a standard temperature, usually 100°F for Saybolt Universal and 122°F for Saybolt Furol. Viscosity indicates how oil behaves when pumped and, more particularly, shows when preheating is required and what temperature must be held.

Flash point represents the temperature at which an oil gives off enough vapor to make an inflammable mixture with air. The results of a flash-point test depend on the apparatus; so this is specified as well as temperature. Flash point measures an oil's volatility and indicates the maximum temperature for safe handling.

Pour point represents the lowest temperature at which an oil flows, under standard conditions. Including pour point as a specification ensures that an oil will not give handling trouble at expected low temperatures.

By *centrifuging a sample* of oil, the amounts of water and sediment present can be determined. These are impurities, and while it is not economical to eliminate them, they should not occur in excessive quantities (not more than 2 percent). Incombustible impurities in oil, from natural salts, from chemicals in refining operations, or from rust and scale picked up in transit, show up as ash. Some ash-producing impurities cause rapid wear of refractories, and some are abrasive to pumps, valves, and burner parts. In the furnace, they may form slag coatings.

All test properties above are covered by ASTM standards, which should be consulted for details of apparatus and methods. A copy of *ASTM Standards of Petroleum Products and Lubricants* can be obtained from the American Society for Testing and Materials at 1916 Race St., Philadelphia 3, Pennsylvania.

Q What rules should be followed in mixing fuel oils?

A Fuel oils have a tendency to deposit sludge in storage; this may be aggravated by mixing oils of different character as when deliveries from two sources go into the same tank. To avoid trouble when oils are mixed, remember that (1) straight-run residuals can be mixed with any straight-run product, and cracked residuals with straight-run residuals; (2) cracked distillates can be added as a third constituent, but (3) cracked residuals cannot be added to straight-run distillates.

OIL BURNERS

Q Describe the functions of an oil burner.

A In addition to proportioning fuel and air and mixing them, oil burners must prepare the fuel for combustion. Two ways (with many variations) are (1) oil may be vaporized or gasified by heating within the burner or (2) oil may be atomized by the burner so vaporization can occur in the combustion space. Vaporizing burners (first group) are limited in range to fuels they can handle and find little use in power plants. If oil is to be vaporized in the combustion space in the instant of time available, it must be broken up into many small particles to expose as much surface as possible to the heat. Atomization is effected in three basic ways: (1) by using steam or air under pressure to break the oil into droplets, (2) by forcing oil under pressure through a nozzle, and (3) by tearing an oil film into drops by centrifugal force. All three methods are used. In addition, a burner must provide good mixing of fuel and air so complete combustion of the oil droplets may ensue.

Q Describe a steam-atomizing burner.

A As a class, steam-atomizing burners possess the ability to burn almost any fuel oil of any viscosity at almost any temperature. Air is less extensively used as an atomizing medium because its operating cost is apt to be high. These burners can be divided into two types:

1. Internal-mixing or premixing oil and steam (Fig. 11.28*a*) or air

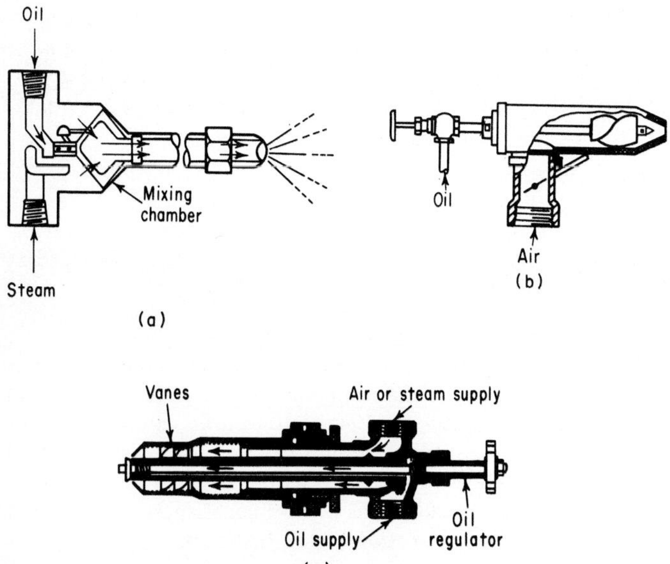

Fig. 11.28 Atomizing oil burners. (*a*) Steam-atomizing burner with mixing done inside; (*b*) air-atomizing burner uses low-pressure air; (*c*) steam-atomizing burner with mixing done externally.

mix inside the body or tip of the burner (Fig. 11.28*b*) before being sprayed into the furnace.

2. External mixing, where oil emerging from the burner is caught by a jet of steam or air (Fig. 11.28*c*).

Steam consumption for atomizing runs from 1 to 5 percent of the steam produced, with the average around 2 percent. The pressure required varies from about 75 to 150 psi.

In the burner of Fig. 11.28*c* oil reaches the tip through a central passage, flow being regulated by the screw spindle. Oil whirls out against a sprayer plate to break up at right angles to the stream of steam, or air, coming out behind it. The atomizing stream surrounds the oil chamber and receives a whirling motion from vanes in its path. When air is used for atomizing, it should be at 10 psi for lighter oils and 20 psi for heavier. Combustion air enters through a register, shown in Fig. 11.29. Vanes or shutters are adjustable to give control of excess air.

Q Explain mechanical atomizing oil burners.

A Good atomization results when oil under a pressure of 75 to 200 psi is discharged through a small orifice, often aided by a slotted disk. The disk gives the oil a whirling motion before it passes on through a hole drilled in the nozzle, where atomization occurs. For a given nozzle opening, atomization depends on pressure, and since pressure and flow are related, the best atomization occurs over a fairly narrow range of burner capacities. To follow the boiler load as steam demand goes up or down, a number of burners may be installed and turned on or off, or burner tips with different nozzle openings may be used.

> **IMPORTANT:** All nozzle openings must be changed to the same size in a given system; never fire with mixed sizes.

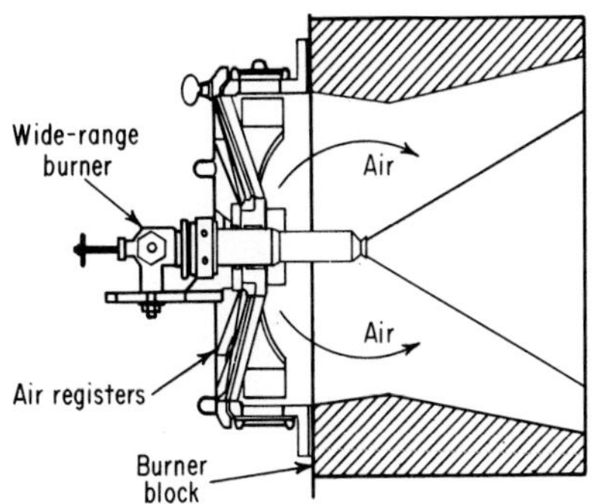

Fig. 11.29 Air registers are used to control excess air in oil burning.

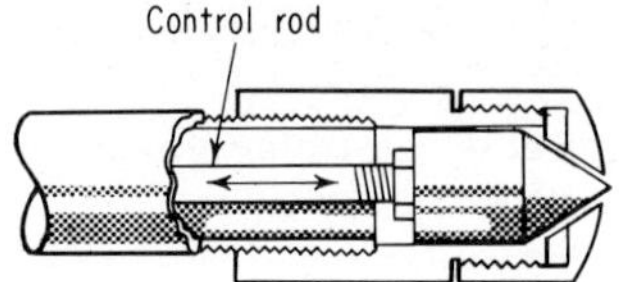

Fig. 11.30 Movable control rod varies tangential slot area, and thus flow of oil to orifice.

There are many burner designs to extend the usual 1.4:1 capacity range of the mechanical-atomizing nozzle. One has a plunger that opens additional tangential holes in the nozzle as oil pressure increases. This gives a 4:1 range. The burner in Fig. 11.30 uses a movable control rod which, through a regulating pin, varies the area of tangential slots in the sprayer plate and the volume of the oil passing the orifice.

The wide-range mechanical atomizers (Fig. 11.31) gives a capacity range of about 15:1 and much higher if needed. By use of either a constant-differential valve or pump, as shown, the difference in pressure between supply and return is held constant. This pump system offers advantages in many plants:

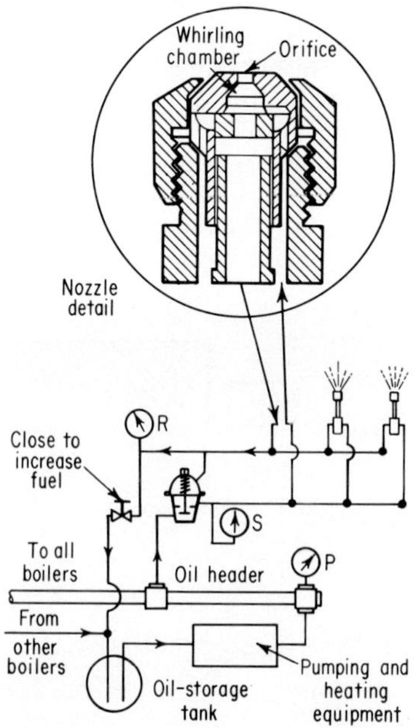

Fig. 11.31 Mechanical atomizer has a wide-range burner tip with whirling chamber.

1. No hot oil is returned to the storage tank or pump suction.

2. Fuel enters the closed circuit at the same rate it is burned, thus simplifying fuel metering and combustion control.

3. The pump may be used to boost pressure on existing oil-burner systems.

Q Explain how a rotary-cup burner operates.

A The horizontal rotary cup atomizes fuel oil by literally tearing it into tiny droplets. A conical or cylindrical cup rotates a high speed (usually about 3500 r/min) if motor-driven. Oil moving along this cup reaches the rim, where centrifugal force flings it into an air stream (Fig. 11.32). This system of atomization requires no oil pressure beyond that needed to bring oil to the cup. But high oil preheat temperatures must be avoided since gasification may develop. The rotary cup can satisfactorily atomize oils of high viscosity (300 seconds Saybolt Universal, or SSU) and has a wide range of about 16:1.

Q Detail some aspects of oil-burner maintenance.

A Make sure that the burner gets uniformly free-flowing oil, clear of sediment that clogs burner nozzles. This means avoiding sludge build-up in storage tanks and keeping strainers in good condition. The preheat temperature must be right for fuel and burner type and must be uniform. Watch for wear caused by abrasion of ash in fuel and for carbon buildup. In rotary-cup burners, worn rims cause poor atomization. If cups are not properly protected after being turned off, carbon forms on the rim. When the burner is shut down, always take out the cup and insert a flame shield.

> **REMEMBER:** Worn or carbonized mechanical-atomizing nozzles give trouble. Always replace worn nozzles and keep them clean.

Q What are low-excess-air oil burners, and what is their purpose?

A There is a decided trend to trim fuel/air ratios on oil burners as close

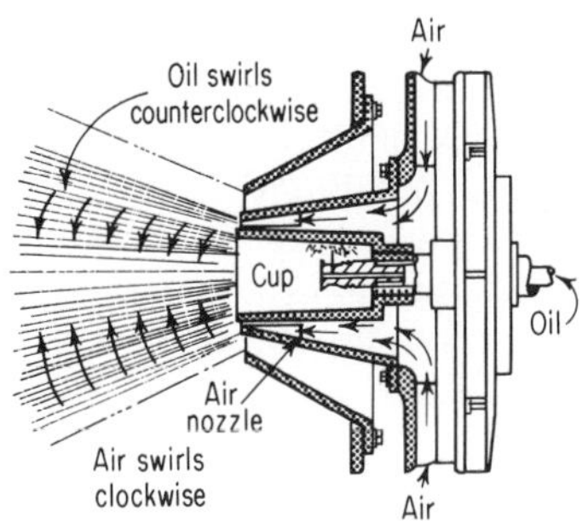

Fig. 11.32 Rotary-cup burner atomizes oil into fine spray by centrifugal force.

to the theoretical combustion equations as possible for fuel-saving purposes and also to control air pollution. Manufacturers have introduced low-excess-air oil burners that improve efficiency while reducing NO_x. These oil burners are designed to operate at 0.5 percent excess air at 70 to 100 percent load and 0.7 percent between 25 and 70 percent load.

The primary reason these burners can operate at low excess air is that they hold the effects of windbox maldistribution to a minimum by using venturi throats that send low turbulent air into the furnace. This flow is thus controlled precisely, which results in the burner's not having to be flooded with excess air to compensate for inefficient mixing.

GASEOUS FUELS

Q Describe some properties of gaseous fuels.

A Of many gaseous fuels, only nautral gas has any commercial importance in steam generation because manufactured gases run too high in cost. By-product gases usually have low heating values and are produced in relatively minor quantities; so they are ordinarily used at the production point and not distributed. Natural gas is colorless and odorless. Composition varies with source, but methane (CH_4) is always the major constituent. Most natural gas contains some ethane (C_2H_6) and a small amount of nitrogen. Gas from some areas, often called sour gas, contains hydrogen sulfide, and organic sulfur vapors. The heating value averages about 1000 Btu/ft^3 (20,000 Btu/lb) but may run much higher. Natural gas is usually sold by the cubic foot but may be sold by the therm (100,000 Btu).

Coal gas and coke-oven gas (manufactured gases) are produced by carbonizing high-volatile bituminous coal in retorts that exclude air and are heated externally by producer gas. Usually a number of by-products result. Cleaned of impurities, these gases are roughly one-half hydrogen and one-third methane, plus small amounts of carbon monoxide, carbon dioxide, nitrogen, oxygen, and illuminants (C_2H_4 and C_6H_6). The heating value runs around 550 Btu/ft^3.

The gas served in a given area may be a mixture of two or more gases or a mixture of natural and manufactured gas. The heating value, usually held to 525 to 550 Btu/ft^3, is often fixed by state or local ordinance.

Q Describe the by-product gases.

A Commercial butane and propane are essentially by-products from the manufacture of natural gasoline and from certain refinery operations. As supplied, propane (C_3H_8) is essentially pure, while butane

(C_4H_{10}) usually contains a small amount of propane. Both have high heating values, are easily liquefied at low pressure, and are widely used as bottled fuels.

Blast-furnace gas, a by-product of iron making, has the lowest heating value of any commercial gas, about 90 Btu/ft^3. It is close to three-quarters nitrogen and carbon dioxide, the only important combustible constituent being carbon monoxide. Raw gas, which usually contains a high concentration of solid impurities, is normally washed before use. But unwashed gas has been successfully burned in boiler furnaces.

Sewage-sludge gas runs about two-thirds methane and one-third carbon dioxide, with small amounts of hydrogen, nitrogen, and usually some hydrogen sulfide. The heating value is about 650 Btu/ft^3. Although used mostly in internal-combustion engines, this gas is also burner-fired. See the table in Fig. 11.33 for properties of fuel gases.

Q Describe and illustrate atmospheric gas burners.

A Burning gas requires no preparation of the fuel, as do other fuels. But proportioning with air, mixing, and burning can be handled in several ways. Also, the fuel's characteristics need to be known for sound selection of equipment and successful operation. Atmospheric burners are used for gas burning, and differ mainly in the way air and fuel mix. The atmospheric burner is popular, as in home gas ranges. The momentum of the incoming low-pressure gas stream is used to draw in, or aspirate, part of the air needed for combustion. A shutter or similar device regulates the amount of air so induced. Gas and air together pass through a tube leading to the burner ports, mixing in the process. The mixture burns at the ports or openings in the burner head (with a blue, nonluminous flame). Secondary air is drawn into the flame from the surrounding atmosphere. Larger counterparts of this general burner type, having ring or sectional burner heads with many ports, are used to fire small boilers and industrial equipment.

A single-port atmospheric burner is shown in Fig. 11.34. A needle valve controls the gas flow through the spud; air is drawn in around the shutter at the end. With burner-port size and shape fixed, the nature of burning depends largely on the amount of primary air, or premix. With premix low, the flame is long and pale blue. It may have a yellow tip, indicating cracking and presence of free carbon.

Operation is usually satisfactory with 30 to 70 percent premix; in some special designs 100 percent primary air is used. This premix range gives a turndown, or capacity, range of about 4:1. Usually premix and capacity ranges are somewhat narrower. Secondary air may be drawn in around the burner, the amount depending on the area of the opening and the draft. The high-pressure burner uses gas at about 20 to 30 psig and air at atmospheric pressure. Another type uses compressed air, with gas at atmospheric pressure.

Fig. 11.33 Properties of fuel gases and their heating values for burning in boilers.

Fuel	Source	Average composition	High heat value, btu/cu ft	Remarks
Blast-furnace gas	By-product of iron making	58% N_2 27% CO, 12% CO_2, 2% H_2, some CH_4	90–100	Good fuel when cleaned—used mainly at source
Butane	By-product of gasoline making, also in casing-head gas	C_4H_{10} (usually has some butylene C_4H_8 and propane C_3H_8)	3200–3260	Liquefies under slight pressure, sold as liquid (bottled gas)
Casing-head gas	Oil wells	Varies, mostly butane, propane	1200–2000	Used mostly in oil field.
Carbureted water gas	Manufactured from coal, enriched with oil vapor	34% H_2, 32% CO, 16% CH_4, 7% N_2, 5% C_2H_4, 4% CO_2, 2% C_6H_6	500–600	Good fuel, but usually costly Part of most city gas
Coke-oven gas	By-product coke ovens	48% H_2, 32% CH_4, 8% N_2, 6% CO, 3% C_2H_4, 2% CO_2, 1% O_2	500–600	Good fuel when cleaned, often used at source
Natural gas	Gas wells	Varies, mostly CH_4, C_2H_6, C_3H_8	950–1150	Ideal fuel, piped to point of use
Oil gas	Manufactured from petroleum	54% H_2, 27% CH_4, 10% CO, 3% N_2, 3% CO_2, 3% C_2H_4	500–550	Used on West Coast, often mixed with coke-oven gas
Producer gas	Manufactured from coal, coke, wood, etc.	51% N_2, 25% CO, 16% H_2, 6% CO_2, 2% CH_4	135–165	Requires cleaning
Propane ...	By-product of gasoline	C_3H_8	2500	Similar to butane
Refinery gas	By-product of petroleum processing	Varies, mostly butane, and propane	1200–2000	Used mainly at refineries
Sewage gas	Sewage-disposal plants	65% CH_4, 30% CO_2, 2% H_2, 3% N_2, traces of O_2, CO, H_2S	600–700	Many disposal plants meet all power needs with this fuel

Q Describe and illustrate refractory gas burners.

A For boiler firing, a slightly different type of burner is widely used. It depends on natural or fan draft to draw in all the air required for combustion; hence draft conditions are important. One design uses multiple gas jets, which discharge into the airstream to cause violent agitation in a short mixing tube or tunnel refractory. In the burner of Fig. 11.35, turbulence vanes impart a swirling motion to the air entering the tunnel.

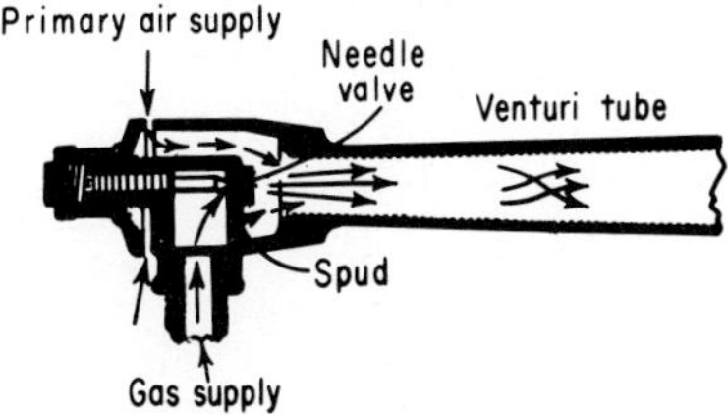

Fig. 11.34 Single-port atmospheric gas burner.

Q What kind of gas burner is used on large steam-generating units?
A Large steam-generating units often use a high-pressure (2 to 25 psi) gas burner of the gas-ring, center-diffusion-tube, or turbulent, design. The gas ring has an annular manifold located between the air register and the furnace wall surrounding the burner opening. Orifices drilled in this ring spray gas angularly across an incoming airstream, controlled in quantity, velocity, and rotation by the resistor.

Q Describe a gas burner where gas and air are mixed at one point and then supplied to several burners.
A Higher burner-head pressure to overcome variable furnace draft, high overload capacity, uniform air-gas mix at all loads, and single-valve control may be had in a system in which the mixture is made at one point and supplied to several burners (Fig. 11.36). In this low-pressure type, gas is at atmospheric pressure while air is at 1 to 2 psi. The heart of the system is the inspirator governor.

COLD-END CORROSION AND FUEL ADDITIVES

Q What does the term *cold-end corrosion* imply?
A Cold-end corrosion is the corrosion that may occur on the tail end of the boiler system, namely, economizer, air heaters, and induced-draft

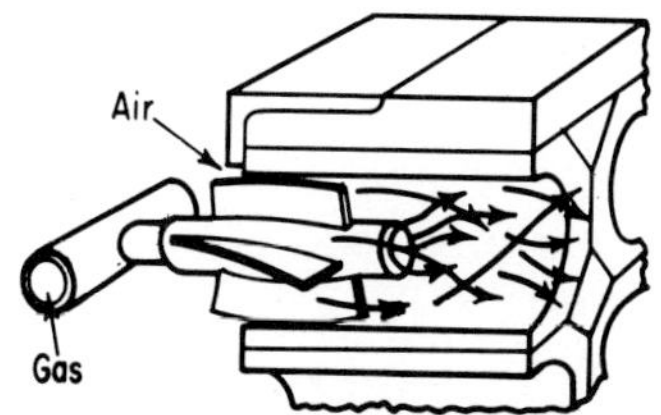

Fig. 11.35 Tunnel gas burner uses vanes to control burning.

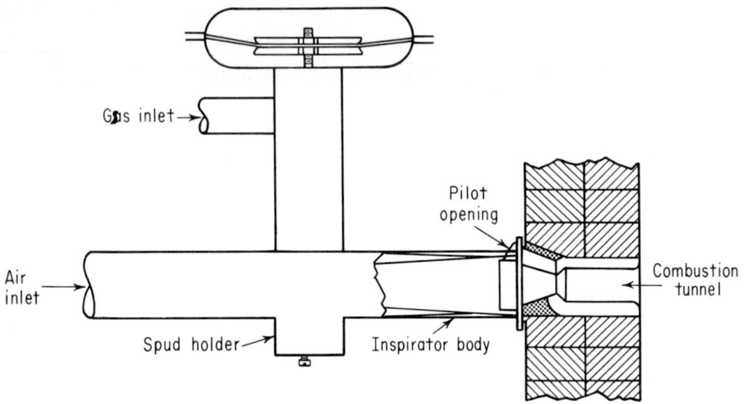

Fig. 11.36 Inspirator governor is used to mix fuel at atmospheric pressure with air at 1 to 2 psi to supply several burners and overcome furnace draft.

fans. The primary cause for this fireside or flue-gas-side corrosion is the sulfuric acid vapors that may be in the flue gas when sulfur dioxide combines with water. This acid condenses in the cooler part of the boiler passages. The acid dew point is the temperature at which sulfuric acid vapor is in equilibrium with liquid sulfuric acid so that the vapor condenses at that temperature.

Q How is cold-end corrosion prevented?

A Prevention of cold-end corrosion can start with design, such as the use of corrosion-resistant boiler passage material and metal-spraying parts with similar material. The usual method is to operate the boiler in a manner so that the back-end temperatures are maintained above the acid dew point in order to prevent acid vapors from condensing in the flue passages. Fuels free of sulfur will, of course, also prevent this type of corrosion.

Q How do fuel additives prevent cold-end corrosion?

A It is reported that additives can prevent the formation of the SO_3 acid molecule by neutralizing this molecule by having it combine with an additive. This raises the melting temperature of the resultant compound, and therefore it passes out the stack instead of depositing on the boiler components. The additives may be in liquid or powder form. Some reportedly provide protective coatings to retard flue-gas-side corrosion. Magnesium oxides are generally used because they have melting points over 5000°F.

Q What causes high-temperature corrosion on the fireside of a boiler?

A This type of corrosion is caused by the presence of vanadium or sodium in the fuel. The vanadium is oxidized in the furnace to V_2O_3. This oxide melts at 1274°F and thus can coat furnace walls, and it is known to be very corrosive under these conditions. The sodium reacts with vanadium to form vanadates. These have low melting points around 995°F, which also corrode metal surfaces on which they adhere.

Additives have been used to raise the vanadium oxide melting points so that they remain solids and thus are swept out of the furnace with the flue gas, especially by the use of soot blowers that remove solid particles from the tube passages.

High-temperature fouling on the flue-gas side can be most pronounced in the superheater sections of the boiler. Here sulfates and vanadium pentoxides adhere to surfaces. Magnesium oxide chemicals are being promoted as a fuel additive, sometimes in combination with aluminum oxide as one method to prevent high-temperature fouling of tube passages. Both of these chemicals raise the melting point of the chemical compound that forms with the additives.

POLLUTION CONTROL

Q What federal legislation has encouraged air-pollution abatement as a national policy?

A The National Environment Policy Act of 1969 has significantly affected designers and operators of power plants who must meet pollution standards considered necessary to clean up the environment by establishing acceptable threshold limits.

Q What two methods are used to lower air pollution in fuel burning?

A Broadly speaking, the two methods used are:

1. Burning clean or treated fuels that have the sulfur or other contaminant removed prior to burning to acceptable levels. This can be done if the fuel has a low ash content.

2. On-site removal of the air pollutants as they are formed in the burning process. This requires the owner to install additional equipment in order to remove the objectionable pollutant before it is discharged into the environment.

Q Name the type of stack emissions that may be regulated by federal or state regulations.

A Acidic flue-gas limits, particulate content of the flue gas, and plume opacity. Boilers generally have SO_2 limits established as well as NO_x stack discharge regulations, and these may be federal, state, or local threshold limit requirements. Figure 11.37 shows federal emission standards for coal-fired boilers of about 250,000 lb/h and over capacity. Coal-fired power plants may even be affected as respects limits by the

Fig. 11.37 Federal emission standards can be complex. Particulate, opacity, SO_2, and different coal-content limits per million Btu heat input are shown. Coal-fired steam-generation units, 250 MBth/h heat input. (*Courtesy of Nalco Chemical Co.*)

	Built before April 1972	Built 1972–1979	Built after 1979
Particulate	0.6 lb/MBtu	0.1 lb/MBtu	0.03 lb/MBtu
Opacity	40%	20% (80% for 6 min/h)	20% (27% for 6 min/h)
SO_2	Exempt	1.2 lb/MBtu	1.2 lb/MBtu and 90% reduction
NO_x		0.7 lb/MBtu	0.6 lb/MBtu
Anthracite	Exempt	0.7 lb/MBtu	0.6 lb/MBtu
Bituminous	Exempt	0.7 lb/MBtu	0.5 lb/MBtu
Subbituminous	Exempt	Exempt	0.6 lb/MBtu
Lignite*	Exempt		

MBtu = million Btu.

* If more than 25 percent of the lignite was mined in North Dakota, South Dakota, or Montana and is burned in a slag tap unit, the standard is 0.8 lb/MBtu.

number of coal-burning plants in the area, instead of just on a boiler basis.

Q Define particulate.
A These are fine, solid particles that are dispersed in flue gas as a result of the combustion process.

Q What methods are used to control particulate emission?
A Mechanical separators are used. These depend on centrifugal force to separate out the particles from the gas stream. Baghouses use filters or glass-fabric material to block the particulate matter but not block the flow of gases through the fabric. Wet scrubbers are used to wash out the particles from a gas stream. See Fig. 11.38. Electrostatic precipitators use the principle that an electric charge between two plates will attract particles to the plate out of a passing gas stream. When the plates are rapped, the deposited particles are dropped into a collection bin for disposal.

Q Differentiate between hot and cold precipitators.
A A hot precipitator is installed ahead of the air preheater, thus handling flue gases that have not been cooled, whereas a cold precipitator is installed after the preheater or after any other heat-absorbing medium in the flue-gas stream.

Q What is the advantage in a fuel having a high ash-fusion temperature?
A The ash-fusion temperature is the point at which an ash particle

changes from liquid to solid. A high ash-fusion temperature means that the particle will not remain molten for a long time after burning, thus not adhering as much to boiler parts in the burning and flue-gas zones but rather leaving the boiler as a solid particle.

Q What can be an operating and maintenance problem with fly-ash-removal systems?

A Fly-ash plugging can short out or misalign discharge electrodes in precipitators. Fly-ash plugging in a baghouse section or hopper will cause fly-ash reentrainment and thus result in restricted flue-gas flow, which will also affect boiler draft condition as a result.

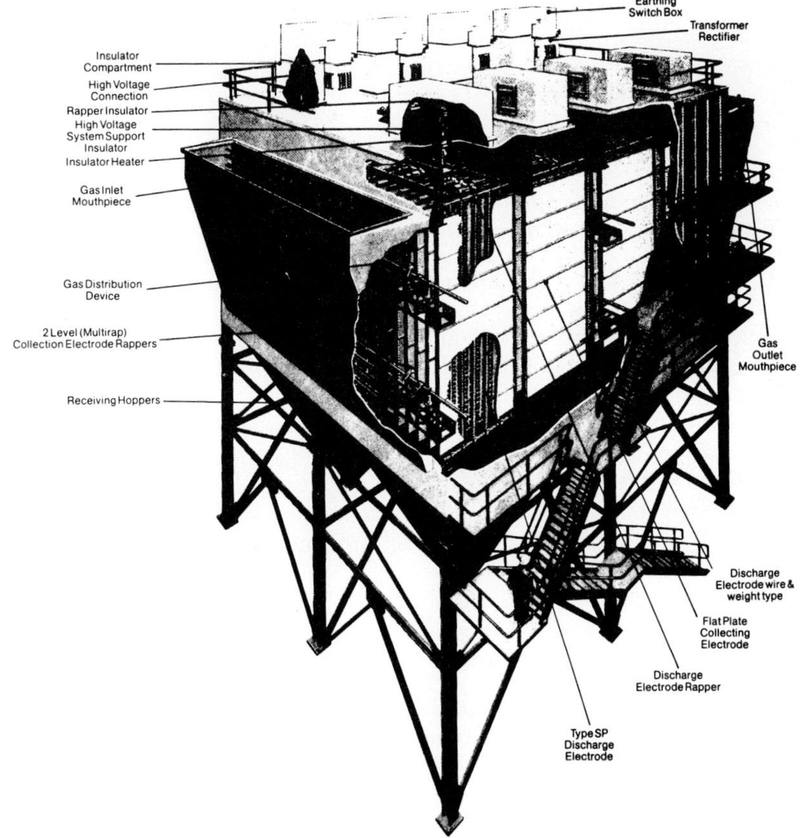

Fig. 11.38 Electrostatic precipitator removes solid particles from flue-gas stream by the use of electrically charged plates. (*Courtesy of Sturtevant Corp.*)

Q What causes fly-ash plugging?

A Fly ash contains silicas, ferrous and nonferrous oxides, and sodium, magnesium, and calcium compounds that make the fly ash very hygroscopic, or moisture-absorbent, at temperatures below the sulfur dioxide/sulfur trioxide dew point. Fly ash will pick up water vapor below this dew-point temperature and form cakes and solidify into chunks or blocks. If the temperature is below the dew-point temperature of the flue gas, condensation in the passages may also cause cement-type clinkers to form.

Q How is fly-ash plugging avoided?

A By preheating the lower portion of the fly-ash-collection hoppers during start-up to 150°F above the ambient temperature. This will prevent condensation and will also raise the hopper temperature above the flue-gas dew-point temperatures. When the flue gas is introduced to a heated hopper, the fly ash will thus not encounter any moisture to combine with and will not produce fly-ash plugging. Usually electric heaters are used to heat the hoppers.

Q What does the term FGD mean in pollution control?

A This is the term applied to the various methods used in removing sulfur from flue gas; it stands for *flue-gas desulfurization.*

Q What are some of the FGD systems being used or under study?

A The removal of SO_2 from flue gas is receiving great attention because of the so-called acid rain threat to the environment, which is being blamed on coal-firing plants. Among the FGD systems under investigation are:

1. Scrubbing the flue gas through limestone beds
2. Fluidized-bed burning by limestone injection
3. Scrubbing the flue gas through magnesia beds
4. Scrubbing with sodium but regenerating the sodium
5. Scrubbing with sodium but throwing the resultant products away in approved landfill areas
6. Catalytic oxidation of the SO_2
7. Carbon absorption
8. Manganese oxide absorption

Q Name the two types of opacity.

A Opacity is a term which represents the percentage of visible light that is not transmitted. The two types are:

1. *Stack* opacity, the opacity measured within the confines of the stack.
2. *Plume* opacity, determined at a distance above the stack from one to three stack diameters.

Q What factors may affect plumes?
A Water plumes appear when the atmospheric temperature drops to about 40°F so that water condenses before the plume is dispersed into the air. Sulfuric acid vapors with concentrations of 5 to 10 ppm of sulfuric acid vapor may also create plumes with the sulfuric acid in them, which can be harmful to the environment.

Q How can operators control emission?
A By following the maintenance, inspection, and operating instructions of the emission-control manufacturers on their installed systems so that jurisdictional requirements are complied with on such items as permissible threshold limits on SO_2, NO_x, noise levels, solid disposal requirements, and similar regulations that involve the environment. This requires paying attention to on-line monitoring systems as well as extracting periodic samples for manual testing on ppm or similar threshold readings.

Q Can oil- or gas-burning boilers be readily adopted to coal burning?
A All boiler manufacturers caution that the substitution of one fuel for another in a boiler must consider what the original design of the furnace may have been, because certain fuels such as coal require larger furnaces than do oil- or gas-burning furnaces. In addition, flue-gas passages for coal-burning boilers require more liberal spacing so that particulate matter may pass through. The problem of slag fly-ash formation needs to be reviewed. Combustion-air requirements differ between coal-burning boilers and oil and gas boilers. Auxiliary equipment to bring the fuel to the boiler and prepare it for combustion is a major consideration. Each conversion project will dictate the appropriate fuel-burning technology that will have to be adopted in order to obtain proper and efficient heat absorption in the boiler. Pollution-control requirements of the jurisdiction may differ between the various fuels.

It is considered good practice to refer any fuel-conversion job to the manufacturer of the boiler or to a designer experienced in analyzing fuel-burning criteria for particular sizes and types of boilers.

CHAPTER **TWELVE**

Combustion Controls and Burner Flame-Safeguard Systems

Fireside boiler explosions and implosions can occur and are quite destructive to property as well as endangering personnel working in power plants or boiler rooms. This chapter is devoted to reviewing the causes of furnace-side boiler explosions and what is or has been done to prevent this type of accident. Chapter 10 reviewed control theories and the types of controls available for boilers. This review was primarily on operating controls and how these controls coordinated combustion or fuel feed with load, pressure, air requirements, water feed, low-water protection, and similar variables on the steam and water side of the boiler.

COMBUSTION CONTROLS

Q Describe the basic functions of a combustion control system.

A Combustion controls in modern boiler systems incorporate operating and safety controls as follows:

1. Maintain proper fuel and air feed for combustion as load on the boiler system varies.
2. Incorporate necessary interlock system on pressure, water feed, draft, and similar associated boiler flows in order to maintain load and safe operating conditions.
3. Provide alarm systems to notify the operator when unsafe operating conditions are being approached or have been reached.
4. Provide safety trips so that if established safe operating conditions are exceeded, the trips will secure the boiler system or safely shut it down.

Q What combustion control is generally used on fire-tube and smaller water-tube boilers?

A On-off or two-position controls are used on commercial and smaller industrial installations, with the chief advantage being in the lower cost of on-off controls compared with more sophisticated controls. Under this scheme of control, when pressure drops to a preset value, fuel and air are automatically fed into the boiler at a predetermined rate until set pressure is reached, when the burner shuts off. The fuel/air ratio stays at one setting.

Q Explain positioning combustion controls.
A Positioning controls track steam pressure and simultaneously adjust or position fuel and air flows to a predetermined alignment, either through a common jackshaft and cam valve arrangement, which is called single-point positioning, or through adjustable cam positioners, which are called a parallel positioning system. In single-point positioning, fuel/air ratio adjustments are set up through a manipulation of the cam valve and linkage angles, and these are set up prior to boiler operation. This arrangement does not permit fuel/air ratio adjustment unless a trim mechanism is used. Parallel positioning permits controlling fuel and air separately through individual actuators and drives, thus making it possible to manipulate the air/fuel ratio by manual or automatic adjustments. The primary advantage of parallel positioning control is the ability to independently adjust fuel and air through the use of hand or automatic stations. The method is generally not suitable for controlling fuel firing in suspension, because an improper manipulation of the fuel or air ratio could permit an unsafe firing condition to develop. Parallel positioning control is thus used more on stoker firing, because a large reserve of unburned fuel exists on the grate and as a result, fuel/air ratios are not as critical as with fuel fired in suspension.

Q Describe the metering combustion controls.
A In this type of control, the combustion process is regulated by feeding the burner with metered amounts of fuel and air. This permits using feedback signals to the combustion controller to match steam flow and fuel-air flows. The advantage of metered control is that compensation can be made for such variable boiler conditions as slagging, changes in fuel quality, combustion air conditions, and similar variables that may affect burning in the furnace.

Q Why is excess air burning required on fuels fired in suspension?
A Excess air is required to ensure complete fuel and air mixing for combustion to proceed to the theoretical chemical amount. This may cause some loss in efficiency, and for this reason, trimming excess air to the barest minimum is being used to improve boiler efficiency. Care must be used, however, to prevent the amount of O_2 at the burner from dropping below the theoretical stoichiometric value during load changes in order to prevent a boiler furnace-side gas explosion.

FURNACE EXPLOSIONS

Q What is a furnace explosion?

A The ignition and almost instantaneous combustion of explosive or highly inflammable gas, vapor, or dust accumulated in a boiler setting. Often it is of greater expansive force than the boiler setting can withstand. In minor explosions, called puffs, flarebacks, or blowbacks, flames may blow suddenly for a distance of many feet from all firing and observation doors. Thus anyone in the flame path may be seriously or fatally burned. Such minor explosions indicate dangerous conditions, even if no real damage is done. Heavier explosions may shatter gas baffles, bulge setting walls, loosen refractory, blow brick tops off boiler settings through roofs, blow the sidewalls out from under the boiler, break connecting piping, and even demolish boiler housings.

Q What conditions are necessary to cause a furnace explosion?

A Usually three (Fig. 12.1): (1) accumulation of unburned fuel, (2) air and fuel in an explosive mixture, and (3) a source of ignition, such as hot furnace walls, improper ignition timing, faulty torch, and dangerous light-off procedures on manually started boiler combustion systems.

Q Name the common types of furnace explosions.

A Gas explosions and coal-dust explosions; each results from the presence of unburned fuel and its delayed ignition. Furnace explosions are of the primary and secondary types. For example, a quantity of unburned fuel in the primary combustion chamber, or furnace, may distill off a large volume of gas during a period of interrupted ignition. If this gas does not ignite promptly, it may fill the furnace and circulate back to a secondary pass.

Continued ignition delay may cause the volatile content of the unburned fuel to be exhausted and the major part of the gas to pocket in the second or third pass of the setting. When the diluted gas mixture remaining in the furnace ignites (because the burner is relighted or for some other reason), a furnace explosion of minor intensity occurs in the primary chamber. The blast of flame may then follow through to ignite

Fig. 12.1 Three conditions determine if fuel explosions may occur. (1) Unburned fuel; (2) an explosive air-fuel mixture; (3) a source of ignition.

the gas in the succeeding passes, resulting in a secondary, and more violent, explosion.

Q How does ignition temperature affect the burning of fuel and the possibility of having unburned fuel in the furnace?

A Research has indicated ignition temperature in a furnace may vary, depending on fuel-air mixtures, temperature of the furnace, and rate of chemical reaction that takes place. The spontaneous ignition temperature of a fuel is defined as that temperature at which a combustible mixture at atmospheric pressure is both self-sustaining and rapidly self-propagating. The combustion reactions in a furnace can be extremely complex, involving aerodynamics of air and fuel flow, thermodynamics, and chemical kinetic considerations. It is known that combustion reaction rates increase with increase in temperatures; therefore, high ignition temperatures will avoid delayed ignitions and possible furnace explosions. Liquid and solid fuels must be vaporized or dried in the furnace before the fuel temperature reaches that of spontaneous ignition. Therefore, ignition may be affected by furnace temperatures, degree of fuel dryness, fuel properties, and similar combustion variables.

Q Does coal have a variable ignition temperature?

A Ignition temperature may vary because of volatile content, size of coal being burned, and whether it is burning in air or oxygen. See Fig. 12.2, which shows ignition temperatures for coals of different volatile content.

Q What are some typical minimum ignition temperatures for gaseous fuels?

A See Fig. 12.3, which provides minimum ignition temperatures of gaseous fuels in air.

Q Does the NFPA have any igniter classification system?

A The National Fire Protection Association's committee on boiler-furnace explosion has defined classes of igniters and has made recommendations on their applications and capacities as follows:

Class 1 igniters are a continuous type able to ignite fuel and support ignition under a large range of burner light-off or operating conditions. Location and capacity are such that they provide sufficient ignition energy (generally in excess of 10 percent of full-load burner input) to raise the associated burner fuel-air mixture above the minimum ignition temperature.

Class 2 igniters are an intermittent type capable of igniting fuel input under prescribed light-off conditions and supporting ignition under low loads or certain adverse operating conditions. The capacity of

Fig. 12.2 Ignition temperature of a coal, °F.

Volatile content of dry coal, ash-free basis, %	Coal size, U.S. standard sieve	Ignition temperature, °F
	In oxygen	
18.0	200	1067
	325	995
30.2	200	932
	325	977
42.5	200	1085
	325	1067
53.0	200	995
	325	996
	In air	
18.0	200	1598
	325	1553
30.2	200	1553
	325	1517
42.5	200	1535
	325	1517
53.0	200	1436
	325	1400

such igniters is generally in the range of 4 to 10 percent of full-load burner input.

Class 3 igniters are the interrupted type applied particularly to gas and oil burners. They are small igniters used to ignite the burner under prescribed light-off conditions. Their capacity generally does not exceed 4 percent of the full-load burner fuel input. These igniters are turned off before the trial time for ignition of the main burner has expired, and they are not used to support ignition or extend the burner operating range.

Q Should fuel ignition be with excess air at all times?

Fig. 12.3 Ignition temperatures of gaseous fuels in air, °F.

Fuel	Temperature, °F
Carbon monoxide	1128
Methane	1170
Hydrogen	1065
Ethane	882
Propane	898

A An ignition system must light off a burner with an excess of air in order to avoid smoke formation. The excess air should be kept to a minimum in the burner region, as the ignition energy required to heat a large quantity of air substantially increases the minimum ignition energy. Once a safe light-off procedure is developed, all the prerequisites of an interlock system such as damper position and windbox pressure should be observed.

If changes to fuel-preparation equipment, fuel, or operation are anticipated, the ignition system should be reevaluated. Typical operating changes which could result in a requirement for additional ignition energy would be operation of gas recirculation fans below specified loads, increase in the burner turndown ratio below recommended limits, utilization of oil with a different viscosity (requiring higher fuel preheat temperature for proper atomization), or operation at high windbox pressures at low loads.

Q How do flammability limits affect flame propagation?
A Figure 12.4 provides upper and lower flammability points for some volatile gases. It is important for boiler operation during start-up or operation that the fuel-air mixture is well within the limits of flammability, because the low resultant flame if not within this range may be easily upset by fuel and air changes and thus will result in an unstable flame.

Q Why is it important to purge a furnace before light-off of fuel?
A For safe operation, the furnace should be purged before light-off in order to reduce the flammability limits well below those at ambient temperature of the fuel being burned. The purging of a furnace will make the air/fuel ratio lean so that ignition may not occur. Figure 12.5 shows a flammability curve for natural gas. Note the lean and fuel-rich

Fig. 12.4 Upper and lower flammability limits of combustibles at different temperatures.

Initial gas temperature		Percent fuel in mixture with air					
		Hydrogen		Carbon monoxide		Methane	
°C	°F	Lower limits	Upper limits	Lower limits	Upper limits	Lower limits	Upper limits
17	62.6	9.4	71.5	16.3	70.0	6.3	12.9
100	212	8.8	73.5	14.8	71.5	5.95	13.7
200	392	7.9	76.0	13.5	73.0	5.50	14.6
300	572	7.1	79.0	12.4	75.0	5.10	15.5
400	752	6.3	81.5	11.4	77.5	4.80	16.6

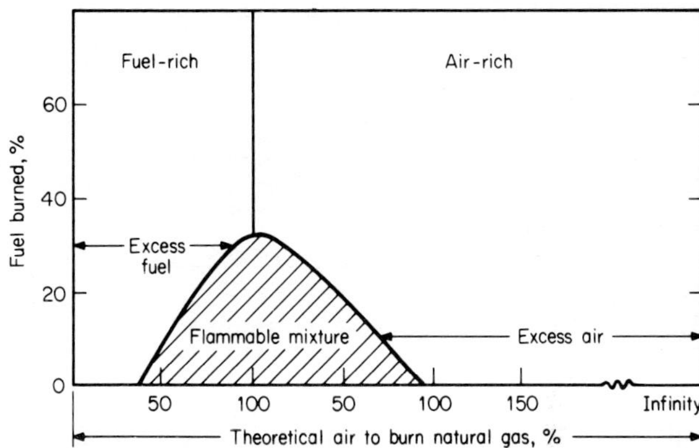

Fig. 12.5 Fuel-rich and air-rich zones of natural gas at ambient temperature.

zones where no ignition would occur. This chart is for ambient temperatures of natural gas. It would be different at higher temperatures. Safe boiler and furnace operation during light-off or during low loads can be best achieved by immediately lighting off or igniting the fuel that enters the furnace in order to avoid the accumulation of unburned fuel in the furnace and then having ignition of this fuel under uncontrolled combustion that actually may be of explosive strength.

Q Define the term smoke-point burning.
A The smoke point is the burning of a fuel with such low excess air that smoke from incomplete combustion begins to appear (see Fig. 12.6). This is done to gain maximum efficiency of burning; however, for safety reasons the boiler should actually be operated above the smoke point at all times.

Q What are some of the contributing factors that cause furnace explosions?
A Statistics show there are nine major causes:

1. Flame failure due to liquids or inert gases entering the boiler fuel system
2. Insufficient purge before the first burner is lighted
3. Human error
4. Faulty automatic fuel regulating controls
5. Fuel shutoff valve leakage
6. Unbalanced fuel/air ratio
7. Faulty fuel supply systems
8. Loss of furnace draft
9. Faulty pilot igniters

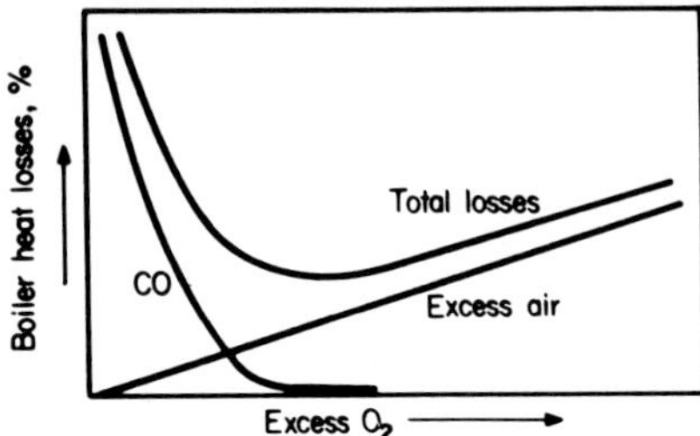

Fig. 12.6 Smoke-point burning aims for maximizing boiler efficiency with low excess-air burning.

Most of these causes, such as insufficient purge, leaking fuel valves, or improper pilots, have long been known. Liquids or inert gases in fuel systems (as cause of explosions) have increased in frequency as more plants have begun using mixtures of process and natural gas for boiler fuel. Plant process upsets, or equipment failures, allow inert or excessively rich gases to enter the fuel system. Many installations are being equipped to burn both gas and liquid fuels. But each fuel has its own peculiar safety problems.

Faulty automatic fuel-regulating controls can immediately create a dangerous condition in a modern high-efficiency boiler because of lower temperatures of fireboxes, mixed fuels, high-capacity burners, and rapid load swings.

Faulty oil or coal fuel-burner systems also cause explosions. But most explosions arc on gas-fired units, usually caused by incorrect limits or stops on the combustion controls, or a control-component failure or malfunction.

Forced or induced draft or both in modern boilers are necessary to achieve complete combustion of the fuel being burned. Failure of fan drives, dampers, or damper controls causes an accumulation of unburned gases in the firebox. Failure or partial failure of combustion air may starve the burner flames, creating conditions for firebox explosions. Human error is a factor in many furnace explosions. Faulty ignition includes inadequate torches, torches held in the wrong position, or pilot lights incorrectly adjusted.

Q How does human error contribute to furnace explosions?

A Human error causes are primarily due to incompetent or poorly trained operators.

> EXAMPLE: Forgetting to purge the firebox before light-off; trying to light one burner from an adjacent burner or, worse still, from hot refractory; incorrectly adjusting combustion and safety controls;

also, not recognizing a dangerous condition because of not understanding the equipment.

Operating today's complex, modern boilers is too difficult for an untrained operator. The only answer is well-trained personnel, backed up by adequate safety controls. The quality and quantity of safety control system maintenance is as important as any other phase of its design and use. The annual cost of maintaining the control system should be from 1 to 5 percent of the equipment's original cost. If the system is selected on cost alone, maintenance can go much higher. Worse still, the reliability of the system will suffer.

Q List the operating precautions needed to prevent furnace explosions.

A Seven basic precautions are:

1. Check the operation of the boiler periodically.

2. If a burner goes out accidentally, shut off the igniter and fuel supply and thoroughly scavenge the furnaces and gas passes before again attempting ignition. Always, always determine and remedy the cause of the stoppage.

3. Keep burners and all allied equipment clean.

4. On boiler using both forced- and induced-draft fans, test the interlock periodically.

5. Do not attempt to secure excessively high CO_2 by using too rich a fuel/air ratio or by an inadequate secondary air supply.

6. Keep the temperatures and pressures for preheated air, drying air, fuel oil, etc., at the right levels.

7. Never allow an unstable flame condition to continue uncorrected.

Q Do modern, compact boilers firing fuel in suspension have a greater potential for having a furnace explosion?

A In general they do, unless the boiler is equipped with well-maintained flame-safeguard systems with a periodic test program on these controls. There are four main reasons for more explosions in modern boilers:

1. Boilers are larger, calling for higher burner capacity and more efficient boiler operation.

2. Boiler capacity has been increased while furnace size has been held to a minimum. Thus firing conditions are more critical.

3. Many boilers of the waterwall type are now being used, and the firebox temperatures are lower.

4. More installations are using mixtures of natural and plant process gases.

The steam- and water-pressure parts of boilers are required by state law to have periodic standard inspections. But few standards for periodic legal inspections have been developed for fireboxes and their controls. This has resulted in a wide difference of opinion as to what is considered necessary to prevent furnace explosions. The maintenance and inspection of boiler combustion safety controls are often left to the discretion of the boiler owner. Thus there is an increase in firebox explosions.

The current trend toward automatic boiler control, with fewer operating personnel, has created a strong need for reliable safety controls. Legal requirements, such as exist for the construction and inspection of the boiler pressure parts, have not been adopted as widely for the construction, installation, and inspection of controls needed to prevent furnace explosions.

Q Are furnace explosions more common with suspension firing?
A While furnace explosions may occur with any type of fuel or firing, they occur far more frequently when fuels burn in suspension than in solid form on grates. This means that the risk is greater with boilers fired with pulverized coal, oil, gas, and waste gas.

Q What characteristic of coal is responsible for furnace explosions?
A During light-load periods, a cold furnace has little igniting tendency, and with pulverized coal low in volatiles, the flame may be unstable, thus may go out. The drop in steam pressure will cause the automatic controls to increase coal feed, filling the furnace with an unburned mixture of pulverized coal and oxygen. A dust explosion may result, or the coal may settle to distill off the volatiles, resulting in a gas explosion before the burner comes on again, or when it does come on. Use of a smaller burner for low-load periods, or burning higher-volatile coal, will prevent explosions of this type.

Good starting pilots are necessary to ignite pulverized coal reliably. A relatively large, stable pilot flame is needed. Either gas or oil may be used, but oil is preferable because its flame is hotter and more stable. Under strong draft conditions, it is possible to pull a gas flame away from the pilot burner.

Pulverized-coal-burning equipment should be mechanically or electrically interlocked so the units can be placed in operation only in the following order: (1) induced-draft fan to purge furnace and passes, (2) forced-draft fan, (3) primary-air blower, (4) coal pulverizer, and (5) coal feeder. Where natural draft is used instead of induced draft, make provision to ensure the wide opening of the flue damper if the forced-draft fan should stop. Failure of any one of the units should automatically shut down all the equipment, following it in the above *reverse* order. An alarm to warn of any interruption in the flow of coal to the pulverizer feed is valuable.

Q How can a burner-igniter malfunction cause an explosion?
A Burners using an auxiliary pilot to ignite a main burner always present the hazard of the pilot going out accidentally before the main burner goes on. By then the entire setting may be filled with fuel. Fire departments and insurance company requirements usually stipulate that pilots be equipped with automatic shutoffs to stop the fuel flow if the flame on the pilot goes out. Pilot-proving flame detectors are now becoming mandatory on larger boilers and also on smaller gas-fired units.

Q Name three types of gas pilot ignition for industrial boilers.
A 1. Interrupted gas-electric ignition. This type uses a pilot for seconds only. The burner fires after ignition without the pilot. Use is mostly for firing residual fuel oil, but also for natural gas, depending upon gas-line valving and vent line to prevent leakage of gas into the furnace during off-firing cycle.

2. Intermittent pilot. This type uses a pilot to ignite fuel and continues to burn during the firing cycle. The burner and pilot go off simultaneously.

3. Continuous pilot. Once ignited, this type of pilot burns continuously whether the burner is firing or is off. Thus protection is provided against unburned gases entering the furnace. This pilot will ignite and burn off any leakage of gas that may enter the furnace.

Q Can fan and damper malfunction contribute to furnace explosions?
A Accidental stoppage of the induced-draft fan, with continued operation of the forced-draft fan, results in fuel being fed to the furnace faster than combustion products leave. Mechanical or electrical interlocks prevent this condition and also prevent the possibility of starting the forced-draft fan unless the induced unit is running. A tightly closed damper creates a situation similar to the stoppage of an induced-draft fan. Cutting back dampers, so there is always an opening at the bottom and top edges (even in the closed position), allows gas movement at all times.

Mechanical and electrical interlocks play a vital role in operating equipment in proper sequence. They immediately shut down a unit if any of the components in a combustion system is not operating within set limits.

BURNER CONTROLS

Q How may burner firing controls be classified as to manual, semiautomatic, automatic, and degree of operator participation on larger industrial and utility-type boilers?

A Here is one classification:

Manual control. An operator watches the burner and boiler, then adjusts the burner manually. Adjustments on operating conditions are made by the operator. But to properly coordinate the start-up and shutdown of burner equipment, good communication is required between the operator and the control room. Today manual supervision and control are found mostly on older boilers.

Manual control with lighter-flame-proving system. This system provides a semiautomatic lighter control, including a flame-proving and interlock system. Starting is from a control board with firing and purge interlocks first satisfied. Then the lighter will be started and proved continuously, thus providing a "go" signal for the introduction of main fuel to that burner. The National Fire Protection Association (NFPA) now recommends that flame detectors or other means be provided to prove igniters in service. For this limited system, the fuel equipment should be operated similarly to manual control.

Manual control with lighter- and main-flame-proving systems (Fig. 12.7). Many industrial boilers with gas or oil firing or both in multiple burners are protected by a burner interlock and safety system, using individual supervisory cocks and individual burner trip valves from main-flame detection. The lighter is initiated from a local panel with manual operation of the air registers and supervisory cocks. If a short ignition-time delay occurs, the individual burner valve will trip on the loss of the main flame. Normally no other monitoring is provided at each burner, except by the operator's judgment. This is a local manual system

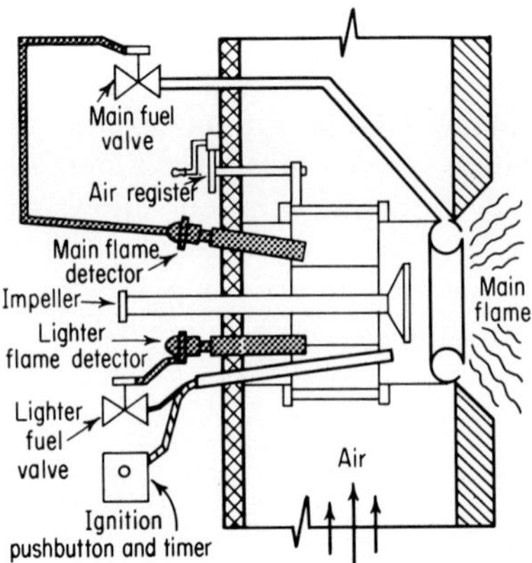

Fig. 12.7 Manual burner control has lighter and main-flame proving system as backup.

because such systems have limited, or no, burner interlocks and trips, except from lighter and main-flame detectors, but also no, or limited, cross interlocks between burners.

Remote manual sequence control. For remote manual operation and for proper evaluation, this system should use instrumentation systems and position switches in the control room. Besides flame detection, such a system needs various burner permissives, interlocks, and trips that sense the position of fuel valve and air registers, in addition to main-flame detection. With this system, operators still participate in the operation of the fuel equipment. They control each sequence of the burner-operating procedure from the control room, and no steps are taken except by their initiation.

Automatic sequence control (Fig. 12.8). This system automates the sequence control, thus permitting initiation of burner equipment from a single push button or switch control. Automation then replaces the operator in control of the operating sequences. The operator still initiates the demand for each fuel-equipment unit and thus must monitor the operating sequence, as indicated by signal lights and instrumentation signals, as the start-up process proceeds to completion of service on automatic control. This category has been widely applied to gas-, oil-, and coal-burning systems.

Fuel management (same equipment as in Fig. 12.8). This system will permit fuel equipment to be placed in service without supervision by the operator. The system will recognize the level of fuel demand to the boiler, sense the operating range of the fuel equipment in service, make a decision concerning the need for starting up or shutting down the next increment of fuel equipment, and then select the next incre-

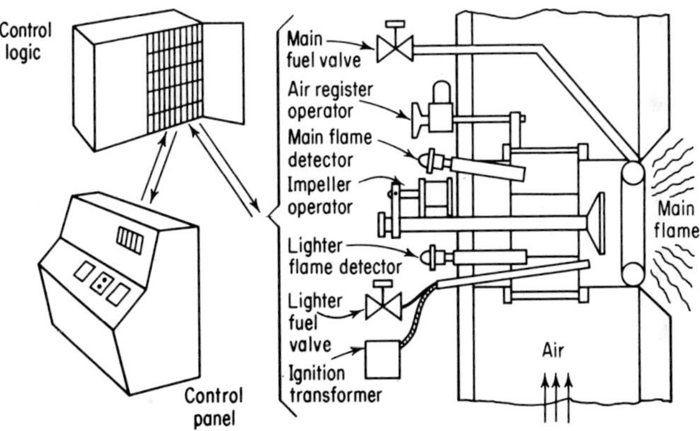

Fig. 12.8 Automatic-sequence-control firing system.

ment based on the firing pattern of burners in service. Demands for the start-up or shutdown of fuel-preparation and burning equipment can be initiated by this system without the immediate knowledge of the operator.

Q Some large-city ordinances cover operator functions with manual and automatic boiler controls. How do they relate to the degree of automation?

A The ordinances usually refer to the terms *manual* and *automatic*. There are four classifications, two manual and two automatic:

1. Manual. A boiler which is purged, started, modulated, and stopped manually.
2. Supervised manual. A boiler which is purged and started manually, modulated automatically, and stopped manually.
3. Automatic nonrecycling. A boiler which, when actuated manually by a push button, is purged, started, modulated, and stopped automatically, but does not recycle automatically.
4. Automatic recycling. A boiler which is purged, started, modulated, and stopped automatically, and which recycles on a preset pressure or temperature range automatically.

Q Explain how a purge interlock system functions.

A Interlocks are used for purging a boiler before lighting off or for shutting down equipment in case of fan failure. The purge interlock (Fig. 12.9) is actuated by a differential pressure that is proportional to the airflow through the boiler. This is usually the differential across the boiler tubes. When the boiler is out of service, no differential is applied to the interlock measuring element (diaphragm), thus contact M2 is open. Contact R3 (closed) may be auxiliary fingers on a fan-motor breaker (on whichever, or both, of the induced-draft or forced-draft fans will be used for purging), or it may be a relay actuated by contact M1 of a fan interlock. With contact M2 open and the top contact of R3 closed, the time-delay relay R1 is deenergized and contact R1 is open, thus opening the fuel-starting circuit to prevent the supply of fuel. Also, the red signal light (if furnished) is energized.

Before fuel may be supplied to the furnace, the fan motors must be started to open the top contact of P3 and close the bottom contact, and the airflow through the furnace must be increased to about 60 percent of capacity. This high rate of airflow differential closes contact M2 to complete the circuit form L1 through M2 and the R1 coil to L2. The amber signal light (if installed) is energized. The time-delay relay R1 allows the high rate of flow through the furnace for several minutes to purge the boiler of any explosive mixture before contact R1 is made to energize the fuel-starting circuit to permit fuel supply. When contact R2 is made, it sustains itself through one of its own contacts and the lower contact of R3. The airflow can then be reduced for lighting off. Upon a

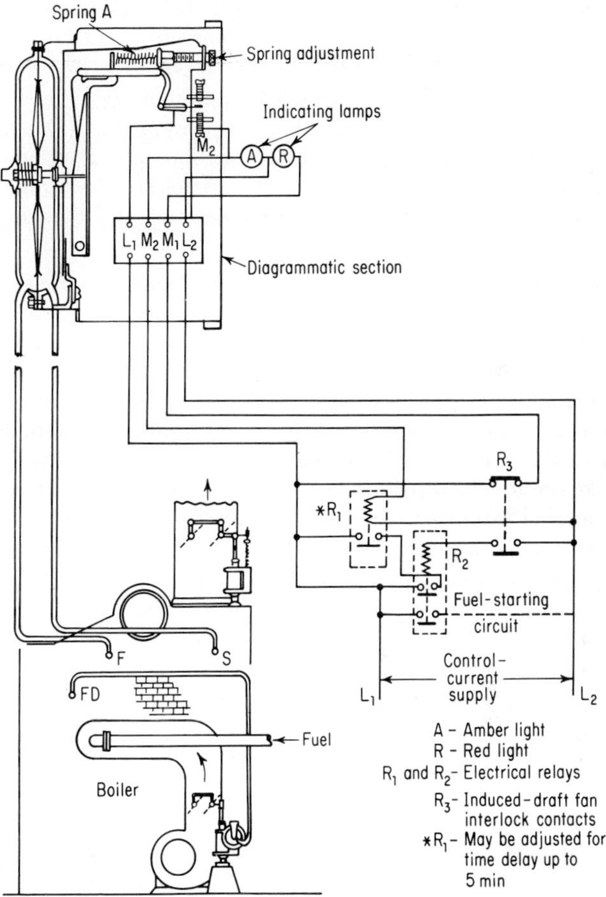

Fig. 12.9 Purge interlock wiring diagram.

boiler shutdown, the lower contact of R3 will open to deenergize the relay R2 and bring about the same conditions as mentioned above when the boiler is out of service.

BURNER FLAME-SAFEGUARD SYSTEMS

Q What is meant by a flame-safeguard system?

A A flame-safeguard system is an arrangement of flame-detection systems, interlocks, and relays which will sense the presence of a proper flame in a furnace and cause fuel to be shut off to the furnace if a hazardous (improper flame or combustion) condition develops. Modern combustion controls are closely interlocked with flame-safeguard systems and also pressure-limit switches, low-water fuel cutoffs, and other safety controls that will stop the energy input to a boiler when a dangerous condition develops. Thus it becomes obvious that a modern flame-safeguard system performs actually two functions: (1) senses the presence of a good flame or proper combustion and (2) programs the operation of a burner system so that motors, blowers, ignition, and fuel valves are energized only when they are needed, and then in proper sequence.

Q What is the primary function of a flame-safeguard system?

A Boilers are always prone to two possible types of explosions.

1. Boiler explosions. These are caused by the release of accumulated energy in the form of pressure. Then, because of structural weakness or malfunctioning of pressure and temperature controls, the vessel (or part of it) can explode from forces that are normally contained by the boiler structural elements. This type of explosion is sometimes referred to as a "steam-type" explosion.

2. Furnace explosions (combustion explosions). These are caused by the sudden ignition of accumulated fuel and air in the fireside of the boiler. This can also lead to devastating property damage and loss of life. The flame-safeguard system used on different boilers firing different types of fuel is designed to sense, and sometimes anticipate, this accumulation of unburned fuel and air in the fireside of the boiler. It safely shuts off the firing equipment, with purging of the furnace usually following in a time sequence so as to drive the unburned fuel-air mixture out of the furnace.

Q What two control parameters are usually considered in a flame-failure system to prevent furnace explosions?

A 1. Control the input composition so that it cannot accumulate to an explosive batch or mixture (Fig. 12.10). This is called *input control.*

2. Ignite in proper sequence all combustible combinations of fuel and air as they enter the furnace. This is called *ignition control.*

Input control depends on a time factor of how much fuel can be put into a furnace before combustion must ensue so no dangerous batch is present for a delayed light-off leading to an explosion. The factor is called the *grace period,* which is the time needed to build up an explosive charge. It varies with the design of the furnace. A small, fast-heating flash boiler might have a grace period of about 1 s, while a large, commercial-type furnace might have 4 s, or more, for ignition or input cutoff.

For safety, ignition protection and shutdown of fuel and air into the furnace must be done before the end of the grace period. As a rule of thumb, a fuel-air mixture greater in volume than 30 percent of the volume of the combustion chamber may be a damaging mixture.

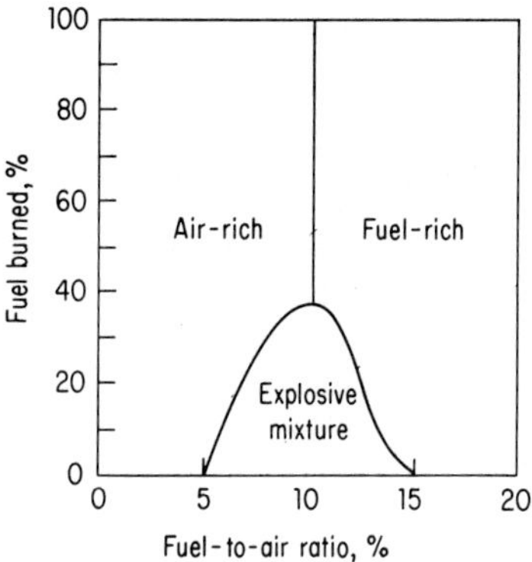

Fig. 12.10 Methane-air mixture showing flammability or explosive-mixture range.

Q What are some of the flame characteristics of a fire that can be used to monitor a flame?

A Many flame-monitoring devices are based on the following physical principles of a flame.

1. A flame produces an ionized zone, meaning that it can conduct a current through it. The amperage is low, as the resistance through the flame is high, being on the order of 250,000 to 150,000 ohms. The currents are in microamperes when a voltage is impressed across a flame, but electronic devices can amplify this current and make it a control signal for proving a flame. Conductivity flame rod detectors use the principle of a conducting flame for flame-detection monitoring.

2. A flame can rectify an alternating current. This is done by making one electrode across a flame larger than the other, thus making electrons flow through a flame much more readily in one direction than in the opposite direction. When an alternating voltage is applied to these electrodes, the resulting current is, in effect, an intermittent direct current.

3. Radiation of light is a known phenomenon of any fire. A flame radiates energy in the form of waves which produce heat and light. Three types of radiation from a flame are:

a. Visible light that can be seen by the human eye. The wavelengths of visible radiation extend only from 0.4 to 0.8 μm. Visible radiation for flame detection has the limitation of low intensity. This intensity varies with different fuels, burner types, and methods of mixing. Refractory-material radiation approaches visible radiation, making flame detection difficult. Visible-radiation flame detection is found to be more suitable for oil flames than for gas flames. Visible-radiation flame detection is by means of the oxide of the metallic element cadmium. When this metallic element is exposed to visible light, it emits electrons with the strength of the visible light. Thus, if a cadmium phototube is designed in an appropriate electronic circuit, electricity will flow through the circuit when the cadmium is exposed to sufficient light. This electricity can be used to trigger relay circuits for flame detection.

b. Infrared radiation covers most of the useful band of wavelengths and also covers most of the radiation strength. Infrared detectors are suitable for gas and oil flames. Because hot refractories also radiate infrared, scanners must avoid hot refractories. Lead sulfide cells are used in photocells to sense infrared radiation. Unlike cadmium phototubes, they do not emit electrons but have the property of having their electrical resistance reduced while exposed to infrared radiation. The greater the strength of the radiation, the lower the resistance of the lead sulfide. This principle is used for flame-detection purposes when such cells are connected to a designed electronic circuit.

c. Ultraviolet radiation is the most widely used flame detector based on the phenomenon of sensing the strength of ultraviolet radiation in a flame. It is insensitive to visual and infrared radiation and is not affected by hot refractories, as these usually do not give off appreciable ultraviolet radiation. When radiation from a flame passes through the typical quartz viewing window of one of these detectors into the flame-sensing tube, the tube becomes electrically conductive. The strength of the detector signal, or current passed through the sensing tube, depends on the kind of fuel, size and temperature of the flame, and distance between the flame detector and the flame. Figure 12.11 shows some typical wavelengths in a flame and response percentage of total wavelengths of typical flame pickup devices.

Q What are the principal flame detectors used on boilers?

A The type of flame detector depends on the fuel used, the type of burner, and the size and arrangement of the boiler. Flame detectors vary from those used on small domestic boilers to those on large boilers. Types of detectors and the task each does are:

1. Stack switch, heat-sensing
2. Rectifying flame rod, heat-sensing
3. Rectifying phototube, visible-light-sensing
4. Lead sulfide photocell, infrared-light-sensing
5. Cadmium sulfide photocell, visible-light-sensing
6. Ultraviolet flame-detector tube, ultraviolet-light-sensing
7. Visible-light detectors using fiber-optic light guides

Q Name some limitations of flame-sensing devices.

A While each of the flame-sensing devices can offer a substantial amount of protection if properly installed, they are all subject to certain

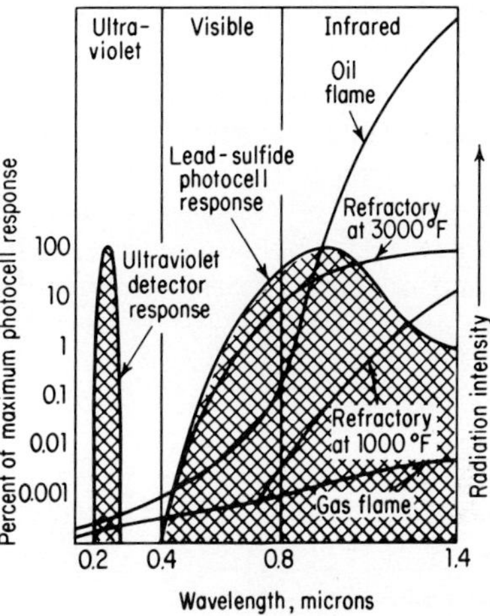

Fig. 12.11 Percentage wavelength light response vs. wavelength for various flame-detection systems.

limitations that must be allowed for. For example, the lead sulfide cell and the photoelectric cell are subject to the following limitations (there is some variation between types):

1. Discrimination between burners. With more than four burners in one firebox, it becomes difficult to locate the sensing cell where it will not be actuated by the flame of an adjacent burner when the burner on which it is mounted has been extinguished.

2. Ambient temperature of sensing cell. High ambient temperatures of the sensing cells (which are easily obtained at the locations where they must be mounted) can result in erratic signals, false signals, and a short life. Air and water cooling have been used to prevent or minimize this limitation.

The flame rod is subject to (1) short service life at high flame temperatures, (2) fouling of insulators causing short circuits and shutdowns, (3) difficulty in providing sufficient rod area to ground the flame, and (4) limitation generally to pilots and small burners. Figure 12.12 gives some comparisons of the different flame safeguard systems.

Q List some precautions to observe when installing flame rods.

A When selecting a mounting location, observe the following precautions:

1. Make sure that the flame rod will check the pilot flame at the desired point.

2. The unit must be clear of the fire door opening radius.

COMPARISON OF FLAME SAFEGUARDS

Principle of Flame Detection	Rectification		Infrared	Visible Light	Ultravision
Type of Detector	Rectifying Flame Rod	Rectifying Phototube	Lead Sulfide Photocell	Cadmium Sulfide Photocell	Ultravision Detector Tube
Advantages					
Same detector for gas or oil flame			▨		▨
Can pinpoint flame in three dimensions	▨				
Viewing angle can be orificed to pinpoint flame in two dimensions		▨	▨	▨	▨
Not affected by hot refractory	▨			▨	▨
Checks own components prior to each start	▨	▨	▨	▨	▨
Can use ordinary TW plastic-covered wire for general application, no shielding needed	▨	▨		▨	▨
No installation problem because of size			▨	▨	
Disadvantages					
Difficult to sight at best ignition point			▨		
Exposure to hot refractory may reduce sensitivity to flame flicker and require orificing			▨		
Flame rod subject to rapid deterioration and warpage under high tenperatures	▨				
Not sensitive to extremely hard premixed gas flow			▨	▨	
Temperature limit too low for some applications	▨	▨	▨	▨	▨
Shimmering of hot gases in front of hot refractory may simulate flame			▨		
Hot refractory background may cause flame simulation		▨			
Electric ignition spark may simulate flame					▨

Fig. 12.12 Advantages and disadvantages of the different flame-safeguard systems.

3. Locate the unit so that drafts will not blow the pilot flame away from the rod.

4. Install the unit so the flame rod is horizontal or angled upward. Use extra support for rods over 12 in long.

5. If the flame rod is to supervise a gas pilot for an oil burner installation, the rod must be located far enough from the oil flame to prevent oil spray from impinging and burning on the surface of the rod.

6. A horizontal or inclined flame rod should enter the pilot flame from the side.

7. Protect the lead wires from excessive radiant or reflected heat. For temperatures under 125°F, use No. 14 wire with thermoplastic insulation. Install a 2-ft flexible lead to the head of the unit to permit easy removal from the combustion chamber.

8. On an oil installation, protect the flame rod insulator from oil and soot deposits which may cause nuisance shutdowns of the burner.

Q Illustrate a typical gas-pilot ignition and main oil-burning flame-sensing installation.

A The rectified impedance system (Fig. 12.13) operates on the principle that either a flame or a photocell sighted at a flame is capable of conducting, as well as rectifying, an alternating current. This alternating current is applied to either a flame electrode inserted in the flame or a photocell sighted at the flame. The resultant rectified current, which can be produced only when a flame is present, is in turn detected by the relay.

The actual flame-detecting units consist of flame-electrode and photocell rectifier assemblies. The flame-electrode type is generally used for nonluminous flames, such as gas flames, whereas the photocell type is usually used for luminous flames, such as oil flames. The system shown in Fig. 12.14 provides gas-pilot flame supervision on start, and oil main-flame supervision on run, for an oil-fired burner. The flame-electrode rectifier is used to supervise the gas-pilot flame (constant pilot) during start-up only, while the photocell rectifier is used to supervise the oil main flame. If the oil-fired burner is ignited by an intermittent pilot, the flame-electrode rectifier is not needed.

Q Why may ultraviolet scanners not be quite suitable for heavy oil and pulverized coal flames?

A The ultraviolet-scanner flame detector also picks up unburned and partially burned hydrocarbon rings or skirts around the main flame because of the ability of these particles to absorb ultraviolet radiation from the root of the flame. This can create nuisance trips or false signals of flame failures as the strength of the ultraviolet radiation varies.

Q What kinds of detectors have been developed for pulverized-coal-burning units?

A There are two:

1. *Infrared detector.* Because ultraviolet detectors are unsuitable for pulverized-coal or heavy-oil flames, developers have found a way to

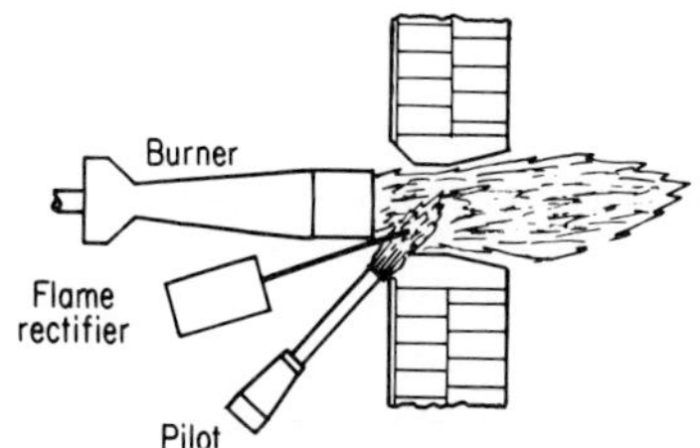

Fig. 12.13 Correct installation is important for a flame-rod-type flame-guard system in order to prove pilot and the main flame at all times.

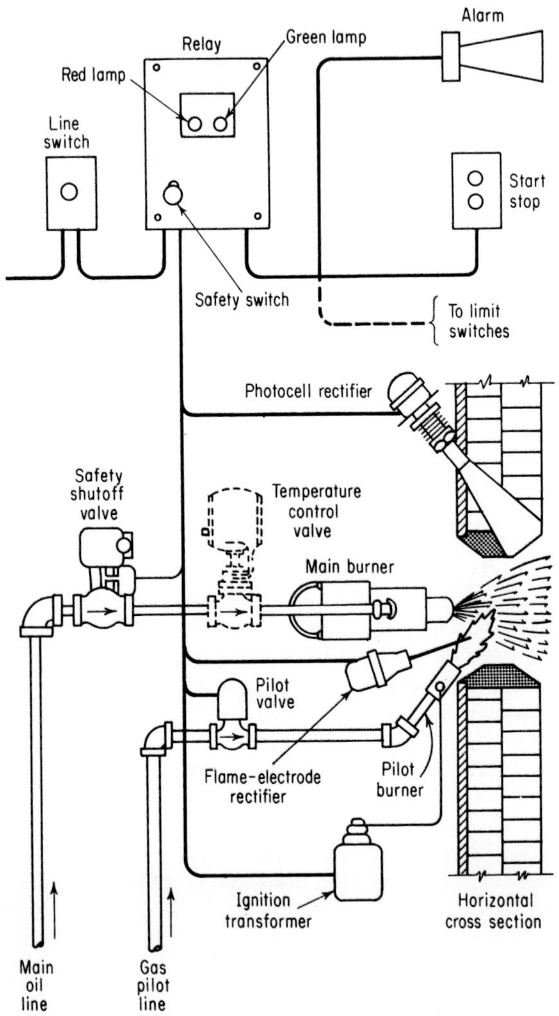

Fig. 12.14 Combustion safeguard system used on oil-fired unit is based on rectified impedance principle.

use an infrared detector to discriminate between adjacent flames. What has come out of research in this area is the so-called *flame-flicker method,* where the high-frequency (ac) component of the radiation intensity is used to discriminate between flames. These detectors generally operate in the visible or near-infrared spectrum with a frequency range of 50 to 600 Hz. Various methods of signal processing using the ac-signal component can now help scanners differentiate between (1) flames of different fuels, (2) flames of the same fuel, and (3) flame and hot refractory.

The flicker method works equally well with all types of waste fuels, including black liquor, hog fuel, tail gases, trash, etc. Ultraviolet detectors generally cannot be used with these fuels for the same reasons as with coal and heavy oil.

2. *Visible-light scanner.* The visible-light scanner is designed to monitor the characteristic frequencies and intensity levels of visible light emitted from the combustion of fossil fuels. Light is transmitted from the windbox by a fiber-optic light guide and converted to an electrical signal by a photodiode. Both the flame intensity and the frequency of the flame pulsations are amplified by electronic circuitry at the boiler side and converted to a current signal suitable for transmission to remote equipment. The sensitivity to visible light follows a response curve similar to that of the human eye (Fig. 12.15). Its spectral sensitivity is limited only by the primary sensor, the photodiode. Both the fiberoptic light guide and photodiode are blind to ultraviolet radiation.

Q Explain the difference between wall-mounted burner arrangements and the flame-monitoring logic in comparison with the flame-scanning practice on tangential fired units.

A Different boiler and burner arrangements dictate to some extent the flame-monitoring system to be found. Wall-fired units are monitored on an individual-burner basis—generally two main-flame scanners and one igniter-flame scanner are adequate. Should one of the main-flame scanners indicate no flame, the igniter is generally removed from service along with the main burner.

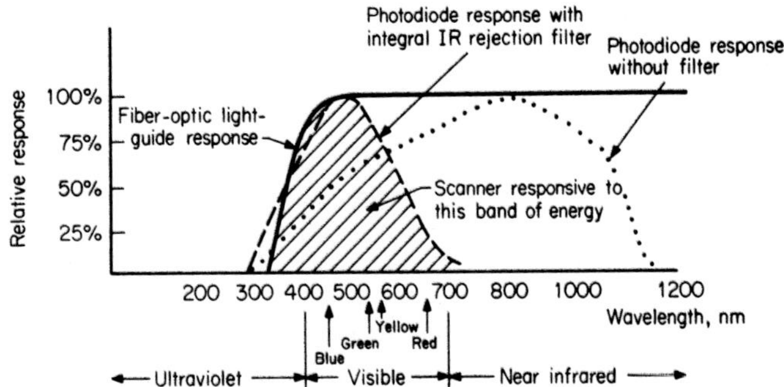

Fig. 12.15 Wavelength response range of visible-light scanner using fiber-optic light guide and photodiode. (*Courtesy of Combustion Engineering, Inc.*)

Shutdown logic for tangentially fired boilers is similar to that for wall-fired units until a single fireball is present in the furnace, typically at 25 to 30 percent of rating. After this point, flame safety logic is transferred to fireball monitoring and flame-failure protection is arranged on a furnace basis. If three out of four scanners indicate no flame on any given burner level, and if loss of ignition energy from associated igniters is confirmed, fuel flow to that elevation of burners is shut off. If all elevations indicate loss of flame, the whole unit is tripped.

Q What is meant by register starting as stipulated in National Fire Prevention Association (NFPA) No. 85-B for gas firing and NFPA No. 85-T for oil firing?
A Open-register start-up procedure for gas firing is outlined briefly thus:

1. Set all, or most, burner-air registers in the normal firing position. Then purge the furnace and boiler setting, using not less than 25 percent of full-load airflow for 5 min.
2. Throughout the start-up period, maintain the same register settings and the same total airflow used for purge.
3. Set fuel header pressure at a value which will provide a burner fuel flow compatible with the burner airflow.
4. Light burners one at a time as increased heat input is required, keeping the burner fuel header pressure and register settings at their initial settings. As each new burner is lit off, close the burner register to light-off position. Since the furnace is air-rich, additional burners may be cut in with no increase in airflow until the fuel flow approaches 25 percent (or whatever airflow rate was used during the purge).

Figure 12.16 is a typical interlock block diagram for unit start-up based on NFPA guidelines. The NFPA has developed requirements for the various interlocks required in the start-up and shutdown of oil-, gas-, and coal-fired boilers, for both single and multiple burner units. Figure 12.17 is a start-up logic diagram for a pulverized-coal unit.

Q Explain the main functions of a burner-management control system for firing pulverized coal on a multiburner system.
A The following basic functions are applied by one burner-control manufacturer:

1. Require and supervise an adequate boiler purge after a master fuel trip before allowing the first fire to be placed in the furnace.
2. Provide for individual or group (as required by the boiler characteristics) igniter lighting and shutdown from a remotely located control panel.
3. Continually supervise igniter operation from flame detected, igniter valve fully opened, and igniter fully extended and coupled (as applicable) for the lighter.

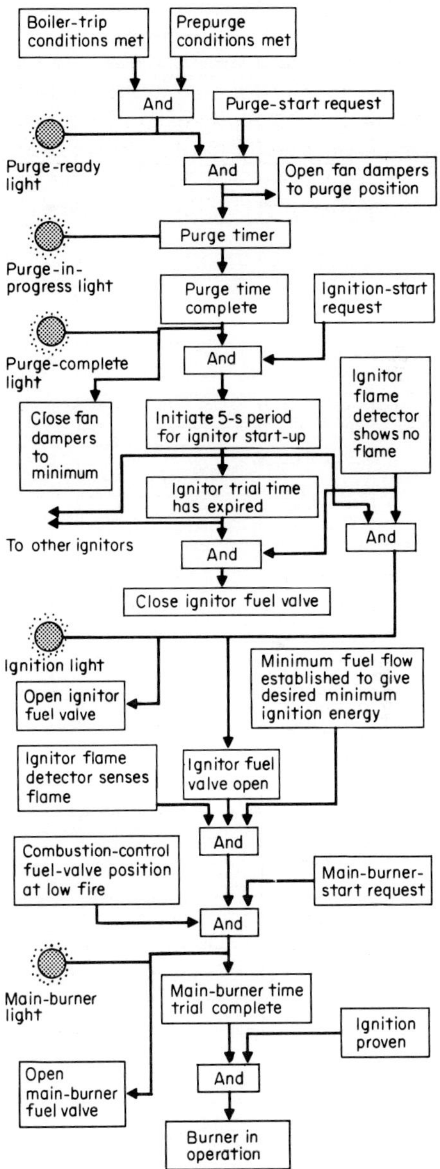

Fig. 12.16 NFPA unit start-up diagram stresses purging, proving of pilot or ignition, main-flame proving, and interlocks to fans and dampers.

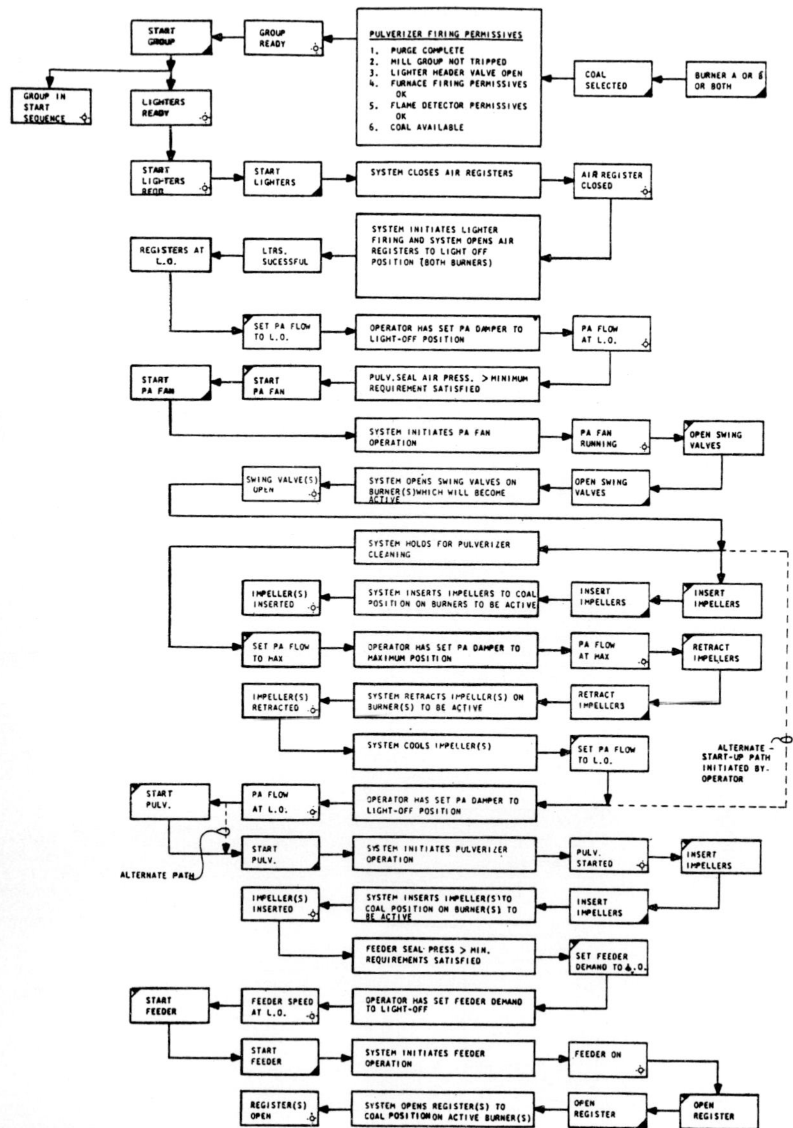

Fig. 12.17 Start-up logic diagram for a pulverized-coal-burning system. (*Courtesy of Bailey Controls Co.*)

4. Automatically shut down an igniter when it fails to light properly, or when there is loss of flame.

5. Provide a properly supervised start sequence for each pulverizer and feeder, including all associated equipment such as dampers (hot and cold air).

6. Provide sequence lights on the operator's control panel insert showing all major steps in each pulverizer-feeder start-up and shutdown sequence.

7. Provide a properly supervised shutdown sequence for each feeder and pulverizer, including all associated equipment such as dampers (hot and cold air).

8. Continuously monitor certain predetermined mill (pulverizer and feeder) operating conditions, and if these conditions are out of limits of safe operation, initiate a mill trip for the respective mill.

In order to accomplish the above functions, the system must monitor the operator's actions and the status of the equipment being controlled as detailed below and illustrated in Fig. 12.17.

a. Boiler must be properly purged and the boiler trip reset in order to start the lighters.

b. Lighters must be in service to start the primary air fan.

c. The primary air fan must be started before the coal-burner coal valves can be opened.

d. The primary air fan must be started, the coal valves opened, and primary airflow at minimum before the pulverizers are started.

e. The pulverizer must be operating and coal feed demand set at minimum to enable starting the coal feeder.

Q What controls are needed to prevent furnace explosion on large boilers firing fuel in suspension?

A These four basic controls:

1. An automatic burner-control system properly programmed and composed of reliable hardware and sensitive discriminating flame detectors

2. A fully integrated combustion-control system to maintain the correct fuel and air ratios under all operating conditions

3. A complete safety interlock system including furnace purge, fan failure, fuel-supply failure, and furnace overpressure interlocks

4. Communication equipment, such as furnace television, for viewing operating conditions, and combustible analyzers and interlocks to trip the fuel when approaching hazardous conditions

Q Outline some causes of furnace explosions on large boilers and corrective actions to be taken by the controls.

A See the table in Fig. 12.18 for gas-fired boilers. Similar causes are found for oil-fired units.

Fig. 12.18 Some firebox-explosion causes and preventive measures.

Causes of firebox explosions	Ways to prevent them
1. Faulty automatic fuel and regulating controls Leaking valve on fuel line Unbalanced air/fuel ratios Faulty fuel injection systems Faulty fans and dampers	1. Provide automatic boiler shutdown for: Low fuel gas pressure High fuel gas pressure Loss of instrument air Loss of fans Loss of dependable power supply Low water
2. Insufficient or no purge	2. Ensure adequate purge by performing these functions: Close all pilot gas valves Close all burner gas valves Close main gas valve Purge with airflow Purge with timed airflow
3. Flame failures Faulty pilot light	3. Ensure a safe light-off by taking these steps: Make sure burner fuel valve is closed Light and check pilot flame-failure detector Extinguish pilot and check burner flame-failure detector
4. Since faulty regulating controls, unbalanced fuel/air ratios, faulty fuel systems, and faulty fans and dampers also can result in flame failure, the inclusion of No. 3 in the safety system offers a double safeguard against these conditions and also a secondary source of protection against the failure of any of the components.	

Q How do furnace television and combustible analyzers aid in preventing furnace explosions?

A Furnace television can provide the operators with the visual picture they once had when stationed at the burners and can be invaluable when placing burners in service. Combustible analyzers can be used to check the effectiveness of a furnace purge and to sound an alarm or trip the unit when approaching unsafe combustible levels.

Q Name four common types of combustion or oxygen analyzers.

A (1) Paramagnetic, (2) catalytic combustion, (3) electrochemical, and (4) inferential thermal conductivity. These four methods, unless specifically designated, are primarily suited to gas analysis. Many of them can be extended to determine dissolved oxygen in liquids by use of an auxiliary gas stream to scrub the liquid and carry the oxygen to the analyzer.

Q Describe a continuous, automatic check system of a flame safety circuit.

A Since it is possible to check the flame sensor by manually cutting off the flame, the same sort of check has been developed to do it automatically. Thus the flame safety circuit is automatically and continuously checked (including the sensor) by simulating flame-out. See Fig. 12.19. Here the flame failure is simulated by means of a swinging shutter that intermittently interrupts the line of sight of an optical flame detector. Each time the sensor's view is blocked, a flame-out is simulated long enough to prove out the sensor but not long enough to actually shut down the system. Should the sensor "see" flame while its view is blocked, the system immediately shuts down on malfunction. But if the sensor detects the intermittent flame-out, its "no-go" signal (of 1 s duration) is too brief to shut down the system, since the logic network has a time delay to prevent nuisance shutdowns. Thus, the system is a constantly repeated check, not only on the flame sensor but on the logic network as well as on the flame-failure circuit.

Q Do fire insurance groups have approval requirements for automatic boilers?

A The two main fire groups that provide insurance and establish

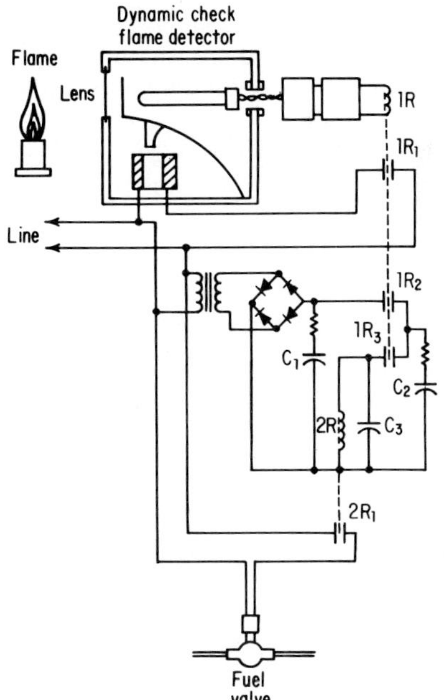

Fig. 12.19 Automatic continuous check system duplicates periodic simulated flame-out.

requirements are the FM (Factory Mutual) and IRI (Industrial Risk Insurers) groups. They work closely with other societies or groups, such as the NFPA, ANSI, and ASME. A typical requirement for automatic gas-fired boilers is given below. See also Figs. 12.20 and 12.21 for fuel-valve train. Similar requirements exist for oil-fired automatic boilers. These standards are usually applied to boilers of the fire-tube and smaller water-tube type that are found in commercial and industrial plants, but the requirements do not apply to resident-size boilers. Here are some of the combustion-control requirements of the IRI for an automatic gas-fired boiler installation:

1. A low-water cutout switch should be provided for all boilers and should be interlocked to shut down the combustion equipment immediately when a low-water condition exists.

2. Excess-pressure cutout switches should be provided for steam boilers. This control should be in addition to operating pressure controls.

3. Excess-temperature cutout switches should be provided for water boilers. This control should be in addition to operating temperature controls.

4. Spark-ignited interrupted-flame gas pilot should be provided. It should have a stable flame over the entire operating range of the main burner. The proved pilot should be adequate and properly located to light the main burner reliably.

5. Approved nonrecycling flame-failure control equipment should be provided for supervision of the pilot burner as well as the main burner.

6. Preignition purging should be provided for the combustion chamber, boiler passes, and breeching. The airflow should be supervised during the entire purge period. Purging should be accomplished

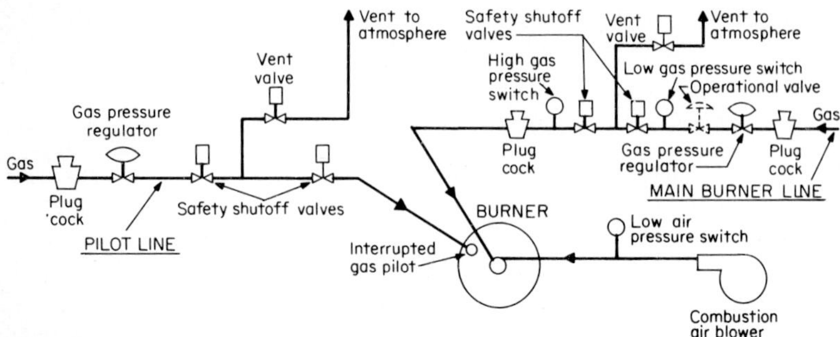

Fig. 12.20 Typical arrangement of IRI-recommended valves for automatic gas-fired single-burner boiler-furnace.

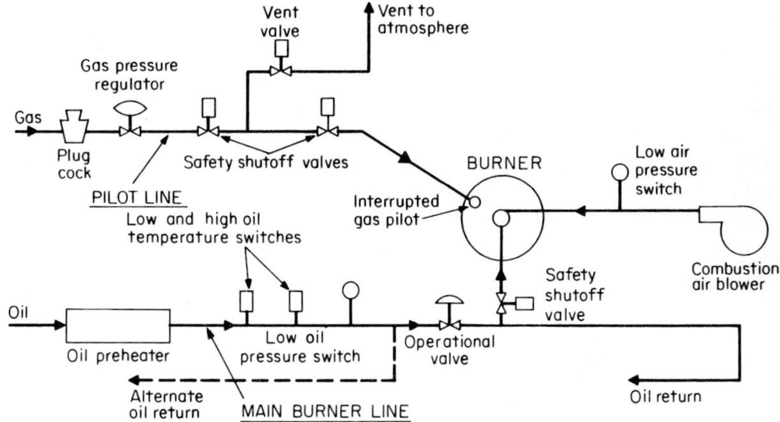

Fig. 12.21 Typical arrangement of IRI-recommended valves for automatic oil-fired single-burner boiler-furnace.

by airflows of at least 60 percent of maximum airflow. This purging should continue for a sufficient time to assure a minimum of four air changes of this volume.

7. Timed trial-for-ignition should be provided for both the pilot and main burner. The main burner trial-for-ignition period should be as short as practical but should not exceed 15 s. The pilot burner trial-for-ignition period should not exceed 10 s.

8. Fuel pressure supervision should be provided by approved pressure switches interlocked to accomplish a nonrecycling safety shutdown in the event of either high or low fuel gas pressure.

9. Forced- and induced-draft fans (including combustion air blowers) should be supervised to assure safe minimum airflow through the combustion chamber and boiler passes. This should be accomplished by both approved airflow proving switches and motor starter interlocks to provide a nonrecycling safety shutdown.

10. Suitable limit switches should be provided and interlocked to assure that the fuel-air proportioning dampers and burner controls are in the low-fire start position before the burner can be fired. Stack dampers should preferably be fully open during the purge and light-off periods.

11. Fuel should be of a type free from all residue and other foreign materials. Suitable and adequate fuel-cleaning equipment such as filters, strainers, drip legs, or water separators should be provided where necessary.

12. An approved motor-driven reset safety shutoff valve should be provided in the main gas line to the burner. An approved safety shutoff valve should be provided downstream from the motor-driven reset

valve. A normally open, fully ported, electrically operated valve should be provided in a vent line connected between the two safety shutoff valves. The vent pipe should be run to outside atmosphere. A manually operated lubricated plug cock should be located downstream of both safety shutoff valves to permit leakage testing of the valves. The gas pilot line should be equipped with dual approved safety shutoff valves and vent valve arranged similarly to those for the main burner if the pilot is rated in excess of 120,000 Btu/h.

13. Alarms indicating interruption to the safety combustion-control circuit should be provided and transmitted to a suitable point where there is constant attendance.

Q Are state laws being passed to include furnace explosion protection in the existing boiler codes?

A New York State has adopted a law for automatic heating boilers, including requirements on combustion safeguards.

> EXAMPLE: Some of the following are now in the state code:
>
> 1. *Gas-fired boilers*. (*a*) Pilot has to be proved, whether manual or automatic, before the main gas valve is permitted to open, either manually or automatically, by completing an electric circuit. (*b*) A timed trial for the ignition period is established based on the input rating of the burner. For instance, for input rating of 400,000 to 5,000,000 Btu/h per combustion chamber, the trial for the ignition period for the pilot of automatically fired boilers cannot exceed 15 s. And the main burner trial for ignition also cannot exceed 15 s. (*c*) The burner flame-failure controls must shut off the fuel within a stipulated time, again depending on the fuel input of the burner. For a burner rated with an input of 400,000 Btu/h or more, the electric circuit to the main fuel valve must be automatically deenergized within 4 s after flame failure. And the deenergized valve must automatically close within the next 5 s.
>
> 2. *Oil-fired boilers*. Similar provisions have been adopted, with requirements on response time for controls to shut off the burner based on fuel input in gallons per hour, instead of Btu per hour. The flame must be continuously supervised by the controls.

No doubt many other states will incorporate requirements on safety combustion controls in the future as furnace explosions continue on their destructive path.

IMPLOSIONS

Q Why may a balanced-draft coal-fired steam generator be exposed to an implosion?

A Balance-draft units are being expanded in size requiring greater induced-draft and forced-draft fan capacities. The addition of pollution-control equipment on retrofitted older units also requires increased draft conditions. Because the implosion problem arose recently, most older boiler furnace structural elements were designed for only −7 in water draft conditions. Negative pressure can occur above this value with resultant structural damage to the furnace parts of the boiler. There is a proposal to require these large boilers to be designed to −35 in negative water pressure or draft by adding suitable structural reinforcements.

Implosions can occur from an induced-draft fan operating with no forced-draft fan in service. This may occur from control failures, operator mistake, and similar causes. A master fuel trip (MFT) can also cause a rapid decay in furnace pressure, similar to a flame-out. This produces instant negative pressures. Therefore, coordination of starting sequence, monitoring purging cycles, and similar operator alertness is required to prevent large negative furnace pressures in order to avoid furnace implosions.

CHAPTER **THIRTEEN**

General Operating Procedures and Practices

There are certain general operating procedures that must be followed for all types of boiler plants. These will assist in observing whether boiler equipment is operating safely and efficiently and will also provide a measure of reliability against unexpected forced outages. This chapter will review general operating practices in order to provide a reasonable assurance of boiler-plant operation. However, because of the variety of boiler systems, readers are encouraged to also review their plants and the associated equipment manufacturer's operation and maintenance instructions.

Q What is an operator's first duty when taking over a watch?
A First observe the water level in the gage glasses of all boilers. Blow down the water column and gage glass on each boiler and observe the return of water back into the glass. Then check the water level with gage cocks, if installed. Check the operating pressure and note if it is within the rated pressure of the boiler.

Q Describe a test on an HP boiler which will indicate that the lower gage glass is obstructed even though the gage glass is half full of water.
A Open the try cocks on the water column. If all show steam, it means the bottom connection is obstructed, permitting steam from the top connection to condense in the gage glass. The boiler should be shut down immediately and inspected for possible dry-firing damage. Naturally, the bottom connection of the water column and gage glass should be cleaned of all obstructions before the boiler is returned to service.

Q What should be done if water is not visible in the gage glass because of failure of the feedwater supply?
A Immediately:

1. Shut off the fuel to the burners and secure the burners.
2. Check the water level by trying the try cocks and water-column

drain. If definite low water is indicated below the gage-glass level, close the main steam valve and feedwater valve.

3. If the boiler is equipped with one, open the superheater drain.

4. Continue operating forced-draft and induced-draft fans until the boiler cools gradually.

5. Let the pressure reduce gradually, and when the furnace area is sufficiently cooled, check for leaking tubes and other signs of overheating damage. On FT boilers, look for cracked or warped tube sheets, broken and leaking stay bolts in the water legs. On SM boilers, check for cracked or leaking furnace-to-tube sheet welds. On CI boilers, look for cracked sections. On steel boilers, check for leaking joints on longitudinal or circumferential welds or riveted joints.

6. If no leakage is evident, give the boiler a hydrostatic test of 1½ times the allowable working pressure. Then again check for leakage at all critical parts of the boiler. If leakage is observed during the initial check or during the hydrostatic test, notify the authorized boiler inspector immediately for inspection of the boiler and advice on permissible repairs.

Q What attention should be given the water level in a boiler when lighting off a cold boiler and also when cutting out a boiler that has been steaming?

A Before lighting off a cold boiler, have the water level about 1 in from the bottom in the gage glass. As the boiler heats, the water expands and the level reaches half a glass or over by the time the boiler is ready to cut in. When the fires are extinguished, before cutting out a boiler steaming on line with other boilers, bring the water level to at least half a glass. Heat stored in hot brick walls generates steam for some time. The steam stop valve on the boiler must remain open while steam is generated with the fires out or the safety valves will lift. During this time steam flows to the main header and water drops in the glass; so feed the boiler from time to time.

When the water level in the glass stops dropping, the boiler is ready to cut out by closing the main stop. If the steam stop valve is not closed then, the water will rise in the glass because the boiler stops generating steam and condenses the steam from other boilers connected to it.

Q What are some causes of water bobbing up and down in a boiler gage glass? How is it corrected?

A This condition is usually from foaming, especially in smaller boilers where contaminated condensate returns to the boiler. Foaming can also be from excessive water hardness, high water density in the boiler, or from impurities forming scum on the boiler-water surface. In severe cases water boils violently and carries over into steam lines.

If the boiler is equipped with a surface blow, open it to blow scum off the water's surface.

> CAUTION: Keep your eyes on the gage glass while blowing and your hand on the valve. This habit keeps you from walking away from the boiler and forgetting to shut off the valve.

You may have to raise the water in the glass and give additional blows to stop foaming. If the test shows that the boiler-water concentration is high, raise the water to near the top of the glass and open the bottom blow valve. Repeat feeding and blowing until the water tests right and the foaming stops.

> CAUTION: Cut out fires while giving bottom blow to a WT boiler. Make a thorough check to find the source of the contamination. Grease extractors may be faulty or loaded with oil. The water should be tested by qualified boiler-water specialists for purity. Follow their advice on how to correct the condition.

Q What should be done if an unusually high feedwater pressure is necessary to maintain the water in the boiler?
A Check the feedwater valves and lines to make sure that a valve has not broken off its seat or that there is not some obstruction in the line itself. Some methods of feedwater treatment have been known to deposit chemicals inside the feedwater line, making it impossible to get water into the boiler. Also look for leaks due to cracked or corroded piping of the feedwater (or condensate line on heating boilers) especially if it is buried anywhere in the system.

Q Explain the value of boiler operating checklists.
A Checklists whether posted or in log form serve as reminders of items that require daily, weekly, monthly, and yearly observation and review to assure that everything is functioning normal. Checklists can be used for record keeping and also as a means of training new people to the plant setup and the routine checks required of the equipment under their care and control. All checklists should be designed to the conditions existing at a particular location, i.e., from simple one-boiler plants to a large power-plant complex, considering the variables involved in each installation.

Q Are any sample checklists available?
A Most good chief engineers have checklists. Here is one for three 60,000 lb/h WT boilers used to provide steam for heat and process at 250 psi, oil-fired.

BOILER DAILY CHECKS

1. Check units for proper pressure-gage readings.
2. Check sight gage for proper water level on all boilers.
3. Check damper motor operation by moving manual modulating control in front of boiler on all operating boilers.
4. Observe movement of feedwater modulating valve stem on all operating boilers.
5. Manually operate blowdown solenoid and check for positive shutoff.
6. Test low-water cutoff for proper operation (blow down both water columns). Do not cycle boiler unless production machines are not running.
7. Observe and record oil temperature and pressure (oil burning only).
8. Observe and record air pressure (oil burners).
9. Observe and record pilot and main gas pressure.
10. Check lube-oil level for compressor (oil burners only).
11. Turn Cuno strainer six turns.
12. Blow down front and rear of boiler as recommended by water-treatment consultant.
13. Test water as recommended by water consultant.
14. Observe and record stack temperature. (Manually operate on low fire according to instructions; then switch to automatic. Boiler should go to high fire and record stack temperature when unit is stabilizing.)
15. Inspect burner linkage to fuel and air.
16. Blow soot from boiler tubes (oil burners only).
17. Test feedwater from boiler for current chemical levels.
18. Mix feedwater chemicals and adjust chemical feed pumps to drain chemical tanks during an operating day whether it is 8 or 24 h.
19. Change steam flow recorder charts, take readings, and check fuel- and water-consumption reports.

DAILY AUXILIARY CHECKS

1. Chemical feed tanks:
   a. Check water-conditioning instructions for chemicals to be fed—check chemical levels; fill if required.
   b. Examine piping for leaks, gages for proper readings, and indicate pen or closed valves, etc. Enter on repair log if necessary.
   c. Examine pumps for proper operation, pump stroke, and leaks. Adjust and note for repair as required.

2. Water softeners and dealkalyzers:
   a. Check meter settings and regenerate when meter reads 0. Reset meter.
   b. Check unit for leaks, power plugs connected.
   c. Check brine level, and fill as required.
3. Air compressor:
   a. Check pressure gages for proper operating range of compressor.
   b. Check lubricator for proper lubrication rate.
   c. Check cooling water for proper temperature. Should be very warm but not excessively hot. Adjust temperature controller as required.
   d. Check oil-separator sight glass for indication of oil buildup. Clean trap as required.
4. Centrifugal blowdown separation:
   a. Check temperature gage.
   b. Check overall system.
5. Continuous blowdown meters:
   a. Check temperature of unit and assure operation (unplug orifices as required).
   b. Open valve on bypass and blow out lines.
6. Condensate receiver tank (plant condensate):
   a. Check water-level variation during start and stop of transfer pump.
   b. Check temperature and pressure of water in tank.
   c. Check for leaks.
7. Feedwater pumps and transfer pump:
   a. Check for excessive water leaking through packing—adjust as required.
   b. Check for excessive vibration.
   c. Observe discharge pressure of operating pumps.
   d. Check for leaks in piping and valves.
   e. Listen for operation of pressure relief valves on all operating pumps.
8. Deaerator/feedwater heater tank (heating system and makeup water):
   a. Check water level in sight glass.
   b. Manually operate makeup water valve.
   c. Check water temperature and pressure in main heating and storage section.
   d. Observe pressure variation while in boiler room checking other equipment.
9. Header:
   a. Check pressure gage.
   b. Check for leaks and partially open valves.

10. Sample cooler:
    a. Unit checked during daily water tests.
11. Control air compressor and air drier:
    a. Operate unit by shutting off air from main air receiver.
    b. Check pressure gages.
12. Electrical control panel:
    a. Check all disconnects and selection switches for proper positions.
    b. Make sure all disconnects are free to trip.
13. Drains:
    a. Check all floor drains for accumulations of trash.
14. General:
    a. Clean all areas as required.

BOILER WEEKLY CHECKS

1. Remove and clean oil nozzle and swirler.
2. Drain sludge from oil filter (Cuno) (oil burners only).
3. Clean lube-oil strainer (oil burners only).
4. Check all belts on boiler, oil pumps, water pumps, and compressors.
5. Clean scanner cell glass and sight tube.

BOILER MONTHLY CHECKS

1. Shut off power to water feed pumps. Observe for normal low-water cutoff. (Not less than ½ in in gage glass.) If operation not proper, contact boiler manufacturer service or replace control.

2. Remove and clean fuel-oil strainers.

3. Check air line from front of boiler to rear sight glass. Must be free for full flow at all times.

4. Observe all gaskets on boiler heads, access hole, handholes, and oil heaters. Replace or repair when required.

BOILER YEARLY OR SEMIANNUAL CHECKS

1. Clean fireside of boilers (when stack temperature increases 75 to 100° over normal).

2. Inspect and wash coat refractory.

3. Take extra care when resealing all door and boiler gaskets. Inspect.

4. Inspect all electronic tubes in flame-safeguard relay. It is suggested that they be either tested thoroughly or replaced each year.

5. Remove and clean low-water-control bowls and floats. Replace gaskets.

6. Remove and clean oil heater assembly. Replace gaskets.

7. Clean lube-oil tank.

8. Lubricate all motors (see manual for instructions). Consult the manuals for other maintenance items on your particular boiler(s).

Q What is the purpose of blowdown?
A The object of blowdowns is to eliminate scale and mud that may settle on the bottom of the boiler in bottom blowdown procedures, and also to maintain total dissolved solids within the limits prescribed by water-treatment instructions. Accumulation of dissolved solids may vary with load conditions, the operating pressure, rate of condensate return, and the quality of any makeup water. Blowdown from the bottom is at selected safe periods, while continuous blowdown is from the steam drum.

Q What is the effect of indiscriminate or improperly controlled blowdown procedures?
A If not controlled properly, continuous or "top" blowdown may waste energy and thus increase operating costs as follows:

1. Using fuel to heat makeup water lost from blowdown.

2. Using more internal chemical water treatment to counteract the introduction of makeup water.

3. Adding pumping costs in introducing makeup water.

Automatic blowdown equipment is available that measures total dissolved solids continuously and initiates blowdown only if set points are exceeded.

Q When should boilers be blown down (bottom blow)?
A Only during minimum steaming periods. The reason is that circulation in some boilers is very sensitive. Thus blowing down during maximum steaming conditions could upset the circulation so badly that some parts, especially tubes, might be seriously damaged.

> CAUTION: Blowdown valves on waterwalls serve primarily as drain valves. *Never* blow down waterwalls when the boiler is in operation. If difficulties arise that require blowing down waterwalls, do this only under banked conditions (low steaming) or only in accordance with your boiler manufacturer's instructions.

Q What is the proper sequence in opening and closing bottom blowoff valves?

A On a boiler equipped with a blowoff valve cock or a quick-opening valve in the same blowoff connection, always open the cock or the quick-opening valve first, the blowoff valve second. To close, always close the blowoff valve first and the cock or quick-opening valve second

> CAUTION: Open and close the blowoff valves and cocks slowly to reduce shocks as much as possible.

Q Should any special precautions be taken in blowing down a boiler where the water-gage glass is not in view of the operator blowing down the boiler?
A Yes. Always station another person in sight of the water-gage glass to signal to the operator blowing down the boiler. Boiler operators must never start blowing down a boiler and then leave it to do some other job.

> CAUTION: Never take your hands off the blowoff valve while it is open.

Q On packaged boilers, list the causes and hazards of the fire going out while the burner is running.
A (1) Oil supply suddenly runs out, (2) dirty or clogged oil line, (3) metering valve clogged, (4) piping or check valve clogged, (5) water is in the oil, (6) suction line springs a leak, (7) low fire set too low, (8) magnetic oil valve burned out, (9) oil is too cold, and (10) fire burns away from the nozzle. The potential hazard of each malfunction is a furnace explosion.

Q On packaged boilers, list the causes of the burner continually going off and on.
A (1) Fluctuating water level tripping low-water cutoff, (2) loose connections or defective control, (3) controls not connected properly or not properly adjusted, (4) partial electrical ground, (5) intermittent low voltage, and (6) combustion control opens the circuit when the burner goes to low fire. The potential hazards of each are overpressure and dry firing.

Q On a packaged boiler, what would prevent the burner from shutting off?
A (1) Limit controls set too high, (2) grounded control circuit wire, and (3) relays not falling out an opening of the circuit. The potential hazards of each are overpressure and dry firing.

Q On a packaged boiler, what would cause fire puffs when starting?
A (1) Starting draft poor, (2) firebox incorrect, (3) wrong burner nozzle, (4) lean fire, (5) insufficient gas pilot or excessive gas pilot, and (6) water in the oil. The potential hazard of each is a furnace explosion.

Q On a packaged boiler, what would cause improper combustion?

A 1. Fire puffs or fluctuations from the fire burning away from the nozzle (incorrect firebox, nozzle, or atomizing cup not located properly); poor draft; excessive oil temperature; water in the oil: unsteady oil pressure.

2. Carbon in the firebox from the wrong shape fire; improper firebox construction; poor draft; forcing burner; damaged atomizing cup; nozzle or cup off-center; carbonized nozzle; carbonized atomizing cup; firebox too small for the load.

3. Fire smoking from improper burner adjustment; burner too small; firebox too small; insufficient air or too much oil; poor draft; carbon in the firebox; fire room starved for air; lean fire burning at low temperatures in a large firebox.

The potential hazard of each is a furnace explosion.

Q On a packaged boiler, what would cause the burner to start and light but then lock out?

A 1. Combustion control is slow in making contact from sooted element; element is worn out; element is in cold-air stream or incorrectly located; element is at the end of long gas travel; auxiliary air damper opens too slowly on high-low or modulated linkage; fire is too lean; poor draft.

2. Oil supply fails or is insufficient.
3. Magnetic oil valve is defective.
4. Insufficient time for ignition.
5. Insufficient time on safety lockout switch.
6. Smothered fire due to poor draft or too little air.
7. Control contact broken or dirty.

The potential hazard of each is a furnace explosion.

Q On a packaged boiler, what would cause oil leaks?

A 1. From the furnace front, atomizing cup does not extend the proper distance beyond the nozzle.

2. From the firebox or checker work, carbon formation on the nozzle or in the firebox; poor atomization; impingement of oil on an exposed boiler surface; electric oil valve fails to shut off tightly the instant the burner is off.

3. From the rear of the burner shaft, clogged atomizing cup; leaking electric oil valve of badly leveled burner.

4. Loose pipe connections, packing glands, porous casting, defective gaskets.

5. Lubricating oil, overflow from oil filler cup caused by filling when burner is running.

The potential hazards are fire and a furnace explosion.

Q On a packaged boiler, what malfunctions would prevent carrying the maximum load?
A Oil pressure is failing, the draft is limiting maximum firing, refractory shields too great a heating surface, the burner is too small, the boiler is too small or short-circuited, the firebox is too small or improperly constructed. The potential hazards are dry firing and furnace explosions.

Q What safety precautions should be observed in operating and maintaining an automatic burner system?
A 1. Always close off all manual fuel valves before working on a burner or disconnect the wiring to automatic fuel valves or do both.

2. Never stand in *front* of a burner or boiler when starting up.

3. Never manually push in relays unless the manufacturer's instructions so advise.

4. Never permanently block in relays with rubber bands, sticks, or other devices.

5. Never change the safety switch timing of a flame supervisory control. If the system is locking out, correct the cause, *not* the symptom.

6. Never install jumper wires, and never bypass any safety interlock switches.

7. Before starting a burner, visually inspect every combustion chamber to make sure there is no accumulation of combustibles.

8. Regard every system lockout as a safety lockout until proved otherwise by competent personnel.

Q List the precautions needed with refractories on oil-fired boilers.
A Refractories (brickwork) are subject to damage from many causes, such as improperly adjusted fires and vanadium-contaminated oil. Impingement and the resulting carbon buildup are common. The flame should travel down the furnace on SM boilers without touching either the furnace or the refractory. Long periods of operation on low firing often cause refractory damage. The very small flame reduces the combustion chamber temperature and causes poor combustion.

Carbon builds up and intense heat is directed at the surface of the refractory, thus causing spalling (facing breaks off). Unless the oil has been specifically treated, there is no relief from the damages caused by vanadium in the oil. Moisture trapped in the refractory may develop steam, which ruptures the surface.

Q What size draft-air opening is needed for a packaged boiler installed in a closed machine room?
A There should be a fixed opening for fresh air, an average area of 2 ft^2 for each 100 boiler HP. Opening windows is not the answer because they are often closed in cold weather. Then the boiler starves for air.

Remember that for each boiler horsepower, about 10 ft^3/min of air is needed.

Q Explain a safe soot-blowing method to prevent furnace explosions.
A The boiler manufacturer provides safe soot-blowing instructions. Always follow them to prevent explosions, puffs, or blowbacks. Each boiler requires a different method, depending on its gas passage design, the type of fuel being burned, and differences in operating instructions. Here are five points to remember:

1. If possible, the boiler to be cleaned should be operated at maximum design load. If soot is blown at low boiler loads, clouds of soot may enter the gas stream to form explosive mixtures. These may ignite from hot furnace walls or from smoldering soot fires. So *all* tube surfaces and *gas* passes should be blown with the gas flow at, or near, maximum.
2. On oil- or pulverized-fuel-fired boilers, use soot blowers with burners operating at the highest possible burning rate. Check the burners during soot blowing for stability. High and stable burning rates prevent flames from being blown out by small puffs or agitation of the gas flow during soot blowing.
3. Before operating soot blowers, always increase the furnace draft to help purge combustible gas pockets.
4. If heating surfaces must be cleaned with soot blowers while burners are out of service, before blowing, make sure that the boiler is cold and that there is no smoldering soot.
5. The normal soot-blowing sequence is to follow the gas flow through the boiler. Thus the first soot blower to operate is that nearest the burners; then each unit in turn along the gas passes to keep from blowing soot on cleaned surfaces. But if back passes are heavily deposited, they may plug unless blown first.

REMEMBER: When units are banked, soot blowers are operated in reverse order, from rear to front.

Q How do tubes usually deteriorate from soot blowers?
A Steam cutting from an improperly aligned or improperly operating soot blower badly erodes tube metal. Idle soot blowers on an idle boiler with leaking soot blower valves can cause condensed steam drippage on the tubes. The drips produce external tube corrosion, especially if sulfur is present in the fuel-gas deposits in the tubes. So carefully examine soot-blower elements each time the boiler is out of service. Soot-blower elements often warp, or a loose bearing may shift, causing misalignment, and thus erode the tubing.

Q What action is necessary when a tube suddenly leaks during boiler operation?

A Immediately extinguish the fires by cutting off the fuel. Then use all measures to protect the firing equipment. To speed cooling the unit, let the airflow remain the same. During this time, if the boiler is so equipped, partially open the superheater outlet header drain, then shut the boiler nonreturn valve.

> CAUTION: Large quantities of water may be discharged from the ruptured tube; so the feedwater supply to other units may be depleted as the automatic feed valves open on the damaged unit. Start additional feed pumps. Maintain the flow of feedwater to the damaged unit so as to keep water in the gage glass. Continue doing so until all the pressure is off and the furnace is sufficiently cooled to prevent overheating of pressure parts. To provide the necessary fast-cooling rate, adjust the airflow. Secure the boiler and find the damaged or leaking tube.

> NOTE: Repairs should be approved by an authorized boiler inspector. After repairing the tube, hydrostatic-test to 1½ times the allowable pressure and check for other weak tubes caused by corrosion, pitting, or dry firing. Replace the tubes as needed until a hydrostatic test shows no leaks. At times, repairs may be permitted for ordinary leaking tubes caused by corrosion or pitting, especially if done by a reliable repair firm. Always save the tube, or tubes, removed so the inspector can check for the cause of damage.

Q What is the best way to test the low-water cutout (LWCO) on heating and smaller HP boilers?
A By duplicating an actual low-water condition. Slowly drain the boiler (through the blowdown line) while under pressure. If a heating boiler does not have proper drains for doing so, be sure to correct this condition. Many operators drain only the float chamber of the cutout for this test. But the float chamber drain is only for blowing sediment out of the float chamber. Usually the float will drop when this drain is opened because of the sudden rush of water from the float chamber.

Tests often show that draining the float chamber will indicate that the cutout performs satisfactorily, but when proper testing is done by draining the boiler, the cutout fails to work. Weekly checking of the LWCO by draining the float chamber (or electrode chamber) is good practice. But at the beginning of each heating season and once a year on other boilers, duplicate an actual low-water condition.

> CAUTION: Watch the water level in the gage glass when making this test.

Q Why do safety valves have to be periodically tested, and what is the usual testing frequency followed?

A If a safety valve is not blown or tested periodically, there is a risk of the valve's sticking in a closed or binding position, because of the buildup of foreign deposits around the seat. Safety-valve testing can be manually performed by lifting the code-required test lever and letting the steam or water blow to a safe point of discharge. The other method is to raise the boiler pressure to the stamped popping point of the valve. Testing of safety or relief valves will vary in frequency because of plant operating requirements, spare boiler capacity available, pressure of the boiler system, seasonal needs of boiler equipment, and similar considerations. In general, the following is usually recommended:

Low-Pressure Boilers During heating season, manually lift valve once per week. Annually pressure-test under competent supervisory personnel.

High-Pressure Boilers

1. Boilers operating below 900 psi—manually test once a month. Pressure-test once a year during annual internal routine.

2. Boilers operating above 900 psi—manually test every 6 months with pressure testing annually.

An alternative to boiler pressure testing is to remove the valve for testing in a recognized safety-valve repair and overhaul shop.

Q What is the proper procedure for warming up and cooling down large industrial boilers?

A Calculating the maximum permissible rate of boiler temperature change is difficult. Most manufacturers suggest that the warming up and cooling down rates of water in the boiler should not exceed 80°F/h, with the superheater elements kept properly cooled by steam flow through them. Although some boilers have been repeatedly subject to much more rapid temperature changes without apparent damage, this conservative rate will ensure longer life of parts. The curve in Fig. 13.1 shows this rate in terms of drum gage pressure and saturation temperature against time.

For rapid cooling, always open the superheater outlet header drains enough to maintain a cooling steam flow through the superheater elements. Then circulate air through the setting at about one-forth capacity or greater, depending upon the emergency. Be sure to feed hot water to the boiler to maintain the proper water level until all pressure is off the unit. Follow the instructions of the boiler manufacturer.

Q What precaution must be observed in starting boilers equipped with nondrainable superheaters?

A Nondrainable superheaters, like drainable superheaters, are always prone to overheating damage due to lack of steam or water flow while the firesides of the tubes are hot. Thus start with a low fire that will limit the hot gases to a temperature below that which might reduce the

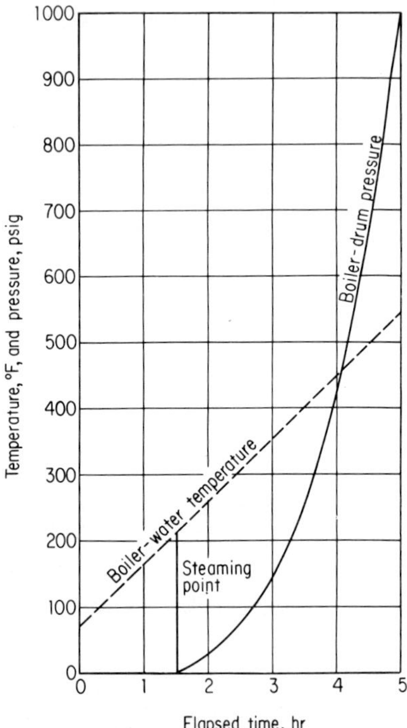

Fig. 13.1 Recommended rate of warming up or cooling down boilers depends on the pressure and temperature rating of the drum at operating conditions.

allowable stress for the pressure being carried on the boiler. This temperature is usually considered to be 900°F for steel tubes and 950°F for alloy tubes. Thermocouples should be installed on superheater tubes to pick up the temperature of the gases.

Q Why is it necessary to drain as much water as possible from a superheater before a boiler is fired?

A Carry-over or condensed water in the superheater will include solids and other impurities which, if not drained prior to firing, may be carried over to a turbine to affect blades, or the impurities may deposit out when the water is evaporated upon firing. Dropped-out deposits will cause scaling of the superheater, which eventually can cause superheater overheating and tube damage. Water deposits in the superheater may also result in uneven heating of the superheater header, which can cause thermal distortion and cracking.

Q What are some methods of removing water from nondrainable superheaters?
A This may depend on the size of the boiler. For smaller boilers, each element may be blown out with compressed air during shutdown procedures. This requires removing handhole caps and piping compressed air through flexible hoses to the superheater elements.

Another method used on smaller boilers is to blow the water out with steam by means of drains or bleeder connections.

Most larger plants boil out the water during the starting routine in getting a boiler ready for operation. The firing rate has to be controlled so that safe tube temperatures are maintained, as previously mentioned.

Q Why must superheater safety valves be set to blow at a lower pressure than boiler safety valves?
A The superheater safety valves should blow first in order to assure a flow of cooling steam through the superheater elements. If the boiler safety valves blow first, the steam flow to the superheaters is seriously reduced, thus causing the superheater to be "starved" of cooling steam, and damage to the elements could result in the form of overheating.

Q What is the procedure for a boiler equipped with reheaters?
A Depending on design, the following procedure is recommended on reheaters on start-up:

1. For those reheaters with drains to the atmosphere, all vents and drains on the inlet and outlet piping must be opened before the boiler is fired so that residual moisture in the reheater is driven off as boiler pressure is raised. Metal temperature of reheaters must be monitored during this period and should not exceed 900°F.

2. For these reheaters with drains to the condenser, the same procedure is followed on opening drains on inlet and outlet piping, except that the drains are directed to the condenser. Metal temperature below 900°F must also be maintained during this start-up period.

Q What is carry-over, and how is it caused?
A Carry-over is entrained moisture and associated solids passing from a boiler with the steam. These slugs of moisture cause erratic superheat and mechanical troubles with engines and turbines. Carry-over also deposits solids in superheaters and on turbine blades. And it may even spoil materials in process. But the main causes of carry-over are priming, foaming, or both.

Q Explain priming of boiler water.
A Priming is the spouting or surging of water into the steam outlet. It is caused by too high a water level, uneven fire distribution, load swings, too high a steaming rate, or even faulty boiler design. Remedies

range from redesigning the boiler or steam drum to installing steam purifiers, lowering the water level, improving firing distribution, or reducing the boiler load. Chemical antifoams also help.

Q What causes foaming?
A Foaming is the formation of small, stable, noncoalescing bubbles through the boiler water. Water film around each steam bubble generated at the heating surface is stabilized by an increase in dissolved and suspended solids in the boiler. Thus the bubble skin becomes tough and doesn't permit coalescence or break readily when the bubble emerges. The resulting expansion of boiler water permits carry-over and priming. The main causes of carry-over are excessive dissolved and suspended solids, high alkalinity, and the presence of oil and various organics that react with alkalinity. Steam washers and mechanical separators in boiler drums effectively control carry-over within reasonable and tolerable limits. A notable exception is silica, which at operating pressures above 600 psi passes over with steam as a vapor. Proper water treatment, including the right amount of blowdown, is the key to maintaining these limits.

Q How would you handle a foaming boiler?
A (1) Reduce the firing rate. (2) Open the surface blow valve. (3) Check the water in the boiler for visual signs of oil. (4) If carry-over is endangering machinery, secure the boiler until the cause of foaming is remedied. (5) Check the feedwater equipment for oil or other contamination. (6) Use the bottom blow, if only to get fresh feed into the boiler.

Q How would you handle a priming boiler?
A (1) Reduce the firing rate. (2) Open the surface blow valve. (3) If necessary, increase blowdown, giving a series of short blows at the lowest firing rate, or secure the fires. (4) Take water samples and test them for either chlorides or dissolved solids. (5) If carry-over is endangering steam machines, secure the boiler until the cause of priming is remedied.

Q When you notice unusual burning of paint on boiler fronts, doors, or other external parts, what may be wrong?
A Refractory or insulation troubles, overload, or more dangerous, low water.

Q When you notice an unusual burning paint odor, especially from piping, or unusually higher steam temperatures, what may be wrong?
A Low water.

Q What should you look for on the first cold day?
A Vapor revealing steam leaks.

Q Name two sources for superheater and reheater contamination while a boiler is operating.
A During boiler operation there are only two flow paths for superheater and reheater contamination for drum-type units. These are solids carry-over from the steam drum and solids introduced from desuperheater spray water. Abnormal amounts of preboiler corrosion products, such as iron oxide and copper oxide, introduced into the steam system while desuperheating can cause deposits to form in the superheater with the possibility of overheating, with damage occurring as a result. When boiler feedwater is used as the desuperheater, it is important to introduce solid chemicals into the water-treatment system downstream from any take-off that is to be directed to the desuperheater. Some plants use condensate for desuperheating to avoid this contamination source of the superheaters.

Q What precaution must be observed before closing a new boiler or a boiler that has been opened for cleaning or repair?
A Make sure all tools, pipes, welding rods, rags, and other such items are removed from drums, headers, furnaces, and tubes. At times, a mirror and flashlight must be used to check headers that are otherwise not accessible for inspection of foreign material. Bent tubes that cannot be looked at from end to end (such as superheater tubes) should be thoroughly flushed, one tube at a time. On new boilers, or where work was done in a tube area, drop rubber balls, and even steel balls, so as to make sure the tubes are free of any obstruction. Water and air can be used to push through the rubber balls.

Q What precautions must be taken when filling an empty boiler that is hooked up in battery with other boilers?
A Remember that when filling, boiler pressure is near zero but the feed-line pressure is high. This causes a large quantity of water to flow with the feed valve only partly open. So make sure that steaming boilers supplied by the feed pumps are not robbed of feedwater. Also, always fill boilers with hot, deaerated or treated water. This procedure permits checking the rise of water level by the temperature rise of drums and headers. It also permits determining whether blowoff or other valves are leaking because such leakage warms up the piping.

Q Why is it necessary to clean the waterside of a new boiler or of a boiler suspected of being contaminated with oil?
A Oxide films formed on steel used for boilers, and oil and grease often applied to metal surfaces to prevent rapid rusting while the boiler is fabricated or erected, must be removed.

Acid cleaning of boilers is often used to remove metallic oxide. Caustic soda and soda ash are the old standbys for cleaning oil from the waterside of boilers. One pound of each chemical is added for every 1000 lb of water required to fill the unit. After the boiler is filled with steam, drum vents open, and a light fire is started and maintained until the vents issue steam. After closing the vents, pressure is built up to 25 psig and held while boiling for 24 h. Some operators blow down to half a gage glass after about 4 h of boiling. But 24 h after boiling, the solution is dumped, and the unit refilled with fresh hot water with vents open. After this flushing water is dumped, a thorough internal inspection is made. If necessary, a hose is used for spraying hot water for final cleaning.

Q Explain the acid method of cleaning boilers.
A Acid cleaning is often done to remove metallic oxide film or scale after the oil and grease boil-out operation. But this should be done only by experienced personnel using recommended dosages for the unit, followed by neutralizing agents. Leaky tubes, especially at rolled joints in older boilers, often develop after improper acid cleaning.

The solvents used for chemical cleaning are varied. Some use hydrochloric acid, others phosphoric acid. The usual procedure is to again fill the boiler until the solution overflows at the air vent (acid is added outside the boiler). The solution is *allowed* to soak the boiler from 4 to 6 h, followed by refilling with a neutralizing agent. If hydrochloric acid is used for soaking, a weak solution of phosphoric acid is used. After draining, fresh warm water is used for flushing; then the boiler is immediately filled with an alkaline solution and boiled again for several hours. This solution is drained, the boiler flushed again, and then refilled with normal service water, with proper feedwater treatment started immediately.

Q Should new piping on a boiler installation also be cleaned?
A Yes. Whenever a new system is started up, not only should the boiler be boiled out, but this process should be applied to the complete system. Every piece of pipe has foreign matter deposited inside. Unless it is cleaned out properly by personnel familiar with boiling out and flushing piping so as to avoid undue strains on piping, trouble may develop.

> IMPORTANT: Elbows, bends, and other dead pockets in piping are dangerous spots and must be thoroughly cleaned.

Q What precautions must be observed in acid cleaning of boilers equipped with a superheater and other such bent tubes?
A Make sure all traces of acid are thoroughly cleaned out of U bends. This is critical in the neutralizing and flushing stage after soaking the tubes with an acid solution. Compressed air may have to be used to

force the solution out of dead pockets. If not, the acid solution may not be completely cleared, and thinning of tubes will result. Acid cleaning of riveted boilers can be dangerous because the acid may settle under the butt straps or lapped plates and eat up these holding elements. Thus riveted boilers are usually not acid-cleaned.

Q What areas of a boiler may become susceptible to corrosion failures from improper acid cleaning or repetitive acid cleaning?
A Acid cleaning of boilers may lead to problems if the following are not considered: (1) chemical analysis of the scale, (2) the age of the boiler, and (3) construction details of the boiler. Welds with surface imperfections are a likely spot for acid attack, leading to possible stress-corrosion cracks.

Areas subject to high local stress and repetitive application of stress may be affected by acid treatment. Tubes that have been repeatedly rolled and acid-cleaned may develop tube-roll leakage. Then with further rolling impossible, new tubes may be required. But the prevention of scale in a boiler is still the best method of keeping a boiler clean.

Q Describe the procedure to be followed in starting a VT boiler so as to prevent overheating the upper segments of tubes that are not surrounded by water.
A Two methods can be used to prevent the upper segment of the tubes from becoming overheated until steam is raised:

1. Start the boiler with a low fire until steam is raised.
2. Fill the boiler with water, operate at this level until the water temperature is near steaming, then adjust the level by blowdown.

Q Should standby or reserve boilers not in use be full of water or empty?
A Whether kept "wet" or "dry" depends. If the unit is subject to freezing, keep it dry. But if the unit will be out of service for only a few days, keep it filled with water. A boiler out of service for over 24 h is subject to corrosion, and storing dry or filling completely with water does minimize corrosion attack. Make sure there is no suspended matter in the drum or shell when storing. And never completely empty a hot boiler because the solids will bake on the interior surfaces. When removing the boiler from service, be sure to blow down frequently to remove mud and suspended solids. And cool down gradually. Then when the boiler and setting have cooled so that residual sludge will not bake onto the metal, open the boiler.

If a boiler is to be stored dry, there is always the danger of dry firing the boiler. For example, a sudden cold spell requiring additional heat often causes someone to fire the empty boiler. Guard against damage by securing the fuel lines, tagging the switches, pulling the fuses, etc.

Q What is the effect of laying up heating boilers improperly during summer outages?

A Cast-iron and steel boilers are susceptible to corrosion damage during summer outages. Soot, if not cleaned, will form sulfuric acid in damp or sweating basements. Water, if not treated, will corrode the inside of the boiler. Burning trash intermittently in a boiler can cause dry firing if the water level is not watched and can also cause acid-type attacks in the fireside when the boiler is idle.

Q Can corrosion attack be minimized on heating boilers during summer lay-up?

A Yes. The boiler should be flushed and cleaned on the waterside. The fireside should be cleaned of all soot deposits. Then lay up the boiler either wet or dry.

Q Explain the wet-storage method.

A This method is best when freezing is not a problem and if the unit will not be needed for at least a month. After it is prepared for storage, fill the boiler to water level with deaerated water. If no deaerated water is used, open a top vent. Then build a light fire to boil the water for 8 h so dissolved gases are driven to the atmosphere. Use 1½ lb of sodium sulfite for each 1000 gal of water stored in the boiler to protect against oxygen. The concentration should be about 75 ppm. Use caustic soda to obtain alkalinity of 375 ppm. Keep the water temperature as low as possible and test the water weekly.

Q Explain the dry-storage method.

A The big problem with the dry method is keeping the insides dry. Air-blast with independent outside hot air after draining. When dry, place shallow pans of quicklime inside, then close all openings tight. Place trays between tubes and one in each steam drum (WT type) and bottom of shell (FT type). Open the boiler every 30 days, and if the quicklime is saturated with water, replace it (or whatever material is used).

Q How and why is a hydrostatic test made on a boiler?

A Fill the boiler with 70°F (or warmer water) until the water comes out of the highest vent. Remove the safety valves and blank their connections. Don't apply gags to the spindle because this will bend them and bind the valve. Apply the gag only to the disk of the valve. If the boiler is in battery, close both stop valves and open the drip valve between the stop valves. Leakage at the drip valve indicates that the first stop valve near the boiler is leaking. Stop the test and insert a blank in the line. Raise the pressure slowly to 1½ times the SV setting and hold this pressure until the areas to be checked can be examined. Then apply the hammer test for weakness.

The purpose of the hydrostatic test is to see that all welds, joints, and tube connections are tight. It is *not* a proof test. This hydrostatic test is also used on new construction, when repairs are made to a boiler (such as welding or new tubes installed), or to determine the exact source of leakage or defect suspected in some part of a boiler. It is used on heating as well as power boilers.

For example, on CI boilers for checking cracked sections, a hydrostatic test is often used. Such a test is mandatory after a major repair for checking the soundness of the repair before returning the boiler into service.

Q What action should operating engineers take if they discover a defect in the boiler or equipment under their charge?
A As most states have laws on permissible repairs, the defect should be reported to the authorized boiler inspector, who can make an inspection of the defect and advise what repairs are permissible. Routine repairs, such as corroded and leaking tubes may normally be repaired during emergencies without waiting for the inspector. But always save the old tubes that were removed for inspection. If the boiler is insured, contact the company. If it is not insured and is in a code state, contact the state inspector.

Q Why are log sheets useful on heating boilers and power boilers of smaller size?
A Log sheets are a forced reminder to check certain components of a boiler to prevent trouble from developing later and to note if proper operation is taking place. Keeping a log serves the same function as a police officer's checking doors, windows, locks, etc. See Figs. 13.2 and 13.3 for typical boiler log sheets.

Q How are utility boilers prepared for any long-term storage or idleness?
A Central-station-type boilers are usually laid up with a pressurized inert gas such as nitrogen. When the boiler is returned to service, and especially if any kind of internal work is to be done on the boiler before it is filled with water, it is imperative to purge the steam and water parts thoroughly of the nitrogen with air in order to prevent a person from being affected in the boiler by insufficient air.

Q How may the fireside of a boiler be affected by the wet-layup method?
A A boiler filled completely with water may act as a condenser for any humid air that may be present on the fireside of the boiler. If the fireside surfaces are not cleaned of soot and ash, there is always the possibility of sulfur products combining with the moisture from the air to form sulfuric acid. This acid will then attack the fireside surfaces of the

Fig. 13.2 Sample of a low-pressure heating boiler log. (*Courtesy of Royal Insurance Co.*)

LOW – PRESSURE HEATING BOILER LOG WEEKLY READINGS

BOILER NO.	YEAR	DATE OF LAST INTERNAL CLEANING	STATE CERTIFICATE DUE

Record Tests each week by checking spaces – Instructions are printed below.

TO TEST:

SAFETY OR RELIEF VALVE -- pull try - lever to open position with pressure on boiler. Release lever to allow valve to snap closed. If lever does not lift valve to discharge steam or water, call your service organization immediately. Do not operate boiler unattended until safety valve functions properly. Watch Pressure Gage On Boiler.

WATER COLUMN OR GAGE GLASS – open drain quickly to void small quantity of water. Water level should return quickly when valve is closed.

LOW WATER FUEL CUT - OFF – drain float chamber while burner is running. This should interrupt the circuit and stop burner. A more positive test is to lower water in boiler to bottom of sight glass; burner should shut off. Do not leave boiler during either test. If burner does not stop when water is lowered in water column, call your burner service organization immediately. Do not operate boiler unattended until low water cut - off functions properly. Watch Water Level.

PUMP AND RETURN SYSTEM -- check pump for proper operation, leaky packing. Examine traps, check valves, makeup float valves, and condition of condensate tank. Check if pump goes on properly when water reaches low level, and if it cuts off properly when water reaches high level. If it does not, call your service organization immediately and do not operate boiler unattended. Watch Water Level.

BURNER OPERATION -- if flame is dirty, a burner starts with a puff, phone your service man at once! If burner is equipped with flame failure protection, this device can be checked weekly. Ask your service organization how this test can be made so as to assure you that proper flame protection is in good operating condition.

	SAFETY VALVE TESTED					WATER COLUMN GAGE GLASS DRAINED					LOW – WATER FUEL CUT – OFF TESTED					PUMP AND RETURN SYSTEM CHECKED					BURNER OPERATION CHECKED				
	1	2	3	4	5	1	2	3	4	5	1	2	3	4	5	1	2	3	4	5	1	2	3	4	5
JAN.																									
FEB.																									
MAR.																									

APRIL																									
MAY																									
JUNE																									
JULY																									
AUG.																									
SEPT.																									
OCT.																									
NOV.																									
DEC.																									

SERVICING:

LOW WATER FUEL CUT - OFF -- The low water fuel cut - off should be dismantled for a complete overhaul by a competent serviceman at least annually. The internal and external mechanism, including linkage, contacts, mercury bolbs, floats, and wiring should be carefully checked for defects. See manufacturer's instructions.

OIL OR GAS BURNER AND CONTROLS -- The oil or gas burner and all operating and protective controls should be thoroughly checked at least once every three months by a competent service organization. See manufacturer's instructions. Record service dates.

SERVICE DATES			

Fig. 13.3 Sample of a high-pressure boiler log. (*Courtesy of Royal Insurance Co.*)

HIGH-PRESSURE POWER BOILER LOG READINGS TAKEN EACH 8 HR. WATCH

Important Remarks

Continued safe operation of a boiler requires regular maintenance and testing of the boiler and its operating and protective controls. The minimum tests and checks outlined below are recommended to determine whether or not the boiler and controls are in good operationg condition. Therefore, test periodically and correct malfunctions. Should any check or test indicate that the device being tested or observed is not in good operating condition, it should be repaired immediately. Record repairs or changes under "remarks" so that a complete record will be available for review at any time. Caution: **Never operate a boiler unattended if the safety valve does not function or if the water control or low water fuel cut-off does not work properly!**

To Test

WATER COLUMN AND GAGE GLASS—open drain valve quickly and flush water from glass and column. When drain is closed, water level should recover promptly.

LOW WATER FUEL CUT-OFF AND WATER LEVEL CONTROL—drain float chamber when firing equipment is operating. Proper operation of the control should shut off the firing equipment and start feed pump. If controls are of probe or other type that require lowering of water level in boiler to test, DO **NOT** lower water level to point below bottom of gage glass.

BOILER NO. __________

WEEK BEGINNING __________

	MON.			TUES.			WED.			THUR.			FRI.			SAT.			SUN.		
CHECK OR TEST OR RECORD EACH 8 HR WATCH	8 to 4	4 to 12	12 to 8	8 to 4	4 to 12	12 to 8	8 to 4	4 to 12	12 to 8	8 to 4	4 to 12	12 to 8	8 to 4	4 to 12	12 to 8	8 to 4	4 to 12	12 to 8	8 to 4	4 to 12	12 to 8
WATER LEVEL-PROPER																					
STEAM PRESSURE, PSI																					
FEED PUMP PRESSURE, PSI																					
FEED WATER TEMPERATURE, °F																					
CONDENSATE TEMPERATURE, °F																					

FLUE GAS TEMPERATURE, °F																					
LOW WATER CUT-OFF-TESTED																					
WATER LEVEL CONTROL-TESTED																					
WATER GAGE GLASS-CLEAN																					
FEED PUMP IN GOOD REPAIR																					
CONDENSATE TANK & FLOAT TESTED																					
BURNER OPERATION-NORMAL																					
FUEL SUPPLY-ADEQUATE																					
FLAME FAILURE SAFEGUARD																					
WATER TREATMENT-TESTED																					
BOILER BLOWDOWN-PERFORMED																					
MONTHLY-SAFETY VALVE TESTED																					
OPERATOR'S INITIAL																					

REMARKS:

boiler, resulting in thinning of pressure parts. Generally, long-term wet layups should be avoided.

Q What other factors may affect tubes from the fireside?

A The abrasive action of fly ash in coal-fired boilers can thin boiler tubes. Waste gases to be fired in boilers may have abrasive dusts in them which can also cut tubes as the abrasive particles sweep the tube banks at high velocity. Leaking soot blowers that use steam can cause fireside corrosion on the tubes affected by the leaks. Misaligned or improperly placed soot blowers, or soot blowers frozen in one position, can cut tubes by the steam-jet action of the steam hitting tubes in a concentrated manner for too long a time.

Q What minimum training or knowledge should be stressed for boiler operators in nonlicensed areas?

A This outline covers the minimum proposed by licensed personnel:

1. Trace and sketch a boiler-fuel system and associated valves, strainers, gages, and controls.
2. Trace and sketch main and auxiliary steam systems and condensate return and boiler feed systems, including valves, gages, controls, and interlocks.
3. Inspect and test boiler casings and settings.
4. Inspect firesides and watersides for leakage, corrosion, cracking, bulging, blistering, and other conditions that weaken the boiler.
5. Clean, inspect, and test oil and gas burners and registers.
6. Check fuel tanks and, if necessary, clean strainers, lines, and valves.
7. Line up, recirculate, check for leaks, and light off a fuel oil system. Check combustion safeguards.
8. Observe and record fuel system pressures and temperatures. Check and adjust control settings.
9. Inspect and operate soot blowers properly.
10. Light off, fire, and bank fires in a coal-burning boiler (if installed).
11. Inspect and regulate stoker and pulverizer operation in coal-fired boiler. Check and adjust controls and interlocks.
12. Start, regulate, and secure forced-draft and induced-draft fans. Check for proper operation, controls, and interlocks.
13. Inspect breechings, uptakes, and stack.
14. Determine draft, windbox, and furnace pressures, and furnace and stack temperatures. Adjust the draft where needed.
15. Take and analyze flue gases for CO_2, CO, and O_2. Interpret the results and make the necessary corrections.
16. Blow down a boiler, both bottom and surface blows.

17. Blow down gage glasses and water columns.

18. Test high- and low-water alarms and low-water cutoff. Dismantle and clean alarms and low-water fuel cutoff.

19. Regulate feed pumps and change-over pumps and adjust feedwater governors.

20. Regulate boiler-water level.

21. Cut in, adjust, and secure feedwater-level regulator.

22. Test SVs.

23. Put a boiler on line.

24. Warm up and cut in steam lines.

25. Take a boiler off line, pull fires (for coal-burning boilers), and secure the boiler.

26. Conduct boiler-water analysis, interpret results, treat the feedwater, and adjust continuous blow according to a water-treatment specialist's advice, if required.

27. Shift combustion control from manual to automatic and back again. Check the safety controls in doing this.

28. Prepare boiler logs and operating records.

29. Cut in and out superheaters properly.

30. Adjust feedwater heater pressures and temperatures.

31. Renew and repack gage glasses.

32. Remove, regasket, and replace manhole and handhole plates.

33. Inspect, repair, and set SVs within code limits.

34. Clean the fireside and waterside of the boiler.

35. Make refractory and other furnace repairs.

36. Conduct a hydrostatic test.

37. Lay up a boiler.

38. Know how to remove and replace tubes in the boiler, superheater, economizer, airheater, and feedwater heater.

39. Adjust soot blower and lances.

40. Inspect, clean, and repair boiler gages, instruments, and controls.

In addition to the above, an operator should know something about (1) boiler and steam system design, (2) construction, (3) operation, (4) maintenance, (5) state or city laws, (6) insurance company requirements and inspections, (7) emergency measures, (8) safety in a boiler plant, and (9) ASME Boiler and Pressure Vessel Codes.

Q Should safety valves be removed from a boiler when a hydrostatic test or acid cleaning is to be performed on a boiler?

A Because a full hydrostatic test is at 1½ times the allowable or operating pressure of the boiler, the safety valves would lift and thus not permit a full hydrostatic test to be made. The preferred method is to remove the safety valves and close the opening with a suitable blank or flange for the pressure.

> IMPORTANT: The flange or blank should be removed and the safety valves reinstalled right after the hydrostatic test so that the boiler is not placed back in service without this important safety device.

Since the acid cleaning solution can damage the internal parts of safety valves, the safety valve should be removed and the opening closed with a suitable blank or flange to avoid safety-valve damage. Again, the safety valve should be reinstalled as soon as the boiler is prepared for normal operation.

In both of the above situations, tagging in a highly visible manner should caution operating people not to fire the boiler until the safety valves have been reinstalled.

Q In power-plant work, what does the term *heat rate* mean?

A In a generating plant, the heat rate is an indication of overall station efficiency, or unit power-generation efficiency. It is expressed as the number of Btu required to generate one kilowatthour of electricity. Heat rates are further classified into gross and net. The gross heat rate is the Btu required to produce one kilowatthour at the generator power terminals. The net heat rate is the gross minus the auxiliary power that is consumed by forced- and induced-draft fans, boiler feed pumps, etc., to generate the power in the station.

Heat rates are a measure of efficiency and can also be used to monitor performance and, if this drops, to determine what maintenance, cleaning, or adjustments will have to be made to restore the previous attained efficient heat rate. Generally, bench marks are established when equipment is new or after a thorough overhaul, and by conducting periodic heat-rate tests or gathering data and calculating the heat rate, performance monitoring for changes can be made. If deviation from a bench mark above a certain percentage is reached, the unit(s) can be scheduled for maintenance to restore bench-mark conditions if possible.

Most utilities have performance-monitoring programs on their generating units. Some of this has been computerized so that net heat rates are readily available per unit and even on a system basis.

Part of this program can be an estimate of the future increase in operating costs at different heat rates. These figures can then be used in comparison with the maintenance costs that may be incurred to correct a declining heat rate.

Q What items generally require measurements for data?

A A full performance test is covered by ASME performance test codes for boiler-turbogenerator output and input criteria and is beyond the scope of this book. In general, kilowatt output is measured, and then it is necessary to determine losses in the generator, turbine, and boiler in order to note where the Btu input went. For a large utility unit the following is measured with calibrated test instruments:

1. Feedwater and total stage heater drain flow
2. Main steam pressure and temperature
3. HP turbine exhaust pressure and temperature
4. IP turbine inlet pressure and temperature
5. IP turbine exhaust pressure and temperature
6. LP turbine exhaust pressure and temperature
7. Condenser circulating water inlet and outlet temperatures
8. Boiler flue-gas analysis including combustible content
9. Boiler fuel analysis
10. Air heater inlet and outlet gas and air temperatures
11. Pump and fan speeds

CHAPTER **FOURTEEN**

Boiler-Water Treatment

Modern boilers are designed for maximum output with the smallest possible size of equipment. Furnace volumes to steam output capacity have decreased, which has resulted in heat-release rates varying from 70,000 to as high as 200,000 Btu/(ft^2 · h). These high heat-release rates are possible because of technological improvements in material, heat-transfer calculations, burner design, combustion-control systems, and boiler controls. Modern boilers tend to be quick steamers provided that heating surfaces are kept free of scale and deposits to a far greater degree than on older boilers. Clean surfaces require water treatment to limit deposits, corrosion, and other water and steam problems to be reviewed in this chapter. Among these are pH, clarification, filtration, softening, ion exchange, demineralization, deaeration, and precipitation.

PURPOSE OF TREATMENT

Q What is the purpose of water treatment?

A Water treatment is more essential as pressures and temperatures of a system rise. The degree of treatment will also vary for this reason. However, even heating boilers may require treatment of the makeup water or the condensate. Water treatment has as its major objectives the following:

1. To prevent boiler corrosion on the waterside.
2. To prevent pressure-part failures from overheating such as may occur from scale. This will improve unit availability and reduce costly downtime in production or lack of use.
3. To improve efficiency by maintaining clean heat-transfer surfaces.
4. To avoid costly and unexpected failures due to corrosion, pitting, gouging, cracking, and similar pressure-part deterioration.

Q What determines the degree of water quality needed?
A The degree or extent of water quality control depends on the operating pressure, process-steam needs, the amount of condensate returned or not returned, which means high makeup water must be introduced into the boiler system, the construction and the material present in the boiler and possibly in connected steam-using equipment such as turbines, and similar considerations. For this reason most equipment manufacturers of steam-using or -generating apparatus specify desired steam and water quality for their equipment. If these are not adhered to, equipment reliability may be affected by the poor steam or water conditions coming out of the boiler system.

WATER ANALYSIS

Q Explain why boiler-water treatment must be considered for heating and smaller commercial boiler installations.
A Usually the most damaging agent in boiler handling is either the lack of, or the improper use of, feedwater treatment. Today most sources of water are surface waters, and to make the water palatable, it is oxidized. Thus untreated boiler water ends up by being high in oxygen content, which must be eliminated so it does not combine with boiler metal. Many water sources also have high calcium or silicon contents, making treatment necessary to eliminate hard-scale forming.

Other chemical alkaline agents are found in water, and all require qualified analysis. An important part of feedwater treatment is blowdown (at times continuous), because most water treatments will cause the dangerous elements to precipitate as solids. And the amount of solids in boiler water affects the quality of the steam. Each solid particle in suspension acts as the nucleus of a drop of moisture when the boiler is operated at high levels of steam production. Thus the higher the concentration of solids, the wetter the steam may be. But high concentrations of solids can also occur from overtreatment, ending up as insulation on heating surfaces, thus promoting boiler failure.

Q Is pure water found in a natural state?
A No, not chemically pure water. See Fig. 14.1 for impurities found in boiler water. It always contains other substances. Nature's nearest approach to pure water is rainwater, yet this is unsuited for boiler consumption because of the foreign matter it picks up in the air. Because most of these foreign particles are harmful, they are called impurities. Their appearance in rainwater can easily be accounted for when we remember that water is the best solvent known. In fact, it is often called the universal solvent because, given time, it can dissolve almost any substance.

Fig. 14.1 Impurities usually found in boiler water.

Common name	Chemical term	Manifestations	Symbol
Lime	Calcium bicarbonate	Soft scale, dissolved in carbonic acid	$Ca(HCO_3)^2$
	Calcium carbonate		$CaCO_3$
Magnesia	Magnesia carbonate	Chalky scale	$MgCO_3$
Silica	Silicon dioxide	Brittle, hard, light color	SiO_2
Gypsum (plaster of Paris)	Calcium sulfate and water in crystal form	Hard, smooth scale	$CaSO_4 + H_2O$
Magnesium chloride	Magnesium chloride	Forms hydrochloric acid	$MgCl_2$
Epsom salts	Magnesium sulfate and water	Corrosive	$MgSO_4 + 7H_2O$
Table salt	Sodium chloride	Causes foaming	$NaCl$
Glauber's salts	Sodium sulfate	Causes foaming	$Na_2SO_4 + 10H_2O$
Soda ash	Sodium carbonate	Causes foaming	Na_2CO_3
Baking soda	Sodium bicarbonate	Causes foaming	$NaHCO_3$
Gases:			
Oxygen	Oxygen	Accelerates corrosion	O_2
Carbon dioxide	Carbon dioxide	Forms acid	CO_2
Chlorine	Chlorine	Forms acid	Cl_2
Acids:			
Sulfuric	Sulfuric	Corrosive	H_2SO_4
Hydrochloric	Muriatic acid	Corrosive	HCl
Alkalies:			
Sodium hydroxide	Caustic soda	Foaming, corrosive, stress corrosion	$NaOH$
Magnesium hydroxide	Magnesium hydroxide	Foaming, corrosion	$Mg(OH)_2$
Organics:			
Leafs, mud, etc.	Same as common name	Foaming, deposits	Varies

Q How do impurities get into boiler water?

A Besides the natural contaminants in raw water from any source, there are air, oil, and several other contaminants from plant equipment. Cooling water from rivers, lakes, or oceans can leak into a steam-condensing system from these sources: (1) condensers, including main or auxiliary condensers, water-cooled air-ejector condensers, distilling condensers, or water-cooled gland-exhaust condensers; (2) evaporators; (3) condensate and drain coolers; (4) leaky feed, suction, and drain lines.

Q How does air contaminate feedwater?

A Air is neither an element nor a compound, but a mixture of several gases. These gases are soluble in water to varying degrees. Thus water has a tendency to absorb and carry these gases (when the two make contact) along with it. But using a closed-feed system reduces these points of contact to a minimum. In addition to air already being present in raw feedwater, air may enter at (1) feed tanks through vent pipes, (2) pump packing and leaky packings in heating system piping and valves, and (3) open hot wells and surge tanks. But some gases in air have no effect whatever on the system or its operation.

Q How does oil get into feedwater?

A Oils and greases have a damaging effect on a boiler and offer high resistance to heat transfer. The higher the operating temperature and pressure, the more the boiler suffers from the effects of oil in the water. Fortunately, high-pressure boilers are usually used with turbines where the dangers of oil contamination are considerably reduced. Reciprocating steam engines and pumps use cylinder oil that is more detrimental than turbine oil. Oil contamination may come from (1) careless, ignorant, and excessive use of oils; (2) leaky fuel-oil heater coils; (3) dirty or improperly operating grease extractors, filters in hot wells, etc.

Q In what state may impurities be in boiler water?

A No matter what the chemical characteristics of the impurity may be, three states are possible:

1. If the impurity is a soluble solid, it appears in a dissolved state or in *solution* with the water.
2. If the solid is not soluble in water, it is not in solution, but in a state of *suspension.*
3. Those impurities of a gaseous nature that are partially soluble are in an *absorbed* state in the water.

Q Explain what minerals may be dissolved in raw water.

A Analysis indicates dissolved minerals of the following type may be present in natural waters: calcium sulfate, called gypsum; calcium carbonate, called limestone; magnesium sulfate, called epsom salts;

magnesium carbonate, called dolomite; silica, or sand; sodium chloride, or table salt; sodium sulfate, or Glauber's salt; and traces of iron, manganese, aluminum, fluorides, and other dissolved minerals depending on geographical location. Some surface waters may also be acidic from wastes coming from mines and other industrial processes. In other cases, the water may be alkaline from wastes that find their way into natural waters.

Q How is water hardness defined?
A Hardness of water is a result of calcium and magnesium compounds reacting with soap to form curds, and thus it is difficult to wash with this type of water. The amounts of calcium and magnesium compounds can vary from 2 to 3 to over 500 ppm. The calcium and magnesium compounds are relatively insoluble in water, thus precipitating out to cause scale and deposits to form in steam generators. Water with this type of hardness must thus be treated to avoid these types of scaling problems. The hardness of the water depends on the degree of calcium and magnesium compounds present in the water.

Q How can the state of the impurities affect the treatment of the water?
A The impurity classification determines the effects the impurities produce and the required methods of treatment. For example, if it is known that a certain impurity will remain in solution regardless of the temperature or concentration, then obviously it will not drop out of solution and thus not be a scale producer.

Most impurities are found in a dissolved state or in solution with the water. The temperature of the water has a marked effect on solubility. Some impurities become less soluble as temperatures rise and start to precipitate, or form scale in the water. Calcium suflate and calcium hydroxide are such impurities. The mere fact that some impurities stay soluble at higher temperatures and do not precipitate does not mean that they should be ignored. They may have a tendency to corrode metals, they may act as moisture carry-overs, or they may cause embrittlement.

Q How can a dissolved impurity leave a solution and become a solid?
A 1. By a temperature increase reducing the solubility of the solid in the water.

2. By exceeding the saturation point of the dissolved impurity in the water. Water can hold only a limited amount of the dissolved impurity; so the concentration is important.

3. By chemical changes of the impurity with heat, causing it to break down and form insoluble substances.

Q List the impurities that produce hard and soft scale and corrosion in boilers.
A Hard scale is caused by calcium sulfate, calcium silicate, magnesium silicate, and silica. Soft scale is caused by calcium bicarbonate, calcium carbonate, calcium hydroxide, magnesium bicarbonate, iron carbonate, and iron oxide. Corrosion is caused by oxygen, carbon dioxide, magnesium chloride, hydrogen sulfite, magnesium sulfate, calcium chloride, magnesium nitrate, calcium nitrate, sodium chloride, and certain oils and organic matter.

Q What is the first step in any water-treatment program?
A Analyze the water, as chemically pure water is rare. Few water supplies are suitable for domestic or industrial use without treatment. And the chemical compositions of different water supplies vary greatly. So it is impractical to prescribe any one "ideal" treatment. Every engineer should know how to make routine water tests such as are used in the daily operation of a water-treatment system.

Remember that boiler-water treatment involves more than just scale prevention. Important are the kind and control of chemical treatment, regulating the preheating and pretreating of makeup, and regulation and control of boiler blowdown to prevent a buildup of boiler-water solids. Supervision and control of this program calls for a qualified operator or a firm specializing in boiler-chemical service. Prevention of feedwater trouble usually costs far less than the repair of neglected equipment.

Q What is meant by a water analysis?
A Analyzing a water sample is the process of finding out how much of the various impurities and other chemical substances are present in the water. The results are usually expressed in parts per million (ppm) and tabulated as shown in Fig. 14.2. Parts per million is a measure of proportion by weight, such as one pound in a million pounds. Grains per gallon (gpg) is another way of expressing the amount of a substance present. One grain per gallon equals 17.1 parts per million. Individual impurities, with the possible exception of hardness, are rarely reported in grains per gallon. But quantities in parts per million are often converted to grains per gallon to aid in calculating the capacity of water-treating equipment.

Water usually contains a wide variety of dissolved compounds, each of which breaks down into its respective ions when dissolved. Thus atoms of the combination molecule separate. While chemists can measure the amount of each cation and anion in the solution, it is impossible to chemically analyze the amount of each individual compound. We can only assume that various ions recombine in certain ways upon evaporation or precipitation. The ionic weight of ions such as calcium, sodium,

REPORT OF WATER ANALYSIS			Parts per million	Equivalents per million
Date Collected 10/30		Silica as SiO_2	5	
Source well		Iron as Fe_2O_3	1.2	
Date analyzed 11/4		Calcium as Ca	62	
Total dissolved mineral solids	ppm	Magnesium as Mg	31	
Organic matter	none ppm	Sodium and potassium as Na	38	
Suspended solids	5 ppm	Bicarbonate as HCO_3	250	
Chloroform, extractable (oil, etc.)	none ppm	Carbonate as CO_3	0	
pH	7.7	Hydroxide as OH	0	
Phenolphthalein alkalinity as $CaCO_3$	0 ppm	Chloride as Cl	11	
Methyl orange alkalinity as $CaCO_3$	205 ppm	Sulfate as SO_4	138	
Hydroxide alkalinity as $CaCO_3$	0 ppm	Nitrate as NO_3	0	
Hardness as $CaCO_3$	282 ppm	Carbon dioxide as CO_2	10	
Specific conductance	micromhos	Turbidity	5	
		Physical characteristics of sample	Clear when drawn	

Fig. 14.2 Water-analysis report form asks for complete information on impurities present.

sulfate, and chloride can be determined. But this doesn't tell us how much individual calcium sulfate, sodium sulfate, or sodium chloride originally went into the solution.

Since analysis measures only ions, never compounds, it is logical to use the ionic form. Usually no attempt is made to indicate hypothetical compounds. If the analyses do show them, they are figured by a chemist (from their ionic weights) by using certain arbitrary assumptions.

The combined form of analysis is best described as a rule-of-thumb attempt to put a water-analysis report into shape for practical interpretation. By adding a column for parts per million expressed as calcium carbonate or equivalents per million (epm), it is possible to reduce all ions present to a common denominator. This makes it easy to calculate the amounts of reacting chemicals and the size of the equipment and to check the accuracy of the analysis.

Expressing an ion such as calcium in terms of its calcium carbonate equivalent is done easily by comparing equivalent weights. Calcium carbonate's weight of 50.1 is 2.5 times greater than calcium's at 20.1. So the calcium carbonate equivalent of calcium in our analysis is 62×2.5, or 155 ppm. Tables giving conversion factors for all ions are available.

This method of expression is accepted as the common standard basis of reporting a water analysis. It is very useful for in-plant treatment control.

An equivalent per million is a unit chemical weight equivalent per million units of solution. To get the equivalent per million, divide concentration in parts per million by the equivalent weight of the element or radical. Equivalents are very helpful in calculating amounts of reacting chemicals from equations. For example, our sample analysis has 250 ppm of bicarbonate. Dividing by 61 (the equivalent weight of bicarbonate), we get 4.10 epm. When we know the equivalents per million, it is easy to find how much 90 percent calcium hydroxide will match and destroy the bicarbonate. Multiply the equivalents per million of carbonate by the equivalent weight of lime. This amounts to 4.10 × 41.1, or 168.5 ppm of 90 percent calcium hydroxide. Since 120 ppm equals 1 lb/1000 gal, our 168.5 ppm becomes 1.4 lb/1000 gal. The analysis can be evaluated also by totaling positive and negative ions separately. As shown in Fig. 14.2, the two columns always must be in complete balance.

Q What is the meaning of pH in water chemistry?
A It is a number between 0 and 14 indicating the degree of acidity or alkalinity. The pH scale resembles a thermometer scale, but the pH scale indicates intensity of acidity or alkalinity. The midpoint of the pH scale is 7, and a solution with this pH is neutral. Numbers below 7 denote acidity; those above, alkalinity. Since pH is a logarithmic function, solutions having a pH of 6.0, 5.0, or 4.0 are 10, 100, or 1,000 times more acid than one with a pH of 7.0.

Q Define the term alkalinity.
A Alkaline substances like sodium hydroxide (NaOH), also called caustic soda, are termed bases by chemists because they form hydroxide ions in water. For example, caustic soda ionizes in water to form Na^+ and OH^- ions. The alkalinity is a measure of the temporary hardness of a water, and is measured by chemical tests. Alkalinity represents the concentration of carbonates, bicarbonates, hydroxides, silicates, and phosphates expressed as equivalent calcium carbonate in ppm.

Q Why must all boilers receive systematic feedwater treatment?
A To eliminate the problems caused by scale, corrosion, carry-over, and such potential trouble as embrittlement. We know that all water, no matter what its source, contains many impurities. Some are of a very corrosive nature, while others have the characteristic of producing adherent deposits (scale) on the waterside of the boiler. Thus continued use of such water, if untreated, reduces the efficiency of any plant. But more important, the corroded metal may become so weakened that leaks

develop. Worse still, a boiler explosion with complete destruction of the plant and hazard to lives may result.

Q Differentiate how carbonate, sulfate, silicate, and iron deposits may be identified in a boiler system.
A Deposits are seldom composed of one deposited mineral alone but are a mixture of minerals, especially in untreated boiler water. However, the following is offered as a guide: (1) A carbonate deposit is usually granular and quite often of a porous nature. Calcium carbonate, the most prevalent of these deposits, is identified as such by dropping the scale in an acid solution. Bubbles of carbon dioxide will effervesce from the scale. (2) Sulfate deposits are harder and denser than carbonate scale. It is a brittle scale which does not effervesce when dropped into an acid solution. (3) Silica deposits are the hardest of the scales, almost resembling porcelain. The scale is lightly colored and is not soluble in hydrochloric acid. (4) Iron deposits are dark-colored and, being magnetic, are usually picked up by a magnet. Iron deposits are soluble in hot acid solutions, with a dark brown solution being formed.

Q Define the term *selected silica carry-over.*
A Silica can be present in the steam as the result of general boiler-water carry-over, or it can leave the boiler as a gas or in volatile form with the steam. When it goes out of the boiler in volatile form, it is considered to be selectively carried over. At pressures above 400 psi, there is an increased silica volatile carry-over with the steam depending on the proportionate amount of silica in the boiler water.

Q How may other carry-overs affect the boiler and steam-using machinery?
A Carry-over will affect efficiency and cause possible wet-steam erosion of turbine blades. Dissolved or suspended solids will deposit out in superheaters and on turbine blades as the steam travels through a turbine at lower and lower pressure. Deposits in superheaters can cause overheating tube failures. Deposits on turbine blades can cause a loss of efficiency due to the nozzle passages being affected. High-velocity scale particles may also erode blade material.

Q How is carry-over prevented?
A Carry-over is prevented by (1) keeping solids concentration in the boiler water at low levels; (2) avoiding high water levels in drum boilers; (3) avoiding overloading the boiler; (4) avoiding sudden load changes on the boiler; (5) avoiding contaminated condensate; (6) using chemical additives called antifoam agents that help in preventing carry-over due to high concentrations of impurities in the boiler water.

Q What is the source of corrosion in a boiler system?

A Corrosion is generally caused by a low pH or acidic water and the presence of dissolved oxygen and carbon dioxide. High temperature and stresses on boiler pressure-containing parts accelerate the corrosion mechanism. Condensate-system corrosion is usually the result of being contaminated with oxygen and carbon dioxide. Ammonia- or sulfur-bearing gases are known to attack copper alloys, such as may be present in steam-turbine condensers. These copper compounds may contaminate the steam-condensate system. Corrosion must be avoided as much as possible, because it thins metals that confine pressure, and rupture may occur if the thinning weakens the metal. The corrosion product may also deposit on heat-absorbing surfaces to the extent that it acts as an insulator, thus causing metal-overheating ruptures on the affected pressure-containing part.

PRESSURE-PART FAILURES FROM WATER PROBLEMS

Q How does scale weaken a boiler tube?
A As the temperature of the metal rises, the allowable stress of the material for that temperature decreases until it cannot carry the same pressure. Figure 14.3*a* shows the temperature gradient through a clean tube wall and also the film effect on the gas side and waterside. Thus the hot side of the tube wall runs at about 300°F, which is safe. In Fig. 14.3*b* scale on the waterside of the tube has caused the hot side to rise to 700°F, which may be near the point of failure for the grade of steel used.

Q What is the effect of flame impingement on a tube?
A Flame impingement on tubes may result from such things as burner

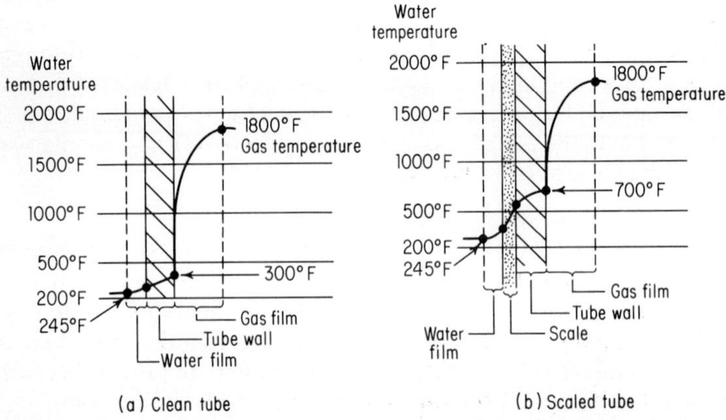

Fig. 14.3 Heat transfer and temperatures of tube metal are affected by scale.

misalignment, damaged baffles, and improper combustion conditions (for example, too long a flame due to overfiring, poor draft). The high heat input to the affected tube area may exceed the heat-transfer capability of the metal or water-steam circuit and cause failure by exceeding stress limits permitted at elevated temperatures for the pressure. See Section I, ASME codes stress tables. Discoloration of tubes on the fireside of tubes with a known clean boiler may be signs of a flame-impingement problem and should be investigated for correction.

Q Define the term *steam blanketing.*
A Steam blanketing is the formation of a semistatic (not moving) layer of steam in tubes. It is most likely to occur in tubes which are parallel or inclined only slightly from the horizontal in natural-circulation boilers. Usually, the volume of steam in these tubes is relatively small and no problems occur with normal firing rates. However, at *high firing rates* and with substantial volume of steam present, the insulating effect of any steam blanket that forms may cause overheating of the tube metal. Also, any subsequent splashing of water onto the "dry hot" metal surface under the steam blanket may remove the protective magnetic iron oxide film and promote corrosion or oxidation. The boiler manufacturer should be contacted for advice in order to avoid steam blanketing. Reduced firing rates in the interval will prevent steam blanketing. At times, correction in rate of water circulation through headers is needed.

Q What causes film boiling on a boiler tube?
A In film boiling, a thin film or layer of steam forms on the heated tube surface because an excessive number of steam bubbles are not being adequately swept away by the water (lack of circulation). Film boiling differs from normal nucleate boiling in which individual steam bubbles form on the metal surface and are carried away by the boiler water. As with steam blanketing, the film may lead to overheating failure because of its insulating effect. It is also primarily caused by overfiring and requires the operator to consult with the boiler manufacturer to determine how it can be avoided.

Q Define the term *phopshate hide-out* as it affects boiler-water treatment.
A Phosphate hide-out is a condition in which phosphate residual is reduced in a boiler because of a deposition on tube surfaces. This temporary deposition or drop-out of normally water-soluble chemicals during normal operation is referred to as hide-out. When the steaming rate is appreciably reduced, or when the boiler is taken off the line, the deposited chemicals redissolve in the boiler water. A number of

substances exhibit this phenomenon, but sodium phosphate is of most interest in boiler operation and maintenance.

At temperatures above 117°C (242.6°F) trisodium phosphate exhibits a retrograde solubility. It is this property that makes hide-out possible (Fig. 14.4). Phosphate hide-out is undesirable because it may increase tube wall temperatures and cause failures in the same manner as other deposits. Further, it makes chemical control of the boiler especially difficult, and it is a good indication of other trouble. Analysis of deposits that show phosphate may be an indication of phosphate drop-out at elevated temperatures. This is a job for an experienced water-treatment chemist to solve and offer solutions.

Q Explain caustic corrosion.

A Caustic corrosion, or caustic gouging, refers to metal attack due to overconcentration of caustic (sodium hydroxide, NaOH). Such overconcentration of caustic in selected boiler areas may be the result of high film concentration, accumulations under deposits, or crevice accumulations. Caustic soda may attack steel by dissolving the magnetic iron oxide film on the surface or by direct reaction with the metal itself. Failure of this type can be spotted because it will occur in an "eaten-up" type of area of boiler metal.

Two chemical reactions are involved:

$$Fe_3O_4 + 4NaOH \rightarrow 2NaFeO_2 + Na_2FeO_2 + 2H_2O \quad (1)$$

$$2NaOH + Fe \rightarrow Na_2FeO_2 + H_2 \quad (2)$$

The control of caustic concentration requires frequent testing of boiler

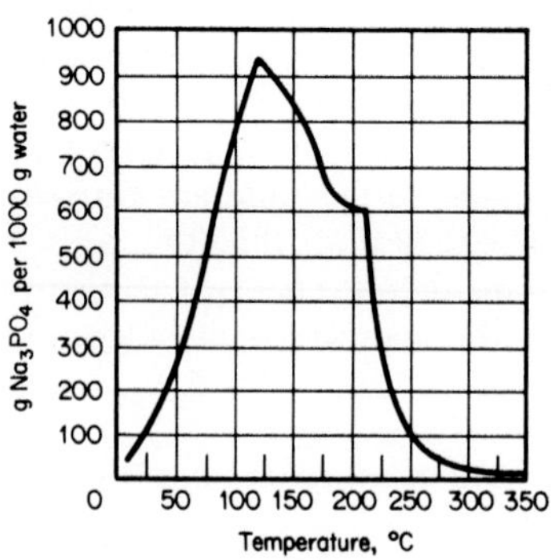

Fig. 14.4 Solubility curve for trisodium phosphate shows retrograde solubility above 117°C. This property makes phosphate hide-out possible and makes chemical control difficult. (*Courtesy of Power magazine.*)

water by the operator, as well as adjustments in water treatment in order to avoid it.

Q How may hydrogen damage occur on very high pressure boilers?
A Hydrogen damage is the result of the diffusion of atomic hydrogen into the boiler metal. Hydrogen may be present in boiler water because of the corrosion reaction (also see equation 2 above):

$$3Fe + 4H_2O_4 \rightarrow Fe_3O_4 + 8H$$

The hydrogen penetrates into the steel and reacts with the carbon in the steel to form methane.

Methane is formed as follows:

$$Fe_4C + H_2 \rightarrow 4Fe + CH_4$$

The presence of these gases within the steel will cause internal stresses due to gas pressure between crystal boundaries. The methane gas exerts pressure on the crystal structure of the metal. The result may be cracking and, ultimately, metal failure. This type of attack is more pronounced on high-pressure units and not in low-pressure boilers.

Q Why is silica a special problem for boilers operating over 850 psi pressure?
A Silica presents a special problem in high-pressure boilers because it may vaporize with the steam and cause deposits in turbines by condensing on blades which are at the lower pressures and temperatures of a staged turbine. At constant pressure, the distribution ratio of the concentration of silica in the steam compared with the concentration of silica in the boiler water is fairly constant. As pressure increases, this ratio also increases. At very high pressures, silica readily volatilizes and may be present in the steam at rather high concentrations (goes over as a gas with the steam). To prevent silica deposition in turbines, silica concentration in the *steam* should be kept below 0.02 ppm. Allowable silica in the *boiler water* which corresponds to 0.02 ppm of silica in the steam is shown in Fig. 14.5 as a function of pressure.

Q Explain caustic cracking or embrittlement.
A Caustic cracking is a form of continuous intergranular cracking originating at the surface of the metal, caused by the attack of caustic on the "cement" between highly stressed metal grains. Because resultant metal fractures have a brittle appearance, this form of attack is commonly referred to as caustic embrittlement.

To get embrittlement of boiler steel, there are three conditions which must be present at the same time:

1. The metal must be under stress near its yield point.

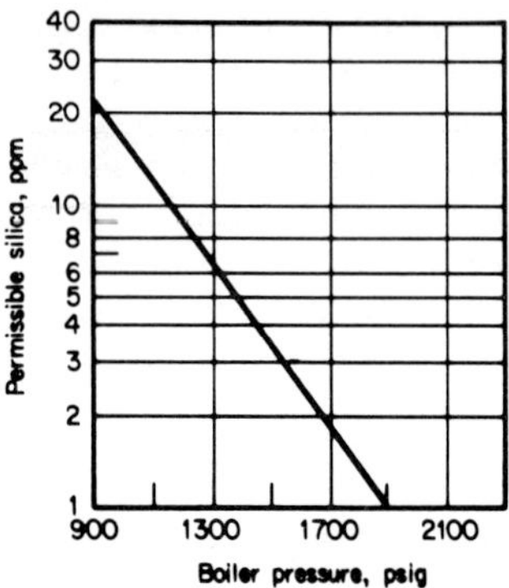

Fig. 14.5 Allowable silica decreases in boiler water as pressure increases to avoid volatilization and deposition in turbines. Curve corresponds to 0.02 ppm of silica in *steam*. (*Courtesy of Power magazine.*)

2. The boiler water must contain caustic. (The presence of some silica along with caustic accelerates attack.)

3. There must be some mechanism (a crevice, seam leak, etc.) permitting the boiler-water solids to concentrate on the stressed metal.

If any *one* of these conditions is absent, embrittlement can't occur.

The contributing factors of metal stress and crevices or leaks are mostly found in boilers with older riveted construction. While modern welded construction minimizes the possibility of embrittlement in the drum itself, embrittlement of highly stressed tube ends where they are rolled into drums and headers is always a possibility. Even blowdown valves have experienced this form of cracking.

Q Define the term *tube exfoliation*.

A Tube wastage with the alloy tube material usually specified for high-pressure service occurs at high steam temperatures above 950°F. The wastage is in the form of metallic oxide scales that peel off during boiler thermal transient operating periods or because of cyclic service. The metallic oxide exfoliations from the tubes find their way into superheaters and reheaters, and then into turbines as high-velocity particles. This has caused erosion of valve parts, first-stage nozzles, blades, shrouding, and similar turbine parts in the steam path. Manufacturers of boiler tubes have developed a high-chrome treatment for use on internal surfaces of tubes that are exposed to steam temperatures above 950°F in an attempt to stop tube exfoliation.

Q Why is water treatment more critical on compact, fast-steaming packaged boilers?

A These units don't require higher-quality feedwater than conventional boilers run at equivalent ratings, but clean boiler surfaces are more critical when operating at high evaporation rates. Scale builds up at a more rapid pace, and their smaller steam space promotes priming and foaming. Therefore, treatment must be carefully selected and controlled to produce high-quality feedwater. Don't compare packaged boilers with older steam generators run at much lower ratings. These older units sometimes do get by with minimum attention to feedwater treatment.

Q What are major treatment problems with heating boilers?
A Corrosion and pitting. Scale is not a problem because the same feedwater is used continuously, and initial treatment usually lasts throughout the heating season.

Q Are many treatment methods available?
A Yes. But they all come under three broad classifications: mechanical, heat, or chemical treatment. Mechanical treatment includes filtration and boiler blowdown. Heat treatment includes makeup distillation and steam purification. But distillation is limited to boilers with small amounts of makeup. Steam purifiers are used where the process demands very dry steam. Chemical treatment, both internal and external, is most widely used. External treatment adjusts raw-water analysis before the water enters the boiler. Internal treatment adjusts the boiler-water analysis by feeding chemicals directly to the boiler.

EXTERNAL TREATMENT

Q Define the term *external treatment.*
A The water added to a boiler in order to replace water lost is termed feedwater. So is the water initially used to fill a boiler. This water needs treatment before it is pumped into a boiler to avoid boiler problems as a result of scale, corrosion, and similar effects detrimental to safe operation. External treatment is used to reduce or remove impurities from the water *outside* of the boiler, especially where the quality of the makeup water is such that large amounts of impurities are present that could not be tolerated by the boiler system if introduced into the boiler directly.

Q What are the recommended limits on boiler water?
A The American Boiler Manufacturers Association publishes permitted limits in order to obtain steam of reasonably good quality. The limits are shown in Fig. 14.6.

Q How are suspended solids removed from water by mechanical means?

Fig. 14.6 American Boiler Manufacturers' Association boiler-water-limit recommendations for WT boilers and associated steam purity.

Drum pressure, psig	Range total dissolved solids,* boiler water, ppm (max)	Range total alkalinity, boiler water, ppm†	Silica (SiO_2), boiler water, ppm¶	Suspended solids, boiler water, ppm (max)	Specific conductance, micromhos/cm, boiler water¶	Range total dissolved solids, steam, ppm (max)†§
For WT drum-type boilers						
0–300	700–3500	140–700	150	15	7000	0.2–1.0
301–450	600–3000	120–600	90	10	6000	0.2–1.0
451–600	500–2500	100–500	40	8	5000	0.2–1.0
601–750	400–2000	80–400	30	6	4000	0.2–1.0
751–900	300–1500	60–300	20	4	3000	0.2–1.0
901–1000	250–1250	50–250	8	2	2000	0.2–1.0
1001–1800	100	‡	2	1	150	0.1
1801–2350	50	As above	1	NA	100	0.1
2351–2600	25	As above		NA		0.05
2601–2900	15	As above		NA		0.05
For once-through boilers						
1400 and above	0.05	NA		NA		

* Actual values within the range reflect the total dissolved solids in the feedwater. Higher values are for high solids; lower values are for low solids in the feedwater.

† Actual values within the range are directly proportional to the actual value of total dissolved solids of boiler water. Higher values are for the high solids; lower values are for low solids in the boiler water.

‡ Dictated by boiler-water treatment.

§ These values are exclusive of silica.

¶ Not in ABMA recommendations.

A Water for final use is filtered by passing it through fine strainers or other porous media to remove suspended solids mechanically. The degree of filtration depends on the fineness of the filtering media and the type of filter aid used, such as cartridge filters. Since fine solids, in the size range from 1 to 75 μm, are readily removed, it is important to know the size of the particles in the water.

Suspended solids are also removed by (1) settling, (2) coagulation, and (3) chlorination before the fine filtering takes place. Settling ponds take longer than do fabricated units; however, settling ponds or tanks are used to settle the heavier suspended solids in raw water. The term *clarification* is also used to denote the removal of suspended matter and even color from raw-water supplies.

Q What can assist clarification?

A When certain chemicals are added in the clarifier station, which then agglomerate suspended solids, the detention time for the settling of the impurities is considerably reduced. Chemicals called *coagulants* are added to the water, and the resultant *coagulation* clumps together the divided colloidal impurities in the water into masses which settle rapidly, and/or which can be more easily filtered out of the water. The most common coagulants used are ferric sulfate, ferric chloride, aluminum sulfate, and sodium aluminate. Another set of chemicals which

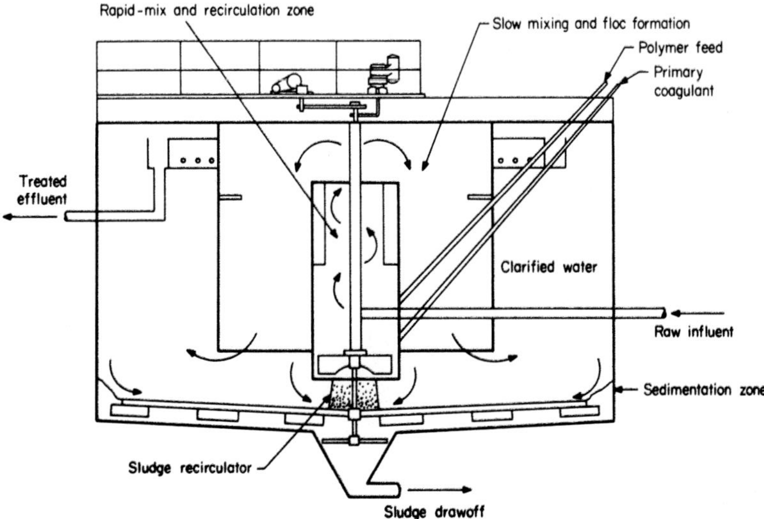

Fig. 14.7 Clarifier station uses coagulants to speed up suspended-solid removal from water. (*Courtesy of Power magazine.*)

promotes coagulation by chemical means are a group of polymers called *polyelectrolytes*. These interact with colloidal particles to form flocs—hence the term flocculants—and these form rather rapidly into larger flocs. They are particularly helpful in forming flocs in cold water. Figure 14.7 shows a clarifier station for removing suspended solids.

Q What is the purpose of chlorination in treating raw water?
A Chlorine is added to destroy any suspended organic matter in the water.

Q Describe the filtration method of fine-suspended-solid removal from water.
A Filtration places a porous barrier in the way of flowing water so that the suspended solid in the water adheres to the barrier. Granular media such as sand, gravel, and even crushed anthracite coal are used individually or in combination. Filtering can be performed by gravity flow or by pressure as shown in Fig. 14.8. Most commercial filters provide means, such as backwash, in order to remove the impurities and also reclaim the filtering medium.

DISSOLVED SOLIDS IN WATER

Q How are dissolved solids in raw water removed?
A There are various methods to remove dissolved solids, and the method to be used will depend on the pressure of the boiler system, process needs, and connected steam-using machinery requirements, to mention a few. Broadly speaking, the external treatment systems will or

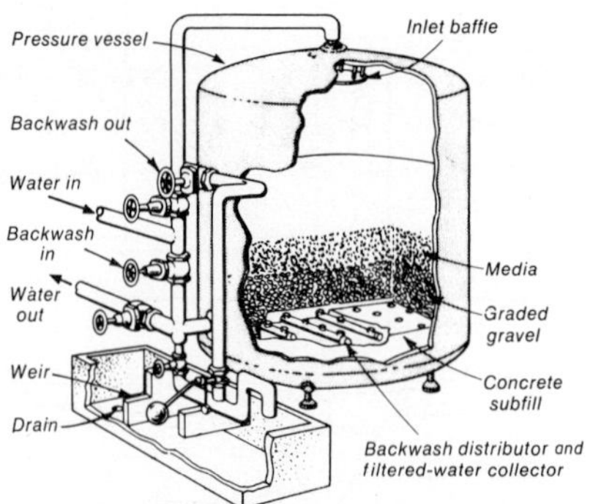

Fig. 14.8 Pressure filters are used to expedite fine-suspended-solid filtering. (*Courtesy of Power magazine.*)

can be: (1) lime soda or phosphate softening, (2) evaporation, or (3) ion-exchange processes such as hot-lime zeolite and demineralizers. Figure 14.9 provides a summary of external treatments that may be applied.

The table in Fig. 14.9 shows what to expect in hardness reduction when using the treatments shown. The quality of the makeup water needed depends mostly on boiler design, operating pressure, and temperature. The degree of treatment called for also depends on boiler capacity, quality of condensate returns available, and raw-water analysis. A large plant with a high percent of makeup usually justifies more elaborate external treatment than does a small installation.

Q Do evaporators using heat remove solids from feedwater?

A Yes. Dissolved and suspended solids are removed from water effectively by the purely thermal process of vaporizing water with heat. The evaporator is simply a pressure tank filled with steam coils and a moisture separator on the outlet. Vapor is condensed in a separate heat exchanger to form pure water. Heating steam flows inside the tubes, and makeup water enters the shell. The makeup water is vaporized and then condensed to be used in the boiler. The impurities remain in the evaporator and must be cleaned out periodically.

Q How does a deaerator reduce the oxygen content of feedwater?

A Oxygen can be a dissolved gas in water that doesn't react chemically with the water and becomes less and less soluble as the water temperature increases. Thus it is easily removed by bringing the water to the boiling point corresponding to its operating pressure. Pressure and vacuum designs are used. Deaerators are used to heat water for boiler feed. But if water is used for cooling or other purposes where heating is not needed, vacuum units are used. Steam deaerators break up water into a spray or film and then sweep the steam across and through it to force out dissolved gases like oxygen or carbon dioxide. The oxygen content can be reduced below 0.005 cm^3/liter, almost the limit of sample testing by chemical means. As carbon dioxide is removed, the increase in pH also gives an indication of deaeration efficiency. Figure 14.10 shows a typical deaerator.

Q What is meant by ion exchange in water treatment?

A Impurities that dissolve in water dissociate to form positively and negatively charged particles known as ions. These impurities, or compounds, are called electrolytes. The positive ions are named cations because they migrate to the negative electrode (cathode) in an electrolytic cell. Negative particles are then anions, since they are attracted to the anode. These ions exist throughout the solution and act almost independently. For example, magnesium sulfate ($MgSO_4$) dissociates in solution to form positive magnesium ions and negative sulfate ions.

Fig. 14.9 External treatments remove impurities from makeup water before they enter the boiler.

Method of treatment	Average analysis of treated water				
	Hardness, ppm as $CaCO_3$	Alkalinity, ppm as $CaCO_3$	CO_2 in steam (potential)	Dissolved solids	Silica
Cold lime-soda	30–85	40–100	Medium-high	Reduced	Reduced
Hot lime-soda	17–25	35–50	Medium-low	Reduced	Reduced
Hot lime-soda phosphate	1–3	35–50	Medium-low	Reduced	Reduced
Hot lime-zeolite	0–2	20–25	Low	Reduced	Reduced
Sodium-cation exchanger	0–2	Unchanged	Low to high	Unchanged	Unchanged
Anion dealkalizer	0–2	15–35	Low	Unchanged	Unchanged
Split-stream dealkalizer	0–2	10–30	Low	Reduced	Unchanged
Demineralizer	0–2	0–2	0–5 ppm	0–5 ppm	Below 0.15 ppm
Evaporator	0–2	0–2	0–5 ppm	0–5 ppm	Below 0.15 ppm

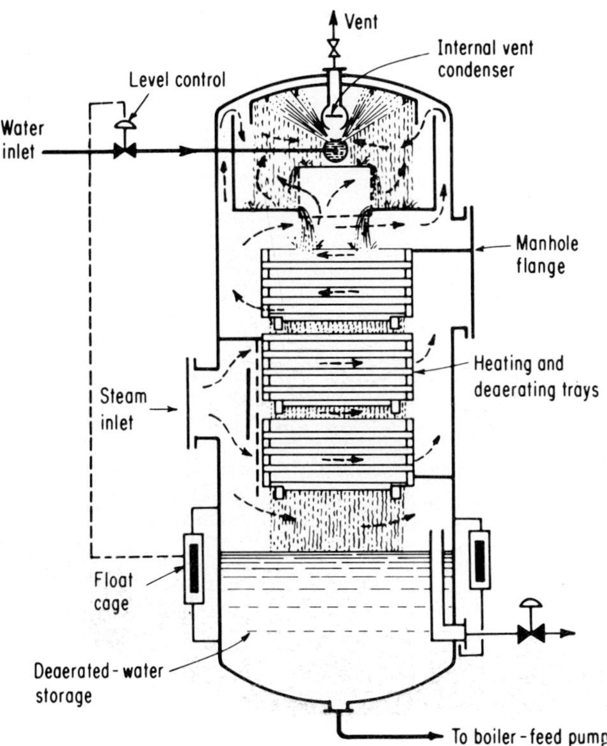

Fig. 14.10 Deaerator of the tray-spray type.

Ion-exchange material has the ability to exchange one ion for another, hold it temporarily in chemical combination, and give it up to a strong regenerating solution. The chart in Fig. 14.11 lists the capacity of the ion exchangers commonly used in water treatment.

Q Describe the cold, hot, and zeolite softening processes for water treatment.

A Four troublesome impurities in water—calcium, magnesium, silica, and oxygen—are removed by chemical reaction with combinations of lime, soda ash, and caustic. When carried out at room temperature, these reactions are termed a *cold process*. When the water is heated well above room temperature, it is called a *hot process*. The hot process is used for treating boiler feedwater. The water is heated by spraying it into the upper steam space of the softener unit. The inlet flow actuates a proportioning device to control the amount of lime and soda ash fed to the heating and mixing zone. Chemical reactions take place almost

Fig. 14.11 Ion-exchange materials in common use.

Ion-exchange materials	Flow rate, gpm/ft^2	Regeneration		Typical capacity, kg/ft^3
		Chemical	lb/ft^3	
Cation exchangers:				
Sodium cycle:				
Natural greensand	5.0	NaCl	1.25	2.8
Synthetic gel	6.0	NaCl	5.0	10.0
Sulfonated coal	6–8	NaCl	3.15	7.0
Styrene resin	8–10	NaCl	10.0	25.0
Hydrogen cycle:				
Sulfonated coal	6–8	H_2SO_4	2.0	8.0
Styrene resin	8–10	H_2SO_4	5.0	11.0
		H_2SO_4	11.0	25.0
		HCl	10.0	30.0
Anion exchangers:				
Weakly basic (aliphatic amine)	6.0	Na_2CO_3	4.2	18.0
Weakly basic (phenolic)	6.0	Na_2CO_3	4.2	18.0
Weakly basic (styrene)	6.0	NaOH	3.25	20.0
Strongly basic (type I)	6.0	NaOH	4.0	11.0
Strongly basic (type II)	6.0	NaOH	4.5	14.0

instantly. Sedimentation proceeds at a rapid pace because of the elevated temperature. The sludge-collecting cone at the bottom of the unit receives the precipitates, then periodically discharges them to the sewer. Clarified water leaves the settling tank and is fed to the anthracite filters for polishing.

Hot-process equipment can treat large flows in relatively small units. Chemical dosages are much lower than required by cold-process units. Treatment jobs handled by the hot-lime soda softener include softening by removing calcium and magnesium, silica removal with magnesium salts, oxygen removal, and removing turbidity.

Ion-exchange resins are used at high temperatures for the hot-lime zeolite method. Effluent hardness is zero compared with 27 ppm from a hot-lime-soda unit. The curve in Fig. 14.12*a* shows that 20 ppm excess soda ash is needed to produce this residual hardness. The resulting alkalinity is in the range of 45 to 50 ppm. Alkalinity from hot-lime-zeolite runs from 12 to 25 ppm. The curve in Fig. 14.12*b* shows the residual carbonate alkalinity and noncarbonate hardness from lime treatment. This residual hardness is removed by a hot zeolite. It is less costly from the chemical angle to remove this hardness with salt in a cation exchanger than with soda ash in the hot-process settling tank.

Q What are the basic types of ion-exchange techniques used in water treatment?

A The three most used are: (1) Water softening, or the removal of

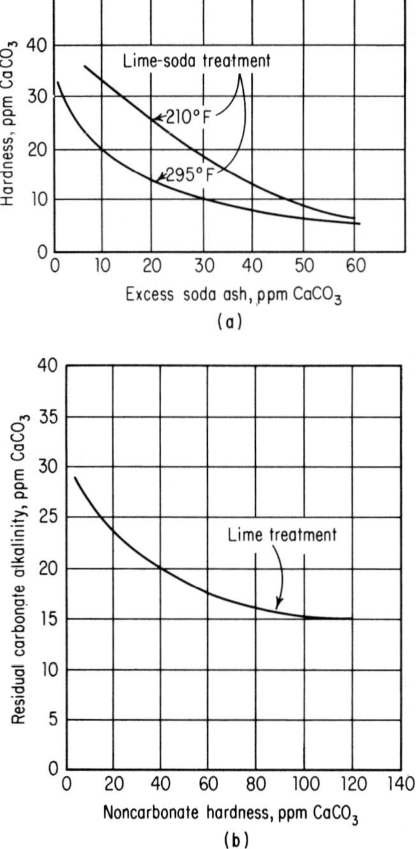

Fig. 14.12 (*a*) Residual hardness depends on excess carbonate that remains in treated water. (*b*) Lime treatment leaves a higher residual hardness, with alkalinity of 15 to 25 ppm.

calcium and magnesium, or hardness. Water softening is usually used for lower-pressure boilers, defined as those below 600 psi, in order to avoid scale formation on heat-transfer surfaces. The softening of water is accomplished with a cation exchanger, usually the sodium form of a strongly acidic, sulfonic-acid cation-exchange resin. (2) Dealkalization, or the removal of bicarbonates and carbonates. Usually at pressures above 900 psi, and depending on the water available, the water is dealkalized as well as softened in order to eliminate any corrosion possibilities due to the liberation of CO_2 formed by the thermal decomposition of bicarbonates. (3) Deionization, or demineralization to remove all electrolytes. At pressures above 900 psi, dissolved solids, CO_2, and silica should be removed to eliminate the possibility of scaling and to prevent carry-over. For demineralizers, both a cation- and an anion-exchange resin are required.

Figure 14.13 shows some of the impurities removed by ion-exchange methods.

Q How are ion exchangers revitalized or regenerated?

A Ion-exchange materials have only a limited capacity for removing ions from water. When their capacity is used up, they are regenerated. This is essentially reversing the ion-exchange process. In the case of cation exchangers operating on the sodium cycle, salt (NaCl) is added to replenish the sodium capacity or acid is added to replenish the hydrogen capacity. Anion exchangers are usually regenerated with caustic (NaOH) or ammonium hydroxide (NH_4OH) to replenish the hydroxide ions. Soda ash (Na_2CO_3) and salt (NaCl) are also used in some cases.

Regeneration involves taking the unit off the line and treating it with a concentrated solution of the regenerant. The ion-exchange material then gives up the ions previously removed from the water, and these ions are rinsed out of the unit. The ion-exchange material is then ready for further service.

Q Explain how a demineralizer works.

A In a demineralizer, water to be treated is passed through both cation- and anion-exchange materials, with the cation-exchange process operating on a hydrogen cycle while the anion exchanger operates on the hydroxide cycle, which replaces hydroxides for all the anions. The final effluent from this process consists of hydrogen ions, hydroxide ions, or water. Among the demineralized forms are the mixed-bed process where the anion- and cation-exchange materials are intimately mixed in one unit. Multibed arrangements use combinations of cation-exchange beds, weak- and strong-based anion-exchange beds, and degasifiers. Figure 14.14 shows a typical water-treatment flow for a multibed arrangement.

Q How can the degradation of an ion exchanger be recognized?

A The following must be watched for signs of degradation.

1. Flow short-circuiting by resin fines and particulates: backwash should be applied thoroughly to loosen bed and remove solids. Consider short runs.

2. Broken or blocked distributor/collector: remove resin to check.

3. Resin contamination due to valve leak: sample water both upstream and downstream; if in doubt, replace valve.

Fig. 14.13 Ion-exchange methods for removing various boiler-water impurities.

Impurity	Composition	Treatment method
Alkalinity	Bicarbonate HCO_3, carbonate CO_3, hydrate OH	Lime-soda softening, hydrogen-cycle cation exchange, chloride anion exchange, chemical neutralization
Ammonia	NH_3	Hydrogen-cycle cation exchange, deaeration, chlorination
Carbon dioxide	CO_2	Aeration, deaeration, chemical neutralization, strongly basic anion exchange
Chloride	Cl^-	Demineralization
Conductivity	Ionized solids	Demineralization
Fluoride	F^-	Alum coagulation, anion exchange
Free mineral acidity	H_2SO_4, HCl, HNO_3, etc.	Anion exchange, chemical neutralization
Hardness	Calcium magnesium as $CaCO_3$	Lime-soda softening, hot-phosphate softening, sodium-cycle cation exchange, hydrogen-cycle cation exchange
Hydrogen sulfide	H_2S	Aeration, strongly basic anion exchange
Iron	Fe^{2+} or Fe^{3+}	Aeration, coagulation, lime softening, cation exchange
Manganese	Mn^{2+}	Aeration, coagulation, lime softening, cation exchange
Nitrate	NO_3^-	Demineralization, biological treatment
Oxygen	O_2	Deaeration
pH	Hydrogen ions	Chemical additives, ion exchange
Silica	SiO_2	Hot-process softening (with magnesium), strongly basic anion exchange
Sulfate	SO_4^-	Demineralization
Turbidity	Suspended solids	Coagulation, sedimentation, filtration

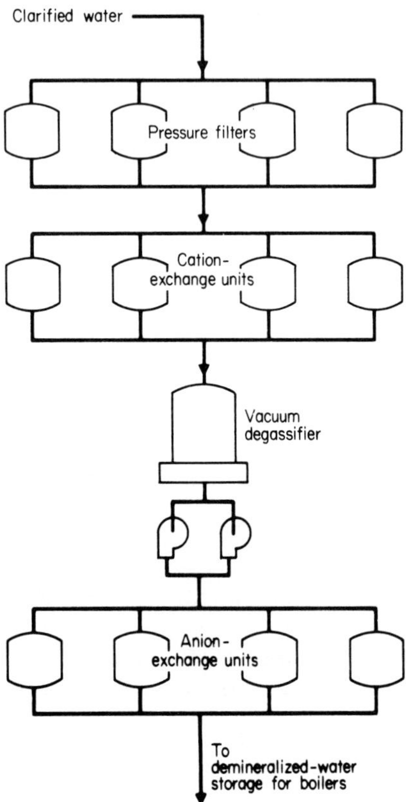

Fig. 14.14 Multibed demineralizer is used to prepare boiler water. Degasifier removes dissolved O_2 and reduces CO_2.

4. Drop in water temperature: cut service-flow rate to compensate.

5. Resin fouled with iron: analyze adequacy of pretreatment, consider using inhibited hydrochloric acid.

6. Increase in dissolved-solids level: consider shorter runs.

7. Channeling due to air blockage: check pressurization; backwash thoroughly.

8. Oil fouling: check pump seals; wash with nonionic detergent.

Anion exchangers:

9. Resin fouled by organics: treat with 10 percent brine.

10. Resin fouled by silica: regenerate with double dose of caustic (possibly at 120°F, with resin bed preheated).

11. Poisoned or degraded resin: check pretreatment: consider replacement with more stable resin.

12. Cation leakage: check regeneration procedure, water analysis.

Cation exchangers:

13. Fouling by calcium sulfate: backwash thoroughly and double regenerant dose. Check regenerant temperature, ratio of calcium to total cations to determine if lower acid concentration is dictated.

14. Resin decross linkage: check water for excess chlorine content; consider replacing bed with more highly cross-linked resin.

CONDENSATE POLISHING

Q Why must condensate systems receive attention in controlling boiler-water contaminants?

A Condensate that is recycled into the boiler feed system may have various impurities that can affect steam generators by corrosion and by depositing out metallic oxides in the steam generator and connected steam-using machinery. The most prevalent reason for corrosion of metal parts is the presence of oxygen and carbonic acid in the condensate. Oxygen attack will be in the form of pitting, while carbonic acid attack is characterized by uniform loss of metal. Carbonic acid is formed when carbon dioxide dissolves in water. The carbon dioxide is formed by the breakdown of bicarbonates and carbonates in the boiler.

Q What treatment or condensate polishing is employed to control impurities in condensate return systems?

A This will be determined by boiler pressure, connected equipment, and process applications. Polishing practices may use treatment by chemical means, oxygen scavengers, deaerators, ion-exchange systems, and even filters to remove suspended metallic oxides in condensate returns.

Q How is oxygen attack usually controlled in condensate systems?

A Deaerators are effective in removing oxygen in the feedwater. Sodium sulfite is used to remove the oxygen by chemical means. Neutralizing amines are added to maintain the condensate pH at about 8.5. Some professional water-treatment concerns also use filming amines on condensate systems. These amines provide a protective barrier between the metal condensate piping and the oxygen in the condensate.

Q How is carbonic acid attack controlled?

A Carbonic acid control in condensate systems is achieved by dealkalizing the condensate in order to remove as much as possible of the bicarbonates or by neutralizing the carbonic acid formed in the condensate. Amines are generally used to control carbonic acid attacks.

Q How does the term *condensate polisher* apply to condensate treatment?
A Metallic oxides formed as corrosion products must be removed. Condensate polishers are used to physically remove suspended metallic oxides from the condensate. Resin-type filters are used. The resin acts as both a filter and an ion-exchange medium. Electromagnetic filters have also been used to remove iron oxides from the condensate.

Q How do electromagnetic filters remove iron oxides?
A Electromagnetic filters usually consist of cylinders filled with iron balls. Strong magnetic fields are set up in the spaces between the iron balls when an electric current is applied. Ferromagnetic impurities are attracted to the region with the strongest magnetic fields. Cleaning or removing of the impurities can be accomplished by demagnetizing the matrix and backwashing the filter.

INTERNAL TREATMENT

Q Define the term *internal treatment.*
A Internal treatment of boiler water is by chemical means and usually complements external treatment by adjustments of the entering feedwater for hardness, oxygen, silica, acidity, and similar contamination to final desired boiler-water permissible concentrations. The treatment is applied to operating boilers by means of special chemical feed pumps.

Q Explain what water problems are prevented by a good internal-treatment program.
A The purpose of internal treatment is:

1. For the chemicals to react with any feedwater hardness introduced into the boiler and prevent it from precipitating on boiler metal parts as scale.
2. To condition suspended matter in the boiler water such as sludge or iron oxide with chemicals so that it does not adhere to metal.
3. To prevent foam carry-over by providing antifoam protection.
4. To eliminate oxygen from the water and also to provide enough alkalinity to prevent corrosion.

Q What chemicals are used to prevent some of the above conditions?
A 1. To remove hardness and soften the water: soda ash, caustic, and various types of sodium phosphates.

2. To prevent foam carry-over, certain organic materials are used as antifoam agents.

3. To condition sludge, various organic materials are also used, such as tannin, lignin, or alginates.

4. To eliminate or scavenge oxygen, sodium sulfite and hydrazine are used. To prevent corrosion, various combinations of polyphosphates and organics are used. These also prevent scale in feedwater systems. For preventing condensate source corrosion, volatile neutralizing amines and filming inhibitors are employed.

Q What are organic materials?

A Organic materials such as tannins, lignins, starches, and seaweed derivatives are also used to keep boiler sludge fluid. Organic colloids such as sodium mannuronate and sodium alginate react with calcium and magnesium salts to form a floc that entangles precipitates.

Q Define the term *precision phosphate control*, *coordinated phosphate control*, and *congruent control* as applied to high-pressure boilers.

A Low-pressure boiler phosphate-treatment programs were originally extended to high-pressure boilers to prevent scale. It was soon discovered that hardness passing through external treatment apparatus was relatively high in magnesium content, and magnesium phosphate was formed. To prevent the formation of this insoluble compound and to keep silica from volatilizing, a system of treatment termed precision control was developed. The following summarizes phosphate-treatment development.

Precision Phosphate Control Precision phosphate control is based on maintaining 2 to 4 ppm of phosphate and 15 to 50 ppm of hydrate alkalinity in the boiler. The phosphate residual is maintained to prevent calcium scale, and the hydrate alkalinity is present to render boiler sludge less adherent and to minimize silica volatilization. With time, use of precision control has been found unsatisfactory in many high-pressure boilers because of possible high alkalinity and resultant caustic gouging. The free caustic formed in precision control systems becomes concentrated in films and under deposits, which eventually causes caustic gouging of metal.

Coordinated Phosphate-pH Control A treatment termed coordinated phosphate-pH control has been developed to prevent caustic gouging. In this system, free caustic in the boilers is eliminated by maintaining an equilibrium between sodium and phosphate. This equilibrium is shown on the curve in Fig. 14.15, which relates pH at 25°C to phosphate residual. The reactions involved are:

$$NaH_2PO_4 + 2NaOH \rightarrow Na_3PO_4 + 2H_2O \quad (1)$$

$$Na_2HPO_4 + NaOH \rightarrow Na_3PO_4 + H_2O \quad (2)$$

Coordinated control is based on maintaining a ratio of 3.0 Na/1.0 PO_4. If

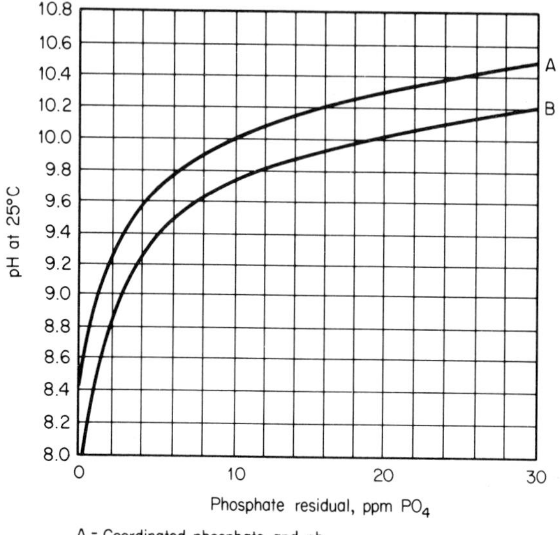

Fig. 14.15 Equilibrium curves between sodium and phosphate residual in boiler water for elimination of free caustic. Phosphate and/or caustic materials are added to remain below the curves. A = coordinated phosphate and pH; B = congruent. (*Courtesy of Power magazine.*)

various phosphate materials and/or caustic are added to the boiler, this ratio can be maintained and "free caustic" eliminated. The proper amount of control depends upon boiler pressure.

Congruent Control Although coordinated control has been developed to prevent caustic gouging, experience began to show that it is not always effective in this regard. A safer system termed congruent control has been developed. Congruent control is based on the same principles as coordinated control but is more restrictive. The equilibrium maintained in a congruent controlled system is shown in Fig. 14.15. This equilibrium is based on maintaining a ratio of 2.6 Na/1.0 PO_4, instead of 3.0 Na/1.0 PO_4.

Effect of Makeup Water The quality of makeup water that is used in a high-pressure operation is also affected by the pretreatment system employed, and the water to be introduced to the boiler is generally consistent. Because of this, it is possible to accurately determine both the type and ratio of various phosphates and/or caustic needed to maintain control within the desired limits. Once the proper ratio has been determined, adjustments may be required only because of variations in operations or load. These adjustments should be based on persistent, conscientious chemical testing at all times for either coordinated phosphate, pH, or congruent control. Operators should be alert to the need for steady, periodic testing and control to avoid problems due to swinging loads.

Q Why is hydrazine used for oxygen control on high-pressure boilers?
A Hydrazine (N_2H_4) is the oxygen scavenger most often used for high-pressure boilers. Its advantages over sulfite, which is used in most low-pressure units, are that it adds no dissolved solids to the boiler. Hydrazine scavenges oxygen according to the following reaction:

$$N_2H_4 + O_2 \rightarrow N_2 + 2H_2O$$

A major disadvantage to the use of sulfite in high-pressure boilers is that it may break down to corrosive H_2S and SO_2. The following reactions are involved:

$$Na_2SO_3 + H_2O + \text{heat} \rightarrow 2NaOH + SO_2 \quad (1)$$
$$4Na_2SO_3 + 2H_2O + \text{heat} \rightarrow 3Na_2SO_4 + 2NaOH + H_2S \quad (2)$$

Hydrazine also has an advantage in that it promotes the formation of protective metal oxides on metal surfaces. In the case of iron surfaces, hydrazine reduces ferric oxide (rust) to magnetite in the following reaction:

$$6Fe_2O_3 + N_2H_4 \rightarrow 4Fe_3O_4 + N_2 + 2H_2O$$

The formation of magnetite stifles the oxidation of the metal surface. In the case of copper, hydrazine reduces cupric oxide to the protective stable cuprous oxide:

$$N_2H_4 + 4CuO \rightarrow 2Cu_2O + 2H_2O + N_2$$

A hydrazine residual of 0.01 to 0.1 ppm is usually maintained in the feedwater. Because the hydrazine breakdown to ammonia (NH_3) begins at approximately 400°F, it is difficult to keep higher residuals at high pressures. The breakdown to ammonia results in a lower hydrazine retention time in the boiler and thus a reduction in the time available for all the hydrazine reactions.

A catalyst which accelerates the reduction reactions but not the degradation reaction offers a solution to the problem of low retention time. An organic catalyst has been developed. It provides the desired acceleration of reduction reactions without the disadvantages associated with the heavy-metal catalysts. Figure 14.16 shows the rate of oxygen removal of a catalyzed hydrazine compared with that of normal hydrazine.

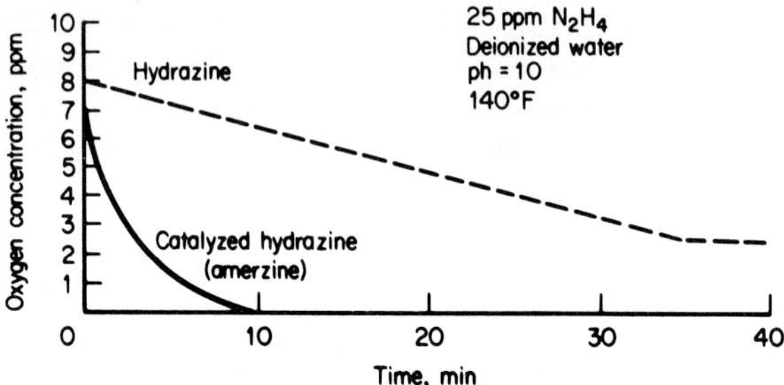

Fig. 14.16 Oxygen removal by N_2H_4 (hydrazine). Catalyzed hydrazine reacts much more quickly and therefore more completely with oxygen than normal hydrazine. This offers a solution to the problem of low retention time. Curves shown are for a catalyzed hydrazine, Amerzine, marketed by the Drew Co. (*Courtesy of Power magazine.*)

Q Define volatile treatment.

A Another method of treatment is termed *volatile treatment.* This system is based on the use of hydrazine and neutralizing amines or ammonia. Major advantages claimed of volatile treatment are that it adds no solids to the boiler and it affords good preboiler protection. Boiler pH is controlled between 8.5 and 9.0, and the hydrazine residual in the feedwater is controlled between 0.01 and 0.10 ppm.

Q What is the purpose in using chelants in water treatment?

A For boilers operating under 1000 psi, chelants have been used to control ion oxide deposits in boiler systems. The chelants react with metal ions to form soluble compounds which can then be flushed with the feedwater. The degree of combining is dependent on the system environment and the reactivity of the chelate to the metal ion that may be encountered. The chelate agents' reaction is considered to be applicable for soluble metal ions, and not for insoluble metals. For the latter, dispersant polymer agents have been developed. The polymer is adsorbed onto the iron or insoluble metal oxide, thus controlling insoluble oxide deposits by altering their charge characteristics and preventing agglomeration. When used this way, the polymers are called *dispersants.*

Q How is pH controlled on high-pressure boilers?

A To reduce the corrosion of preboiler equipment such as feedwater heaters, feed pumps, and feed lines, feedwater is treated to control the pH level and the amount of free carbon dioxide and oxygen gases.

Maintaining the condensate pH between 8.0 and 9.5 is generally accomplished by one of two methods. The first is the addition of neutralizing amines (ammonia, morpholine, cyclohexylamine, and hydrazine) which neutralize the acids present in the condensate. The alternative method, the injection of filming amines, is used in situations where high carbon dioxide content in the steam is causing increased corrosion rates. The filming amines are a waxlike substance which coats the tube walls, protecting them from the corrosive condensate. Their major drawback is that control of the amine feed rate is critical to ensure coating without causing flow restrictions.

Q What is the difference between erosion and corrosion as it affects wearing of boiler parts?
A Corrosion is an electrochemical attack, whereas erosion is a mechanical action causing wear by abrasion. Improperly adjusted soot blowers, for example, can cause high-velocity steam to cut or abrade a tube. Dirty or gritty fly ash traveling through a boiler at high speed can cause tube thickness to be reduced to dangerous levels by this sandblasting effect.

Q Define the term *chemical precipitation.*
A As chemicals are added in internal treatment to react with dissolved minerals in the water, the products formed are insoluble in the water and deposit out as precipitates. The soda ash softening process is an example, as is the lime process where precipitation takes place.

Lime Reaction Calcium hydroxide or hydrated lime reacts with soluble calcium and magnesium bicarbonates to form insoluble precipitates.

Soda Ash Reaction Soda ash reacts with calcium sulfate and calcium chloride to form calcium carbonate, a sludge out of solution.

BLOWDOWN

Q What is the purpose of blowdown?
A Sludge, concentrated dissolved and suspended solids, can accumulate in the boiler from natural water and as a result of water treatment. Blowdown or draining the bottom water from a drum or header is required in order to remove these deposited sludges from the boiler. By regulating the amount of blowdown, the amount of solids suspended in the boiler can be controlled.

Q How is blowdown rate determined?
A Generally by the desired or permissible boiler-water concentration, which depends on the pressure of the system, alkalinity, silica, and iron oxide limits that are or must be maintained.

The stairstep curve in Fig. 14.17 indicates the total boiler-water concentrations vs. the drum pressure, according to the American Boiler Manufacturers Association (ABMA). To use, assume a 650 psi operating pressure and 85 ppm feedwater concentration. To find the permissible boiler concentration and percent blowdown required, extend the drum pressure (dashed line, 650 psi) horizontally to the *pressure-vs.-concentration* transfer line (stairstep curve). Now read vertically down to *boiler-water concentration*, which is 2000 ppm. Extend the feedwater concentration horizontally to boiler-water concentration; then read the percent blowdown required on the diagonal lines, which is 4¼ percent.

> **CAUTION:** This curve represents the total concentrations and does not designate the chemical constituents or physical properties of the solids. Thus do not assume that adhering to these concentrations will result in trouble-free boiler operation.

Q What is intermittent blowdown, and why is it necessary?

A Intermittent blowdown is taken from the bottom of the mud drum, waterwall headers, or lowest point in the circulation system. The blowoff valve is opened manually to remove accumulated sludge, about

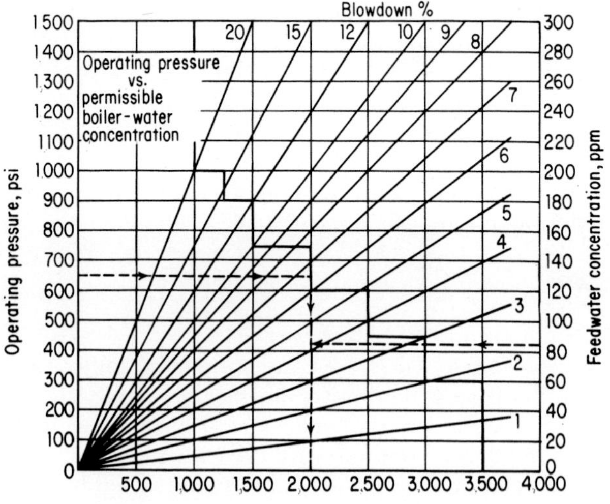

Fig. 14.17 Blowdown percentage is determined from ABMA recommended boiler-water concentration limits and boiler pressure.

every 4 to 8 h, or when the boiler is idle or on a low-steaming rate. But hot water is wasted, and control of concentrations is irregular.

Q Why is the continuous surface blowdown method used?
A Continuous surface blowdown automatically keeps the boiler water within desired limits. Continuously removing a small stream of boiler water keeps the concentration relatively constant. Savings by transferring heat in the blowdown to incoming makeup (Fig. 14.18) often pay for the investment.

Q What impurity in water requires critical attention on very HP boilers?
A Silica, which in HP water volatilizes and passes over with steam to the turbine. As the steam expands and drops in pressure and temperature, the silica condenses and deposits on the turbine blades, cutting the machine's efficiency. Vaporous carry-over involving the volatilization of matter (hot gases) increases most rapidly around 2000 psi. Above this range, mechanical carry-over predominates, especially above the critical pressure (3206.2 psi), as there is no longer a liquid-to-steam phase. Steam quality on these units is no longer measured by a calorimeter in

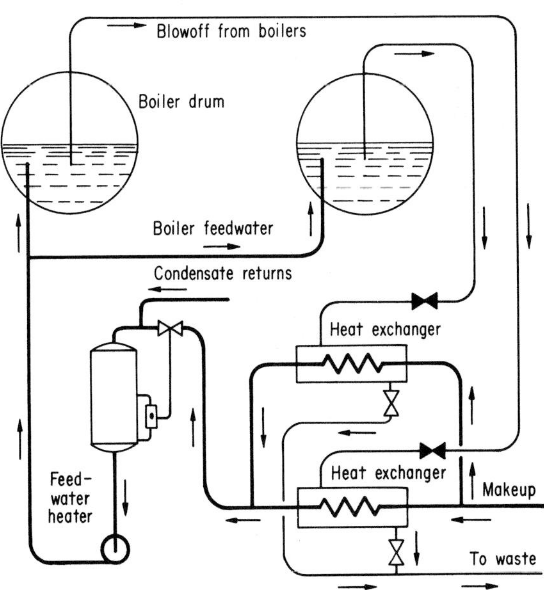

Fig. 14.18 Continuous blowdown heat is removed and recovered in feedwater heat exchangers.

percent of moisture but in parts per billion by the use of flame photometers.

Q What precautions must be taken when starting a systematic water-treatment program for a boiler that has long been neglected?
A When steaming for a few days after initiating the new chemical treatment, if possible shut down the boiler for an internal inspection. Often the new chemical treatment loosens up so much scale after the first few days, especially in HRT and SM boilers, that the scale piles up on the lower shell plates and the lower row of tubes. Then the boiler plate and tubes may bag from overheating. Scale may loosen too rapidly for removal by the usual blowdowns.

Q Is water treatment being automated?
A Yes. Today there are very sophisticated automatic controls on the market. They should be investigated and used where possible.

WATER-TREATMENT TESTING

Q What are some common water-treatment tests?
A Titration, colorimetric, hardness, chloride, alkalinity titration, etc. Complete instructions on water testing are contained in (1) the manual on industrial water published by the American Society for Testing and Materials (ASTM) and (2) *Standard Methods for the Examination of Water, Sewerage, and Industrial Wastes*, published jointly by the American Public Health Association and the American Water Works Association.

Q What is the purpose of an electrical conductivity test of boiler water?
A To measure the extent to which dissolved substances are concentrated in the boiler water. This test then helps in controlling carry-over of dissolved solids, which condense in lines or equipment such as turbine blades, into the steam system.

Q How is boiler-water analysis made by ion chromatography?
A Ion chromatography, first introduced in 1975 by Small et al. of the Dow Chemical Company, can also be used for boiler-water analysis. This technique is based on an ion-exchange separation followed by eluant suppression and measurement of effluent conductivity. It is used for rapid determinations of ions with ionization constants greater than 1×10^{-7}, such as chloride, sulfate, phosphate, calcium, and ammonium. Applications have been developed for trace determinations in steam condensate, feedwater, boiler blowdown, reactor coolant, radioactive waste, stack-gas scrubbers (see Fig. 14.19).

For anion analysis, the eluant (usually a mixture of dilute solutions

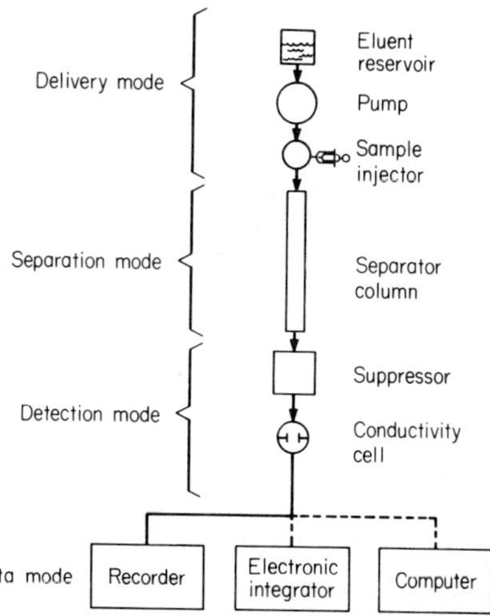

Fig. 14.19 Typical Dionex Ion Chromatograph configuration. Ion chromatography is used to analyze the ions in a water sample. A liquid sample is introduced at the top of the analytical column. An eluant is pumped through the Ion Chromatograph and causes the ionic species to move through the column(s) at rates determined by their affinity for the resin bed. The differential rate of migration separates the ions into discrete bands. The detection is accomplished via suppressed conductivity, electrochemical, ultraviolet, or fluorescence. The identification and quantitation is done by comparing retention times and peak heights of the sample with those of a standard solution. Data reduction may be done manually or by using electronic integration. (*Courtesy of Dionex Corp.*)

of sodium carbonate and sodium bicarbonate), used in a separator column containing an anion resin in the HCO_3^-/CO_3^- form, causes the sample ions to split into distinct physical bands. The retention time of each anion band is governed by the affinity of that ion, the eluant used, its flow rate, and the length of column. The effluent from the separator column is then fed to a resin-filled suppressor column in which two reactions occur.

The resin, which is highly acidic, exchanges hydrogen ions for sodium ions, forming carbonic acid. The analysis of cations utilizes the same physical design. However, the separator column contains a cation resin, and the suppressor column contains an anion resin in hydroxide form. The acidic eluant is neutralized by the suppressor resin, and the cations are detected as hydroxides.

Since electrical conductivity is a function of ionic concentration, conductivity is measured and the value obtained is converted into concentration units.

Ion chromatography is a powerful analytical tool capable of characterization and quantitization of chemical ions in water samples found in power generation. The advantages of this technique are (1) high sensitivity, (2) freedom from matrix interferences, (3) specificity in analyzing similar types of ions, (4) multiple ion determinations in a single chromatographic run, and (5) time efficiency.

Q What is the purpose of a sodium analyzer?

A Sodium analyzers are installed to measure steam purity going to a turbogenerator. Presence of sodium in steam may cause stress-corrosion cracking. The sodium monitor can detect to ppb levels sodium contamination of the steam. Sodium monitors are also useful, because the amount of sodium detected is also an indication of the amount of dissolved solids being transmitted by the steam. Water treatment can then be adjusted to reduce the amount of dissolved solids to prescribed limits for the plant.

Q What is an effective way of checking on the results of a water-treatment program?

Q Water treatment can be effectively measured by the final outcome, namely, a clean boiler free of scale, pitting, gouging, etc., and no carry-over problems. Thus, internal inspections and sometimes cutting out sections of tubes are absolutely necessary in order to check on the soundness of a water-treatment program. This is even more important when treatment changes are made.

High-pressure WT boilers (in general, those operating at over 600 psig, but in particular those operating at or over 850 psig) are susceptible to many of the same water-treatment problems as lower-pressure units but, in addition, have special problems of their own. Improper or poorly controlled water treatment can result in tube overheating, fluid-film concentration, and those problems unique to high-pressure operation, hydrogen damage and silica volatilization (vaporization of silica with steam). Water chemistry and treatment of boiler water is a job for specialists experienced in correcting problems. However, by careful observation, an experienced operator can assist specialists by calling problems to their attention, so that they can apply the water-treatment service that may be needed to correct the conditions noted during inspections or during a failure investigation.

CHAPTER **FIFTEEN**

Integrating Boiler-Turbine Operations

Steam-using machinery can be immediately affected by boiler problems, and boiler operation can be affected by machines using steam, today mostly turbines, if these machines are tripped suddenly, experience heavy load swings, are cycled on and off, and have auxiliaries such as feedwater heaters that use turbine extraction steam. This chapter will review some of the problems in integrating turbine-boiler operation, with the main emphasis being on generating plants.

Q Explain modular boiler-turbine arrangements and integrated steam and generating systems.

A Figure 15.1 shows a modular boiler-turbogenerator system. Note that each boiler supplies its own turbogenerator. There is no interconnection from one boiler to the other. If a turbogenerator has to be shut down, the connected boiler cannot supply steam to the other turbogenerators. While the plant has three boilers and three turbogenerators, they are arranged to operate as separate units, the same as if they were in another plant.

In contrast, Fig. 15.2 shows an integrated steam system involving three boilers generating 1250 psi that supplies steam to a 15,000-kW turbogenerator, and five boilers that produce 600-psi steam for two generators and also seven mechanical-drive turbines. Process steam at 160 psi is obtained from the exhaust steam of the mechanical-drive turbines and also from the extraction lines from the 600-psi turbogenerators. Note also that the six recovery boilers help to burn black liquor and recover chemicals used in the digesters, besides producing steam for power and process. The need for close coordination between boiler operation and steam turbogenerators should be apparent, as should the effect to the plant process if there are any major disturbances on boiler-associated auxiliaries and the equipment using the steam. The diagram shown is for a paper mill. The paper industry, as is indicated, has

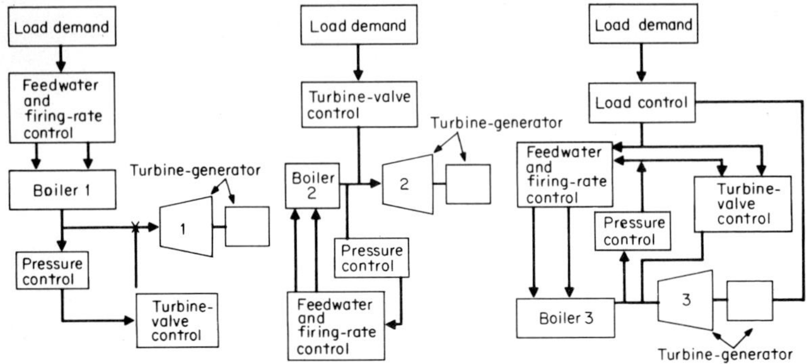

Fig. 15.1 Modular boiler-turbogenerator arrangements permit independent operation from each other.

practiced cogeneration, or the simultaneous generation of electricity and production of process steam, from the beginning of electric generation. The process steam is obtained as a by-product of electricity generation and thus is generally economically justifiable.

TURBINE AND BOILER THERMAL CYCLING

Q Define the term low-cycle fatigue.

A Low-cycle or thermal fatigue is primarily due to uneven heating of turbine parts during the time when rapid temperature changes are occurring on the unit. These may occur from rapid cold start-up or shutdown cycles. Low-cycle fatigue is also affected by the geometric layout of turbine components and the selection of material in the original design such as yield point, ductility, corrosion resistance, and similar material properties.

Q What turbine components are most affected by rapid thermal cycling?

A Plastic strain may occur on the massive, thick sections of the turbine, where the parts are exposed to hot incoming steam. These would include valve bodies (Fig. 15.3), high-pressure and reheat turbine shells or casings, and the corresponding rotors in these shells.

Q How does peaking or cycling service affect the larger utility-type turbines?

A Most machines in the past were designed primarily for base-load service or steady operation. As the units become older and are replaced by more efficient units, the older units are delegated to "backup"

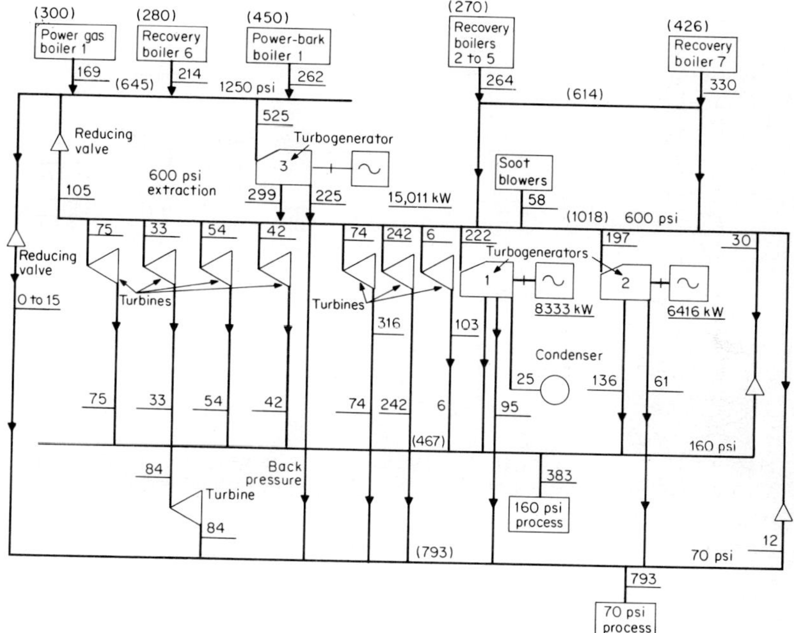

Fig. 15.2 Power-plant steam balance. An industrial power plant has an integrated boiler-turbogenerator and process steam flow arrangement. Numbers in parentheses indicate rating in thousands of pounds per hour. Numbers without parentheses indicate design flow in thousands of pounds per hour.

peaking or cyclic service by owners. However, this type of service can create cyclic-type failures, mostly from thermally induced up and down transient stress conditions. For example, shell cracking, steam chests, and casing warping incidents have occurred because of severe, repetitive thermal stresses on old base-load units that were converted to peaking units. In addition, rotor and bore surface cracks due to temperature swing have occurred to rotors by repetitive stresses caused by improper start-up procedures, heavy load swings, and similar operating practices that can make a machine go through too rapid temperature swings.

Cracking of large-capacity rotors due to the large diameters of the new larger units occurs because of the temperature gradient that exists between the surface of a rotor and the bore. Cracking of shells and steam chests is primarily due to the change from a heavy casting to a thinner casting with insufficient radius provided in between to even out the thermal gradient that is established. Severe restraints on heavy castings

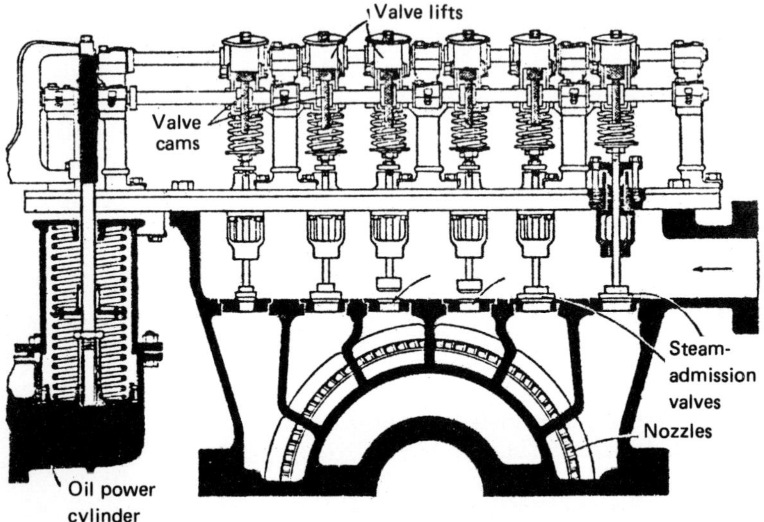

Fig. 15.3 Heavy steam chests that are integral to turbine shells, and as shown with cam lift steam-admission valves, may be exposed to unequal heating and cooling and thus may experience low-cycle fatigue cracks.

which restrict expansion and contraction between transition sections can also cause cracking.

Q Define ihe term *ramp rate.*
A Ramp :ate is applied to turbines during cold and hot starts. It is the degrees Fahrenheit per hour rise that metal surface temperatures are exposed to when bringing a turbine up to rated conditions of temperature and load.

Q How is the thermal rate of rise measured on turbine components?
A On larger machines, thermocouples are installed on the inner and outer surfaces of the shell to note temperature gradients that may exist between these two surfaces. The differential temperature is carefully detailed as respects permitted rate of rise and cooling by most turbine manufacturers. The rotor surface at the present state of instrument development is considered to be at the same temperature as the nearest shell surface being measured by thermocouples.

Q How can operators keep track of the effects of thermal cycling a turbine?
A Major turbine manufacturers can analyze the effect of thermal

cycling on their units. Generally, the analysis includes the following for rotor surface thermal cracking:

1. The amount of thermal fatigue damage produced per cycle, and hence the number of cycles to cracking of a turbine rotor, depends on the following:

a. The magnitude of the nominal thermal stresses produced

b. The magnitude of the stress (strain) concentration factors (e.g., the rotor surface geometry)

c. The low-cycle fatigue strength of the rotor material

2. Manufacturers can calculate the induced stresses due to temperature changes and can also calculate the number of stretch and contraction cycles needed to produce fatigue cracks from so-called low-cycle fatigue due to temperature swings.

3. The major turbine manufacturers provide turbine operators with life-cycle curves for their machines. These curves show the operator how much of the life of a machine is being expended due to heating and cooling rates and temperature swings. See Fig. 15.4. By utilizing these curves as a guide, operators can:

a. Select the proper temperature ramp rate for their machines.

b. Review the different past and future modes of operation of the machine and also review as well operation of other units in the station in order to minimize life expenditure from thermal swings on the machines. Turbine-cooling operations must also be controlled the same way as heating cycles. In addition to causing fatigue damage, cooling stresses the steam side of the turbine in tension. Cracks that may exist tend to open more and progress deeper into the metal thickness.

Close control of cycle life expenditure is more critical on old base-load units that are now used in cycling service, such as peaking. Repeated thermal stresses due to the high rate of start-up and shutdowns in addition to load swings can produce rotor surface cracks in a relatively short time. If these cracks go undetected for a considerable amount of time, further damage will continue as the cracks progress, and this can result in expensive repairs. Periodic NDT inspections should supplement good operating practices in limiting thermal low-cycle fatigue failures.

Q Which thermal transients can affect a boiler-turbine system the most?

A Thermal transients of the following type: (1) cold start, (2) hot start, and (3) major load swings. Each of these can cause thermal transient conditions. Figure 15.5 illustrates a machine's initial and reheat steam temperatures and the corresponding first-stage shell temperature with

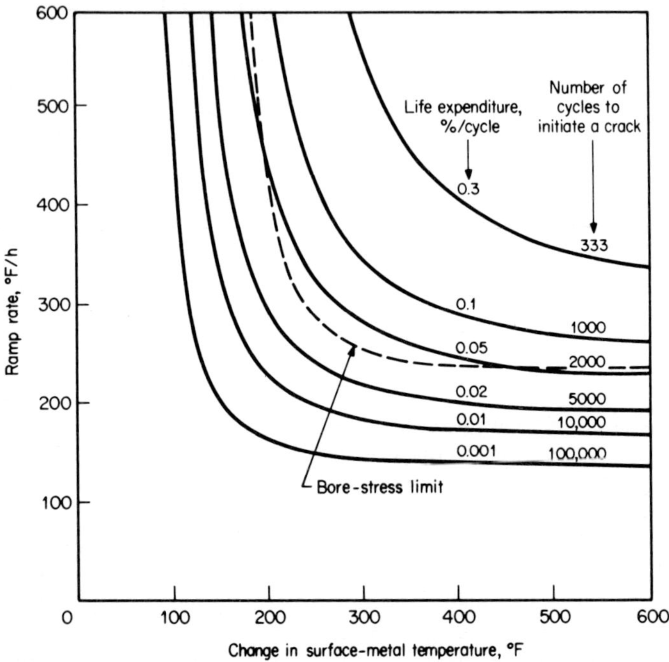

Fig. 15.4 Cyclic life expenditure curves show number of cycles to crack initiation for various surface-temperature changes and ramp rates. These are furnished by turbine manufacturers. The area above the dashed line shows bore-stress limits for this unit. (*Courtesy of Power magazine.*)

loading to 100 percent rated. Note there is a 300°F difference between inlet steam and shell temperature at 15 percent load, and as the machine is loaded, the gap narrows to 40°F. Figure 15.6 is a heating curve for shell, rotor, and bore from a cold start with corresponding rotor surface stresses and bore stresses. Note the large difference between the bore and surface stresses.

Q How can the mismatch that exists between turbine metal temperatures and the main steam-supply temperatures on start-up be avoided?

A Prewarming of the turbine with low-pressure steam while the rotor is turned slowly by a turning gear is the method usually employed on large turbogenerators to reduce the mismatch between turbine metal parts temperatures and the steam supply. This is applicable primarily to cold starts. To stay within the prescribed ramp-rate limits of the turbine manufacturer, the steam supplied to the turbine must be increased in

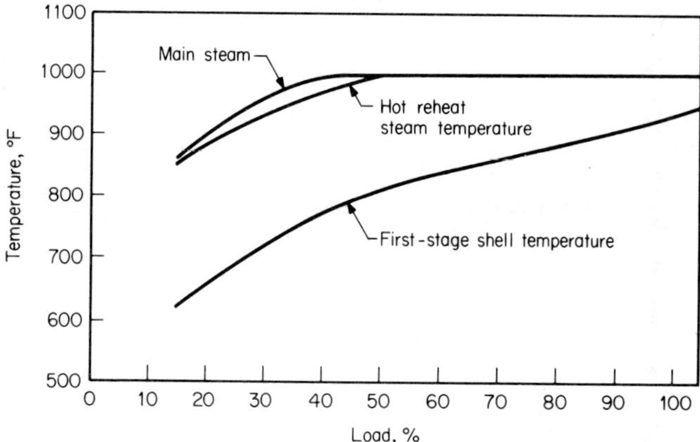

Fig. 15.5 Difference between steam and first-stage shell temperature on a large turbine cold start to 100 percent load. (*Courtesy of Westinghouse Electric Corp.*)

temperature at a controlled rate to reach operating steam temperatures. If the controlled rate is not maintained, the turbine cycling life will be used up more rapidly, and thermal induced cracks may develop in the turbine.

Q What is the usual temperature rate rise increase allowed on high-pressure drum-type boilers?

A For drum-type boilers, limitations on allowable rates of temperature increase are measured by the saturation temperature of the steam at pressure buildup. For the usual 2400-psi drum-type boiler used in utilities, the saturation temperature increase recommended by one prominent boiler manufacturer is 300°F per hour. Figure 15.7 shows the pressure stress and thermal stress produced on the drum from this rate of temperature rise. The alternating stresses resulting from this temperature rise, even when considering drum stress-concentration factors, produce an allowable cyclic life of about 18,000 cold starts without any thermal-gradient effects.

Q Where do most rotor surface cracks occur on turbines subject to thermal cycling?

A The most common locations are the surface areas with sudden geometric changes in the rotor surfaces, such as sharp shaft radiuses where shaft diameters change, in grooved areas for seals, and other similar areas where some form of stress concentration may exist.

Q What is the effect on a turbine's temperature and the boiler supplying it with steam if load is suddenly lost?

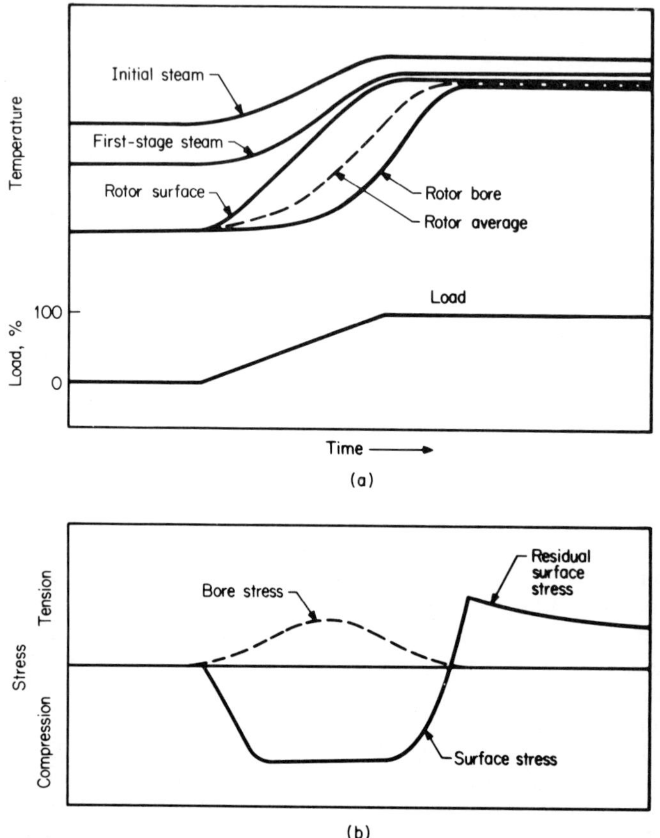

Fig. 15.6 Cold-start factors. (*a*) Steam and rotor bore temperatures; (*b*) rotor surface and bore stresses. (*Courtesy of Westinghouse Electric Corp.*)

A If a large utility-type turbine loses full load and remains in service carrying auxiliaries or just spinning with no load, it will be subjected to sudden, fairly large changes in temperature, particularly in the region of the high-pressure admission and first-stage shell. Three factors cause the changes—the temperature drop in throttling through the control valves, the characteristics of the partial arc control which most large turbines have, and the inevitable decrease in steam temperature at the greatly reduced flow. The first two of these factors result in increasingly large temperature changes at the higher rated pressures.

Figure 15.8 shows temperatures in the first-stage region as a function of load. This particular curve is for a four-valve, 3500-psi

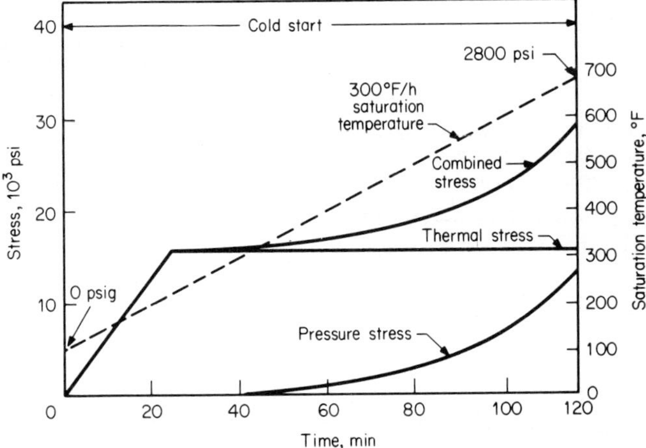

Fig. 15.7 Stress-combination chart for a boiler drum during cold start-up. (*Courtesy of Combustion Engineering, Inc.*)

turbine with a supercritical boiler that provides steam of constant temperature and pressure at all loads.

It should be noted that from full load to 10 percent load the first-stage steam temperature will change approximately 300°F. Since this occurs literally in seconds, the resulting thermal stresses in massive turbine parts can be quite high.

The stress may actually be considerably greater. The steam temperature from a modern boiler may drop more than 150°F at light load and thereby aggravate the situation.

The boiler firing rate must be reduced rapidly with loss of load to prevent pressure buildup and the operation of safety valves. If either gas

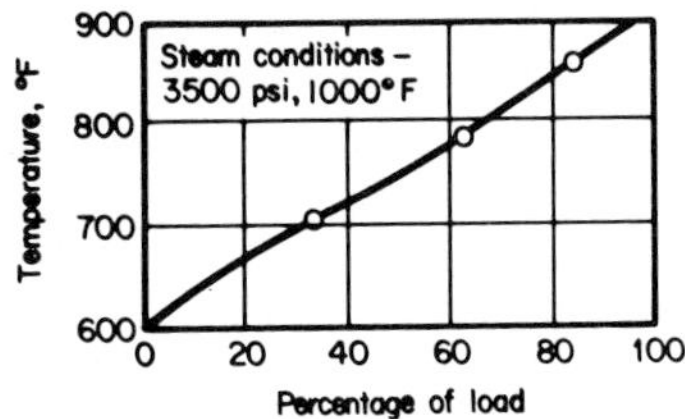

Fig. 15.8 Constant inlet or throttle pressure and temperature. First-stage temperatures can vary considerably with percentage of rated load carried by the turbine. (*Courtesy of General Electric Co.*)

or oil fuel is used, it may be possible to reduce the firing rate and maintain ignition. This is the most favorable course of action so far as the turbine is concerned because it will result in the smallest drop in steam temperatures. Where solid fuel is used, it is a common practice to trip the fires on loss of load. For drum boilers there will probably be enough stored energy to permit carrying auxiliaries while the furnace is purged for 5 min. Then the fire can be relighted. Operation without fires for 5 min will result in a larger drop in steam temperature but may be acceptable in an emergency.

Q Explain the difference between partial-arc and full-arc steam admission.

A Flow control of steam to a turbine is accomplished by area control or throttling control. In area control, also called *partial-arc* control or sequential valve control, the flow area of the steam to the turbine is adjusted to control the flow with minimum steam throttling. Steam is admitted at full pressure to a part of the arc of the first HP turbine stage. A control valve controls the steam flow to each segment of the arc. As the valves operate in sequence, at any point only one valve (or group of valves) is partially open, throttling, and consequent thermodynamic losses are only a part of the total flow.

Partial-arc turbines require a first control stage operating on the impulse principle which permits partial-arc operation under the influence of individual control valves. The remaining stages may use reaction blading. In *full arc control*, all the steam is throttled by inlet valves to give an admission pressure to the full annulus of the HP turbine which will pass the flow required. Full-arc admission turbines dispense with the control stage and instead can use all reaction stages with the inlet pressure being uniform around the arc of the first stage. In the absence of the control stage, more stages can be incorporated into the design of a given bearing span, and this can lead to some full-load efficiency advantages for this type of HP turbine.

Full-arc admission provides a more even warming of the HP turbine because of the admission all around the annulus of the turbine inlet. However, partial-arc design is capable of better efficiencies at part loads because of lower throttling losses. Modern partial-arc turbines use full arc for turbine warmup and include the control capability to change automatically from full to partial arc control, even when operating under load.

Q What is the advantage of variable-pressure operation in limiting thermal-stress effects on large turbogenerators?

A Variable-pressure operation is another way of reducing the change in first-stage shell temperature as the load is changed. If the main steam pressure is reduced as load is reduced, the advantage of decreasing the first-stage shell temperature change for a given load change can often be achieved without any sacrifice in heat rate. In fact, the light-load heat rate can often be actually improved by variable-pressure operation. Variable-pressure operation requires close coordination between boiler and turbine operation in order to achieve desired results.

Q Explain the advantage of having a stop-valve bypass system for turbines with integral steam chests and turbine shells.
A When the throttle valve is opened to admit steam to the steam chest, a large temperature gradient will exist between the hot steam chest and the cold turbine shell, because the control valves have not yet opened to admit steam to the turbine. Under these conditions, the stop valve is fully opened, admitting hot steam to the steam chest. The control valves are closed, so that no heating steam is admitted to the turbine shells. Even with the control valves slightly opened, the steam throttles to low temperature across the control valves so that the shell temperature remains lower than the chest temperature, and mismatch thermal stresses between the chest and the shell are produced. The installation of a stop-valve bypass will cause the steam throttling to occur at the stop valve, so that "cool" steam is admitted to both the chest and the shell, thus preventing large temperature mismatches between the chest and the shell. This is especially useful for peaking units where frequent start-stops are experienced on steam chests and shells. Older machines can have the steam-chest design modified to avoid large castings being connected to thin shells.

Q How may previously used base-load units be modified for peaking service in order to avoid low-cycle fatigue?
A 1. Continue to prewarm the high-temperature rotors on cold starts. The thermal stresses thus can be minimized by prewarming on turning gear, and then at two levels of speed before reaching synchronous speed. Another advantage of prewarming is that the rotor's metal temperature is raised above the transition temperature of some steels, and thus as the rotor is brought up in speed, the chance for brittle cracking is reduced.

2. Change to full-arc admission, which would reduce the metal temperature gradient around the annulus of the turbine inlet, or at least have it more even around the inlet shell portion of the turbine.

3. If the peaking unit is on a variable swinging load, use variable-pressure operation to help reduce the change in first-stage pressure as the load is changed.

4. Modify some parts of the turbine, such as heavy steam-chest castings in order to minimize thermal gradients.

Q How does partial-load operation affect the air heater and flue-gas cleanup equipment?

A Partial or cycling loads may produce flue-gas temperatures below the acid dew point that can affect the following boiler components in the flue-gas passages: (1) air-heater plugging and corrosion, (2) fabric filter-bag binding, (3) precipitator fouling and corrosion, (4) scrubber scaling. All these conditions require increased maintenance as a result.

Q How are turbine manufacturers assisting operators in preventing low-cycle fatigue failures from thermal effects?
A Major turbine manufacturers provide their new machines with instruments, monitoring devices, and controllers in order to limit cyclic stresses from thermal gradients. Some of these devices are capable of calculating the stresses on the HP and IP rotors as the machine is warmed up and accelerated to rated speed. Supervisory instruments monitor and display rotor vibration, eccentricity, cylinder expansion, differential expansion between rotor and cylinder, rotor position to a fixed reference point, and position of governor valves. The operator must learn the acceptable limits in observing these supervisory instruments and must take corrective action if these limits are exceeded. If this is done, turbine damage from low-cycle fatigue can be minimized. Manufacturers are also working closely with operators in tracking life-cycle expenditure due to cyclic service. NDT examination methods are used to detect incipient crack formations or the growth of any existing defects as a result of cycling service. Based on the results of these examinations and analyzing machine base data, a periodic inspection program is established to keep track of any developments that may take place because of the cyclic service. A similar program is followed on steam generators, but as indicated, rotating machinery is more prone to failure from cyclic service.

WET STEAM AND WATER INDUCTION

Q Name the sources from which water may enter a steam turbine.
A Wet steam is an old problem on steam-using power equipment. Many machines have been severely damaged from such elementary causes as operators failing to drain condensate out of casings properly, and/or steam chests and steam lines. When steam pressure drops below unit design pressure, load should be reduced to prevent blades downstream from the inlet from starting to have steam condense in the intermediate stages and thus cause water impingement on the rest of the blading. Wet steam also erodes and eats up blades.

The most frequently cited causes for water admission to turbogenerators are (1) backflow of feedwater from or through extraction lines, (2) water coming over from the boiler for various reasons, (3) flooded drain lines, (4) defective spray attemperators on superheat control systems, and (5) excessive leakage into the turbine from defective steam seal regulators. Some operators use the term water *injection* when water enters the steam turbine directly from the boiler, such as water slugging. *Induction* of water is applied generally where water reenters the steam turbine from undrained extraction lines, such as may occur from a badly leaking feedwater heater.

Q What is the effect of water admission into a hot steam turbine?
A Considerable and expensive damage may result. These are the most common:

1. Excessive thrust forces (5 to 10 times normal in severe cases)
2. Damaged buckets
3. Possible shell cracking due to abrupt quenching and high thermal stressing
4. Excessive vibration due to transient rotor distortion
5. Steam leaks due to permanent distortion of shell joint flanges
6. Hard rubs of packing teeth on rotors with possible rotor damage
7. Internal rubbing due to distortion of wheels or diaphragms

CARRY-OVER DAMAGE

In the event of water carry-over from the boiler, the primary area of concern is the turbine thrust bearing. The greater density of water as compared with steam prevents its proper acceleration and direction by the turbine nozzles. The relative velocity of the water is therefore backward against the buckets instead of through them, and a large pressure drop across the buckets develops. The actual amount of thrust increase will usually be from 5 to 10 times normal.

As the point of water entry moves downstream from the turbine inlet, the length of the buckets in the stage first reached by the water increases and so does the likelihood of damaged or broken buckets, primarily due to impact. The risk is greatest for the longest buckets, and in the last several stages of condensing turbines, cracked tie wires or covers and, in extreme cases, broken buckets can be the result of water induction from an extraction line, even with conservatively designed buckets with the best operating history.

Q Explain why tubes may leak in feedwater heaters, and thus possibly cause water induction into the turbine through extraction lines.
A Most feedwater heaters have the boiler water going through tubes at a higher pressure than the shell pressure, which generally has low-pressure steam coming to it from the turbine through extraction lines. A tube leak will thus increase the pressure on the shell if drains and other alarms fail to alert the operators to control the filling of the heater with feedwater. A sudden, large tube rupture will permit water to flow back through the extraction lines to the turbine. It is normal practice to have check valves in extraction lines to prevent this backflow, but these must

be periodically exercised so that they remain functional. Tubes in feedwater heaters have leaked because of flow-induced vibration of the tubes that affected the tubes in the intermediate support area from the constant rubbing that results from vibration. Proper venting of noncondensable gases is necessary, because some of these gases form acids, such as CO_2 to carbonic acid. The acid concentrates in the slower-flow region of the heater (see Fig. 15.9), such as where the tubes enter the tube sheet, and can attack the metal until leakage and rupture occur. Internal vent pipes are recommended usually to release the large amounts of corrosive gases.

Q What controls and alarms are usually provided on feedwater heaters to detect tube leakage? How often should these be tested?

A Most modern feedwater heaters have high- and low-level controls, alarms, and even interlocks to warn operators of high water in the shell of the heater. These sensors are mounted external to the shell with pipe connections similar to low-water fuel cutoffs on boilers. The sensors have drains, but some are piped back to the shell—an installation mistake, because now the level control cannot be tested by lowering the water in the pipe column. Sludge can accumulate, as can noncondensable gases, in the drain leg under these conditions, which may make the level control inoperative. Major turbine manufacturers recommend testing these heater level controls and alarms at least once per month for this reason. Extraction lines are usually equipped with check valves for prevention of not only backflow of water but also flashing steam from the connected heater. Since utility-type turbines have uncontrolled extraction (no turbine control valves on the extraction opening), backflow of steam in an uncontrolled manner can cause overspeeding damage as the steam from the extraction heater expands through the turbine to the condensor. Turbine manufacturers recommend daily testing of air-operated positive-assist check valves in order to make sure they are free to move as intended.

Q How are thermocouples used for water-induction detection?

A Thermocouples can quickly detect temperature changes and are being recommended to be installed in pairs at each shell location that may have a steam inlet or extraction line. One thermocouple is placed on the top and one near the bottom of the pipe penetration. Water induction will usually be detected by the lower thermocouple by an abrupt change in temperature compared with the top thermocouple. By suitable alarms, an operator will thus immediately know where water induction is taking place and will quickly isolate this line from the turbine in order to minimize damage. Without thermocouples and with inoperative heater alarms, first signs of turbine trouble may be high increase in vibration, fluctuating speed control, steam-line hammering, and similar manifestations of turbine distress.

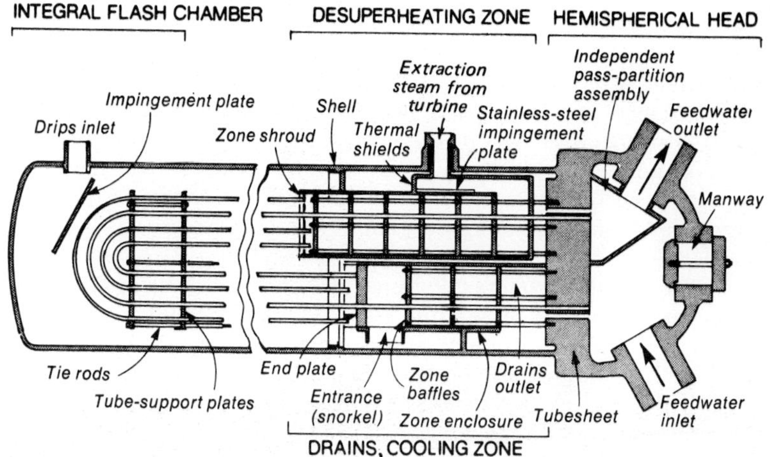

Fig. 15.9 A tube failure in a feedwater heater may cause water to back up through the extraction steam line and possibly affect the turbine.

Q How can water entering a steam turbine be minimized?

A In addition to the testing of check valves on extraction lines, there are a number of plant design and operating considerations that can minimize the likelihood of a water-entry incident. A key first step is to recognize that the boiler is not necessarily the only hazard but that other connected piping systems can be contributors to the total problem. Other steps include proper installation and maintenance of drum level controls, installation of suitable automatic controls for attemperating and desuperheating sprays, provision of a stop or blocking valve in series with any desuperheater or attemperator spray control valve, proper pitching and draining of main steam leads and extraction piping, providing dependable level controls for heaters, and providing for check valve and stop valve isolation of heaters in the event of high water level in the heater.

Piping systems directly connected to the turbine, including steam sealing supply systems, should be specifically examined at periodic intervals to ensure that hazardous conditions cannot exist under abnormal conditions of operation. Particular attention should be given to the sizing, pitching, and routing of drain and blowdown lines, the location of drain valves with respect to shutoff valves, the flashing potential of heaters and other cavities, the possible need for continuous blowdown or drainage of normally inactive lines (i.e., the steam seal makeup supply), and potential reverse-flow capabilities of drain and blowdown manifolds.

Q Are there any standard recommendations for a proper installation to prevent water induction?

A As respects required drains, high- and low-level alarms, protection from defective attemperators, or superheat control, the ASME has published a recommended practice guide to prevent water damage to turbogenerators. There is a need to coordinate boiler-turbine and auxiliary operation where the installation consists of single boiler-turbogenerator block design. Once-through, supercritical unit startup operations can be complex and may even require computer tracking for proper sequencing, timing, etc., of the whole starting process of placing a unit on the line.

The ASME recommended practice* guide stresses redundancy in preventing water from entering a turbine, by the following statement: "No single failure of equipment should result in water entering the turbine." Of particular importance in the ASME recommended practice guide to prevent water damage are the following sections:

1. Drain requirements on cold reheat piping.

2. Spray-water systems to control superheat are a favorite source of water entering a system, especially at low loads. High- and low-level alarms are needed on drain pots at the low point of cold reheat lines. A block valve to back up the control valve is needed on spray attemperators, as well as a flow indicator on the waterline to show leakage when the attemperator may not be needed.

3. On feedwater heaters that are connected to turbine extraction lines, the ASME standard requires:

a. A high- and low-level alarm

b. An automatic heater drain on a high-water level alarm

c. A nonreturn or check valve on the steam line going to the heater (usually the extraction line from the turbine)

d. Automatic shutoff valves on all sources of water entering the heater shell and tubes so that the unit can be quickly isolated from water sources

STEAM-CONTAMINATION PROBLEMS

Q Detail some of the effects to steam turbines from steam-path deposits.

A Steam-path deposits may cause the following short-term and long-term problems:

1. *Short-term problems*

a. Reduction in turbine efficiency due to steam-path plugging and control-valve sticking

b. Increase in thrust forces by as much as 200 percent, endangering the thrust bearing and lubrication of other bearings

* ASME standard TWDPS-1, *Recommended Practices for the Prevention of Water Damage to Steam Turbines Used for Electric Power Generation*, pt. 1, "Fossil Fueled Plants," July 1972.

c. Reduction in turbine output capacity by as much as 25 percent due to valve plugging and sticking as well as steam-path plugging

d. Potential overspeeding damage from slow, sticking valves and stems

2. *Long-term effects* of carry-over deposits and other harmful contaminants

a. Damage to turbine metal parts in the form of cracking, pitting, and erosion

b. Packing-ring freeze-up resulting in excessive shaft wear and seal leakage

c. Stress-corrosion cracking of disks and shafts

d. Blade and bucket damage due to chemical attack as well as rubbing damage from carry-over deposits overloading thrust bearings

e. Prolonged forced outages and loss of use, with income loss resulting

Q What boiler additives are considered harmful to turbogenerators if present in excessive amounts?

A These are shown as boiler-water additives and steam contaminants as follows, realizing that some contaminants which have potentially harmful effects on the turbine are normal feedwater additives. In this case, the key is maintenance of proper control over concentration rather than complete elimination.

Common *boiler-water* additives

Additive	Possible turbine carry-over problem
Sodium hydroxide	Caustic cracking
Sodium sulfite	Corrosion from acidic sulfur dioxide Stress-corrosion cracking from hydrogen sulfide Plugging deposits from decomposition products (sodium sulfate)
Disodium phosphate Trisodium phosphate	Plugging deposits

Common *steam* contaminants

Contaminant	Possible turbine carry-over problem
Sodium (and other) chlorides	Stress-corrosion cracking Plugging deposits
Sodium (and other) sulfides	Stress-corrosion cracking
Sodium sulfate	Plugging deposits (from decomposition products)
Silica	Deposits (principally with steam pressures 1250 psig and higher)

Q How do transient operating conditions affect impurity concentrations in boiler water or steam?

A Increase in impurity concentrations is usually observed as a result of changes in operating modes. The water-steam chemistry balance may be affected by load changes, condensate polisher exhaustion, introduction of a regenerated polisher, condenser leakage, batch addition of water-treatment chemicals and makeup water, intermittent blowdown, and pH change, just to name a few possible operating transient conditions that must be watched and considered. On-line analyzers can assist operators in noting rapidly developing transient steam chemistry problems.

Q What average steam chemistry limits are usually advocated?

A Establishing steam purity threshold limits is an ongoing program on utility-sized large units. The large turbine manufacturers are requesting lower and lower ppb as they wrestle with high-alloy material-selection problems.

Periodic visual and NDT examinations to catch cracking before it becomes serious from contaminants are still a major reason for dismantled inspections. The Electric Power Research Institute and an ASME Research Committee on Water Impurities in Power Systems is working on establishing common acceptable threshold limits on large turbogenerators. The following average steam chemistry limits for normal operation are given as a guide:

Cation conductivity	0.24 μS/cm
pH	8.55–9.26
Oxygen	11.6 ppb
Sodium	5.0 ppb
Potassium	1.5 ppb
Silica	17.4 ppb
Copper	5.0 ppb
Iron	17.4 ppb
Phosphate	7.0 ppb
Free OH	0.1 ppb

Q Explain the importance of the "Wilson" line.

A The Wilson line is where the steam enters a dry-wet transition as it expands through the turbine. Chemical compounds precipitate and deposit out when their concentration in superheated steam exceeds the solubility range for the pressure and temperature involved. Solubility in dry steam decreases as the steam expands, with the deposition most likely to occur on turbine components near the dry-wet transition steam-path line. High chemical-impurity concentrations can build up in these regions, leading to material attack on turbine components and possible blade and similar failures. Fatigue failures are a major source of turbine outages, with many occurring in this dry-to-wet-steam transition area.

Q How may stress-corrosion cracking be minimized in steam turbines?

A Most manufacturers recommend the following be adhered to:

1. Maintain steam contaminants as low as possible, but at least to the level recommended by the turbine manufacturer.

2. Avoid caustic contamination of the turbine. Maintain recommended coordinated phosphate boiler-water program in order to maintain a balance in the pH and phosphate-ion concentration.

3. It is essential to watch condensate and makeup demineralizers carefully. Failure to recharge the resins punctually, and properly, may lead to sodium or chlorine "breakthrough" and to contamination of the boiler water with unacceptable levels of sodium or chloride ions. In the case of mixed-resin beds, extreme care must be used to assure thorough separation of the resins for regeneration. If separation is poor, the remixed bed can discharge free sodium ions, which then can combine with hydroxide ions to form caustic.

4. It is necessary to permit only treated condensate in the steam path. Condenser-tube leaks into the hot well allow untreated water to enter the condensate system. If this condensate is used for (1) attemperating sprays to control boiler superheat and reheat temperatures, or (2) turbine exhaust-hood water sprays to control temperatures in the low-pressure exhaust sections during low-load operation, before being treated in the demineralizer or boiler, it can introduce contaminants into the turbine.

5. Try to prevent turbine contamination during cleaning operations. The use of acidic or caustic chemicals to remove deposits or scale from the steam side of power-plant components can contaminate turbine parts. To illustrate: Hot chemical discharge to the condenser from some other part of the system can easily cause vaporous contamination of low-pressure turbine parts. During cleaning operations involving the boiler, steam piping, feedwater heaters, condenser, and other related components, keep the turbine properly isolated to prevent contamination of steam-path parts. Example: When chemically cleaning the condenser or feedwater heaters in the condenser neck, place a vaporproof membrane in the condenser neck to isolate the turbine.

6. It is necessary not to wash turbines with caustic solutions, even if they are dilute. Reason: When evaporation occurs, as the turbine heats up, the caustic solutions produced can become sufficiently concentrated to cause stress-corrosion cracking.

7. Proper instrumentation on feedwater systems is important. Con-

ductivity and pH measurements are not enough to ensure good control of steam chemistry. Sodium, oxygen, hydrazine, and silicia levels must be measured, too.

Maintain feedwater instrumentation in accordance with the manufacturer's instructions. Poor maintenance can result in erroneous information, which, in turn, can cause delays, or overreaction on the part of operators, in correcting the feedwater treatment.

Q Why are stress-corrosion problems more prevalent on the larger-size turbogenerator units?

A Higher pressures, temperatures, and unit sizes require stronger and lighter material, which, however, is stressed more to maintain a low material per unit of rating ratio. This has permitted building larger rated units at a lower cost per kilowatt rating. The materials required to limit the size needed per kilowatt are more prone to contaminant attack that can lead to stress-corrosion cracking. Thus, tighter steam chemistry limits are required, which may not have been the case on machines built smaller in output but large in physical size and with different materials.

Q How can corrosive pitting affect turbine blades?

A Corrosive pitting is usually in the form of small pits and can occur on nozzles, blades, and even rotor wheels or disks. It is believed that the presence of both chloride contamination and oxygen with moisture accelerates pitting attack. It is most often noted in the low-pressure sections of large turbines, because moisture from the condenser hotwell can circulate into the low-pressure turbine through the condenser. Condenser-tube leakage is a favorite source of chlorides entering the condensate.

Pitting can roughen surfaces, which can affect efficiency of steam flow to produce the desired work per blade. Pits may cause stress concentration and are also a source of contaminant deposit. This can cause stress-corrosion cracks to develop. Deposits found in cracked sections of blades showed chlorides to be frequently present. Because of different solubilities, various contaminants precipitate out in different parts of the turbine. For example, sodium hydroxide has been found in deposits in the HP, IP, and LP sections of the turbine, while sodium chloride is only near the Wilson line. This line may shift with load. Where contaminants are found, some form of material attack may occur on turbine components.

Q What three approaches are being used to minimize turbine steam chemistry attack on turbine components?

A Turbine manufacturers are (1) reviewing mechanical design by modifying blade geometry in order to further reduce stress and vibration or flutter of blades in order to improve fatigue life; (2) specifying to the operators better steam chemistry control for their machines, which may differ from those required by boiler manufacturers for their units, thus producing a need for boiler-turbine operations to be coordinated in order to control the contamination levels needed on the entire boiler-turbine system; (3) striving to develop new turbine material to be used on the most highly stressed parts that are prone to steam chemistry attack. One such material is titanium for blades, which is being researched as a possible solution to blade fatigue cracking failures as a result of stress corrosion. Part of the problem with titanium may be its vulnerability to alkaline attack, thus requiring close control of pH. It is hoped the search for a suitable alloy will result in a solution to this problem.

SOLID-PARTICLE EROSION

Q What is solid-particle erosion, and what causes it?

A Solid-particle erosion manifests itself by abrasive cutting of turbine components by foreign particles carried into the turbine at high velocity with steam. The cause of the erosion damage is the solid particles which exfoliate as metal oxide from boiler parts and are carried into the turbine by the steam. The metal oxide may come from other pieces of equipment in the boiler-turbine loop, such as condensers and feedwater heaters. The metallic oxides build up over a period of time and spall off during rapid load changes or thermal transient conditions. Exfoliation of metallic oxides is most prevalent from boiler superheater and reheater tubes, and the associated high-pressure and -temperature piping. The variables that may affect the extent of exfoliation include such items as particle size and shape, steam density and velocity, and the geometric arrangement of turbine components in the steam path.

Q Where in the turbine may one find solid-particle erosion?

A Reheat-type turbines experience solid-particle erosion on the following sections of the turbine on high-pressure admission and after the reheater: first-stage nozzles, first-stage buckets, covers, and tenons.

The effect of the erosion is to reduce the efficiency of the turbine, and as thinning progresses, the affected parts will have to be replaced in order to prevent a piece from breaking off and going through the turbine to cause additional blade, cover, tenon, and bucket damage.

Q How may erosion damage be detected in operation?

A It is necessary to log data recommended by the turbine manufacturer, and by observing the changes taking place, erosion can be assumed to be occurring. For example, an increase in first-stage shell pressure and steam flow and lower power output under identical steam conditions and valve openings may be a sign of eroded blades and nozzles. Most utilities today take pictures of affected parts during dismantled inspections, and compare subsequent pictures to note if serious additional erosion has taken place. The rate of erosion per year or per month can be calculated, and blade replacement can then be planned for as blade cross-sectional area reaches the minimum required.

Q How may solid-particle erosion be prevented?
A Most experts recommend:

1. Not operating for long periods of time boilers known to be shedding metallic oxides from superheaters and reheaters.
2. Minimizing frequent cycling of the boiler due to rapid load changes, temperature changes, and frequent starts and stops.
3. Chemically cleaning the boiler, but under experienced personnel familiar with the procedures in removing all traces of acid from such components as nondrainable superheaters and reheaters.

IMPURITY MONITORING

Q Why are stringent requirements on sampling and analytical techniques in controlling steam contaminants to a turbine usually specified?
A The average steam chemistry limits, as shown previously, require parts per billion control limits. The range of impurities to be monitored is very large as pressure and temperature of the boiler-turbine system increase to the supercritical and 1000°F range. This requires accurate instrumentation to achieve such fine control, and also good sampling methods by experienced personnel.

Q Why is steam sampling for contaminants now recommended in addition to feedwater, blowdown, and condensate monitoring on large utility units?
A All major turbine manufacturers are specifying steam chemistry limits. It is difficult to predict turbine steam chemistry limits by sampling only feedwater, blowdown, and condensate, and as a result steam chemistry monitoring is also being used for finer turbine-contamination control.

Q Why are continuous-reading or printout instruments on steam and water contaminations important to an operator of a boiler-turbine system?
A Continuous-reading instruments provide rapid detection of the increase in impurities, and this permits corrective actions to be taken in order to limit the ingress into the boiler-turbine system.

Q What on-line continuous analyzers are available?

A This is a rapidly expanding field, which has been prompted by the need for good sensors in the automatic control being applied to industrial development for all sorts of process applications. Generally, pH, oxygen, conductivity, cation conductivity, and sodium and chloride analyzers are used for corrosion and deposit control of turbine steam.

Q What is recommended for analysis by sampling methods?
A One turbine manufacturer supplements the continuous chemical monitoring detailed in the previous answer by sampling and analyzing by liquid-ion chromatography for Na, Cl, NH_4, K, F, and SO_4. Around the boiler-turbine loop, grab sampling is recommended in order to identify the source of the impurities. These could be the boiler, condensers, condensate polishing system, makeup water, attempering sprays, and similar sources of contamination intrusion possibilities due to control or equipment malfunction.

Q Who should participate in assisting operators in controlling boiler-turbine steam chemistry desired limits?
A The boiler manufacturer, turbine manufacturer, plant's top management, water-treatment specialists, station chemists, and test instrument personnel should all be involved in drawing up impurity source controls, sampling methods, permissible limits, chemical analysis, and similar areas of known variables that should be agreed to before guidance is provided operating personnel. Training of all involved personnel in the control of the possible contaminants should be a continuous effort by plant management. In this way, plant reliability will be improved, and the forced outage rate will benefit.

Q What care must be exercised on turbines located in process plants?
A Where steam is used for process application and power generation, such as paper mills, refineries, sugar mills, and chemical complexes, there is always the possibility of contaminating steam and condensates from leakage in process equipment using steam. Especially harmful are acid or alkaline contaminants. The condensate will carry this through the boiler and from the boilers into the turbines. The impurities in the steam can attack the internal parts of the turbine, which can cause damage through corrosion and erosion. Suitable on-line monitoring instruments or detectors are one way of quickly detecting steam contamination from process-equipment malfunction.

CHAPTER **SIXTEEN**

Maintenance, Inspection, and Repair

This chapter will review the objectives of maintenance and inspections, which generally are to prevent and/or minimize forced outages, prevent serious damage to equipment, and thus reduce downtime and permit scheduled outages at appropriate production periods that will have a minor effect on plant productivity. Maintenance, repairs, and even unscheduled outages that interrupt normal operation can occur for various reasons, because materials fail, controls do not operate, or operating errors are made. Repairs may then be necessary, and this chapter will review some requirements on acceptable repairs to heating and power boilers.

Q What is the main purpose of a planned preventive-maintenance program?
A The main purpose of planned programs is to test, inspect, adjust, repair, and replace equipment before it fails unexpectedly during production runs and thus not only causes more expensive repairs but also adds the cost of equipment downtime, which generally may be substantial for large, critical pieces of equipment such as boilers and steam-using machinery. A planned maintenance program avoids forced outages and permits scheduled downtime for repairs, adjustments, and similar work that will correct a deteriorating condition before it progresses to failure.

Q How do operators contribute to maintenance programs?
A Operators are the main contributors to the maintenance required, because they are the best people to detect any early symptoms of operating or malfunctioning as they try to meet the plant needs for steam, hot water, or energy. During their daily rounds of repetitive operations, they will note items that need attention. These will be supplemented by review of logs to note changes in operating data or even calculations. Figure 16.1 is a typical listing of some boiler

Fig. 16.1 Common noted boiler deficiencies.

Symptoms	Probable cause	Remedial action
Inoperable low-water fuel cutoff	Sludge buildup from lack of testing	Dismantle cutoff, clean internal parts, establish testing program at more frequent intervals.
Water gage cocks frozen	Boiler compounds or rust binding cocks	Clean and free the cocks; test them daily.
Water returns slowly to gage glass when draining gage glass	Obstructed gage glass connection	Rod out pipe connections to gage glass; drain gage glass daily to remove sediments.
Safety valve is inoperable	Sludge buildup is binding valve	Free or replace valve, and test valve more often to keep mechanism free.
Leakage at handhold shell opening	Loose or defective gasket, corroded plate surfaces	Secure boiler, clean metal surfaces under gasket seating area, replace gasket, and reinstall handhold plate.
Oil leakage at base of oil burner	Oil pump seal leak or pipe joint leak	Clean up spilled oil to prevent possible fire; trace and locate leak and repair as necessary.
Oil noted on inside surfaces of boiler along water line	Leaking oil heater or similar heat-exchanger equipment using steam; also steam-using machinery	Clean oil out of boiler. Trace for source of leak and correct. Consider conductivity meter and alarm in condensate return to warn early of contamination taking place.
Eroded tubes by soot blower	Misaligned soot blower, condensate not drained from soot blower	Align soot blower, correct leaking valve to soot blower. Drain condensate before blowing.
Cracks at furnace to tube sheet weld (SM)	Probable thermal shocking and thermal cycling of boiler	Review operation, including the possible use of water preheater. If on-off burner control, consider modulating control.
Water column connection plugged	Inadequate column blowdown procedure, lack of clean-out and inspection plugs on column connections	Install clean-out plugs; rod out column connections; blow down column daily.

Fig. 16.1 Common noted boiler deficiencies (*continued*).

Symptoms	Probable cause	Remedial action
Heavy scale noted on internal surfaces	Precipitated solubles	Clean out scale. Review scale content; adjust water treatment and blowdown procedures.

NOTE: The above list is not all-inclusive but is designed to alert operators not to become complacent in operation and maintenance practices.

deficiencies that will require attention. Therefore, the operator should always be requested to participate in any preventive-maintenance program in addition to supervisory personnel.

Q Can a planned preventive-maintenance program be assigned a monetary value?

A Some are easy to convert to monetary value, while others are difficult because they involve comparing maintenance costs against potential savings. Fuel savings are usually the easiest way to justify a maintenance procedure.

EXAMPLE: Should a boiler be cleaned of scale? SM boiler, 20,000 lb/h rating, generates on the average 16,000 lb/h, operates two shifts, 5 days per week, saturated steam at 150 psi. A year ago, when new, the boiler had an efficiency of 84 percent. The average efficiency since then is 80.5 percent. Condensate return averages 150°F to the boiler. If oil with a heating value of 160,000 Btu/gal is used, how much extra oil was burned owing to the decrease in efficiency?

Steam generated per year

$$= 16{,}000(16)(5)(52) = 66{,}560{,}000 \text{ lb}$$

Enthalpy at 150 psi

$$= 1195.6 \text{ Btu/lb}$$

Enthalpy of 150°F liquid

$$= 117.89 \text{ Btu/lb}$$

Btu generated per lb

$$= 1195.6 - 117.89 = 1077.71 \text{ Btu/lb}$$

Total Btu generated

$$= 1077.71(66{,}560{,}000) = 71{,}732{,}377{,}000 \text{ Btu}$$

Fuel used at 84% efficiency

$$= \frac{71{,}732{,}377{,}000}{0.84(160{,}000)} = 533{,}723 \text{ gal/year}$$

Fuel used at 80.5% efficiency

$$= \frac{71{,}732{,}377{,}000}{0.805(160{,}000)} = 556{,}928 \text{ gal/year}$$

Fuel saved with clean surfaces $= 556{,}928 - 553{,}723 = 3205$ gal/year

Of course, in addition to fuel savings, clean boiler surfaces will also eliminate the possibility of overheating metal parts of the boiler, which can result in a forced outage with its indirect costs, including loss of use income.

Q Can checklists assist maintenance personnel during turnaround maintenance periods?

A Checklists are invaluable in reminding personnel of what has to be done during the outage time. Each plant should list items that need attention during the outage. This list may come from known complaints of operating problems, manufacturer's recommendation on what to check, or the plant's previous adverse experience on a particular item that may have been neglected in the past. Here is a checklist used by a plant on their fuel-burning equipment.

1. *Oil-burning equipment*
   a. Blow out and flush all oil, air, and steam lines in the fuel system.
   b. Have oil guns cleaned.
   c. Properly store oil guns when not in use.
   d. Check out the oil supply, atomizing steam or air supply, pump, heaters if used, reducing stations, control valves, and interlocks.
   e. Mount oil guns to check tip alignment and proper lengths of protrusion beyond diffuser cone in furnace. Refer to appropriate instruction manual, supplied with burners.
   f. See that proper oil and atomizing medium pressures are set up. Refer to appropriate set of instructions, supplied with burners.
2. *Gas-burning equipment*
   a. Blow out all gas lines before admitting gas to burners.
   b. See that vents are provided to permit scavenging of air from entire system. Remember to have all gas lines filled with nothing but gas right up to the burner cocks before attempting to light off.
   c. Check that all gas control equipment is ready for operation.
   d. Check that all safety devices are installed and working (interlocks, gas system safety valves, low-gas-pressure trips, etc.).
3. *Pilot igniters*
   a. Check that the correct igniter instruction manual is on hand for the particular type of igniter supplied.
   b. Check inside furnace for proper clearances around igniter horns. Check that horns have proper number and arrangement of wedges. Refer to igniter drawings supplied with unit.
   c. Check sparkplugs for spark.
   d. Blow out or flush all piping. Check and clean oil strainers if used.

e. Check tubing lines to horn differential pressure switch, if used, for correct installation and cleanliness.

WARNING: Do not blow through these lines into the differential pressure switch; it is fragile.

f. On oil-fired igniters, check spray nozzle assembly against drawing.

g. Check installation, rotation, and lubrication of igniter windbox fans, where supplied.

h. Check igniter windbox dampers. These are normally left wide open, at least in preliminary operation.

i. Set up pilot igniters for operation in accordance with appropriate instruction manual.

4. *Dampers*

a. Check all dampers for freedom of movement and for tightness of closing.

b. Check marks on the ends of every damper shaft to show position of each damper blade.

c. Check setup of control drives and linkage for proper operation of power-operated dampers.

d. With induced-draft and forced-draft fans running, check for leakage around damper shafts.

e. Check operation of interlock system on air-heater shutoff dampers where provided.

Q What maintenance is generally recommended on low-water fuel cutoffs?

A The inspection departments of boiler and machinery insurance companies recommend the following procedure be followed on automatically fired boilers equipped with low-water fuel cutoffs:

1. In accordance with the instruction tag or plate attached to each device, a blowdown should be effected on each high-pressure device at least once a shift, and on each low-pressure device at least once a week.

2. At least once a year all types of devices should be completely dismantled for inspection and cleaning, following any special instructions in this regard as furnished by the manufacturer. All water-equalizing piping should be cleaned of all scale or sediment accumulation.

3. At least every 6 months, all electrical components of probe- or electrode-type controls (such as relays, condensers, vacuum tubes, and resisters) should be carefully inspected for positive and correct electrical function; in addition, probe rods should be checked for scale films.

4. At least once every 6 months, on *all* types of electrical devices, inspect all high-temperature wiring connected to the device for condition of insulation and tight wiring connections. If the device employs

mercury switches in transparent glass envelopes, they should also be checked for "cloudiness," fine-line cracks, or "lazy" mercury action. Any of these symptoms indicates a dangerous condition and dictates immediate replacement of the mercury switch; in addition a thorough investigation to determine cause (overload, short, or ground) should be made to prevent recurrence.

5. Any additionl care or maintenance instructions as prescribed by the manufacturer of a particular device should also be followed, but the foregoing minimum maintenance instructions should be followed.

Q Can stack-temperature increases be an indication of fireside tube fouling?

A Stack temperatures at similar boiler operating conditions are an effective means for monitoring boiler-tube cleanliness. It is necessary first to establish baseline stack temperatures when the boiler has been cleaned in order to note changes. Similar boiler operating conditions must also be used, because stack temperatures usually increase with higher firing rates, caused by load conditions as an example. Stack-temperature measurements and logging can also be used to warn of a low-water condition developing on a boiler. It is by comparing readings on well-placed sensors that much maintenance work in a power plant is decided upon, and smaller plants should follow a similar practice in analyzing instrument readings.

Q Illustrate by an example how steam leaks on an auxiliary piece of equipment may add to fuel costs from lack of maintenance.

A A 600-hp steam turbine driving a boiler feed pump has a normal heat rate of 25 lb/(hp · h) at full load. If steam costs $5/1000 lb, and by test it is found the turbine was using 26 lb/(hp · h) because of shaft leakage, what would be the cost of the extra steam needed if the unit operated at full load all year?

$$\text{Steam lost} = 1(24)(365)(600) = 5{,}256{,}000 \text{ lb/year}$$

$$\text{Cost of lost steam} = \frac{5{,}256{,}000}{1000} \times 5 = \$26{,}280 \text{ per year}$$

This is a considerable sum and should easily justify keeping the packing or seals on the turbine reasonably leak-free.

INSPECTIONS

Q Describe the broad objectives of daily, weekly, monthly, and yearly inspections.

A The objective is to keep equipment operating in an economical and safe manner, and to note when deviations from prescribed performance are not being attained. Inspections are thus for maintaining reliable operating conditions, maintaining efficiency, and preventing unplanned outages and casualties.

LEGAL INSPECTIONS

Q Various states and municipalities have laws pertaining to inspection of boilers. What is their purpose?
A The main purpose is to protect the life and limb of employees and the public, as well as protect from property damage. The legislation sets up standards for design, installation, and inspection. These are usually ASME or NB rules, made legal requirements. Most of the laws provide for periodic inspection of boilers coming within their scope, either by state, municipal, or insurance company inspectors. The owner or operator must arrange for these inspections at required, stipulated intervals. For power boilers, the requirement is usually a yearly internal inspection and an external inspection 6 months later. Low-pressure heating-boiler inspections may be on a yearly or biannual basis, depending on the jurisdiction.

Q What inspectors make the legal inspections and reports (to a jurisdiction) that a boiler is safe or unsafe to operate, or that it requires repairs before it can be operated?
A Three types of inspectors are:

1. State, province, or city inspectors who see that all provisions of the boiler and pressure-vessel law, and all the rules and regulations of the jurisdiction, are observed. Any order of these inspectors must be complied with, unless the owner or operator petitions (and is granted) relief or exception.

2. Insurance company inspectors qualified to make ASME code inspections. If commissioned under the law of the jurisdiction where the unit is located, they can also make the required periodic reinspection. As commissioned inspectors, they require compliance with all the provisions of the law and rules and regulations of the authorities. In addition, they may recommend changes that will prolong the life of the boiler or pressure vessel.

3. Owner-user inspectors are employed by a company to inspect unfired pressure vessels for direct use and not for resale by such a company. They must also be qualified under the rules of any state or municipality which has adopted the code. Most states do not permit this group of inspectors to serve in lieu of state or insurance company inspectors.

Q What is the relationship between insurance company boiler and pressure-vessel inspectors and state laws?
A The relationship between authorized insurance company boiler

and pressure-vessel inspectors and the enforcement officials of the various states is very close. Upon passing a required examination, the inspector is issued a certificate of competency to inspect boilers and pressure vessels by the jurisdiction in which the examination was held. Generally a person having received such a certificate then may apply for an NB commission, provided NB examinations were used by the jurisdiction. Holders of such a commission are considered as being legal representatives of the jurisdiction in which they perform inspection work.

Q Explain how the relationship between the commissioned insurance inspector and the legal jurisdiction is implemented.
A When a boiler is insured (usually by a company licensed to write boiler and machinery insurance in that state), the engineering department of the insurance company sends the state legal jurisdiction a notice that the boiler is insured. If the company has commissioned inspectors in their employ (and most companies writing boiler and machinery insurance do), the legal jurisdiction does not schedule an inspection on the boiler for which a notice was received. Why? Because it will be inspected and reported to them on a formal state report when the inspection comes due by the commissioned insurance company inspectors.

The owner must prepare the boiler for the internal inspection. On some LP boilers, internals (inspections) may be scheduled only when it is deemed advisable, depending on the state law. In others, the boiler must be drained and opened for inspection whether it is of the HP, LP, steam, or hot-water-heating type.

The commissioned inspector makes the inspection, and, if conditions are satisfactory, a report is filed to the state, requesting the state to renew the certificate on the boiler. In some jurisdictions, the insurance company issues the certificate directly. If repairs are needed or if conditions need correction, certificates are withheld until the violation is removed. If this is not done, the inspector notifies the state of the violation and requests that no certificate be issued until the violation is removed. The state or legal jurisdiction can then use its police power to enforce the requirements.

When the insurance on a boiler is canceled or not renewed, the insurance company notifies the legal jurisdiction and gives the reasons for it. Many legal jurisdictions employ a numbering system on boilers. Commissioned inspectors are given state numbers to assign to boilers, and these are often stamped on the boilers.

Some legal jurisdictions have a fee for renewing certificates. The owner has to pay this when the legal jurisdiction receives a satisfactory report from the commissioned inspector when the boiler is inspected. In some states, the commissioned inspector has to collect this and send it to the state with the report. Dangerous conditions on a boiler can be reported to the state over the phone, if necessary, by the commissioned inspector. The owner will then have the full pressure of the legal jurisdiction to correct the condition or take the boiler out of service. Most insurance companies also have provisions for immediate suspension of insurance by the inspector under these adverse conditions.

Q What assistance should the owner or operating engineer give the legal boiler inspector during inspections?
A It is the responsibility of the owner to prepare the boiler for the required legal internal inspection. All openings must be removed. All scale and mud must be removed so metal surfaces are exposed for inspection. Firesides must be cleaned of soot so tubes can be inspected for corrosion, thinning, erosion, and evidence of overheating.

Inspectors must be given all the help they need. Point out any known defects. Station someone immediately outside the boiler when they are making their internal inspection. If the boiler is in battery with others, make sure that all steam, water, and blowoff valves are locked and cannot be opened. Make provision for the hydrostatic test if inspectors deem it advisable. In general, assist in every way to make their examinations thorough and complete.

INTERNAL INSPECTIONS

Q How is the waterside of WT boilers cleaned for internal inspection?
A After opening the waterside, wash out the shells and tubes with a spray of HP water. Be sure that all tubes are washed thoroughly. If it is a header-type boiler, either sectional or solid, remove all handhole caps necessary for washing and examining the tubes for scale.

> CAUTION: Wash immediately after draining the water from the boiler. This prevents any heat contained in the brick setting or furnace walls from baking the soft, muddy scale to the shell and tubes.

Examine the inner shell and tubes for scale. If it is heavier than *eggshell* thickness, remove it by hand or mechanical means. If the right feedwater treatment is used, the thin scale should peel off easily by running a hand scraper lightly over the metal. Then the tubes should be turbined. If the boiler has been neglected and is badly scaled, the drums will have to be hand-chipped. Chipping is a long and tough job. If time is important, several concerns remove all scale completely by acid cleaning, but this should not be done on riveted boilers.

Q What precaution must be observed in turbining the tubes?

A Turbining tubes can cause local tube wear or nicking if the turbining tool is forced through a tube or held in one position too long.

Q Is it safe to use portable lamps and extension cords inside boiler drums, shells, or headers?
A Only if low-voltage lamps, 32 V or less, supplied by transformers or batteries, are used to avoid electrical shocks in case a lamp or bulb breaks and creates a current flow through the boiler shell. Never use extension cords without proper waterproof fittings. Make all connections outside the boiler. And light bulbs should have explosionproof guards. Fittings, sockets, and lamp guards must be grounded.

Q What is the purpose of the internal inspection, and what areas of the boiler should receive the most attention?
A To check on the structural soundness of the pressure-containing parts, and to note any conditions that can affect its strength to confine the pressure. Wear, deterioration, corrosion, scale, oil, cracks, grooving, thinning, and other such weakening conditions require attention. Most boilers develop their own areas of trouble spots, depending on design, operating conditions, and maintenance practices. Check all exposed metal surfaces inside the boiler for effectiveness of water treatment and scale solvents, also for oil or other substances that enter with feedwater. Oil or scale on heating surfaces weakens the metal, causing bagging or rupture. Corrosion areas next to a seam are more serious than in a solid plate away from seams. Thinning on a joint is dangerous because the strength of a joint is less than that of a solid sheet.

Check for evidence of grooving and cracks along longitudinal seams of shells and drums. Carefully look for internal grooving in fillets of unstayed heads. Inspect stays and stay bolts for even tension, fastened ends for cracks where stays or stay bolts are punched or drilled for rivets or bolts. Access holes and other openings are subject to corrosion thinning and cracks. See that openings to water column connections, dry pipes, and SVs are free of obstructions such as mud and scale.

Ligaments between tube holes in heads (of all types of boilers) often crack, then leak and weaken the boiler. Also, on both WT and FT boilers the beading and flaring on tube ends need checking for erosion and corrosion, cracks and thinning. Welded nozzle and other such openings require inspections for weld washout, cracks, and evidence of deterioration of the joints.

Q How can the presence of oil be determined in the steam drum?
A Oil is usually hard to detect, especially if only a very small amount is present. Run the back of the finger along the waterline. If stained and the stain cannot be washed off with soap and water, oil is entering with

the feedwater and is being distributed throughout the boiler by circulation. Oil, being lighter than water, rises gradually and forms a scum along the water level. The real danger comes from tiny solid particles sticking to the oil before it adheres to the drum sides. Then gradually this weighted oil settles to the heating surfaces, causing tubes to blister or completely fail.

Q What precautions should be taken before entering a boiler shell?
A If the blowdown enters a common line with other boilers in operation, be sure that all valves on the blowdown line to the open boiler are closed. If other boilers are operating on the same steam header, *both* stop valves must be closed and the drip valve between them must be open. Any other valves on lines under pressure leading to the boiler must be checked. The engineer in charge and the operator on duty must be told that someone is going inside. A responsible person should be stationed at the access hole or entrance doors while anyone is inside the boiler.

Q What precaution must be observed before closing a new boiler or a boiler that has been opened for cleaning or repair?
A Make sure all tools, pipes, welding rods, rags, and other such items are removed from drums, headers, furnaces, and tubes. At times, a mirror and flashlight must be used to check headers that are otherwise not accessible for inspection of foreign material. Bent tubes that cannot be looked at from end to end (such as superheater tubes) should be thoroughly flushed, one tube at a time. On new boilers, or where work was done in a tube area, drop rubber balls, and even steel balls, to make sure the tubes are free of obstruction. Water and air can be used to push through the rubber balls.

Q Name some shortcomings of internal inspection on today's modern HP steam generators as compared with older boilers.
A As boiler size and capacity increase, the possibility of a forced outage, especially one resulting from a tube failure or explosion, takes on greater significance. The length of outage and the cost of repairs are proportionate to the size of the boiler. Thus every effort must be made to prevent failure of pressure parts by adequate inspection and maintenance. Visual inspections are still required to get as close to the parts of the boiler, both internal and external, as is practicable.

On older-type installations this was possible because a greater percentage of the surfaces of the pressure parts was accessible because of design. And more openings were provided in the setting. But with newer boilers this is not often possible. Thus inspection of large boilers should also include a review of instrumented readings so as to pinpoint areas of trouble. For example, an increase in pressure drop across a bank of tubes may indicate tube fouling in that bank, and is quicker than looking at the tubes.

The internal surfaces of most tubes are not accessible for close examination. Tube samples should be taken at each required inspection from areas not accessible and sectioned for close examination to determine the effectiveness of feedwater treatment, need for chemical cleaning, or development of internal or external corrosion. This is particularly important on newly installed boilers within the first year or two of operation.

Q Provide a checklist for drum internals.

A This is what one plant uses:

1. *Steam drum internals* checks

a. Check drier screens for fit-up, tightness.

b. Check for loose or missing bolts. See that there are washers used under bolts and nuts.

c. Check all seal welding for soundness and completeness.

d. Check that all U bolts and J bolts are double-nutted and tight.

e. Check all vents and drain holes in baffles.

f. Have steam-sampling connections checked for tightness and proper alignment of the sampling holes.

g. Check all internal piping (chemical feed, sampling, blowdown, feedwater) for tightness of joints, freedom from obstructions, and proper tightness of joints, freedom from obstructions, and proper alignment of distribution or intake holes. See that small drain holes on underside of long horizontal piping runs are open.

h. Check for obstructions in safety-valve nozzles, water-column nozzles, vent nozzles, pressure-gage nozzles, downcomer nozzles, etc.

i. Remove hoods from downcomers and check screens over nozzles for fit, cleanliness, soundness.

j. Have primary separators cleaned with wire brush; install and check for tightness and alignment.

k. Make sure that all foreign material is cleaned from the drum.

2. *Drum water-level* connections checks

a. Check to see that elevation setting and installation are correct on drum water column, gage glass, level recorder, regulator, and any remote-reading water-level indicators.

b. Check all connections from drum to water-level devices for tightness and freedom from obstruction. See that the lower (water) connection to each of these slopes downward toward the drum with no dips or pockets in it.

FIRESIDE INSPECTIONS

Q What checks are made on the fireside of boilers?

A Carefully inspect the plate and tube surfaces that are exposed to the fire. Look for places that might become deformed by bulging or blistering during operation. Solids in the waterside of lower generating tubes cause blisters when sludge settles in tubes and water cannot carry away heat. The boiler must be taken out of service until the defective part or parts have been properly repaired. Blistered tubes usually must be cut out and replaced with new ones.

Lap-joint boilers are apt to crack where plates lap in a longitudinal or straight seam. If there is evidence of leakage or trouble at this point, remove the rivets and examine the plate carefully if cracks exist in the seam. Cracks in shell plates are usually dangerous, except fire cracks that run from the edge of the plate into the rivet holes of girth seams. Usually, a limited number of such fire cracks are not very serious.

Test stay bolts by tapping one end of each bolt with a hammer. For best results, hold a hammer or heavy tool at the opposite end while tapping. A broken bolt is indicated by a hollow sound.

Tubes in HRT boilers deteriorate faster at the ends toward the fire. Tapping the outer surface with a light hammer shows if there is serious thinness. Tubes of VT boilers usually thin at the upper ends when exposed to the products of combustion. Lack of water cooling is the cause. Tubes subject to strong draft often thin from erosion caused by impingement of fuel and ash particles. Soot blowers, improperly used, will also thin the tubes. A leaky tube spraying hot water on nearby sooty tubes will corrode them seriously from an acid condition. Short tubes or nipples joining drums or headers lodge fuel and ash, then cause corrosion if moisture is present. First clean, then thoroughly examine all such plates.

Baffles in WT boilers often move out of place. Then combustion gas, short-circuiting through baffles, raises the temperature on portions of the boiler, causing trouble. Heat localization from improper or defective burners, or operation causing a blowpipe effect, must be corrected to prevent overheating.

Tubes in the vicinity of soot blowers should be inspected closely for any signs of metal loss due to erosion or steam cutting. Direct impingement of steam on tubes may result from the shifting of soot blower elements, from the growth on soot blower elements which may move the nozzle out of line with the spaces between the tubes or from defective or worn nozzles. Any tubes found to be affected by such impingement should be closely examined and the thickness of the remaining metal determined. If the thinning is serious, proper repair should be made before returning the boiler to service.

Waterwall tubes at air ports should be examined closely for loss of metal. The wall thickness of the tubes in this area should be determined periodically by using a thickness-measuring instrument or, if considered necessary, by removing tube samples.

Finned waterwall tubes should be checked for evidence of cracking at or adjacent to the fins. It may be necessary to sand or grit-blast questionable areas and use liquid dye penetrant to find the extent of the cracks if any are observed. All waterwall tubes should be carefully examined for any evidence of corrosion, erosion, or other defects. Any defects found should be repaired prior to returning the boiler to service.

Q What internal and fireside inspections are made on superheaters and economizers?

A Because of limited access to tubes, it is common practice to cut out a section of the tubes to check for scale, pitting, and grooving. External inspections are made for evidence of erosion, and the following checklist is used by one plant.

1. Check assembly spacing for excessive misalignment which might produce problems in operation.
2. Superheater assemblies should be cleaned of all scaffolding, chokers, wires, and other debris.
3. Check all thermocouples for continuity with a potentiometer and see that all are properly identified.
4. Check desuperheater and/or attemperator control valve drive units and linkage for proper operation and limits.
5. Make removable thermocouple probe for use in front of the superheater in each furnace when starting up. These should be installed at about arch level in each furnace.
6. Check installation of O_2-sampling lines.
7. Check gas-sampling probe and thermocouple lance.

EXTERNAL FITTINGS

Q What inspections are needed on the external fitting of boilers?

A Safety valves are the most important attachments on a boiler. There should be no rust, scale, or foreign matter in casings to hinder free operation. The best way to test the setting and freedom of SVs is by popping the valve with pressure. If this cannot be done, test by try levers. Inspect the discharge pipe to make sure it is secure. Operators have been killed because a valve discharging into the boiler room fills the space with steam in a few seconds. The opening in the discharge line must not be plugged.

Pressure gages have to be removed to test by comparing with a standard test gage. Blow out the pipe leading to the pressure gage. Make sure water-column connections are free by removing plugs, or the tees. Examine the condition of the water-column and gage-glass attachments.

Examine the supports of the boiler structure. Make sure that ash and soot won't bind the boiler structure to produce excessive strains from expansion under operating conditions. Look also for evidence of corrosion from soot on structural supports. Check the blowdown valves to see that they work freely and are packed and that external piping and fittings are not corroded or damaged.

Q What inspections are generally made on air-pollution-control equipment?

A Look for plugging of bags, filters, and precipitator passages where the particulate matter is removed from the flue gas. Evidence of the corrosion and its effect on the components is essential to prevent an unexpected operating problem later. Damper freedom to move as intended is also an important check. Electrostatic precipitators need special attention on the following:

1. Are electrodes in place and not broken?
2. Are the hoppers plugged?
3. Are the insulators clean and not cracked or broken?
4. Are the rapping mechanisms free and operable?

Q List some problems on electrostatic precipitators and the reasons for them.

A Wire breakage from transformers can be caused by fly-ash erosion. Excessive hopper buildup can cause moisture condensation, which in turn causes agglomeration, caking, and fly-ash bridging. Buildup to the electrode frame may cause transformer problems in that bank of electrodes. Rapper not operating properly causes contaminant buildup and diminished electrode capacity, which will affect efficiency of collection. Moisture condensation from lack of heat in the precipitator can cause insulator problems and electrical tracking, with electrical arcing.

Q How may the outage of a flue-gas desulfurization (FGD) system affect operation?

A If the FGD system cannot be bypassed, a forced outage on the rest of the equipment may be necessary.

USE OF NDT IN INSPECTION

Q How is nondestructive testing equipment being used in boiler inspection to locate potential areas of failure?

A There are five major nondestructive tests being used: ultrasonics, radiography, magnetic particle, dye penetrant, and eddy current. Ultrasonic equipment is now portable for field use and is extensively used for plate and tube thickness checks. These instruments become useful as tracing instruments for determining causes of failure of a repetitive nature. For example, tube failures in waterwalls are a common problem. After one tube failure, adjacent tubes can be checked ultrasonically and thinned tubes replaced prior to failures.

A similar practice is followed on tubes subject to fly-ash erosion. The thickness of the tubes in a suspected area is checked by ultrasonic equipment, and those found thinned are replaced during normal outages. Plate thickness around access hole and handhole openings, water legs, shells, and heads is checked ultrasonically for thickness in order to determine allowable pressure.

Flaw detection, such as checking for laminations, cracks, porosity underneath plate surfaces, or welds on inaccessible visual parts, is playing a more important role. Pulse-echo instruments are now available for field testing to do flaw detection.

Radiography, so important in new construction, is extensively used in field testing. Welded repairs on HP boilers are tested by x-ray or other radiographic equipment.

Magnetic-particle inspection finds its chief use in surface crack detection. Its main use is on piping and joints of boilers.

Eddy-current testing is finding its chief use in nonmagnetic-tube searching for defects, such as condensers and heat exchangers connected to a boiler.

Nondestructive testing of nuclear reactors in service will supplement also the traditional visual internal inspection which is limited because of the radiation hazard.

TUBE FAILURES

Q Which boiler component is most susceptible to corrosion attack?
A Corrosion failures in boiler tubing represent one of the major causes of unscheduled outages, especially on larger, high-pressure boilers. In addition to repair costs, these forced outages have high indirect costs due to interruption in production or due to the extra expense of buying replacement power that is applicable to generating plants.

Q Differentiate between uniform and localized corrosion attack.
A Uniform corrosion corrodes metal at the same rate over its entire surface. Uniform corrosion is the most common form and is also easiest to measure as respects depth and area of effect. Unexpected failures can usually be avoided by regular NDT and visual inspections.

Localized corrosion can be caused by many types of corrosion and is thus more difficult to control and prevent. Among these are (see Fig. 16.2) the following:

1. *General attack* makes the entire surface recede uniformly, as Fig. 16.2*a* shows. If a dense corrosion product adheres tightly, like iron oxide in a boiler, it can protect the metal against further corrosion even though the layer is permeable to water. The iron ions cannot diffuse easily into the water. However, constant removal of all or part of the iron oxide means trouble—either accelerated attack or pitting.

2. *Pitting* is a deep attack on small areas of the metal surface. As shown in Fig. 16.2*b*, the pit bottom acts as an anode to a nearby cathodic area, such as metal relatively protected by corrosion deposits, paint films, or plating deposits. Pitting attack is independent of the crystallographic and grain structure of the metal.

3. *Concentration cell* corrosion's net result resembles pitting (Fig. 16.2*c*), but the anode and cathode and their corrosion-engendering current are caused more by local differences in the composition of the electrolyte solution than in the metal. For example, crevices or surface dirt hinder diffusion of, say, oxygen to the solution under an obstruction. Once the original oxygen has chemically combined with the metal in a mild attack, the lack of uniformity in the solution then sets up an electrolytic battery that corrodes the metal.

4. In a *selective attack* (Fig. 16.2*d*), one metal component is preferentially leached out of the matrix, leaving a skeleton of the more resistant components.

5. *Intergranular attack* (Fig. 16.2*e*) might be called "selective path" attack. The thin zone between metal grains is a borderland inhabited by many kinds of foreign substances left over after solidification. This and other factors may promote corrosion along or near grain boundaries.

6. *Stress corrosion.* Under tensile stress, a metal can be much more seriously attacked by a mild corrosive. Figure 16.2*f* and *g* shows the two chief cases. Compressive load has no effect, only tensile. Failure is more rapid than that caused by the summation of separate effects. Stress-corrosion cracking occurs only under specific conditions not always predictable, although several tests provide a general clue.

In the past, certain brasses and especially stainless steels have been susceptible to stress-corrosion cracking. Heat treatment has helped with the brasses and also in the case of stainless steel.

7. *Hydrogen damage*, shown in Fig. 16.2*h*, causes steel to lose its ductility and strength because of tiny cracks resulting from the internal pressure of hydrogen or methane gas. The methane gas is a "corrosion product" formed in tiny amounts when hydrogen diffuses through the steel grains and combines with carbon on the way. The gas then collects

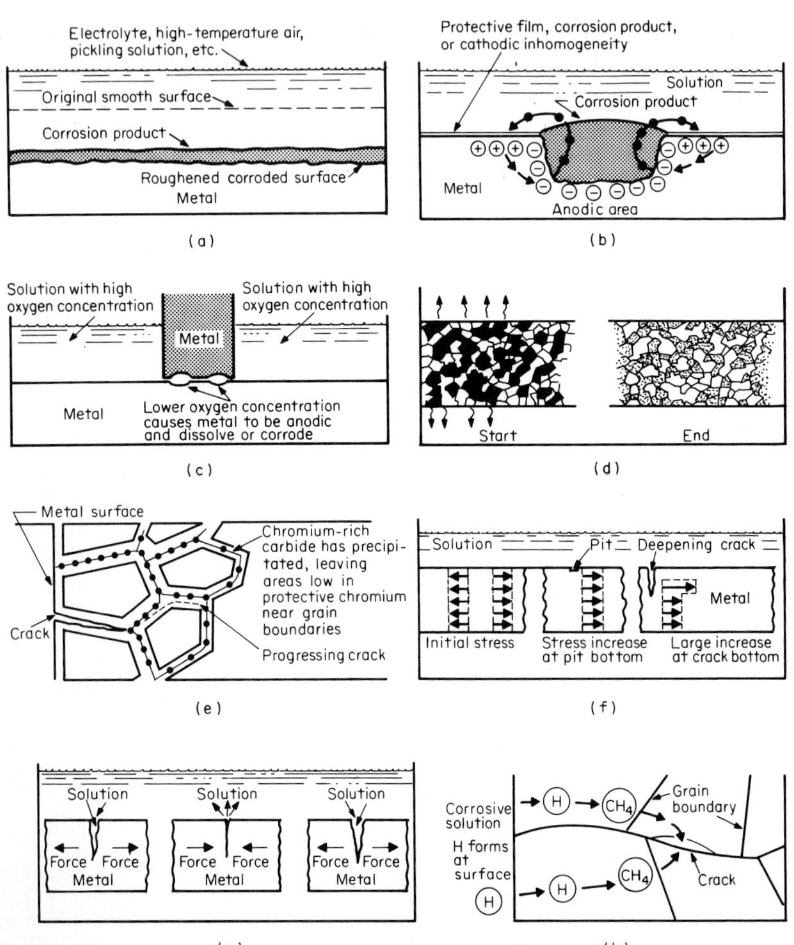

Fig. 16.2 Types of corrosion. (*a*) General attack depletes a surface uniformly. The corrodent can be a liquid or a low- or high-temperature gas. The corrosion product can dissolve or be rinsed away, or it can blow or spall off the surface. (*b*) Pitting sometimes attacks metals highly resistant to general attack. In some cases, the corrosion products slow down an initially rapid pitting. In rare cases, the pit may meet a resistant inclusion. (*c*) Concentration-cell corrosion results from differences in the corrosive medium rather than in the metal. Differences in aeration at various water levels and crevices in metal are two causes. (*d*) Selective attack, formerly common as dezincification in copper-alloy condenser tubes, removes one component and leaves a weakened structure. It is often combated by alloying. (*e*) Intergranular attack in some stainless steels results from the loss of the protective alloying element near grain boundaries. The depleted boundary areas are then easily attacked by the corrosive agent. (*f*) Stress-corrosion cracking, shown here at three different stages of

at small voids and discontinuities along the grain boundary, building up enormous pressures in microscopic regions. Microfissures result, followed by major cracks and failure.

Q Explain the term *fretting corrosion.*
A Fretting corrosion occurs as a result of vibration with two metals coming in contact by the relative motion, which destroys the protective oxide coating on the metals, exposing clean metal to the corroding media. Superheater and similar tubes that are braced by ties to limit vibration are susceptible to this type of attack.

Q Outline some of the reasons tubes fail in boilers.
A A boiler tube fails because some condition has occurred that either weakens the metal to such an extent that it cannot withstand the stress to which it is normally subjected or imposes forces on the tube that exceed tube design limits. The nature of the fracture, the size and shape of the opening, and the appearance of the edges, as well as the condition of the internal surface, will usually give evidence of the cause. The causes of tube failures can be divided into the aforementioned two general classes as follows:

1. *Weakens tube*
   a. Defective material
   b. Defective workmanship
   c. Overheating as a result of
      (1) Internal deposits
      (2) High coolant temperature (superheater tubes)
      (3) No coolant
   d. Corrosion
      (1) Internal
      (2) External
   e. Erosion
2. *Imposed loads*
   a. Differential thermal expansion
   b. External loads
   c. Strain at attachments
   d. Notch effect (propagation of cracks from attachments)
   e. Fatigue

---

time, occurs at a constant minimum tensile stress and under specific conditions. Stress concentration plays a big part in this. (*g*) Corrosion fatigue, shown here in sequence of stress application, results in a tearing apart of metal as the cracked bottom material is corroded and then subjected to alternate stresses. (*h*) Hydrogen damage is important in boilers. Why hydrogen produced by corrosion sometimes causes this damage and does not at other times is not yet completely understood in research. (*Courtesy of Power magazine.*)

Pressure-part failure is the greatest single cause of failure and the one most difficult to eliminate. Causes of pressure-part failure include overheating, weld failures, corrosion, erosion, stress cracks, vibration fatigue, mechanical damage, runaway sootblowers, and falling slag in some boilers such as recovery boilers. Proper operation and maintenance procedures, improved design and fabrication, and thorough inspections help minimize pressure-part failures. The biggest single cause of pressure-part failures is overheating.

Q Explain the *exfoliation* mechanism on high-pressure superheater and reheater tubes.

A It is believed that the magnetite layer of superheater and reheater tubes spalls off during start-up and shutdown boiler procedures. It is theorized that this occurs because of the difference in the coefficient of expansion between the magnetite or oxide layer and the tube metal. Rapid temperature changes cause the spalling of the oxide to be swept by the steam to turbines, and there cause rapid erosion of blades.

Q Define the term *caustic embrittlement.*

A See Fig. 16.3 showing a riveted boiler cracked from rivet hole to rivet hole. Caustic embrittlement is a form of stress-corrosion cracking, but it is the sodium hydroxide or caustic solution that affects certain steels that are under stress and causes them to crack suddenly in a brittle manner. Stressed rivet holes on older boilers were a place for the caustic to concentrate and attack the metal, especially if slight leaks were present in the riveted seam. The boiler water flashing into steam in the leaks would coat or leave the caustic chemicals on the boiler metal. Today it is known that welding a structure will avoid the caustic

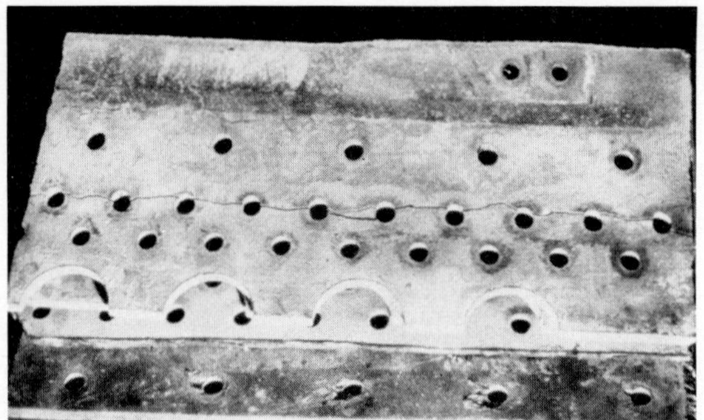

Fig. 16.3 Riveted boiler experienced rivet hole to rivet hole cracking from caustic embrittlement. (*Courtesy of Hartford Steam Boiler Inspection and Insurance Company.*)

concentrations previously experienced on riveted boilers; therefore, caustic embrittlement is not as pronounced a problem as it was years ago. Metallurgists also have developed well-defined limits for sodium hydroxide concentrations at different temperatures of the water. For example, welded steel is considered acceptable at all NaOH concentrations below 120°F.

WELD PROBLEMS

Q What are some of the reasons weld defects occur?

A The variables in welding different types of metals with an assortment of welding rods for various thicknesses and configuration of plate, tubes, and similar boiler components create problems in making satisfactory welds. This is one of the reasons the ASME has tried to provide step-by-step procedural methods in qualifying a welding method to be applied and has also studied how to qualify the operators who are to perform the welding in the procedure. If these requirements are carefully followed, the chances for weld failures are minimized. Most weld problems arise because of the following:

1. There is no worked-out welding procedure involving such factors as fit-up details, arc characteristic or strength, filler metal to be used, compatibility of filler metal with base metal, and preheat and postheat requirements.

2. The welder may not have the experience and knowledge to weld the particular joint involved. Operator qualification for the material, the type of welding process, the thickness range, and similar variables that the ASME code lists for operator qualification will help in eliminating weld defects due to lack of welding know-how.

3. There is insufficient information on or not enough consideration of the long-term effects to welds from cycling, environmental conditions, and the effects of chemical attack on welds.

Q What are the usual weld defects noted?

A Some of the most common are:

1. *Cracking*, defined as linear separation of metal due to stress. The term fissure is also used; it generally means very small cracks not considered dangerous yet in the service life of the component, but the cause of fissure should be investigated.

2. *Lack of penetration* in welding. This is a common defect. It means there is insufficient penetration of the weld through the entire thickness of the joint to effect a complete fusion of the two pieces being welded together.

3. *Incomplete fusion*, or failure to fuse together the entire joint.

The lack of fusion may be below the surface and thus not visible to the naked eye. NDT may detect lack of fusion below the surface. The lack of fusion throughout the joint means there is insufficient metal in the joint to resist the load that may be imposed. It is also a stress-concentration point as well as a place for boiler-compound hide-out.

4. *Porosity*, a form of incomplete fusion. In this case, small holes are formed in the weld by gas bubbles. The ASME code has charts that give permissible range of porosity as respects hole diameters and number of holes within a defined area of the weld, and these should be consulted.

5. *Undercuts* of welds. See Fig. 16.4. These can act as stress raisers and are also a good place for hostile contaminants to concentrate and attack the weld and base metal.

6. *Heat-affected zone stresses*, sometimes called residual stresses. These stresses are caused by metallurgical effects in the welding process as well as the heating and cooling effect of the weld and base metal. Postweld heat treatment is generally required on most, but not all, steels to avoid heat-affected weld stresses. As a complement to postweld heat treatment, preheating the joint to be welded will also assist in reducing the temperature gradient between base and weld metal. The ASME code lists preweld and postweld requirements for the different materials used in boiler and pressure vessel manufacture.

Q Why may backing rings on tube welds create problems?
A Backing rings placed on the inside of the tubes may cause fluid-flow disturbances similar to those caused by an orifice plate. The downstream side of the flow can be a good place for chemical hide-out and eventually may cause the concentrated contaminants to attack the weld joint. Most major boiler manufacturers recommend not using backing rings in joining tubes together by welding. The original purpose of the backing ring, of course, was to assure good joint-penetration welds on the ID portion of the tube.

Q Name some causes of heating-boiler failures.
A Heating-boiler explosions (also fireside explosions) have become very pronounced. Here are some of the causes:

1. More unattended automatic operation of boilers, with complete reliance on automatic controls for overpressure and fireside explosion prevention. Though controls can malfunction in many ways, their installation can lead to a false sense of security.

2. Failure to test SRVs on a consistent, regular basis.

3. Failure to maintain boiler and auxiliaries properly. The latter includes reserve boiler feed and low-water fuel cutoff. Maintenance is often neglected on water treatment, cleaning, and checking of controls.

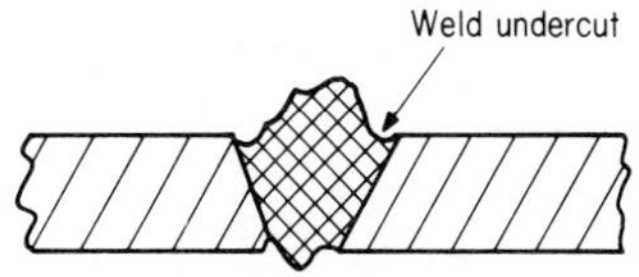

Fig. 16.4 Weld undercut serves as stress raiser and a place for contaminants to deposit.

4. As automatic boilers become more complex in control arrangement, tampering with controls or blocking the safety controls may lead to a failure.

5. The higher firing rates with suspended fuel on today's more compact boilers can quickly lead to dry firing or to improper fuel/air ratios that trigger fireside explosions, again if safety controls don't work fast enough.

Q Name the usual causes of failures on large steam generators.
A Recent figures show that 5 percent of failures were caused by internal corrosion, while only 2 percent were caused by external corrosion. Seven percent of tube failures had been caused by erosion which involved baffles, high-velocity gases, and soot blowers. The largest amount, or 50 percent of the failures, involved overheating resulting from scale, poor circulation, and some flame impingement. Most of these resulted in tube failures.

Superheater tube failure from deposits in carry-over and from overheating the tube material during start-up was common. Overheating is sometimes caused when the boilers are placed in service and steam generation is increased too rapidly. These failures indicate that failure prevention also depends on the method of operating the boiler, instrumentation, and feedwater treatment.

Q What is the most common type of boiler accident occurring today on heating and packaged boilers?
A Low water, usually resulting in overheating, loosening of tubes, collapse of furnaces and, in some cases, complete destruction of the boiler. In some classes of boilers, a low-water condition can set the stage for a disastrous explosion.

Q Is the preceding answer true for both FT and WT boilers?
A Yes. The only difference is in the parts of the boiler that are affected most. In WT boilers, damage due to low water is heavily concentrated on tubes. On FT boilers, while the highest percentage of failures is also on tubes, a large percentage includes furnaces, tube sheets, water legs, and even the shell portions (see Fig. 16.5). Cast-iron boilers generally

Fig. 16.5 Shell rupture on FT boiler can cause extensive property damage. (*Courtesy of Hartford Steam Boiler Inspection and Insurance Company.*)

crack because of low water, unless a sudden inrush of cold water occurs. In that case, sections can blow out in the form of an explosion.

Q Why is low water such a pronounced cause of failure in packaged boilers?

A The main reason is the compactness of many designs, having little reserve water storage capacity. With a high firing rate of fuels in suspension, it does not take long to develop low water in the event that water feed is interrupted. Today's greater knowledge of heat transfer and flow calculations results in "tight" designs. They leave little margin for the unpredictable effects caused by flow interruptions, scale buildup, or too rapid starting or other such shock factors. And the automatic nonattended boiler (especially heating boiler) can lull an owner into a false sense of security.

Q What parts of WT, FT, and CI boilers have the highest frequency of accidents?

A This varies from year to year, but generally follows this pattern:

1. Water-tube boilers, generating tubes, superheater tubes, waterwall tubes, and coil tubes on coil-type boilers.

2. Fire-tube boilers, tube failures; tube sheets on SM boilers, usually because of ligament cracking, on furnace-to-tube sheet weld

cracking. Fire-tube boilers over 15 psi also have shell and drum accidents, usually around access hole and handhole openings.

3. Cast-iron boilers, cracking of sections.

Q What usually causes bulging of tubes in a WT boiler?

A 1. Excessive localized scale deposits causing overheating of the tube metal.

2. Low water or improper water circulation, such as a partially plugged tube, leading to localized overheating.

3. Localized concentration of heat, such as may occur from adjacent tubes being bridged with slag or fly ash, thus throwing excessive heat on other tubes.

Q What causes tube ends and tube sheets to crack on FT boilers?

A 1. Waterside corrosion (pitting) or scale coating the tubes or both.

2. Turning the boiler feedwater system off too soon on a weekend or evening before letting the boiler cool down properly to about 200°F. Then the hot rear refractory boils off steam, reduces the water level, and exposes the rear tubes to refractory heat.

3. Unusually heavy periodic steam draw periods when the boiler is delivering heavier-than-normal amounts of steam, causing unusually heavy feedwater inputs. These tend to shock the boiler with the cooler feedwater. This can happen in the mornings when drawing heavily on steam to heat up vats, tanks, etc. Low-water-cutoff operation at this time could be intermittent, causing hot-and-cold shocking of the boiler. Obviously if the low-water cutoff does not function during this draw on steam, dry firing will affect the boiler.

Q What causes one or more bulges in tubes in a WT boiler?

A Usually a piece of scale breaks away from inside the boiler and settles in a tube, causing localized overheating of tube metal (Fig. 16.6). Flame or heat impingement and sludge deposits are also causes. If no evidence of scale is found after a tube bulges and leaks, it may have been blown free with the tube failure.

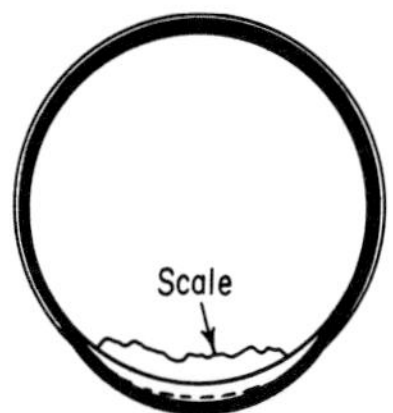

Fig. 16.6 Local bulging on WT boiler tube is caused by scale.

Q Is flame impingement on boiler parts harmful?
A Direct flame impingement causes local overheating of metal because water circulation may not be fast enough to keep the metal within safe limits. On CI units, flame impingement can create uneven expansion stresses that may crack a section. On WT boilers, steam pockets can form in a tube, leading to tube bulging or rupturing.

Q What welding problems have been experienced in nuclear generating plants?
A Intergranular stress-corrosion cracking has been experienced in welded stainless steel, type 304, areas of the steam generators, and on some external piping. No danger existed to the public, but because of the radioactivity surrounding the heat exchangers, even a minor crack repair takes a long time. It is believed the weld failures occurred because of sensitization of the stainless steel. Because stainless steels consist of 18 percent chromium, 8 percent nickel, any welding process may precipitate out chrome carbide particles and thus reduce the chromium content of the metal. This can make the stainless steel more susceptible to corrosive attack. The problem is being attacked by applying a corrosion-resistant cladding over the welds and also by altering heat-treatment methods of stainless steel to avoid sensitization.

REPAIRS

Q On what basis are repairs *allowed* on boilers?
A Repairs permitted are based on restoring the affected part or parts to as near the original strength as possible. They are governed by code requirements for new construction or by NB rules on permissible repairs, where the state has adopted NB rules for repairs.

Q Must all repairs be approved by an authorized inspector?
A Yes, if the strength of the vessel has been impaired in any way requiring repairs involving code enforcement and interpretation. Repairs not affecting the strength of the boiler, or of a minor or routine nature, may not require approval. But the inspector should be consulted on the problem if there is any doubt about the safety of the boiler. Crack repairs, welding, tube replacement, SV replacement, and similar repairs or changes require approval. The best rule to follow on any structural repairs or changes on a boiler is to immediately contact an authorized boiler inspector.

Q Is a qualfied welder permitted to make welded repairs on any part of a boiler?
A Not necessarily. A welder who is qualified to make some welds may

not be qualified for welding (1) the particular thickness of plate, (2) the type of material, (3) in the position of welding to be used, or (4) the method of welding required.

Q Can welded repairs be made on any kind of metal found on boilers?
A The repairs that may be made are limited to steels having known weldable quality; repairs are further limited to carbon steels having a carbon content of not more than 0.35 percent and low-alloy steels having a carbon content of not more than 0.25 percent. The welding of high-alloy material and nonferrous material should be done in accordance with the requirements of the ASME Boiler and Pressure Vessel Code.

Q Can a section of thinned tube or bulged tube in an FT or WT boiler be cut out and a new section of tube welded in?
A Reending or piecing tubes or pipes in either FT or WT boilers is permitted, provided the thickness of the tube or pipe away from the defective section has not been reduced by more than 10 percent from that required by the ASME code for the pressure to be carried. In all cases the requirements of the ASME power boiler code should be met on tubes. An authorized boiler inspector must approve the repair.

Q What is meant by a window patch?
A See Fig. 16.7. This type of patch may be used to seal a hole cut in a waterwall tube to provide access for welding the back side of a circumferential joint, or to replace a small, sharp bag. Window patches should comply with the provisions of the ASME power boiler code. The patch should be cut from a tube of the same size, material, and thickness as the one being repaired.

Q What general requirements does the NB stress in repair procedures?
A That the repair is approved by an authorized inspector of the jurisdiction. All welding done must be to the ASME standards of an approved written procedure that the repairer must have for the approval

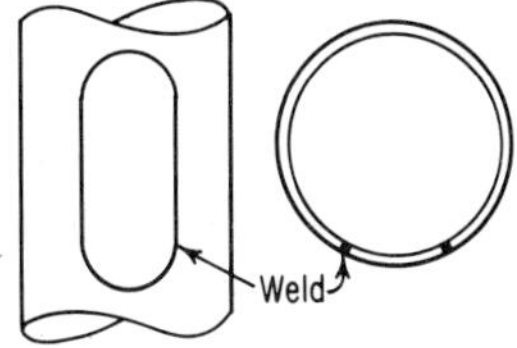

Fig. 16.7 Window patches are applied to tubes that are locally corroded, or where a piece is cut out to check on deposits or corrosion within the tube.

and review of the authorized inspector. The repair organization must have their welding procedures comply with the requirements of Section IX of the ASME code for the type of welding that is to be performed in the repair.

The repair organization must employ ASME qualified welders for the welding process to be employed, again following boiler code welding requirements. The record of such welder qualification should be certified and be available for the review of the authorized inspector.

Q Can repairs be made first and then approved by an authorized inspector?
A NB and jurisdictional rules require that no repairs be initiated without the review and authorization of the authorized inspector. For minor repairs, he or she may give approval over the phone, but the responsibility of the repairer is to first obtain approval from the jurisdictional or authorized inspector in order to avoid possible difficulties. For example, welding may not be accepted if not first approved by the authorized inspector. The welds may have to be redone or proved to be of good quality by NDT methods, all subject to the approval of the jurisdictional authorities.

Q Do repairs require NDT examinations?
A The original ASME code construction rules apply. If the joint, area, seal weld, etc., required NDT examination in the applicable section of the ASME code for new construction, the repair must also receive the same type of NDT examination. At times, the authorized inspector or the repair contractor may request or perform NDT examinations to prove the quality of a repair even though the code may not require it. Alternate NDT examinations may also be agreed to where the repair procedure prohibits following the original NDT methods that were used during construction.

Q Differentiate between repairs and alterations as respects NB rules.
A Per NB rules, "an alteration is a change in any item described on the original Manufacturer's Data Report, which affects the pressure containing capability of the boiler." Rerating a boiler to higher capacity is also considered an alteration. The NB requires alterations to be made by an ASME stamp holder for the type of boiler to be altered, usually either an S stamp for power boilers, an H stamp for heating boilers, or a PP stamp for pressure piping within the code boiler limits. Similar rules exist in having the appropriate ASME nuclear code stamps for the alteration to be made on nuclear components. The NB now has an R-1 form for repairs and alterations that must be completed and forwarded to the jurisdiction and NB if the boiler is altered from original design. The NB has a boiler data file on all NB registered boilers, and the alteration

form properly completed will retain the current data on these registered boilers.

Q Must a repair organization on boilers have certificates or credentials to show they are capable of making code repairs?
A The NB issues an R stamp to qualified boiler and pressure vessels repair firms upon application and approval by the NB. The NB rules on repairs have been accepted by ANSI as a national standard. The NB rules permit ASME stamp holders to make repairs on the equipment for which the stamp was issued. Finally, some jurisdictions that have their own equivalent rules require licensed repairers to complete a jurisdictional repair form, usually titled "Welded Repair Form." In all of the above, authorized inspectors must be consulted before any repairs or alterations are made and must approve the repairs by signing the forms with the repairer's authorized representative.

Q Can a broken stay bolt, as indicated by a leaking telltale hole, be repaired by welding?
A No. Repairs are not permitted because the leaking telltale hole indicates the stay bolt is cracked inside the boiler.

WARNING: Install a new stay bolt immediately.

Stay bolt heads cannot be welded to stop leakage around the heads. The heads should be recalked. If leakage persists, it may be an indication of a corroded sheet on the boiler. The old stay bolt must be removed, the sheet examined for corrosion and thinning, and if satisfactory, new stay bolts must be installed. If the sheets are corroded more than 50 percent of the original thickness, the defective section must be cut out and a flush-welded patch installed.

Threaded stays may be replaced by welded-in stays, provided that in the judgment of the qualified boiler inspector, the plate adjacent to the stay bolt has not been materially weakened by deterioration or wasting away. Stress relieving other than thermal may be used as provided in NB rules for welding.

Q What are the NB rules for repairing cracks on boilers?
A Two conditions are covered:

1. *Unstayed areas*. Cracks in unstayed shells, drums, or headers of boilers or pressure vessels may be repaired by welding, provided the complete repair is radiographed and stress-relieved per ASME code rules as required. Cracks of any length in unstayed *furnaces* may be welded, provided the welds are thermally stress-relieved. Welds applied from both sides of the plate should be used wherever possible. Welds applied from one side only should be subject to the approval of the authorized inspector. Field repair of cracks at the knuckle or the turn

of the flange of the furnace opening is prohibited unless approved by the enforcement authority.

2. *Stayed areas.* Cracks of any length in stayed areas may be repaired by fusion welding, but multiple or star cracks (Fig. 16.8) radiating from rivet or stayed holes should not be welded without first removing the rivet or stay bolt and checking the plate for the extent of the cracks and for evidence of corrosion.

Q What is a lap crack, and where would you expect to find it?
A Lap cracks are fatigue-type cracks (Fig. 16.9) in the longitudinal joints of lap-seam-riveted boilers. They develop because a lap seam is not part of a true cylinder. Thus a bending action is created. The crack progresses until the plate cannot hold the force created by internal pressure, resulting in an explosion.

WARNING: Lap cracks cannot be repaired.

Q What are fire cracks, and can they be repaired?
A Cracks caused by radiant heat usually around circumferential riveted seams (Fig. 6.10) of thick plates. They are caused by the comparatively greater difference in expansion on the boiler plate between the waterside and fireside surfaces. Look for them along the circumferential riveted seams of HRT boilers on the lower seam exposed to radiant heat. Fire cracks can be repaired by welding as long as they are not of the star crack type. But first check with a commissioned boiler inspector.

Q What causes CI sections in CI boilers to develop cracks besides those due to low water?
A Three possible causes are:

1. Tie rods holding sections together are drawn up too tightly, resulting in cracked CI sections as the boiler expands and contracts during heating and cooling. Check the strain on the tie rods with a torque wrench and compare it with the boiler manufacturer's recommendation. Slack off on the nuts if the strain is excessive.

WARNING: Do not draw up on the nuts if leakage develops from the fireside, but renew the asbestos seals between the sections.

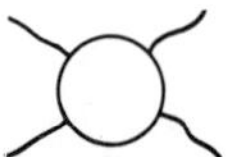

Fig. 16.8 Multiple or star cracks can emanate from rivet or stay-bolt holes because of stress.

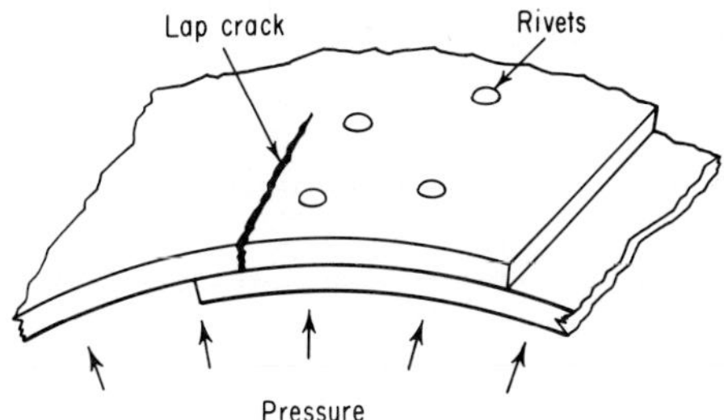

Fig. 16.9 Lap crack on a riveted joint is caused by repetitive bending stresses.

2. Sediment or scale buildup on the waterside will cause localized overheating. Cast-iron boilers need cleaning and flushing like most other boilers.

3. Feeding water of 60°F or cooler suddenly and rapidly into a CI boiler is a bad practice. Always feed makeup through the return lines so it mixes with the warm water returning to the boiler. This is a code requirement.

Q What is the usual problem with handhole and access hole openings, and what potential danger exists on these openings?

A Handhole and access hole openings are commonly oval in shape, with the cover fitted inside the boiler (Fig. 16.11). A woven asbestos gasket seals this joint. The gasket softens when heated so the nut against the dog must be tightened more as pressure is brought up on the boiler. Unless it is tightened gradually as pressure builds, the gasket may leak or be blown out, scalding personnel.

Persistent gasket leakage results in the boiler plate being thinned as a result of corrosion, which can only lead to expensive repairs. Always correct gasket leakage as soon as possible.

Q If a corroded tube sheet is to be built up by welding, what must be done to the tubes in the corroded area?

A Corroded areas of tube sheets (Fig. 16.12) may be built up by welding where tubes act as stays. But all tubes in such corroded areas

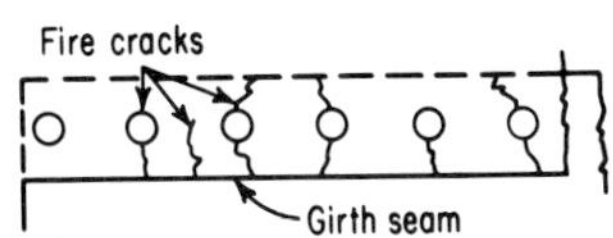

Fig. 16.10 Fire cracks appear on riveted girth seams because the thick plates cause uneven expansion.

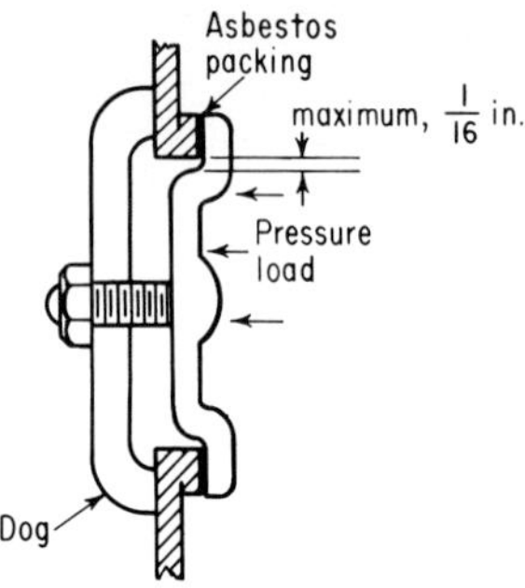

Fig. 16.11 Handhole opening, cover plate, and packing or gasket is a common place for leaks to develop.

must be removed before welding is applied. After welding, the tube hole should be reamed before new tubes are installed.

> NOTE: Stayed surfaces and tube sheets must also conform to the following requirements before welding. Corroded areas in stayed surfaces may be built up by fusion welding, provided the remaining plate has an average thickness of not less than 50 percent of the original thickness, and also if the areas so affected are not corroded enough to impair the safety of the boiler. Before repairing, approval by an authorized inspector is necessary.

Q What can boiler owners do to ensure that they have taken all possible steps to prevent boiler failure?
A First, purchase the best equipment available for a given service. Second, see that the boiler is properly installed and equipped with all necessary code-approved appurtenances and safety devices. Third, make sure the boiler is inspected regularly by a commissioned inspec-

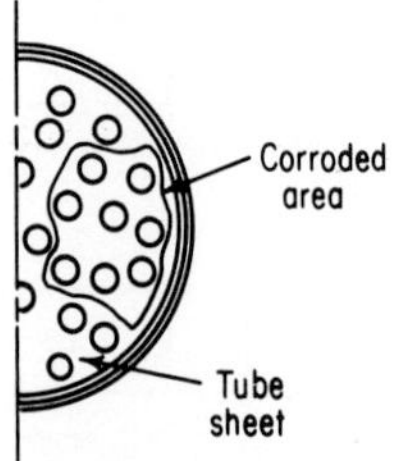

Fig. 16.12 Corroded areas of tube sheets may be repaired by welding, but tubes in the area must be removed.

tor. Only then will all the legal requirements be covered. Keep a log book check system and a set of preventive maintenance and testing procedures. See that such checks and testing procedures are followed and that the results are always recorded. Immediately correct any malfunctions found during any check or test. And never operate the boiler unattended until the proper repairs or replacements have been made.

Q Many state laws require full-time or part-time operator attendance on automatic-fired boilers. Define the meaning of *attendance*.
A Water-tube boilers with capacities over 20,000 lb/h and those fired with pulverized coal should have full-time operator attendance. High-pressure FT boilers should have full-time or part-time operator atten-

Fig. 16.13 Repairs to boilers after flooding. This checklist should be followed prior to firing any boiler that has been exposed to flood water.

Safety valves	Clean, lubricate, inspect, and test the valves. Inspect discharge pipe for mud obstructions.
Low-water cutoffs	Clean, rewire if necessary, renew probes and mercoid switches if necessary, inspect and test.
Limit controls	Clean, inspect, and test. Renew if necessary.
Flame-safeguard devices	Clean, inspect, and test. Renew if necessary.
Electric motors, transformers, relays, and wiring	Electric motors should be cleaned, baked out, and tested before energizing. Conduit, BX, wiring, and relays should be checked by a qualified electrician. If defective, renew. Failure to follow safe procedures regarding electrical facilities could lead to electrocution and/or electrical failures including burnouts and fires.
Refractory, brickwork, and insulation	After inspection and repair, all of these should be dried out as much as possible before firing. Initial firing should be light and in short intervals in order to allow moisture to escape gradually. Rapid firing could build up steam pockets with resultant pushing out of refractory.
Internal inspection	Boilers should be thoroughly inspected internally and any accumulation of slime and mud removed.
Starting up	As soon as the boiler is started, the following should be done: (1) test flame-safeguard devices, (2) test low-water cutoffs, (3) test the safety valves, (4) test the limit controls.

NOTE: To check on the function of safety devices, boiler logs should be used after the boiler becomes operational.

dance. *Full-time attendance* means the presence of a competent operator who never leaves the boiler room for more than 20 min. *Part-time attendance* means the presence of an operator who may leave the boiler room for more than 20 min, leaving the boiler operating unattended but making periodic checks at least every 2 h. A boiler is considered operated unattended when it is operated for more than 2 h without a competent operator checking it for proper operation. Check your own state and city laws.

Q Can flooding affect a boiler, and what should be checked prior to making an attempt to fire the unit?
A The severity of the flooding will dictate the actions required; however, secure a boiler that is to be exposed to flooding. Take all necessary steps to remove as much electrical equipment to high ground as possible. An emergency diesel generator will be invaluable in restoring power to the boiler room after the floodwater recedes. See Fig. 16.13 on items that may need attention after a boiler is flooded.

CHAPTER **SEVENTEEN**

Boiler-Room Management

Energy and facilities system management is concentrating again on improving efficiency of operation since fuel costs have risen dramatically and also because of emphasis on the fact that the earth's fossil-fuel supply is limited. This trend has created a recognition of the need to conserve fuel for this generation's benefit and for those to come. Owners of boiler plants consider the operation primarily in terms of costs and the benefits received in the form of heat and power. Operators and engineers realize there is a skill and conscious labor involved in producing this heat and power. It is essential, however, to consider the economics of boiler-room operation in order to maintain reasonable costs of operation within acceptable limits from the investor or owner's viewpoint. This chapter is devoted to a review of boiler-room economics and management.

MAIN MANAGEMENT FUNCTIONS

Q What are the main elements of boiler-room management that require engineering attention and review?

A Boiler-room operation and related costs usually must consider the following:

1. Purchase of equipment and supplies needed to keep the equipment operating efficiently and reliably.
2. Training and supervision of the staff required to operate the plant.
3. Managing and controlling the chemical process of combustion in its varied forms from fuel purchase to burner control.
4. Controlling and supplying load demands as reliably as possible.
5. Controlling the impurities in the boiler water and supplying as much as possible contaminant-free steam for process or power.

6. Managing routine cleaning, inspection, and periodic repairs of heat-transfer equipment and the auxiliaries involved.

7. Conducting routine performance tests to check on efficiency of operation, and also safety tests such as low water, loss of flame, loss of power, and similar tests for emergency training of staff, and to note if safety devices perform as intended.

8. Record keeping on output, fuel consumption, and supplies, and calculating efficiency of performance from recorded data are all boiler-room management functions.

Q Which of the above items require daily testing?

A To be used as guides in the operation of the boiler, the combustion process requires daily test for checking on the efficiency of burning, as does the control of the impurities in the boiler water. Both can immediately affect boiler operation, including efficiency and reliability.

Q How have on-line flue-gas analyzers affected combustion control and boiler efficiency?

A Before the advent of on-line flue-gas analyzers, the only flue-gas characteristics that boiler operators could use continuously were opacity and temperature. They also looked at the appearance of the flame and used a lot of other subjective judgments to decide when the boiler was operating at the best excess-air level. If they let the level fall too low, unburned hydrocarbons in the convection passages led to danger of explosion. And more recently, even if they starved the flame only slightly, smoke and other pollutants made them and their company liable to EPA penalties.

Many boiler operators, particularly those in industrial plants, have been in the habit of pumping lots of excess air into their boilers—that is, until their management realized how many Btu were going up the stack.

All boilers need some excess air to ensure proper combustion of the fuel. Coal-fired boilers need more air than oil- or gas-fired boilers. Excess-air needs of oil-fired boilers depend on the design of the burners and atomizers, and on the cleanliness of the burner nozzles. Low-excess-air burners, which can operate efficiently at lower excess-air levels, are now available for liquid and gaseous fuels.

At least nine flue-gas parameters change over a wide range as the amount of air supplied to the furnace is varied from a dangerously low to a wastefully high level. They are opacity, temperature, unburned hydrocarbons, O_2, CO, CO_2, NO_x, SO_2, and SO_3. Several of these parameters can be and are used to determine the optimum level of air for safe and efficient boiler operation.

A variety of instrumentation has become available that has made it possible to control the level of excess air far more precisely than was previously possible. Information on flue-gas composition can now be made available to the operator. For larger boilers, where the energy loss incurred with only small amounts of excess air can quickly run into thousands of dollars, automatic control of combustion based on flue-gas measurement is now a practical reality.

Q How may energy tracking assist boiler-room management?
A By knowing where the flow of energy goes, the boiler-room manager can advise plant managers accurately what amount of the generated energy is consumed by each department in an industrial enterprise; and as respects generating plants, a heat-rate calculation can be made on the Btu required to make one kilowatt of electricity. By measuring the flow of energy to the different parts of a plant, it is possible to cost account each department with a portion of the plant's energy costs. In addition, heavy swinging plant loads may be traced to a particular department's method of operation that could be adjusted to avoid gross swings that affect boiler-room operation. For example, a series of batch operations using steam could be scheduled to go on per batch instead of on a total department basis of five to six reactor vessels being turned on at once.

Q What is needed to perform energy tracking?
A The most important requirement is to have instrumentation or monitoring devices installed in order that the flow can be measured and analyzed. One value of a flow analysis is that it can be used to determine when maintenance may be needed on a piece of equipment by an increase in steam consumption that is evident only if flows are measured and analyzed. Monitoring of energy flows will also assist in optimizing energy usage by comparison. Adjustments are made in control, methods of operation, and similar variables, and the effect on energy flow as a result can be determined if in-place flow measurements are available. Inefficient operating procedures can be quickly identified also if energy flows are monitored and analyzed.

It can be seen that management at all levels of production must be willing to invest in energy-tracking devices and then to use these measurements as a method to track their energy consumption as part of their manufacturing costs.

Q Illustrate how fuel-consumption tracking can be used to note a decline in equipment performance, and thus result in an investment decision to replace the equipment.
A Smaller plants can track boiler-output decline by comparing fuel consumption.

EXAMPLE: A 600-hp boiler was noted to burn 611,000 gal/year of No. 6 fuel oil. A new boiler was guaranteed to burn less than 500,000 gal/year for the same output. The existing boiler was 20 years old.

An analysis indicated a new boiler would cost \$250,000, and with fuel costing \$6/1,000,000 Btu with the No. 6 oil used averaging 158,000 Btu/gal, the following payback period would result:

$$\text{Btu saved} = (611{,}000 - 500{,}000)158{,}000 = 17{,}538{,}000{,}000$$

$$\text{Fuel-cost saving} = \frac{17{,}538{,}000{,}000}{1{,}000{,}000} \times 6 = \$105{,}228$$

$$\text{Payback on fuel saving alone} = \frac{250{,}000}{105{,}228} = 2.34 \text{ years}$$

This would indicate that the boiler had a very low efficiency and it was economical to replace it even considering interest expense on the \$250,000 investment needed.

Q What charges are generally applied to the costs of supplying a unit of power?

A The charges applied are

1. *Fixed charges.* These would include interest on invested capital, depreciation, taxes, and insurance.

2. *Production costs.* These would include cost of fuel, labor costs, supplies, such as lubricating oil, office paper, charts and graphs, and similar expenses, and maintenance and repair costs to keep equipment operating efficiently.

Q Illustrate by an example fixed and production costs.

A The following figures were provided on a generating plant in a paper mill. It is required to compute the cost of power in cents per kilowatthour as delivered at the switchboard.

Station investment cost	\$6,000,000
Fixed charges	15% of investment costs
Average year-round load	30,000 kW
Annual labor costs	\$250,000
Btu/kW · h heat rate	12,500
Heating value of fuel (average)	13,700 Btu/lb
Cost of coal fuel	\$22 per ton
Maintenance and repair costs	0.025 cent/kW · h
Miscellaneous supplies	0.006 cent/kW · h

$$\text{kW} \cdot \text{h generated per year} = 30{,}000(24)(365) = 262{,}800{,}000 \text{ kW} \cdot \text{h}$$

$$\text{Fixed charges/kW} \cdot \text{h} = \frac{6{,}000{,}000(0.15)(100)}{262{,}800{,}000} = 0.342 \text{ cent/kW} \cdot \text{h}$$

Production costs. Fuel costs and labor costs can be added first before the cost per kilowatthour is calculated as follows:

$$\text{Fuel costs per year} = \frac{262{,}800{,}000(12{,}500)}{13{,}700(2000)} \times 22 = \$2{,}637{,}591.20$$

$$\text{Fuel cost plus labor cost} = \$2{,}637{,}591.20 + 250{,}000 = \$2{,}887{,}591.20$$

$$\text{Fuel cost plus labor cost per kW}\cdot\text{h} = \frac{2{,}887{,}591.20(100)}{262{,}800{,}000} = 1.1 \text{ cents}$$

$$\begin{aligned}\text{Total costs per kW}\cdot\text{h} &= \text{fixed charges plus production costs}\\ &= 0.342 + 1.100 + 0.025 + 0.006\\ &= 1.473 \text{ cents/kW}\cdot\text{h} \qquad \textit{Ans.}\end{aligned}$$

ORGANIZATIONAL STRUCTURES

Q Why are organization charts essential in a boiler plant?

A A clearly identifiable organizational chart will advise the staff of their areas of responsibility and who has the authority in these areas of boiler-plant operation. For example, fuel preparation, water treatment, burner controls, feed pumps, induced-draft and forced-draft fans, and steam and condensate lines are only a few plant functional areas that continuously require attention and care. The organization chart should identify operating personnel, maintenance and testing personnel, supervisors and emergency contacts, including outside personnel, such as servicing engineers of manufacturers and utility contacts on power, water, sewers, and similar areas.

Q Are job descriptions used in boiler-room operations?

A The size of the plant and the division of work necessary to maintain an efficient plant will determine the extent of any written job descriptions. In general the following procedures are usually stressed:

1. The supervising staff in each plant should prepare a job description for each job classification required for the proper operation of the boilers and auxiliary equipment.
2. The job descriptions should be most thorough and cover all duties to be performed by each classification. A copy of the job description should be given each employee covered by the particular classification.
3. It should be understood that these descriptions are flexible and may have to be altered from time to time as the equipment or process changes.
4. The instructions as to the responsibilities of each individual and the group as a whole should be very complete and thorough, but concise. The responsibilities of each during normal start-up and opera-

tion, also normal shutdown and emergency shutdown, should be given in detail.

Q What parameters should a boiler-room supervisor stress to an operator working in a control room?
A Basically, operating problems are deviations from normal conditions. The operator must first be trained to recognize good normal operation conditions, and once this is accomplished, the operator can easily recognize departures from these standards.

The control room of a modern boiler room is filled with recorders and indicators which convey important information to the operators, remote manual controls which save time and effort, and automatic controls which perform part of their duties. Accurate instruments are the most useful devices for detecting operating problems.

Operators should learn the meaning of each instrument reading and should know the desired operating range when normal conditions prevail. They should know the function of each control and the effect that adjustments will have on the boiler operation. Operators must familiarize themselves with the physical location of each control valve, each control drive, and each component of the boiler unit. They must study the plant layout and process until they have a mental picture of where each component fits into the boiler system.

Operators must learn the operating characteristics of the units under their control so that they know what readings will be affected when the load on the units is increased or decreased. Then they can anticipate problems that may be expected as the result of changes of the load at which the units are operating.

Drilling for operating problems should be supplemented with a troubleshooting chart, listing the problem, the manifestations to be expected, and what must be done to correct the problem. Supervisory people should also stress in the same way, but perhaps with more emphasis, emergency conditions that may arise and immediately endanger plant equipment if not properly diagnosed and corrected promptly.

EQUIPMENT RECORD CARDS AND OPERATION RECORDS

Boiler-room management includes keeping accurate records on purchases, age of equipment, suppliers of parts, maintenance performed, and repairs made on the many pieces of equipment found in a boiler operation.

Q Briefly describe the equipment record card system.
A The equipment history is a case history of the equipment from installation to the present. To start a basic maintenance program, it is

necessary to catalog the various pieces of equipment involved, to maintain records of spare parts, to maintain and dispense the manufacturer's operating and maintenance instructions, and to keep accurate, up-to-date maintenance records. The catalog or parts breakdown gives immediate access to data for ordering new parts for both preventive maintenance and unscheduled maintenance and provides a cross reference to other similar equipment in the plant, thereby allowing the user to keep a proper maintenance equipment inventory.

Q Why should every boiler plant keep log sheets, and what should be recorded?

A A log sheet should record all important operating data, such as pressures and temperatures, and it should also record procedures such as the times the soot blowers and water columns are blown and when blowdown valves were operated. A continuous record of operating data and important procedures carried out is then at hand when needed. It is also important to log when testing safety appurtenances.

Any irregular operation or event should be recorded in a separate book, with a description of the irregularity and the corrective measures taken. In this book all orders should be written and initialed by the operator.

Firemen or shift engineers reporting for duty should read the notations made by the previous watches. Then they will know what the past operation has been, what orders have been issued for future operation, and what trouble spots to keep their eyes on. They should

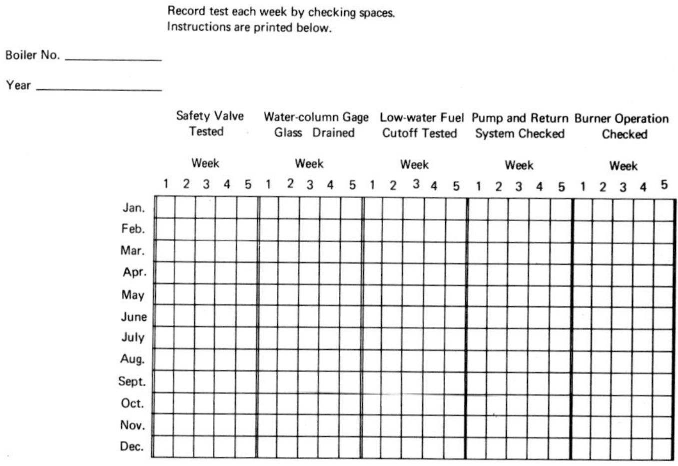

Record test each week by checking spaces.
Instructions are printed below.

Boiler No. ____________

Year ____________

	Safety Valve Tested					Water-column Gage Glass Drained					Low-water Fuel Cutoff Tested					Pump and Return System Checked					Burner Operation Checked				
	Week					Week					Week					Week					Week				
	1	2	3	4	5	1	2	3	4	5	1	2	3	4	5	1	2	3	4	5	1	2	3	4	5
Jan.																									
Feb.																									
Mar.																									
Apr.																									
May																									
June																									
July																									
Aug.																									
Sept.																									
Oct.																									
Nov.																									
Dec.																									

Fig. 17.1 Low-pressure boiler heating logs for weekly readings are available from boiler insurance companies.

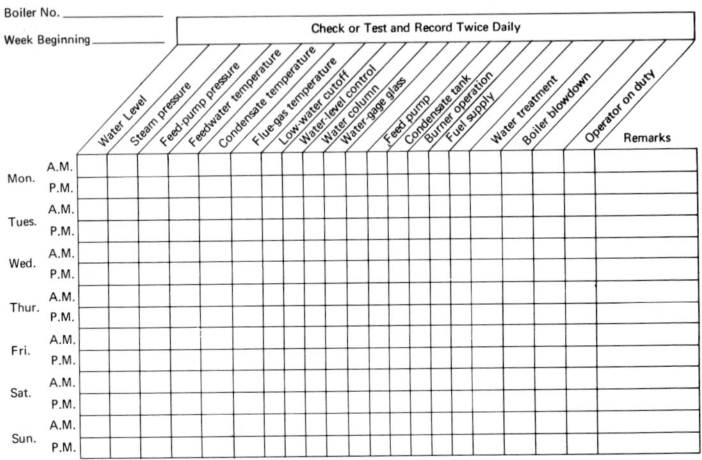

Fig. 17.2 High-pressure power-boiler log for twice-daily readings.

then initial those items for which they are responsible so as to indicate that they are familiar with the situation.

In larger plants (especially) complete records should be used to calculate the overall daily performance. Figures 17.1 and 17.2 are samples of some log sheets supplied by some insurance companies for small low-pressure plants and industrial HP boilers. Many plants design their own log sheets to record important data pertaining to their specific plant details and layout.

IDENTIFYING PIPING

Q Should power plant piping be identified by color coding or labeling?

A Yes. The ASME has on sale ASA Standard A-13.1, which suggests identifying colors for piping to make the operator's life less confusing, especially in an unfamiliar plant.

Q Describe an efficient way to become familiar with a new plant.

A Trace out every important piping system and make a sketch of all the valves and equipment in each system. Do the same with electrical systems. Operators should know their plant well enough to be able to go to key valves and controls in the dark during an emergency. Make all new operators trace out and sketch each system. Keep a file of literature pertaining to machinery, instrumentation, and equipment in the plant, and study them. Also set up a filing system for blueprints of important equipment in case they are needed in a hurry. Set up an inventory of critical components so they are always on hand. This includes boiler tubes and components of equipment that would shut down the plant if not in stock and needed because of failure.

OPERATOR-MAINTENANCE INTERFACE

Q Differentiate between operator and maintenance personnel functions as found in larger plants.

A There are certain items which need daily inspection by someone other than the operator, especially in the control system. Items such as oxygen recorders and density meters need regular cleaning; some equipment needs regular adjustment, and occasionally, recorders need calibration or similar checks. Routine inspections should be made by the operators when making their rounds to check equipment, lubrication, and cooling water. The operator is probably the primary source of information that maintenance is needed. The operator should not, however, be relied on for all information related to preventive maintenance, as this information should be accumulated by maintenance personnel. Preventive maintenance can be worked out only from experience with the equipment in the particular plant where it is used. A start can be made from consultation with the manufacturer, but the ultimate system should be set up based upon actual plant experience.

Log sheets can be simple but should contain pertinent data for future record so that this information may be used in maintenance scheduling and planning.

Q Does scheduled and predictive maintenance follow the same maintenance programs?

A Maintenance programs exist in most plants. Two schools of thought have prevailed in years past concerning maintenance. One school urged preventive maintenance (PM), with scheduled downtimes for routine cleaning, repairs, etc. The other school opted for extended operation, with maintenance called for *only* when equipment showed clear signs of impending malfunction. The latter approach wastes energy, because equipment is operating at reduced efficiency part of the time. But PM can be expensive, especially on boilers.

Predictive maintenance skirts both these pitfalls: It slashes costs and saves energy. The approach requires taking key measurements, along with a sure knowledge of how the equipment should be operating for the conditions existing. For example, a surface condenser is operating at 5 in Hg abs with a steam flow of 34,000 lb/h. Under the conditions of temperature and flow available, however, the condenser should be operating at 2.62 in Hg abs. If it were, steam flow to the turbine exhausting to it would be only 31,744 lb/h. This excess steam is unnecessary waste. Here's why: Analysis shows that roughly half the loss can be traced to tube fouling; the other half is due to either excessive air leakage or improper operation of the vacuum jet system.

This is predictive maintenance. In essence, it's very simple: The process is carefully monitored; when performance begins to tail off, the deterioration is recognized and probable causes pinpointed. Maintenance is then needed, but it can be scheduled in between the extremes of routine and emergency conditions.

Q In what manner are power plants tracked for efficient operation?
A The heat rate is the most accurate method used to determine the efficient use of energy, because it is tracking the Btu required to make a kilowatt of electricity. As a result of higher fuel costs, greater emphasis is again being placed on tracking the heat rate, including the formalization of programs in maintenance scheduling, better operational controls, and more frequent equipment performance tests in attempts to optimize station heat rates. Management is using heat rates as an effective tool in evaluating station performances in comparison with other stations or systems. Modular boiler-turbine generators can be compared on a unit basis by using the heat rate per units as a measurement criterion. Incentives for good, economical operation can thus be established by comparing heat-rate results.

Q Define the term *real-time recording.*
A A real-time system is one that permits control transactions based on events as they occur, and not after reviewing a chart, for example. Real-time recorders are thus continuously recording data as events occur, such as throttle temperature and pressure or hot and cold reheat temperatures. Real-time recording permits adjustments to be quickly made in a process to gain maximum yield. It is a term applied in the utilities more frequently as a way of maintaining a good heat rate. The real-time feedback permits operators and managers of power plants to adjust control settings immediately to achieve the design heat rate.

Q How are computers being applied to improve energy-management systems?
A Computers can be used to store many measurements for energy-management purposes. These measurements can be fed to the computer to make calculations quickly for optimum control settings needed to gain the best or most efficient results. The data gathered can also be printed on a daily basis as a permanent record or as management information. It depends on what has been programmed into the computer and the complexity of the plant. For example, Fig. 17.3 shows a

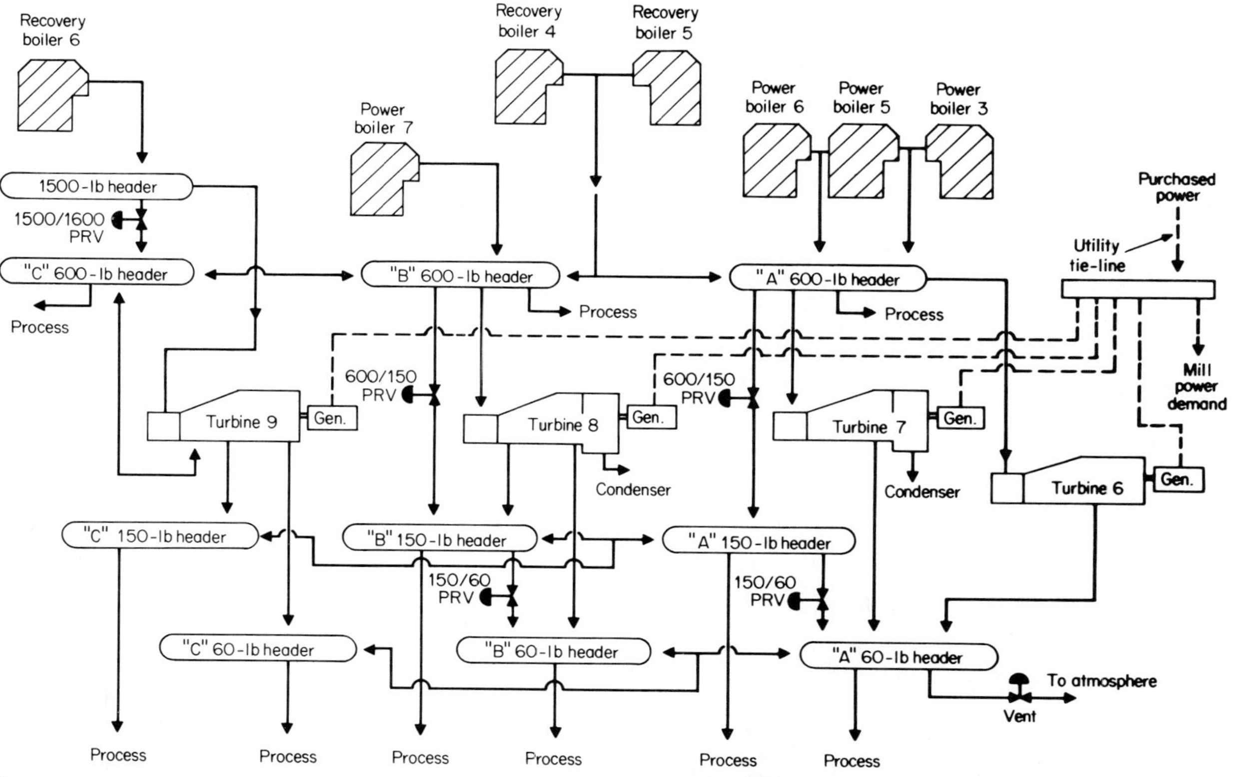

Fig. 17.3 Paper mills can have complex steam and electric integrated systems that require constant adjustments as loads for steam and power vary. Computers can assist in optimizing energy management of complex systems. (*Courtesy of Pulp & Paper magazine.*)

paper-mill steam and electric system. The complexity of matching electric power needs with process steam needs through pressure-reducing valves (PRVs) and extraction steam is apparent. Purchased power must be limited to avoid additional energy and demand charges from the utility. It is possible for a computer control system to keep generated and purchased power within desired limits, while also satisfying process needs, all at optimum levels of heat utilization.

Q How may leak-reduction management improve the conservation of fuel resources?
A Leak reduction should be a power-plant management goal, because minimizing the amount of energy losses from leaks is also a method of reducing fuel costs. Many a power-plant manager has effectively reduced steam generation needs by a program of leak detection and correction. The result quite often was idling a boiler that previously was forced to supply demand. Leak tracking can also be used as a tool for improved maintenance. For example, if it is found a certain packing or gasket continuously develops leaks, this may indicate that a different grade may be needed.

SAFETY

Q Why is it necessary to promote safety tags around power-plant equipment?
A In order to avoid having equipment start accidentally while a plant operator or others are working on the equipment. During maintenance, when equipment is out of service, a safety tag system should be used. This system should include an inspection by a responsible person to ensure that the equipment is safe to enter. Also tags should be installed on all operating buttons, electrical switches, operating valves, etc., to notify personnel that the equipment should not be operated until all work is completed and the equipment is returned to a safe condition. These tags should be removed only by the person responsible for completion of the work.

The following general precautions should be observed:

1. Before entering or checking any rotating equipment, ensure that the drive unit is tagged out and the circuit breakers are open.

2. Before entering pressure parts, ensure that all valves which possibly could lead to water, steam, or plant sewer connections are tagged out and locked closed.

3. All pumps and other rotating equipment should be turned over by hand before power is applied to ensure that they are not bound internally.

4. Any undesirable condition found should be reported to manage-

ment so that immediate steps can be taken to correct these conditions. When an unusual condition is found, whether involving repairs, feedwater conditions, corrosion, or operating problems, that is beyond the experience of the operators, experts in that field should be called for consultation and correction made.

Q What methods are used to protect plant workers from excessive noise?

A Exposure to excessive noise can cause loss of hearing, which can be either partial or full. Length of exposure is also a consideration on the effect of noise to the human body. Methods used to protect against excessive noise include:

1. Reduce the intensity of the noise source by employing different methods to do the same job or operation.
2. Place a barrier or shield between the source and the plant worker, such as acoustic soundproof barriers.
3. When engineered control methods fail, require that personal protective devices such as earmuffs or plugs be worn.

The permissible noise exposure limit can be found in CFR Title 29, Chapter XVII, Part 1910. This is available from the Government Printing Office, Washington, D.C. 20402. Additional information may be found in *Threshold Limit Values for Chemical Substances and Physical Agents in the Workroom Environment* published by the American Conference of Governmental Industrial Hygienists, 6500 Glenway Ave., Bldg. D-5, Cincinnati, Ohio 45211.

Q Describe the hazards in welding operations around power-plant equipment.

A Sparks from welding are always a possible ignition source for combustibles; therefore, the welding area must be kept free of combustibles. A common danger is sparks dropping to areas below, which may injure a person as well as be a source for igniting combustibles. Openings must be temporarily closed to avoid sparks dropping below. A fire watch should be established.

Welding fumes have received great attention. Many welding, cutting, and similar processes produce fumes and gases which if inhaled too long are dangerous to health. Fumes are considered solid particles which originate from consumable welding rods, from the base metal and the coatings that may be present on consumables or base metal. Gases are emitted during the welding operation. The effects of overexposure to these fumes and gases can be irritation of the eyes, skin, and respiratory system. Fumes can also cause nausea, headaches, and dizziness. Proper ventilation is essential in order to avoid some of these symptoms. Air sampling can be used to verify that adequate ventilation exists.

Q Briefly describe personal protective equipment.

A Personal protective equipment is the last defense against injury to eyes, head, feet, and hands. Personal protective wear protects these vital organs by providing a barrier—safety glasses for eyes, gloves for hands, hard hats for heads, and special steel insert barriers in footwear. Most plants today require workers and visitors to wear personal protective equipment for specific jobs or for certain plant locations where a specific hazard may exist. Management should consistently enforce the rules that protective wear must be worn; otherwise a lax practice will undermine the original purpose—to reduce the injury rate in a plant. In addition, supervisors should set a good example by wearing the prescribed personal protective equipment at all plant locations designated as requiring it.

Q What is the main reason for fires in long-term coal-storage areas?

A Spontaneous combustion is the primary reason for fires in long-term coal-storage areas. The fire potential is reduced by keeping oxygen out and minimizing the coal surface area exposed to air by grading and compacting the coal pile in 1-ft layers.

Q Describe the difference between an open-head and closed-head sprinkler system.

A The closed-head system employs fusible links, and when the sprinkler is activated, water will flow only out of those sprinkler heads that have opened, whereas the open-head system permits water to flow at once through all openings. The open-head system is also called a deluge system.

Q Where are wet- and dry-pipe systems used?

A Wet-pipe systems always have water in the pipes; therefore, they are suitable for areas not subject to freezing, while the dry-pipe system is usually employed in areas subject to possible freezing.

Q What is the advantage of a deluge system?

A A deluge system sprays water ahead of the fire, thus making it difficult for the fire to spread. Deluge systems, however, require a greater water supply to maintain the flow at capacity.

Q Why should a coal conveyor be stopped in case of a coal fire on the conveyor?

A In case of a fire in a conveyor, stopping the conveyor and associated fans, dust collectors, etc., will prevent burning material from being transferred downstream, passing through bins or chutes and causing sparks to ignite dust collectors and similarly increasing the potential for spreading the fire. If a closed or deluge system is installed for conveyor protection, most fire insurance organizations require an interlock to shut the conveyor down when the fire-protection system is actuated.

Q Why is it essential to eliminate tramp metal from coal-handling systems?
A A metal can set off a spark as it strikes another metal, and thus serve as a source of ignition for a dust explosion or fire. Magnetic separators are usually installed at strategic points of the coal conveying, crushing, and dumping locations. The NFPA has many specific requirements on how to prevent coal-dust explosions and fires, as do most fire insurance organizations.

Q Why are water-cooled furnace-wall-equipped boilers, firing coal, more susceptible to furnace explosions than other furnace-wall boilers?
A Boilers with water-cooled furnaces remain at a relatively low temperature in comparison with furnaces of the refractory or air-cooled types. With loss of flame in the furnace, the refractory and air-cooled wall furnace will usually ignite the coal as it enters the furnace because of the high retained heat, whereas the water-cooled furnace wall will not do so because of the lower wall temperature. Pulverized coal may thus accumulate on loss of flame, to be ignited downstream from some other hot source. The accumulation of fuel and subsequent ignition can be very explosive and destructive. Modern boiler units require flame scanning and pilot ignition of the coal to avoid some of the above on water-cooled furnace-wall-equipped boilers.

Q How do simulators assist in training operators?
A Large complex plants require central control room displays and elaborate controls to integrate the programmed starting, operation, and shutting down of a huge complex. Simulators of plant equipment will help to train personnel in an orderly manner before they are placed on actual duty. Simulators can provide a realistic environment that permits trainees the opportunity to acquire the skills, knowledge, and confidence that are essential in the operation of large plants. Simulators have the hardware, including computers, so that the interplay between a condition (simulated) and an operator or control action that might occur in the plant can be duplicated and repeated for training purposes. This enables the operator to face problems and offer solutions on the simulator without actually endangering real process equipment while obtaining the necessary experience.

Q Why is there increased emphasis on unit availability on large utility boilers and turbogenerators requiring management involvement?
A There are several reasons, among which are:

1. The large sizes involved in utility power generation make the

cost of buying replacement power very high. Typically, a 600-MW unit may require $600,000 dollars per day expenditure to replace the power lost if the unit has a forced outage. Figure 17.4 shows forced outages as compiled by the North American Electric Reliability Council for the 1971–1980 period.

2. Managers of utility systems know that if availability can be improved, more load can be handled by existing plant, thus saving the cost of building additional facilities even though load growth will eventually require more capacity to be built.

3. Improvement in availability can be assigned a dollar value. Often cited is the fact that a 1 percent improvement in a 500-MW unit could represent a 1 million dollar saving per year. Similar comparisons are possible on small boiler and generating plants.

Q How is modern monitoring equipment assisting plant owners in improving the reliability of power equipment?
A Sensors have been and are being developed not only for control purposes but also to automatically start and load large complex trains of power-plant equipment. Extensive use of microprocessors will program the operation so that allowable or design limits for temperature, pressure, vibration, expansion, vacuum, and similar criteria will not be exceeded. Steam-purity sensors are being developed so that steam conditions can be continuously displayed and analyzed.

WORK AND SHIFT SCHEDULES

Q What is the purpose of shift and work schedules in power-plant operation?
A To provide an equitable distribution of overtime and weekend work for the operating crew. The following questions and answers will illustrate this point.

Q Set up a power-plant rotating shift schedule for a five-person crew for 24 h per day, 7 days per week, 365 days per year, covering a 10-week period.
A The rotating-shift schedule in Fig. 17.5 shows that at the end of the 10-week period the schedule is repeated. If the pay period is 2 weeks, this schedule will cover five pay periods. During each 2-week pay period, each worker works 80 h.

This schedule is useful in plants where the operators do some maintenance work because here 8 days of maintenance out of every 50 days of work are provided. Thus overtime is not necessary.

Here is how during the 10-week period every person on every crew will receive exactly the same treatment adding up to 50 days worked

Fig. 17.4 North American Electric Reliability Council data—1971 to 1980—on utility equipment outages and gigowatthours of power lost from the outages.

Item	Forced full		Forced partial		Total	
	No.	GWh lost	No.	GWh lost	No.	GWh lost
Waterwalls	14,540	291,101	2,022	16,282	16,562	307,383
Superheater	9,922	142,677	1,873	5,168	11,795	147,844
Pulverizer	502	4,941	74,105	89,443	74,607	94,384
Boiler feed pump	1,700	21,219	14,210	70,003	15,910	91,222
Boiler, general	3,863	77,839	6,589	13,111	10,452	90,950
Reheater—first	3,944	80,997	785	3,071	4,729	84,068
Vibration of turbogenerator	2,758	71,074	2,253	9,709	5,011	80,783
Buckets or blades	187	53,324	1,747	17,979	1,934	71,303
Fresh-water heater leak	1,090	16,885	11,544	14,217	12,634	64,102
Economizer	4,783	59,860	564	3,142	5,347	63,002
Induced-draft fan	977	14,548	12,184	38,078	13,161	52,626
Forced-draft fan	785	10,420	9,455	38,678	10,240	49,098
Lube-oil system, turbogenerator	911	41,782	1,099	4,786	2,010	46,568
Generating tubes	1,691	42,448	409	1,626	2,100	44,074
Stator windings, etc.	187	37,767	226	2,910	413	40,677
Furnace slagging	549	17,694	14,487	22,835	15,036	40,529
Main turbine, general	999	37,399	1,101	3,090	2,100	40,489
Control, turbine, and reheat stop valves	1,937	28,285	2,268	9,251	4,205	37,537
Total of all forced outages for the decade						
All fossil codes	92,223	1,681,331	338,546	809,457	430,769	2,490,788
All nuclear codes	4,178	313,457	14,645	90,752	18,823	404,209

Fig. 17.5 Five-person rotating shift schedule.

Hours	S	S	M	T	W	T	F	S	S	M	T	W	T	F
12–8	B	B	B	B	B	D	D	D	D	B	B	B	B	B
8–4	E	E	E	D	D	E	E	E	E	E	D	D	E	E
4–12	C	C	A	A	A	A	A	A	A	A	A	A	C	C
Maint.			D	C	C	C				C	C	C	D	
12–8	D	D	D	D	D	A	A	A	A	D	D	D	D	D
8–4	B	B	B	A	A	B	B	B	B	B	A	A	B	B
4–12	E	E	C	C	C	C	C	C	C	C	C	C	E	E
Maint.			A	E	E	E				E	E	E	A	
12–8	A	A	A	A	A	C	C	C	C	A	A	A	A	A
8–4	D	D	D	C	C	D	D	D	D	D	C	C	D	D
4–12	B	B	E	E	E	E	E	E	E	E	E	E	B	B
Maint.			C	B	B	B				B	B	B	C	
12–8	C	C	C	C	C	E	E	E	E	C	C	C	C	C
8–4	A	A	A	E	E	A	A	A	A	A	E	E	A	A
4–12	D	D	B	B	B	B	B	B	B	B	B	B	D	D
Maint.			E	D	D	D				D	D	D	E	
12–8	E	E	E	E	E	B	B	B	B	E	E	E	E	E
8–4	C	C	C	B	B	C	C	C	C	C	B	B	C	C
4–12	A	A	D	D	D	D	D	D	D	D	D	D	A	A
Maint.			B	A	A	A				A	A	A	B	

NOTE: A = first person, B = second person, C = third person, D = fourth person, E = fifth person.

and 20 days off: 14 days on a 12-to-8 shift, 14 days on an 8-to-4 shift, 14 days on a 4-to-12 shift, 8 days on maintenance, 6 weekends worked, 4 weekends off, two 4-day weekends off, one 3-day weekend off.

Q Prepare an 8-week permanent shift schedule for four people, each averaging 42 h per week.

A In Fig. 17.6 schedule *A* suitably divides the operating hours of four people with each averaging 42 h per week. There are twenty-one 8-h shifts in a week here. Thus with a four-person crew, three people will work 40 h a week and one will work 48 h. This is rotated progressively among the workers in an 8-week cycle. With this schedule, each person will have a complete rotation of schedule in 8 weeks and will have two Saturdays and two Sundays off. Thus, each person works 6 consecutive days (except for the swing-shift worker, who will have 2 days off).

The swing-shift worker's week will consist of 2 days on the first shift, 24 h off, 2 days on the second shift, 24 h off, then 2 days on the third shift; finally 48 h off, after which the schedule starts over. Some plants

Fig. 17.6 Eight-week shift schedule with three people on permanent shifts and one person on rotating shift.

8-week shift—schedule A																												
Shifts	M	T	W	T	F	S	S	M	T	W	T	F	S	S	M	T	W	T	F	S	S	M	T	W	T	F	S	S
Third shift	R	R	C	C	C	C	C	C	R	R	C	C	C	C	C	C	R	R	C	C	C	C	C	C	R	R	C	C
First shift	A	A	A	R	R	A	A	A	A	A	A	R	R	A	A	A	A	A	A	R	R	A	A	A	A	A	A	R
Second shift	B	B	B	B	B	R	R	B	B	B	B	B	B	R	R	B	B	B	B	B	B	R	R	B	B	B	B	B
Off	C	C	R	A	A	B	B	R	C	C	R	A	A	B	B	R	C	C	R	A	A	B	B	R	C	C	R	A
Third	C	C	C	C	R	R	C	C	C	C	C	C	R	R	C	C	C	C	C	C	R	R	C	C	C	C	C	C
First	R	A	A	A	A	A	A	R	R	A	A	A	A	A	A	R	R	A	A	A	A	A	A	R	R	A	A	A
Second	B	R	R	B	B	B	B	B	B	R	R	B	B	B	B	B	B	R	R	B	B	B	B	B	B	R	R	B
Off	A	B	B	R	C	C	R	A	A	B	B	R	C	C	R	A	A	B	B	R	C	C	R	A	A	B	B	R

8-week shift—schedule B																												
	M	T	W	T	F	S	S	M	T	W	T	F	S	S	M	T	W	T	F	S	S	M	T	W	T	F	S	S
Third shift	R	R	C	C	C	C	C	C	R	R	C	C	C	C	C	C	R	R	C	C	C	C	C	C	R	R	C	C
First shift	A	A	R	R	A	A	A	A	A	A	R	R	A	A	A	A	A	A	R	R	A	A	A	A	A	A	R	R
Second shift	B	B	B	B	R	R	B	B	B	B	B	B	R	R	B	B	B	B	B	B	R	R	B	B	B	B	B	B
Off	C	C	A	A	B	B	R	R	C	C	A	A	B	B	R	R	C	C	A	A	B	B	R	R	C	C	A	A
Third	C	C	C	C	R	R	C	C	C	C	C	C	R	R	C	C	C	C	C	C	R	R	C	C	C	C	C	C
First	A	A	A	A	A	A	R	R	A	A	A	A	A	A	R	R	A	A	A	A	A	A	R	R	A	A	A	A
Second	R	R	B	B	B	B	B	B	R	R	B	B	B	B	B	B	R	R	B	B	B	B	B	B	R	R	B	B
Off	B	B	R	R	C	C	A	A	B	B	R	R	C	C	A	A	B	B	R	R	C	C	C	A	A	B	B	R	R

NOTE: A = first shift (8 A.M to 4 P.M), B = second shift (4 P.M to midnight), C = third shift (midnight to 8 A.M), R = relief shift.

find it preferable to assign one person in a group of four to work this swing shift. The reason is that it isn't easy to find many people who can adapt to changing shifts, but it is usually possible to find at least one in the group who does not mind the swing shift. In this case, if the labor contract calls for time and a half for the sixth consecutive workday and double time for the seventh day, it is possible that the swing-shift worker will have to be paid the premium rate; if so, use schedule *B* in Fig. 17.6.

In schedule *B* in Fig. 17.6 the swing shift is completed in six consecutive days. Some contracts call for time and a half on Saturdays and double time on Sundays. This would apply to the swing-shift worker as well, and schedule *A* in Fig. 17.6 would then be preferable. Most contracts call for shift premiums. Using schedule *B* in Fig. 17.6, the swing-shift worker will have 24 h off at every shift change and 48 h off at the end of the workweek. The swing-shift worker's week starts with the third shift and is completed at the end of the second shift. The other shift workers work the same weekly hours in either schedule.

During the nonheating season, if experienced, licensed workers are needed for the plant, most will have to be used for maintenance. If the spring and fall schedule is on a 5-day basis, the minimum crew is two workers, each working 9 h a day. But if three workers are used, one can do maintenance work for 6 h each day and operate the two middle hours of the day, thus keeping the other two on 8-h shifts. But if the plant can be shut down 2 h in the middle of the day, two workers can do the work.

The charts are only for illustrative purposes on how shift schedules can be conveniently developed by an appropriate layout of weekly and monthly charts.

Q How can energy costs be minimized in the operation of energy-producing equipment?

A Energy costs will continue to be significant in commercial, industrial, and power-generation enterprises. It is necessary in optimizing a balanced energy program to maintain an incentive-type administrative program that keeps track of deteriorating performance, plans maintenance to restore the equipment to targeted design conditions, provides the capital to replace old and inefficient equipment, and maintains an operating atmosphere that encourages efficient and reliable operation, including the maintenance of a safe working place for the plant personnel and the public.

APPENDIX A

Commonly Applied English-Metric Conversions

Quantity	Multiply quantity by	To obtain
	Lengths	
centimeters	0.03281	feet
centimeters	0.3937	inches
fathom	6.0	feet
feet	30.48	centimeters
feet	0.3048	meters
inches	2.540	centimeters
meters	100.0	centimeters
meters	3.281	feet
meters	39.37	inches
meters	1.094	yards
mils	0.001	inches
mils	0.00254	centimeters
yards	91.44	centimeters
yards	0.9144	meters
square centimeters	0.155	square inches
square centimeters	0.001076	square feet
square inches	6.452	square centimeters
square inches	645.2	square millimeters
square inches	$1.273 \times (10)^{6}$	circular mils
	Volumes	
barrels (U.S. liquid)	31.5	gallons
barrels (oil)	42.0	gallons (oil)
cubic centimeters	0.06102	cubic inches
cubic centimeters	$2.642 \times (10)^{-4}$	gallons (U.S. liquid)
cubic centimeters	$1.057 \times (10)^{-3}$	quarts (U.S. liquid)
cubic feet	28.32	liters
cubic feet	29.92	quarts (U.S. liquid)
cubic feet	1728	cubic inches
cubic feet	28,320	cubic centimeters
cubic inches	$4.329 \times (10)^{-3}$	gallons (U.S. liquid)
cubic inches	0.01732	quarts
cubic inches	0.01639	liters
gallons	3785	cubic centimeters

Quantity	Multiply quantity by	To obtain
	Volumes	
gallons	0.1337	cubic feet
gallons	231	cubic inches
gallons	$3.785 \times (10)^{-3}$	cubic meters
gallons	3.785	liters
liters	1000.0	cubic centimeters
liters	61.02	cubic inches
liters	0.2642	gallons (U.S. liquid)
liters	1.057	quarts (U.S. liquid)
pints	28.87	cubic inches
pints	0.125	gallons
pints	0.4732	liters
quarts (liquid)	57.75	cubic inches
quarts	946.4	cubic centimeters
quarts	0.25	gallons
quarts	0.9643	liters
	Weights	
grams	$2.205 \times (10)^{-3}$	pounds
grams	0.001	kilograms
kilograms	2.205	pounds
kilograms	1000	grams
ounces	28.35	grams
ounces	0.0625	pounds
pounds	16	ounces
pounds	453.5924	grams
	Force	
newtons	0.22481	pounds
pounds	4.44822	newtons
	Pressure	
atmospheres	29.92	inches of mercury
atmospheres	1.0333	kilograms per square centimeter
atmospheres	14.7	psi
bars (metric unit)	0.9869	atmospheres
bars (metric unit)	14.50	psi
bars (metric unit)	1.020	kilograms per square centimeter
bars (metric unit)	10	megapascals (MPa)
feet of water	0.4335	psi
feet of water	0.03048	kilograms per square centimeter
feet of water	0.8826	inches of mercury
inches of mercury	0.03342	atmospheres
inches of mercury	0.03453	kilograms per square centimeter
inches of mercury	0.4912	psi
kilograms per square centimeter	28.96	inches of mercury
kilograms per square centimeter	14.22	psi
kilograms per square centimeter	0.9678	atmospheres
kilograms per square centimeter	0.981	bars
pascals	1.0	newtons per square meter
pascals	10^{-5}	bars
pascals	$145.036 \times (10)^{-6}$	psi

Quantity	Multiply quantity by	To obtain
	Pressure	
megapascals (MPa)	1.0	newtons per square millimeter
megapascals (MPa)	10	bars
megapascals (MPa)	145.038	psi
psi	0.06804	atmospheres
psi	2.307	feet of water
psi	2.036	inches of mercury
psi	0.0690	bars
psi	$6.8948 \times (10)^{-3}$	megapascals (MPa)
psi	0.07032	kilograms per square centimeter
	Power and heat	
Btu	778.3	foot-pounds
Btu	252.0	gram-calories
Btu	1054.8	joules
Btu/h	$3.929 \times (10)^{-4}$	horsepower
foot-pounds	$1.286 \times (10)^{-3}$	Btu
foot-pounds	0.3238	gram-calories
foot-pounds	1.356	joules
foot-pounds per minute	$3.030 \times (10)^{-5}$	horsepower
gram-calories	$3.9685 \times (10)^{-3}$	Btu
horsepower	0.7457	kilowatts
horsepower	33,000	foot-pounds per minute
horsepower	42.44	Btu/min
horsepower (boiler)	33.479	Btu/h
horsepower (boiler)	9.803	kilowatts
horsepower-hour	2546	Btu
horsepower-hour	$2.684 \times (10)^{6}$	joules
joules	$9.480 \times (10)^{-4}$	Btu
joules	0.7376	ft-lb
kilowatts	56.92	Btu/min
kilowatts	737.6	foot-pounds per second
kilowatts	1.341	horsepower
kilowatthours	3413	Btu/h
kilowatthours	$3.6 \times (10)^{6}$	joules
megawatts	1000	kilowatts
therm	1,000,000	Btu
	Temperature	
	$°F = \frac{9}{5}(°C) + 32 \quad °C = \frac{5}{9}(°F - 32)$	

How to use the conversion list: Convert 200 psi into bars, MPa, and kg/cm^2:

$$200 \times 0.0690 = 13.80 \text{ bars}$$
$$200 \times 6.8948 \times (10)^{-3} = 1.37896 \text{ MPa}$$
$$200 \times 0.07032 = 14.064 \text{ kg/cm}^2$$

Appendix **B**

Transposing Equations and Decimal Carrying

TRANSPOSING EQUATIONS

A brief review of transposing equations is presented for those readers who are preparing to take operator or inspector examinations in order to quickly transpose some equations illustrated in the text. Transposing can be performed in two ways, namely, (1) algebraically and (2) by number substitution and reduction. An illustrative example will be used to demonstrate the two methods.

Assume:

$$P = \frac{StE}{R + 0.6t}$$

Given $S = 15{,}000$
$t = 0.5$
$E = 1$
$P = 700$

What is R?

Solving Algebraically This method first solves for R, and then the known quantities are substituted in order to obtain the answer. This is done as follows for this problem:

$$P = \frac{StE}{R + 0.6t}$$

By multiplying both sides of the equation by $(R + 0.6t)$, we get

$$P(R + 0.6t) = StE$$

By dividing both sides by P, we get

$$R + 0.6t = \frac{StE}{P}$$

By shifting the $0.6t$ factor to the other side of the equal sign, so that R is all by itself, we get

$$R = \frac{StE}{P} - 0.6t$$

This gives us R algebraically. Now we can substitute the given numbers to solve for R, namely,

$$R = \frac{15{,}000(0.5)(1)}{700} - 0.6(0.5)$$

Solving:

$$R = 10.714 - 0.3$$
$$R = 10.414 \qquad \textit{Ans.}$$

Number-Substitution Method

$$P = \frac{StE}{R + 0.6t}$$

Substitute the numbers given:

$$700 = \frac{15{,}000(0.5)(1)}{R + 0.6(0.5)}$$

Reducing:

$$700 = \frac{7500}{R + 0.3}$$

Clearing:

$$700(R + 0.3) = 7500$$

Reducing:

$$700R + 210 = 7500$$

Transposing:

$$700R = 7500 - 210$$
$$700R = 7290$$
$$R = 10.414 \qquad \textit{Ans.}$$

Either method will give the correct answer, and it is suggested readers use the method they are most familiar with.

Examples of Transposing Equations Examples of transposing equations are given below with numbers provided to correspond with the solution to the equation for practice purposes. (See Chapter 8 for equations and what they mean.)

1. *Tube equation for WT boilers*. Check by using $P = 2455$, $S = 14{,}400$, $t = 0.188$, $e = 0$, $D = 2.25$.

$$P = S\frac{2t - 0.01D - 2e}{D - (t - 0.005D - e)}$$

$$t = \frac{PD}{2S + P} + 0.005D + e$$

$$D = \frac{(2S + P)(t - e)}{1.005P + 0.01S}$$

$$S = \frac{P}{2}\left(\frac{1.005D - t + e}{t - 0.005D - e}\right)$$

2. *Tube equation for FT boilers*. Check by using $P = 233$, $t = 0.115$, $D = 3.0$.

$$P = 14{,}000\left(\frac{t - 0.065}{D}\right)$$

$$D = 14{,}000\left(\frac{t - 0.065}{P}\right)$$

$$t = \frac{PD + 910}{14{,}000}$$

3. *Shell equation (a)*. Check by using $P = 125$, $R = 33$, $S = 13{,}750$, $E = 0.866$, $t = 0.4375$.

$$P = \frac{0.8SEt}{R + 0.6t}$$

$$t = \frac{PR}{0.8SE - 0.6P}$$

$$R = \frac{(0.8SE - 0.6P)t}{P}$$

$$E = \frac{P(R + 0.6t)}{0.8St}$$

$$S = \frac{P(R + 0.6t)}{0.8Et}$$

4. *Shell equation (b).* Check by using $P = 908.8$, $S = 17{,}500$, $E = 1.0$, $t = 1.469$, $R = 27.406$, $y = 0.4$, $C = 0$.

$$P = \frac{SE(t - C)}{R + (1 - y)(t - C)}$$

$$t = \frac{PR}{SE - (1 - y)P} + C$$

$$S = \frac{P[R + (1 - y)(t - C)]}{E(t - C)}$$

$$R = \frac{(t - C)[SE - (1 - y)P]}{P}$$

$$E = \frac{P[R + (1 - y)(t - C)]}{S(t - C)}$$

5. *Ligament efficiency.* Check by using $p = 29.8125$, $n = 5$, $d = 3.28125$, $E = 0.448$.

$$E = \frac{p - nd}{p}$$

$$p = \frac{nd}{1 - E}$$

$$n = \frac{p(1 - E)}{d}$$

6. *Semiellipsoidal head.* Check by using $P = 513.5$, $t = 0.75$, $L = 36$, $E = 0.9$, $S = 13{,}750$.

$$t = \frac{PL}{2SE - 0.2P}$$

$$P = \frac{2SEt}{L + 0.2t}$$

$$S = \frac{P(L + 0.2t)}{2Et}$$

$$E = \frac{P(L + 0.2t)}{2St}$$

$$L = \frac{t(2SE - 0.2P)}{P}$$

7. *Flat heads.* Check by using $S = 13{,}750$, $C = 0.417$, $P = 289$, $d = 16$.

$$t = d\sqrt{\frac{CP}{S}}$$

$$d = \frac{t}{\sqrt{\frac{CP}{S}}}$$

$$S = \frac{CPd^2}{t^2}$$

$$P = \frac{St^2}{Cd^2}$$

8. *Furances.* Check by using $t = 0.4375$, $L = 42$, $D = 26$, $P = 174.3$.

$$P = \frac{51.5(300t - 1.03L)}{D}$$

$$D = \frac{51.5(300t - 1.03L)}{P}$$

$$t = \frac{PD + 53.045L}{15{,}450}$$

$$L = \frac{15{,}450t - PD}{53.045}$$

NUMBER OF PLACES TO CARRY DECIMALS

Carry out any number to a decimal that reflects the extent to which measurements and specifications are made in practice or by code requirements.

As a guide, the following is suggested:

1. *Pressure.* Nearest tenth of a pound (one place). This means you have to calculate to two places in order to round out to the nearest tenth. For example:

102.18 = 102.2 psi
102.14 = 102.1 psi
102.15 = 102.2 psi

2. *Length.* Nearest thousandth of an inch (three places). Calculate four places as follows:

26.6491 = 26.649 in
26.6495 = 26.650 in
26.6497 = 26.650 in

3. *Stress.* Nearest pound per square inch. Calculate to the nearest tenth of a pound as follows:

13,749.6 = 13,750 lb/in^2
13,750.2 = 13.750 lb/in^2
13,749.1 = 13,749 lb/in^2
13,750.5 = 13,751 lb/in^2

It can be seen that you carry a decimal one step further than your anticipated answer in order to be able to "round" it out to that number.

APPENDIX **C**

Tabulation of Water-Treatment Methods

Source of water	Purpose of treatment	Type	Method used
Raw water	To remove dissolved and suspended solids, gases, and organic matter	External	Aeration: to oxidize iron, manganese, and H_2S; remove CO_2
		External	Suspended solids: Settling Coagulation Chlorination Filtration
		External	Dissolved solids: Sodium-cycle softening Hot lime zeolite softening Demineralization and evaporation (under 1000 psi boilers)
Condensate return	To remove oil, dissolved contamination	External	Condensate polishing systems: deep-bed demineralizers
Feedwater	To control pH, CO_2, O_2 to minimize corrosion in preboiler equipment		Control of O_2:
		External	Deaerator heater
		Internal	Chemical scavengers—sodium, sulfite, hydrazine
		Internal	pH control: adding neutralizing amines By injection of filming amines

Source of water	Purpose of treatment	Type	Method used
Boiler water (natural-circulation units)	Scale prevention; control pH	Internal	Phosphate-hydroxide, coordinated phosphate, chelant for boilers under 1000 psi, volatile for over 1000-psi units

Selected Bibliography

ASME Boiler and Pressure Vessel Codes—Sections I–VI, IX, XI, American Society of Mechanical Engineers, New York, N.Y.

Fundamentals of Welding, American Welding Society, Miami, Fla.

National Board Inspection Code, National Board of Boiler and Pressure Vessel Inspectors, Columbus, Ohio.

National Fire Protection Codes, National Fire Protection Association, Boston, Mass.

Power magazine reprints, New York, N.Y.—Steam Generation, Fuels and Firing, Nondestructive Testing, Air Pollution, Fuel Handling and Storage, Combustion Control, Computers, Pumps, Water Treatment, Valves, Corrosion, Nuclear Power Reactors.

Power Piping Code, ANSI B31.1, American National Standards Institute, New York, N.Y.

Recommended Practices for NDT Personnel Qualifications and Certification, American Society for Nondestructive Testing, Evanston, Ill.

State, County and City Synopsis of Boiler and Pressure Vessel Laws on Design, Installation and Reinspection Requirements, Uniform Boiler and Pressure Vessel Laws Society, Oceanside, N.Y.

Index

D1550349

CHICAGO PUBLIC LIBRARY
HAROLD WASHINGTON LIBRARY CENTER
R0019596474

Oak Ridge Associated Universities (ORAU) is a private, not for profit association of 45 colleges and universities. Established in 1946, it was one of the first university-based, science-related, corporate management groups. It conducts programs of research, education, information, and training for a variety of private and governmental organizations. ORAU is noted for its cooperative programs and for its influence on the development of science in the South.

SPONSORING INSTITUTIONS

University of Alabama
University of Alabama in Birmingham
University of Arkansas
Auburn University
Catholic University of America
Clemson University
Duke University
Emory University
Fisk University
University of Florida
Florida State University
University of Georgia
Georgia Institute of Technology
University of Houston
University of Kentucky
Louisiana State University
University of Louisville
University of Maryland
Meharry Medical College
Memphis State University
University of Miami
University of Mississippi
Mississippi State University
University of New Orleans
University of North Carolina
North Carolina State University
North Texas State University
University of Oklahoma
University of Puerto Rico
Rice University
University of South Carolina
Southern Methodist University
University of Tennessee
Texas A&M University
University of Texas at Austin
Texas Christian University
Texas Woman's University
Tulane University
Tuskegee Institute
Vanderbilt University
Virginia Commonwealth University
University of Virginia
Virginia Polytechnic Institute and
State University
West Virginia University
College of William and Mary

FUTURE STRATEGIES FOR ENERGY DEVELOPMENT

A QUESTION OF SCALE

Proceedings of a conference at Oak Ridge, Tennessee
October 20 and 21, 1976
Sponsored by
Oak Ridge Associated Universities

Oak Ridge Associated Universities
Oak Ridge, Tennessee

ORAU-130

International Standard Book Number: 0-930780-01-9
Library of Congress Catalog Card Number: 77-87589
Printed in the United States of America

Manuscript editor: Irene Kiefer
Manuscript editor for Prof. Haefele's paper: Jeannette Lindsay
Book design: Evelyn G. Bradford
Conference graphics: Marilyn A. Schuette

Contents

Acknowledgments

The staff of Oak Ridge Associated Universities gratefully acknowledges the generous financial support for the conference provided by the Rockefeller Foundation, the Xerox Corporation, and the ORAU Board of Directors. The Association wishes to thank the moderators at the conference: Walter R. Hibbard of Virginia Polytechnic Institute and State University; Chester L. Cooper, ORAU Institute of Energy Analysis; William Fulkerson of the Oak Ridge National Laboratory; and Frank M. Potter of the staff of the Subcommittee on Energy and Power, Interstate and Foreign Commerce Committee, U.S. House of Representatives. The efforts of Alvin M. Weinberg, director of the ORAU Institute of Energy Analysis, William E. Felling, assistant director of ORAU, and Robert A. Potter, head of the ORAU Information Services Department, were instrumental in the success of the conference. "Energy Strategy: The Road Not Taken?" by Amory Lovins originally appeared in *Foreign Affairs* and is reprinted by permission.

Preface

Having recognized the potential importance of the emerging debate over the scale of future energy systems, Oak Ridge Associated Universities convened an international conference—*Energy Strategies for the Future.* The conference, which was held in the fall of 1976, addressed time and spatial dimensions of scale in terms of social, economic, and technological considerations.

The debate over appropriate scale has now become a major controversy. In its least interesting form, the argument over scale is no more than polemical exchanges between those on the one hand who believe that *small* represents all that is good for man and society and the advocates of *big* on the other hand for whom growth is the *sine qua non* of civilization. But such oversimplifications should not cloud the fundamental importance of the debate, because the resolution holds profound implications for future energy systems and the nature of the societies they will serve. As one participant asked at the conference about the future, "Is it tribal? Is it feudal? Or is it a world of multinational energy corporations?"

The conference also anticipated important aspects of our emerging national energy strategy, and the sometimes spirited debate presaged the reactions to the Carter administration's energy initiatives. Let us hope that the differing viewpoints assembled in this volume enrich the national debate at this critical juncture and that they increase the probability for global acceptance of solutions to the energy challenge.

Philip L. Johnson
Executive Director
Oak Ridge Associated Universities

Introduction

Norman Metzger
Washington, D.C.

Norman Metzger's most recent book, Energy, The Continuing Crisis, *was published in mid-1977. He lives in Washington, D.C., where he has been on the staff of the National Academy of Sciences since 1974. After earning a bachelor's degree from Brooklyn College, he spent three years as a researcher in organic chemistry at Sloan-Kettering Institute. In 1962, he began writing for the American Chemical Society, where he became well-known for the radio show* Men and Molecules; *his first book carried the same title and was based upon his experiences as writer-producer of the show. Metzger, who was a Sloan-Rockefeller Fellow at Columbia University in 1968-1969, is also a prolific filmwriter and contributor to scientific yearbooks and other publications. In 1972, he joined the staff of the American Association for the Advancement of Science, where he wrote and produced an unusual set of audiotape programs about the energy crisis under the title* Energy: A Dialogue.

"If there is a comprehensive energy problem, it is a problem of choice and value in a world of finite capabilities," a political scientist, Lynton Caldwell, has reminded us.[1] The maturation of the energy crisis has been one of learning exactly what those choices and values are and the costs and rewards of choosing particular ones. The arguments of environmental protection versus energy supply needs, of price mechanisms versus regulatory mechanisms for controlling energy use, of the links between Gross National Product and the growth of energy demand, of the risks and benefits of new technologies—these and other binary issues are expressions of society's problem, if not agony, in choosing how it wants to use finite supplies of energy.

The argument that small is beautiful and big is bad is possibly the most powerful expression of the social roots of our energy situation. For in the arguments that we can do more with less (albeit carefully defining what more is) and that energy sources should be shaped by their actual uses, we are really being asked what our values are. And we are being asked to make choices: Do we indeed wish to change the historical nature of energy supply and distribution? Do we wish, for example, to decentralize our energy supplies and reduce their scale, shaping energy sources according to their final use, rather than to availability of massive primary energy sources? Do we wish to turn as rapidly as possible to renewable energy sources fitted to small-scale systems? Can we reduce our energy consumption without inviting a stagnant, even depressed economy? Do energy supply and GNP march in lock step, or are the causal relations more apparent than actual? At what cost do we abandon current programs intended to foster massive new supply technologies, such as the fission breeder or fusion—all intended to supply electricity?

The view that the energy situation forces social decisions also implies that it has no technical solutions, and certainly none are offered in this volume. What is offered is an examination of alternative futures, and costs, risks, and benefits inherent in each. It attempts to lay before society the facts and sophisticated intuitions that are available as an aid to making choices. Is small truly beautiful or is big necessary? What are the features of centralized systems—the creation of 1,000-megawatt-plus electrical power plants, for example—that have made them apparently so successful (depending, of course, on the measures used) in advanced, industrialized countries? Or, to look at it differently, why has society chosen to put the bulk of its governmental energy investments into the creation of even more massive centralized systems—the breeder reactor and nuclear fusion?

But the first reality in approaching the social choices we have in dealing with energy is not in understanding the nature of energy supply systems and effects of various levels of demand, but in understanding clearly the assumptions involved in making a choice. For in all cases we are dealing with the future, and, as Niels Bohr reportedly remarked, "It is very difficult to make predictions, especially about the future."

Part of the problem in making choices is our own cleverness. We can no longer take the historical route of testing new concepts and technologies by trial and error, because the trials involved are too massive and the possible errors terribly costly. One does not want to deploy several hundred breeder reactors, only to find out that the plutonium produced simply cannot be dealt with. Nor does one want to put a stop to nuclear fusion power, and then learn that the consequent lack of electricity has seriously damaged our economy. Nor does one want to develop various forms of solar energy using a rapid timetable only to find out that there are unforeseen environmental effects and very high costs to the consumer of electricity. The latter possibilities are quite real, given that both the problems and costs of new technologies, such as commercial nuclear power, have increased as we became better acquainted with them.

"The experiment," C. S. Holling comments in this volume, "should not, ideally, destroy the experimenter—or, at the least, someone must be left to learn from it. Nor should the experiment cause irreversible changes in its environment . . . we are now ignoring those minimum conditions. Our trials are capable of producing errors larger and more costly than society can afford."

We try to close the gap between what we need to know and what we can actually test in various ways: with computer models, with small-scale, even full-scale tests that still do not include the elements of time, of actual operating conditions, of human foibles. All our efforts to find out what will happen appear inadequate, and therefore our choices are invariably difficult, and never completely defensible, and often partly wrong.

In decisions there is no surety, only the certainty of risk and the inevitability of surprise. As Kenneth Boulding emphasizes, "every decision is a choice among alternative images of the future present in the mind of the decision-maker. The images themselves . . . may be more or less realistic; they are created, however, by the learning experiences of the decision-maker, and these in turn are affected profoundly by the psychological makeup of the decision-maker, by the organizational milieu and by the 'image-climate' in which the decision-maker is placed."

Part of the problem facing the decision-maker, although not always apparent, is that decisions rely on images that are themselves abstractions, adumbrations of a world too complex for the mind to rationalize. "All image formation . . . involves abstractions," Boulding says, "a simplification of the real world to the point where our mind can contain it." The danger is in forgetting that these images are abstractions; in thinking of scenarios, models, myths, or metaphors as portrayals of a real world.

Holling and Boulding, both of whom are concerned in their essays with the pitfalls and tactics in making choices, see that the correction of errors is intrinsic to the forming of images and, more particularly, in attempting actions based on them.

"Myths," Holling explains, "are a way in which mankind captures some essence of experience or wisdom in a simple and elegant form. Myths are necessary to guide us in our actions, and they protect us from that larger reality of the frightening unknown . . . But myths are only a partial representation of reality. Through their acceptance, man becomes bold enough to accumulate new knowledge that eventually exposes their incomplete nature. That introduces confusion; old myths are dusted off, and new ones developed." There is also, after all, the real world. Thus, Boulding writes, "error has some chance of being detected, whereas truth has the real world on its side."

Related to this is some analysis of the pathology of decisions, what makes bad decisions so. That is ultimately decided politically, for, quoting Boulding again, "there are processes—of election, of impeachment, even of revolution, and ultimately the verdict of historians—by which we reach evaluations that are more than the private evaluations of individual persons and that are widely, if not universally, shared."

A part of perceiving the worth of decisions is understanding the terrain in which they are made. Holling attempts that analysis. We can assume, and historically have, the condition of what Holling describes as Nature Benign, one that "comfortably and logically leads to big is necessary." Holling also points out that "trials and mistakes of any scale can be made in this world, and the system will recover once the disturbance is removed." The reverse of this is Nature Ephemeral, which is "in harmony with the persistent traditions of anarchism and of small is beautiful. It is in complete opposition to big is necessary, since any effort to increase size, to homogenize, will lead to collapse as the underlying instability of nature becomes obvious." These are absolute portraits, and the errors are obvious; for nature is not always benign, nor is it invariably unstable. Therefore, we also have Nature Perverse/Tolerant that "can accommodate either small is beautiful or big is necessary." And even while we are trying to find our way on a decisional landscape, the very topography changes; we are given a kaleidoscopic world that is difficult to understand and mathematically confounding. Therefore, the final myth is Nature Resilient, one that recognizes that the "dynamic pattern of the variables of the system and of its basic stability structure lies at the heart of coping with the unknown."

A myth that nature is resilient has the "property that allows a system to absorb *and utilize* (or even benefit from) change," according to Holling. "Such a myth explicitly recognizes the unknown and the ability to survive and benefit from 'failures."

All well and good, but in what way does the myth of Nature Resilient mirror reality and in what ways does it delude? The bulk of the essays in this book—those of Lovins, Haefele and Sassin, Craig, Weinberg, and Lee—take up that question.

The issues common to all the essays are those of how much energy we need to enable a quality of life that remains somewhat indefinable and by what sort of infrastructure that energy shall be supplied. That is, will energy be supplied by large-scale, centralized systems, assuming an ever-increasing proportion of primary fuels converted into electricity, the massive availabilities of these primary fuels (whether coal, uranium isotopes, or tritium), and long-distance distribution networks? Or, alternatively, shall there be a transition to a decentralized energy economy, one in which energy sources are shaped to needs, distribution distances tend to be very short, and the fuels are renewable?

Of course, the fact that our energy systems are now centralized is a product of social choice. For example, W. Haefele and W. Sassin point out that the centralized infrastructure reflects the availability of cheap oil and gas, fuels that are (or were) cheap through a combination of economic and political policies, national and international. If we were to opt for something different, it would be a result, as Alvin Weinberg remarks, of society's responses to some difficult questions.

Thus, Alvin Weinberg notes, in asking about centralization versus decentralization, small is beautiful versus big is bad, of soft versus hard paths, we are asking what is "the ideal of a good society: Can we have a good society in which the primary means of production, in this case energy, are in the hands of a small, remote elite? Is true human freedom possible in a society so technologically centralized? . . . But we would turn the question around. Can human freedom prevail in a society that is short of energy?"

Amory Lovins seems to see the menace in Weinberg's last question as still worth the risk, that what is to be gained is worth the gamble and indeed that the gamble is very small. "While soft technologies can match any settlement pattern," he says, "their diversity reflecting our own pluralism, centralized energy sources encourage industrial clustering and urbanization. While soft technologies give everyone the costs and benefits of the energy system he chooses, centralized systems allocate benefits to suburbanites and social costs to politically weaker rural agrarians. Siting big energy systems pits central authority against local autonomy in an increasingly divisive and wasteful form of centrifugal politics that is already proving one of the most potent constraints on expansion."

Inherent in the debate over small and big is whether what we have now is what people wanted or simply the product of factors that have led to a result that is both unnecessary and ultimately inimical to the best interests of all men. Lovins answers that in his very definition of "hard" and "soft" energy paths for the United States in the next half-century: "The first [i.e., hard] path resembles present federal policy and is essentially an extrapolation of the recent past. It relies on rapid expansion of centralized high technologies to increase supplies of energy, especially in the form of electricity. The second [i.e., soft] path combines a prompt and serious commitment to the efficient use of energy, rapid development of renewable energy sources matched in scale and in energy quality to end-use needs, and special transitional fossil-fuel technologies."

Lovins argues that rich and (artificially) cheap energy supplies have produced systems in which end-use patterns are shaped more by the sources of supply than what people actually need, with the result that "we are using premium fuels and electricity for many tasks for which their high-energy quality is superfluous, wasteful and expensive, and a hard path would make this inelegant practice even more common."

Much of Lovins' essay is invested in showing why the continuance of the historical modes of supplying energy are—or certainly will be— inimical to the interests of society, whether those interests are measured in economic costs, in environmental lesions, in the "quality of life." Rather than that bleak future, we can turn to the goal of using energy more efficiently and making the transition to the soft ways of supplying energy. Such an approach relies on technologies that are diverse, renewable, flexible, of a scale that is consonant with the way they are actually used, and of a quality fitted to their use.

It is a specific case of E. F. Schumacher's "Buddhist economics" in which, as Schumacher describes it, "production from local resources for local needs is the most rational way of economic life, while dependence on imports from afar and the consequent need to produce for export to unknown and distant peoples is highly uneconomic and justifiable only in exceptional cases and only on small scale."[2]

Paul Craig agrees with Lovins that "our thinking in the United States about energy technologies is at present far too strongly oriented toward massive approaches to energy supply." Given the perhaps unbearable costs of new energy systems of huge scale, along with their increased technological complexity and potentially serious environmental effects, ways must be sought to reduce energy demand without affecting quality of life. A new era for urban mass transit is one, and Craig carefully examines it. Measured in terms of energy outgo, investments in mass transit, he argues, will use financial capital more efficiently than new systems for supplying energy. And, possibly, the amount of capital needed may be lower in absolute terms. Craig's conclusion is that "plausible estimates of long-term energy savings that might derive from the most expensive form of mass transit system suggest that investment in such systems may be, from a purely energy point of view, comparable to investment in new energy supply systems when measured in terms of efficiency in the use of capital."

Of course, the question again is whether what people do is a reflection of technology or rather, that technology is successful because it satisfies social needs and wishes. If the latter, then new technology, no matter how elegant and economic, can have a hard road. Craig acknowledges the problem: "Causal relationships between mass transit and population density are not well known." Studies of the Bay Area Rapid Transit (BART) system and of an Illinois commuter line show that construction tends to follow rail lines. But the success of transportation systems depends axiomatically on population density, and "we do not know how much population density increase can actually be attributed to a mass transit system and how much to other factors." In all, while mass transit has the potential, it is also "clear that far more extensive analysis will be required in order to assess just how large the opportunities for energy conservation really are."

The urgency of increasing energy efficiencies makes the case for Craig's analysis, while the very fact that urban mass transit has had a difficult time in this country reminds us that there are other elements to consider than just good technology. Thus, the car—its availability is constant and immediate, requires no queuing, and is *still* cheap to run. The problem seems to be in closing the gap between long-term good sense and immediate satisfactions and necessities.

Nevertheless, both Craig and Lovins seem to be disagreeing with Napoleon's comment that history is destiny; that we can indeed, once we are persuaded by the argument, make a deliberate social choice of embarking upon a different way. Haefele and Sassin, as well as Lee, try to examine history before deciding on destiny, while Weinberg attempts to gauge, within "asymptotic settings," the ability of different styles of energy power systems to satisfy probable demands.

Their conclusions are relatively uniform: Haefele and Sassin, for example, see lower energy use in the rural regions of Third World countries as an expression of poverty and not as a crude model for soft paths; Lee argues the universal economic benefits of centralized electrical power, and Weinberg, calling on a study by the ORAU Institute for Energy Analysis of the effects of a nuclear moratorium beginning in 1985, states that "our findings on the whole give little consolation to those who urge decentralization."

Haefele and Sassin see no evidence—evidence in terms of what is actually happening versus what is wanted—that people are freely choosing low-scale, limited distribution, decentralized, labor-intensive energy technologies. Where these conditions exist, it is because of a lack of capital, not desire. They acknowledge, and indeed it is the refrain to their paper, that there are "contrasting views of the future." But in contrast to some projections that various *national* rates of growth in energy consumption may begin to taper off, Haefele and Sassin see increasing per capita energy use on a *global* scale. The driving forces are continued economic growth of the developed countries, further economic improvements in the less-developed countries, and, underlying these two, the continued growth in world population. As standards of living improve universally, as urbanization continues, as capital becomes more available, the demand for more energy will intensify, swept along by the global demand for something better. The gaps between the rich and poor nations are now enormous—gaps not simply of resources, but also of the wherewithal to exploit them or to buy them, of industrial infrastructure, of an educated, entrepreneurial class. Whether these gaps can be closed is problematical, but we should plan our agendas, including those of energy supply, as if they will. In an interdependent world, Haefele and Sassin argue, if stability is to be maintained then the differences in the qualities of lives must be narrowed. However, that will take time, with the transition certainly not possible before 2000.

Urbanization has been the dominant demographic trend in the 20th century, and therefore its effects on energy consumption cannot be waved away. Indeed, its effects are massive, if not controlling. Centralization of the population, Haefele and Sassin note, is followed by high per capita energy consumption that must be met by centralized energy supply systems, often relying on massive power plants, long transportation routes, and distribution lines. The picture seems to be the same no matter the country or its comparative economic condition. Haefele and Sassin site metropolitan areas in India and Germany with virtually the same energy densities, but per capita consumption in India is still lower, with the equivalents of energy densities due to the higher densities of Indian urban populations.

If the urban Indian were to improve his standard of living, satisfying his hopes and (we like to think) ours, then total energy demand would go up sharply, pressing even more heavily on energy supply systems, and making the possibilities of turning to the soft paths even less plausible.

Moreover, Haefele and Sassin argue, the nature of future metropolitan areas, assuming that urbanization continues, will determine the nature of the energy technologies that can be applied. If United Nations projections are accepted, then future patterns of urbanization in developing countries will follow that of the industrialized countries, with the population clustering about a fairly small number of metropolitan areas. Such a pattern, Haefele and Sassin emphasize, aside from its social and physical aspects, also has technological implications, including the manner of supplying energy and distributing it.

Thomas Lee, in his analysis of the benefits of centralized supplying of electricity, does not rely, as Haefele and Sassin did, on analyses of population and developmental pressures. Rather, he examines the apparent reasons for the historical pattern of the nature of electricity supply, the modulating role of capital, and the relationship of the existing system to what people pay for electricity. The basic point, as Lee states it, is "what is the optimum way to supply the electricity still needed after all reasonable efforts are made to conserve?" A slightly different way to put the question is whether a different structure for the electrical power industry—one consisting of small, neighborhood-level units—would serve the public better. Lee thinks not, because ". . . there is economy of scale in both component and plant construction. By taking advantage of this through centralization, the industry has been able to continually reduce the price of electricity in the past 20 years . . . The analysis also shows that while we may have already reached the optimum plant size for the present technology, there is no reason to think that decentralizing the electric power systems will be beneficial to the public."

Weinberg approaches future energy strategies from an "asymptotic" perspective, looking infinitely into the future to the time when fossil fuels are gone and we face alternative futures relying either on the energy of the sun or upon the fertile energy of atoms freed by breeder reactors. In effect, Weinberg is examining Holling's Nature Resilient. He moves from one miniequilibrium to another, leaving the understandable and familiar terrain where fossil fuels are still available to voyage into that murky region where the current stopgaps—such as relying on coal in the absence of nuclear—are no more. It is a region in which "the great economic discrepancies between poor and rich have been eliminated."

This is, of course, a mythical world, a metaphorical one, one set up for an exercise; and we should keep in mind the warnings of both Boulding and Holling on the uses and limits of such games. Nevertheless, as Weinberg points out, looking at an asymptotic world in which the choice is between the sun and the breeder raises interesting questions and forces some disturbing conclusions. Creating such a world "brings out most clearly what may be the most essential choice: between a stable world in which all have a relatively large per capita energy but which places great pressure on the environment, and an unstable world in which the average per capita demand is very low . . . but the environmental pressures are much smaller."

Weinberg finds that an "all-solar future is almost surely a low-energy future, unless man is prepared to pay a much larger share of his total income for energy than he now pays." If it is to be nuclear, then it will have to be a world in which currently estimated accident rates are much lower, in which socially acceptable systems exist to deal with some ten tons of plutonium each day. According to Weinberg, "we can hardly escape the impression that the price nuclear energy demands, if it is indeed to become the dominant energy system, may be an attention to detail, and a dedication of the nuclear cadre that goes much beyond what other technologies have demanded."

Weinberg cannot reconcile a "cornucopian world," in which global per capita energy use rises to about the West German level today, with decentralized systems for delivering energy. If the cornucopian world is to be possible, then our energy path will indeed be hard, consisting of technologies that are centralized, predominantly nonsolar, heavily electric, with breeder reactors the main source of the heat needed to raise steam for electrical turbines.

Lovins also asks another question, tangential to that of the fundamental choice between hard and soft paths: "Since the energy needed today to produce a unit of GNP varies more than 100-fold depending on what good or service is being produced, and since GNP in turn hardly measures social welfare, why must energy and welfare march forever in lock step?" A reasonable question, and one that has been and is being asked in any number of energy studies. Usually, the question is "simplified" to that of rate of growth in energy consumption as related to the rate of growth in GNP, with the usual disclaimer, as stated by Lovins, that GNP is a measure of the whole system but tells us nothing of what is happening within.

The first inclination is to say that there is a strong

relationship, after noting the parallelism in the curves of past growths in energy consumption and GNP. But do these relationships reflect causal links or simply common ones? Sam H. Schurr and Joel Darmstadter approach the issue by disassembling these historical curves and then analyzing them. That is not simple, for, as they note, "there is a striking absence of solid information on the relationship between energy and economic growth."

But they do find that the nature of the relationship between energy and GNP has changed radically in the past several decades, reflecting structural changes that are both social and economic. For example, the nature of primary energy sources changed radically in the 20th century—from coal to oil and gas. Uses changed—to electricity and gasoline. These interwove with transformations of industry, agriculture, transportation, manner of life. There were changes in the efficiencies of labor and in the use of capital, in patterns of employment, and, coming full circle, in the use of energy as evidenced by the fact, Schurr and Darmstadter note, that "over much of the time, energy consumption per unit of national output has persistently declined."

Overall, they conclude that "despite the similarity of movement in energy consumption and employment in the aggregate and over the long term, it seems to us questionable to assert a rigid linkage between these two factors, at least not without probing underlying elements in the relationship."

International comparisons are also instructive. The question is what accounts for the high per capita energy consumptions in the United States compared to that of other advanced industrialized countries. In answering that question, Schurr and Darmstadter point out, we can achieve a "better understanding of the combined effects on the growth of future energy consumption of structural changes in the economy, of relative price increases in energy, and of changes in energy-use technology in response to rising energy prices."

There are also events and elements that are not easily counted, but, as Schurr and Darmstadter note, still affect the question of energy and economic growth. These include, for example, public policies on energy conservation. In what way do prices—should prices—support legislative goals? What of changing social attitudes, of people living closer to their jobs or a greater interest in urban living? What do these quite possible trends, as well as more extreme ones such as a mass return to nature movement, portend for energy consumption and economic growth?

The discussions in this book come to no final resolution; no questions are answered with certainty, no difficulties waved away. The intent was to examine what we can do, and where, according to men who have thought about it a good deal, our best opportunities are for promoting the general welfare. As obvious as it is, it's sometimes forgotten that energy is a tool for doing something, with the opinion of what that something is differing individually, regionally, nationally, and globally. But, one hopes, whatever disagreements on the scale of the tool, there is agreement on its uses, that in common there is shared the goal that Ruben Nelson described in his "The Illusions of Urban Man."

". . . we are after a world of multitudinous forms which reflect and reinforce our deepest and truest insights into what it is to live together as persons. It is not enough that we want to be men and women who are fit to live with, or that we want a world that is fit to live in. What we must learn to do is embody our best intentions and aspirations in our physical, organizational, and linguistic forms. Only so will we be able to move towards a world which sustains rather than distorts life . . ."[3]

REFERENCES

1. Caldwell, Lynton K. Energy and the Structure of Social Institutions. Human Ecology, Vol. 4, No. 1, 1976. p. 32.

2. Schumacher, E. F. Small Is Beautiful: A Study of Economics As If People Mattered. Harper & Row, Publishers, New York. 1973.

3. Nelson, Ruben F. W. The Illusions of Urban Man. Published by Macmillan Co. of Canada, Ltd., for the Ministry of State for Urban Affairs. 1976. p. 61.

Kenneth E. Boulding
Institute of Behavioral Science
University of Colorado
Boulder, Colorado

Kenneth E. Boulding, director of research on general social and economic dynamics at the University of Colorado's Institute of Behavioral Science, was professor of economics at the University of Michigan from 1949 to 1967, directing its Center for Research on Conflict Resolution for three years. Holder of 20 honorary degrees, the Oxford graduate was a member of the faculty of the University of Edinburgh, Scotland, early in his career and later was associated with Colgate, Fisk and McGill universities and Iowa State College. He also served as a Danforth Visiting Professor at the International Christian University in Tokyo.

As author-editor of some 25 books, Mr. Boulding has published such titles as Beyond Economics, *which was nominated in 1970 for a National Book Award, and* Sonnets from the Interior Life and Other Autobiographical Verse. *He is past-president of the Society for General Systems Research and the Peace Research Society.*

At Oxford he took his bachelor's degree in the natural sciences and the master's in the School of Philosophy, Politics, and Economics. Mr. Boulding, who was born in Liverpool, became an American citizen in 1948.

Determinants of Energy Strategies

Strategy is a decision about decisions. Shaving is a decision; being clean shaven is a strategy. Getting up in the morning is a decision; taking a job is a strategy that affects our decision about getting up in the morning. Building a nuclear plant might be the result of a decision; a nuclear moratorium would be a strategy.

Energy strategy, like all others, almost certainly for good rather than for ill, is not the decision of any single person. It emerges as a result of the interaction of a very large number of the 4 billion decision-makers of the human race, some of whom of course are more powerful, that is, affect more people more strongly, than are others. The dynamic processes by which the role of energy in society is changed may be affected in a small way even by this conference. They may be affected more by the decisions of the powerful, like the people who make the decisions for the Organization of Petroleum Exporting Countries. They are also affected by the political and social climate, and by the fashions of opinion or behavior that affect the decisions of the powerless, who collectively make far more difference to the world than do the decisions of the powerful. Power, one suspects, is 99 percent illusion, though of course the 1 percent that is not illusion does matter.

Behind all strategies lie the conditions of decision. Every decision is a choice among alternative images of the future present in the mind of the decision-maker. The images themselves (the "agenda" of choice) may be more or less realistic; they are created, however, by the learning experiences of the decision-maker, and these in turn are affected profoundly by the psychological makeup of the decision-maker, by the organizational milieu, and by the "image-climate" in which the decision-maker is placed. Once the agenda of choice is present in the mind of the decision-maker, the selection process by which one of the items or elements of the agenda is chosen emerges out of the valuation processes of the decision-maker. The agenda-formation and the valuation processes constantly interact; an alternative future, for instance, that may be quite realistic may be rejected because it is painful in terms of the values of the decision-maker, or one that is unrealistic may be accepted because it conforms to the values.

All thinking is at least partly wishful, and values always affect perceptions; up to a point this may not matter, but beyond a certain point this process becomes pathological. Out of these very complex processes finally emerges a value ordering of a specific agenda; the agenda item with the highest value is selected, and a decision is made. This is what the economists call "maximizing behavior," which simply says that all people do what they think is best at the time, a proposition so empty that it is almost impossible to deny. To put content into it we must examine the processes by which both agendas and value orderings are formed in the mind of a decision-maker. These processes are extremely complex and hard to discover.

A critical question, which is not easy even to formulate, is what makes a decision a "bad" decision. This is the problem of the evaluation and the pathology of decisions. In economics, there seem to be no bad decisions. Everybody chooses among a set of realistic alternations according to a structure of value orderings, beautifully rendered by indifference curves, about which there can be no further question. This type of analysis is a good place to begin. It is a poor place to end, however, simply because it is hard to deny the possibility of bad decisions, for everyone is conscious of having made them. The definition of a bad decision, however, is not easy. It is not merely a decision with unexpected consequences, for our agenda may have been faulty and a decision can still turn out well. Furthermore, almost all decisions have unexpected consequences, for all decisions are made about imaginary futures, not about real ones, and it is only for rather trivial, very shortrun decisions that the imagined futures turn out to be real. The definition of a bad decision in terms of subsequent regret has merit, though it runs into the difficulty that it does not distinguish between regret about unanticipated consequences and regret about the values that determined the original decision. Furthermore, although the minimizing of regret, as Marschak[1] has suggested, is a useful formal principle in dealing with decision-making under uncertainty (and all nontrivial decisions are under uncertainty!), it suffers from the formality of the general principle of maximizing behavior if we cannot specify a learning process by which regret about past decisions leads to less regret about present and future decisions.

An even more difficult question in the evaluation of decisions is that of decisions that other people regret—people, that is, who are not the decision-makers. Almost every decision affects other people besides the decision-maker, and these effects may be evaluated by the subject as being for the better or for the worse. The "power" of a decision-maker indeed may be defined by the number of other people who are affected by the decisions, or by the sum of the estimates of "betterness" or "worseness" that these decisions evoke in others, if these could be summed. The political process is the means by which joint estimates of the quality of powerful decisions are made over time. We may never reach consensus in these matters; nevertheless, there are processes—of election, of impeachment, even of revolution, and ultimately of the verdict of historians—by which we reach evaluations that are more than the private evaluations of individual persons and that are widely, even if not universally, shared.

Another possible approach to the pathology of decisions is to look at the environment of the decision-maker to see if there is anything in this structure that is likely to lead to bad decisions rather than good ones. One does not have to be a determinist (which I am not) to argue that the decision-maker's environment affects the probability that one decision will be made rather than another. This environment, as noted earlier, has a number of facets. There is first the internal psychological environment of the mind of the decision-maker—the genetic predispositions; the conscious and unconscious trauma from past experience; the general quality of disposition towards openness or secrecy, benevolence or malevolence, trust or mistrust, responsibility or irresponsibility; or any of the whole dictionaryful of dimensions of the human disposition. This internal environment is partly a biological artifact through the genetic and somatic constructions that build the body, but it is mainly a social artifact through the information inputs that the individual has received from birth. It is n-dimensional to the point where exact description becomes impossible, but without some attempt at describing it, psychohistory at least would be impossible. In systems that have a small number of powerful decision-makers, their internal environments cannot be neglected. Where there are large numbers of not very powerful decision-makers, something like a law of large numbers may come to the rescue, and their internal environments may be regarded as random disturbances that cancel out in the mass, as in the economic theory of perfect markets.

A second facet of the decision-maker's environment is the "image climate." This consists of the images of the world and of values present in the minds of the persons who constitute the culture within which the decision-maker has been formed and within which he or she operates. The image climate has a large effect on the learning processes by which the internal psychological environment is so largely formed.

This accounts for the fact that within a given culture or subculture of interacting persons, there is convergence in the internal environments so that the members of such a culture become more alike.

A very important part of the image climate consists of the metaphors by which images of complex systems are presented to the mind and transmitted to other minds. The development of images of complex systems, such as the world economy, the international system, or even a large organization, is very difficult. The real world is more complex than any image that we can form of it; this must be so, because the human nervous system itself is part of the real world. All image-formation therefore involves abstraction, that is, a simplification of the structure of the real world to the point where our minds can contain it.

Abstraction takes two forms: models and metaphors. A model is a structure of specified parts, and relationships among the parts, that has something approaching a one-to-one relationship with the reality of which it is a model; a map is perhaps the simplest example; a mathematical model is a map-description in various abstract dimensions. A metaphor is a verbal mapping of the unfamiliar on the familiar; it says that something we do not know is "like" something that we do know. The construction and understanding of models is an advanced skill, largely confined to the scientific community. Metaphors can be understood even by the least sophisticated people and are the main stock in trade of preachers, politicians, lobbyists for both virtuous and mercenary causes, advertising men, prophets, persuaders, and visionaries. The danger is that while models may be more accurate, metaphors are more persuasive and often imply models that are not accurate. The sliced pie so beloved of distribution theorists, the balance of power, the domino theory, the eagle, the lion and the bear that are the totems of nations, the lifeboat ethic, the nuclear priesthood—all are metaphors and all imply models that may not be very accurate; nevertheless they have great powers of persuasion.

Fortunately, or perhaps for some people unfortunately, there is a real world. This constantly exerts an impact on the image climate, and even on metaphors, simply because of a fundamental asymmetry—that error has some chance of being detected, whereas truth has the real world on its side. Consequently, over the long pull our images of the world have a better chance of moving towards truth than they do of moving towards error, even though there are plenty of examples of the latter, thanks to the persuasive powers of human communication of all kinds in words, music, buildings, regalia, art, and all the other instruments of deception. But all persuasion, one hopes, will ultimately fall victim to truth, though it does seem to take an unconscionably long time.

Another facet of the decision-maker's environment, and a frequent source of bad decisions, is organizational structure. This particularly applies to powerful decision-makers, for power is largely a function of position in a hierarchical organizational structure. Even nonhierarchical structures, however, like a committee with a weak chairman, are capable of generating errors of agenda and of values, of which the supposed "risky-shift" phenomenon[2] is a possible example. According to this phenomenon, a committee will make riskier decisions than individuals acting on their own, simply because everyone on the committee has an alibi and does not have to take full responsibility. Private decisions by single persons, the consequences of which impinge only on the decision-maker or his immediate circle, are probably the least subject to error and produce the most rapid feedback and learning process when they are in error. Unfortunately, this class of decision is denied to the powerful. In the mass, however, it may account for the important element of truth in Adam Smith's "invisible hand" metaphor, and the remarkable capacity of societies to correct, by innumerable small decisions of its ordinary members, for the follies, perversions, and extravagances of the powerful. This indeed is the great political virtue of the market mechanism, that it makes political decisions virtually unnecessary and relies on the cumulative impact of the small decisions of little people, which are much less likely to be in error than the large decisions of the great.

The decisions of powerful people, indeed, are almost doomed to be bad. Hierarchy corrupts communications, for a person rises in hierarchy by pleasing a superior, not by telling the truth. Consequently, the information reaching the top of a hierarchy has been filtered out at all the intermediate levels until it is likely to contain a great deal of error. Usually, it is only the bottom ranks of a hierarchy that are in direct touch with the real world—the foot soldiers, the parish priests, the students, the sales clerks. The powerful people at the top always have to rely on indirect and filtered information, which organizations are usually designed to corrupt. To some extent, this tendency for hierarchy to corrupt information can be offset by informal meetings, staff organization that bypasses the hierarchy, or the king's going out and wandering incognito through the slums, but even these devices tend themselves to become hierarchical. Nevertheless, no large organization is possible without hierarchy, simply because of the principle that whereas a small number of people can all communicate with each other, as the number rises the number of pairs of potential communicators increases almost exponentially, so that beyond a certain magic number (around seven) communications increasingly break down unless they are organized hierarchically. There must, however, be important other economies of scale in terms of division of labor, indivisibility of equipment, control of the market or the political environment, and so on, that offset the diseconomies of hierarachy; otherwise large organizations, like the dinosaur, would fall of their own weight and communications corruption. From the point of view of the probability of bad, or even fatal, decisions, however, each form of organization is likely to have an optimum scale, above which the probability of bad, size-reducing, or even catastrophic decisions rises sharply.

Another aspect of this problem is the mechanism by which powerful roles get filled. If there are only a few powerful roles in a society, some selective process is needed to distribute them at any one time among the persons of the population. These processes include hereditary roles as in a hereditary monarchy and roles filled by promotion, election, or chance. I have suggested that there is a "dismal theorem" of political science—that most of the skills enabling individuals to rise to power unfit them for exercising it. If this is true—and in the absence of strong random elements in the selection of powerful role occupants it is all too likely to be true, as in the famous Peter principle[3]—then the wide distribution of power is a certain safeguard against the almost inevitably bad decisions of the powerful. There is a certain dilemma here—wide distribution of power may protect against very bad decisions, at the cost of assuring there will be no exceptionally good ones. The cost of egalitarianism, especially in regard to power, may be a safe mediocrity.

If the evaluation of decisions is difficult, the evaluation of strategies seems even more difficult, because a strategy is precisely a decision about the evaluation of future decisions. Nevertheless, the "theory of bad decisions" that we have been outlining, while it is not the whole problem of strategy, is a very important part of it, for one of the major objectives of strategy is precisely to outline certain principles and organizational structures that will make for better decisions, rather than worse, in the future. Thus, a constitution is a political strategy designed to protect a society from bad political decisions in the future, partly by organizational structures like "checks and balances" and also by the removal of certain potential alternatives from future political agendas. Much bad decision comes from a failure of agenda—too large agendas lead to information overload and a "clutching-at-straws" pattern of decision; too restricted agendas lead to a failure to perceive alternatives that would be valued more highly than any on the immediate menu. Some things should be removed from agendas—these are the "no-no's" of morality and law—and some should be deliberately added, like the environmental impact statements that force decision-makers to consider agendas they might otherwise have overlooked.

Perhaps the greatest problem of strategy is that it contemplates a much longer future than do most decisions, and the problem of the validity of images of the future becomes increasingly difficult—except under some unusual conditions of future convergence—as the time horizon expands and we look further and further into the future. The images of the future that form the agendas of decisions are obtained mainly by projecting the perceived patterns of the past, though often in very complex ways. We may use simple "celestial mechanics," as when we decide to jump out of the path of a moving vehicle that is bearing down on us. More often, we use complex combinations of images of space-time invariances, as when we decide to go to the post office and draw on our past experience that gives us an internal map of the town and knowledge of our capacity to follow a certain route to where we think the post office is located. All images of the future, however, involve some uncertainty; the approaching vehicle may accelerate or swerve, the post office may have moved, we may have an accident on the way, and so on. Uncertainty rises as we take longer time spans, mainly because the longer the time span we take, the larger becomes the probability that any uncertain event will happen *within* the time span. However, we cannot know when, and the "when" matters enormously in determining the actual pattern of the future. We are virtually certain that San Francisco will be destroyed by an earthquake in x years unless we find out how to lubricate faults, and that Washington will be destroyed by a nuclear weapon in y years unless we learn how to live at stable peace. We just do not know what x and y are, and there is no way in which we can know.

The specter of irreducible uncertainty broods over all models of the future, and strategy of all kinds must take account of this. Indeed, it is possible that a new source of bad decisions has been generated in recent years—the illusions of certainty that come with belief in sophisticated, computerized models of the future that do not take sufficient account of the irreducible uncertainties. The greater the uncertainty, the more important it becomes that, as new knowledge accumulates, strategy should involve liquidity, noncommitment, holding options open, and widening the agendas of future decisions. The most catastrophic situation can well be the illusion of certainty, in which case the strategy becomes to zero in on disaster. The Vietnam War may well have been an example of this.

How, then, do we apply these principles to the problems of energy strategy? We must start with some kind of image of the "real world" and the role of energy in it, with the full realization that any image that we present will have errors and uncertainties in it. What follows is my own image, offered for critique and possible testing.

The role of energy, both in the biosphere and in society, is dominated by the fact that energy is essential to any process of production. Production is a process by which a genotype becomes a phenotype, by which an egg becomes a chicken or a blueprint becomes a house. All processes of production start off with a structure of know-how that is the genotype. This know-how is able to capture energy for two major functions. One is to sustain the appropriate temperature at which the production processes go on, whether this is the 98.6° of the human body or the $1{,}000^{\circ}$ of the potter's kiln. The other is to perform mechanical and chemical work by which materials are first selected, then transported, transformed, and rearranged into the improbable shapes of the phenotypes. From the point of view of the productive process, energy is not just ergs; it is ergs when and where and in the form in which they are wanted.

In the biosphere, production is "epigenesis," the process by which biological artifacts such as rabbits are made from fertilized eggs. In the social system, production is the process by which human artifacts, from the flint arrowhead to the space lab, are made from fertilized ideas. When these artifacts enter into exchange or gift relationships, they become commodities and enter into the field of economics. They become commodities, however, only because they have a larger social aspect as creators of utility. That is, they must be things people want, whether to satisfy biological needs such as food or to satisfy social desires for adornment, ceremony, and status. The production of all human artifacts, like that of biological artifacts, begins with an idea, combined with know-how, which is the genotype. If this is to be realized in production, it must be able to direct energy towards sustaining temperatures and transporting and transforming materials into the appropriate shapes. The absence of either energy or materials can limit the process of production. We see this on the moon, where evolution stopped about three and a half million years ago, mainly because of the lack of an essential material, water, and only resumed sporadically in this decade as the result of the invasion of human beings and human artifacts from the earth. We see this limitation operating in paleolithic societies, where energy input into human processes of production consisted of little more than human muscles, with the result that only stone, skins, bone, and wood were available as materials and the output of human artifacts was very small. With the ability to draw on accumulated energy reserves in the form of wood fires, the materials horizon expanded to include metals and pottery, and the quantity and variety of human artifacts increased by orders of magnitude.

Material objects are not the only human artifacts. Organizations likewise originate in some social genetic structure (an idea) and are formed through communication, persuasion, threat, exchange, and so on. They involve energy in bringing people together and in making the material artifacts that facilitate their operation, from temples and ritual objects to factories and machines. Individual human beings are partly biological artifacts made by biological processes from fertilized human eggs and partly social artifacts insofar as they acquire language, knowledge, skill, and behavior patterns from the other humans who surround them in the course of life. Persons, organizations, and material artifacts interact constantly in a complex social ecosystem and interact beyond this with the larger ecosystem that includes biological species and physical environments. Evolution—whether in the biosphere, the sociosphere, or both together—consists of this ecological interaction under conditions of constantly changing parameters: genetic mutation, climatic or physical changes, or social mutation. Human artifacts are part of this process of evolution; the automobile is just as much a species as a horse and just as "natural," just as much produced as a result of the evolutionary rise of know-how. The main difference between human and biological artifacts is that human artifacts are multiparental and can draw on the whole range of human know-how and social "genes," whereas biological artifacts can draw only on the gene pool of their own species. This is why social evolution is so rapid.

Human history from the very beginning has been characterized by something that can be called "development." In some of its aspects, this becomes "progress," which is a movement from states of the world evaluated as "worse" to those evaluated as "better." The process has also resulted in a very substantial expansion of the human "niche," with a fairly steady increase in the human population, barring a few wars and plagues, that has carried it now to over 4 billion. The process has resulted also in a continual expansion of the "sociomass"—the total mass of human artifacts—from a few flint knives and camp sites to cities, throughways, farms, factories, plantations, and so on. In addition there are unintended human artifacts like carbon dioxide in the atmosphere, pollution of air and waters, soil depletion, strip mines, and so on; most of these are perceived as "bads"—something that when increased makes things worse rather than better. Development, like the evolutionary process of which it is a part, is essentially a process in know-how, or genetic information structure. The know-how cannot be realized in the production of a product (phenotype) if there is a deficiency in energy or materials. Nevertheless, one of the remarkable things about human history is the way in which the expansion of know-how has constantly increased the availability both of energy and materials. As a result, the expansion of know-how has constantly pushed back the energy or materials frontiers that might limit it.

Furthermore, an expansion of the energy frontier seems to push back the materials frontier, and an expansion of the materials frontier often opens up new sources of energy or new ways of storing or transmitting energy. Thus, fire led to metals, metals led to tools, tools led to new materials and to coal, coal led to further expansion of metals, better tools led to oil and gas, oil and gas led to plastics, copper led eventually to electricity, and electricity led to aluminum and titanium, and so on in an extraordinary arabesque of interacting processes through time. Then the expansion of the energy and materials frontier feeds back in turn to the learning process, the essential increase in knowledge and know-how that is the central pattern of both biological evolution and human development. At some point in the expansion of energy and materials sources we were able to turn trees into paper, metals into type and later into computers; all facilitated the growth of human knowledge. Organizations likewise play a part in this complex feedback; material improvements like the telephone, the telegraph, and the typewriter made possible the "organizational revolution" after 1870—a massive reduction in the diseconomies of hierarchy and of scale that permitted the rise of large-scale organizations such as General Motors, the Pentagon, and the Soviet Union. Large organizations then create new methods of getting and using energy and materials that would be impossible without them, and so we go on, in a never-ending process of evolutionary disequilibrium.

In the last hundred or two hundred years, we have seen an enormous expansion of production of human artifacts, mainly because of rising knowledge, but also partly as a result of the discovery of fossil fuels, especially of gas and oil, and the discovery of new sources of materials. Petroleum and natural gas are extraordinarily cheap, convenient, storable, and transportable sources of energy. They are also an important source of new materials in the petrochemical industry. The discovery of electricity and its applications, which also took place in the nineteenth century, did not open up a new source of energy. It did, however, permit its transportation and especially its subdivision, facilitating its utilization in many convenient forms of light, power, and information transfer. Electricity, however, has a severe handicap in that it cannot be stored, especially in quantity, except at considerable expense. In addition, its transportation is both costly and clumsy. Without fossil fuels, the modern world would not have come into existence. If the geological history of the earth had been such that there had been no coal, petroleum, or natural gas, the world today would still look very much as it did in the 1700's, though we probably would have radios and television powered pretty expensively by wood-fired power plants. We would almost certainly not have airplanes or automobiles, and the horse would still be producing high rates of infant mortality from fly-induced diarrhea.

We are now facing a "hundred years' crisis" in which petroleum and natural gas will be exhausted. Coal will last a little longer, but it is dirty, inconvenient, and dangerous.

Existing nuclear technology will not last much longer than gas and oil, for uranium-235 will soon be exhausted. If we go to the breeder reactor, we can burn all the uranium, perhaps expanding the fuel supply by 5,000 years or more. If we go to the fusion, especially the deuterium reaction, we will expand the fuel supply to half a million years. All this is rather problematic. The direct use of solar energy on which we have always relied can undoubtedly be improved and expanded. Solar collectors and insulation can save half our heating and cooling bills, which account for 10 percent of our energy use. Solar power stations seem very expensive, dangerous, and hard to maintain. Nuclear reactors, especially the breeder, have a low probability of very large catastrophes, rather like dams. No substitute for oil and gas seems to be on the horizon. Most of the proposed substitutes produce only electricity, which is expensive, hard to transport, almost impossible to store in large quantities, and inefficient, except for the more delicate tasks. Paradoxically enough, it is the electricity produced by big power plants that makes the small beautiful.

It is easy to get rather pessimistic about the outlook for the next hundred years. World per capita real income may even have peaked in 1973. It will almost certainly peak before the end of this century. The population explosion is probably increasing the number of poor people faster than the number of rich. Distributional strains in a world of declining per capita real income may become very large. The more complex a society becomes, the more vulnerable it may get. Development of the poor countries may be frustrated, in spite of the spread of know-how, if production is limited by lack of energy or materials.

In the rich countries, up until very recently we have simply taken energy for granted, because it has been granted. Consequently, we have not really had to have an energy policy or strategy. Public reaction to energy scarcities is hard to predict. The antinuclear movement is quite strong, especially among the virtuous middle class. On the other hand, when the chips are down, if the choice is having nuclear power or putting the lights out, nuclear power will probably win. There is a certain case for delaying it, especially in the United States, which could do without it for 25 years, simply because we have so much coal. The economic difference between coal and uranium-235 as a source of electricity is so small that it is within the error of estimate of the Gross National Product. Nuclear power, however, may be much more important for the development of the poor countries, where alternative fuels are available only at rapidly rising expense and where energy shortages would severely limit economic development. The consequences of this for the international system are not pleasant to contemplate in terms of nuclear weapons proliferation, but we may have to learn to live with it. Fortunately, at least 99 percent of human potential for doing harm is not realized, and even this percentage can be raised. We have enormous overkill capacity in table knives and in arsenic, with its half-life of forever.

Ideological considerations may determine whether energy is produced mainly by public or by private institutions. One is not sure that this makes much difference, as long as the institutions are large. We are shifting from a period in which the emphasis has been on supply to one in which conservation and improving the efficiency of the consumption of energy are likely to take on increasing importance. Conservation, however, is a strategy involving many small units (households for instance), and it is harder to do than the strategy of building large power plants. It can be done with suitable tax, price, and information systems. We do need to work on this, though we should not have any illusions that conservation will be popular. Waste, after all, is a rich man's good.

Cartels like OPEC can raise the prices of their product and divert economic rent to the owners of energy resources, but historically cartels have been a bit unstable. There is at least some probability, hard to estimate but not negligible, that OPEC will produce a temporary oil surplus, which will in turn produce strong temptations for the weaker members to break away and undercut the price. If this led to a collapse of the cartel—a process that has happened often in the past—and a return to competitive conditions in the world oil market, there would undoubtedly be shortrun gains, both for the very poor countries without oil and for the rich countries. Unless there were offsetting tax policies, however, this might mean that oil would stay cheap until the last drop, perhaps leading to a catastrophic crisis for which nobody had prepared.

The main justification for at least a modest longrun optimism in regard to all these future scarcities is that all price elasticities are likely to be much higher in the longer run than they are in short periods. A fourfold rise in the overall relative price of energy would have spectacular effects on increasing the total supply. Many sources of energy that now look rather exotic, simply because oil and gas have been so fantastically cheap, would burgeon into important industries under the stimulus of high energy prices. The fact that there are quite a number of different unexploited sources, especially of energy income directly or indirectly derived from the solar flux—wind, water, tides, and ocean temperatures—as well as geothermal sources suggests that the longrun elasticity of supply may be quite high.

The demand for energy likewise will probably have a much greater elasticity in the longer run than it does in the short run. A quadrupling of the average price of energy, for instance, would produce large changes in architectural design, in automobiles, in the use of solar heaters and panels, even in clothing to be worn at lower temperatures, in large numbers of industrial processes, and in the general commodity mix away from the high energy-using commodities. There are a great many things we have not had to do because energy has been so cheap, things that can be and will be done as energy becomes more expensive. The fact that energy cost still is a fairly small part of the total Gross National Product means that even a quite substantial rise in its price, representing a roughly equal increase in the resources devoted to producing it, could be offset by a relatively small improvement in the efficiency of its use. Thus, if the energy industry constituted 5 percent of the GNP, with the rest being 95 percent, a quadrupling of the price of energy would raise the energy industry to 20 percent of the GNP, assuming perfectly inelastic demand; with the conservation that would be stimulated this rise might be only 15 percent. The rest of the economy would then be 85 percent of the GNP, and a general rise in resource efficiency of only (1-95/85), or 11.8 percent, would offset the 400 percent rise in the cost of energy.

While elasticity is a genuine source of longrun optimism, it would be unwise to deduce, as some of the more optimistic economists seem to think, that the elasticity of supply and demand will take care of all our problems. While it seems highly probable that at the present low price of energy, both supply and demand over the next generation or so will be highly elastic, we may then easily move into a price range where inelasticity sets in; if this is short of what would be required for continued world expansion of per capita real income, we could be in for a very rough time. Just to point out that the limits to growth are not absolute, and are modified by the elasticities of supply and demand, does not mean that they cannot become very real. On the supply side, the new technologies may simply not be enough to replace that vast elasticity of oil and gas supply that we have enjoyed so briefly. On the demand side, conservation is a one-shot operation; once it has been accomplished, demand may become very inelastic, and low-energy life styles may not prove to be appetizing. If we move into a price range in which both demand and supply are inelastic, however, a worsening of the supply situation (a movement of the supply curve to the left) can produce both very large changes in the relative price of energy and only small adjustments in the quantity. Sharp changes in the relative price structure, however, inevitably produce sharp redistributions of income, both among occupational groups and among income classes. These not only increase social and political tension, but are also likely to produce foolish policies in the effort to offset them.

It is a curious fact that while economists have had a modest success in persuading political decision-makers to follow their advice in regard to employment and fiscal policy, they have had no success whatever in influencing price policy. The United States' reactions to the OPEC cartel are an almost classic example of what not to do. We have kept oil relatively cheap by quotas and a two-price policy. If we had put a substantial tariff on all imported oil, we would have captured some of the economic rent that now goes to the OPEC oil producers; if we had put an equivalent tax on domestic oil production, we would have encouraged conservation; and if we had allowed the price of oil to rise, we would have encouraged exploration. It is a sound general principle that if we have something that is plentiful now, but is going to be scarce later, there is a strong case for anticipating the future price rise and so produce an early stimulation of the long-run elasticities. The sensible policy is first to tax it heavily, which will raise the consumer price and encourage conservation, and perhaps in addition to subsidize new discoveries.

Can we then apply some of the principles of good and bad decisions outlined in the first part of this paper to the problem of energy strategy? Strategy involves thinking about how to improve the probability of good decisions and diminish the probability of bad ones. The problem of the psychology of the decision-maker we must neglect—not because it is unimportant, but because it is intractable to strategy. Possibly an improvement in parent education would yield us more sane adults in the next generation. For the most part, however, psychology must be included in the probabilistic part of our image of the future; just as we must include the probability of the hundred-year flood or the hundred-year nuclear disaster, so we must include the probability of the hundred-year lunatic in a powerful position. Hitler perhaps was a thousand-year lunatic. The great danger here is that we overlearn from the actual occurrence of improbable events and overestimate their probability after they have happened.

The problem of the image climate is likewise rather intractable to strategy, but, like psychology, it should be part of the background. We must be on the lookout, for instance, for the rise of taboos. A sense of cleanness or uncleanness is an important constituent of most people's images of the world. This is particularly important in the case of nuclear energy. Because of its dramatic use as a weapon against civilians (it is hard to visit the atomic museum in Hiroshima without being ashamed of the human race), nuclear energy had a curse on it from the start. In addition, it produces uncleanness in the shape of nuclear wastes, especially plutonium, whose radiations are deadly, invisible, and imperceptible except to priests with Geiger counters and other sacred objects—properties that they share with evil spirits. Plutonium is easily perceived as more unclean even than pigs, or liquor, or semen, or any of the other numerous objects that have driven mankind into taboos. Taboos on nuclear energy, therefore, are by no means out of the question. Taboos, furthermore, are very hard to deal with, because they involve an ineradicable mixture of rationality and irrationality. They may contribute to bad decisions, but they may also be a strategy for survival.

Decision-makers, especially powerful ones, rarely consider the impact of their decisions on the image climate. The decision, or absence of decision, by which the nuclear and military industries concluded that the problem of nuclear waste could be left to the future may have led to a change in the image climate very adverse to them. The most important element in the ongoing dynamics of the social system is the perceptions of legitimacy. Without internal legitimacy, morale—the sense that what one is doing is "okay"—no person or no organization can function. The withdrawal of external legitimacy, the perception on the part of other persons in the environment of an individual or an organization that what is going on is okay, makes the environment so unfavorable that it becomes extremely difficult to function, and the loss of external legitimacy often leads to the loss of internal legitimacy. These considerations should be in the background of any strategic decision. Deceit, the irresponsible exercise of power, secrecy, and many other factors can have powerful effects in eroding legitimacy.

Organization is fairly susceptible to strategic decision. Part of the perquisites of power is the ability to set up organizational structures. Government and any large organization can do this, and the question, therefore, of how organizational structures create bad decisions is highly relevant. One of the critical questions here is the scale of organization and the advantages or disadvantages of centralization and decentralization. There are no easy answers to this question; a great deal depends on the particular circumstances and environment and on the image climate of organizational decision-makers. The fact that there seems to be a centralization-decentralization cycle in most large organizations suggests that this is one of the insoluble problems—that everything we do is wrong, so we try a succession of wrong things one after the other.

Small is not necessarily beautiful nor big necessarily necessary. The small can be mean, and the large can be generous. On the other hand, a policy dominated by large organizations can easily miss the advantages of small scale. We should not overlook the household as a very significant unit in both energy consumption and possibly in production from wind, solar energy, and geothermal heat pumps. On the other hand, we are probably quite a long way from the idyllic utopia of household self-sufficiency. Large organizations will be with us for a long time, and we must learn to run them better. It is perhaps significant that Schumacher,[4] the author of "Small Is Beautiful," was an economist for a very large organization (the British Coal Board) and that his observations on how to run a large organization are very cogent!

There is no litmus paper for good strategies, though perhaps there is for very sour ones. The most one can hope for is that the agendas of strategists will be widened, and that the worst mistakes can be avoided. But even if that is all we can do, it is worth doing.

REFERENCES

1. Marschak, Jacob. The Role of Liquidity Under Complete and Incomplete Information. American Economic Review/Supplement, Vol. 39, May 1949. p. 182-195.

2. See, for example: Myers, David G., and Helmut Lamm. The Group Polarization Phenomenon. Psychological Bulletin, Vol. 83, No. 4, 1976. p. 602-627. This article contains an excellent bibliography.

3. Peter, Laurence, and Raymond Hull. The Peter Principle. Bantam Books, New York. 1970.

4. Schumacher, E. F. Small Is Beautiful: A Study of Economics As If People Mattered. Harper & Row, Publishers, New York. 1973.

C. S. Holling
Institute of Resource Ecology
University of British Columbia
Vancouver, B.C.

Mr. Holling was one of the first ecologists to apply the techniques of systems analysis to ecosystem problems. Following a year as a consultant to the Ford Foundation helping develop the foundation's ecology program, he returned to the University of British Columbia to create the Institute of Animal Resource Ecology.

The Institute's research emphasized the use of simulation models simple enough to be used by decision makers unfamiliar with computer techniques. The success of this work led to Mr. Holling's appointment to head the first group of systems ecologists at the International Institute for Applied Systems Analysis (IIASA).

Mr. Holling earned his bachelor's degree in biology and his master's in zoology at the University of Toronto. He received his doctorate in zoology at the University of British Columbia. In 1966, he received the George Mercer Award from the Ecological Society of America. He is a Fellow of the Royal Society of Canada, and received the Gold Medal of the Entomological Society of Canada for 1969-70.

Mr. Holling has been a visiting professor at the University of California, Berkeley; he is an affiliate professor at the University of Washington and the University of Idaho. He is currently a consultant for Environment Canada.

He is editor of the International Journal of Environmental Sciences *and the* Journal of Theoretical Population Biology. *He has published widely in scientific journals.*

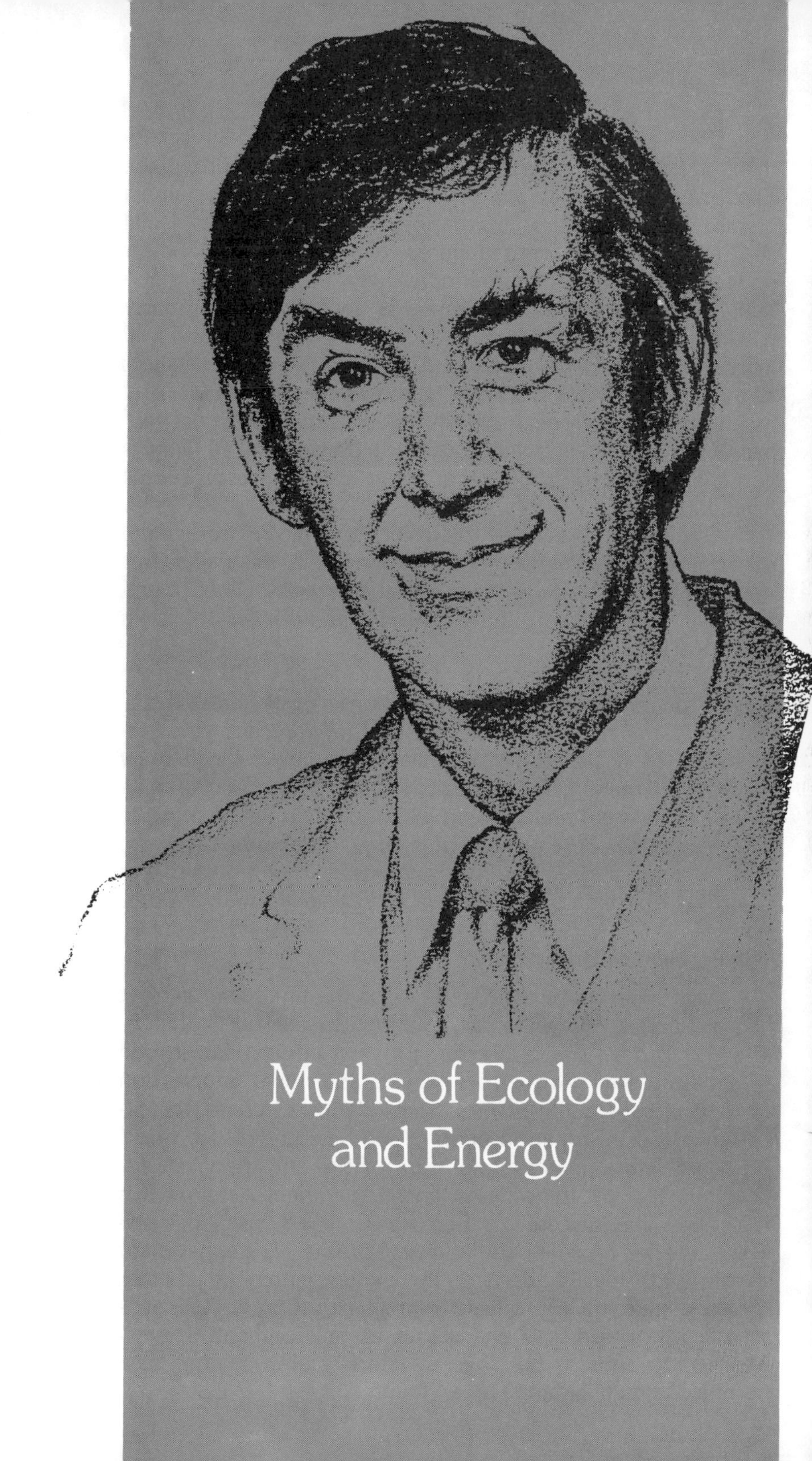
Myths of Ecology
and Energy

The key issue for the design of an energy policy is how to cope with the uncertain, the unexpected, the unknown. The price and availability of imported oil are uncertain in precisely the terms that international relations are. The social and environmental consequences of emphasizing nuclear power, coal, solar or geothermal power, or a mix are uncertain. Even the ultimate objective is uncertain. Community wisdom has held that increasing the energy per capita will increase the standard of living. And yet at least one analysis suggests[1] the reverse has been true over the recent past. And in his espousal of small and diverse solutions, Schumacher[2] has raised the heresy again, suggesting that quality of living is perhaps not improved by unending growth in Gross National Product.

Originally, I intended to explore specific environmental constraints of alternative energy proposals in this paper. But the central constraint is their very uncertainty. That, therefore, will be my theme—how can we deal with the unknown and uncertain.

After exploring aspects of uncertainty, I will treat the subject under three topics: the response of systems to disturbance, the role of hierarchies of function, and the consequence of aggregations of scale. Throughout I will draw examples largely from ecological systems—in part to reap any benefits from analogy and in part to help define possible environmental consequences of energy developments. From that, I point a direction that suggests neither that small is beautiful nor that big is necessary, but something of both.

The Problem of Uncertainty

Man has always lived in a sea of the unknown and yet has prospered. His customary method to deal with the unknown has been trial and error. Existing information is used to set up a trial. Any errors provide additional information to modify subsequent efforts. Such "failures" create the experience and information upon which new knowledge feeds. Both prehistoric man's exploration of fire and the scientist's development of hypotheses and experiments are in this tradition. The success of this time-honored method, however, depends on some minimum conditions. The experiment should not, ideally, destroy the experimenter—or at least someone must be left to learn from it. Nor should the experiment cause irreversible changes in its environment. The experimenter should be able to start again, having been humbled and enlightened by a "failure."

We are now ignoring those minimum conditions. Our trials are capable of producing errors larger and more costly than society can afford. This leads to the dilemma of "hypotheticality" posed by Haefele,[3] who argues that the design of energy policies is locked in a world of hypothesis because we dare not conduct the trials necessary to test and refine our understanding.

One of the reactions to this dilemma is, paradoxically, the generation of a plethora of alternative hypotheses that are, in principle, untestable. They appear in a variety of guises—as the hidden or explicit assumptions behind scenarios, alternative forecasts, exploratory calculations, or "world" models. Such exercises can be salutary by forcing questions that otherwise would not be raised, but they can be just as pernicious by tempting us to use the exercise to make decisions.

Consider, for example, even such a noble task as the prediction of environmental constraints. Haefele and Sassin[4] have made a preliminary and yet effective projection of the constraints to an energy policy that emphasizes a shift from oil to coal. Accumulation of carbon dioxide in the atmosphere suggests that global climate would be affected so much that new technologies would be essential within 60 years. New technologies, however, are not created overnight. Other analyses suggest a lead time of the same duration. Hence the pressure to design and implement a new technology now. But once such a decision was made, the magnitude of the effort would foreclose the options of withdrawing from it if other, unexpected problems arose. And they would arise.

The problem, therefore, is not only one of inadequate knowledge of environmental constraints or of social consequences. There is a more fundamental issue, suggested by the theme set by Schumacher[2]—small is beautiful versus big is necessary. Both have the flavor of myths. Myths are a way in which mankind captures some essence of experience or wisdom in a simple and elegant form. Myths are necessary to guide us in our actions, and they protect us from that larger reality of the frightening unknown. Myths may be as all-embracing as "there is a God" or as narrow as "the free market model will correct all inequities."

But myths are only a partial representation of reality. Through their acceptance, man becomes bold enough to accumulate new knowledge that eventually exposes their incomplete nature. That introduces confusion; old myths are dusted off, and new ones developed. In many ways, the present emergence of a multitude of hypotheses is a process of evolving new myths. But let us call them myths and not pretend they are recipes for action.

Moreover, it might be useful to structure consciously alternative myths and trace their logical consequences in relation to dealing with the unexpected. The way systems respond to unexpected events is determined by their stability properties. Hence the myths might usefully be myths of stability.

Stability of Systems

Initially, it is less important how variables respond to disturbances than whether a system is stable at all and under what conditions. These are questions of structural stability, and there is little interest in kinds of equilibria (fixed points, limit cycles, nodes, etc.) and even less for the details of a trajectory as it evolves after a disturbance. If we don't strain the analogy too far, these structural properties of stability can be imagined as forming a topographic surface representing some aspect of the time-independent forces of a system.

One of the simplest and most common images would be a surface with a valley shaped like a bowl, within which a ball moved, partially as a consequence of its own acceleration and direction and partially as a consequence of the forces exerted by the bowl and by gravity. If the bowl were infinitely large, or events beyond its rim meaningless, then this would be an example of global stability. That is, no matter how far from the bottom the ball was displaced, eventually it would return and come to rest there again.

Now this myth of global stability has been pervasive. It underlies much of the presumptions of economics, with its focus on equilibrium and near equilibrium conditions. After all, the dominant property seems to be the final condition of rest or equilibrium, and any other behavior is transient and seems trivial. Moreover, equilibrium mathematics is easy!

It is also a comforting myth, for it represents a benign and infinitely forgiving Nature. Trials and mistakes of any scale can be made in this world, and the system will recover once the disturbance is removed. Since there are no penalties to size, only benefits of scale, this myth of Nature Benign comfortably and logically leads to big is necessary.

An opposing myth is that of instability, where the

imagined surface is now dominated by a smoothly convex hill rather than a bowl. The top of the hill represents an unstable equilibrium, for if the ball is only slightly displaced from this point, it will roll away. Ultimately, complete instability of this kind leads to all variables becoming zero—to extinction. Since we know that systems persist, this myth of instability would seem to be impossible and uninteresting. Yet it has a rich tradition in philosophy and science.

Persistence is conceivable if we introduce true spatial dimensions. Imagine a mosaic of spatial elements or sites, each occupied by such an unstable system, each out of phase with the other and contributing excess individuals by movement to neighboring sites. The result could be as much as was actually observed in a classic experiment by Huffaker et al.[5] They examined the interaction between populations of a plant-eating mite and a predatory mite. In the relatively small enclosures used, when there was unimpeded movement throughout the experimental universe (a homogeneous world, therefore), the system was unstable and the populations became extinct. When barriers were introduced to impede dispersal between parts of the universe, small-scale heterogeneity was introduced and the interaction persisted. Thus populations in one small area that suffered extinction were reestablished by invasion from other populations that happened to be at the peak of their numbers.

This myth of ephemeral nature leads to a concentration on issues of spatial heterogeneity, diversity over space, and of fine-scaled, local autonomy. Nature Ephemeral is in harmony with the persistent traditions of anarchism and of small is beautiful. It is in complete opposition to big is necessary, since any effort to increase size, to homogenize, will lead to collapse as the underlying instability of nature becomes obvious.

Both these myths of stability and instability used to be espoused by two separate schools within ecology. One school showed that many population processes, such as predation, parasitism, and competition, have important regulatory properties. Such processes have a density-dependent, or negative feedback, influence such that if numbers become excessively high, there will be a proportionate increase in mortality to damp the upswing. The reverse will occur if numbers become too low. That, essentially, is the myth of the single equilibrium and global stability. The opponents of this school saw evidence in nature of what Huffaker demonstrated in the laboratory. That is, populations are ephemeral, chance events profoundly influence numbers, and invasion-reinvasion is the rule.

The familiar Lotka-Volterra equations demonstrate a stability condition precisely midway between these two myths. These equations are coupled differential equations representing the interactions between two populations—a predator and prey, for example, or two competitors—in a highly simplified form. They result in neutral stability as if the topography were represented as a flat plane. It is the stability of a frictionless pendulum. But the relationships represented in the Lotka-Volterra equations are not only simple, they are simplistic. Adding any kind of realistic negative feedback (or density dependence) results in the myth of Nature Benign and global stability. Adding any kind of realistic positive feedback, or explicit lags, results in the Myth of Nature Ephemeral and instability.

Yet both kinds of addition have been demonstrated to exist.[6] The component structure of key population processes leads to classes of behavior that can be formally defined[7] and that have mixed stabilizing and destabilizing properties. When all of these are included, a different myth emerges.

The dominant feature of this myth is the appearance of more than one equilibrium state.[6] Each equilibrium state or attractor is separated from its neighbors so that two or more basins of attraction or domains of stability are formed. As long as variables remain within one basin of attraction, they will tend to the same attractor. If, however, variables happen to be close to the boundaries of these basins, then an incremental disturbance could shift the variables into another basin, thereby causing radically altered behavior. Or, returning to the theme of trial-and-error learning, incremental trials could seem not to be causing significant error and yet could accumulate until one more trial "flips" the system into a different stability region. Consider eutrophication, where progressive phosphate loading of aquatic systems can be accommodated only up to a point. Then algal blooms are triggered, and subsequent rotting of the plant material can lead to anaerobic conditions.

Rather than the image of hill or bowl, a better analogy for this myth would be a mesa with a depression at its top. As long as the ball is in the depression, the system appears qualitatively stable. If the ball is tipped over the edge of the mesa, it will move to a different position, one of which could well represent extinction. So, if we perceived reality as being the myth of Nature Benign, this would seem to represent a rather perverse Nature. On the other hand, if we perceived reality as being the myth of Nature Ephemeral, it would seem to represent a quite tolerant Nature—more stable than we thought.

The burden of evidence suggests that the multiequilibria world of Nature Perverse/Tolerant is common—and not only for ecological systems. Like all good myths, it can be usefully captured by some simple differential equations, as long as some effort is made to represent relationships simply and not simplistically. Such models have been proposed for ecological systems,[8] institutional systems,[9] and societal systems.[10] Even these simple structures exhibit rich topologies with the dominant feature of multiple stability regions.

Moreover, many empirical studies of specific systems suggest a multiequilibria structure and sharp behavioral shifts between equilibria. A preliminary survey[6,11] shows numerous examples in the ecological, water resource, engineering, and anthropological literature. We know enough about the functional form of ecological processes at least to know that their qualitative properties can produce multiple equilibria. This makes topological approaches in ecology so important,[12] since they emphasize equilibria manifolds and conditions that trigger sharp shifts of behavior. Simple examples are the catastrophe manifolds, familiar now even to readers of *Newsweek.*

The myth of Nature Perverse/Tolerant can accommodate either small is beautiful or big is necessary. The small-is-beautiful theme operates much as before, with perhaps a more formal sense of optimal scale. It is still contrary to big is necessary. In a spatial world with multiequilibria systems, a distinctive spatial scale can be identified. Steele,[13] for example, in a pioneering analysis of spatial attributes of plankton in the ocean, shows that there is a distinctive size to a patch of plankton that results from a balance between the kind of stability described by this third myth, and diffusion forces that tend to dissipate the patch. This primary scale is measured in kilometers, and he points out the essential constraints this places on magnitude of intrusion of pollutants or of thermal discharge from power plants. Terrestrial systems show the same scale definitions—insect defoliators, grazing ungulates, plant colonizers. Again, the minimum viable "patch" size or scale is set by the balance between magnitude and extent of dispersal and the endogenous stabilizing and destabilizing forces.

But note that ecological and environmental scales can be very small (the soil bacteria's world) or very large (the scale of global climatic systems). Are the big ones ugly and the small ones beautiful? There is an inconsistency, and we will turn to that later when we consider the role of hierarchies in buffering the unknown.

For now, the principal point is that this multiple equilibria myth can also be accommodated by big is necessary; one only has to be a little more cautious. Thus, if boundaries exist separating "desirable" from "undesirable" stability regions (that is, if separatrices exist), then the task is to carefully control the variables to keep them well away from the dangerous separatrix. To do this effectively, big might well be necessary as the only way to gain sufficient knowledge of the separatrix, to monitor the distance to it, and to institute control procedures to maximize that distance. Haefele and Buerk[10] suggested this in their search for a resilient energy policy that can absorb unexpected events. They start with a brilliant articulation of a simple societal model that clearly has two equilibria—one representing high population and low energy per capita and one the reverse. That representation has great value as a myth—as one partial representation of one view of reality that serves as a paradigm but not as a blueprint for action. But they go on to use it as such by treating this powerful myth as a model of reality. To use it in this way (even with the careful and responsible qualifications they emphasize) is to destroy its value as myth, as hypotheses with untestable consequences.

The goal of maximizing the distance from an undesirable stability region is exactly in the highly responsible tradition of engineering for safety, of nuclear safeguards, of environmental and health standards. It demands and presumes knowledge. It works beautifully if the system is simple and known—say, the design of a bolt for an aircraft. Then the stress limits can be clearly defined, these limits can be treated as if they are static, and the bolt can be crafted so that normal or even abnormal stresses can be absorbed. The goal is to minimize the probability of failure. And in that, the approach has succeeded. The probability of failure of nuclear plants is extremely small. But in parallel with that achievement is a high cost of failure—the very issue that now makes trial-and-error methods of dealing with the unknown so dangerous. Far from being resilient solutions, they seem to be the opposite, when applied to large systems that are only partially known.

For this reason, I did not address myself to the issue of this symposium—"environmental constraints to energy developments." To be able to identify environmental constraints presumes sufficient knowledge, presumes the possibility of testing by hypotheses, presumes a myth of Nature Tolerant and Static. It reinforces the trap of hypotheticality. It emphasizes a fail-safe design at the price of a safe-fail one.

One final myth is therefore needed to make a Compleat Mythology. The three myths discussed so far have been described in three steps of increasing reality and comprehensiveness and implicitly assumed constancy and determinacy. That is, the topography of hills, valleys, and mesas was fixed and immutable, and the ball followed a predetermined path to some constant condition. But ecological—and, for that matter, economic, institutional, and social—systems are neither static nor completely determined. Variability and change are the rule and provide the next step toward reality.

Stochastic events dominate some ecosystems. Fire, rather than being a disaster, is the source of maintenance of some grassland ecosystems. Shifting patterns of drought determine the structure of some savannah systems in Africa. The internal variables themselves can move, through endogenous mechanisms, from one equilibrium to another. Such seems to be the lesson from our recent studies of forest insect pests. There, periodic outbreaks can be triggered by stochastic events, by spatial events, or by the growth of the forest itself. Populations then increase explosively from a lower to a higher equilibrium. While the higher one is stable for the insects, it is unstable for the forest. Consequently, large swings and movements among stability regions contribute to forest renewal and the maintenance of diversity.

Hence for natural ecosystems, residence in one stability region, far from separatrices, is not necessarily the rule. Locally, a system may even move to an extinction region, and stochastic or deterministic events will reinstate it. Hence the ball on our surface is moving continuously, and the stability boundaries are continually being tested.

But the topography itself is not static. In the mathematical representations, changing the values of parameters can change the stability landscape. Hence some regions in parameter space can be identified that indeed represent global stability, others that represent instability, and still others that show various multiequilibria configurations. Now, parameters in equations are presumed to be constant only as a convenience and first approximation. They can better be viewed as variables that normally change so slowly that, within limits, they can be presumed to be constant or, at most, stochastic variables.

The values of these parameters at any moment define the form of the topographic landscape. They assume these values in nature as a consequence of long- and short-term adaptation to variation—both genotypic and phenotypic. In essence, natural selection produces a balanced set of parameters whose value is, in part, the consequence of the historic variation of the system. Changing the pattern of variability can thus change this balance. Moreover, our mathematical models show that if key parameters are arbitrarily moved, often there is little change in the topography until a certain point, when the topography suddenly shifts: Stability regions implode, new regions of stability and instability appear. There are, in short, separatrices in parameter space as well as in state variable space.

This dynamic pattern of the variables of the system and of its basic stability structure lies at the heart of coping with the unknown. However much we may be sure of the stability landscape of a physical system, rarely will we know the societal or ecological stability landscape in great detail. Policies often attempt to reduce variability within these partially understood systems, either as a goal in itself or as an effort to meet standards of safety, health, or environment. That changed variability in turn may itself shift the balance of natural, cultural, or psychological selection so that stability regions will contract. Paradoxically, success in maximizing the distance from a dangerous stability boundary may cause collapse because the boundary may evolve to meet the variables. That is, if surprise, change, and the unexpected are reduced, systems of organisms, of people, and of institutions can "forget" the existence of limits until it is too late.

This final myth is the myth of Nature Resilient. Originally, resilience was defined[6] as the property that allowed a system to absorb change and still persist. But if that is followed with the myth of Nature Tolerant and Static in mind, resilience becomes, as Haefele has translated it, "Schlagabsorptionsfaehigkeit" or "strike absorption capability." Using that as a goal leads to avoidance of separatrices, followed, in partially known systems, by their possible collapse. A better definition, therefore, is that resilience is a property that allows a system to absorb *and utilize* (or even benefit from) change. Thus, it can lead to radically different kinds of approaches—for example, the dynamic management of environmental standards[14] or of environmental impact assessment.[15]

Such a myth explicitly recognizes the unknown and the ability to survive and benefit from "failures." It certainly would allow for a world that can accommodate trial-and-error learning. Since it is a myth of a natural, evolved system, we might profit from looking deeper into the organization of such systems, as Simon[16] has done in an important essay on hierarchical organization.

Hierarchies of Function versus Aggregations of Scale

Hierarchy, as defined in theories of complex systems, is not the linear rank structure of some human organizations. Rather, it can be visualized as the roots of a tree that branch into distinct levels of function. Each level consists of a number of components with different functions whose integration produces the higher level. Hierarchies seem to be universal within a large class of evolved systems—physical, chemical, biological, or social—because they have properties that facilitate the evolution and survival of complexity. If that is true, then it touches precisely the issue of energy, society, and environment.

The familiar biological hierarchy runs from organic compounds to macromolecules to cells to tissues and organs to whole organisms. This hierarchy can be extended to include at one end a microhierarchy of molecules, atoms, and so on, and at the other an ecological hierarchy. The individual organisms form populations. These populations are organized into distinct functions—primary energy acquisition (the plants or primary producers), herbivory, predation, and decomposition. These populations are connected by energy transfer, mass transfer, and competition; together they form ecosystems. We could, as some have done, continue the hierarchy to communities, biogeocoenoses, and biomes. But beyond the ecosystem, the concept becomes fuzzy and is, at the moment, more an exercise in classification than understanding.

Simon[16] shows that certain attributes are found in all hierarchies. Each level has a distinctive time scale and, we might add, spatial scale. The higher the level in the hierarchy, the slower the time scale and the larger the spatial scale. Hence, to return to an earlier point, the parameters of one level are, in reality, slow variables of the next level.

Each level exists as the consequence of interactions between components of the lower level. That interaction results in a buffering of the extremes and easing the demand for adaptation at the lower level. Thus components need not be infinitely adaptable.

Communication between components and levels is simple compared to the interactions within components. As long as that simple communication is maintained, the details of operations within the components can shift, change, and adapt without threatening the whole. In ecological systems, for example, there is a great diversity of populations performing any one function, such as primary production. Part of that diversity is due to heterogeneity—solar energy, while diffuse, is not homogeneous. Hence "sun-loving" trees and shade-tolerant plants live together. Part of that diversity is due to uncertainty. Any particular function represents a role that at different times can be performed by different actors (species) that happen to be those available and best suited for the moment.

All these attributes make for flexibility and resilience. Additionally, the simple mathematics of hierarchical structures allow rapid evolution and the absorption and utilization of unexpected events. Simon's example of the two Swiss watchmakers demonstrates this.

One watchmaker assembles his watch as a sequence of subassemblies—a hierarchical approach. The other has no such approach, but builds from the basic elements. Each watchmaker is frequently interrupted by phone calls, and each interruption causes an assembly to fall apart if it has not yet reached a stable configuration. If the interruptions are frequent enough, the second watchmaker, having always to start from scratch, might never succeed in making a watch. The first watchmaker, however, having a number of organized and stable levels of assembly, is less sensitive to interruption. The probability of surprise (of failure) is the same for each. The cost of surprise (of failure) is very different.

Let me now draw some temporary conclusions from this treatment of stability and hierarchies. First, in terms of function, the value attached to big versus small is meaningless. Higher levels in a hierarchy are slower and larger in scale than are lower levels. We could even say that embedding in a larger system buffers the operation of smaller components. Thus, the reason why there is a flea on the frog on the bump on the log in the hole at the bottom of the sea is because there is a sea, a set of holes, a number of logs, and other objects, etc., etc. Equally, we could reverse the argument. The higher, larger level exists as a consequence of the interactions between the smaller components. In short, small is necessary for big to be.

Second, missing levels in a hierarchy will force bigger steps. And bigger steps will take a longer time. An example might be the one mentioned in my introduction—that is, the 60-year lead time to develop technologies that avoid excessive accumulation of carbon dioxide in the atmosphere. Bigger steps also presume the greatest knowledge and require the greatest investment. Hence, once initiated, they are more likely to persist even in the face of obvious inadequacy. Finally, bigger steps will produce a larger cost if failure does occur. To avoid that, the logical effort will be to minimize the probability of the unexpected, of surprise, or of failure.

For example, another solution to the watchmaker's problem of interruption might be for a number of watchmakers to join together, pool their resources, occupy a large building, and hire a secretary to handle the phone calls. This would control the disturbance and interruptions, and both watch-building strategies would succeed. Without the interruptions, there is not that much to gain by maintaining very many steps in a hierarchy of subassemblies. The hierarchically organized watchmaker could well begin a process of aggregation of scale, reducing the number of steps and having larger subassemblies at each step. That might increase efficiency in use of time and produce economies of scale, but it is totally dependent on complete and invariant control of disturbance. If the secretary were sick for one day, production would halt.

This is the same conclusion I reached in the discussion of the stability myths, and the circle of argument is now closed. The engineering-for-safety, fail-safe approach, while appropriate for totally known systems, is not so for uncertain, heterogeneous, and partially known ones. Whither nuclear safeguards, then, and the proliferation of nuclear power plants? The best safeguard system conceived by the mind of man can be deceived by another mind. That line of argument leads to the proposals for nuclear parks, floating energy islands that serve whole hemispheres of the globe via pipelines carrying hydrogen. It is not likely that we need treat these proposals as anything more than bad science fiction. But they do suggest an alternative that might even be of the same scale. Is it not possible, by formally restructuring a hierarchical organization to energy production, supply, and use, to have a global energy system? Then trial-and-error learning would become possible again, and the fears of the unknown could, temporarily, hide behind the myth of Nature Resilient.

REFERENCES

1. Watt, K. E. F., Y. L. Hunter, J. E. Flory, P. J. Hunter, and N. J. Mosman. The Long-Term Implications and Constraints of Alternate Energy Policies. In: Part I, Middle and Long Term Energy Policies and Alternatives. Hearings before the Subcommittee on Energy and Power of the Committee on Interstate and Foreign Commerce, 94th Congress. March 25-26, 1976. Serial No. 94063. U.S. Government Printing Office, Washington, D.C. 1976. p. 303-362.

2. Schumacher, E. F. Small Is Beautiful: A Study of Economics As If People Mattered. Harper & Row, Publishers, New York. 1973.

3. Haefele, W. Hypotheticality and the New Challenges: The Pathfinder Role of Nuclear Energy. Minerva, 10:303-323, 1974.

4. Haefele, W., and W. Sassin. Energy Strategies. IIASA Research Report RR-76-8. International Institute for Applied Systems Analysis, Laxenburg, Austria. 1976.

5. Huffaker, C. D., K. P. Shea, and S. S. Herman. Experimental Studies in Predation. Complex Dispersion and Levels of Food in an Acarine Predator-Prey Interaction. Hilgardia, 34:305-330, 1963.

6. Holling, C. S. Resilience and Stability of Ecological Systems. Annual Review of Ecology and Systematics, 4:1-23, 1973.

7. Holling, C. S., and S. Buckingham. A Behavioral Model of Predator-Prey Functional Responses. Behavioral Science, 3:183-195, 1976.

8. Bazykin, A. D. Volterra's System and the Michaelis-Menten Equation (in Russian). In: V. A. Ratner, ed. Problems in Mathematical Genetics. U.S.S.R. Academy of Sciences, Novosibirsk. 1974. (Available in English as Structure and Dynamic Stability of Model Predator-Prey Systems. Institute of Resource Ecology Research Report R-3-R. University of British Columbia, Vancouver, B.C. 1975.)

9. Holling, C. S., C. C. Huang, and I. Vertinsky. Technological Change and Resource Flow Alignments: An Investigation of Systems Growth Under Alternative Funding/Feedback. In: R. Trappl, ed.

Progress in Cybernetics and Systems Research. Austrian Institute for Cybernetics, Vienna. (In press.)

10. Haefele, W., and R. Buerk. An Attempt of Long-Range Macroeconomic Modeling in View of Structural and Technological Change. IIASA Research Memorandum RM-76-32. International Institute for Applied Systems Analysis, Laxenburg, Austria. 1976.

11. Holling, C. S. Resilience and Stability of Ecosystems. In: E. Jantsch and C. H. Waddington, ed. Evolution and Consciousness: Human Systems in Transition. Addison-Wesley Publishing Co. Reading, Mass. 1976. p. 73-92.

12. Jones, D. D. The Application of Catastrophe Theory to Ecological Systems. In: G. S. Innis, ed. Simulation in Systems Ecology. Simulation Council, Logan, Utah. (In press.)

13. Steele, J. H. Spatial Heterogeneity and Population Stability. Nature, 248:83, 1974.

14. Fiering, M. B., and C. S. Holling. Management and Standards for Perturbed Ecosystems. Agro-Ecosystems, 1:301-321, 1974.

15. Hilborn, R., C. S. Holling, and C. J. Walters. Managing the Unknown: Approaches to Ecological Policy Design. In: J. J. Reisa, ed. Biological Analysis of Environmental Impacts. Council on Environmental Quality, Washington, D.C. (In press.)

16. Simon, H. A. The Organization of Complex Systems. In: Howard H. Pattee, ed. Hierarchy Theory. George Braziller, Inc., New York. 1973. p. 1-28.

ACKNOWLEDGMENTS

This paper owes its origins to a group of individuals at the Institute of Resource Ecology, University of British Columbia. Their mix of practical experience in renewable resource systems, policy analysis, and theory color much of the writing. I would particularly like to thank Dixon Jones and Bill Clark for their continuing concern, criticisms, and contributions.

Amory B. Lovins
Friends of the Earth, Inc.

After resigning a Junior Research Fellowship of Merton College, Oxford in 1971, Mr. Lovins became British Representative of Friends of the Earth, Inc., a nonprofit conservation group. A consultant physicist (mainly in the United States) since 1965, he now concentrates on energy and resource strategy. His current or recent clients include the Organization for Economic Cooperation and Development, several UN agencies, the International Federation of Institutes for Advanced Study, the MIT Workshop on Alternative Energy Strategies, and other organizations in several countries. He is active in international energy affairs and has testified before parliamentary and congressional committees.

Mr. Lovins has written, lectured, and broadcast extensively on energy policy, particularly as it relates to the environment. Mr. Lovins has studied at Harvard and at Oxford's Magdalen College and Merton College, where he received the master of arts in 1971. As a physicist his research interests included nuclear magnetic resonance, low noise r-f electronics, and r-f physics of pure noble metals. He has been a British resident since 1967.

At Mr. Lovins' request, reprints of his paper "Energy Strategy: The Road Not Taken?" were distributed to the participants in the Oak Ridge conference, and since his paper for the conference relies upon familiarity with it, it was included in this volume.

Energy Strategy: The Road Not Taken?

Where are America's formal or de facto energy policies leading us? Where might we choose to go instead? How can we find out?

Addressing these questions can reveal deeper questions—and a few answers—that are easy to grasp, yet rich in insight and in international relevance. This paper will seek to explore such basic concepts in energy strategy by outlining and contrasting two energy paths that the United States might follow over the next 50 years—long enough for the full implications of change to start to emerge. The first path resembles present federal policy and is essentially an extrapolation of the recent past. It relies on rapid expansion of centralized high technologies to increase supplies of energy, especially in the form of electricity. The second path combines a prompt and serious commitment to efficient use of energy, rapid development of renewable energy sources matched in scale and in energy quality to end-use needs, and special transitional fossil-fuel technologies. This path, a whole greater than the sum of its parts, diverges radically from incremental past practices to pursue long-term goals.

Both paths, as will be argued, present difficult—but very different—problems. The first path is convincingly familiar, but the economic and sociopolitical problems lying ahead loom large, and eventually, perhaps, insuperable. The second path, though it represents a shift in direction, offers many social, economic and geopolitical advantages, including virtual elimination of nuclear proliferation from the world. It is important to recognize that the two paths are mutually exclusive. Because commitments to the first may foreclose the second, we must soon choose one or the other—before failure to stop nuclear proliferation has foreclosed both.[1]

Most official proposals for future U.S. energy policy embody the twin goals of sustaining growth in energy consumption (assumed to be closely and causally linked to GNP and to social welfare) and of minimizing oil imports. The usual proposed solution is rapid expansion of three sectors: coal (mainly strip-mined, then made into electricity and synthetic fluid fuels); oil and gas (increasingly from Arctic and offshore wells); and nuclear fission (eventually in fast breeder reactors). All domestic resources, even naval oil reserves, are squeezed hard—in a policy which David Brower calls "Strength Through Exhaustion." Conservation, usually induced by price rather than by policy, is conceded to be necessary but it is given a priority more rhetorical than real. "Unconventional" energy supply is relegated to a minor role, its significant contribution postponed until past 2000. Emphasis is overwhelmingly on the short term. Long-term sustainability is vaguely assumed to be ensured by some eventual combination of fission breeders, fusion breeders, and solar electricity. Meanwhile, aggressive subsidies and regulations are used to hold down energy prices well below economic and prevailing international levels so that growth will not be seriously constrained.

Even over the next ten years (1976-85), the supply enterprise typically proposed in such projections is impressive. Oil and gas extraction shifts dramatically to offshore and Alaskan sources, with nearly 900 new oil wells offshore of the contiguous 48 states alone. Some 170 new coal mines open, extracting about 200 million tons per year each from eastern underground and strip mines, plus 120 million from western stripping. The nuclear fuel cycle requires over 100 new uranium mines, a new enrichment plant, some 40 fuel fabrication plants, three fuel reprocessing plants. The electrical supply system, more than doubling, draws on some 180 new 800-megawatt coal-fired stations, over one hundred and forty 1,000-megawatt nuclear reactors, 60 conventional and over 100 pumped-storage hydroelectric plants, and over 350 gas turbines. Work begins on new industries to make synthetic fuels from coal and oil shale. At peak, just building (not operating) all these new facilities directly requires nearly 100,000 engineers, over 420,000 craftspeople, and over 140,000 laborers. Total indirect labor requirements are twice as great.[2]

This ten-year spurt is only the beginning. The year 2000 finds us with 450 to 800 reactors (including perhaps 80 fast breeders, each loaded with 2.5 metric tons of plutonium), 500 to 800 huge coal-fired power stations, 1,000 to 1,600 new coal mines and some 15 million electric automobiles. Massive electrification—which, according to one expert, is "the most important attempt to modify the infrastructure of industrial society since the railroad"[3]—is largely responsible for the release of waste heat sufficient to warm the entire freshwater runoff of the contiguous 48 states by 34–49° F.[4] Mining coal and uranium, increasingly in the arid

West, entails inverting thousands of communities and millions of acres, often with little hope of effective restoration. The commitment to a long-term coal economy many times the scale of today's makes the doubling of atmospheric carbon dioxide concentration early in the next century virtually unavoidable, with the prospect then or soon thereafter of substantial and perhaps irreversible changes in global climate.[5] Only the exact date of such changes is in question.

The main ingredients of such an energy future are roughly sketched in Figure I. For the period up to 2000, this sketch is a composite of recent projections published by the Energy Research and Development Administration (ERDA), Federal Energy Administration (FEA), Department of the Interior, Exxon, and Edison Electric Institute. Minor and relatively constant sources, such as hydroelectricity, are omitted; the nuclear component represents nuclear heat, which is roughly three times the resulting nuclear electric output; fuel imports are aggregated with domestic production. Beyond 2000, the usual cutoff date of present projections, the picture has been extrapolated to the year 2025—exactly how is not important here—in order to show its long-term implications more clearly.[6]

The flaws in this type of energy policy have been pointed out by critics in and out of government. For example, despite the intensive electrification—consuming more than half the total fuel input in 2000 and more thereafter—

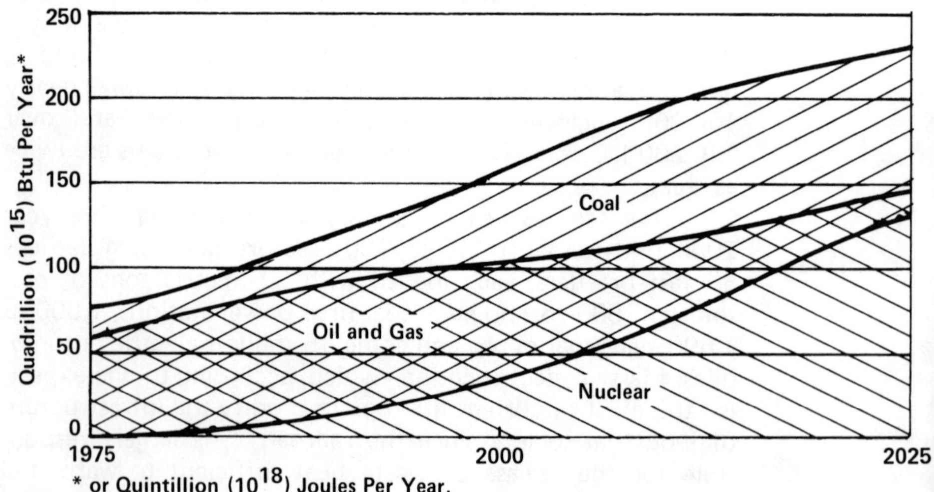

FIGURE 1. An Illustrative, Schematic Future for U.S. Gross Primary Energy Use

we are still short of gaseous and liquid fuels, acutely so from the 1980s on, because of slow and incomplete substitution of electricity for the two-thirds of fuel use that is now direct. Despite enhanced recovery of resources in the ground, shortages steadily deepen in natural gas—on which plastics and nitrogen fertilizers depend—and, later, in fuel for the transport sector (half our oil now runs cars). Worse, at least half the energy growth never reaches the consumer because it is lost earlier in elaborate conversions in an increasingly inefficient fuel chain dominated by electricity generation (which wastes about two-thirds of the fuel) and coal conversion (which wastes about one-third). Thus in Britain since 1900, primary energy—the input to the fuel chain—has doubled while energy at the point of end use—the car, furnace or machine whose function it fuels—has increased by only a half, or by a third per capita; the other half of the growth went to fuel the fuel industries, which are the largest energy consumers.

Among the most intractable barriers to implementing Figure I is its capital cost. In the 1960s, the total investment to increase a consumer's delivered energy supplies by the equivalent of one barrel of oil per day (about 67 kilowatts of heat) was a few thousand of today's dollars—of which, in an oil system, the wellhead investment in the Persian Gulf was and still is only a few hundred dollars. (The rest is transport, refining, marketing and distribution.) The capital intensity of much new coal supply is still in this range. But such cheaply won resources can no longer stretch our domestic production of fluid fuels or electricity; and Figure I relies mainly on these, not on coal burned directly, so it must bear the full burden of increased capital intensity.

That burden is formidable. For the North Sea oilfields coming into production soon, the investment in the whole system is roughly $10,000 to deliver an extra barrel per day (constant 1976 dollars throughout); for U.S. frontier (Arctic and offshore) oil and gas in the 1980s it will be generally in the range from $10,000 to $25,000; for synthetic gaseous and liquid fuels made from coal, from $20,000 to $50,000 per daily barrel.

The scale of these capital costs is generally recognized in the industries concerned. What is less widely appreciated—partly because capital costs of electrical capacity are normally calculated per installed (not delivered) kilowatt and partly because whole-system costs are rarely computed—is that capital cost is many times greater for new systems that make electricity than for those that burn fuels directly. For coal-electric capacity ordered today, a reasonable estimate would be about $150,000 for the delivered equivalent of one barrel of oil per day; for nuclear-electric capacity ordered today, about $200,000–$300,000. Thus, the capital cost per delivered kilowatt of electrical energy emerges as roughly 100 times that of the traditional direct-fuel technologies on which our society has been built.[7]

The capital intensity of coal conversion and, even more, of large electrical stations and distribution networks is so great that many analysts, such as the strategic planners of the Shell Group in London, have concluded that no major country outside the Persian Gulf can afford these centralized high technologies on a truly large scale, large enough to run a country. They are looking, in Monte Canfield's phrase, like future technologies whose time has passed.

Relying heavily on such technologies, President Ford's 1976–85 energy program turns out to cost over $1 trillion (in 1976 dollars) in initial investment, of which about 70 to 80 percent would be for new rather than replacement plants.[8] The latter figure corresponds to about three-fourths of cumulative net private domestic investment (NPDI) over the decade (assuming that NPDI remains 7 percent of Gross National Product and that GNP achieves real growth of 3.5 percent per year despite the adverse effects of the energy program on other investments). In contrast, the energy sector has recently required only one-fourth of NPDI. Diverting to the energy sector not only this hefty share of discretionary investment but also about two-thirds of all the rest would deprive other sectors which have their own cost-escalation problems and their own vocal constituencies. A powerful political response could be expected. And this capital burden is not temporary; further up the curves of Figure I it tends to increase, and much of what might have been thought to be increased national wealth must be plowed back into the care and feeding of the energy system. Such long-lead-time, long-payback-time investments might also be highly inflationary.

Of the $1 trillion-plus just cited, three-fourths would be for electrification. About 18 percent of the total investment could be saved just by reducing the assumed average 1976–85 electrical growth rate from 6.5 to 5.5 percent per year.[9] Not surprisingly, the combination of disproportionate and rapidly increasing capital intensity, long lead times, and economic responses is already proving awkward to the electric utility industry, despite the protection of a 20 percent taxpayer subsidy on new power stations.[10] "Probably no industry," observes Bankers Trust Company, "has come closer to the edge of financial disaster." Both here and abroad an effective feedback loop is observable: large capital programs → poor cash flow → higher electricity prices → reduced demand growth → worse cash flow → increased bond flotation → increased debt-to-equity ratio, worse coverage, and less attractive bonds → poor bond sales → worse cash flow → higher electricity prices → reduced (even negative) demand growth and political pressure on utility regulators → overcapacity, credit pressure, and higher cost of money → worse cash flow, etc. This "spiral of impossibility," as Mason Willrich has called it, is exacerbated by most utilities' failure to base historic prices on the long-run cost of new supply: thus some must now tell their customers that the current-dollar cost of a kilowatt-hour will treble by 1985, and that two-thirds of that increase will be capital charges for new plants. Moreover, experience abroad suggests that even a national treasury cannot long afford electrification: a New York State-like position is quickly reached, or too little money is left over to finance the energy *uses,* or both.

Summarizing a similar situation in Britain, Walter Patterson concludes: "Official statements identify an anticipated 'energy gap' which can be filled only with nuclear electricity; the data do not support any such conclusion, either as regards the 'gap' or as regards the capability of filling it with nuclear electricity." We have sketched one form of the latter argument; let us now consider the former.

Despite the steeply rising capital intensity of new energy supply, forecasts of energy demand made as recently as 1972 by such bodies as the Federal Power Commission and the Department of the Interior wholly ignored both price elasticity of demand and energy conservation. The Chase Manhattan Bank in 1973 saw virtually no scope for conservation save by minor curtailments: the efficiency with which energy produced economic outputs was assumed to be optimal already. In 1976, some analysts still predict economic calamity if the United States does not continue to consume twice the combined energy total for Africa, the rest of North and South America, and Asia except Japan. But what have more careful studies taught us about the scope for doing better with the energy we have? Since we can't keep the bathtub filled because the hot water keeps running out, do we really (as Malcolm MacEwen asks) need a bigger water heater, or could we do better with a cheap, low-technology plug?

There are two ways, divided by a somewhat fuzzy line, to do more with less energy. First, we can plug leaks and use thriftier technologies to produce exactly the same output of goods and services—and bads and nuisances—as before, substituting other resources (capital, design, management, care, etc.) for some of the energy we formerly used. When measures of this type use today's technologies, are advantageous today by conventional economic criteria, and have no significant effect on life-styles, they are called "technical fixes."

In addition, or instead, we can make and use a smaller quantity or a different mix of the outputs themselves, thus to some degree changing (or reflecting ulterior changes in) our life-styles. We might do this because of changes in personal values, rationing by price or otherwise, mandatory curtailments, or gentler inducements. Such "social changes" include car-pooling, smaller cars, mass transit, bicycles, walking, opening windows, dressing to suit the weather, and extensively recycling materials. Technical fixes, on the other hand, include thermal insulation, heat-pumps (devices like air conditioners which move heat around—often in either direction—rather than making it from scratch), more efficient furnaces and car engines, less overlighting and overventilation in commercial buildings, and recuperators for waste heat in industrial processes. Hundreds of technical and semi-technical analyses of both kinds of conservation have been done; in the last two years especially, much analytic progress has been made.

Theoretical analysis suggests that in the long term, technical fixes *alone* in the United States could probably improve energy efficiency by a factor of at least three or four.[11] A recent review of specific practical measures cogently argues that with only those technical fixes that could be implemented by about the turn of the century, we could nearly double the efficiency with which we use energy.[12] If that is correct, we could have steadily increasing economic activity with approximately constant primary energy use for the next few decades, thus stretching our present energy supplies rather than having to add massively to them. One careful comparison shows that *after* correcting for differences of climate, hydroelectric capacity, etc., Americans would still use about a third less energy than they do now if they were as efficient as the Swedes (who see much room for improvement in their own efficiency).[13] U.S. per capita energy intensity, too, is about twice that of West Germany in space heating, four times in transport.[14] Much of the difference is attributable to technical fixes.

Some technical fixes are already under way in the United States. Many factories have cut tens of percent off their fuel cost per unit output, often with practically no capital investment. New 1976 cars average 27 percent better mileage than 1974 models. And there is overwhelming evidence that technical fixes are generally much cheaper than increasing energy supply, quicker, safer, of more lasting benefit. They are also better for secure, broadly based employment using existing skills. Most energy conservation measures and the shifts of consumption which they occasion are relatively labor-intensive. Even making more energy-efficient home appliances is about twice as good for jobs as is building power stations: the latter is practically the least labor-intensive major investment in the whole economy.

The capital savings of conservation are particularly impressive. In the terms used above, the investments needed to *save* the equivalent of an extra barrel of oil per day are often zero to $3,500, generally under $8,000, and at most about $25,000—far less than the amounts needed to increase most kinds of energy supply. Indeed, to use energy efficiently in new buildings, especially commercial ones, the additional capital cost is often *negative*: savings on heating and cooling equipment more than pay for the other modifications.

To take one major area of potential saving, technical fixes in new buildings can save 50 percent or more in office buildings and 80 percent or more in some new houses.[15] A recent American Institute of Architects study concludes that, by 1990, improved design of new buildings and modification of old ones could save a third of our current *total* national energy use—and save money too. The payback time would be only half that of the alternative investment in increased energy supply, so the same capital could be used twice over.

A second major area lies in "cogeneration," or the generating of electricity as a by-product of the process steam normally produced in many industries. A Dow study chaired by Paul McCracken reports that by 1985 U.S. industry could meet approximately half its own electricity needs (compared to about a seventh today) by this means. Such cogeneration would save $20–50 billion in investment, save fuel equivalent to 2–3 million barrels of oil per day, obviate the need for more than 50 large reactors, and (with flattened utility rates) yield at least 20 percent pretax return on marginal investment while reducing the price of electricity to consumers.[16] Another measure of the potential is that cogeneration provides about 4 percent of electricity today in the United States but about 12 percent in West Germany. Cogeneration and more efficient use of electricity could together reduce our use of electricity by a third and our central-station generation by 60 percent.[17] Like district heating (distribution of waste heat as hot water via insulated pipes to heat buildings), U.S. cogeneration is held back only by institutional barriers. Yet these are smaller than those that were overcome when the present utility industry was established.

So great is the scope for technical fixes now that we could spend several hundred billion dollars on them initially plus several hundred million dollars per day—and still save money compared with increasing the supply! And we would still have the fuel (without the environmental and geopolitical problems of getting and using it). The barriers to far more efficient use of energy are not technical, nor in any fundamental sense economic. So why do we stand here confronted, as Pogo said, by insurmountable opportunities?

The answer—apart from poor information and ideological antipathy and rigidity—is a wide array of institutional barriers, including more than 3,000 conflicting and often obsolete building codes, an innovation-resistant building industry, lack of mechanisms to ease the transition from kinds of work that we no longer need to kinds we do need, opposition by strong unions to schemes that would transfer jobs from their members to larger numbers of less "skilled" workers, promotional utility rate structures, fee structures giving building engineers a fixed percentage of prices of heating and cooling equipment they install, inappropriate tax and mortgage policies, conflicting signals to consumers, misallocation of conservation's costs and benefits (builders vs. buyers, landlords vs. tenants, etc.), imperfect access to capital markets, fragmentation of government responsibility, etc.

Though economic answers are not always right answers, properly using the markets we have may be the greatest single step we could take toward a sustainable, humane energy future. The sound economic principles we need to apply include flat (even inverted) utility rate structures rather than discounts for large users, pricing energy according to what extra supplies will cost in the long run ("long-run marginal-cost pricing"), removing subsidies, assessing the total costs of energy-using purchases over their whole operating lifetimes ("life-cycle costing"), counting the costs of complete energy systems including all support and distribution systems, properly assessing and charging environmental costs, valuing assets by what it would cost to replace them, discounting appropriately, and encouraging competition through antitrust enforcement (including at least horizontal divestiture of giant energy corporations).

Such practicing of the market principles we preach could go very far to help us use energy efficiently and get it from sustainable sources. But just as clearly, there are things the market cannot do, like reforming building codes or utility practices. And whatever our means, there is room for differences of opinion about how far we can achieve the great theoretical potential for technical fixes. How far might we instead choose, or be driven to, some of the "social changes" mentioned earlier?

There is no definitive answer to this question—though it is arguable that if we are not clever enough to overcome the institutional barriers to implementing technical fixes, we shall certainly not be clever enough to overcome the more familiar but more formidable barriers to increasing energy supplies. My own view of the evidence is, first, that we are adaptable enough to use technical fixes *alone* to double, in the next few decades, the amount of social benefit we wring from each unit of end-use energy; and second, that value changes which could either replace or supplement those technical changes are also occurring rapidly. If either of these views is right, or if both are partly right, we should be able to double end-use efficiency by the turn of the century or shortly thereafter, with minor or no changes in life-styles or values save increasing comfort for modestly increasing numbers. Then over the period 2010–40, we should be able to shrink per capita primary energy use to perhaps a third or a quarter of today's.[18] (The former would put us at the per capita level of the wasteful, but hardly troglodytic, French.) Even in the case of fourfold shrinkage, the resulting society could be instantly recognizable to a visitor from the 1960s and need in no sense be a pastoralist's utopia—though that option would remain open to those who may desire it.

The long-term mix of technical fixes with structural and value changes in work, leisure, agriculture and industry will require much trial and error. It will take many years to make up our diverse minds. It will not be easy—merely easier than not doing it. Meanwhile it is easy only to see what not to do.

If one assumes that by resolute technical fixes and modest social innovation we can double our end-use efficiency by shortly after 2000, then we could be twice as affluent as now with today's level of energy use, or as affluent as now while using only half the end-use energy we use today. Or we might be somewhere in between—significantly more affluent (and equitable) than today but with less end-use energy.

Many analysts now regard modest, zero or negative growth in our rate of energy use as a realistic long-term goal. Present annual U.S. primary energy demand is about 75 quadrillion Btu ("quads"), and most official projections for 2000 envisage growth to 130—170 quads. However, recent work at the Institute for Energy Analysis, Oak Ridge, under the direction of Dr. Alvin Weinberg, suggests that standard projections of energy demand are far too high because they do not take account of changes in demographic and economic trends. In June 1976 the Institute considered that with a conservation program far more modest than that contemplated in this article, the likely range of U.S. primary energy demand in the year 2000 would be about 101—126 quads, with the lower end of the range more probable and end-use energy being about 60—65 quads. And, at the further end of the spectrum, projections for 2000 being considered by the "Demand Panel" of a major U.S. National Research Council study, as of mid-1976, ranged as low as about 54 quads of fuels (plus 16 of solar energy).

As the basis for a coherent alternative to the path shown in Figure I earlier, a primary energy demand of about 95 quads for 2000 is sketched in Figure 2. Total energy demand would gradually decline thereafter as inefficient buildings, machines, cars and energy systems are slowly modified or replaced. Let us now explore the other ingredients of such a path—starting with the "soft" supply technologies which, spurned in Figure I as insignificant, now assume great importance.

There exists today a body of energy technologies that have certain specific features in common and that offer great technical, economic and political attractions, yet for which there is no generic term. For lack of a more satisfactory term, I shall call them "soft" technologies: a textural description, intended to mean not vague, mushy, speculative or ephemeral, but rather flexible, resilient, sustainable and benign. Energy paths dependent on soft technologies, illustrated in Figure 2, will be called "soft" energy paths, as the

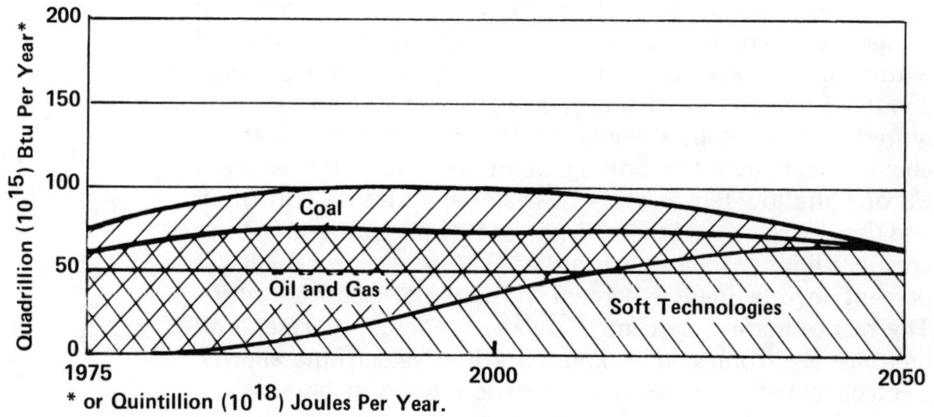

FIGURE 2. An Alternate Illustrative Future for U.S. Gross Primary Energy Use

"hard" technologies sketched in Section II constitute a "hard" path (in both senses). The distinction between hard and soft energy paths rests not on how much energy is used, but on the technical and sociopolitical *structure* of the energy system, thus focusing our attention on consequent and crucial political differences.

In Figure 2, then, the social structure is significantly shaped by the rapid deployment of soft technologies. These are defined by five characteristics:

- They rely on renewable energy flows that are always there whether we use them or not, such as sun and wind and vegetation: on energy income, not on depletable energy capital.
- They are diverse, so that energy supply is an aggregate of very many individually modest contributions, each designed for maximum effectiveness in particular circumstances.
- They are flexible and relatively low-technology—which does not mean unsophisticated, but rather, easy to understand and use without esoteric skills, accessible rather than arcane.
- They are matched in *scale* and in geographic distribution to end-use needs, taking advantage of the free distribution of most natural energy flows.
- They are matched in *energy quality* to end-use needs: a key feature that deserves immediate explanation.

People do not want electricity or oil, nor such economic abstractions as "residential services," but rather comfortable rooms, light, vehicular motion, food, tables, and other real things. Such end-use needs can be classified by the physical nature of the task to be done. In the United States today, about 58 percent of all energy at the point of end use is required as heat, split roughly equally between temperatures above and below the boiling point of water. (In Western Europe the low-temperature heat alone is often a half of all end-use energy.) Another 38 percent of all U.S. end-use energy provides mechanical motion: 31 percent in vehicles, 3 percent in pipelines, 4 percent in industrial electric motors. The rest, a mere 4 percent of delivered energy, represents *all* lighting, electronics, telecommunications, electrometallurgy, electrochemistry, arc-welding, electric motors in home appliances and in railways, and similar end uses which now *require* electricity.

Some 8 percent of all our energy end use, then, requires electricity for purposes other than low-temperature heating and cooling. Yet, since we actually use electricity for many such low-grade purposes, it now meets 13 percent of our end-use needs—and its generation consumes 29 percent of our fossil fuels. A hard energy path would increase this 13 percent figure to 20—40 percent (depending on assumptions) by the year 2000, and far more thereafter. But this is wasteful because the laws of physics require, broadly speaking, that a power station change three units of fuel into two units of almost useless waste heat plus one unit of electricity. This electricity can do more difficult kinds of work than can the original fuel, but unless this extra quality and versatility are used to advantage, the costly process of upgrading the fuel—and losing two-thirds of it—is all for naught.

Plainly we are using premium fuels and electricity for many tasks for which their high energy quality is superfluous, wasteful and expensive, and a hard path would make this inelegant practice even more common. Where we want only to create temperature differences of tens of degrees, we should meet the need with sources whose potential is tens or hundreds of degrees, not with a flame temperature of thousands or a nuclear temperature of millions—like cutting butter with a chainsaw.

For some applications, electricity is appropriate and indispensable: electronics, smelting, subways, most lighting, some kinds of mechanical work, and a few more. But these uses are already oversupplied, and for the other, dominant uses remaining in our energy economy this special form of energy cannot give us our money's worth (in many parts of the United States today it already costs $50–120 per barrel-equivalent). Indeed, in probably no industrial country today can additional supplies of electricity be used to thermodynamic advantage which would justify their high cost in money and fuels.

So limited are the U.S. end uses that really require electricity that by applying careful technical fixes to them we could reduce their 8 percent total to about 5 percent (mainly by reducing commercial overlighting), whereupon we could probably cover all those needs with present U.S. hydroelectric capacity plus the cogeneration capacity available in the mid-to-late 1980s.[19] Thus an affluent industrial economy could advantageously operate with no central power stations at all! In practice we would not necessarily want to go that far, at least not for a long time; but the possibility illustrates how far we are from supplying energy only in the quality needed for the task at hand.

A feature of soft technologies as essential as their fitting end-use needs (for a different reason) is their appropriate scale, which can achieve important types of economies not available to larger, more centralized systems. This is done in five ways, of which the first is reducing and sharing overheads. Roughly half your electricity bill is fixed distribution costs to pay the overheads of a sprawling energy system: transmission lines, transformers, cables, meters and people to read them, planners, headquarters, billing computers, interoffice memos, advertising agencies. For electrical and some fossil-fuel systems, distribution accounts for more than half of total capital cost, and administration for a significant fraction of total operating cost. Local or domestic energy systems can reduce or even eliminate these infrastructure costs. The resulting savings can far outweigh the extra costs of the dispersed maintenance infrastructure that the small systems require, particularly where that infrastructure already exists or can be shared (e.g., plumbers fixing solar heaters as well as sinks).

Small scale brings further savings by virtually eliminating distribution losses, which are cumulative and pervasive in centralized energy systems (particularly those using high-quality energy). Small systems also avoid direct diseconomies of scale, such as the frequent unreliability of large units and the related need to provide instant "spinning reserve" capacity on electrical grids to replace large stations that suddenly fail. Small systems with short lead times greatly reduce exposure to interest, escalation and mistimed demand forecasts—major indirect diseconomies of large scale.

The fifth type of economy available to small systems arises from mass production. Consider, as Henrik Harboe suggests, the 100-odd million cars in this country. In round numbers, each car probably has an average cost of less than $4,000 and a shaft power over 100 kilowatts (134 horsepower). Presumably a good engineer could build a generator and upgrade an automobile engine to a reliable, 35-percent-efficient diesel at no greater total cost, yielding a mass-produced diesel generator unit costing less than $40 per kW. In contrast, the motive capacity in our central power stations—currently totaling about 1/40 as much as in our cars—costs perhaps ten times more per kW, partly because it is not mass-produced. It is not surprising that at least one foreign car maker hopes to go into the wind-machine and heat-pump business. Such a market can be entered incrementally, without the billions of dollars' investment required for, say, liquefying natural gas or gasifying coal. It may require a production philosophy oriented toward technical simplicity, low replacement cost, slow obsolescence, high reliability, high volume and low markup; but these are familiar concepts in mass production. Industrial resistance would presumably melt when—as with pollution-abatement equipment—the scope for profit was perceived.

This is not to say that all energy systems need be at domestic scale. For example, the medium scale of urban neighborhoods and rural villages offers fine prospects for solar collectors—especially for adding collectors to existing buildings of which some (perhaps with large flat roofs) can take excess collector area while others cannot take any. They could be joined via communal heat storage systems, saving on labor cost and on heat losses. The costly craftwork of remodeling existing systems—"backfitting" idiosyncratic houses with individual collectors—could thereby be greatly reduced. Despite these advantages, medium-scale solar technologies are currently receiving little attention apart from a condominium-village project in Vermont sponsored by the Department of Housing and Urban Development and the 100-dwelling-unit Mejannes-le-Clap project in France.

The schemes that dominate ERDA's solar research budget—such as making electricity from huge collectors in the desert, or from temperature differences in the oceans, or from Brooklyn Bridge-like satellites in outer space—do not satisfy our criteria, for they are ingenious high-technology ways to supply energy in a form and at a scale inappropriate to most end-use needs. Not all solar technologies are soft. Nor, for the same reason, is nuclear fusion a soft technology.[20] But many genuine soft technologies are now available and are now economic. What are some of them?

Solar heating and, imminently, cooling head the list. They are incrementally cheaper than electric heating, and far more inflation-proof, practically anywhere in the world.[21] In the United States (with fairly high average sunlight levels), they are cheaper than present electric heating virtually anywhere, cheaper than oil heat in many parts, and cheaper than gas and coal in some. Even in the least favorable parts of the continental United States, far more sunlight falls on a typical building than is required to heat and cool it without supplement; whether this is considered economic depends on how the accounts are done.[22] The difference in solar input between the most and least favorable parts of the lower 49 states is generally less than twofold, and in cold regions, the long heating season can improve solar economics.

Ingenious ways of backfitting existing urban and rural buildings (even large commercial ones) or their neighborhoods with efficient and exceedingly reliable solar collectors are being rapidly developed in both the private and public sectors. In some recent projects, the lead time from ordering to operation has been only a few months. Good solar hardware, often modular, is going into pilot or full-scale production over the next few years, and will increasingly be integrated into buildings as a multipurpose structural element, thereby sharing costs. Such firms as Philips, Honeywell, Revere, Pittsburgh Plate Glass, and Owens-Illinois, plus many dozens of smaller firms, are applying their talents, with rapid and accelerating effect, to reducing unit costs and improving performance. Some novel types of very simple collectors with far lower costs also show promise in current experiments. Indeed, solar hardware per se is necessary only for backfitting existing buildings. If we build new buildings properly in the first place, they can use "passive" solar collectors—large south windows or glass-covered black south walls—rather than special collectors. If we did this to all new houses in the next 12 years, we would save about as much energy as we expect to recover from the Alaskan North Slope.[23]

Secondly, exciting developments in the conversion of agricultural, forestry and urban wastes to methanol and other liquid and gaseous fuels now offer practical, economically interesting technologies sufficient to run an efficient U.S. transport sector.[24] Some bacterial and enzymatic routes under study look even more promising, but presently proved processes already offer sizable contributions without the inevitable climatic constraints of fossil-fuel combustion. Organic conversion technologies must be sensitively integrated with agriculture and forestry so as not to deplete the soil; most current methods seem suitable in this respect, though they may change the farmer's priorities by making his whole yield of biomass (vegetable matter) salable.

The required scale of organic conversion can be estimated. Each year the U.S. beer and wine industry, for example, microbiologically produces 5 percent as many gallons (not all alcohol, of course) as the U.S. oil industry produces gasoline. Gasoline has 1.5–2 times the fuel value of alcohol per gallon. Thus a conversion industry roughly 10 to 14 times the scale (in gallons of fluid output per year) of our cellars and breweries would produce roughly one-third of the present gasoline requirements of the United States; if one assumes a transport sector with three times today's average efficiency—a reasonable estimate for early in the next century—then the whole of the transport needs could be met by organic conversion. The scale of effort required does not seem unreasonable, since it would replace in function half our refinery capacity.

Additional soft technologies include wind-hydraulic systems (especially those with a vertical axis), which already seem likely in many design studies to compete with nuclear power in much of North America and Western Europe. But wind is not restricted to making electricity: it can heat, pump, heat-pump, or compress air. Solar process heat, too, is coming along rapidly as we learn to use the 5,800°C. potential of sunlight (much hotter than a boiler). Finally, high- and low-temperature solar collectors, organic converters, and wind machines can form symbiotic hybrid combinations more attractive than the separate components.

Energy storage is often said to be a major problem of energy-income technologies. But this "problem" is largely an artifact of trying to recentralize, upgrade and redistribute inherently diffuse energy flows. Directly storing sunlight or wind—or, for that matter, electricity from any source—is indeed difficult on a large scale. But it is easy if done on a scale and in an energy quality matched to most end-use needs. Daily, even seasonal, storage of low- and medium-temperature heat at the point of use is straightforward with water tanks, rock beds, or perhaps fusible salts. Neighborhood heat storage is even cheaper. In industry, wind-generated compressed air can easily (and, with due care, safely) be stored to operate machinery: the technology is simple, cheap, reliable and highly developed. (Some cities even used to supply compressed air as a standard utility.) Installing pipes to distribute hot water (or compressed air) tends to be considerably cheaper than installing equivalent electric distribution capacity. Hydroelectricity is stored behind dams, and organic conversion yields readily stored liquid and gaseous fuels. On the whole, therefore, energy storage is much less of a problem in a soft energy economy than in a hard one.

Recent research suggests that a largely or wholly solar economy can be constructed in the United States with straightforward soft technologies that are now demonstrated and now economic or nearly economic.[25] Such a conceptual exercise does not require "exotic" methods such as sea-thermal, hot-dry-rock geothermal, cheap (perhaps organic) photovoltaic, or solar-thermal electric systems. If developed, as some probably will be, these technologies could be convenient, but they are in no way essential for an industrial society operating solely on energy income.

Figure 2 shows a plausible and realistic growth pattern, based on several detailed assessments, for soft technologies given aggressive support. The useful output from these technologies would overtake, starting in the 1990s, the output of nuclear electricity shown in even the most sanguine federal estimates. For illustration, Figure 2 shows soft technologies meeting virtually all energy needs in 2025, reflecting a judgment that a completely soft supply mix is practicable in the long run with or without the 2000–25 energy shrinkage shown. Though most technologists who have thought seriously about the matter will concede it conceptually, some may be uneasy about the details. Obviously the sketched curve is not definitive, for although the general direction of the soft path must be shaped soon, the details of the energy economy in 2025 would not be committed in this century. To a large extent, therefore, it is enough to ask yourself whether Figure 1 or 2 seems preferable in the 1975–2000 period.

A simple comparison may help. Roughly half, perhaps more, of the gross primary energy being produced in the hard path in 2025 is lost in conversions. A further appreciable fraction is lost in distribution. Delivered end-use energy is thus not vastly greater than in the soft path, where conversion and distribution losses have been all but eliminated.

(What is lost can often be used locally for heating, and is renewable, not depletable.) But the soft path makes each unit of end-use energy perform several times as much social function as it would have done in the hard path; so in a conventional sense, social welfare in the soft path in 2025 is substantially greater than in the hard path at the same date.

To fuse into a coherent strategy the benefits of energy efficiency and of soft technologies, we need one further ingredient: transitional technologies that use fossil fuels briefly and sparingly to build a bridge to the energy-income economy of 2025, conserving those fuels—especially oil and gas—for petrochemicals (ammonia, plastics, etc.) and leaving as much as possible in the ground for emergency use only.

Some transitional technologies have already been mentioned under the heading of conservation—specifically, cogenerating electricity from existing industrial steam and using existing waste heat for district heating. Given such measures, increased end-use efficiency, and the rapid development of biomass alcohol as a portable liquid fuel, the principal short- and medium-term problem becomes, not a shortage of electricity or of portable liquid fuels, but a shortage of clean sources of heat. It is above all the sophisticated use of coal, chiefly at modest scale, that needs development. Technical measures to permit the highly efficient use of this widely available fuel would be the most valuable transitional technologies.

Neglected for so many years, coal technology is now experiencing a virtual revolution. We are developing supercritical gas extraction, flash hydrogenation, flash pyrolysis, panel-bed filters and similar ways to use coal cleanly at essentially any scale and to cream off valuable liquids and gases as premium fuels before burning the rest. These methods largely avoid the costs, complexity, inflexibility, technical risks, long lead times, large scale, and tar formation of the traditional processes that now dominate our research.

Perhaps the most exciting current development is the so-called fluidized-bed system for burning coal (or virtually any other combustible material). Fluidized beds are simple, versatile devices that add the fuel a little at a time to a much larger mass of small, inert, red-hot particles—sand or ceramic pellets—kept suspended as an agitated fluid by a stream of air continuously blown up through it from below. The efficiency of combustion, of other chemical reactions (such as sulfur removal), and of heat transfer is remarkably high because of the turbulent mixing and large surface area of the particles. Fluidized beds have long been used as chemical reactors and for burning trash, but are now ready to be commercially applied to raising steam and operating turbines. In one system currently available from Stal-Laval Turbin AB of Sweden, eight off-the-shelf 70-megawatt gas turbines powered by fluidized-bed combusters, together with district-heating networks and heat pumps, would heat as many houses as a $1 billion-plus coal gasification plant, but would use only two-fifths as much coal, cost a half to two-thirds as much to build, and burn more cleanly than a normal power station with the best modern scrubbers.[26]

Fluidized-bed boilers and turbines can power giant industrial complexes, especially for cogeneration, and are relatively easy to backfit into old municipal power stations. Scaled down, a fluidized bed can be a tiny household device—clean, strikingly simple and flexible—that can replace an ordinary furnace or grate and can recover combustion heat with an efficiency over 80 percent.[27] At medium scale, such technologies offer versatile boiler backfits and improve heat recovery in flues. With only minor modifications they can burn practically any fuel. It is essential to commercialize all these systems now—not to waste a decade on highly instrumented but noncommercial pilot plants constrained to a narrow, even obsolete design philosophy.[28]

Transitional technologies can be built at appropriate scale so that soft technologies can be plugged into the system later. For example, if district heating uses hot water tanks on a neighborhood scale, those tanks can in the long run be heated by neighborhood solar collectors, wind-driven heat pumps, a factory, a pyrolyzer, a geothermal well, or whatever else becomes locally available—offering flexibility that is not possible at today's excessive scale.

Both transitional and soft technologies are worthwhile industrial investments that can recycle moribund capacity and underused skills, stimulate exports, and give engaging problems to innovative technologists. Though neither glamorous nor militarily useful, these technologies are socially effective—especially in poor countries that need such scale, versatility and simplicity even more than we do.

Properly used, coal, conservation, and soft technologies together can squeeze the "oil and gas" wedge in Figure 2 from both sides—so far that most of the frontier extraction and medium-term imports of oil and gas become unnecessary and our conventional resources are greatly stretched. Coal can fill the real gaps in our fuel economy with only a temporary and modest (less than twofold at peak) expansion of mining, not requiring the enormous infrastructure and social impacts implied by the scale of coal use in Figure 1.

In sum, Figure 2 outlines a prompt redirection of effort at the margin that lets us use fossil fuels intelligently to buy the time we need to change over to living on our energy income. The innovations required, both technical and social, compete directly and immediately with the incremental actions that constitute a hard energy path: fluidized beds vs. large coal gasification plants and coal-electric stations, efficient cars vs. offshore oil, roof insulation vs. Arctic gas, cogeneration vs. nuclear power. These two directions of development are mutually exclusive: the pattern of commitments of resources and time required for the hard energy path and the pervasive infrastructure which it accretes gradually make the soft path less and less attainable. That is, our two sets of choices compete not only in what they accomplish, but also in what they allow us to contemplate later. Figure 1 obscures this constriction of options, for it peers myopically forward, one power station at a time, extrapolating trend into destiny by self-fulfilling prophecy with no end clearly in sight. Figure 2, in contrast, works backward from a strategic goal, asks what we must do when in order to get there, and thus reveals the potential for a radically different path that would be invisible to anyone working forward in time by incremental ad-hocracy.

Both the soft and the hard paths bring us, each in its own way and at broadly similar rates, to the era beyond oil and gas. But the rates of internal adaptation meanwhile are different. As we have seen, the soft path relies on smaller, far simpler supply systems entailing vastly shorter development and construction time, and on smaller, less sophisticated management systems. Even converting the urban clusters of a whole country to district heating should take only 30–40 years. Furthermore, the soft path relies mainly on small, standard, easy-to-make components and on technical resources dispersed in many organizations of diverse sizes and habits; thus everyone can get into the act, unimpeded by centralized bureaucracies, and can compete for a market share through ingenuity and local adaptation. Besides having much lower and more stable operating costs than the hard path, the soft path appears to have lower initial cost because of its technical simplicity, small unit size, very low overheads, scope for mass production, virtual elimination of distribution losses and of interfuel conversion losses, low exposure to escalation and interest, and prompt incremental construction (so that new capacity is built only when and where it is needed).[29]

The actual costs of whole systems, however, are not the same as perceived costs: solar investments are borne by the householder, electric investments by a utility that can float low-interest bonds and amortize over 30 years. During the transitional era, we should therefore consider ways to broaden householders' access to capital markets. For example, the utility could finance the solar investment (leaving its execution to the householder's discretion), then be repaid in installments corresponding to the householder's saving. The householder would thus minimize his own—and society's—long-term costs. The utility would have to raise several times less capital than it would without such a scheme—for otherwise it would have to build new electric or synthetic-gas capacity at even higher cost—and would turn over its money at least twice as quickly, thus retaining an attractive rate of return on capital. The utility would also avoid social obsolescence and use its existing infrastructure. Such incentives have already led several U.S. gas utilities to use such a capital-transfer scheme to finance roof insulation.

Next, the two paths differ even more in risks than in costs. The hard path entails serious environmental risks, many of which are poorly understood and some of which have probably not yet been thought of. Perhaps the most awkward risk is that late in this century, when it is too late to do much about it, we may well find climatic constraints on coal combustion about to become acute in a few more decades: for it now takes us only that long, not centuries or millennia, to approach such outer limits. The soft path, by minimizing all fossil-fuel combustion, hedges our bets. Its environmental impacts are relatively small, tractable and reversible.[30]

The hard path, further, relies on a very few high technologies whose success is by no means assured. The soft path distributes the technical risk among very many diverse low technologies, most of which are already known to work well. They do need sound engineering—a solar collector or heat pump can be worthless if badly designed—but the engineering is of an altogether different and more forgiving order than the hard path requires, and the cost of failure is much lower both in potential consequences and in number of people affected. The soft path also minimizes the economic risks to capital in case of error, accident or sabotage; the hard path effectively maximizes those risks by relying on vulnerable high-technology devices each costing more than the endowment of Harvard University. Finally, the soft path appears generally more flexible—and thus robust. Its technical diversity, adaptability, and geographic dispersion make it resilient and offer a good prospect of stability under a wide range of conditions, foreseen or not. The hard path, however, is brittle; it must fail, with widespread and serious disruption, if any of its exacting technical and social conditions is not satisfied continuously and indefinitely.

The soft path has novel and important international implications. Just as improvements in end-use efficiency can be used at home (via innovative financing and neighborhood self-help schemes) to lessen first the disproportionate burden of energy waste on the poor, so can soft technologies and reduced pressure on oil markets especially benefit the poor abroad. Soft technologies are ideally suited for rural villagers and urban poor alike, directly helping the more than two billion people who have no electric outlet nor anything to plug into it but who need ways to heat, cook, light and pump. Soft technologies do not carry with them inappropriate cultural patterns or values; they capitalize on poor countries' most abundant resources (including such protein-poor plants as cassava, eminently suited to making fuel alcohols), helping to redress the severe energy imbalance between temperate and tropical regions; they can often be made locally from local materials and do not require a technical elite to maintain them; they resist technological dependence and commercial monopoly; they conform to modern concepts of agriculturally based eco-development from the bottom up, particularly in the rural villages.

Even more crucial, unilateral adoption of a soft energy path by the United States can go a long way to control nuclear proliferation—perhaps to eliminate it entirely. Many nuclear advocates have missed this point: believing that there is no alternative to nuclear power, they say that if the United States does not export nuclear technology, others will, so we might as well get the business and try to use it as a lever to slow the inevitable spread of nuclear weapons to nations and subnational groups in other regions. Yet the genie is not wholly out of the bottle yet—thousands of reactors are planned for a few decades hence, tens of thousands thereafter—and the cork sits unnoticed in our hands.

Perhaps the most important opportunity available to

us stems from the fact that for at least the next five or ten years, while nuclear dependence and commitments are still reversible, all countries will continue to rely on the United States for the technical, the economic, and especially the *political* support they need to justify their own nuclear programs. Technical and economic dependence is intricate and pervasive; political dependence is far more important but has been almost ignored, so we do not yet realize the power of the American example in an essentially imitative world where public and private divisions over nuclear policy are already deep and grow deeper daily.

The fact is that in almost all countries the domestic political base to support nuclear power is not solid but shaky. However great their nuclear ambitions, other countries must still borrow that political support from the United States. Few are succeeding. Nuclear expansion is all but halted by grass-roots opposition in Japan and the Netherlands; has been severely impeded in West Germany, France, Switzerland, Italy and Austria; has been slowed and may soon be stopped in Sweden; has been rejected in Norway and (so far) Australia and New Zealand, as well as in two Canadian Provinces; faces an uncertain prospect in Denmark and many American states; has been widely questioned in Britain, Canada and the U.S.S.R.[31]; and has been opposed in Spain, Brazil, India, Thailand and elsewhere.

Consider the impact of three prompt, clear U.S. statements:

- The United States will phase out its nuclear power program[32] and its support of others' nuclear power programs.
- The United States will redirect those resources into the tasks of a soft energy path and will freely help any other interested countries to do the same, seeking to adapt the same broad principles to others' needs and to learn from shared experience.
- The United States will start to treat nonproliferation, control of civilian fission technology, and strategic arms reduction as interrelated parts of the same problem with intertwined solutions.

I believe that such a universal, nondiscriminatory package of policies would be politically irresistible to North and South, East and West alike. It would offer perhaps our best chance of transcending the hypocrisy that has stalled arms control: by no longer artificially divorcing civilian from military nuclear technology, we would recognize officially the real driving forces behind proliferation; and we would no longer exhort others not to acquire bombs while claiming that we ourselves feel more secure with bombs than without them.

Nobody can be certain that such a package of policies, going far beyond a mere moratorium, would work. The question has received far too little thought, and political judgments differ. My own, based on the past nine years' residence in the midst of the European nuclear debate, is that nuclear power could not flourish there if the United States did not want it to.[33] In giving up the export market that our own reactor designs have dominated, we would be demonstrating a desire for peace, not profit, thus allaying legitimate European commercial suspicions. Those who believe such a move would be seized upon gleefully by, say, French exporters are seriously misjudging French nuclear politics. Skeptics, too, have yet to present a more promising alternative—a credible set of technical and political measures for meticulously restricting to peaceful purposes extremely large amounts of bomb materials which, once generated, will persist for the foreseeable lifetime of our species.

I am confident that the United States can still turn off the technology that it originated and deployed. By rebottling that genie we could move to energy and foreign policies that our grandchildren can live with. No more important step could be taken toward revitalizing the American dream.

Perhaps the most profound difference between the soft and hard paths is their domestic sociopolitical impact. Both paths, like any 50-year energy path, entail significant social change. But the kinds of social change needed for a hard path are apt to be much less pleasant, less plausible, less compatible with social diversity and personal freedom of choice, and less consistent with traditional values than are the social changes that could make a soft path work.

It is often said that, on the contrary, a soft path must be repressive; and coercive paths to energy conservation and soft technologies can indeed be imagined. But coercion is not necessary and would signal a major failure of imagination, given the policy instruments available to achieve a given technical end. Why use penal legislation to encourage insulation when tax incentives and education (leading to the sophisticated public understanding now being achieved in Canada and parts of Europe) will do? Policy tools need not harm life-styles or liberties if chosen with reasonable sensitivity.

In contrast to the soft path's dependence on pluralistic

consumer choice in deploying a myriad of small devices and refinements, the hard path depends on difficult, large-scale projects requiring a major social commitment under centralized management. We have noted in Section III the extraordinary capital intensity of centralized, electrified high technologies. Their similarly heavy demands on other scarce resources—skills, labor, materials, special sites—likewise cannot be met by market allocation, but require compulsory diversion from whatever priorities are backed by the weakest constituencies. Quasi-warpowers legislation to this end has already been seriously proposed. The hard path, sometimes portrayed as the bastion of free enterprise and free markets, would instead be a world of subsidies, $100-billion bailouts, oligopolies, regulations, nationalization, eminent domain, corporate statism.

Such dirigiste autarchy is the first of many distortions of the political fabric. While soft technologies can match any settlement pattern, their diversity reflecting our own pluralism, centralized energy sources encourage industrial clustering and urbanization. While soft technologies give everyone the costs and benefits of the energy system he chooses, centralized systems allocate benefits to surburbanites and social costs to politically weaker rural agrarians. Siting big energy systems pits central authority against local autonomy in an increasingly divisive and wasteful form of centrifugal politics that is already proving one of the most potent constraints on expansion.

In an electrical world, your lifeline comes not from an understandable neighborhood technology run by people you know who are at your own social level, but rather from an alien, remote, and perhaps humiliatingly uncontrollable technology run by a faraway, bureaucratized, technical elite who have probably never heard of you. Decisions about who shall have how much energy at what price also become centralized—a politically dangerous trend because it divides those who use energy from those who supply and regulate it.

The scale and complexity of centralized grids not only make them politically inaccessible to the poor and weak, but also increase the likelihood and size of malfunctions, mistakes and deliberate disruptions. A small fault or a few discontented people become able to turn off a country. Even a single rifleman can probably black out a typical city instantaneously. Societies may therefore be tempted to discourage disruption through stringent controls akin to a garrison state. In times of social stress, when grids become a likely target for dissidents, the sector may be paramilitarized and further isolated from grass-roots politics.

If the technology used, like nuclear power, is subject to technical surprises and unique psychological handicaps, prudence or public clamor may require generic shutdowns in case of an unexpected type of malfunction: one may have to choose between turning off a country and persisting in potentially unsafe operation. Indeed, though many in the $100-billion quasi-civilian nuclear industry agree that it could be politically destroyed if a major accident occurred soon, few have considered the economic or political implications of putting at risk such a large fraction of societal capital. How far would governments go to protect against a threat—even a purely political threat—a basketful of such delicate, costly and essential eggs? Already in individual nuclear plants, the cost of a shutdown—often many dollars a second—weighs heavily, perhaps too heavily, in operating and safety decisions.

Any demanding high technology tends to develop influential and dedicated constituencies of those who link its commercial success with both the public welfare and their own. Such sincerely held beliefs, peer pressures, and the harsh demands that the work itself places on time and energy all tend to discourage such people from acquiring a similarly thorough knowledge of alternative policies and the need to discuss them. Moreover, the money and talent invested in an electrical program tend to give it disproportionate influence in the counsels of government, often directly through staff-swapping between policy- and mission-oriented agencies. This incestuous position, now well developed in most industrial countries, distorts both social and energy priorities in a lasting way that resists political remedy.

For all these reasons, if nuclear power were clean, safe, economic, assured of ample fuel, and socially benign per se, it would still be unattractive because of the political implications of the kind of energy economy it would lock us into. But fission technology also has unique sociopolitical side-effects arising from the impact of human fallibility and malice on the persistently toxic and explosive materials in the fuel cycle. For example, discouraging nuclear violence and coercion requires some abrogation of civil liberties[34]; guarding long-lived wastes against geological or social contingencies implies some form of hierarchical social rigidity or homogeneity to insulate the technological priesthood from social turbulence; and making political decisions about nuclear hazards which are compulsory, remote from social experience, disputed, unknown, or unknowable, may tempt governments to bypass democratic decision in favor of elitist technocracy.[35]

Even now, the inability of our political institutions to cope with nuclear hazard is straining both their competence and their perceived legitimacy. There is no scientific basis for calculating the likelihood or the maximum long-term effects of nuclear mishaps, or for guaranteeing that those effects will not exceed a particular level; we know only that all precautions are, for fundamental reasons, inherently imperfect in essentially unknown degree. Reducing that imperfection would require much social engineering whose success would be speculative. Technical success in reducing the hazards would not reduce, and might enhance, the need for such social engineering. The most attractive political feature of soft technologies and conservation—the alternatives that will let us avoid these decisions and their high political costs—may be that, like motherhood, everyone is in favor of them.

Civilization in this country, according to some, would be inconceivable if we used only, say, half as much electricity as now. But that is what we did use in 1963, when we were at least half as civilized as now. What would life be like at the per capita levels of primary energy that we had in 1910 (about the present British level) but with doubled efficiency of energy use and with the important but not very energy-intensive amenities we lacked in 1910, such as telecommunications and modern medicine? Could it not be at least as agreeable as life today? Since the energy needed today to produce a unit of GNP varies more than 100-fold depending on what good or service is being produced, and since GNP in turn hardly measures social welfare, why must energy and welfare march forever in lockstep? Such questions today can be neither answered nor ignored.

Underlying energy choices are real but tacit choices of personal values. Those that make a high-energy society work are all too apparent. Those that could sustain life-styles of elegant frugality are not new; they are in the attic and could be dusted off and recycled. Such values as thrift, simplicity, diversity, neighborliness, humility and craftsmanship—perhaps most closely preserved in politically conservative communities—are already, as we see from the ballot box and the census, embodied in a substantial social movement, camouflaged by its very pervasiveness. Offered the choice freely and equitably, many people would choose, as Herman

Daly puts it, "growth in things that really count rather than in things that are merely countable": choose not to transform, in Duane Elgin's phrase, "a rational concern for material well-being into an obsessive concern for unconscionable levels of material consumption."

Indeed, we are learning that many of the things we had taken to be the benefits of affluence are really remedial costs, incurred in the pursuit of benefits that might be obtainable in other ways without those costs. Thus much of our prized personal mobility is really involuntary traffic made necessary by the settlement patterns which cars create. Is that traffic a cost or a benefit?

Pricked by such doubts, our inflated craving for consumer ephemerals is giving way to a search for both personal and public purpose, to reexamination of the legitimacy of the industrial ethic. In the new age of scarcity, our ingenious strivings to substitute abstract (therefore limitless) wants for concrete (therefore reasonably bounded) needs no longer seem so virtuous. But where we used to accept unquestioningly the facile (and often self-serving) argument that traditional economic growth and distributional equity are inseparable, new moral and humane stirrings now are nudging us. We can now ask whether we are not already so wealthy that further growth, far from being essential to addressing our equity problems, is instead an excuse not to mobilize the compassion and commitment that could solve the same problems with or without the growth.

Finally, as national purpose and trust in institutions diminish, governments, striving to halt the drift, seek ever more outward control. We are becoming more uneasily aware of the nascent risk of what a Stanford Research Institute group has called ". . . 'friendly fascism'—a managed society which rules by a faceless and widely dispersed complex of warfare-welfare-industrial-communications-police bureaucracies with a technocratic ideology." In the sphere of politics as of personal values, could many strands of observable social change be converging on a profound cultural transformation whose implications we can only vaguely sense: one in which energy policy, as an integrating principle, could be catalytic?[36]

It is not my purpose here to resolve such questions—only to stress their relevance. Though fuzzy and unscientific, they are the beginning and end of any energy policy. Making values explicit is essential to preserving a society in which diversity of values can flourish.

Some people suppose that a soft energy path entails mainly social problems, a hard path mainly technical problems, so that since in the past we have been better at solving the technical problems, that is the kind we should prefer to incur now. But the hard path, too, involves difficult social problems. We can no longer escape them; we must choose which kinds of social problems we want. The most important, difficult, and neglected questions of energy strategy are not mainly technical or economic but rather social and ethical. They will pose a supreme challenge to the adaptability of democratic institutions and to the vitality of our spiritual life.

These choices may seem abstract, but they are sharp, imminent and practical. We stand at a crossroads: without decisive action our options will slip away. Delay in energy conservation lets wasteful use run on so far that the logistical problems of catching up become insuperable. Delay in widely deploying diverse soft technologies pushes them so far into the future that there is no longer a credible fossil-fuel bridge to them: they must be well under way before the worst part of the oil-and-gas decline. Delay in building the fossil-fuel bridge makes it too tenuous: what the sophisticated coal technologies can give us, in particular, will no longer mesh with our pattern of transitional needs as oil and gas dwindle.

Yet these kinds of delay are exactly what we can expect if we continue to devote so much money, time, skill, fuel and political will to the hard technologies that are so demanding of them. Enterprises like nuclear power are not only unnecessary but a positive encumbrance for they prevent us, through logistical competition and cultural incompatibility, from pursuing the tasks of a soft path at a high enough priority to make them work together properly. A hard path can make the attainment of a soft path prohibitively difficult, both by starving its components into garbled and incoherent fragments and by changing social structures and values in a way that makes the innovations of a soft path more painful to envisage and to achieve. As a nation, therefore, we must choose one path before they diverge much further. Indeed, one of the infinite variations on a soft path seems inevitable, either smoothly by choice now or disruptively by necessity later; and I fear that if we do not soon make the choice, growing tensions between rich and poor countries may destroy the conditions that now make smooth attainment of a soft path possible.

These conditions will not be repeated. Some people think we can use oil and gas to bridge to a coal and fission economy, then use that later, if we wish, to bridge to similarly costly technologies in the hazy future. But what if the bridge we are now on is the last one? Our past major transitions in energy supply were smooth because we subsidized them with cheap fossil fuels. Now our new energy supplies are ten or a hundred times more capital-intensive and will stay that way. If our future capital is generated by economic activity fueled by synthetic gas at $25 a barrel-equivalent, nuclear electricity at $60–120 a barrel-equivalent, and the like, and if the energy sector itself requires much of that capital just to maintain itself, will capital still be as cheap and plentiful as it is now, or will we have fallen into a "capital trap"? Wherever we make our present transition to, once we arrive we may be stuck there for a long time. Thus if neither the soft nor the hard path were preferable on cost or other grounds, we would still be wise to use our remaining cheap fossil fuels—sparingly—to finance a transition as nearly as possible straight to our ultimate energy-income sources. We shall not have another chance to get there.

REFERENCES

1. In this essay the proportions assigned to the components of the two paths are only indicative and illustrative. More exact computations, now being done by several groups in the United States and abroad (notably the interim [autumn 1976] and forthcoming final [1976–1977] reports of the energy study of the Union of Concerned Scientists, Cambridge, Mass.), involve a level of technical detail which, though an essential next step, may deflect attention from fundamental concepts. This article will accordingly seek technical realism without rigorous precision or completeness. Its aim is to try to bring some modest synthesis to the enormous flux and ferment of current energy thinking around the world. Much of the credit (though none of the final responsibility) must go to the many energy strategists whose insight and excitement they have generously shared and whose ideas I have shamelessly recycled without explicit citation. Only the limitations of space keep me from acknowledging by name the 70-odd contributors, in many countries, who come especially to mind.

2. The foregoing data are from M. Carasso et al., The Energy Supply Planning Model, PB-245 382 and PB-245 383, National Technical Information Service (Springfield, Va.), Bechtel Corp. report to the National Science Foundation (NSF), August 1975. The figures

assume the production goals of the 1975 State of the Union Message. Indirect labor requirements are calculated by C. W. Bullard and D. A. Pilati, CAC Document 178 (September 1975), Center for Advanced Computation, Univ. of Illinois at Urbana-Champaign.

3. I. C. Bupp and R. Treitel, "The Economics of Nuclear Power: De Omnibus Dubitandum," 1976 (available from Professor Bupp, Harvard Business School).

4. Computation concerning waste heat and projections to 2000 are based on data in the 1975 Energy Research and Development Administration Plan (ERDA-48).

5. B. Bolin, "Energy and Climate," Secretariat for Future Studies (Fack, S-103 10 Stockholm); S. H. Schneider and R. D. Dennett, Ambio 4, 2:65–74 (1975); S. H. Schneider, The Genesis Strategy, New York: Plenum, 1976; W. W. Kellogg and S. H. Schneider, Science 186:1163–72 (1974).

6. Figure 1 shows only nonagricultural energy. Yet the sunlight participating in photosynthesis in our harvested crops is comparable to our total use of nonagricultural energy, while the sunlight falling on all U.S. croplands and grazing lands is about 25 times the nonagricultural energy. By any measure, sunlight is the largest single energy input to the U.S. economy today.

7. The capital costs for frontier fluids and for electrical systems can be readily calculated from the data base of the Bechtel model (footnote 2 above). The electrical examples are worked out in my "Scale, Centralization and Electrification in Energy Systems," Future Strategies for Energy Development symposium, Oak Ridge Associated Universities, October 20–21, 1976.

8. The Bechtel model, using 1974 dollars and assuming ordering in early 1974, estimates direct construction costs totaling $559 billion, including work that is in progress but not yet commissioned in 1985. Interest, design and administration—but not land, nor escalation beyond the GNP inflation rate—bring the total to $743 billion. Including the cost of land, and correcting to a 1976 ordering date and 1976 dollars, is estimated by M. Carasso to yield over $1 trillion.

9. M. Carasso et al., op. cit.

10. E. Kahn et al., "Investment Planning in the Energy Sector," LBL-4479, Lawrence Berkeley Laboratory, Berkeley, Calif., March 1, 1976.

11. American Institute of Physics Conference Proceedings No. 25, Efficient Use of Energy, New York: AIP, 1975; summarized in Physics Today, August 1975.

12. M. Ross and R. H. Williams, "Assessing the Potential for Fuel Conservation," forthcoming in Technology Review; see also L. Schipper, Annual Review of Energy I:455–518 (1976).

13. L. Schipper and A. J. Lichtenberg, "Efficient Energy Use and Well-Being: The Swedish Example," LBL-4430 and ERG-76-09, Lawrence Berkeley Laboratory, April 1976.

14. R. L. Goen and R. K. White, "Comparison of Energy Consumption Between West Germany and the United States," Stanford Research Institute, Menlo Park, Calif., June 1975.

15. A. D. Little, Inc., "An Impact Assessment of ASHRAE Standard 90–75," report to FEA, C-78309, December 1975; J. E. Snell et al. (National Bureau of Standards), "Energy Conservation in Office Buildings: Some United States Examples," International CIB Symposium on Energy Conservation in the Built Environment (Building Research Establishment, Garston, Watford, England), April 1976; Owens-Corning-Fiberglas, "The Arkansas Story," 1975.

16. P. W. McCracken et al., Industrial Energy Center Study, Dow Chemical Co. et al., report to NSF, PB-243, 824, National Technical Information Service (Springfield, Va.), June 1975. Extensive cogeneration studies for FEA are in progress at Thermo-Electron Corp., Waltham, Mass. A pathfinding June 1976 study by R. H. Williams (Center for Environmental Studies, Princeton University) for the N.J. Cabinet Energy Committee argues that the Dow report substantially underestimates cogeneration potential.

17. Ross and Williams, op. cit.

18. A calculation for Canada supports this view: A. B. Lovins, Conserver Society Notes (Science Council of Canada, Ottawa), May/June 1976, pp. 3–16. Technical fixes already approved in principle by the Canadian Cabinet should hold approximately constant until 1990 the energy required for the transport, commercial and house-heating sectors; sustaining similar measures to 2025 is estimated to shrink per capita primary energy to about half today's level. Plausible social changes are estimated to yield a further halving. The Canadian and U.S. energy systems have rather similar structures.

19. The scale of potential conservation in this area is given in Ross

and Williams, op. cit.; the scale of potential cogeneration capacity is from McCracken et al., op. cit.

20. Assuming (which is still not certain) that controlled nuclear fusion works, it will almost certainly be more difficult, complex and costly—though safer and perhaps more permanently fueled—than fast breeder reactors. See W. D. Metz, Science 192:1320–23 (1976), 193:38–40, 76 (1976), and 193:307–309 (1976). But for three reasons we ought not to pursue fusion. First, it generally produces copious fast neutrons that can and probably would be used to make bomb materials. Second, if it turns out to be rather "dirty," as most fusion experts expect, we shall probably use it anyway, whereas if it is clean, we shall so overuse it that the resulting heat release will alter global climate: we should prefer energy sources that give us enough for our needs while denying us the excesses of concentrated energy with which we might do mischief to the earth or to each other. Third, fusion is a clever way to do something we don't really want to do, namely to find *yet another* complex, costly, large-scale, centralized, high-technology way to make electricity—all of which goes in the wrong direction.

21. Partly or wholly solar heating is attractive and is being demonstrated even in cloudy countries approaching the latitude of Anchorage, such as Denmark and the Netherlands (International CIB Symposium, op. cit.) and Britain (Solar Energy: A U.K. Assessment, International Solar Energy Society, London, May 1976).

22. Solar heating cost is traditionally computed microeconomically for a consumer whose alternative fuels are not priced at long-run marginal cost. Another method would be to compare the total cost (capital and life-cycle) of the solar system with the total cost of the other complete systems that would otherwise have to be used in the long run to heat the same space. On that basis, 100 percent solar heating, even with twice the capital cost of two-thirds or three-fourths solar heating, is almost always advantageous.

23. R. W. Bliss, Bulletin of the Atomic Scientists, March 1976, pp. 32–40.

24. A. D. Poole and R. H. Williams, Bulletin of the Atomic Scientists, May 1976, pp. 48–58.

25. For examples, see the Canadian computations in A. B. Lovins, Conserver Society Notes, op. cit.; Bent Sørensen's Danish estimates in Science 189:255–60 (1975); and the estimates by the Union of Concerned Scientists, footnote 1 above.

26. The system and its conceptual framework are described in several papers by H. Harboe, Managing Director, Stal-Laval (G.B.) Ltd., London: "District Heating and Power Generation," November 14, 1975; "Advances in Coal Combustion and Its Applications," February 20, 1976; "Pressurized Fluidized Bed Combustion with Special Reference to Open Gas Turbines" (with C. W. Maude), May 1976. See also K. D. Kiang et al., "Fluidized-Bed Combustion of Coals," GFERC/IC-75/2 (CONF-750586), ERDA, May 1975.

27. Small devices were pioneered by the late Professor Douglas Elliott. His associated firm, Fluidfire Development, Ltd. (Netherton, Dudley, W. Midlands, England), has sold many dozens of units for industrial heat treatment or heat recuperation. Field tests of domestic packaged fluidized-bed boilers are in progress in the Netherlands and planned in Montana.

28. Already Enköping, Sweden, is evaluating bids from several confident vendors for a 15-megawatt fluidized-bed boiler to add to its district heating system. New reviews at the Institute for Energy Analysis and elsewhere confirm fluidized beds' promise of rapid benefits without massive research programs.

29. Estimates of the total capital cost of "soft" systems are necessarily less well developed than those for the "hard" systems. For 100-percent solar space heating, one of the high-priority soft technologies, mid-1980s estimates are about \$50,000–\$60,000 (1976 dollars) of investment per daily oil-barrel-equivalent in the United States, \$100,000 in Scandinavia. All solar cost estimates, however, depend sensitively on collector and building design, both under rapid development. In most new buildings, passive solar systems with negligible or negative marginal capital costs should suffice. For biomass conversion, the 1974 FEA Solar Task Force estimated capital costs of \$10,000–\$30,000 per daily barrel equivalent—toward the lower part of this range for most agricultural projects. Currently available wind-electric systems require total-system investment as high as about \$200,000 per delivered daily barrel, with much improvement in store. As for transitional technologies, the Stal-Laval fluidized-bed gas-turbine system, complete with district-heating network and heat-pumps (coefficient of performance = 2), would cost about \$30,000 per delivered daily barrel equivalent. See Lovins, op. cit., footnote 7.

30. See A. B. Lovins, "Long-Term Constraints on Human Activity," Environmental Conservation 3, 1:3–14 (1976) (Geneva); "Some Limits to Energy Conversion," Limits to Growth 1975 Conference (The Woodlands, Texas), October 20, 1975 (to be published in conference papers). The environmental and social impacts of solar technologies are being assessed in a study coordinated by J. W. Benson (ERDA Solar Division), to be completed autumn 1976.

31. Recent private reports indicate the Soviet scientific community is deeply split over the wisdom of nuclear expansion. See also Nucleonics Week, May 13, 1976, pp. 12–13.

32. Current overcapacity, capacity under construction, and the potential for rapid conservation and cogeneration make this a relatively painless course, whether nuclear generation is merely frozen or phased out altogether. For an illustration (the case of California), see R. Doctor et al., Sierra Club Bulletin, May 1976, pp. 4 ff. I believe the same is true abroad. See Introduction to Non-Nuclear Futures by A. B. Lovins and J. H. Price, Cambridge, Mass.: FOE/Ballinger, 1975.

33. See Nucleonics Week, May 6, 1976, p. 7, and I. C. Bupp and J.-C. Derian, "Nuclear Reactor Safety: The Twilight of Probability," December 1975. Bupp, after a detailed study of European nuclear politics, shares this assessment.

34. R. Ayres, 10 Harvard Civil Rights-Civil Liberties Law Review, Spring 1975, pp. 369–443; J. H. Barton, "Intensified Nuclear Safeguards and Civil Liberties," report to USNRC, Stanford Law School, October 21, 1975.

35. H. P. Green, 43 George Washington Law Review, March 1975, pp. 791–807.

36. W. W. Harman, An Incomplete Guide to the Future, Stanford Alumni Association, 1976.

Scale, Centralization and Electrification in Energy Systems.

Amory B. Lovins, London, England
Consultant physicist and British representative,
Friends of the Earth, Inc.

In what has come to seem an extended metaphor for energy strategy, Robert Frost once said:

Two roads diverged in a yellow wood,
And sorry I could not travel both
And be one traveler, long I stood
And looked down one as far as I could
To where it bent in the undergrowth;

Then took the other, as just as fair,
And having perhaps the better claim,
Because it was grassy and wanted wear;
Though as for that the passing there
Had worn them really about the same,

And both that morning equally lay
In leaves no step had trodden black.
Oh, I kept the first for another day!
Yet knowing how way leads on to way,
I doubted if I should ever come back.

I shall be telling this with a sigh
Somewhere ages and ages hence:
Two roads diverged in a wood, and I—
I took the one less traveled by,
And that has made all the difference.

Today, as the conventional road we gaze down bends ever more sharply into more obscure and brambly undergrowth, the sense of divergent choice of which Frost spoke is growing on many of us in the energy policy community. Three tangled issues—scale, centralization, and electrification—are emerging as key elements of that choice.

This symposium is apparently the first coherent and public attempt by "establishment" energy policy professionals to address these fundamental issues. This seems odd: after all, hundreds of billions of dollars have already been committed to large-scale central electrification on the assumption that its logic is obvious and unassailable. But it is the fate of all knowledge to begin as heresy and end as superstition (Huxley). For many analysts, overtaken by events that suddenly reversed traditional assumptions, the notion that scale issues are worth discussing has only lately emerged from heresy into respectability. Thence arises our challenge, and opportunity, to break new intellectual ground.

In this paper, I shall informally explore the implications of choice in scale, centralization, and electrification. If I seem at times to be presenting advocacy as well as analysis, it is because the results of the analysis have so impressed me, not because of preconceived ideological adherence to any particular position. Big high technologies probably have their place, and as a former high technologist I would not deny them that place. I merely conclude that it is a limited place and that they have long since saturated and overreached it.

I am likewise impressed by the way the arguments for matching energy supply in both scale and quality to end use fit naturally and powerfully into a coherent energy strategy. I have explored this thesis in a recent paper[1] that outlines and contrasts, not as precise recommendations but as a qualitative vehicle for ideas, two energy paths that the United States (and, by analogy, other countries) might follow over the next 50 years. The first, or "hard," path is high-energy, primary-supply-oriented, high-technology, centralized, increasingly electrified, and reliant chiefly on depletable resources (coal and uranium). The second, or "soft," path is high-efficiency (in end use), end-use-oriented, relatively low-technology and decentralized, electrified only where essential, based on renewable resources, and fission-free. I argue that though both paths entail difficult problems, the soft path offers major advantages in costs, rates, risks, equity, geopolitics (including control of nuclear proliferation), and domestic sociopolitics. Finally, I suggest that though each path is only indicative and embraces an infinite spectrum of variations on a theme, there is such a deep conceptual dichotomy between the two paths that the incremental resource and cultural commitments of the hard path, as "way leads on to way," are rapidly making a soft path less attainable in principle, and in practice, without prompt redirection, will soon foreclose it. This exclusivity makes choice urgent.

Since my earlier paper has already developed many of the arguments appropriate to this paper, I shall incorporate relevant sections here by cross-reference rather than repeating them. The relatively nontechnical treatment in my earlier paper will be complemented here with numbers, details, and qualifications that were omitted or excised there for reasons of both space and readership. The methodology used in this paper is discussed in Appendix I.

Underlying both papers, and indeed this symposium, is the contrast between two or more conflicting perceptions of what the energy problem "really" is. It will probably take at least another century of 20/20 hindsight to decide with universal certainty which perception(s) will have led to more appropriate actions. Indeed, I am uncertain whether our attempts to influence the future are likely to have the effect we intend rather than the opposite one—though that doesn't mean we shouldn't try. Accordingly, however persuasive or appealing a particular view of the world may seem to each of us, it is premature to conclude that other views are "wrong." We must accept them at face value as a basis for discussion in good faith, appreciating that identical facts seen by different people can give rise to completely different perceptions and to different views of what facts are important. Some relevant facts as I see them are set out elsewhere.[1-4] As a basis for mutual understanding, however, it may be useful for me to summarize a few of my underlying opinions—not on every aspect of social philosophy, geopolitics, ethics, international development, peace, and other parts of the whole universe of perceptions that must support any consistent view of our energy future, but at least on a few basic values. Briefly, then, I think that:

1) We are more endangered by too much energy too soon than by too little too late, for we understand too little the wise use of power.

2) Through technical changes only, the United States can roughly double its end-use energy efficiency* by about the turn of the century, then roughly redouble it over the ensuing half-century or so (the pattern implied by Figure 2 of reference 1 being by no means the lowest that might be realistically considered).†

3) Through voluntary value and structural changes

* *People who believe that the Cooper-Weinberg projections[6] of demographic and economic growth are too low should reflect that improvements in energy efficiency occur most rapidly at the margin. The more new houses, machines, etc., are built, the faster the old, less energy-efficient stocks can be diluted or replaced. Thus people who postulate rapid population and economic growth and correspondingly rapid energy growth are trying to have it both ways.*

† *Many people still do not know the difference between improved end-use efficiency and curtailment of end-use functions. For example, Americans for Energy Independence, who advocate "maximum feasible production of all forms of energy," are reportedly preparing a conservation program consisting largely of curtailment—presumably to induce the public to identify conservation with the taste of cod-liver oil.*

only, we could do approximately the same thing, with at least equal benefits to equity, liberty, comfort, employment, happiness, and national capacity to cope with domestic and world problems.

4) Many of these value changes are desirable or essential, and many are already occurring rapidly (Harman's transformationalist theory[5] has considerable merit).

5) In practice, we can and should use some combination of technical and structural changes to increase end-use efficiency at least three- or four-fold over the next three or four generations.

6) Over that period our concepts of social welfare are likely to change profoundly, and using Gross National Product (GNP) to measure it will savor of a category mistake.

7) Accordingly, the energy problem is not how to expand supplies to meet the postulated extrapolative needs of a dynamic society, but rather how to accomplish social goals elegantly with a minimum of energy and effort, meanwhile taking care to preserve a social fabric in which diverse values can thrive.

8) The technical, economic, and social problems of fission technology[3,7] are so intractable, and technical efforts to palliate those problems are politically so dangerous, that we should abandon the technology with due deliberate speed.

9) Many other energy technologies[1,2,4] are exceedingly unattractive and should be developed and deployed sparingly or not at all—for example nuclear fusion, large coal-fired power stations and coal-conversion plants, many current coal-mining technologies, urban-sited liquefied natural gas terminals, much Arctic petroleum extraction, most "unconventional" hydrocarbons, and many "exotic" large-scale solar technologies such as solar satellites and monocultural biomass plantations.

10) Ordinary people are qualified and responsible to make these and other energy choices through the democratic political process, and on the social and ethical issues central to such choices the opinion of any technical expert is entitled to no special weight.

11) Technical experts contributing to public discussion of such choices should call special attention to uncertainties and professional disagreements, lest they give the impression that disputed questions are actually resolved.

12) Issues of material growth are inseparable from the more important issues of distributional equity, both within and among nations—indeed, high growth in overdeveloped

countries is inimical to development in poor countries.

13) The self-reliant ecodevelopment concepts inherent in the New Economic Order approach—and inconsistent with hard energy technologies—are commendable and practicable, provided we overcome the danger of not being imaginative enough.

14) Though the potential for growth in the social, cultural, and spiritual spheres is unlimited, resource-crunching material growth is inherently limited, and, in countries as affluent as the United States, should be not merely stabilized but returned to sustainable levels at which the net marginal utility of economic activity is clearly positive.

15) People are more important than goods; hence energy, technology, and the economy are means, not ends; hence economic rationality is a narrow and often defective test of the wisdom of broad social choices.

16) Though humanity and human institutions are not perfectable, legitimacy and the nearest we can get to wisdom both flow, as Jefferson believed, from the people.

17) Pragmatic Hamiltonian concepts of central governance by a cynical elite are unworthy of our people and are ultimately tyrannical.

18) Since sustainability is more important than the momentary advantage of any generation or group, long-term discount rates should be zero or even slightly negative, reinforcing a frugal (though not penurious) ethic of husbanding.

19) Nature knows best, whereas we know little of the natural systems and cycles on which we depend; hence we must take care to preserve safety margins whose importance we do not yet understand, and to design for resilience and flexibility.*

20) The national interests of the United States lie less in traditional geopolitical balancing acts than in striving to attain a just and equitable, therefore peaceful, world order, even at the expense of temporary commercial advantage.

Perhaps these elements of a credo will help readers to see what lies behind my views on energy strategy, just as other perceptions lie behind different views. I must again stress, however, that my views on energy and on other matters coevolve; neither can be said to be derived from, or reshaped to justify, the other.

One further introductory comment. I consider the

* *Since, as Niels Bohr remarked, "It is difficult to make predictions, especially about the future."*

structural, sociopolitical, and value implications of energy paths to be paramount both in theoretical importance and in practical impact on political acceptability. These implications are highlighted by the soft/hard dichotomy[1], which is definitive and crucial. Whether an energy path is high or low is derivative and less interesting. In energy strategy as in system dynamics, the *structure* of the system is far more important to its behavior than the exact values assumed for its coefficients.

Some commentators seem to have missed this point. Alvin Weinberg, for example, has proposed that a strongly nuclear-electric future, modified by "technical fixes" (such as clustered and perhaps underground siting, and administration by a technological priesthood socially isolated from the public), should be acceptable to me and to other nuclear critics if at the same time the demand for energy is greatly reduced by increasing end-use efficiency. Let us suppose—though I do not think it is true—that the problems of fission are amenable to such technical fixes without untoward social side effects. Let us further suppose that a strong-conservation-plus-strong-nuclear policy is internally consistent, though I doubt, as a matter of empirical sociology, that these two constituencies overlap much. Yet even on these heroic assumptions, the argument misses the point: that an essentially nuclear-electric future, even scaled down, *is still a hard energy path*[1] in structure and in sociopolitical implications. The structural effects of pervasive centralized electrification matter far more than whether its enormous extent is doubled or halved. Thus the arbitrary assumption[6] that even a nuclear moratorium would still mean a half-electric economy by about 2000—i.e., that it only means building big fossil-fueled power stations instead of big nuclear power stations—has kept Alvin Weinberg and his group at the Institute for Energy Analysis from addressing the most interesting policy questions. The breadth of the soft-path concepts, whose attractions[1] seem to me a key element in the nuclear debate, requires far more than monovariate analysis. Suppressing these concepts by an invariant ex cathedra assumption of centralized electrification reminds me of the popular lapel button that says:

TECHNOLOGY IS THE ANSWER!
(But what was the question?)

I sympathize with the difficulties that the Institute for Energy Analysis faces in addressing the deeper and harder questions on which I hope to make a modest start here, for the issues of scale, centralization, and electrification are so inextricably intertwined that it is not clear how to grasp any one of them. The economies available at small scale, for example, are in principle separable from the issue of thermodynamic matching of sources to end-use needs. Yet in practice, electrification of the type currently proposed tends to be centralized and large-scale, thus adding structural issues to those of thermodynamic appropriateness and inherent capital intensity.[1] In this paper, I have been unable to treat the three issues separately for more than a few paragraphs at a time. I shall instead outline some basic concepts of scale and energy quality, then amplify my earlier summary comments[1] on comparative whole-system costs, and finally address the crucial sociopolitical and value problems of soft vs. hard energy paths.

Scale

Since energy is but a means to social ends, and since energy is useful only insofar as it performs specific tasks relevant to those ends, anyone seeking to perform tasks with the least possible energy and trouble should start with an inventory of tasks. Specifically, to design a rational energy system, one should first assess the unit scale, type, quality, and degree of geographical clustering of energy needs as a function of space and time. Such an elementary data base does not exist today anywhere in the world. So far as I know, no country even has a usable inventory of the scale of individual end-use energy needs. One must instead fall back on naive order-of-magnitude observations—simple but instructive.

For example, the heating load of houses is typically measured in thousands of watts (W); the peak power output of a car is of order 100,000 W; the total input to one of our society's largest integrated industrial facilities—certain primary metals and isotope-enrichment plants—is on the order of 1,000 megawatts (MW) with current designs (which might arguably be different were it not for large dams and power grids). With isolated exceptions, mainly pathological cases associated with large dams or coal complexes, virtually all our end-use systems are probably smaller than about 100 MW (the inefficient World Trade Center is wired for a peak electrical load of 80 MW), and great many—probably the majority—are clustered in domestic, commercial, and industrial units smaller than about 1 MW. Most of the end-use devices important to our daily lives require of order 0.1 to 1,000 W and are clustered within living or working units requiring of order 1,000 to 100,000 W. Most production processes of practical interest can be (and long have been) carried out in units of roughly that scale.

Thus it is not obvious prima facie that energy must be converted in blocks of the order of hundreds of thousands to tens of thousands of megawatts. The arguments usually articulated for such large-scale include reduced unit capital cost, increased reliability, central high-volume delivery of primary fuel, localization and hence simplified management of residuals and other side effects, ability to use and finance the best high technologies available, ease of attracting and supporting the specialized maintenance cadre that giant systems require, ease of substituting primary fuels without retrofitting numerous small conversion systems, and convenience for the end user (who need merely pay for the delivered energy he purchases as a service, not involve himself in the details of its conversion). These contentions are not devoid of merit. Big systems do have some real advantages—though advantages are often subjective, and one person's benefit can be another's cost. But I shall suggest below that many of the advantages claimed for large-scale may be doubtful, illusory, tautological, or outweighed by less tangible and less quantifiable but perhaps more important disadvantages and diseconomies.

Some diseconomies of large scale are starting to be widely appreciated. For example, it is now well known that large electrical components, notably turbogenerators, often lose in reliability—hence in contributions to grid operating costs, standby capacity costs, grid instability, and lost revenues—what their size gains in unit capital cost. These effects are so common that there is mounting evidence[8] that most types of power stations can have lower busbar costs in sizes of the order of hundreds rather than of thousands of megawatts.

Dispersed generation near load centers is well known to improve system integration and stability. According to one recent study[9]

> For one small system, and with assumptions that we believe are realistic, it was found that kW of dispersed generation was equivalent from the standpoint of reserve requirements to 2.5 kW of central generation. The reliability of supply within the network

> was determined by means of an index related to the LOLP [loss-of-load probability]. The reason why the dispersed device can be so effective is that it protects the load in its vicinity against generation *as well as* transmission and distribution outages.

Thus the "dispersion credit" traditionally assigned to local supply—e.g., from battery banks or fuel cells associated with distribution facilities—may be far too low, since it reflects only the cost, on the order of $100/kilowatt (kW), of saved transmission facilities. Unfortunately, publication of a study by the Electric Power Research Institute with this conclusion has been suspended, so few recent data are available. The extreme capital intensity of all electrical facilities offers ample reason to explore carefully the reliability implications of more centralization. The classical literature of this subject seems to me sketchy and unpersuasive and its quantification primitive.

A further diseconomy of large scale is so obvious that it is often forgotten. The past few decades' military experience in Europe and Indochina has taught us that central energy systems reliant on a few large facilities are far more vulnerable, and harder to restore when damaged, than dispersed systems.* This concept has led to design criteria used today in Israel and, I am told, the People's Republic of China.

Small energy systems suited to particular niches can mimic the strategy of ecosystem development, adapting and hybridizing in constant coevolution with a broad front of technical and social change. Large systems tend to evolve more linearly like single specialized species (dinosaurs?) with less genotypic diversity and greater phenotypic fragility. Large systems also accrete costly, specialized infrastructure that strongly influences future lines of development. Thus unamortized natural-gas pipelines, the third largest U.S. industry, provide a strong incentive to make synthetic pipeline-quality gas even at a price an order of magnitude higher; building a grid dependent on 1,000-MW blocks of electricity discourages a future shift to smaller-scale or reduced electrification. Small systems, in contrast, tend to depend more on infrastructure installed at the point of end use, thus increasing the user's ability to adapt. For example, resistive heaters and electric heat pumps are not very adaptable, so an all-

* *My statement[1] that soft technologies are not militarily useful referred to direct applications, not reduced vulnerability. In the latter sense they are extremely useful.*

electric house is hard to heat except with electricity from some source. A domestic or district heating system based on circulating hot water, however, can use virtually any heat source at any scale without significant change to the domestic plumbing, adapting to as wide a range as solar collectors, solar/heat-pump hybrids, and combined-heat-and-power district stations. Thus if a transitional technology such as coal-fired fluidized-bed gas turbines with district heating[1] is deployed first in those urban areas where solar backfits will be slowest and least convenient, with district heating clustered in holding tanks of neighborhood scale, then that interim heat distribution system (coupled to existing domestic plumbing) can be adapted later to whatever soft source of heat becomes available in each neighborhood. Infrastructure at or near the point of end use can be designed for this sort of piggybacking, whereas large-scale distribution infrastructure enables one only to choose one or another kind of enormous central power station, gas plant, etc.

Large-scale systems already in place (not at the margin) can be useful transitional tools on a regional scale. For example, my studies of a soft energy path for Denmark[3] came to conclusions similar to Sørensen's,[10] though I used more organic conversion, less electricity, and no hydrogen. But the 50-to-75-year transition, even though no more awkward than under present policy, could be made far easier within a Scandinavian hegemony that shared surplus Norwegian hydroelectricity in return for food and other exports from Denmark and Sweden. Current studies within the European Economic Community[11] may reveal scope for other such regional policies—not that they are an argument for building more big systems than we have now.

I have already outlined[1] some of the potential technical and economic advantages of small scale:[12]

1) Virtual elimination of the capital costs, operation and maintenance costs, and losses of the distribution infrastructure (see below).

2) Scope for greatly reducing capital cost by mass production if desired.

3) Elimination of direct diseconomies of scale, such as the need for spinning reserve on electrical grids.*

4) Major reductions in indirect diseconomies of scale that arises from the long lead times of large systems: for ex-

* *Trying to judge the true cost of prompt spinning reserve leads one into dense definitional thickets; but it is a big number, and someone must pay for it.*

ample, exposure to interest and escalation during construction, to mistimed demand forecasts, and to wage pressures by a large number of strongly unionized crafts well aware (as in the Trans-Alaska Pipeline project) of the high cost of delay.

The very conditions that make the indirect diseconomies of large scale important make them hard to quantify. Nonetheless, some utility managers are realizing that interest, escalation, delays owing to greater complexity, and the effects of forecasting errors can make a single large plant of capacity C more costly than N smaller plants of capacity C/N with shorter lead times. It is an inherent engineering feature of soft technologies,* moreover, that their lead times are qualitatively shorter than those of conventional big systems.[1] Whether in development, demonstration, or deployment, a small and technically simple system such as a rooftop solar collector, requiring months and thousands of dollars, is quicker and has lower indirect costs than one such as a fast breeder reactor, requiring several stages of scaling-up and billions of dollars (and perhaps a decade) per stage. Even a relatively adventurous soft or semisoft technology, such as a large wind machine, requires less than 2 years to design and build from scratch. The complete process for large solar heating systems today can be less than 2 months.

An intangible but important advantage of small scale is its scope for reducing public alienation (hence opposition, costly in these escalatory times) by building a greater sense of participation. A giant energy facility is arcane, remote, unfamiliar, and so impressive as to be threatening. A small system has an obvious relevance to everyday life because it is both physically and conceptually closer to the end-use task. For example, the New Alchemy Institute's biologically sophisticated Ark project in Prince Edward Island has taken an impressive hold on the popular imagination all over the

* *As noted elsewhere,[1] "soft" is not a wholly satisfactory term, though already widely used. Some alternatives, like "supple" (vs. "brittle") or "pluralistic" (vs. "monolithic"), are not definitive enough. "Convivial" (vs. "radical-monopolistic"[13]) is, alas, too obscure. "Mature"—i.e., surprise-free and straightforward rather than speculatively straining technology to and beyond its limits—is apt, but some people may reverse Commoner's original sense.[14] "Proven" (vs. "unproven") likewise shifts meaning with speaker. Perhaps the best term, "appropriate," which comes to us from modern development economics, lacks concreteness for some people and is perhaps broader in application than we want. Terms stressing only smallness, renewability, etc. are of course too narrow to embrace the five defining characteristics of "soft technology,"[1] All suggestions for a better term will be welcome.*

Province. It has tapped a hitherto unsuspected reservoir of intelligent interest and initiative, not only because it is easy to visit and understand,* but also because it is a farm that grows fish and tomatoes—a readily assimilated extension of everyday agrarian life. It makes more sense and seems more relevant to the Islanders than a CANDU in New Brunswick or Ontario, just as Biharis could justifiably suppose that having methane digesters and solar cookers in their villages would help them more directly than building more reactors in California.

In short, because small, soft technologies[1] use equitably distributed natural energies to meet perceived human needs directly and comprehensibly, rather than being oriented toward abstract economic services for remote and anonymous consumers, they are, in Illich's sense,[13] "convivial." Of course, there may be Economic People who want as little as possible to do with their own life-support systems, and are content to pay their utility bills without a murmur, gobble precooked plastic food as the television exhorts, and eagerly turn every aspect of life into a prepackaged component of GNP. Yet I do not seem to meet such people. It seems to me rather that most people want to understand their own systems and feel responsible for their own destinies, not be mere economic cogs. That is why, for example, even the most narrowly materialistic people complain about impersonal and shoddy "service,"† arrogantly paternalistic utilities, incompetent auto mechanics, outrageous bills for simple repairs, and petty bureaucracy. That is why many increasingly cherish the small corner shop, feel guilty rather than satisfied at eating hamburgers that are a by-product of petrochemicals manufacture, and try to persuade the dentist to explain exactly what is wrong with that tooth. That is why creative personal activities of all kinds are fluorishing—from gardening and canning to weaving and do-it-yourself carpentry (and, arguably, citizens' band radio). Perhaps I just meet a skewed sample of people, but I do not think so. When typ-

* *And makes manifest our interdependence with the natural world, reintegrating us into it and enhancing our sense of wholeness: a special strength of combined innovation in energy and agricultural systems.*

† *A recent survey showed 40 percent of consumer purchases resulting in complaints (mostly in vain). The costs of correction, if any, were of course passed through to the consumer via Galbraith's "cost-plus economy," and the only incentive to do better—a powerful one, but very slow—would be public withdrawal of legitimacy from the institutions responsible.*

ical all-American slurbians tell me with obvious pride that they have just changed their own fuses, made their own preserves from their own fruit, sewed their own clothes, or insulated their own attics, I sense that they have done these things not only because it pays but because it is satisfying and fulfilling. They are proud to recount it because it symbolizes a small triumph of quality over mediocrity and of individualism over the System. The emotions that such involvement releases—a psychological aspect of scale—are powerful, lasting, and contagious. They must not be ignored.

Of course this striving for personal participation and understanding occurs often at a neighborhood or community level (though seldom at a city or regional one, since by definition one cannot then know one's neighbors). There are often good reasons to share even simple energy systems among, say 10 to 1000 people. For example, neighborhood solar heating systems (for individual or cluster housing) can clearly offer substantial economies over single-house systems through freer collector siting and configuration, reduced craftwork, reduced surface-to-volume ratio in storage tanks, more favorable ratio of variable to fixed costs, and perhaps even a bit of user diversity (different people using, say, hot water at different times, though this would be a very small term). Many transitional energy systems, too, might have to be on a substantial scale: It is even too early to rule out entirely the possibility of fairly large-scale coal conversion to make industrial gas for a transitional ring-main in the Midwest, until fluidized-bed backfits (particularly for industrial packaged boilers, with or without cogeneration) can take up the slack.[1] Most such boiler backfits will also have to match existing industrial scale, whether it is optimal in the long run or not.[12] But all such transitional technologies, unlike those now planned, would be designed at appropriate scale so that they and their infrastructure can later mesh with smaller, softer technologies. Thus the scale issues in the soft-hard debate are relevant to larger transitional systems too.

Energy Quality

I am not aware of any definitive or detailed survey of the thermodynamic structure of end use in any country. But to obtain round numbers for the United States, we can start with the Ross and Williams revision[15] for 1973 of the somewhat problematical Stanford Research Institute data base[16] for 1968, as shown in Table 1.

If we ignore fuels used as materials and consider only

TABLE 1
U. S. Primary Energy Consumption by End Use, 1973
(quads, or 10^{15} BTU/year)

Sector	Direct fuel	Electricity	Total fuel
Residential	7.89	1.97	14.07
Space heat	6.16	0.32	7.19
Water heat	1.26	0.28	2.13
Air conditioning		0.32	1.00
Refrigeration		0.38	1.18
Cooking	0.38	0.08	0.63
Lighting		0.26	0.82
Clothes drying	0.09	0.08	0.34
Other electrical		0.25	0.78
Commercial	6.65	1.74	12.06
Space heat	4.28		4.28
Water heat	0.61		0.61
Air conditioning	0.35	0.41	1.63
Refrigeration		0.28	0.87
Cooking	0.15		0.15
Lighting		1.05	3.26
Asphalt and road oil	1.26		1.26
Industrial	21.44	2.96	29.65
Process steam	10.54		10.54
Electricity production	0.33		0.33
Direct heat	6.58	0.18	7.09
Electric drive		2.34	6.48
Electrolysis		0.34	0.94
Other electrical		0.10	0.28
Feedstocks	3.99		3.99
Transportation	18.91	0.02	18.96
Automobiles	9.81		9.81
Trucks	3.90		3.90
Aircraft	1.29		1.29
Rail	0.58	0.02	0.63
Pipelines	1.81		1.81
Ships	0.26		0.26
Buses	0.16		0.16
Other	1.10		1.10
Grand total	54.89	6.69	74.74
(% of grand total)	(73)	(9)	(100)
Grand total excluding petrochemical feedstocks, reductant coke, asphalt, and road oil	49.64	6.69	69.49
(% of grand total so modified)	(71)	(10)	(100)

Source: M. Ross and R. H. Williams, "Assessing the Potential for Fuel Conservation," Institute for Public Policy Alternatives, State University of New York at Albany, July 1975.

the separable problem of fuels used for their energy content, we can now construct an illustrative table of the *approximate* distribution of end-use enthalpy according to the type of work to be done, as shown in Table 2. Note that conservation laws are satisfied: The horizontal sum of the grand total row is 69.49 x 10^{15} British thermal units (BTU) or 69.49 quads, which is the primary energy use (as fuels) shown in Table 1. The upper row of percentages at the bottom of Table 2 is equivalent, in round numbers, to that given on page 78 of reference 1. There the obligatorily electrical component is disaggregated into roughly half industrial electric drive and half other applications. The temperature distribution of heat requirements mentioned there pessimistically assumes classification by terminal temperature, not total temperature change (see Appendix II).

What does Table 2 mean? Of course it takes no account of end-use effectiveness or efficiency—of how much of a stove's heat, or the primary energy used to produce it, gets into the pot, or of the ability of 0.32 quads of electricity applied to residential air conditioners to move about 0.61 quads of unwanted heat[17] into the volume of air partly cooled by a neighbor's air conditioner. But the coefficient of performance (COP) attainable from electrically driven heat pumps such as refrigerators may not be as important as we would like to suppose: Heat pumps can be driven by any source of mechanical work (including wind-hydraulic drive), and refrigeration, as Robert Williams has pointed out, can be advantageously integrated into the heating and cooling system of a house (solar or otherwise) rather than restricted to a special plug-in appliance. Even the undoubted efficiency of electricity for driving motors pales a bit when we recall that mechanical work in fixed devices is about 7 percent of all end use; that fossil fuels are also very high-grade forms of energy and are often used at very respectable First Law efficiencies; and that the First Law efficiency of the electric motors studied by Goldstein and Rosenfeld[18] has fallen from 71 percent to 49 percent for a *single manufacturer* since 1940 (and to 30 percent for another manufacturer).

There is room for infinite argument about the significance of the losses at power stations shown in Table 2, which I have included for clarity and completeness. But however one takes account of the low entropy attained by partitioning entropy (and losing enthalpy) at power stations, it is clear that the specialized applications that can take greatest advantage of electricity's high quality are but a small

TABLE 2

U. S. Deliveries of Enthalpy by End Use, 1973
(quads, or 10^{15} Btu/year)

Sector	Heating and Cooling $\Delta T < 100°C$	Heating and Cooling $\Delta T \geq 100°C$	Portable liquids	Misc. mech. work	Obligatorily electrical	Lost at power stations
Residential	8.72	0.62			0.52	4.21
Space heat	6.48					0.71
Water heat	1.54					0.59
Air conditioning	0.32					0.68
Refrigeration	0.38					0.80
Cooking		0.46				0.17
Lighting					0.26	0.56
Clothes drying		0.16			0.01*†	0.17
Other electrical					0.25†	0.53
Commercial	5.93	0.15			1.05	3.67
Space heat	4.28					
Water heat	0.61					
Air conditioning	0.76					0.87
Refrigeration	0.28					0.59
Cooking		0.15				
Lighting					1.05	2.21
Industrial	4.84	12.46			2.78	5.58
Process heat	4.84‡	12.46‡				0.33
Electric drive					2.34†	4.14
Electrolysis and other electrical					0.44	0.33††
Transportation			17.10	1.81	0.02	0.03
Autos, trucks, buses, air, ships			15.42			
Rail			0.58		0.02	0.03
Pipelines				1.81		
Other			1.10			
Grand total	19.49	13.23	17.10	1.81	4.37	13.49
(% of 56.33 quads**)	(35)	(23)	(30)	(3)	(8)	-
(% of 69.49 quads)	(28)	(19)	(25)	(3)	(6)	(19)

* *This is a rough estimate of drive (as opposed to heat) requirements.*

† *All these applications are assumed to be obligatorily electrical even though some are likely in practice to be readily substitutable—e.g., by compressed air.*

‡ *For the temperature distribution of process heat, see Appendix II.*

†† *The 0.33 quads of fossil fuel used to generate electricity at industrial sites is arbitrarily assumed to provide its output to "other electrical"; the exact mix is unkown.*

** *The total delivered enthalpy shown in Table 1 is 49.64 + 6.69 = 56.33 quads.*

NOTE: for comparison, the approximate distribution of Canadian end-use energy by thermodynamic category is (see Figure 1): 42 percent heat below 100° C, 20 percent heat 100° to 260° C, 5 to 7 percent heat above 260° C, 25 percent portable liquids, 1 percent feedstocks, and about 5 to 7 percent obligatorily electrical. This is strikingly similar to the U.S. distribution, with slightly more low-grade heat and slightly less vehicle fuel.

Source: From Table 1 and Appendix II.

fraction of all end uses. The reasons for restricting electricity as nearly as possible to those few advantageous end uses—cost, First and Second Law efficiency, elegance, and sociopolitics—will be treated below. Meanwhile, we need merely note that approaching the energy problem from the viewpoint of end-use structure, rather than of the forms of energy most readily made from the most abundant domestic fuels, gives a very different impression of what kinds of energy we need.*

To fix ideas, let us consider the approximate end-use structure of a structurally similar country, Canada. As in the United States, end use is mostly heat, most of that at modest temperatures, and nearly all the rest liquids for transport (Figure 1). The end uses requiring electricity for purposes other than making small temperature differences are so few

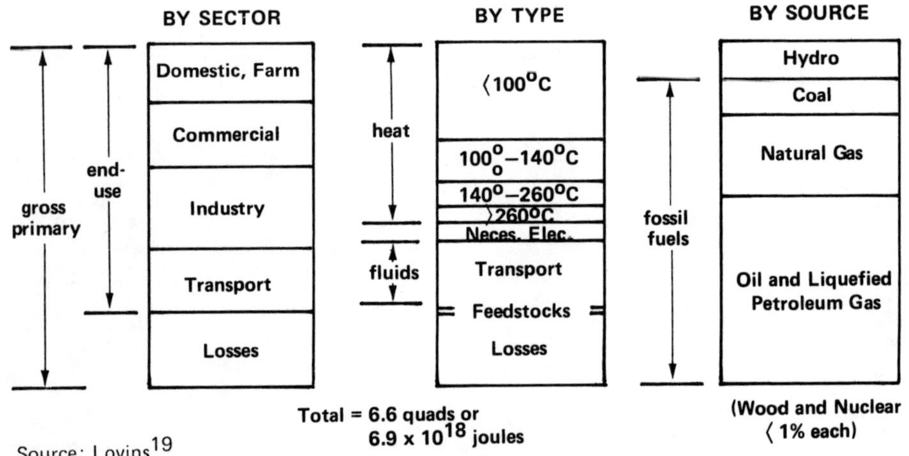

Source: Lovins[19]

FIGURE 1 Canadian Energy Use, 1973 (Population 22 Million)

that all of them, plus all high-temperature heat needs, could be covered (on an aggregated basis, but a fairly realistic one) by present Canadian Hydroelectric capacity. It is thus not obvious that additional electricity can be used to advantage: the same situation I have found in each of about 14 other countries in the past year.

A study[19] sponsored by the Science Council of Canada takes this logic one step further. The Canadian Cabinet has

* *Similar methodological lessons—being demand-oriented, matching quality to end-use needs, etc.—are just starting to be learned in other resource problems such as water. A systematic survey of what we have learned the hard way in energy policy should be useful elsewhere.*

approved in principle, and the Government is now implementing, technical fixes that are officially estimated to hold roughly constant, over the next 15 years, the primary energy needed to heat houses and to run the transport and commercial sectors, despite normally continuing secular growth. But whereas official calculations run only to 1990, capital stocks turn over so slowly that only half the housing stock, for example, is expected to be replaced by 2025. The technical fixes thus have large long-term effects that are invisible in short-term calculations. My estimates, considered by Canadian officials to be technically conservative, suggest[19] that technical fixes similar to those already approved in Canada, combined with a standard growth projection (with modest exceptions), will yield in 50 years a halving of today's per capita primary energy. (Applying essentially the same technical fixes to the per capita activity levels of 1960—perhaps a rough surrogate for a luxurious version of a "Conserver Society"—would yield a further halving—i.e., to a per capita level roughly one-fourth today's U.S. level, or about that of New Zealand.)

The middle bar in Figure 2 shows the approximate estimated end-use structure for the higher of these two estimates, allowing for population growth to 1.74 times the 1976 level and probably exaggerating the needs for high-

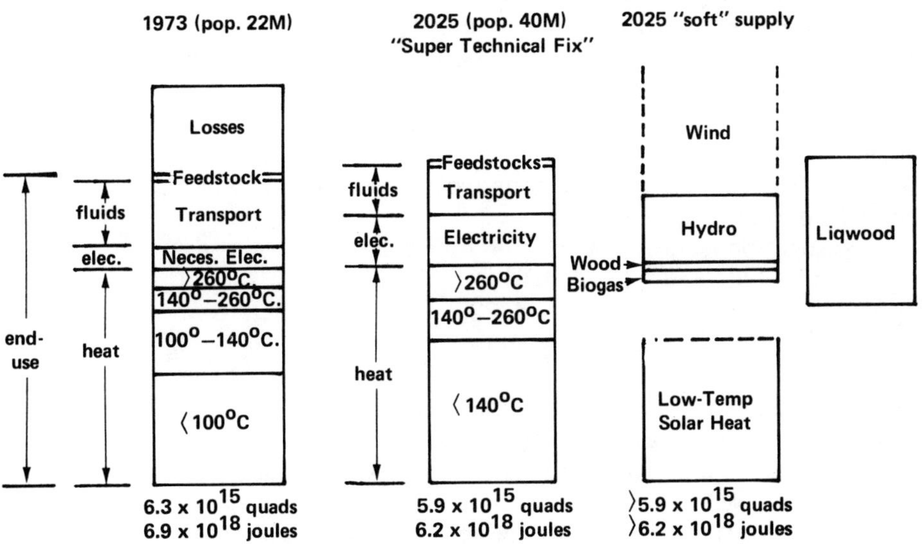

Source: Lovins[19]

FIGURE 2 Projections of Canadian Energy Use

quality energy. The Figure also includes some elements of supply, to show graphically in a back-of-the-envelope fashion how the concept of end-use matching[1] can be applied. The hydro block is the grandiose and controversial minimum hydroelectric capacity already firmly committed for 1985 (not counting the James Bay project, whose future is uncertain), and substantially exceeds the electrical requirements in 2025 for all purposes other than low-grade heating and cooling. Figure 2 assumes that 50 years is long enough to supply essentially all space heating, plus perhaps some medium-temperature process heat, with solar collection incorporating seasonal storage. (This is technically and economically feasible under Canadian conditions,[19] and the main uncertainty is in the deployment rate; people who think it will take longer than 50 years are welcome to change 2025 to a later date and stretch out the transition.) The box labeled "liqwood" is the Canadian Forestry Service's name and estimate (2.55 quads/year) for the conservatively sustainable net yield of fuel alcohols from forestry if pulp and paper production is held at the present level. The estimate may prove too high on further ecological study, but is far larger than is needed to run the transport sector, even today. Finally, wind is an unknown but very large number (far off-scale): for example, if instead of flooding the Phase One (LaGrande Complex) watershed of James Bay we built currently commercial Canadian wind machines there, we would obtain about 13,000 MW average output instead of 8,300 MW at broadly comparable unit cost, from an area that is not particularly windy.[19] Further, wind can make electricity (or hydrogen) if desired, but can also pump heat, compress air, and otherwise fit anywhere into the end-use structure.

Inspection suggests that the hydroelectricity already committed, plus any two of the three "soft" sectors, should suffice to match the calculated 2025 end-use structure *in both quantity and type.* In practice, one would instead use a mix of sources. The next steps in this exercise—not yet done for Canada—will be to disaggregate by regions and work backwards[1] through the fossil-fueled transitional period, then to compare the total costs, risks, impacts, etc., of the policy with those of present policy. For two reasons, I expect the results will be attractive. First, pairing off each soft technology with the hard technology that one would otherwise have to use in the long run to do the same thing reveals, as I shall outline below, that the soft technologies are cheaper; the same is probably true for transitional technologies, for similar reasons; so they should all still be cheaper when added up. Second, this exercise has been done for several countries less well placed than Canada, such as Denmark[3,10] and Japan[3]. In each case, the soft path appears to be quicker and cheaper than present policy, with no worse dependence on imported fossil fuels during the execution of the soft path than without it. (This is because end-use efficiency and soft technologies can each grow faster than hard technologies.)

The traditional objection to such a soft path is that its supply technologies would be far more costly than conventional hard technologies. Let us now examine this thesis.

Costs

Before we plunge into the numbers, a few caveats are in order. First, the cost calculations below are approximate, illustrative, incomplete, even naive. Their style will be more familiar to a physicist than to an economist. Though I consider the calculations adequate to support the tentative conclusions drawn, they need refinement, which I earnestly hope readers will attempt.

Second, I shall concentrate not on life-cycle costs but on initial capital costs, for the following reasons:

1) Though most thoughtful analysts agree that most or all of our fossil fuels are currently too cheap relative to the real costs of depleting them (future depletion is heavily discounted), trying to estimate what fuels should or will cost decades from now is entirely guesswork, since long-term market-clearing prices depend heavily on future taxes, subsidies, negotiable economic rents, etc., not on ideal free-market equilibria.

2) Since any soft technology, given ordinarily good engineering, will have an operation and maintenance cost lower than the *sum* of fuel cost and operation and maintenance cost for a conventional technology (which, unlike the soft one, requires fuel), the soft technology will have a lower life-cycle cost than the hard technology if the soft one has a lower (or even a slightly higher*) capital cost.

3) Since virtually all significant energy technologies

* *At a 10 percent/year discount rate the present value of 40 years of fuel costs for a reactor ordered today is approximately 42 percent of its initial capital cost, assuming no further rise in fuel costs.*

at the margin are capital intensive, their capital cost is an increasingly dominant term in life-cycle cost, and, to first order, provides the long-term signals that one would otherwise seek from life-cycle cost.

4) Capital cost is an engineering number that can be calculated, even for future production, with far more confidence than future fuel costs, for it is relatively resistant to the interfering effects of tax, price regulation, and oligopoly.

5) The drastic capital-cost escalation that has befallen some adventurous hard technologies in the past few years is unlikely to afflict soft technologies to any remotely similar degree, since they are technically more mature,[14] far simpler, and more acceptable to the public.

6) Engineering calculations provide information that is unobtainable in principle from any kind of economic analysis (e.g., no amount of regressions of historical elasticities can tell you that you can build a heat pump).

A third caveat, as argued elsewhere[1], is that it is essential to compare the costs of alternative *complete* energy systems needed *in the long run* to perform a given *end-use* task. Otherwise one cannot avoid long-term misallocations of resources. Marginal-cost pricing of fuels is not enough, as one must compare the costs of different ways of performing a *function,* not just of delivering a fuel. (Thus capital cost per delivered daily barrel of enthalpy, for example, must be complemented by consideration of how efficiently that delivered enthalpy, of whatever quality it may be, will be translated into end-use function: an example is considered later for space heating.) Though a few analysts are starting to look at whole-system costs of performing a function, no institution today does so. Vertically integrated major oil companies, uniquely, perceive whole-system costs because they own all the pieces, but they only see costs per daily barrel or per barrel of output, not per unit of function. Likewise, utilities are not concerned with end-use function, but only seek the cheapest source of large blocks of busbar electricity to feed into the grid.

Fourth, a serious question can be raised whether economic calculations are particularly relevant today. Of course, sophisticated managers realize that they should base investment decisions on sensitivity to altered circumstances rather than on tiny marginal-cost differences in the base case. But are even enormous marginal-cost differences that important? Is cheap energy necessary—or even desirable? (Paul Ehrlich argues[20] that plentiful cheap energy can be considered less a benefit than a *cost* because of the very expensive damage it lets us do—or its conversion unavoidably does—to essential life-support systems.) As Alan Poole cogently argues:[21]

> The ultimate condemnation of a project is that it is "uneconomic," which is to say it costs 5 mills per kWh more than another option. By comparison, human labor in a dirt-poor preindustrial society costs roughly 2,000 mills/0.5 kWh, 4,000 mills per kWh. In this light I find it odd that so little work has been done on the long-term importance of the cost of energy. Does it ***really*** matter if, say, a solar option costs twice as much as a nuclear option? In the conventional wisdom such a price discrepancy would dismiss the solar option ***automatically*** even if we were rather concerned about the long-term implications of nuclear power. A pretty good hand-waving case can be made that the effect on economic growth is not very important. A number of European countries appear to have had real energy costs substantially higher than ours yet their per capita GNP is comparable.

Moreover, as Kneese points out,[22] the usual cost-benefit or cost-cost comparisons break down when some costs are not quantifiable (and may hence seem less real than benefits) or are transferred to other times and places (which makes the theoretical basis of cost-benefit comparison invalid). In my opinion, the noneconomic aspects of, say, nuclear power are vastly more important, both theoretically and politically, than its disputed and probably unknowable economic properties. Even Keynes admonished us not to "overstate the importance of the economic problem, or sacrifice to its supposed necessities other matters of greater and more permanent significance."[23]

With these substantial caveats in mind, let us now try to compute a few capital costs, first for central electric systems and then for others, both hard and soft. I shall use the data base of the Bechtel Energy Supply Planning Model,[24] expressed in third-quarter 1974 dollars for ordering early in 1974. The data base, like any, could be improved, but is probably the most detailed, authoritative, and up-to-date available, and is used extensively in Federal and private studies today. It is being continuously updated.

The Bechtel estimate of direct construction cost—architect-engineer's passthrough—for a 1.1-gigawatt(GW) light-water reactor is $418/kW installed.* To this must be

* *Many current data—e.g., the Rancho Seco 2 projections,[28]—strongly suggest that this figure is already 10 to 25 percent too low.*

added 40 percent for indirect or owner's cost—interest during construction, design, and administration—for a total of \$585/kW installed. This is comparable to the \$520/kW (also 1974 dollars) assumed by the Brookhaven data base,[25] which relies on the out-of-date (July 1973) and poorly documented A. D. Little study for Northeast Utilities.

Associated with a marginal GW of generating capacity in a typical U.S. program is about 140 miles of transmission circuit[26] with updated nominal characteristics, shown in Table 3. From these data, total capital cost of transmission (direct plus owner's costs, 1974 dollars) per marginal kilowatt of generation is about \$69/kW. (The corresponding 1973 and 1985 average costs are about \$87/kW and \$80/kW, respectively, internally consistent within rounding errors.) The Brookhaven estimate of \$75/kW (1974 dollars) is fairly close. Note, however, that the current-dollar escalation rate averaged about 11 percent/year during 1966-1972, when GNP inflation was relatively slow and steady, and that regional cost variations are about 3.5-fold.[27] I shall take as authoritative the Bechtel data,[26] which are based on extensive field experience and checked with utility and Federal Power Commission data.

The Bechtel estimate[24,26] of capital cost for distribution assumes a nominal 160-MVA-input facility that feeds 29 MW, 21 MW, and 70.2 MW to industrial 34.5 kV, commercial 13.2-kV, and domestic 220-V customers, respectively, at an average power factor such that MVA x 0.823 = MW. Direct construction cost (1974 dollars)[26] is \$34.5 million with all lines aerial, \$55 million with all lines under-

TABLE 3

Nominal Characteristics for Transmission Circuits

kV	MVA/facility In	MVA/facility Out	GW-miles/ facility*	Million dollars/ facility†	Approximate modal split‡ 1973	1985	Marginal 1973-85‡
230	600	530	125	95	0.47	0.32	0.185
345	1,200	1,060	300	120	0.23	0.27	0.30
500	2,600	2,450	600	188	0.23	0.29	0.335
765	4,000	3,800	1,250	228	0.05	0.09	0.12
400DC	1,400	1,370	1,200	317	0.02	0.04	0.055

* GW transmitted; to express in terms of GW generating capacity, divide by 0.791.
† Direct construction cost, third-quarter 1974 dollars; for owner's costs add 20 percent.
‡ Fraction of total energy transmitted per mode (not fraction of facilities per mode).
‡ My computation; generating capacity 415 GW in 1973 and 899 GW[24] in 1985.
Source: Carasso and Gallagher[19]

ground, and \$41 million, or \$256,000/MVA, for a nominal split (about 80 percent underground) considered realistic at the margin. Indirect costs[24] are 35 percent. Total distribution capital cost is thus about \$420/kW (1974 dollars). Since the Brookhaven estimate is only \$145/kW, I have queried the Bechtel estimate with special care and am assured that internal review has confirmed it as realistic.* Bechtel believes[26] that lower utility estimates reflect less stringent undergrounding assumptions and special internal accounting procedures that reduce apparent overheads and cost of money.

The Bechtel estimates[24], also used by Brookhaven, for marginal capital costs of nuclear fuel-cycle facilities yield a total cost of about \$61/kW installed. I believe this figure is too low by a substantial factor[28] (probably at least two), in view of recent escalation and altered regulatory requirements. Nevertheless, I shall use it.

A fairly up-to-date estimate of the approximate capital cost of the initial core—a front-end cost even though it can be credited later against fuel-cycle costs—is shown in Table 4

TABLE 4

Capital Cost of Initial Core of 1-GW Pressurized Water Reactor*

	Millions of dollars	
Direct cost		
400,023 lb U_3C_8 @ \$42/lb	\$16.8 x 3 =	\$50.4
153,838 kg U to convert @ \$3.50/kg	0.5 x 3 =	1.6
135,068 kg U to enrich @ \$67.25/kgSW,		
27,001 kg U at 2.63 percent (74,685 SWU)	5.0 x 3 =	15.1
28,994 kg U to fabricate @ \$90/kg	2.6 x 3 =	7.8
		\$74.9/GW
Interest compounded at 12 percent/year with 3-year lead time for U_3O_8 production & conversion, 1.5 years for enrichment, 0.6 years for fabrication		\$24.4/GW
Total cost (excluding insurance, property taxes, other minor overheads)		\$99.3/GW

* Initial design enrichment 2.63 percent, tails assay 0.3 percent, once-through fuel cycle, values for annual core (including standard process losses).

Source: Harding[28]

in 1976 dollars for 1976 ordering.[28] I shall thus take the initial core cost as \$100/kW installed (1976 dollars). Total

* *The original error-band estimate[24] is -10 percent, + 20 percent, with an exponential scaling factor 0.7.*

cost per *installed* kilowatt is thus \$585 + 69 + 420 + 61 = \$1,135 in 1974 dollars plus \$100 in 1976 dollars. The empirical capacity factor of the system to mid-1976 is about 0.58. Some analysts argue that this will improve on a "learning curve"; others question whether this theory is applicable in the circumstances peculiar to nuclear engineering with possibly declining quality control, and note that the infirmities of light water reactors (LWR's) appear from outage analysis to be proceeding from the pediatric to the geriatric with scarcely a pause. In any event, since the lifetime average capacity factor assumed in the Atomic Energy Commission Document WASH-1139(74), page 23 is 0.57, since performance appears to be deteriorating with increasing unit size, and since international experience is ambiguous (West Germany vs. Japan), I shall use a generally acceptable 0.55 for lack of a better value. Moreover, since I have not been able to find definitive data on whether the corresponding capacity factor for marginal transmission and distribution equipment should be higher or lower than for the power station, I shall use 0.55 for all the components. The total cost per kilowatt *sent out* is thus \$2064 in 1974 dollars plus \$182 in 1976 dollars. Adding a surcharge of 12 percent for transmission and distribution losses (the Bechtel data[24] show average losses of 8.4 percent and 8.7 percent, respectively) yields a total cost per kilowatt *delivered* of \$2,311 in 1974 dollars plus \$200 in 1976 dollars.

We now have the difficult task of putting everything in 1976 dollars for 1976 ordering. One crude approximation, certainly too low, would be to inflate all the items in 1974 dollars by a standard index for general construction, such as Handy-Whitman. This might yield a total of about \$3,068/kW delivered. A better approximation would be inflate the nuclear plant component at the observed rate carefully derived by Bupp et al.[29]* —on the order of 15 percent/year in current or 7 percent/year in constant dollars—yielding \$1,575 for the station, \$200 for the initial core, and (assuming, somewhat arbitrarily, a 1974-1976 inflator of about 25 percent based on the Marshall & Stevens Equipment Cost Index[25]) \$1,245 for total transmission and distribu-

* *The undocumented escalation rates used in The Atomic Energy Commission publication WASH-1345 (1974) are virtually useless. More recent work by Bupp shows better correlation coefficients for regressions with LWR costs escalating at 20 percent/year in constant dollars (26 in current dollars), which would yield a whole-system cost of \$3,496 per kilowatt delivered, not the \$3,179 shown here.*

tion and $155/kW for fuel-cycle facilities. On this basis, the grand total in 1976 dollars, ordering in 1976, would be $3,179/kW delivered, or about 213,000/daily barrel delivered.

Of course, all these data can be greatly refined and improved—though it may not be worth working hard on second and third significant figures in view of the major unknowns in future escalation rates. The main defect in the approach illustrated seems to me rather to be that one should take account both of generating mix (for different parts of the load-duration curve) and of user diversity. On the former point, clearly gas turbines, combined-cycle plants, etc., are less capital-intensive than baseload stations. Since transmission and distribution (especially the latter, which siting near load centers would not greatly reduce) cost about as much as a nuclear station, however, the sensitivity of total investment to plant mix is reduced. Moreover, detailed data on capital costs of complete electrical networks, including a realistic generating mix, appear not to be available. The only recent number I can find is an estimate[30] of $2000 capital cost per average kW of capacity to deliver to consolidated Edison customers in New York City (where land and underground cable costs are high). This number appears to refer to capital cost per kilowatt installed, not sent out or delivered, and should presumably therefore be roughly doubled to reflect cost per kilowatt delivered. Moreover, most long-term projections of generation mix assume that nuclear capacity will extend far beyond the present baseload level and that much of the current intermediate-load and gas-turbine territory will be occupied by very capital-intensive pumped storage schemes.[31]

As for user diversity—the fact that not all users will demand electricity at once, so they can share capacity—it is a real issue on which better data are needed. It is not particularly relevant, though, to the specific application I shall examine below, space heating, since European experience clearly shows that the winter peak heating load determines the system's simultaneous maximum demand with negligible user diversity. For industrial process heat (which, from resistance heaters, accounts for 60 percent of the marginal electricity use in the Energy Research and Development Administration's (ERDA) 1975 "intensive electrification" *scenario,*[32] or more electricity than the U.S. used for everything in 1975) one can again probably assume a good approximation to base-loading, so the question does not arise. For intermittent applications such as home appliances, it clearly does arise and should be taken into account—though the terms to which a large correction factor would apply seem small.

It is also important to note what the above estimate of over $200,000/daily barrel does *not* include: marginal capital investment in land, reserve margin and spinning reserve,[31] taxpayer-supported regulation and security services, Federal R&D, future services (waste management and decommissioning), and end-use devices. No allowance is made for escalation after 1976 beyond the GNP inflation rate—presumably a highly conservative assumption—nor for the energy that must be debited against station output to operate the nuclear fuel cycle (a debit of about 6 to 8.5 percent for most LWR's).[34] A realistic calculation including all these terms except end-use devices, and still excluding all externalities and dynamic net-energy considerations,[3] would yield a nuclear capital cost nearer $300,000/daily barrel, or close to $5,000/kW delivered. The numbers at this point obviously become fuzzy because proper data are not available and, for some terms, may never become available. Nonetheless the estimate of over $3,000/kW seems unrealistically low.

For comparison with this figure, the Bechtel data[24] for 800-MW coal-electric plants, averaging over the various types of coal and of coal mines and delivery systems in a typical U.S. program,[24] are about $355/kW installed (1974 dollars, direct plus owner's costs). About $106/kW for the fuel cycle and $120/kW for scrubbers must be added.[24] Assuming 1974-1976 inflation for the plant[29] and scrubbers of about 13 percent/year in current dollars, a 1974-1976 inflator of about 25 percent as assumed above for transmission and distribution (T&D) equipment and of 20 percent for the fuel cycle, a 12 percent surcharge for T&D losses as above, and a predicted effective capacity factor of 62 percent[33] for all components yields a total system cost of $2,476/kW delivered, or about $166,000/daily barrel delivered enthalpy. The omitted terms are analogous to, but presumably smaller than, those for nuclear power as summarized above.

Neither of these calculations is, strictly speaking, a whole-system cost, not only because of their noted omissions, but also because neither calculates the cost of supplying a unit of end-use *function* and thus takes proper account of the low entropy of the form of energy being supplied. An example is worked out below.

The inherent capital intensity of the complex devices needed to make large blocks of electricity is impressive. I am even more impressed, however, by the little-known diseconomy of distributing centrally generated electricity in a country whose load density averaged over 1965-1971, ranged from about 0.61 to 230 mW/square meter (m^2) (mean about 30 mW/m^2)[27] and whose time-averaged load density for residential and small light and power consumers ranged from about 1.2 to 500 mW/m^2 (mean about 15 mW/m^2*). (Both sets of data refer *only* to investor-owned lower-48 U.S. utilities, which had about 80 percent of the total U.S. utility business with about 258,000 large customers and 54.2 million small ones—whereas all utilities combined had about 363 million of the latter.) A recent study[27] finds:

> Transmission and distribution costs contribute significantly to the total costs of providing electrical service. In 1974, privately-owned electric utilities in the United States spent about 35% (over $7 billion) of their total capital expenditures for transmission and distribution equipment. The expenditures for operation and maintenance of this equipment were about $3.0 billion, an amount equal to about ½ the total costs of fuel in 1972.
>
> The costs derived from the transmission and distribution (T&D) system have historically comprised about 2/3 the costs of producing and delivering electricity to residential-commercial customers, and over 1/3 the total costs [of] supplying electricity to large industrial customers.

The study concentrates on major terms—high-voltage transmission lines, distribution lines, and operation and maintenance costs of the T&D system. These three terms together account for about 80 percent of total T&D costs. When capital costs are converted into million dollars/kWh charges at a 13.5 percent annual fixed charge rate, the remarkable results of Table 5 emerge.

The authors conclude:[27]

> . . . almost 70% [69.3%] of the costs of power to residential and small light and power customers are related to transmission and distribution. Of this 70%, almost half [49.3% of the total T&D costs] can be attributed to costs of installing transmission and distribution lines, the two items of T&D equipment that exhibited the most significant regional cost variations. For large light and power customers on the other hand, transmission equipment related costs are only 34% of the total cost of power, while gen-

* *Implying that an average 1-GW plant serves a radius of 146 kilometer (km); the actual average haul length is of the order of 343 GW-km/GW.*

eration comprises about 55%. Distribution equipment and operation and maintenance, including billing, comprise the other 11%.

TABLE 5

National Average Costs of U.S. Privately Owned Utilities, 1972

Item	Residential & Commercial (cents/kWh sold)	Industrial
Transmission equipment	0.45	0.43
Distribution equipment	0.58	0.06
Operation & maintenance of T&D	0.50	0.08
Total T&D	1.54	0.57
Estimated cost of generation	0.69	0.69
Estimated total cost (delivered)	2.23	1.26
Average revenue	2.37	1.17

Source: Boughman and Bottaro[27]

To put it differently, for the smaller customers (average load 1.04 kW) who accounted for about 55 percent of the annual energy sales in the whole sample, a dollar spent on electricity was allocated approximately 19 percent to transmission equipment, 24 percent to distribution equipment, 21 percent to operation and maintenance of all that equipment (including metering—a small term—and billing), about 6 percent to profit and to arithmetic discrepancies in the analysis (owing largely to differential escalation of various components)—and *only 29 percent to electricity.* This seems to me an undeniable diseconomy of centralization. I am sure utility customers would be surprised to learn that they are paying 2.2 times as much to have the electricity delivered to them as to generate it. For the large customers (average load 177 kW), the mix is less extreme—generation cost is 1.2 times delivery cost—but still odd, and suggests that there are strong economic incentives for central electrification to encourage centralized use patterns. The wide regional variation in these values—the 1972 T&D cost ranged from 1.0 to 2.3 cents/kWh for small and from 0.36 to 0.82 cents/kWh for large customers—does not alter the startling size of the T&D overheads.* It does not appear from the data presented[27] that truly marginal (rather than recent historic) costs would tell a very different story. It would be interesting to check that, though, and to see whether public utilities (about a

* *Since 1 barrel = 5.8 gigajoule and 1 kWh = 3.6 megajoule, 1 cent/kWh = $16.1/barrel enthalpic equivalent.*

quarter the total size of the private ones surveyed) yield similar results.

Direct fossil-fuel systems are surveyed in detail by Bechtel,[24] with the results (used also in the Brookhaven model[25]) shown in Table 6. The costs are approximate total marginal capital costs (direct plus owner's costs) in 1974 dollars per daily barrel, ordering in early 1974. Since the delivery-system costs are a function of the source, the detailed structure of the distribution system should be calculated separately in each case. In general, the Arctic sources, with low wellhead investment, entail high transport investment (an Arctic gas pipeline ordered now could easily cost much more[35] than the wellhead investment). The exact data (insofar as anyone knows them) are in principle available from the Bechtel data base,[24] but would require very laborious calculations of modal splits. As a crude approximation, therefore, one can only combine weighted averages of sources in each category with weighted average costs of delivery. The result in 1976, dollars, ordering in 1976, is approximately \$2,800 for domestic coal, \$7,200 for the oil sources shown, and \$13,100 for the gas sources shown. Both the latter figures are significantly lowered by the inclusion of

TABLE 6

Total Marginal Capital Costs for Direct Fossil Fuel Systems

System	Exploration/development/ production cost at well-head/minemouth/terminal (1974 dollars/(daily barrel)	Cost of system to process/deliver to consumer*
Domestic coal: weighted average	\$ 1,437*	
Eastern underground	1,150	
Eastern surface	1,980	\$1,380
Western underground	968	
Western surface	1,150	
Crude oil import	110	
Alaskan oil	803	3,697
Offshore lower-48 oil	11,594	
Oil shale	9,823	
Liquefied natural gas import	1,034	
Alaskan gas	3,410	4,698
Offshore lower-48 gas	11,594	
Onshore lower-48 gas	12,496	

* Average weighted by the modal mix (for both sources and means of transport) calculated by the Bechtel model[24] as optimal for President Ford's January 1975 energy program for 1976-1985.

imports (imported oil is only about as capital-intensive as domestic coal, a few thousand dollars/daily barrel). It appears that domestic frontier fluids fall generally in the range $10,000 to $25,000/daily barrel in 1976 dollars.

As for coal synthetics, the Bechtel data[24] relied on by the Brookhaven model[25] are in the range (1974 dollars and ordering) from $14,900/daily barrel for low-BTU gas to $20,700 for pipeline-quality gas and $34,300 for methanol. (Brookhaven estimates[25] $12,500 for hydrogen.) These estimates do not include, however, the fuel cycle, water supply, or delivery system (though the latter might rely on pipe already laid). Broadly comparable estimates are given in the November 1975 Interagency Task Force report on synthetic fuels. September 1975 data[35] suggest that the specific investment (1975 dollars) in a Lurgi plant is of the order of $18,000/daily barrel for Western and $20,000 for Eastern coals, with about a 20 percent saving possible with second-generation (e.g., HYGAS) technology. The total cost for such an advanced plant, including the fuel cycle, water supply, and connections to existing pipelines, would be of the order of $22,000/daily barrel "or more,"[35] roughly comparable to the cost of a tar-sands system. The extraordinary escalation of most synthetic-fuels projects in the past 2 years, however, makes me suspect, with Arthur Squires, that a whole-system capital cost around $40,000/daily barrel in 1976 dollars, ordering now and excluding pipelines, is probably realistic, and might even be low if such rapid escalation continued for long.

Let us turn now to transitional fossil-fuel technologies.[1] Though no detailed cost estimates appear to be available for small-scale coal conversion—by e.g., the flash processes[36] or supercritical gas extraction[37]—both seem technically simpler and more reliable than conventional large-scale conversion, and it would be surprising if they did not have correspondingly lower capital costs and escalation rates. Firmer estimates can be given for the fluidized-bed technologies—for industrial and institutional boiler backfits, cogeneration,* and district heating—on which I believe strong and immediate commercial emphasis is warranted.[1,38] I have unfortunately been unable to find out the tender

* *Robert Williams tells me that the SRI estimate[1] that 29 percent of West German electricity comes from cogeneration is incorrect: apparently some 23 to 25 percent is generated at industrial sites, but only about half is true cogeneration rather than condensing turbines. I regret the error and shall seek to correct it when Foreign Affairs corrects note 25.[10]*

prices for the 25-MW(t) thermal Enköpings Värmeverk fluidized-bed boiler now under construction by a Norwegian-Finnish consortium that is operating a 2-MW version commercially. (I wrongly cited 25-MW/Enköping as 15-MW/Linköping.) A detailed design study by Stal-Laval Turbin AB (Finspong, Sweden) has been developed, however, to the point of cost estimates on which a commercial tender could be based.[39] After distribution losses, the Stal-Laval system[1] would deliver 60 MW plus 105 MW(t) of district heating. If we assume that the electricity is used in heat pumps with COP = 2, total delivered space heat is 225 MW(t). Recent estimates[39] of total construction costs (including interest and standby boilers) are $40 million for the plant and $25 million for the district heating grid;[40] both appear generous.* If all electrical distribution were counted as marginal—which it might not be—it would add about $35 million. Heat pumps at a rather generous $200/kW would add $12 million. The grand total implies a system cost of $33,400/daily barrel delivered heat. To this must be added a few thousand dollars for the fuel cycle and some allowance for capacity factor (though the latter modification would presumably be very small, since the capacity factor should be extremely high, and outages would be covered by the standby boilers already allowed for). I suspect that both these terms would be more than covered by conservatisms in the cost estimates, and that a round number in the vicinity of $30,000/daily barrel (1976 dollars) is realistic.

It would be particularly interesting to have firmer cost estimates for fluidized-bed industrial boiler backfits—first at atmospheric, but then at elevated pressures suitable for later conversion to gas-turbine cogeneration, via either direct or closed cycles.[42] The sharing of investment between electrical and process-steam functions seems attractive:[41] indeed, utility-owned industrial cogeneration by 1985 could be profitably generating 26 to 42 percent more than all U.S. utility sales of electricity in 1975.[41] Better field data for pressurized fluidized beds—generally conceded to be cheaper than those at atmospheric pressure—should become available in the next year. Data for pressurized trash- and wood-waste-burners of this type are also relevant, though I have not yet

* *For example, the Thermo-Electron study[41] estimates that a new conventional industrial gas turbine of this size costs about $170/kW installed without, or $230/kW with, a waste heat boiler. This implies $11.9 or $16.1 million for a conventional 70-MW gas turbine, leaving a large sum for the fluidized bed, coal-handling system, cyclones, etc.*

obtained firm cost figures for such designs as Combustion Power Co.'s 11-MW + $61,000 pounds/hour CPU-400 system [reportedly about $15,000/(input ton-day)].

Let us turn now to the capital costs of soft energy technologies,[1] beginning with wind machines. Unfortunately, all available modern data refer only to wind-electric systems, though mechanical work from wind may be used more cost-effectively for pumping heat or water, compressing air, etc.[1] Moreover, the field is developing very rapidly, and details of many of the most interesting studies are not yet published (e.g., Lockheed-California's estimates for ERDA of low mass-production costs for horizontal-axis machines—the sort of technology that mass-production cost specialists can handle with confidence because it rests on mature engineering). Data on some intriguing systems, such as James Yen's Grumman vortex tower, are also not yet firmly developed.

As a baseline, however, we can take a currently commercial 200-kW vertical-axis Darrieus device made by Dominion Aluminum Fabricating Ltd (Mississauga, Ontario). The first such device is currently being installed by Hydro-Quebec on the Magdalen Islands in the Gulf of St. Lawrence. It has two blades in tensile stress, is 80 feet in diameter, and has a 120-foot shaft atop a 30-foot tower. It is expected to supply about 0.5 GWh to the grid annually at 3.5 to 4.0 cents/kWh from a site with average windspeed of 18 to 19.5 miles/hour. The installed capital cost, turnkey except foundation and grid connection, is $235,000 (1976 Canadian dollars, nearly the same as U.S. dollars). But this cost includes virtually the whole investment in the large extrusion dies. Accordingly, the second unit has a list price of $175,000 FOB factory (exclusive of switchgear), with further marked declines thereafter. The run-on production cost in small lots (tens) is probably $90,000 to $100,000 and the corresponding price about $130,000. Cost estimates for larger orders (or for the 1-MW unit under development) are not yet available. If we take the total installed price for small-lots production as $150,000 (1976 dollars) and assume a capacity factor of about 0.3, characteristic of respectable sites, the capital cost is $2,500 per average kilowatt sent out. Since trunk transmission would be avoided by feeding directly into local lines, distribution loss of about 4% seems reasonable. If we assume that a third of the 600-VAC-and-downwards distribution investment (altogether of the order of $380/kW) is at the margin, the whole-system cost might be about $204,000/daily barrel delivered enthalpy, less than that of an LWR. It could be more reasonable to assume that the entire distribution investment has already been sunk, however, just as one might assume the same for pipelines associated with a new coal-gas plant. This is because the wind system would normally be used as a fuel-saver integrated into the existing fossil-plus-hydro grid. Such supplementary use makes more sense than an attempt to imitate, with full storage, the operation of existing central stations. Of course, it might make even more sense to use the mechanical work directly to pump heat or compress air at the point of end use, and in such a system the storage required would be simple and cheap.[1]

I do not propose here to develop cost estimates for geothermal heating (a more attractive system than geothermal-electric), partly because many geothermal systems would not be truly soft[1] and none in principle is indefinitely renewable. It is interesting, however, that technical advances in telethermics now permit very remote siting of wells from heat loads. For example, I understand[43] that an insulated pipe in Italy has continuously shipped hot water since 1969 for over 100 km at an average thermal efficiency of 98.5 percent.

Estimates of the capital cost of bioconversion vary widely. One consistent set of estimates in 1974 dollars suggested values from less than $10,000/daily barrel output to about $20,000 for crop residues, and up to as much as $30,000—often less—for municipal waste pyrolysis[44] (which can offer substantial economic credits for recovered materials and for saved disposal costs).[45] For most agricultural projects, $13,000 to $20,000/daily barrel (1976 dollars), plus a few thousand dollars for local delivery systems, might be typical, including collection investment.[46] On the other hand, the local distribution might use existing equipment.

Solar process heat,[47] though attractive and starting to be widely studied, is not yet in a state where definitive cost estimates are generally accepted for a range of designs. Good numbers are about to emerge from the studies described in Appendix II. High working temperatures can be obtained not only by various forms of concentrators but also by highly selective surfaces, either on flat nonfocusing plates or in cylindrical configurations that concentrate severalfold. For example, thin-film sputtering technology now suffices to produce selectivities (ratio of visible absorptivity to infrared emissivity) of 50 to 60,[48] at pilot-scale costs of a few dollars per square meter. It is readily calculated that such a surface, suitably insulated and contained in a hard vacuum, is so insensitive to cloudiness that it should provide working temperatures, under load, of the order of 400° to 600°C on a cloudy day in the winter at Scandinavian latitudes.

Fundamental to any discussion of solar space heating as a major long-term energy technology is the cost of seasonal storage,[19] which eliminates backup investment (and associated potential for load management problems) and can increase collector capacity factors. With modern tank design—prefabricated tongue-and-groove concrete slabs assembled modularly in a hole in the ground, then sprayed with insulating foam, lined, and backfilled—installed costs of hot water storage are about 60 to 95 cents per cubic foot or about \$21 to \$34 per cubic meter (1975 dollars).[49] With such a tank, the marginal capital cost of the solar heating system now operating with no backup at Lyngby, Denmark,[50] would have been about \$6,000 rather than about \$8,000—or even less without the high costs of first-of-a-kind design. I suspect that storage costs will drop further as tanks are integrated into buildings: apparently heat-sink tanks up to nearly 5,000 cubic meters are currently used in Japan, partly as seismic stabilizers.[49]

Estimates of the capital cost of solar space heating depend sensitively on both building and collector design. For new buildings, passive systems with negative, zero, or negligible marginal capital cost should suffice.[51] The following discussion thus refers to retrofits requiring hardware. Further, impressive economies are available in principle from mass production of very simple designs. Suitable concepts include Jerry Plunkett's paper-and-phenolic honeycomb sandwich that unrolls like roofing paper, replaces shingles, and is estimated to have a very long lifetime (many decades) and an installed cost of a few dollars per square foot; the Philips tubular collector designed for cloudy northern European latitudes and expected to enter advanced pilot production next year;* grids of flexible plastic tubing laid in poured asphalt on a flat roof; plastic tubes imbedded in a lightweight rollable mat like the "Sunmat" made by Calmac;[52] and the Thomason system.[52] Commercial versions of the last two sell today for about \$3 to \$5 per square foot installed. Their merit

* *Though the envelopes used in current Philips tests use low-pressure sodium-vapor-lamp technology and are thus somewhat larger and heavier than fluorescent-lamp tubes, the latter, at less than 10 cents each, are perhaps the cheapest manufactured commodity per unit mass, and are extruded at about 30 miles per hour. Owens-Illinois uses a similar approach.*

of simplicity merits close attention.[51]

Let us be pessimistic, however, and assume that solar heating is to be done with conventional flat-plate assemblies that are not integrated into the building structure, have high transport and installation costs, and involve the inherent assembly and materials costs of high-quality glazing and sheet-metal work. In late 1974, factory prices for such collectors were \$10 to \$50/m^2, while various versions assembled and delivered, including moderately selective plates, were \$70 to \$120/m^2 for the collector assembly only.[53] Many solar hardware production experts estimate that an installed square meter of high-quality flat-plate collector plus 1 cubic meter of seasonal water storage,* complete with all plumbing, will cost about \$150 (1976 dollars) around 1978-1979 and about \$100 in the mid-1980's (1976 dollars)—the period of interest for comparison with hard technologies ordered today, since the latter have about a 10-year lead time.

Let us apply this \$100/(m^2 + m^3) figure—consistent with estimates of the Office of Technology Assessment and A. D. Little and with detailed analyses by private mass-production specialists—to Denmark. A Danish south wall receives average total insolation[10] of about 125 W/m^2, and modern (double-glazed, slightly selective) flat plates in Danish conditions can achieve a First Law efficiency of about 0.42. Seasonal fluctuations in Denmark do not appear to require—taking account of storage losses—more than about 0.7 m^3 of storage per square meter of collector in a well-insulated house, thus saving about 8 percent of the estimated system cost. (Actually the storage volume and collector area can be traded off against each other.) Completely solar space heating for a typical Danish house (125 m^2 floor) in the mid-1980's should thus cost on of the order of \$118,000/daily barrel delivered enthalpy—about \$3,500 for an average heat load of 2 kW (good insulation) or \$7,000 for a typical 1974 heat load of 4 kW.

The less favorable regions of the United States receive about 180 W/m^2 on an optimally oriented fixed flat plate (about 270 W/m^2 in southwestern deserts), and seasonal fluctuations require even less storage than in Denmark. Typical mid-1980's specific investment for completely solar space heating should thus be of the order of \$50,000 to \$70,000/

* *A 100-m^3 tank cycling over 50°C and discharging over a period of 4 months produces 2 kW. My macroeconomic approach makes it unnecessary to worry about partial solar heating with backup, or combined solar heating and cooling, as tricks to make solar heating economic.*

daily barrel, assuming total insolation of 180 W/m^2. (Extra installation costs of retrofit should be roughly balanced by savings through the reuse of existing hot-water plumbing or hot-air ductwork, though more detailed study of this point would be useful.) All such figures should be treated with caution, not only because they assume fixed technologies in a rapidly changing field, but also because, as Weingart points out,[53] the installed price even of completely conventional, nonsolar domestic water heating systems varied by a factor of *more than two* in a recent Southern California survey of similar apartment buildings, due to variations in local costs of labor and materials and in the skill (and avarice) of the contractors. The few integrated solar heating and cooling projects now built or being built on a large scale are probably a poor guide to future costs. For example, the Los Alamos project,[53] which ran a $150/$m^2$ collector cost up to a $690/$m^2$ installed collector cost, appears to be a typical example, like many ERDA-funded solar demonstration projects, of a highly sophisticated and instrumented, therefore excessively complex and costly, installation.* Simpler systems, such as the Oss houses[54] (where 2/3 solar heating competes even with cheap Groningen gas), may be a better measure of realistic overhead levels.

Subject to the numerous caveats above, the *approximate* marginal capital cost (1976 dollars) of complete systems to deliver a daily barrel of enthalpy (not of end-use function!) to a U.S. consumer in the mid-1980s are summarized in Table 7.[1] I have generally neglected future escalation beyond the GNP inflation rate and considered only those soft technologies that are currently state-of-the-art. Even if one quibbles with details, this table suggests several broad conclusions:

1) Big electrical technologies are pricing themselves out of the capital market. It is no wonder that the internal financing ratio of utilities fell from 0.40 to 0.19 during 1960-1973,[56] nor that institutional investment in utility

* *I am concerned that major aerospace and defense corporations are so heavily involved in ERDA's solar program: not because they do not have technical resources that we need, but rather because they may not be able to keep the hardware simple enough and the overheads low enough. (Just responding to a recent ERDA coal Request for Proposal would have cost $2 million!) I am reminded of the new Boeing-Vertol subway cars for Boston, which reportedly used, in the original design, some 1,300 components per door. This was sweated down, with difficulty, to about 300, and the doors might now work, but clearly the designers have become so sophisticated that they can't design a door any more. Mass production is hardly an answer to this institutional problem.*[53]

TABLE 7
Approximate Marginal Capital Cost For Delivering One Daily Barrel of Enthalpy [about 67 kW(t)] in the U. S. (1976 dollars)

Traditional direct fuel technologies, 1950-1970	$2000-3000
Imported oil or domestic coal, 1970s	2000-3000
Frontier oil and gas, 1980s	10,000-25,000†
Coal synthetics and "exotic" hydrocarbons, 1980s	20,000-40,000†
Central coal-electric with scrubbers, 1980s	170,000†
Light-water reactor, mid-1980s	200,000-300,000†
Fluidized-bed gas turbine/district heating/heat pumps, early 1980s	30,000*
Wind-electric	200,000†
Retrofitted 100 percent solar space heat, mid-1980s	50,000-70,000*
Bioconversion of agricultural/forestry residues, 1980s	13,000-20,000
Pyrolysis of municipal wastes, late 1970s	30,000
Improved end-use efficiency:	
New commercial buildings[55]	-3000*
Common industrial/architectural "leak-plugging"	0-5000*
Most industrial/architectural heat-recovery systems	5000-15,000*
Difficult, extremely thorough building retrofits	25,000*

*Including cost of end-use devices to deliver desired function.
†Delivered in the form of electricity.

debt has fallen off sharply in the past year or two. The Invisible Fist strikes again!*

2) Improved end-use efficiency will long remain the most cost-effective investment.

3) Though all major domestic supply technologies available at the margin (with the exception of direct coal) are relatively capital-intensive, the life-cycle and even the capital costs of soft and transitional technologies, viewed in traditional terms (cost per unit of energy or power), are arguably attractive compared with those of complete hard-technology systems.

4) To make such comparisons specific, one must examine end-uses one by one.

To take up this last point, consider the task of supplying heat at modest temperatures, shown by Table 2 to be the dominant term (34 percent) in U.S. end use, and typically over half of European end use. Let us suppose that my estimate of $50,000 to $70,000/daily barrel for completely solar space heating of an existing building in the mid-1980's (assumed to be very well insulated, which is worthwhile no

* *(As in Adam Smith's "Invisible Hand.")*

matter how it is heated), using relatively costly technologies within the present art, is far too low and that a Danish figure, say of the order of \$100,000/daily barrel, is in fact more appropriate for the United States. Let us also suppose that the long-run alternative is, say, an LWR–powered grid (say, \$213,000/daily barrel), operating a heat pump (say \$200/kW) with COP = 2.5. The whole-system marginal capital cost of this alternative is thus

$$\frac{\$213{,}000/\text{daily barrel}}{2.5} + \frac{(\$200/\text{kW}) \times (67.1\ \text{kW(t)/daily barrel}}{2.5\ \text{kW(t)/kW}} = 90{,}700/\text{daily barrel}$$

suggesting that the solar system will certainly win on life-cycle cost and may even win on initial capital cost. (I think the solar system will also have lower life-cycle cost than an equivalent synthetic-fuel-and-furnace system.)

Numerical quibbles apart, several methodological objections might be raised to this calculation. First, it takes no account of user diversity. Second, it takes no account of the likelihood that part of the winter peak load will in fact be met by lower-capital-cost generation such as gas turbines. Both these points were answered earlier. Third, it does not consider the attractions of, say, an ACES[49] system with the potential for higher effective COP. True; but neither does it come close to a true and full accounting of the nuclear costs, even of the conventional internal costs, nor does it take account of the likelihood that urban and even suburban solar retrofits will in fact be technically simpler than assumed here, or on a neighborhood scale with substantially lower unit costs, or both.

This sort of argument is complex and perhaps can't be resolved. It does, however, leave me with the impression that soft-system economics leave a strong prima facie case for proponents of hard technologies to answer.* But there are broader perspectives in which to view this comparison. For example:

* *To say nothing of the economic comparison between electricity and direct fuel use. For example, if fuel oil in the mid-1980's cost, say, \$30/barrel (1984 dollars), while nuclear electricity cost 5 cents/kWh (1984 dollars) at the busbar and only 8 cents/kWh delivered, then an existing oil furnace of 67 percent First Law efficiency will compete with an existing heat pump with COP* $\leqslant$ *2.87. With capital costs not treated as sunk, the limiting COP would probably rise. It is no wonder that Table 1 shows no electrical water heating or cooking or space heating, and only partial electrical air conditioning, in the cost-conscious commercial sector, despite electrical rate structures favoring large users.*

1) First Law efficiency. I do not mean by this the First Law efficiency of a soft technology like a solar heating system; it is not an interesting criterion, not only for the reasons adduced by Shurcliff,[52] but also because what is "lost" (not converted from flux to function) is renewable, not depletable,* and is either desirable for albedo compensation of (in other soft systems) usable locally as low-grade heat because of small unit scale. I refer rather to the low net First Law efficiency of the electric grid and its fuel cycle—about 0.25 in Britain, for example. A great deal of high-grade fuel is thus used to produce a small amount of delivered enthalpy. The delivered energy is indeed of low entropy; but to compensate for the low First Law efficiency of the grid one would need the equivalent of a reciprocal COP (say, about 3 to 4) in each electrical application in order not, in a real and irreversible sense, to waste precious fuel compared with the First Law efficiency of using the same fuel directly. Analogously, in what significant applications does *electricity* have a high enough First Law efficiency to compensate for its-order-of-magnitude penalty in capital intensity relative to competing systems?

2) Second Law efficiency.[57] This is a more subtle and useful criterion, and makes the same argument emerge even more strongly. Recall that the Second Law efficiency of a typical U.S. domestic gas furnace system in winter is of the order of 0.05, that of a typical New York City office air conditioning system of the order of 0.02, that of a car of the order of 0.12.[15,57] Plainly any process that uses a flame temperature of kilodegrees or a nuclear temperature of teradegrees to produce a steam temperature of hectodegrees to drive turbines and make electricity of very high quality, all to make an end-use temperature change of dekadegrees, is going to have an abysmal Second Law efficiency that heat pumps cannot greatly alter. Worse, it is inelegant—a cardinal sin to a physicist. Natural energy fluxes, on the other hand, are of relatively low thermodynamic potential when used diffusely. The 5,800°K potential of concentrated sunlight, the kilodegree potential of fixed carbon, and the infinite effective temperature of wind-derived mechanical work need only be used for applications requiring them, so preserving high Second Law efficiency through end-use matching.[1]

* *Thus the definition of "primary energy" must shift for soft technologies to show the output of the device, not the gross input of which only some is converted. This approach is used in Figure 2 of reference 1.*

3) Sociopolitical implications. This crucial perspective—perhaps the most important—is considered elsewhere[1] and in the concluding section of this paper.

Are there not countervailing arguments? Of course there are, notably the convenience with which a utility can control electricity and a user use it. I do not know how to measure convenience; nor am I persuaded that it is as important a criterion as some have proposed, especially at the margin. I have not found that gas mantle lights are significantly more trouble to use and maintain than electric lights—in some respects they are more pleasant*—nor that machinery driven by compressed air is in any sense less desirable than electrical machinery. Convenience, however one defines it, can also be dearly bought, and I think many people are getting to the point where they would rather have the trouble (or creative task) of drawing the curtains at night, putting up storm windows, and even adjusting solar plumbing each season than of paying an exorbitant utility bill and living next to a reactor. Badly designed soft technologies can be time consuming; but irrational utility systems can be at least as time consuming in terms of the amount of alienating work that must be done to earn the money to pay the utility bill. Some people think that all soft technologies must be fiddly and entail onerous personal involvement. But to the extent that this is historically true, it is largely because most existing soft hardware is the work of enthusiastic amateur tinkerers who enjoy understanding their own systems and increasing their independence, rather as some people enjoy camping or messing about in boats. I have never seen a respectable engineering argument why soft technologies cannot be as "hands-off" as the domestic systems that we have—and have to fix—today, or even more so. We do, after all, have some solar heating systems that have operated with zero maintenance for 30 years. Likewise, inherently benign soft energy systems ought not to give rise to the hard-to-manage systemic environmental impacts and public nuisances that inherently dirty decentralized technologies like cars have done. The management and maintenance issues are real, and I must not be understood to say that the bad engineering that has plagued heat pumps, reactors, etc., cannot occur in simpler systems. (Many episodes of corroded

* *I do not know whether they are more dangerous. Electric wiring is the number two cause (after smoking) of the approximately 12,000 annual U.S. deaths in 3 million fires.*

solar collectors come to mind.) But I think the impact of soft technologies on personal time and care has been much exaggerated. Generally we would need only to return to the standard of engineering and quality control characteristic of many domestic appliances in the 1930's and 1940's.

Is this argument about the marginal—not historical—convenience and appropriateness of pervasive electrification academic? I think not. Major policy decisions such as whether to proceed with the liquid metal fast breeder reactor program are proposed today on the basis that, say, U.S. electricity "demand" will increase 15-fold, or at least a "conservative" 7.5-fold, by 2020—for what purposes is not made clear. On current ERDA projections, late-1975 U.S. nuclear capacity is to increase 11- to 20-fold by 2000. A cogent new paper,[32] however, argues that on the contrary, no conventional economic or engineering grounds could conceivably justify a nuclear capacity increase above, say, about 6-fold (250 GW) by 2000, and that ERDA projections showing the contrary are spherically senseless. Indeed, it appears[32] that cogeneration in 2000 could provide electricity conservatively equivalent to over two-thirds of the total 1975 electrical demand, directly displacing over 200 GW of projected nuclear capacity.

If existing centralized systems do not now make economic and engineering sense, why were they built? There are at least four rational explanations. First, because objective conditions have changed drastically,[58] and an industry not noted for quick and imaginative responses has been slow to adapt. Second, because centralized energy systems have been built by institutions in no position to ask whether those systems are the best way to perform particular end-use functions—an omission reinforced by our failure to price fuels at long-run marginal cost. Third, at times we have seen powerful institutions deliberately seeking to reinforce their power by constricting consumer choice, as in the classic monopoly tactics of the early electric utilities[59] or the fight against public power and (abroad) private wind machines. Fourth, the long economic shadow cast by large sunk costs has often led us to seek to reinforce past mistakes through subsidies, bailouts, $100-billion slush funds, etc., thus further restricting consumer choice,[60] rather than writing off (or gradually retiring through attrition) ill-conceived infrastructure. Energy decisions are always implemented gradually and incrementally; major shifts take decades. A chief element of strategy, inherent in the soft path,[1] is thus to avoid incremental commitments of resources to major infrastructure that locks us into particular supply patterns for more decades thereafter. We are already stuck with gigantic infrastructure that constrains our choices, and nobody is suggesting we wipe the slate clean. The question is rather what we do at the margin. What made sense when real costs of electricity (both average and marginal) were steadily falling,[58] may need to be reversed when real costs are rapidly rising with no end in sight.

To avoid misinterpretation of my views on electrification, I must make it clear that I am almost—not quite—as unimpressed by the virtues of central-station solar electricity as of central-station fossil or nuclear electricity. Both, in my view, are socially (and economically) unattractive.[1] Some analysts believe, probably on good grounds, that we may well have a large solar-thermal-electric scheme operating in the late 1980's at a cost roughly comparable to the cost of nuclear power at that time (or, according to estimates of the International Institute for Applied Systems Analysis, cheaper than oil-electric power in Central Europe). Such systems would have land requirements of the same order as those of equivalent coal-electric, coal-synthetic, or nuclear-electric fuel cycles,[61] but, unlike them, could coexist with some other land-uses and need cause no really serious scars or impacts.[62] What is not clear to me is why one should want such a system—or, for that matter, equivalent central stations based on the cheap photovoltaic cells (with or without optical concentrators) that seem likely to arrive much sooner. The justification for having cheap photovoltaics would be if they can be used in decentralized fashion—e.g., on houses (either using AC inverters or, preferably, adapting end uses to low-voltage DC). In such circumstances, I would welcome their use where it is thermodynamically advantageous—a criterion subject to both physical and economic tests—or, perhaps, even where it isn't, as that is not the only criterion, and we have already noted that economic rationality is too narrow a test. Dispersed cheap photovoltaics—especially if they can be made cheap relative to direct fuels, not just to central-station electricity—would certainly be useful and should be pursued. I still believe, however, that *no* solar electric technology is necessary to our national energy supply if we take full advantage of the diversity of soft technologies already demonstrated.[1] Interest in solar electricity is unfortunately reinforced by ERDA's arbitrary and inexplicable 1975 decision that only electrified technologies will be considered capable of making a major long-term contribution.[63]

The illogic of the ERDA position is this: If we are running out of oil and gas but do not like coal, it is said, we need nuclear power; but if we are not going to have nuclear power, we need other systems that would do what nuclear stations would have done—namely, deliver gigawatt blocks of electricity. But we should instead be seeking systems that will do what we would have done with the oil and gas if we had had them in the first place. It is the function that interests us, not substituting for reactors. By not structuring the problem in this way, ERDA has so far failed to grasp the immense short- and medium-term opportunities for deploying available technologies for end-use efficiency, cogeneration, fluidized-bed boiler backfits, organic conversion, and extensive solar space heating. The longer this delay, the worse will be our shortages of clean heat and fluid fuels.[1]

Sociopolitics

Section IX of reference 1—the pivotal section of that paper—argues that a hard path inevitably entails structural changes with high political costs: that it nurtures dirigiste autarchy, by-passes traditional market mechanisms, concentrates political and economic power, centralizes human settlements, persistently distorts political structures and social priorities, encourages bureaucratization[64] and alienation, compromises professional ethics, is probably inimical to greater distributional equity within and among nations, inequitably divorces costs from benefits, enhances, vulnerability and the paramilitarization of civilian life, introduces major economic and social risks, reinforces current trends toward centrifugal politics and the decline of federalism, and encourages—even requires—elitist technocracy whose exercise erodes the legitimacy of democratic government. Several of these points deserve amplification or qualification.

In a soft energy path, the technological measure to be achieved can be readily separated from the policy instrument used to encourage it. The former—cogeneration, bioconversion, insulation—is per se relatively neutral, the latter politically charged. It is the latter only that is likely to irritate us if ill-conceived. But I believe the policy tool can be chosen, according to practical and ideological convenience, from such an enormous armamentarium that the choice can fully respect pluralism and voluntarism.

I do not see how the same pluralism can possibly extend to a hard, coarse-grained energy path. The scale and the technical difficulty of its enterprises are so vast that corresponding concentrations of social resources must be efficiently mobilized without substantive regard to diverse opinions and circumstances. Only large corporations, encouraged by large government agencies, using large sums of private and public money to employ large numbers of workers on large areas of land, can possibly get the job done. It is not a task for householders, small businesses, block associations, or town meetings.

Soft technologies are thus inherently, structurally less coercive and more participatory than hard technologies. In a nuclear society, nobody can opt out of nuclear risk. In an electrified society, everyone's life style is shaped by the economic imperatives of the energy system, and, from the viewpoint of the consumer, diversity becomes a vanishing luxury. Like purchasers of model T Fords, the consumer can have anything he wants so long as it's electrified. But in a soft path, each person can choose his own risk-benefit balance and his own energy systems to match his own of caution and involvement. People who do not care to partake of the advantages of district heating will be free to reject them—and, if the system is thoughtfully designed, to change their minds later. People who want to drive big cars or inhabit uninsulated houses will be free to do so—and to pay the social costs. People can choose to live in city centers, remote countryside, or in between, without being told their life style is uneconomic. People can choose to minimize what Robert Socolow calls their "consumer humiliation"—their forced dependence on systems they cannot understand, control, diagnose, repair, or modify—or can continue to depend on traditional utilities, for large grids are already with us and in some degree will persist for a long time. In a soft path, then, dissent and diversity are not just a futile gesture but a basis for political action and a spur to private enterprise. But the monolithic nature, gargantuan scale, exacting requirements, and homogenizing infrastructure of hard technologies does not offer such pluralism. Only our largest conglomerations of resources, shielded by the powers of the state from the vagaries of the economic and political marketplace, can perform such demanding tasks. Analogously, on an international level the risks of the hard path, such as accumulation of carbon dioxide[65] or misuse of strategic materials (see Appendix III), are so formidable that only supranational coercion can reduce them, so abridging national sovereignty and freedom of choice. (People who consider this a good idea in the abstract are often reluctant for it to happen to them at a national or personal level.)

One could argue forever whether the structure of the energy system is a cause, effect, or concomitant of social structure.[66] Obviously, other influences on social structure, such as information flows, water flows, and land tenure, are important (though often related to the energy system). The skein of causation is thoroughly tangled, and I do not wish to appear simplistic. But it seems undeniable that, for example, energy decisions can and do affect the spatial distribution of jobs, hence of settlements,* hence of political power. That is, energy decisions, not only at the point of resource extraction (e.g., the stripping of coal in Wyoming's Powder River basin) but at the point of conversion and use, are unavoidably land-use and regional decisions, just as the railroads once shaped our land-use patterns. One could at least argue, too, that the extreme capital intensity of new energy systems requires end use to be nearly optimized in an economic sense, so that its nature and patterns conform to the needs of the source of supply rather than the other way around. (I favor peak load pricing, for example, and possibly such measures as ripple-signal load control for some applications; but one can hardly pretend that load-management techniques[18] are not intended to alter the patterns of use and in some sense—one of which undersized solar water heaters are often accused—to alter the consumer's life style.)

Centralized energy systems are also inequitable in principle because they separate the energy output from its side-effects, allocating them to different people at opposite ends of the transmission lines, pipelines, or rail lines. The export of these side-effects from Los Angeles and New York to Navajo country, Appalachia, Wyoming, and the Brooks Range (not to mention Venezuela, the Caribbean, Kuwait, and British Columbia) makes the former more habitable and the latter more resentful. That resentment is finding political expression. As the weakest groups in society, such as the native peoples, come to appear to stronger groups as miners' canaries whose fate foretells their own, sympathy for the recipients of the exported side-effects grows.

Throughout the world, central government is trying to promote expansionist energy policies by preempting regula-

* *Nietzsche remarks that cities that have not decided what they want to be tend to grow to unnatural size. Rome in its glory was slight compared to Athens, for in Rome the energy that should have gone into the flower went instead to make the stem and leaves, and hence was of no consequence. It is a sobering perspective for modern planners who move industrial complexes around on a map like chess-pieces on a board.*

tory authority, and in the process is eliciting a strong State (or Provincial) and local response. Washington, Ottawa, Bonn, Paris, and Canberra are coming to be viewed locally as the common enemy. Unholy alliances form. Perhaps Montana might mutter to Massachusetts, "We won't oil your beaches if you won't strip our coal." As Congress—made of State people with no Federal constituency—increasingly molds interregional conflict into a common States'-rights front, decisions gravitate by default to the lower political levels at which consensus is still possible. At those levels, further insults to local autonomy by remote utilities, oil companies, banks, and Federal agencies are intolerable. Thus people in Washington sit drawing reactors and coal complexes on maps, but the exercise has increasingly an air of unreality because it is overtaken by political events at the grassroots. The greater the Federal preemption (as in offshore oil leasing), the greater the homeostatic State response. The more the Federal authorities treat centrifugal politics as a public relations problem, the more likely it becomes that they will not only fail to get their facilities sited, but will also in the process destroy their own legitimacy. To some extent this has already occurred, and I have no doubt that States will soon gain a veto power, at least, over nuclear facilities in their jurisdiction (as pending Federal legislation proposes). On this issue, as in other spheres, the traditional linear right-left political spectrum seems to become cyclic as differently grounded distastes for big government merge across gaps of rhetoric. The resurgence of individual, decentralized citizen effort in politics, as in private life and career, seems to me an important political universal in most industrial nations today.[67]

Big Brother does not like losing his grip. Only last year, for example, some Federal officials were speculating that they might have to seek central regulation of domestic solar technologies, lest mass defections from utility grids damage utility cash-flows and the State and municipal budgets dependent on utility tax revenues. Since utilities are perceived as having too much power and utility regulators too little sensitivity already, a surer recipe for grassroots revolt would be hard to imagine. I think perceptions of the value of dependence on utilities are shifting rapidly as the enterprise reaches such a size that it starts to intrude on life in many traditionally "safe" areas, as in Ontario, or as its vulnerability becomes painfully manifest, as in England, or as general political consciousness rises in step with utility bills. The disillusionment and resentment I see in many industrial countries are akin, perhaps, to that of a citizen of a poor country who is realizing that an energy technology predicted to bring him self-reliance, pride, and the development of his village has actually brought him dependence, a cargo-cult mentality, and the enrichment of urban elites. I believe the recent shift of institutional and individual investment away from utilities reflects not only concern with debt structure and interest coverage but also, more fundamentally, with gradual withdrawal of legitimacy by a fickle public that has already done the same to major oil companies. I believe further that the grounds of this shift among a previously tolerant, even supportive, public are structural, arise essentially from suspicion of centrism, and would not be reversed by nationalization or rechartering that ignored scale.

Living in England, I am keenly aware of the vulnerability of an energy system that permeates the end-use structure, distributing a form of energy that cannot readily be stored in bulk. Its supply relies on hundreds of large and precise machines rotating in exact synchrony across a continent, strung together by a vast web of frail arteries that can be severed with a rifle. Disruption can be instantaneous and pronounced, so much so that English users have gone to very costly lengths to protect themselves, reportedly installing more private than public capacity in some recent years. The vulnerability of the grid is enhanced not only by the guerrilla activity[68] current in Britain but also by the capital intensity of the grid. In the United States, a power station requires roughly a quarter of a million dollars' investment per direct job. Correspondingly low labor intensity and technical specialization mean that political power resides in fewer and fewer workers whose highly developed skills may be matched by coherent purpose and organization. Most power engineers appear to have an admirable sense of dedication to the public good. In Britain, however, a leader of the power engineers remarked recently—and correctly—that "the miners brought the country to its knees in 8 weeks; we could do it in 8 minutes." His union, before starting a recent round of wage negotiations, even proposed a 24-hour initial strike as a token of sincerity. In this sphere as in the military sphere, the nature of a technology has fundamentally altered the power balance between large and small groups,[68] making it less tolerable for potentially hostile or abnormal people to enjoy traditional civil liberties, and generally increasing the level of suspicion and intolerance in society. "Power to the people" now makes us ask:[60] "Power to *which* people?"

Such political costs, prominent in the list of nuclear power issues, correspond to analogous costs of possessing military nuclear weapons: indeed, the latter costs may have helped to sensitize us subliminally to the former. For example, nuclear weapons demand an elite, self-perpetuating cadre that will meticulously and perpetually guard and prepare the weapons, isolated alike from social unrest and from proximate (effective) political accountability, rather like Weinberg's concept of a "technological priesthood" for guarding long-lived nuclear wastes.[3] Nuclear weapons demand a fast-reacting central authority that can make decisive and timely military responses without the niceties of balanced democratic consultation, just as nuclear power tends to weaken political responsibility through centrism. Further, the power of nuclear weapons requires that willingness to use them be periodically demonstrated obliquely—by fighting small nonnuclear wars as theater,[70] not to attain classical military objectives so much as to present an image of united national will. This image may require public dissent at home to be suppressed, or prevented by deception,[70] and thus contributes both to loss of public candor and trust and to the evolution of an executive siege mentality—events with an obvious parallel in the civilian sector. In several fundamental ways, then, the military imperatives of nuclear weapons conflict with the political imperatives of a free society. I suspect that latent public unease with this conflict is an important indirect element in the political climate[71] in which opposition to official siting proposals can flourish.

Akin to this problem, but less visible in public, is the array of pressures on individual high technologists.[1] Intense social pressures within committed institutions tend to discourage deviance from shared values and beliefs, reinforcing them with apparent consensus (deserved or not) even at the expense of personal ethics. It is well known that people in groups, even with high intentions, do things they would never dream of doing individually; and our professions have not yet devised ways to shield the delicate spirit of inquiry and dissent from subtle (or even not so subtle) peer pressures. Even at Oak Ridge National Laboratory I know of instances where the public and private views of fine technologists perforce must differ, so corroding both personal and technical quality. Shifting one's career to a field that one can have good dreams about may not be an attractive option: so long as Federal R&D interest is perceived to be mainly in hard technologies, good technologists will be reluctant to shift to soft ones, thus limiting the long-term ability of those sectors to absorb funding effectively and further reducing their attractions. This is a practical demonstration of the exclusivity of paths.[1]

The big organizations needed to develop big technologies suffer often from fuzzing of responsibility. So many people contribute to a project that nobody is really responsible,[72] and responsibility in the traditional engineering sense may slip through the cracks. But more destructive to technical quality is the tendency of such big organizations, after the exciting surge of initial pioneering, to become routine and deadly serious. For that reason alone—to say nothing of public controversy—fission is now no fun any more; therefore it will not be done well; therefore it has failed.* A subtle but important disadvantage of big technologies (as Freeman Dyson points out) is that they are so big that one cannot play with them, so an essential breath of both fun and creativity is lost. In contrast, soft technologies bring out the ingenious tinkerer, the spirit that made the sophisticated (but understandable) gadgets that fill any good farm museum. The challenge of simplicity, the art of artlessness, have full scope. Human scale relative to the technologist himself lets ingenuity—the source of paper solar collectors, heat cables,[73] and exterior retrofit insulation[74]—expand far beyond the narrowly focused limits attainable in hard technologies.

A further institutional problem from which soft technologies are nearly exempt is control of potentially hazardous developments. Soft technologies are by definition nonviolent. Hard technologies are often Siamese twins of weapons (see e.g., Appendix III, or consider solar-satellite schemes or laser fusion). Worse, subjecting civilian or quasicivilian hard technologies to the necessary political control may require, for informed discussion, the release of information that endangers the public. For example, to provide a basis for intelligent decision-making about LMFBR disassembly accidents, one must at the same time reveal a good deal about how to design atomic bombs. In order not to be faced with the unpalatable choice between endangering the public by releasing information useful to the malicious and, on the other hand, ceding decision-making on a crucial matter to an elite[75] and so sacrificing democratic principles, one may have to head off such technologies as soon as such a choice

* *One can likewise argue that really well-engineered nuclear facilities will be so boring to operate that talented people will seek other careers. Some observers see this as a major long-term problem at facilities like the reprocessing plant being built at Barnwell, South Carolina.*

appears on the horizon. To do this effectively, one may be forced to try to suppress lines of research that are still in the realm more of science than of technology. This in turn raises unpalatable issues of liberty vs license in academic freedom: a controversy now raging in genetic research, which is in some ways analogous to the state of nuclear science in the 1940's. Such encroachment is less likely to be considered or needed at all if we channel our efforts into less exploitative, more adaptive, gentler technologies.*

I have sketched a few examples of the labyrinthine complexity with which energy policy questions (to say nothing of their consequences) ramify into every aspect of our lives. The policy questions that the soft path makes so prominent raise also some more basic structural and philosophical questions that we can neither answer definitively nor ignore: questions of the personal values within which, and the social ends for which, energy is sought. My earlier sketch[1] of a few of these questions--economic growth and its diseconomies, distributional equity, transindustrial values—may need expansion here.

A cursory glance at a graph of U.S. per-capita primary energy since, say, 1880 dispels any notion of a significant correlation with social welfare on a time scale of decades. The graph remains flat for several long periods when welfare by almost any measure improved significantly, then rises steeply during the 1960's out of all proportion to real improvements. Nonetheless the existence of a close correlation--even an ironclad causality—is an article of faith to many who measure our success by the quantity of goods and services consumed, rather than by how far we achieve human satisfaction, joy, and inward growth with a minimum of consumption (the concept central to Schumacher's classic essay[12] on Buddhist economics). The former view is consistent with the industrial-era paradigm in which, as Mumford caustically observes, [76]

> All but one of [the seven deadly] . . . sins, sloth, was transformed into a positive virtue. Greed, avarice, envy, gluttony, luxury, and pride were the driving forces of the new economy . . . Goals and ends capable of working an inner transformation were obsolete: mechanical expansion itself had become the supreme goal.

* *A simple example from E. Robertson: There are three ways to make limestone into a structural material. We can cut it in blocks, or we can bake it at thousands of degrees into Portland cement. Or—most elegantly—we can feed chips of it to a chicken. Twelve hours later it emerges as eggshell, several times as strong as the best Portland cement. Evidently the chicken knows something we don't about ambient-temperature technology.*

As a result of this process, as Diesing notes,[77]

> One cultural element after another has been absorbed into the ever-widening economy, subjected to the test of economic rationality, rationalized, and turned into a commodity or factor of production. So pervasive has this process been that it now seems that anything can be thought of as a commodity and its value measured by a price . . . time, land, capital, labor; also personality itself, . . . art objects, ideas, experiences, enjoyment itself, and even social relations. As these become commodities they are all subject to a process of moral neutralization.

The very language that weaves our thoughts about the world is thus conditioned by an economic paradigm that treats means as ends[78] and places goods above people.[12] But the cult of material acquisitiveness underlying our assiduously promoted image of the good life has its flaws, and we are coming to see them more clearly.

For example, we concentrate our resources not on microefficiency or resilience but on managing our growth and resolving its numerous conflicts and inequities. So we weave a web of bigness and of incomprehensible, unmanageable complexity. But are we sure that our transaction costs do not exceed our productivity? Is it worthwhile, or possible, to go on like this? Miles states a related thesis thus:[79]

> (1) The more energy a society uses, the more interdependent it tends to become, both within itself and in relation to other societies.
>
> (2) The more interdependent a society becomes, the more complex it becomes, and the more man-designed and man-controlled its economic, ecologic, and political systems and subsystems become.
>
> (3) The more complex and interdependent the systems and subsystems, the more vulnerable they become to design failures, since:
>
> (a) No human designers, and this applies especially to the politicians who are responsible for designing the largest human systems, can know or comprehend all the factors that need to be taken into account, and their interrelation, sufficiently to make the current set of systems work well. If complexity and interdependence increase further, the problems will be further compounded.
>
> (b) Those responsible for selecting the designers—the voting public in a society like the United States—are even less informed about the intricacies of the systems than the politicians who represent them. They cannot judge, therefore, which programs or social designers (politicians) to support,

and in consequence they are highly likely to vote for the representatives who promise to support programs that benefit them directly and immediately—a fatal flaw in designing workable complex systems for interrelating enormous numbers of human beings with each other and their environments.

(4) The United States is probably nearing the point (it could even be beyond it) where the complexity of the systems of interdependence exceeds the human capacity to manage them, causing system breakdowns to occur as fast as, or faster than, any combination of problem-solvers can overcome them.

(5) World systems of interdependence are more remote, inefficient, and precarious than national systems, and may have exceeded their sustainable level of complexity. More and more nations are looking for ways of becoming more independent, rather than more interdependent, even if to do so they will have to be satisfied with a lower standard of living. Those social analysts who have asserted that we will inexorably continue to move upward toward higher levels of technology, higher levels of energy consumption, greater volume of international exchange of raw materials, goods, information, culture, and tourists may be in for a surprise. Humanly designed and operated systems have upper limits of complexity, and when they reach those limits of complexity, they simply break down.

Here is another conundrum. Assuming that people seek to escape from repugnant work into blissful leisure, we strove mightily to mechanize, automate, and fragment work. But while that effort at first relieved mindless drudgery, it soon came to deprive people of meaningful roles, even of jobs themselves. As the craftsperson working creatively with tools was displaced by the machine demanding a stultified operator, people, especially those whom age or sex or ability ill-suited for economic roles, were deprived of a share in a visible and widely shared public purpose. We systematically substituted money and energy for people. We redefined work as obtaining a commodity (a job) from a vendor (an employer), substituting earning for an older ethic of serving and caring. Thus alienation in place of fulfillment, inner poverty alongside outward affluence, a pathologically restless and rootless mobility, became the symptoms of a morbid social condition that corroded humane values. Our brilliant success in achieving economic goals has thwarted the human goals of much of our population by declaring their lives to lack value. Might not happy people (in D. T. Suzuki's phrase) instead enjoy life as it is lived rather than trying to turn it into a means of accomplishing something else? Our sense of the whole, in personality and family as in community and country, has been sacrificed to specialization and mobility, the imperatives of efficient production. But is that price worth paying? Soft-technology life near enough to the soil to understand the life process has nurtured every previous culture,[80] and for all its hardships in the past—many of them now avoidable—it has a cultural value. Are we sure that by rejecting that awareness of the life process and turning ourselves into mere cogs of mechanical production we are not losing something essential to the human psyche and to our own mythic coherence?

Again, if we work to buy a car without which we cannot get to work, is the *net* benefit so very substantial? Is not this circularity a measure less of our wealth than of our failure to create fulfillment with an elegant economy of effort and time? Illich[13] cites a number which, whether it is quite correct or not, still conveys an important idea: He says that the average American man drives about 7,500 miles a year in his car, but that the total time it takes him to do this and to earn the money to finance it is about 1,500 hours, which works out to 5 miles an hour, and we can walk nearly that fast.

Again, 3 billion people offer living refutation of a theory that assured us that our growing wealth would automatically enrich the poor without our having to redistribute anything. We are waking up: In the famous Harris poll of late 1975, a solid 61 percent felt it is "morally wrong" for Americans to consume such a disproportionate share of the world's resources, and 50 percent worried that a continuation of such behavior will "turn the rest of the world against us." Yet the theory persists: Appeals for more electricity with which to help the poor still come from utilities that have long charged the poor several times as much as the rich (or profligate) for the same unit of electricity.* Yet we

* *Here equity and narrow economic arguments[27] may conflict. The broader question of energy price is vexing. Unlike many colleagues, I do not consider energy prices an appropriate instrument of distributional equity. Energy should be properly priced, and if poor people cannot then afford it, they should be made less poor in ways that do not entail giving cheap energy to rich people. Of course, we may wish to direct to good social purpose the windfall profits that higher prices may otherwise give to oiligopolies; but that is a separable problem that should not be beyond our wit to solve.*

can now calculate that raising, say, the poorest three-fifths of Americans to the per-capita energy level of the rest would not require massive energy growth, but rather a level well below any current official projections. Even more significantly, we can now see that reducing ourselves to the "primitive" energy level of most Europeans need not reduce our material comforts at all and could markedly improve the quality of life.[15]

As we learn to question the ability of present policy to serve both public and private ends, the legitimacy of those ends themselves comes up for review. Our know-how has far outstripped our know-why; and as we seek to reverse the balance, old political concepts begin to reassert themselves. Grassroots democracy acquires[67] a more concrete meaning. Jefferson and Mao gain a curious affinity. Control of property and land—a cornerstone of free-enterprise democracy—comes to embrace control of the energies essential to life, liberty, and the pursuit of happiness, for to control those energies we must now control the land they lie under or fall upon. In the process we may start to approach Aldo Leopold's land ethic, or even the native American concept* that absolute, monopolistic ownership of land, at first incomprehensible, is in a sense blasphemous.

In our dynamic, pluralistic culture, our decisions will be made at the points of tension between competing value systems. Some of these tensions have long been with us but are now taking on a new significance. For example, for many historical reasons (including the forces that selected those who emigrated here), it is a basic tenet of American social philosophy that economic and social activity are primarily to serve private and individual, not public and communitarian, ends. Public ends are indeed conceived not as abstract moral universals, such as making a good or a just society, but rather as a pragmatic summation of diverse private ends. This value structure, this very individualism that has enabled us to accomplish so much materially, has nurtured the primacy of economic and technological values over spiritual and humanistic ones, and has long kept us from coming to grips with our own Tragedies of the Commons (Garrett Hardin).

* *This is not a new theme. "Moreover the profit of the earth is for all: the king himself is served by the field. He that loveth silver shall not be satisfied with silver; nor he that loveth abundance with increase: this is also vanity. When goods increase, they are increased that eat them: and what good is there to the owners thereof, saving the beholding of them with their eyes? The sleep of a laboring man is sweet, whether he eat little or much: but the abundance of the rich will not suffer him to sleep."—Ecclesiastes 5:9-12.*

As "economic growth and technical achievement, the greatest triumphs of our epoch of history, [show] . . . themselves to be inadequate sources for collective contentment and hope" (Robert Heilbroner), we are starting to appreciate that we have no monopoly on wisdom. If we can suspend our cultural arrogance, we may learn much from other peoples with different values. We may, for example, be surprised and much moved to hear the chiefs of the Micronesian island of Maap, threatened by a Japanese hotel complex, petition us thus:[82]

> Whereas we love our land and the ways in which we live together there in peace, and yet live humbly and still cherish them above all other ways, and are not discontent to be the children of our fathers, it has become apparent to us that we have been persuaded to subscribe to processes that will quickly extinguish all that we hold most dear. We, pilungs and langanpagels, elders and elected officers, Chiefs in Council of and on behalf of all the people of the Eighteen Villages and Fiefs of the Island of Maap . . . declare our love of this place and of the ways passed down to us by the generations. We have inherited from our fathers a land that is lovely and provides for us the fruits of the earth and of the sea. We are few in numbers but have a brave history and are strong in our resolve to preserve these things that are sweet to us and freely to determine the affairs of our island with respect to custom and deference to the law . . . We . . . now urgently and passionately unite . . . to ask the help of . . . [others] in ridding us of this invasion and freeing us, that we do not become servants in our own land, to choose for ourselves the paths that will be good for the people of all the villages of Maap. [W]e meet today under the shadow of . . . change and innovation and . . . we know our people to be roused against these things, as they today do forcefully convey through a petition . . . [W]e are therefore all the more solemnly moved to affirm our united will in the face of the unfamiliar contingencies of this age and the ages to come, so that our home may not be vulnerable to the casual invasions of those who do not know our hearts or the disloyal speculations of those who do.

Such language, so reminiscent of our own national beginnings, must remind us of the power of ideas—republican government in monarchical Europe, or Christianity in Rome—to transform societies with extraordinary speed. The thesis that we stand today on the verge of such a transformation must command attention,[5] and accounts for much of our present failure of public discourse. As Willis Harman[72] points out, different people with similar perceptions of reality may have different interests, and our society has mechanisms for resolving conflicting interests, but we have no mechanisms for reconciling different perceptions of reality, or even for recognizing their divergence as a basis of conflict and misunderstanding. It is perhaps encouraging, then, that the concept of a soft energy path brings a broad convergence that, even as it coincides with many preexisting strands of social change, cuts across traditional lines of political conflict. It offers a potential argument for every constituency: civil rights for liberals, States' rights for conservatives, availability of capital for business people, environmental protection for conservationists, old values for the old, new values for the young, exciting technologies for the secular, spiritual rebirth for the religious. As we realize, in Alwyn Rees's phrase, that when we have come to the edge of an abyss, the only progressive move we can make is to step backwards, we begin to see that we can instead turn around and then step forwards, and that the turning around—the transition to a future unlike anything we have ever known—will be supremely interesting, an unprecedented central project for our species.

Faust, having made a bad bargain by not reading the fine print, brought disaster on innocent bystanders (Gretchen's family). Still, he was eventually redeemed and accepted in heaven because he changed his career, redevoting his talents to bringing soft technologies to rural villagers. That choice of "the road less traveled by" made all the difference to him; and so it can to us, for underlying the structural differences between the soft and hard paths is a difference of perceptions[3] about mankind and his works. Some people, impressed and fascinated by technology's glittering achievements, say that if we will only have faith in human ingenuity (theirs) we shall witness the Second Coming of Prometheus (if we have yet recovered from the First), bringing us undreamed-of freedom and plenty. Other people say we should plan on something more modest, lest we find instead undreamed-of tyrannies and perils; and that even if we had a clean and unlimited energy source, we would lack the discipline to use it wisely. Such people are really saying, first, that energy is not enough to solve the ancient problems of the human spirit, and second, that the technologists who claim they can satisfy Alfven's condition that "no acts of God can be permitted" are guilty of hubris, the human sin of divine arrogance. In choosing our energy path we have an opportunity—perhaps our last—to foster in our society a greater humility, one that springs from an appreciation of the essential frailty of humankind.

ACKNOWLEDGMENTS

A synthetic paper, like a wall, is mostly bricks. I have added a bit of mortar and art to assemble both this paper and reference 1, but the bricks bear other names: notably James Benson, Amasa Bishop, David Brooks, Harvey Brooks, Clark Bullard, Alex Campbell, Monte Canfield, Jr., Meir Carasso, Henrik Casimir, Peter Chapman, Umberto Columbo, David Comey, Charles Correa, Paul Craig, Herman Daly, Ned Dearborn, Peter Dyne, Freeman Dyson, Bent Elbek, Duane Elgin, the late Douglas Elliott, Richard Falk, Bernard Feld, Ian Fells, Jay Forrester, Peter Glaser, Wolf Häfele, Bruce Hannon, Henrik Harboe, Jim Harding, Willis Harman, Denis Hayes, Hazel Henderson, John Holdren, Takao Hoshi, Ivan Illich, Barbara Jackson, David Jhirad, Ann and Henry Kendall, Vladimir Kollontai, Tjalling Koopmans, Lew Kowarski, Gerald Leach, Thomas Long, Måns Lönnroth, Arjun Makhijani, Sicco Mansholt, Bert McInnes, George McRobie, Dennis and Donella Meadows, Niels Meyer, Peter Middleton, Bruce Netschert, Kees Daey Ouwens, Walter Patterson, William Peden, Jerry Plunkett, Alan Poole, Frank Potter, Gareth Price, John Price, William Raup, James Robertson, Marc Ross, Charles Ryan, Ignacy Sachs, Lee Schipper, Walter Schroeder, Fritz Schumacher, Robert Socolow, Bent Sørensen, Arthur Squires, Theodore Taylor, John and Nancy Todd, Eric Jan Tuininga, Frank von Hippel, Alvin Weinberg, Jerome Weingart, Andy Wells, Robert Williams, and Carroll Wilson. These generous (and often inadvertent) contributors are in no way responsible for the use I have made of their bricks. Any loose, crumbly, or missing bricks have my name on them.

Above all I am grateful to Gerald Barney for suggesting these papers, to David Brower for inspiring them, and to William Bundy and Jennifer Seymour Whittaker for nurturing reference 1 far beyond the call of editorial duty. This paper was kindly commissioned by Oak Ridge Associated Universities, who cannot be held accountable for my views.

I must acknowledge in advance the helpful suggestions and criticisms that this exploratory paper will doubtless elicit from intrigued or indignant readers. My permanent forwarding address is c/o Friends of the Earth Ltd, 9 Poland St., London W1V 3DG, England. Lack of lead time has prevented me from putting this draft through my usual process of preliminary peer review, so I expect there will be a lot to improve.

Neither this paper nor reference 1 could have been written without the generous and tolerant hospitality of Ann and Henry Kendall and of the staff of the Appalachian Mountain Club's Pinkham Notch Camp. I am deeply grateful to them all.

APPENDIX I: Methodological Note

Present policy rests on the belief that the future is essentially the past writ large. Quantitative assumptions, such as rates of growth in primary energy demand or in GNP, are extrapolated by so many percent per year—i.e., inherently exponentially—changing according to historically observed patterns (elasticities, saturations, etc.) or not at all. Such models have trouble adapting to a world in which, for example, real electricity prices are rapidly rising rather than slowly falling as they used to.[58] More generally, such models have a certain inflexibility that tends to lock us into a single narrow vision of life styles and development patterns. Extrapolations have fixed structure and no limits, whereas real societies and their objectives evolve over decades—i.e., they have slow variables affecting model structure and objective function in a finite world where the key questions are those of choice and value within consecutively approached constraints. Extrapolations assume essentially a surprise-free future even when written by and for people who spend their working lives coping with surprises. Worst, extrapolations are remote from real policy questions. Decision-makers are seldom called upon to fix a growth rate at *X* or *Y* percent a year, but rather to apply pragmatic judgment to intricate decisions of detail. If the modeler assumes a single value, or only a few values, for each basic variable, the model may elicit a "my number is better than your number" response; if many values, the model unhelpfully offers the policy-maker enough options to keep him undecided indefinitely, so he will simply ignore it. What it does not offer him is a sense of the goals and criteria that would make some numbers preferable to others.

Having diagnosed these defects (set out in a 1974 occasional paper for the Aspen Institute for Humanistic Studies), Monte Canfield, Jr., and colleagues in the Energy Policy Project of the Ford Foundation developed several scenarios that are policy-oriented, descriptive (not prescriptive), exploratory, and adaptive. They are defined not by exogenous numerical assumptions but rather by policy goals and mental models presumed to prevail among governments and their constituencies. A reader can thus identify that scenario that most nearly reflects his own frame of mind, note its consequences, and gain a fuller understanding of the consequences of his own everyday choices. Scenarios are also easier than extrapolations to test for resilience in the face of surprises—important in an era when discontinuities and singularities may matter more than the fragments of secular trend in between.

This scenario technique is valuable and should be extended, just as extrapolation, in skilled hands, has its uses in short-term forward planning (especially when bolstered by multivariate sensitivity analysis) and has rightly had much talent devoted to its refinement. But I believe both techniques must be supplemented by a third method that highlights what can be done in the long term by recommitting resources in the short term. This method is the broad-brush "working-backwards" policy exploration exemplified by reference 1 and by the Canadian,[19] Danish,[3,10] and Japanese[3] studies it cites. Such an exercise should use a time frame long enough to make major supply changes and turn over major capital stocks: I normally use 50 years, about the lifetime of a power station ordered today. I also do not pretend to achieve econometric precision (which often seems spurious anyway), nor try to simulate markets and prices decades ahead, but instead estimate main terms in the spirit of experimental physics, striving not for elaborate sophistication but for transparent simplicity.[1,19] (See my discussion of fossil fuel costs for why I use capital intensity as an a posteriori signal rather than market equilibrium as an a priori mechanism.) I argue elsewhere[1] that working backwards, by turning divergences into convergences, provides insights unavailable to anyone working forwards. One can then work forwards, or both ways at once, to devise transitional tactics.

Studies of soft energy paths or of related topics, using methods akin to those of reference 1, are in progress (generally with very limited resources) in at least the following centers with which I am in touch via the main researchers shown:

USA: Union of Concerned Scientists, Cambridge, MA (Prof. Henry Kendall) Institute for Energy Analysis, Oak Ridge, TN (Dr. Alvin Weinberg) Center for Environmental Studies, Princeton University (Prof. Robert Williams) Energy & Resources Group, Univeristy of California, Berkeley (Prof. John Holdren)

Canada: Science Council of Canada, Ottawa (Dr. Ray Jackson) Ministry of Energy, Mines, & Resources, Ottawa (Dr. Peter Dyne) Energy Probe, University of Toronto

(Mr. Barry Spinner) Premier's Office, Charlottetown, P.E.I. (Mr. Andy Wells)

UK: Friends of the Earth Ltd., London (Mr. Walter Patterson) International Institute for Environment & Development, London (Mr. Gerald Leach) Energy Research Group, The Open University, Milton Keynes, Bucks. (Dr. Peter Chapman)

France: Faculté Sciences Economiques, University of Grenoble (Dr. J.-M. Martin)

Netherlands: A dispersed group (Dr. Eric-Jan Tuininga, TNO, Apeldoorn; and Dr. Kees Daey Ouwens, Rijksuniversiteit Utrecht)

Norway: Råd for Natur- og Miljøfag, Universitetet i Oslo (Dr. Paul Hofseth)

Denmark: Niels Bohr Institutet, København (Dr. Bent Sørensen)

EEC: Technical University of Denmark, Lyngby (Prof. Niels Meyer)

Sweden: Future Studies Group, Stockholm (Dr. Måns Lönnroth)

Austria: International Institute for Applied Systems Analysis (Dr. Wolf Häfele; also active in West Germany and other countries)

Israel: Weizmann Institute of Science (Dr. Tullio Sonnino)

New Zealand: University of Auckland (Dr. Robert Mann)

Finally, energy studies cannot hope to succeed in isolation. The energy problem is intimately related to all the other great issues of our day, and is arguably only a symptom of deeper social disorders that any energy strategist must address. Thus when I suggest[1] that "The most important, difficult, and neglected questions of energy strategy are not mainly technical or economic but rather social and ethical," I am not thinking only of social and ethical *energy* issues. Caldwell sums it up:[66] "The energy crisis is a natural product of the sociotechnical system that it now threatens."

APPENDIX II:
Temperature Spectrum of Process Heat

Nearly one-fourth of all U.S. primary energy is devoted to process heat in industry (Table 1). Its temperature distribution is the main unknown in constructing a rough thermodynamic classification of end-use energy, as in Table 2.

Before 1976, the only attempt of which I am aware to construct a modestly detailed spectrum or histogram for U.S. process heat was an unpublished study by Westinghouse Electric Corp. that gives five temperature categories for 15 process sectors* in 1972. I do not know the conventions or methodologies used. The "fossil fuel uses" (10^{12} Btu) are shown as follows in the summary table:

Industry	230°F	231°-400°F	400°F & steam drive	Elec. Gen.	Space heat	Total
Process	2,363	2,944	13,641	732	1,063	20,743
Other	708	471	590	24	564	2,357
Total	3,071 (13%)	3,415 (15%)	14,231 (62%)	756 (3%)	1,627 (7%)	23,100 (100%)

These figures, however rough, suggest that it would be useful to confirm how much heat is used at low and moderate temperatures in industrial processes.

Two studies nearing completion for the ERDA Solar Division (Mr. William Cherry, project manager), to be published in early 1977 by the National Technical Information Service, are exploring the economic and engineering parameters of solar process heat, and in the process are developing a temperature spectrum of industrial process heat in the United States. The two contractors, working independently at ERDA's request, are InterTechnology Corp. (ITC) (Warrenton, VA) and Battelle Columbus Laboratories (BCL). Mr. Malcolm Fraser of ITC and Dr. Elton Hall of BCL have

* *Food and kindred products; textile mill products; apparel and other textile products; lumber and wood products; paper and allied products; chemical and allied products; petroleum and coal products; rubber and plastic products; stone, clay, and glass products; primary metal industries; fabricated metal products; machinery, except electrical; electrical equipment and supplies; transportation equipment; natural gas processing.*

generously sent me early versions of their data, which are incomplete, preliminary, subject to revision, but nonetheless very interesting.

Mr. Fraser summarized interim ITC results in his Survey of the Applications of Solar Thermal Energy to Industrial Process Heat, given at the Sharing the Sun Conference, International Solar Energy Society, Winnepeg, Aug. 15-20, 1976. He stressed that

> . . . the survey was performed from the point of view of process requirements rather than the point of view of current methods of using heat. Thus, the temperature of major interest for a particular application was the required temperature of the process material rather than the temperature at which the heat is currently provided. Currently much heat of high thermodynamic availability is wasted because it is used for a low-temperature application which could readily be satisfied with lower-temperature heat. Fuels with the capability of a flame temperature of 2,000+°F are burned with the ultimate objective of making hot water, for example. A solar process heat system should be designed to satisfy the needs of a process and not merely to substitute for the current method of providing heat.

The ITC survey is exceedingly detailed. Of the more than 450 four-digit Standard Industrial Classification groups in mining and manufacturing, ITC selected more than 100: mostly those that used over 5,000 GWh(t) of direct fuels in 1972, plus some others required "to obtain a broad-based coverage of industry." This sample uses about 70 percent of all industrial direct fuel. For each industry, ITC constructed process flowsheets showing process, temperature, and amount of nonelectric heat needed for specific applications, then integrated the results into a temperature spectrum (Figure A). The preliminary spectrum here includes some 67 four-digit SIC groups (and over 170 specific processes) using a total of 7.2 quads /year, or about 48 percent of the estimated total use of process heat in industry in 1972. ITC believes the final (larger) data base will not yield radically different curves than this interim sample.

In terms of terminal temperature, about 5 percent of the heat load is below 100° C and about 24 percent below 177° C. But if the heat required to preheat from ambient is also considered—for it "should be possible to visualize a method of utilizing preheat [e.g., from solar sources] in essentially every application"—these figures change to about 28 percent and 42 percent respectively, as shown in the upper curve of Figure A. These proportions are the basis

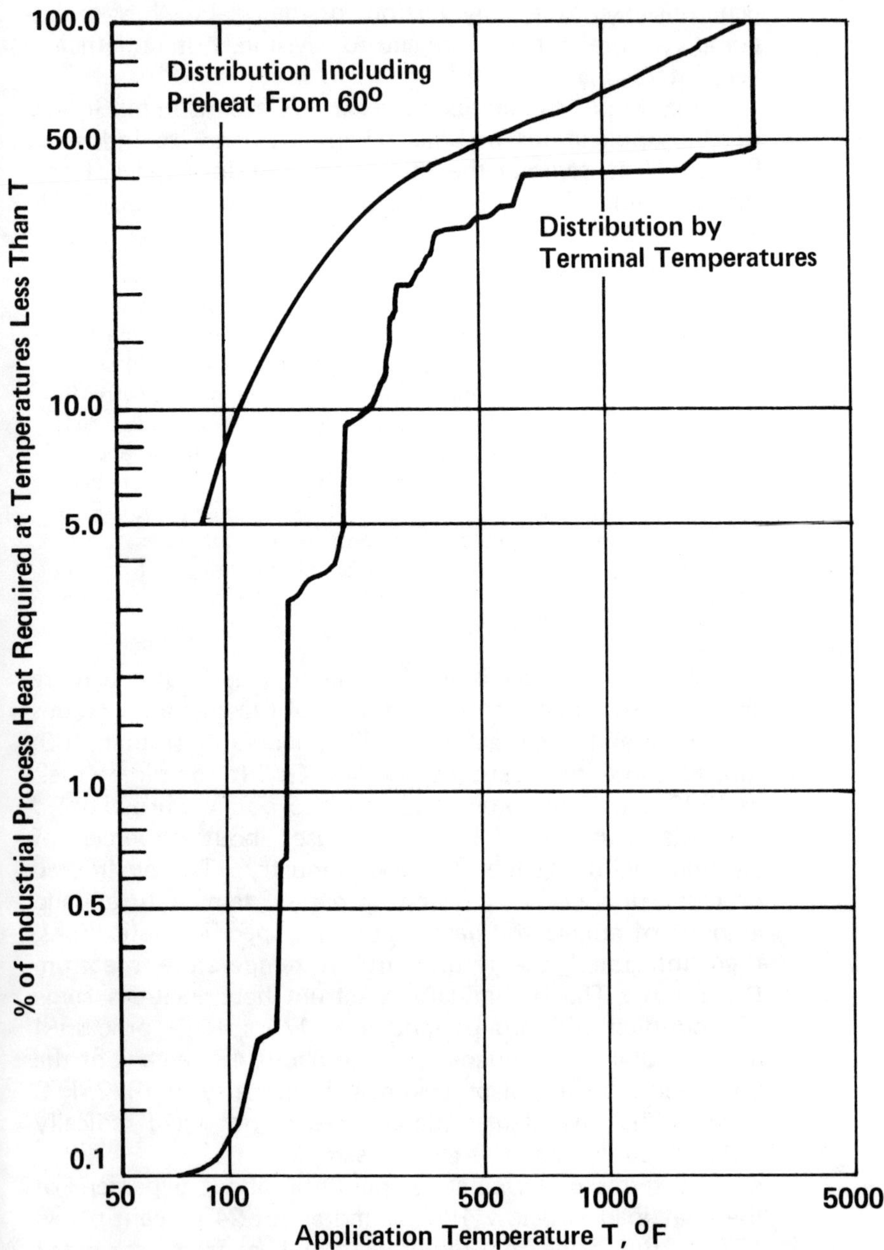

Figure A. Preliminary, Incomplete Data of Inter-Technology Corporation Cumulative Distribution of Process Heat Requirements.

of the process heat classification used in Table 2, whereas preheat is neglected in the temperature distribution mentioned in reference 1.

The BCL data, which use the same process-oriented approach, appear less detailed and at an earlier stage of development. The preliminary results of process analysis of an incomplete sample (7.87 quads), and of a less fine-grained and less precise statistical analysis of a larger sample, are as follows (direct fuel, quads/year, ca. 1975):

Category	Process analysis	Statistical analysis
<100°C (<212°F): Total	0.27 (2%)	0.26 (3%)
Hot water	0.07	0.12
Direct heat or hot air	0.1	0.14
100-177°C (212-350°F): Total	1.34 (17%)	3.25 (32%)
Steam	1.2	2.6
Direct heat	0.14	0.65
>177°C (>350°F): Total	6.35 (81%)	6.53 (65%)
Steam	0.5	0.56
Direct heat	5.85	5.97
Grand Total (rounding errors not corrected)	7.86 (99%)	10.04 (100%)

The sample in the right-hand column reportedly includes about 80 percent of all process heat; the rest is in unexamined small industries with unknown temperature distribution.

The BCL data thus show 2 to 3 percent of process heat; below 100° C and a cumulative total of 19 percent (by process analysis) or 35 percent (by statistical analysis) below 177° C. In view of the inherent uncertainties and limited samples used, the data seem reasonably consistent with the ITC data if the BCL data are assumed to reflect terminal temperatures without regard to preheat from ambient. The greater detail of the ITC data base, however, leads me to prefer it as a basis for Table 2. More useful comparisons will of course be possible once the reports are both published around early 1977.

APPENDIX III:
Supplementary Notes on Nuclear Proliferation

The brevity of reference 1 leaves several points to be qualified or expanded in this more technical companion-piece.

First, a tacit premise of my argument[1] is that civilian nuclear technology is a principal driving force behind the proliferation of nuclear weapons today. This case has been made in detail by other authors.[83-88] I need not reiterate here the role of civilian power reactors in spreading knowledge, hardware, and expectations and in reducing the time and the political leap needed to convert latent to actual proliferation. It is also obvious that the intended or actual possession of civilian nuclear facilities, because of their military potential, is a destabilizing psychological influence that may impel neighboring countries to seek a similar potential.

A more disturbing and little-known aspect of civilian nuclear power programs, however, needs to be spelled out here to combat a widespread misconception that I shared until recently. It is a delicate subject, but I believe the only thing more dangerous than talking about it would be not talking about it.

It is well known that reactor-grade plutonium (exposed to high neutron flux, hence relatively high in concentration of isotopes of high spontaneous fission rate such as plutonium-240) is not ideally suited to making explosives. It is now widely conceded that such plutonium can be used by terrorists, on the basis only of the extensive information in the open literature,[88-91] to make crude bombs of low and unpredictable yield (but still sufficient to wipe out a government, breach the containment of a nuclear facility, etc.). It is also known that a higher yield with rather narrow dispersion can be obtained by using sophisticated techniques unlikely to be available to amateurs. This second fact has so far been largely ignored, since traditional concerns in this area have focused on what terrorists can do. It has been frequently assumed, therefore, that governments wishing to make bombs would either juggle civilian fuel cycles to obtain low burnup (probably evading international safeguards) or, more likely, make "dedicated facilities," probably clandestine, specifically to generate weapons-grade plutonium. However, my considerations of this problem have led me to conclude that the techniques likely to be available even to small, modestly sophisticated governmental weapons programs—essentially the same techniques that such programs would be likely to use anyhow—would permit reactor-grade plutonium to be used directly in bombs of substantial military utility. Yields could be made quite predictable and well into the kiloton range, though for many such applications nuclear testing would be important (if not essential).

It is difficult to make a technically satisfying but unclassified statement on these matters. Never having received classified information, I am not as constrained as those with expert knowledge. Nonetheless, if my conclusions based on the open literature are correct—and they are shared by some professionals—they place a very different complexion on civilian nuclear programs. These programs are expected to accumulate very large inventories of reactor-grade plutonium very rapidly in many countries.[83] If, as I believe, this plutonium can be used directly to make military bombs while still maintaining high performance, then the civilian stocks appear in a different light—a problem much more prompt, and orders of magnitude larger, than the stocks that might otherwise be created covertly in dedicated facilities.

Reference 1 describes a package of initiatives that, if the United States implemented them simultaneously, unilaterally, and universally, would be likely in my opinion to halt virtually all nuclear power programs abroad. The essential aim of these steps would be to foster an international climate of denuclearization[64,92] in which it is socially unacceptable—perceived as a mark of national immaturity—to possess or desire either reactors or bombs. Without such a development I see no way to avoid the consequences of endemic nuclear violence,[71] for the technical and political arrangements that must prevent it once the requisite materials are created must last not merely for the usual lifetime of treaties or nations but for the lifetime of the materials themselves—a period many times as long as the period from the Neolithic to the present. The implausibility of success in such arrangements is illustrated by the inability of both Britain[93] and France[94]—perhaps the countries with the greatest experience of exercising political control—to subject their own original nuclear weapons programs to political control or formal decision before those programs were virtually completed.

My thesis that a halt to foreign nuclear power programs

under the conditions I describe[1] would go very far to defuse proliferation—i.e., would reduce it by several orders of magnitude, and perhaps ultimately to zero—needs several qualifications. First, if laser enrichment turns out to have tractable engineering and chemistry on a crude pilot scale, then any clever technician should be able, in due course, to make a bomb from a few tons of natural uranium. Uranium ore would become a strategic commodity and we would be back to the Baruch plan. Recent preliminary work by Prof. Allan Krass (currently at Princeton) suggests that the technology is very unlikely to be easier than the research-reactor or theft-and-black-market route to strategic materials. I hope this is correct. Diverse opinions, however, persist, and if there is any chance that some form of laser enrichment will be relatively easy, then our best protection would be to have a fission-free economy in which uranium is not ordinarily mined or traded in international commerce. The quantities needed to make a bomb would then no longer be an undetectable fraction detachable from commercial flows. If there turned out to be no worryingly simple form of laser enrichment, we would still have played safe.

Second, there are pockets of technical metastasis that might be relatively impervious to internal and external political pressures resulting from the initiative I suggest.[1] India might be such a place, though I think India is vulnerable to political and economic pressures, especially from within. Nonetheless, India already has bombs and will want to keep them as long as others do. The question is whether India could and would maintain the domestic political, economic, and technical base needed to become an independent nuclear exporter, as well as find qualified client states seeking her exports, all within the political and psychological context I suggest.[1] I consider this implausible and believe that even Indian technical ability would be hard pressed to maintain a substantial domestic program for more than a decade or two in the face of an effective quarantine. Recent demolition of the bizarre idea that nuclear power is a necessary, appropriate, or economically attractive energy source for developing countries[83,95] makes such secondary metastasis seem even less likely.

Third, research reactors would of course still be a poor country's route to bombs—either by making plutonium in moderate amounts and extracting it in a crude several-bombs-per-year reprocessing plant that might be built[96] in a year or two with perhaps $1 million to $3 million, or else by buying a suitable number of fully enriched uranium cores. Appropriate natural-uranium/graphite research or production reactors are relatively easy to build.[97] This could be made much harder through better international controls and intelligence directed to a few strategic materials such as nuclear-grade graphite, uranium, and deuterium. A more important deterrent or disincentive, though, would be the psychological shifts referred to above, since a country that is not going to have a nuclear power program, in a world that is phasing out nuclear technology, can hardly justify building or keeping research reactors.

I have pointed out elsewhere[98] that nobody has yet thought seriously enough about what the terminal phase of a nuclear economy should look like—i.e., how to phase out nuclear programs in an orderly fashion that minimizes residual hazard. There are some nice technical puzzles here. It is not clear whether reprocessing is advisable or not; nor whether, for example, it might even be worth building some low-power-density reactors specially designed to burn transuranic stocks. Such questions should be addressed by the technical community. I believe, however, that though such technical questions are open, there are grounds[98] for confidence that the institutional barriers to reversing our nuclear commitments can be overcome without prolonged or massive disruption. An appropriate and urgently needed first step widely urged by the arms control community[83,85,87,99]—suspension and stringent reexamination[100] of all plans for reprocessing, recycling, and breeding plutonium—would cause insignificant social or economic disruption and is long overdue.[57,87]

The political plausibility of the steps I propose[1] is a matter for political judgment. Consistent with many recent declarations in the nuclear trade press, I believe U.S. abandonment of nuclear power is politically plausible within a few years. I am confident that nuclear power is already politically and economically dead* in the United States, and am only slightly less confident that its demise may be put in a constructive context[1] closely linked with soft energy policies, strategic arms reduction, and nonproliferation. Looking at European nuclear programs from the inside, I believe the same is true. This may surprise American observers who

* *In the sense of a brontosaurus that has had its spine cut, but is so big and has so many ganglia near the tail that it can keep thrashing around for years not knowing it's dead yet.*

listen to, say, French and German rhetoric; but to a French or German observer who only reads Presidential speeches and ERDA press releases, the U.S. program might also appear, erroneously, to be easily brushing off a minority fringe protest and going from strength to strength. It is distance, not familiarity, that gives us in the United States our impression of the political and economic robustness of European electronuclear plans (as recent developments in Britain, Sweden, and Denmark attest). Bupp[101] has even been told at the highest levels of the French LWR program that its chance of actually happening—completing most of the reactors currently planned—is perhaps one in three. Surprisingly, the remarkable recent concern in Congress about nuclear export policy is coming to be mirrored in some high French circles, bringing dominolike pressure to bear—ultimately irresistably, I believe—on West Germany. Of course, the commercial enterprises involved have strong incentives, given their cash-flow problems, to recycle ExImBank or similar loans[83,95,102] via the treasuries of less developed countries into their own revenues; but I believe political revulsion, both public and private, will soon prove stronger.

Our approach so far to nuclear proliferation is all too reminiscent of Paul Ylvasaker's definition of a region as "an area safely larger than the one whose problems we last failed to solve." By treating the civilian and military atoms as distinguishable, and by treating our own possession of bombs as patriotic but others' possession as irresponsible, we have woven a web of hypocrisy[92] and double-think that is as inimical to strategic arms reduction as to control of proliferation.[103] The former, which is the greater and harder problem, cannot be addressed without a new approach to the latter, and the collapse of nuclear power offers us, briefly, the opportunity to do better; but we must stop passing the buck before our clients start passing the bombs.

NOTES

1. A. B. Lovins, Energy Strategy: The Road Not Taken?, Foreign Affairs, 55 (1): 65-96, October 1976. Reprinted without illustrations in Congressional Record 122, (151): Part , pp. (1 October 1976).

2. A. B. Lovins, Long-term Constraints on Human Activity. Environmental Conservation 3(1):3-14, Spring 1976. Geneva.

3. A. B. Lovins and J. H. Price. Non-Nuclear Futures: The Case for an Ethical Energy Strategy. Friends of the Earth. Ballinger, Cambridge, Mass. 1975. A review by Alvin Weinberg is published with a response in Energy Policy, December 1976.

4. A. B. Lovins, World Energy Strategies: Facts, Issues, and Options. Friends of the Earth. Ballinger, Cambridge, Mass., 1975. See also: W. C. Patterson and C. Conroy. Energy Alternatives and Energy Policy: A UK Viewpoint. Memorandum to the Energy Resources Sub-Committee of the Select Committee on Science & Technology, House of Commons, London, May 1976; and P. F. Chapman, Fuels Paradise, Penguin, 1975.

5. W. W. Harman, An Incomplete Guide to the Future. Stanford Alumni Association. 1976.

6. Economic and Environmental Implications of a U.S. Nuclear Moratorium 1985-2010 (draft). Institute for Energy Analysis, Oak Ridge, Tenn. August 1976.

7. W. C. Patterson, Nuclear Power. Pelican. 1976. See also reference 100.

8. W. L. Wilson, Energy World, 26:2, April 1976. Institute of Fuel, London; Weekly Energy Report, 3 (10):10, March 10, 1975, reporting the FEA availability study; Electrical World, p. 43-58 (especially the unavailability graph on page 51), Nov. 15, 1975, reporting the latest Steam Station Survey.

9. J. Peschon, et al., Regulatory Approaches for Handling Reliability and Reserve Margin Issues and Impact on Reserve Margin of Dispersed Storage and Generation Systems. Testimony to California Energy Resources Conservation and Development Commission, July 28, 1976. Systems Control, Inc., Palo Alto, Calif.

10. B. Sørensen, Energy and Resources. Science, 189:255-60, 1975. In reference 1, note 25, the initial page number is incorrectly given as 225; this will be corrected.

11. As noted in Appendix I, Prof. Meyer, head of the Danish Academy of Technical Sciences, is exploring soft energy paths for the Science Policy Committee of the EEC. The main initiative for this project was taken by Prof. H. B. G. Casimir of the Netherlands. No results are yet available.

12. E. F. Schumacher, Small Is Beautiful. Harper & Row, New York. 1973. His followup, Small Is Possible, is in preparation. The concepts have been proven in practice by the Intermediate Technology Development Group (25 Wilton Road, London SW 1, UK), notably in the manufacture of bricks, Portland cement, sugar, and egg-cartons, and are well supported by design studies (now being implemented) for paper, glass, and other products. This experience shows that increased efficiency, low overheads, and low transport costs can produce a wide range of products more cheaply in a small than a large factory. Some of the economies of small scale are rather subtle: For example, The Times noted early this year that British companies lose working time through strikes in direct proportion to their size, perhaps a measure of alienation. For an incisive discussion of scale and equity, see D. H. Meadows, Not Man Apart, 6 (17);1, October 1976, Friends of the Earth, San Francisco, Calif.

13. I. Illich, Tools for Conviviality. 1973. And Energy and Equity., Calder & Boyars Ltd., 18 Brewer St., London W 1, UK. 1974.

14. B. Commoner, The Energy Crisis: The Solar Solution, Sharing the Sun Conference, International Solar Energy Society, Winnipeg, Aug. 16, 1976. See also M. Garner, Electrical Review, p. 390-1, March 28 / April 4, 1975 (England).

15. M. H. Ross, and R. H. Williams. Improved Efficiency: Our Most Underrated Energy Resource. Bulletin of the Atomic Scientists, November 1976. (In press.). Parallel publication in press; Technology Review, February 1977 (cited in reference 1). See also R. Doctor et al., California's Electricity Quandary: III. Slowing the Growth Rate, R-1116-NSF/CSA, 1972; and W. Ahern et al., Energy Alternatives for California: Paths to the Future, R-1793-CSA/RF, 1975, both RAND Corp., Santa Monica, Calif.; and A. C. Sjoerdsma and J. A. Over, ed., Energy Conservation: Ways and Means, publication 19, June 12, 1974, Stichting Toekomstbeeld der Techniek (Prinsessegracht 23, Den Haag).

16. Patterns of Energy Consumption in the United States. Stanford Research Institute. 1972. The Ross and Williams update is in the Technology Review version of their paper.[15]

17. The average COP of currently installed U.S. air conditioners is given as about 1.9 in several recent studies—e.g., R-1641-NSF,

p. 31 and ORNL-NSF-EP-51.

18. D. B. Goldstein, and A. S. Rosenfeld. Conservation and Peak Power Demand. LBL-4438. Lawrence Berkeley Laboratory. 1975. See also their Projecting an Energy Efficient California, LBL-3274, 1975.

19. This study, done in March 1976 by the author in the Ministry of Energy, Mines, and Resources (Ottawa), is fully reported in Exploring Energy-Efficient Futures for Canada, Conserver Society Notes 1, (4); 5-16, May/June 1976. Science Council of Canada, 150 Kent St., Ottawa, Ont. K1P 5P4. The Cabinet-approved technical fixes on which the calculations rest were announced in the House of Commons (Ottawa) by the Hon. Alastair Gillespie, Minister of EMR, on Feb. 29, 1976 and are reviewed, with graphs, in the EMR press packet of that day. The graphs are reproduced in the CS Notes article. The staff of the Office of Energy Conservation kindly made available much in-house EMR material underlying the Government's conservation policy, and offered many helpful comments.

20. P. R. Ehrlich, An Ecologist's Perspective on Nuclear Power. Federation of American Scientists, Public Interest Report 28, (5-6): 3-7, May/June 1975.

21. A. D. Poole, Questions on Long-Term Energy Trajectories. Notes from a talk delivered to the ORAU summer faculty institute, July 1976: typescript, Institute for Energy Analysis, Oak Ridge, Tenn.

22. A. V. Kneese, The Faustian Bargain. Resources 44:1, September 1973. Resources for the Future, Washington, DC.

23. Quoted by E. F. Schumacher in Small Is Beautiful.[12]

24. M. Carasso, et al., The Energy Supply Planning Model. Bechtel Corp. report to NSF, PB-245 382 and PB-245 383. National Technical Information Service, Springfield, Va. August 1975.

25. M. Beller, ed., Sourcebook for Energy Assessment. BNL-50483. December 1975.

26. M. Carasso, and M. Gallagher. Personal communications, 1976. Dr. Carasso has kindly reviewed my use of his group's data[24] to ensure that I was using it correctly, and Dr. Gallagher has generously shared the results of Bechtel's update of the model's data base.

27. M. L. Baughman, and D. J. Bottaro. Electric Power Transmission and Distribution Systems: Costs and Their Allocation. Research Report 6. NSF-RA-N-75-107. Center for Energy Studies, University of Texas, Austin. July 1975. The discussion of T&D costs appears not to take account of T&D losses.

28. J. Harding, (on staff of California Energy Resources Conservation and Development Commission). Personal communication. 1976. Harding is among the best-informed authorities on fuel-cycle economics.

29. I. C. Bupp, et al., Trends in Light Water Reactor Capital Costs in the United States: Causes and Consequences. CPA 74-8. Center for Policy Alternatives, MIT. Dec. 18, 1974. Summarized in Technology Review, 77; (2) 15, 1975. Prof. Bupp told me recently that a check in early 1976 confirmed that the escalation rates described in the earlier study were continuing unabated.

30. J. J. O'Connor, The $2000 Kilowatt. Power p. 9, April 1976.

31. A. B. Lovins, Things That Go Pump in the Night. New Scientist, 58:564-6, May 31, 1973. True to form, the E100 million (plus) Dinorwic scheme was both overrun and obsolete before it was even off the drawing board. The economic details are in the Commons Committee hearings, particularly in the author's evidence.

32. F. von Hippel, and R. H. Williams. Energy Waste and Nuclear Power Growth. Bulletin of the Atomic Scientists, December 1976. A similarly devastating critique of British plans for electrification has been published by P. F. Chapman et al., Energy Research Group, The Open University, Milton Keynes, Bucks., UK. 1976.

33. A detailed 1976 study by Charles Komanoff of the Council on Economic Priorities (New York City) has developed this figure as a fair representation of what such stations would achieve if their capacity factor were unaffected by their merit-order position.

34. ERDA-76-1, Appendix B (1976): the static analysis only, as the dynamic one appears (from the limited information given) to be done incorrectly. The static analysis is nearly a factor of two less favorable to pressurized water reactors than Price's[3] (using a common accounting basis for both): using Price's conventions, the ERDA P_o/P_i is 1.97. Price's Epilogue[3] gives a range of other estimates. It is also important to note that all analyses so far published assume that the design-basis fuel burnup is actually achieved. A large fuel sample analyzed by Harding[28] in

responding to interrogatories in the Jamesport (LILCO) hearings in mid-1976 strongly suggests that only about half to two-thirds of the design-basis burnup is being achieved by mature LWR cores; hence that the reactors are consuming yellowcake and separative work far faster than standard economic calculations assume. Since this issue was raised in a Louisiana licensing hearing, ERDA has reportedly been trying for the first time to survey actual fuel-cycle performance, but the results do not yet appear to be available.

35. H. R. Linden, (President, Institute of Gas Technology). Information submitted for the record, House Committee on Science and Technology, full Committee meeting on loan guarantee provisions of HR 3473, ERDA Appropriations Bill, FY 1976, 18 Sept. 18, 1975. Interesting comparative data on escalation and on electric vs. synthetic conversion of coal are included.

36. See e.g., A. M. Squires, Science 184:340-6, 1974; 189:793-5, 1975, and 191:689-700 1976. also, The City College Clean Fuels Institute, paper at symposium on Clean Fuels from Coal, Institute of Gas Technology, Chicago, June 26, 1975; J. Yerushalmi et al., Science 187:646-8, 1975; and I&EC Process Design & Development, 15:47-53, January 1976.

37. D. F. Williams, Applied Energy 1:215-21, 1975 Applied Science Publishers, England; K. D. Bartle et al., Fuel 54:226-35, October 1975, IPC Science & Technology Press, England. The process exposes lumps of coal, without special heating, to solvent vapor above the critical pressure, e.g. toluene at 100 bar. The vapor penetrates between the leaves of carbon and pulls out intact (usually aromatic) molecules by the interstitial hydrogen without cracking them. The hydrocarbons then fall out of the vapor when it is reexpanded in another container; a substantial fraction of the mass of the coal is extracted in a form (typically benzene, anthracene, etc.) whose spectrum is controllable by the solvent, pressure, and exposure time. Virtually all the hydrogen is removed. Metals may be removed too. (nobody knows yet). The residual char (carbon plus minerals) generally melts in the range 100° to 200° C and can be liquid-fed as an under-boiler fuel. Like the flash processes,[35] supercritical gas extraction offers the intriguing possibility of small units sited in a coal-using factory; premium fuels extracted on the premises could then be used for kiln firing, for example, and the char for cogeneration and process steam.[40] This flexible scale, simplicity, and lack of tar formation seem to me immensely attractive compared to conventional billion dollar conversion plants.

38. See the citations in reference 1 (notes 26-28); A. M. Squires, Applications of Fluidized Beds in Coal Technology, in J. P. Hart-

nett, ed., Alternative Energy Sources, in press, Hemisphere, Washington, D.C.; the brochure Fluidized Combustion, Combustion Systems Ltd, 66-74 Victoria St., London SW 1, which is amply illustrated for a nontechnical audience; H. B. Locke & H. G. Lunn, The Chemical Engineer, p. 667-70, November 1975; and the general review by A. M. Squires, Ambio 3, (1), 1-14, 1974. The address of Stal-Laval (GB) Ltd is 41-7 Strand, London WC 2. The domestic fluidized-bed furnace developed by the late Prof. Elliott should be, from the user's point of view, indistinguishable from an oil furnace. Granular coal is delivered by hose from a tank-truck to a bunker, then gravity-fed to a small packaged furnace with very low emissions. Ash is collected in a snapon container akin to a vacuum-cleaner bag, changed at each fuel delivery. The technical problems of such a system appear to be substantially solved, and plans for a U.S. field test are well advanced despite lack of Federal interest outside The Council on Environmental Quality and The Environmental Protection Agency. (In September 1976 ERDA's coal technologists did not appear to have heard of Fluidfire or Stal-Laval, and in April 1976 a leading industrial coal expert told me that Professor Elliott's devices, which I have seen operating, could not possibly exist.)

39. H. Harboe, (reference 1, note 26), citations there and personal communication, 1976. Pilot tests have given Stal-Laval high confidence of success in avoiding blade erosion by sintered ash, largely because the proposed design, rather than striving (like U.S. schemes) for high turbine inlet temperature, limits it to about 800° C, so that sintering, NO_x formation, and volatilization of metallic impurities are greatly reduced. The resulting system, operating at 15-20 bar, attains a First Law turbine efficiency of only 27 percent, but boosts this to over 70 percent (plus any bonus for heat pumps) by taking district heating off the bottom of the cycle. Failure to consider the implications of this design philosophy appears to account for most U.S. skepticism about the readiness of fluidized-bed gas turbines for prompt commercial application. The problem is made still easier by a closed-cycle air bypass (see Harboe & Maude. reference 1, note 26).[41]

40. A detailed Swedish study shows that the marginal capital cost of distributing electricity for heating Stockholm buildings is nearly twice that of distributing hot water for the same purpose (considering only the capital costs of the distribution networks). The district heating would be done by tunneling tens of meters down in rock, then tunneling up into basements to connect rather than tearing up the streets. The details are given in Utredning om Södermalms framtide värmeförsjörning, met förslag till värmeplan för Södermalm, Stockholms Kommunstyrelse, 1974, and summarized in Neil Muir's informative district-heating paper in the International CIB Symposium (reference 1, note 15).

41. Two very important papers on cogeneration have just appeared: R. H. Williams, The Potential for Electricity Generation as a By-product of Industrial Steam Production in New Jersey, report to the NJ Cabinet Energy Committee, June 21, 1976, Center for Environmental Studies, Princeton University; and S. E. Nydick et al., A Study of Inplant Electric Power Generation in the Chemical, Petroleum Refining, and Paper and Pulp Industries, Thermo Electron Corp. report to FEA, PB 255 658 *executive summary) May 1976, and PB 255 659, National Technical Information Service (Springfield, Va). Prof. Williams has kindly called my attention to an informative prospectus for a Weyerhaeuser co-generation system, available from K. W. Rinard, Eugene Water & Electric Board, Eugene, Oreg. It appears that cogeneration could be so attractive as to compete economically with solar heat for both space (with heat pumps) and process, on certain assumptions about allocation of costs between the dual outputs. It would, however, rely on depletable fuels and thus be transitional only.[1] As the use of electricity was gradually trimmed back to appropriate end uses, cogeneration would become less attractive as a source of electricity relative to installed hydroelectric capacity, and as a source of heat relative to solar process heat, materials recycling, or process redesign (note a, page 40).

42. K. Bammert, and G. Groschup. Status Report on Closed-Cycle Power Plants in the Federal Republic of Germany. ASME paper 78-GT-54 for presentation at the Gas Turbine Conference and Products Show, New Orleans, March 21-5, 1976. I am grateful to Prof. Williams for calling this useful paper to my attention.

43. A. E. Haseler. Personal communication. 1976.

44. Project Independence Blueprint. Solar Task Force Report, FEA. (4118-00012, USGPO). November 1974.

45. Good data should emerge from the 22 projects summarized at page III-D-35, Vol. III, Recommendations for a Synthetic Fuels Commercialization Program, Synfuels Interagency Task Force report to President's Energy Resources Council. November 1975.

46. A. D. Poole, and R. H. Williams, Flower Power: Prospects for photosynthetic energy, Bulletin of the Atomic Scientist 32. (5): 48-58, 1976; and M. J. Antal, Jr., ibid., p. 59-62; J. Marshall et al., E-X-25, Forestry Service, Environment Canada, February 1975; Forest Service, USDA, The Feasibility of Utilizing Forest Residues for Energy and Chemicals, NSF/RA-760013, March 1976; C. G. Golueke and P. H. McGauhey, Ann. Rev. Energy 1:257-77, 1976; K. Sarkanen, Science 191:773-6, 1976 and adjacent papers; D. Hayes, Energy: The Case for Conservation, Paper 4, Worldwatch Institute, Washington, D.C., 1976.

47. E. P. Gyftopoulos, et al., Potential Fuel Effectiveness in Industry, Ballinger, Cambridge, Mass., for The Energy Policy Project, 1974; J. A. Day et al., Industrial Process Heat from Solar Energy, UCRL-76390, May 1975; and a large international literature. Solar crop-drying seems ready for immediate large-scale use.

48. A. B. Meinel, in Briefings Before the Task Force on Energy, Subcommittee on Science, Research, and Development, Committee on Science and Astronautics, USHR, Serial U (92d Congress), USGPO, 1972; and in Solar Energy for the Terrestrial Generation of Electricity, hearing before the Subcommittee on Energy, same Committee, 12 (93d Congress), USGPO, 1973. The α/ϵ of about 50 which Meinel mentions, with an apparent lifetime of order 40 years on the basis of accelerated-lifetime tests, has since been bettered in proprietary work. I understand that wet chemical deposition at much lower cost can yield films with α/ϵ of at least 30, but the lifetime is not yet known.

49. H. C. Fischer, The Annual Cycle Energy System. Energy Division, Oak Ridge National Laboratory. 1975. An illustrated but less detailed account of the ACES system, which looks useful for warm climates, appears in Professional Engineer Magazine, June 1976.

50. T. V. Esbensen, and V. Korsgaard. The Zero Energy House. International CIB Symposium (reference 1, note 15), also T. V. Esbensen, personal communication, 1976.

51. Proceedings of the Passive Solar Heating and Cooling Conference and Workshop. Albuquerque, N. Mex. May, 18-20, 1976. National Technical Information Service, Springfield, Va., 1976.

52. Calmac Corp., Box 710, Englewood, N.J. 07631; Thomason Solar Homes, Inc., 6802 Walker Mill Rd. SE, Washington, D.C. 20027; W. A. Shurcliff, Active-Type Solar Heating Systems for Houses: a Technology in Ferment, Bulletin of the Atomic Scientist 32 (2):30-40 1976, which includes much good advice and the latest fusible-salts reference.

53. J. M. Weingart, Solar Energy Conversion and the Federal Republic of Germany—Some Systems Considerations. Draft WP-75-158. IIASA, Laxenburg, Austria. December 1975. IIASA studies of solar heating and cooling costs (capital and life-cycle) and of available hardware are in progress. See also S. W. Herman and J. S. Cannon, Energy Futures: Industry and the New Technologies, INFORM 25 Broad St., New York City. 1976.

54. H. van Bremen, and J. M. van Heel. Solar Energy in Local Authority One-Family Houses in Holland. International CIB Symposium (reference 1, note 15). Also H. van Bremen, personal

communication, 1976. Interesting comparative data are available for a different climate from the SAGE (solar assisted gas energy) project of the Jet Propulsion Laboratory, Pasadena, and Southern California Gas Co. The field data from a 32-apartment-unit complex will become available late this year from SAGE Project Director, Southern California Gas Co., P.O. Box 3249 Terminal Annex, Los Angeles, Calif.

55. A. D. Little, An Impact Assessment of ASHRAE Standard 90-75, report C-78309 to FEA. December 1975. Summarized in FEA's f. Conservation Paper 43A.

56. D. P. Kamat, et al. Regulatory and Tax Alternatives and the Financing of Electricity Supply. Research Report 7, NSF-RA-N-75-123, Center for Energy Studies, University of Texas, Austin, September 1975. p. 29.

57. Efficient Use of Energy. American Institute of Physics Conference Proceedings 25. American Institute of Physics, New York City. 1975. Gyftopoulos, op cit.[46] and Ross and Williams, op. cit.[15]

58. L. D. Chapman, et al. Electricity Demand: Project Independence and the Clean Air Act. FPC Task Force Report. Oak Ridge National Laboratory, Oak Ridge, Tenn. November 1975. von Hippel and Williams show[32] that the 1975 price of residential electricity would have to more than quadruple to get back to the constant-dollar level of 1940.

59. S. Novick, The Electric Power Industry. Environment 17 (8) p. 7-13, 32-9, November 1975.

60. E. Kahn et al., Investment Planning in the Energy Sector, LBL-4479, Lawrence Berkeley Laboratory, March 1, 1976.

61. This result has been derived both by IIASA and by Dr. Kees Daey Ouwens (Rijks-universiteit Utrecht). I consider both calculations more detailed and persuasive than those of a recent Institute for Energy Analysis study, and hope the discrepancies will be publicly reconciled.

62. I expect this result will appear in the Stanford Research Institute's important study (reference 1, note 30) A Preliminary Social and Environmental Assessment of the ERDA Solar Energy Program 1975-2020, if and when ERDA publishes it (I understand a draft was submitted to ERDA in July 1976).

63. This assumption is stated in terms in ERDA-48 (1975). See the Office of Technology Assessment critique (summarized in Sci-

ence, 190:535-7, 1975) and the author's critique for the President's Council on Environmental Quality (CEQ hearings, Washington, D.C. Sept. 3, 1975; similar hearings on ERDA-76-1 were held in December 1976).

64. R. A. Falk, Rejecting the Faustian Bargain. The Nation, p. 301-5, March 13, 1976.

65. Good beginnings are being made on the strategic implications of the CO_2 problem: see for example the 1976 Institute for Energy Analysis paper Energy and the Climate, or Häfele and Sassin's Energy Strategies (IIASA RR-76-8, March 1976). A high-energy future implies a fossil-fuel burn so large (with or without nuclear power) and increasing so rapidly that the great uncertainty in the timing of climatic effects of CO_2 is compressed, by the steep slope of the curve, into a few decades or so. For general background, see Study of Man's Impact on Climate, Inadvertent Climate Modification, MIT Press, 1971.

66. R. N. Adams, Energy and Structure. University of Texas Press. 1975. Also L. K. Caldwell. Energy and the Structure of Social Institutions. Human Ecology 4 (1), 31-45, 1976.

67. D. E. Lilienthal, Sr. America: The Greatest Underdeveloped Country. Address to a public meeting sponsored by the American Institute of Planners and the American Society of Planning Officials, Washington, D.C., March 22, 1976. The individualistic renascence Lilienthal welcomes could presumably be encouraged, as in the land-grant days, by an Energy Extension Service.

68. B. M. Jenkins, High Technology Terrorism and Surrogate War: The Impact of New Technology on Low-Level Violence. P.-5339. Rand Corp., Santa Monica, Calif. January 1975.

69. M. Flood, Nuclear Sabotage. Bulletin of the Atomic Scientist, in press.

70. J. Schell, The Time of Illusion: VI. Credibility. The New Yorker, July 7, 1975.

71. J. C. Mark, Global Consequences of Nuclear Weaponry, Annual Review of Nuclear Science, 1976. (In press.)

72. A. J. Ackerman, Trans. IEEE (Aerosp. Electr. Syst.) AES-8, 5, 576 1972.

73. A. E. Haseler tells me (1976) that flexible, insulated heat mains for distributing hot water to low-density residential areas can now be delivered on cable drums for rapid installation at very low

cost, making district heating economically attractive even for dispersed suburbs.

74. I. Höglund, and B. Johnsson (Royal Institute of Technology, Stockholm), in their paper at the International CIB Symposium (reference 1, note 15), describe a method they have developed for insulating existing apartment buildings on the outside, so as not to disturb the tenants or make the rooms smaller.

75. H. P. Green, The Risk-Benefit Calculus in Safety Determinations. 43 George Washington Law Review 791-807, March 1975, also response by P. Handler at 808-13; A. B. Lovins, comments in S. M. Barrager et al., The Economic and Social Costs of Coal and Nuclear Electric Generation: A Framework for Assessment and Illustrative Calculations for the Coal and Nuclear Fuel Cycles, Stanford Research Institute MSU-4133 (USGPO, 1976); A. B. Lovins, George Washington Law Rev., August 1977. (In preparation.)

76. L. Mumford, The Transformations of Man, Harper & Row, Publishers, New York City. 1956.

77. P. Diesing. Reason in Society. University of Illinois Press, Urbana. 1962.

78. H. E. Daly, Toward a Steady-State Economy. W. H. Freeman, San Francisco, Calif. 1973. Also, more recent manuscripts; C. Cooper, ed., Growth in America, Woodrow Wilson Center, Washington, D.C., 1976.

79. R. E. Miles, Jr., Awakening from the American Dream: The Social and Political Limits to Growth. Summarized in Washington Star, May 9, 1976.

80. For a specific example of rational economics, see The Return of The Draft Horse. Organic Gardening & Farming, p. 156-64, February 1976.

81. Several passages in this section are drawn from my draft contributions to The Unfinished Agenda, report of the Environmental Agenda Task Force (G.O. Barney, ed.), a project sponsored by the Rockefeller Brothers Fund, 1976. (In press.)

82. R. Wenkam, Introduction to K. Brower, Micronesia: Island Wilderness, Friends of the Earth, San Francisco, Calif. 1976.

83. A. Wohlstetter, et al. Moving Toward Life In A Nuclear Armed Crowd? Report to Arms Control & Disarmament Agency, U.S. State Dept., ACDA/PAB-263. April 22, 1976. Pan Heuristics, 1801 Avenue of the Stars, Suite 1221, Los Angeles, Calif. 90067.

84. See the report of the summer 1976 Aspen Institute of Humanistic Studies workshop on proliferation issues; the rapporteur was Michael Nacht (Harvard).

85. See the report of the summer 1976 Institute for Man and Science (Rensselaerville, N.Y.) workshop on proliferation issues; the rapporteur was Prof. Richard Gardner (Columbia University Law School).

86. Nuclear Energy and National Security. Research and Policy Committee, Committee for Economic Development, New York City. September 1976. also: L. A. Dunn and H. Kahn. Trends in Nuclear Proliferation, 1975-1995. HI-2336-RR/3. May 15, 1976 also: Hudson Institute.

87. H. A. Feiveson, and T. B. Taylor. Alternatives for International Control of Nuclear Power. Bulletin of the Atomic Scientist. December 1976.

88. M. Willrich, and T. B. Taylor. Nuclear Theft: Risks and Safeguards. Ballinger, Cambridge, Mass. 1974.

89. D. B. Hall, The Adaptability of Fissile Materials to Nuclear Explosives. October 1971 MS (out of print). Slightly revised version reprinted at p. 275+ in R. B. Leachman and P. Althoff, ed., Preventing Nuclear Theft: Guidelines for Industry and Government, Praeger, New York, 1972.

90. J. C. Mark, Nuclear Weapons Technology and ensuing discussion, in B. T. Feld et al., ed., Impact of New Technologies on the Arms Race, MIT Press, 1971.

91. P. Jauho and J. Virtamo. The Effect of Peaceful Use of Atomic Energy upon Nuclear Proliferation. Helsinki Arms Control Seminar, Finnish Academy of Arts and Letters. June 1973.

92. R. A. Falk, Reconsidering American Foreign Policy. Foreign Policy, October 1976. Also, A World Order Analysis of Nuclear Proliferation, Princeton University typescript. 1976. See also reference 64.

93. M. Gowing, Independence and Deterrence. 2 vol. Macmillan, London. 1974. This gripping official history of the UK nuclear program stirs one to speculate about what the world would be like if our species had evolved somewhat later when the uranium-235 was gone (a thought experiment suggested by Alvin Weinberg).

94. L. Scheinman, Atomic Policy in France Under the Fourth Republic. Princeton University Press. 1965. As quoted on p. 44-5 of reference 83.

95. Richard J. Barber Associates, LDC Nuclear Power Prospects 1975-1990. ERDA-52, 1975.

96. W. Epstein, The Last Chance, Nuclear Proliferation and Arms Control. Free Press, New York. 1976.

97. J. R. Lamarsh, On the Construction of Plutonium-Producing Reactors by Small and/or Developing Nations. QC 170 Gen., April 30, 1976. Report for the Congressional Research Service, Library of Congress. Reprinted by CRS June 4, 1976.

98. D. R. Brower, ed. Politics As If Survival Mattered: A New-Era Testament for America. In preparation.

99. I understand that in October 1976 the National Council of Churches in Christ adopted a resolution dealing with nuclear exports, but I do not yet have details.

100. Sixth Report: Nuclear Power and the Environment. Royal Commission on Environmental Pollution, Sir Brian Flowers FRS, Chairman. Cmnd. 6618, HM Stationery Office, London. September 1976. See also J. M. Brown, Health, Safety and Social Issues. In the California Nuclear Initiative. Stanford University Institute for Energy Studies. April 1976.

101. I. C. Bupp, and J.-C. Derian. Nuclear Reactor Safety: The Twilight of Probability. December 1975 seminar paper available from Prof. Bupp, Harvard Business School.

102. W. C. Patterson, Exporting Armageddon. New Statesman 92, 2371, p. 264-6 Aug. 1976.

103. G. Kistiakowsky, et al. The Threat of Nuclear War. Transcript of Granada TV program, March 29, 1976.Granada TV, Manchester M60 9EA, England.

Paul P. Craig
Council on Energy and Resources
University of California Systemwide Administration
Berkeley, California
and
Department of Applied Science, University of California
Davis, California

Paul P. Craig was trained as a physicist and has carried out research in a number of areas of solid state and low temperature physics at the Los Alamos Scientific Laboratory and at Brookhaven National Laboratory. For the past five years he has been involved exclusively with energy policy issues at the National Science Foundation and presently at the University of California. At the National Science Foundation he served as deputy director and acting director of the Office of Energy R & D Policy, the office established to support the director of the National Science Foundation in his role as Science Advisor to the President. At the University of California he is director of the Council on Energy and Resources, an all-University Council exploring new mechanisms by which the University can respond to changing societal needs. Mr. Craig received his B.A. degree from Haverford College in 1954 and his Ph.D. from the California Institute of Technology in 1959. He was a Guggenheim Fellow in 1965 and taught at the State University of New York from 1967 until 1971. He is also a Fellow of the American Physical Society. Mr. Craig has been a member of the Environmental Defense Fund's board of trustees and he has been active in the Sierra Club.

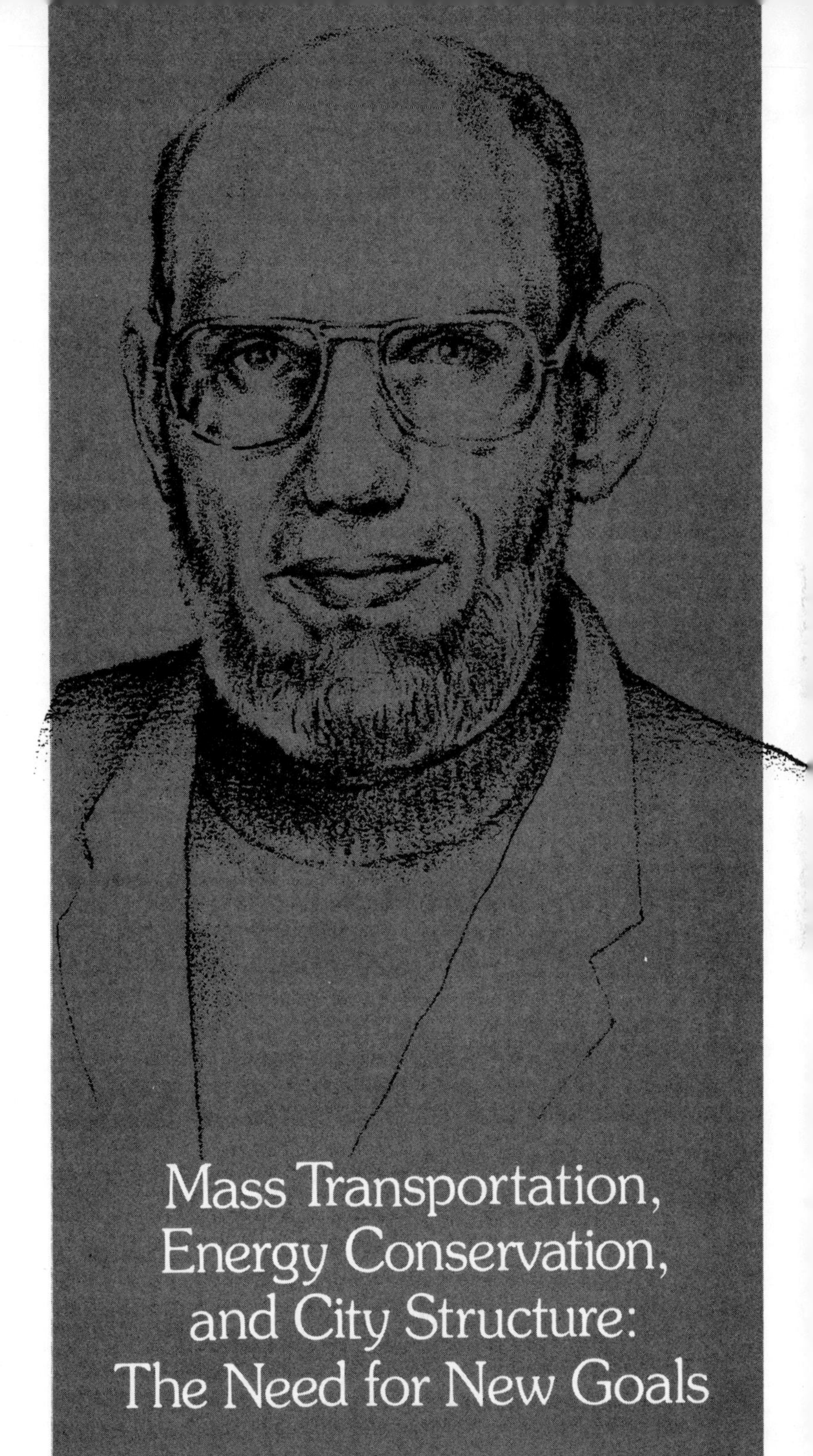

Mass Transportation, Energy Conservation, and City Structure: The Need for New Goals

New energy technologies will prove critical to the nation's evolution over the coming decades, as the world inevitably moves from reliance upon oil and natural gas, through a period of reliance on coal and other transitional energy forms, and finally to reliance upon solar energy—the ultimate renewable energy form. The beginning of the decline of oil and gas use globally will depend upon many factors, but will almost certainly occur within the final decade of this century. This central fact must motivate long-range energy planning, and we must bear it in mind even as we deal with the many short-term and internal issues that can so readily absorb the bulk of our effort.

The transition away from oil and gas will require development of many new approaches to the operation of our society. National recognition of the magnitude of the problems facing us has only begun, and the balance of our present approaches is, at this time, inappropriate to the size and character of the problem.

Our thinking in the United States about energy technologies is at present far too strongly oriented toward massive approaches to energy supply, toward development and implementation of complex and expensive breeder reactors, coal gasification systems, fusion reactors, and the like. Such systems are the lifeblood of the Energy Research and Development Administration (ERDA), the federal agency charged with development of energy alternatives. And these systems will surely be urgently needed. But energy supply is just one element of the energy system. Our approaches to the use of energy require at least equal attention. At the core of a successful approach to a world without oil and gas must lie a vigorous approach to constraining energy use.

Our understanding of how energy is used and what our choices are for modifying this use is presently inadequate. The major innovations required in energy use are not predominantly technological. The key issues relate to the capability of our national system to respond, during a period of extreme uncertainty and ever tighter coupling of nation to nation, to the extreme pressures that will build up on the global energy system as very cheap and abundant energy resources must be rather rapidly replaced by much more expensive ones. The stresses associated with this transition are such that the future must inevitably look very different from the past. Projections that call for "more of the same" are virtually certain to be wrong. And our challenge in planning is to expand our perspective so as to be prepared for the unexpected.

Decisions about which energy systems will be used to meet energy needs in the post-fossil era involve technological and social factors. Until quite recently there has been consensus that virtually any technology that can be developed should be used. As our technological capabilities expand, this viewpoint is likely to be less readily accepted. The national decision not to build supersonic transports is an example. Current debate over recombinant DNA research may lead to decisions not to undertake certain types of research, on grounds that there is too much potential risk associated with the outcome. In the energy area, the most important example is nuclear energy. If we could agree on just how safe nuclear systems need to be, then it is likely that systems meeting these requirements could be built. Yet there are many who feel that the risks associated with nuclear power are so great as to make this technology inappropriate, regardless of the level of safety assurance.

I believe it is presently an open question whether the United States will use nuclear systems as a major element of our long-term energy system.

Any assessment of whether to proceed with a new technological system must include cost as an important element. However, issues of the appropriateness of technically feasible systems transcend cost considerations. Methodologies for benefit/cost analysis have been developed and are being improved. Yet the relation between economics and values is complex and often not definitive. And so decisions on implementation of technological systems will be made on grounds that include economics as one component, but which transcend economics.

These considerations are of particular importance in planning energy systems. Energy is a central driving force in our society. The energy systems we build as we make the transition to the post-fossil fuel era will strongly influence the shape and structure of our society, and thereby the lives of our children and their children. This is true not only of energy supply systems, but also of energy using systems. Energy conservation tends to be thought of (especially in Washington) as a short-term expedient, handy as a means to tide us over until new energy supplies become available. But our patterns of energy use must change permanently.

In this process, transportation plays a key role. For transportation is the primary user of liquid fuels. The structure of cities is affected by their transportation systems. The time scale for the growth and change of cities is measured in decades, perhaps in centuries. Thus, our transportation system planning must have a similarly long-term perspective. Energy analyses presently being made of transportation systems do not do this. Perspectives are typically just a few years. The focus is on short-term intermodal shifts. Change in the structure of cities which may result over decades from new transportation systems is ignored, or given short shrift. I believe this is a mistake. In this paper I will give some arguments that in the long-term mass transit systems may prove to be sound investments from an energy point of view.

The thrust of this paper is to challenge the generally held belief among transportation experts that mass transit systems are poor investments in terms of energy conservation. The paper does not purport to give definitive answers. Its purpose is much more limited—to raise the issues in a broader context than is usual, and to suggest that conventional analyses may be too limited in scope.

The argument is simple. Conventional analysis assumes that the only energy savings resulting from new mass transit systems are those due to intermodal shifts—that is, shifts of persons who previously used automobiles, buses, etc., into the railed transport systems. This point of view is particularly well expressed by Webber[1] and by Lave.[2] But there is another effect that may require many years to operate but may be much larger. This is the change in building types, and hence population density, that results from the presence of a permanent mass transit system. A higher density of buildings leads to two effects—more energy-efficient buildings and less need to travel. But these effects do not occur instantly. Many decades are required for land use patterns to change. And, thus, any calculation that looks only at short-term effects will necessarily miss changes in land use patterns.

Before undertaking this analysis, it is important to describe the institutional context in which approaches to modification of energy demand are embedded. The cornerstone of the discussion is the major role played by massive institutional structures in the United States today. These institutional structures have response mechanisms that differ qualitatively from those of earlier institutions. It is essential to recognize this in developing policy options in the energy area, just as in all other areas of national importance.

The changes in our society that have occurred in the past century have involved introduction of technologies of such power that they only superficially resemble what has gone before. Simultaneous with technological change has been institutional change. Today, many of the directions in which our society moves are dominated by the influence of new institutional structures capable of organizing both human and material resources to accomplish their objectives. One measure of the impact of these structures is the decline in this century of self employment, to be replaced by employment in large organizations.

The situation has been analyzed by Galbraith[3] in many of his books, especially "Economics and the Public Purpose." Small business and the self-employed person fall into what he calls the "market system." This system operates along more or less classical economic lines. The large organizations, on the other hand, belong to the "planning system" and are able through organizational skills and enormous capital bases to modify the competitive market system in critical ways. The relationship between the government and the massive corporations that supply military hardware is one area in which the planning system can be clearly seen. The decision by the government to support Lockheed Corporation when it was in financial straits dramatically illustrates the importance of nonmarket forces in shaping the institutions of the planning system. And the array of government regulations on price, profit, and character of service in, for example, communications (telephone, telegraph, television) and transportation (airline schedules, trucking rates) shows that many of our daily activities are strongly affected by the nonmarket forces so characteristic of the planning system.

These are but a few of the many areas in which noneconomic factors dominate the decision-making process. Decision theorists today endeavor to make economic analysis more broadly inclusive by speaking of the importance of including "externalities" and of "costing out the full energy system." Nevertheless, this remains impossible when the externalities are national security, public health, and the preservation of our environmental heritage.

The importance of social criteria as a complement to economic criteria for national decision making is emphasized by Bell[4] in "The Coming of Post-Industrial Society": "A postindustrial society is one in which there will necessarily be more conscious decision making. The chief problem is the stipulation of social choices that accurately reflect the 'ordering' of preferences by individuals." The role of knowledge is central to Bell's discussion, and much of this knowledge—both organizational and technical—is so complex that its organization and application are certainly not possible without large organizations.

These ideas offer guidance in approaching planning for the use of depleting energy resources and for the replacement of these resources by alternative forms. Two guidelines should be remembered:

- The present trend is toward replacement energy systems that are, for the most part, extremely complex and capital intensive. Their construction and operation will require large organizational structures.
- Decisions on these replacement energy systems will be made on grounds that include direct economics as only one of many elements.

The large organizations to develop and implement replacement energy systems can be either public or private. But whatever their form, they will have to call upon sophisticated technology, large numbers of persons, and major capital resources. These organizations will require longevity, for the replacement systems will operate for decades and require sophisticated attention throughout their lifetimes. (In some cases, such as radioactive waste disposal, the time scales may be measured in millennia.)

While extensive attention to economics of the replacement energy systems is essential, many national decisions include economics as but one factor. National defense and education are but two examples in which noneconomic factors dominate the decision-making process.

New energy supply technologies are, with few exceptions, large and capital intensive. They offer clear and exciting opportunities to large organizations. A new electricity generating plant typically has an output of 1,000 megawatts and costs a billion dollars. Breeder reactors will carry the same or higher price tags. Coal conversion systems presently proposed will produce 250 billion British thermal units (Btu) of gas per day and will also cost about a billion dollars. Some supply technologies are even larger. The Trans Alaska Pipeline is costing many billions of dollars, and its financing was well beyond the capabilities of any single corporation. The dominant theme of new energy supply systems is that the prices of the elements are exceedingly high, as are the potential adverse effects associated with their use.

There are exceptions. Solar energy systems utilize an intrinsically diffuse energy form—sunlight. This characteristic assures that many applications of solar energy will be at a small scale, building heating and cooling are the best example. However, other applications of solar energy such as central electricity generation may well turn out to be of a scale comparable to the systems referred to above. If this is the case, such systems will probably prove appealing to the same large organizations involved with other massive supply systems. Indeed, funding patterns in the federal solar program suggest that this evolution is already under way.

The situation with regard to energy use is exactly the reverse. Energy is used by every person and institution. With few exceptions, energy is used in relatively small units, as compared to the units in which it will be supplied by the new technologies. Further, energy is not an end in itself; it is an intermediate product. It is what the energy does that is important to the energy user, not the energy itself.

Few opportunities exist to conserve energy on a scale comparable to the massive supply technologies. And, accordingly, similar incentives do not exist for large organizations to become involved with energy use to the degree that they are involved with energy supply.

The situation is clearly seen within ERDA. This organization, with responsibility for developing new approaches to the entire national energy posture, concentrates 97 percent of its resources on energy supply technologies, and the bulk of this on massive systems. Neither small energy supply systems nor energy use receive major attention.

This situation does not result from lack of opportunities for technological improvement in energy use. These opportunities are legion, and they are well documented. A study we carried out in California shows that even with present natural gas prices, investments in simple energy-conserving measures in houses justify expenditures of several billion dollars in California alone (in building insulation, storm windows, etc.).[5] Extensive material has been prepared documenting energy-savings potentialities in every sector of the society—residential, commercial, industrial, transportation.[6-9] Much can be done with existing technology, and even more will be possible with new concepts. A good example is the use of microprocessors to optimize building climate conditioning systems and lighting levels, to run industrial processes, and to optimize automobile engine operation.

Forecasts of future U.S. energy needs demonstrate the potentialities for conventional energy conservation.[10] The last major energy forecast carried out prior to the oil embargo of 1973 projected use in the year 2000 of 192 quadrillion Btu (quads).[11] After the embargo, the Ford Foundation Energy Policy Project estimated a year 2000 range of 122 to 165 quads.[12] In fact, technical analysis suggests that very much lower levels of energy use are entirely feasible and are consistent with continued growth in Gross National Product. While these projections are not closely coupled to energy price projections, they are useful in illustrating what technological improvements may be possible. Higher energy prices and constraints due to regulation (building codes, miles-per-gallon standards, appliance efficiency standards) are mechanisms that could speed change.

The diffuse nature of energy use—the fact that it affects us all—makes it far more difficult to focus programs in this area than in energy supply. To change this situation, I believe it will be necessary to identify management programs for energy use comparable in scope to the major energy supply projects. There is an important and intrinsic difference between such energy use programs and the energy supply projects. Energy supply projects have well defined objectives—to provide a certain amount of energy in particular forms. Energy use projects generally have a multitude of effects, and consideration of them must take into account not energy use reduction alone, but the full spectrum of implications for the society. Analysis of programs to modify energy use thus tends to be far more complex than analysis of energy supply.

To make these remarks more precise, I now turn to a specific proposal—that mass transit systems may offer the nation one of the most interesting areas for possible modification of energy use. Mass transit is generally not considered an important area for attention from an energy conservation point of view. Transportation uses about 25 percent of total energy used in the United States. The bulk of it—60 percent–is for automobiles.[13] Airplanes, trucks, and military use consume most of the rest. Mass transit uses just one-half percent of all transportation energy. Options for improving automobile efficiency are very great indeed, and it is in this area that most effort is presently being expanded. A recent interagency report summarizes the situation well.[14] By introducing improved and more efficient engine drive trains and by reducing vehicle weight, projected automobile fuel consumption in the year 2000 can be held to about 80 percent of today's level, even allowing for a 2 percent per year growth rate in vehicle miles traveled. One strategy calls for increasing fuel economy of the new car fleet to about 30 miles per gallon (from 17.1 miles per gallon for the 1976 new car fleet) through use of advanced diesel engines—and later by Stirling or Brayton cycle engines--and with careful attention to vehicle weight.

Growth in the number of vehicle miles traveled has negative implications, which must be taken into account. One of these is the accident rate. According to studies by the Department of Transportation, urban automobile transportation will increase urban fatalities by 49 percent, from 18,840 in 1972 to more than 28,000 in 1990. Automobiles will continue to be responsible for 99 percent of total transportation-associated fatalities in urban areas.[15]

Mass transit systems, I believe, offer an important means for reducing long-term energy consumption in urban regions. This assertion contradicts the results of virtually every energy study made to date. There appears to be virtual consensus that mass transit is, from an energy point of view, a rather poor investment. While head of the Federal Energy Administration, Frank Zarb[16] expressed the conventional view particularly clearly. He chose to evaluate the utility of energy conservation measures in terms of the price to which oil would have to rise to justify the measure. Mass transit, he argued, would require oil prices of $300 per barrel. Since oil presently sells for about $13 per barrel, mass transit would make no sense until oil prices rose by a factor of about 25 above today's prices (which are already a factor of four above preembargo prices).

Zarb's conclusion is justified when mass transit is viewed in the short term—over 5 to 10 years. The short-term situation has been carefully analyzed by Stuntz and Hirst.[17] They note the historic decline in mass transit ridership from nearly 20 percent in 1950 to only about 2.5 percent in 1973. Some experiments in encouraging ridership are explored (Vanpool, Dial-a-Ride, fare reductions, etc.), but none are found to have major potential. The authors conclude that "it is clear that transit cannot contribute substantially to the reduction of petroleum imports during the next 5 (and probably not even the next 10) years because of the extremely low base from which transit operates today: less than 3 percent of urban passenger travel." In an extensive review, the Office of Technology Assessment reached the same conclusion.[18] Stuntz and Hirst do note, however, that "unless transit improvement projects are undertaken now, the long-term potential benefits of transit will never be realized."

For a number of years, rail transit in the United States has been in a stage of secular decline. In the 5 years ending in 1974, only 108 miles of new rail transit were installed in the United States.[15] Total installed urban rail mileage in 1972 was 2,200 miles, and there were 613,000 miles of urban streets and highways. There were 110,000 miles of motorbus routes and just 600 miles of electrified trolley coach. Rail transit is neither a major contributor to urban energy use in transportation, nor is it expanding rapidly. Our interest here, however, is not in the present situation, but rather in the potential for change.

The major new subway systems in the United States are the Bay Area Rapid Transit (BART) in the San Francisco Bay area, and Metro in Washington, D.C. Like energy supply systems, both have experienced major cost overruns. The 71 miles of BART originally estimated at $994 million cost about $1.6 billion,[19] and the almost 100 miles of Metro will cost in excess of $4.5 billion, with the final cost not yet known.[20]

These numbers, while enormous in absolute terms, become less formidable when compared with the $1 billion cost of a typical energy supply unit. The BART system cost less than two modern electricity generating plants, and Metro about as much as four plants. On a per capita basis, the costs are also comparable. The BART system, which services an area with a population of 2.3 million,[21] cost $700 per capita to construct. The average capital cost for all urban rail systems to be installed by 1990 is estimated at $448 per capita, and for bus systems $64 per capita.[15]

Electrical plant capacity in the United States now amounts to about 2 kilowatts per person. At the presently projected growth rate of about 5 percent per year, the system size will double in the next 14 years. (This period is also roughly the time required to install a mass transit system.) The cost per person of doubling the installed electricity generating plant will be about $2,000. (The full cost of the electrical system will be close to twice this number because of requirements for transmission lines and distribution systems.)

Of course, the fact that the capital costs of a mass transit system are less than those of an electricity generating plant tells us little about the comparative merit from an energy point of view. What the comparison does do—and I believe this is critical—is to show that there are major energy-related social expenditures contemplated that are substantially larger, per person, than those of mass transit systems. If it can be shown that mass transit systems are potentially comparable contributors in terms of energy, then a trade-off analysis between energy production and conversion systems and mass transit systems becomes interesting from a public policy viewpoint.

To carry the argument further requires analysis of the energy implications of mass transit. Let's begin by noting that BART's energy requirements for construction and operating are modest. Over a 50-year lifetime, the energy of the system is composed of construction energy (44 percent), traction energy (40 percent), and operation and maintenance energy (16 percent).[22] The total lifetime energy is 0.25 quads, or 2.2 million Btu per person per year in the service region. This is just 2.4 percent of the per person energy used in the United States for transportation. The energy requirement on BART is 5,200 Btu per passenger mile, versus 16,200 Btu per passenger mile for automobiles, taking into account both system construction and operating energy.[22] This is a striking energy reduction. However, mass transit systems tend to have relatively low ridership and to be used predominantly for commuting to and from work. Thus the energy saved per dollar invested in the system resulting *directly* from shifts between various transportation modes is relatively small, giving rise to high costs for the system per Btu saved (i.e., the equivalent of $300 per barrel of oil saved quoted by Frank Zarb).[16] (The ridership problems of the BART system[1,2] are not relevant here. BART still has technical problems of reliability, a very poor bus feeder system, and inadequate parking lots at the stations.)

An analysis of short-term intermodal shifts alone surely represents a poor approximation of the impact of mass transit systems on energy use. This is essentially a "static" analysis. It entirely ignores shifts in population patterns that mass transit systems may induce. To date, there have been few attempts to explore the energy implications of transportation systems associated with population shifts, and the shifts in energy use associated with differing forms for the built environment. Fels and Munson[7] explored major life-style shifts in which work was located near living quarters. Under extreme assumptions, they estimated a 90 percent reduction in energy. Further, no mass transit was considered. A group at Brookhaven Laboratory and the State University of New York at Stony Brook considered a less extreme situation in which land use patterns in Long Island were modified so as to require less automobile transport. Reductions of 15 to 25 percent in total system energy requirements were found.[23] The Regional Plan Association (RPA) has conducted an intensive study of the New York City region.[24] And, finally, the classic study "The Costs of Sprawl" made by the Council on Environmental Quality explores changes in automobile energy usage resulting from changing land use patterns.[25]

At this stage in the analysis it becomes necessary to make some heroic assumptions in order to proceed. The attitude required is exemplified by the mathematical biophysicist Nicolas Rashevsky, who introduces one chapter of his classic work "Mathematical Biophysics" as follows:

> In a very rough way, a quadruped, such as a horse or an elephant, may be considered from a mechanical point of view as a bar supported at its ends. This sets definite limits to the length . . . for a given weight.[26]

The transportation network acts as the backbone to the city, and the character of the transportation network determines many aspects of the city, including the capabilities and directions of growth. The following estimates are rough, but do capture some important aspects of the relationship between urban transportation systems and city growth. These aspects, which have been omitted from previous discussions, are crucial for understanding long-term urban energy use.

The RPA study provides the best starting point for our purposes. RPA explored the total amount of energy used in the counties surrounding New York City. The study discovered that there is a correlation between energy use per capita and population density. At very low densities (100 persons per square mile), energy use was about 190 million Btu per capita. This use dropped to about 120 million Btu per capita at 30,000 persons per square mile. (Energy use rises rapidly at higher densities for reasons that could be specific to Manhattan or else fundamental to extremely high population densities. The average energy use per capita in the entire region is only two-thirds of the U.S. average, primarily as a consequence of the small amount of manufacturing and the high population density.)

As population density increases in the New York metropolitan area from 100 to several thousand persons per square mile, energy use per capita drops by about 50 million Btu. Typical population densities in Standard Metropolitan

Statistical Areas (SMSA's) range from about 1,000 to about 5,000. Within the SMSA's, the percentage of population using public transport of all types for work trips increases roughly linearly with density, from negligible at densities near 1,000 persons per square mile to 20 percent at densities of 5,000 persons per square mile.[27]

Causal relationships between mass transit and population density are not well known. However, there appears to be little doubt that the prospect of a new mass transit system in the San Francisco Bay region has had a major effect on high density construction in the downtown area and has also led to considerable construction in outlying areas, especially at stations.[19] A similar conclusion was drawn from a study of "the Skokie Swift," a short commuter line running from the Village of Skokie, Ill., to Chicago. A survey by the Federal Energy Administration disclosed that developers often identified the Swift as a major factor behind their decision to build near the Swift Terminal area.[28]

The Toronto subway system is generally acknowledged to be one of the most successful new systems, largely because of its effect in encouraging new construction. The *Toronto Daily Star* reported:

> Toronto's subway system draws new building like a magnet, a survey of recent construction shows. Ninety percent of all office building and half of all apartment construction is occurring within a 5-minute walk of subway systems, the study by A. E. LePage Ltd., a real estate firm, shows.[29]

One way to estimate the long-term energy impact of a mass transit system is to use the energy-density relationships found in the RPA study[24] to evaluate the energy impact of density increases assumed brought about by the system. Suppose that over a period of several decades a fully developed and reliable BART system with an adequate feeder system induces an increase in population density level in the San Francisco/Oakland SMSA from its 1970 value of 4,387 persons per square mile[27] to 6,000 persons per square mile. The RPA study would imply a decrease in annual energy use of about 10 million Btu per capita. Most of this shift arises from changes in automobile usage. Residential energy use varies by a considerable amount in apparently similar areas. (Energy use per square foot of residential space is 134,000 Btu in the Bronx, but 191,000 Btu in Queens.) As a result of increasing energy costs and new standards, new construction will be more energy efficient than old, and apartments more efficient than single family dwellings. One analysis[8] shows 76 million Btu per dwelling in 1972, and 44 million Btu per dwelling in 2000.

A decrease of 30 million Btu per capita due to all effects would, if applied to the full 2.3 million people in BART's service area, lead to an annual reduction in energy use of 69 trillion Btu per year. This is about 15 percent of total energy use.

Increases in efficiency of energy use in both buildings and in the transportation system would be complemented by changes in the form of the energy used. Increasingly scarce petroleum used for automobiles would be replaced by electricity to operate mass transit. New buildings could be operated on energy forms other than oil and natural gas—such forms as electricity (using heat pumps for efficiency) and solar energy. The saving of petroleum that results from mass transit systems could, in the long run, prove far more significant than the total number of Btu's saved.

From a capital investment point of view, the mass transit system would then be comparable in payoff to investment in new energy supply systems. The investment required to save 1 million Btu per year would be $35 per million Btu per year saved. The corresponding capital investment figure for an electric generating plant is about $50 per million Btu per year delivered as electricity. Synthetic gas plants cost about $12 per million Btu per year delivered.

The finance of mass transit systems is a complex matter that goes far beyond energy. Ever since 1968, transit operations nationally have operated at a net loss.[21] There are many reasons for this. Transit systems of all forms have traditionally been subsidized heavily in our society, and the existence of an operating loss does not allow any conclusions regarding the societal cost or benefit of such systems. Of more interest are the operating costs per seat mile (excluding capital costs). In the San Francisco Bay area, for all forms of mass transit except cable cars, these range from 3 to 5 cents per seat mile, with BART lying at the lower end.[15,19] This is less than automobile operating costs (excluding capital and roadway costs), which are about 12.7 cents per vehicle mile, or 8 cents per mile with the 1.6 occupants per vehicle typical of urban driving patterns.[14,17]

The arguments given here for energy savings due to mass transit systems are clearly only first approximations. We do not know how much population density increase can actually be attributed to a mass transit system and how much to other factors. The time for new construction is measured in decades, and energy savings will not accrue until many years following construction. On the other hand, the example used is for the most expensive form of mass transit system. Other mass transit systems that cost a small fraction of subway systems may have the same or greater impact on land use patterns. Typical guideway costs (in millions of dollars per mile)[15] are $10 (rail rapid transit), $3.5 (small vehicle systems), $2.5 (private auto).

What the analysis does show, however, is that plausible estimates of long-term energy savings that might derive from the most expensive form of mass transit system suggest that investment in such systems may be, from a purely energy point of view, comparable to investment in new energy supply systems when measured in terms of efficiency in use of capital.

Thinking in terms of long-run growth of the built environment, and the forces affecting that growth, allows planning for technologies that go beyond transportation. An interesting example is district heating, where a single energy conversion system provides both electricity and heat for a complex of buildings. Such systems are very widely used in Europe, especially in Sweden.[30] Fuel cells now under development may lead to highly efficient district heating systems, which would be attractive and feasible on a small scale.[31] By taking into account the new choices that such technologies make possible, planning for city growth could increase the potential for energy conservation well beyond the estimates given here.

The purpose of this discussion has been to suggest that mass transit, when considered on a time scale of decades rather than just a few years, offers the potential for conserving large amounts of energy, especially in the form of oil. It is clear that far more extensive analysis will be required in order to assess just how large the opportunities for energy conservation in mass transit really are. Surely they exceed the estimates generally made today.

A few remarks about the technologies associated with mass transit might also be appropriate. In recent decades, extensive research programs have been devoted to development of new transportation types. Most have been relatively unsuccessful. A striking example is the personal rapid transit prototype installation at Morgantown, W.Va., which never operated properly and ultimately had to be demolished. The BART system has had continuing technical difficulties, many of them associated with efforts to apply high technology too rapidly to a mass transit system. At the other extreme, the Vanpool approach that requires no new technology is relatively successful. Intercity high speed transportation concepts such as the use of magnetic levitation have been explored briefly, but have been abandoned in the United States. They are being pursued elsewhere, however, notably in Japan and Germany.

Although I have concentrated predominantly on heavy rail transit, it is probable that the best approach to the use of mass transit systems to revitalize cities will be dominated by other transport forms such as trolley systems and electrified or diesel buses. Except in extremely high population density areas, these systems are cheaper and offer more flexibility.

There are several areas where technology can improve overall energy efficiency of bus systems. Probably the least important of these is the bus itself. Of much more importance are load factors and attractiveness of the system to potential riders. Experiments in progress—such as the General Motors Cincinnati Project—will use extensive monitoring to provide detailed information to dispatchers of bus location, occupancy, and mechanical state. Simultaneously, riders will be advised—through displays at bus stops or by telephone inquiry—when the next bus can be expected and how full it will be. Sophisticated sensors combined with computer analysis may be able to significantly reduce the delays and inconvenience that deter so many persons from riding any mass transit system regardless of the cost of the ride.[32]

In my view, we should not be unduly disheartened by the difficulties experienced in applying technology to mass transit. In energy supply technologies, prototypes and demonstration units are required to perfect systems. It is not at all unusual for many years to be required to shake down these systems. Some of these difficulties are extreme—an example being the nuclear reprocessing business. In energy supply, these difficulties are accepted as being inevitable concomitants to the complexities of technology. We should take the same viewpoint with respect to energy demand projects. In initial projects, we should expect significant errors in estimation of costs and reliability and should budget accordingly. Investment in complex end-use technologies like mass transit will pay off on a time scale of decades—the same kind of pay-off period that is associated with energy supply technologies. Learning curves apply to both energy supply and energy demand.

In the nineteenth century, the fabric of American life was changed by decisions to expand the railroad system throughout the nation. Even more striking changes were brought about during the 1960's by the decision to develop the Interstate Highway System. The implications of the Interstate Highway were poorly understood by those who planned it. Within Europe, emphasis on rail rather than road transportation network has affected the growth of cities, the attitudes of citizens, and many other aspects of life in ways very different from the United States. Railroad transport is accepted as the norm, substantially reducing the need for automotive transport. The major energy and resource shortages and rising costs we will face within a few decades provide incentive to reexamine the major factors influencing the long-term structure of our society. It is my belief that such restructuring is essential if our standard of living is not to deteriorate.

Land use patterns and the character of the built environment in the United States are continually changing. In many areas, development is anticipated, and debate over goals is intense. A case in point is the Box Elder watershed near Denver, Colo., where virtually every land use issue facing the Rocky Mountain States comes to a focus and where transportation decisions that must be made soon will largely determine the growth patterns of the region.[33]

In San Francisco, citizens realized that freeway expansion was destroying the character of the city, and mounted an effective effort to stop further construction. Downtown San Francisco is now dominated by high rise buildings that would have been inconceivable 30 years ago. The role of the freeways and of BART in influencing these changes is impossible to accurately assess, but is certainly significant.

Large scale investment in new transportation-related facilities has profound positive implications for our society. What I urge is that we expand our efforts to explore these implications. I also urge that we put to work the technological capacity of this nation, especially as it exists in the large organizations such as the aerospace, defense, shipbuilding, and automobile industries.

The United States has virtually unlimited capacity to develop and produce technological devices. In the 1960's, the major thrust of this capability was the space program. The revitalization of American cities through the intelligent application of technology is a far more significant goal in terms of long-term implications for the nation's future.

Energy shortages will recur and grow increasingly severe in the next few decades. Responses based purely upon increasing energy supply will prove inadequate. We need to give equal attention to modifying demand. While much can be accomplished through "technical fixes," changes in social patterns will be required as well. This prospect is exciting, for it allows us to utilize energy as an organizing principle to focus discussion on national objectives.

REFERENCES

1. Webber, Melvin M. The BART Experience—What Have We Learned. Institute of Urban and Regional Development and Institute of Transportation Studies, Monograph No. 26. University of California, Berkeley, Calif. 1976.

2. Lave, Charles A. The Negative Energy Impact of Modern Rail Transit Systems, (Report ITS-I-SP-76-2). Transportation and Energy: Some Current Myths, (Report ITS-I-76-1). School of Social Sciences and Institute of Transportation Studies, University of California, Irvine, Calif. September 1976. Rail Rapid Transit: The Modern Way to Waste Energy. Report submitted to the Transportation Research Board for the Annual Meeting, January 1977.

3. Galbraith, J. K. Economics and the Public Purpose. New American Library, New York. 1976.

4. Bell, Daniel. The Coming of Post-Industrial Society. Basic Books, New York. 1973.

5. Craig, P. P., B. D. Goldstein, R. W. KuKulka, and A. Rosenfeld. Energy Extension for California: Context and Potential Impact. Presented at the Energy Extension Service Workshop, Berkeley, Calif. July 19-24, 1976. Proceedings published by Lawrence Berkeley Laboratory, 1977.

6. See for example: A Time to Choose: America's Energy Future. Report of the Energy Policy Project of the Ford Foundation. Ballinger Publishing Co., Cambridge, Mass. 1974.

7. Williams, R. H., ed. The Energy Conservation Papers. Ballinger Publishing Co., Cambridge, Mass. 1975.

8. Energy Demand Studies: Major Consuming Countries. First report of the Workshop on Alternative Energy Strategies (WAES). MIT Press, Cambridge, Mass. 1976.

9. Lovins, Amory. Exploring Energy-Efficient Futures for Canada. Conserver Society Notes, May-June 1976.

10. Middle and Long Term Energy Policies and Alternatives. Part 7 of appendix to hearings before the Subcommittee on Energy and Power of 94th Congress. The Committee on Interstate and Foreign Commerce, Serial No. 94-77. Washington, D.C. March 1976.

11. DuPree, W. G., Jr., and J. A. West. U. S. Energy Through the Year 2000. U.S. Department of the Interior, Washington, D.C. December 1972.

12. A National Plan for Energy Research, Development and Demonstration. ERDA-48 (1975) and ERDA 76-1 (1976) Energy Research and Development Administration, Washington, D.C.

13. Hirst, Eric. Policies to Reduce Transportation Fuel Use. Science, April 6, 1976.

14. Report by the Federal Task Force on Motor Vehicle Goals Beyond 1980. Available from Assistant Secretary for Systems Development and Technology, U.S. Department of Transportation, Washington, D.C. September 1976.

15. National Transportation Report, 1974. U.S. Department of Transportation, Washington, D.C. July 1975.

16. Zarb, F. Energy Conservation. Briefing document. Federal Energy Administration, Washington, D.C. 1975.

17. Stuntz, M. S., Jr., and Eric Hirst. Energy Conservation Potential for Urban Mass Transit. Energy Conservation Paper No. 34. Federal Energy Administration, Washington, D.C. 1975.

18. Energy, the Economy and Mass Transit. Report OTA-T-15. Office of Technology Assessment, Washington, D.C. December 1975.

19. BART Impact Program—Transportation and Travel Impacts of BART: Interim Service Findings. U.S. Department of Transportation, Washington, D.C. 1976.

20. An Assessment of Community Planning for Mass Transit, Vol. 10, Washington, D.C., Case Study. Report OTA-T-25. Office of Technology Assessment, Washington, D.C. February 1976.

21. Transit Operating Report, 1974. Also: Transit Fact Book, 1975-76 ed. American Public Transit Association, Washington, D.C.

22. Healy, T. J. Energy Requirements of the BART System. Unpublished report. University of Santa Clara, Santa Clara, Calif. November 1973.

23. Carroll, T. Owen et al. Land Use and Energy Utilization, Interim Report. Informal Report BNL 20577. Brookhaven National Laboratory, Upton, N.Y. 1976.

24. Regional Energy Consumption, Second Interim Report, 1974. Regional Plan Association, New York, N.Y., and Resources for the Future, Washington, D.C. (Full report in press.)

25. The Costs of Sprawl. Council on Environmental Quality, Washington, D.C. April 1974.

26. Rashevsky, Nicolas. Mathematical Biophysics. University of Chicago Press, Chicago, Ill. 1948. p. 586.

27. Urban Transportation Fact Book, 1974. American Institute of Planners, Washington, D.C.

28. Hershberger, B. The Skokie Swift: Study of a Transit Line and Development. Federal Energy Administration, Washington, D.C. Draft dated August 13, 1974.

29. Toronto Daily Star. July 20, 1971. Quoted in a brochure prepared by the Toronto Transit Commission, Toronto, Canada.

30. Schipper, Lee. Efficient Energy Use: The Swedish Example. Lawrence Berkeley Laboratory Report ER6-76-09. University of California, Berkeley, Calif. 1976.

31. National Benefits Associated with Commercial Applications of Fuel Cell Powerplants. ERDA 76-54. Energy Research and Development Administration, Washington, D.C. 1976.

32. Naylor, Michael. General Motors Research Laboratories. Private communication.

33. Ingraham, Elizabeth Wright. Lead Time for Assessing Land Use: A Case Study. Science, 194: 17, Oct. 1, 1976.

W. Haefele and W. Sassin
International Institute for Applied Systems Analysis
Laxenburg, Austria

Wolf Haefele is deputy director of the International Institute for Applied Systems Analysis and leads IIASA's research in energy systems, working specifically on problems of energy demand and energy strategies. He came to IIASA in July 1973 from the Karlsruhe Nuclear Research Center of the Federal Republic of Germany.

After receiving his Ph.D. in theoretical physics from the University of Goettingen in 1955, Professor Haefele joined the Karlsruhe Nuclear Research Center, where he was responsible for studies on theoretical reactor physics. He participated in the design of the first German-built reactor FR2 and then spent one year at the Oak Ridge National Laboratory. He returned to Karlsruhe and established the Fast Breeder Project, serving as its leader for the next 12 years until the SNR 300 started and industry assumed the project responsibility. During this period, he was also head of the Institute for Applied Systems Analysis and Reactor Physics.

He was scientific advisor to the West German government on the nonproliferation treaty, and in that capacity participated in the conception and design of the international material safeguards system of the International Atomic Energy Agency (IAEA) in Vienna. Professor Haefele is a member of the Scientific Advisory Committee of the IAEA. He has served as president of the German Nuclear Society and is a Fellow of the American Nuclear Society.

Wolfgang Sassin joined the Energy group of IIASA in July 1975. His work focuses on geographical deployment of modern energy systems and supply and demand patterns. He came to IIASA from the Nuclear Research Center in Karlsruhe. Dr. Sassin received his Ph.D. in solid state physics from the Technical University in Aachen (1969).

From 1964 to 1973 he was with the Nuclear Research Establishment in Julich, F.R.G., working on radiation damage, low temperature electricity transmission, and systems analysis in the fields of energy and the environment.

He is a member of the Environment and Energy project group of the Federal Ministry of the Interior, F.R.G.

Contrasting Views of the Future and Their Influence on our Technological Horizons for Energy

Introduction

It is well known that the same data can be used to support many different conclusions, even when the "truth" can be ascertained by direct observation. Thus, it is not surprising that views of the future, which of necessity are based on tentative projections, often founded on differing estimates of the same values and filtered through differing assumptions and biases, tend to diverge. In this paper, we examine contrasting views of the future as they relate to energy production and use.

We treat five aspects of future energy use—global energy demand; world energy resources; spatial patterns of energy demand and centralization; energy storage and transportation; and capital stocks, investments, and less developed countries. In the last case, we offer a minimodel that may be of use in elucidating the relationships among big and small, capital- and labor-intensive technologies in developed and less developed nations. Clearly, the complexity and diversity of the factors that affect each of these aspects—and there are social, political, environmental, geological, technological, ethical, and other factors, each of them of local, regional, or global significance—and the inherent uncertainty of the future make contrasting views inevitable and resolution of the differences impossible. When estimates of even so basic a consideration as available coal reserves vary by a factor of 10, it is easy to see that estimates of other, less quantifiable, components must also vary and that alternative futures based on each combination of these varying values must differ greatly.

It is to be hoped that the observations, data, and conclusions reported here can help us to think about the future, and about alternative futures, and the implications for energy technology in a reasonable and appropriate way. If this can be done, we may be able to avoid unpleasant surprises.

Global Energy Demand

The present global energy demand totals about 7.5 terawatt years per year (terawatts for short), or an average of roughly 2 kilowatts (kW) for each of the world's 4 billion people. There are three reasons to expect a continued increase in this demand:

- Continuing economic growth in developed countries (DC).

- Development of less developed countries (LDC's) and the consequent increase in per capita consumption of energy.
- Growth of world population.

Economic growth of developed countries has been the focus of intensive studies and debates in the United States, Sweden, and elsewhere.[1,2] Part of this larger issue is the question of energy conservation in these nations, a question whose answer is no longer in doubt: DC's must conserve energy, and there is room for such energy saving. Indeed, their second law efficiencies are at low values, say 5 percent. A certain amount of energy can be saved quickly and easily, at least in principle, by such simple measures as reducing room temperatures and lowering the speed of automobiles. More extensive energy-saving schemes require adjustment of the existing infrastructure through altered patterns of capital investment, labor allocation, and energy use; clearly, this is a more complex undertaking.

A typical estimate of the energy savings to be realized through such simple measures as mentioned above is 5 percent of the total primary energy input, the savings to be achieved through adjustments of the infrastructure, however, are difficult to assess. Estimates range from half a percent per year to not more than 20 percent or so. The longer range potential of energy savings requires complex analysis, which in itself may lead to contrasting views; nevertheless, the desirability of saving energy in DC's is not a matter of controversy, so, there is no point here in going into it in greater depth. The problem of an appropriate economic growth rate of DC's, on the other hand, is quite a different matter, and it will be dealt with in a broader context later in this paper.

The development of LDC's as reflected in an increase in per capita energy consumption is quite a different matter. Figure 1 shows the present distribution of per capita demand. In about 80 countries the per capita consumption of energy is about 0.2 kW, barely enough to run the households: and 72 percent of the world population is below the global average of 2 kW/capita, while 6 percent of the world population consumes more than 7 kW per capita. In an increasingly interdependent world, this distribution will not continue. However, existing economic and social patterns make it unlikely that the LDC's will make a quick transition. Because of this, economic forecasts up to the year 2000 show no major influence on world energy demand due to such development of LDC's. The controversy lies beyond the year 2000, where the forecasts diverge. Since modern energy strategies must cover roughly 50 years if major transitions are to be included,[3-5] it would be inappropriate to restrict forecasts to the year 2000.

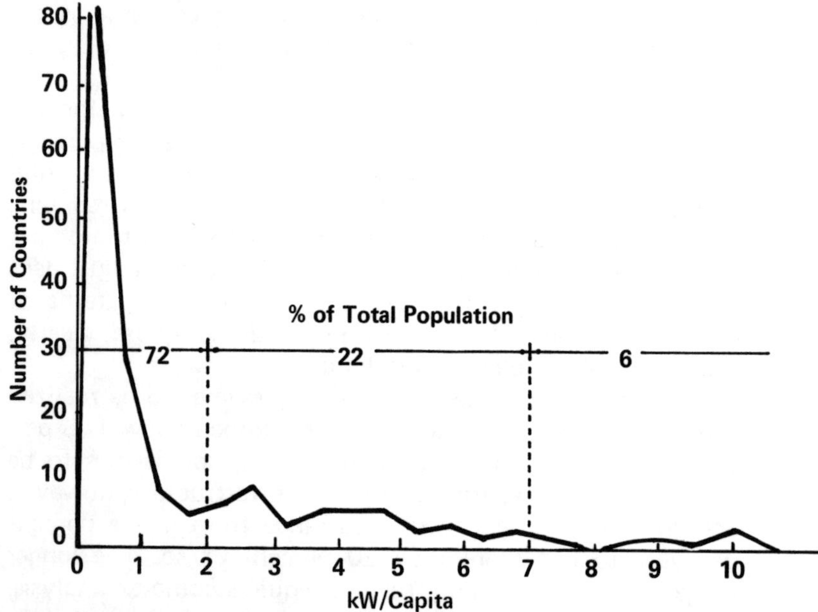

Source: Charpentier, IIASA

FIGURE 1. Distribution of Per Capita Energy Consumption (1971)

Another pertinent question related to the development of LDC's is whether existing economic and social patterns, specifically patterns of world trade, will prevail. The UNCTAD (United Nations Council on Trade and Development) Conferences of 1964-1976 led to the creation within the UN of Group 77 and to a scheme called the New Economic Order. Among other things, this scheme assumes for the LDC's a 25 percent share of the gross world product by the year 2000. Their share today is 12 to 15 percent. If existing economic arrangements prevail—as expressed, for instance, by GATT (General Agreement on Tariffs and Trade)—a target of 25 percent in the year 2000 is probably unrealistic. Nevertheless, one should not underestimate the strength of the political issues that underlie such claims, and the fundamental issue here is whether we will have over the next 50 years a compartmentalized world or an interdependent one. World energy demand will be heavily influenced by the resolution of this issue. (We will come back to such considerations toward the end of this paper.)

The third component of an increasing world energy demand is global population; including population as a factor implicitly assumes that all people do indeed participate in the world economy in one form or another. There are contrasting views on the growth of world population. Figure 2 shows the expected percentage increase, based on achievement of replacement fertility in 1970-75, 2000-05, and 2040-45, for the developed and the developing countries. The upper curves would result in an equilibrium population of about 17 billion, while the lower curves correspond to a total of only 7 billion people. The gap between such projections can be seen in Figure 3, which also reveals that extension of current family sizes leads to even higher figures. The UN Conference held in Bucharest in 1974 assumed equilibrium at 12 billion people.

A straightforward calculation using this last figure and

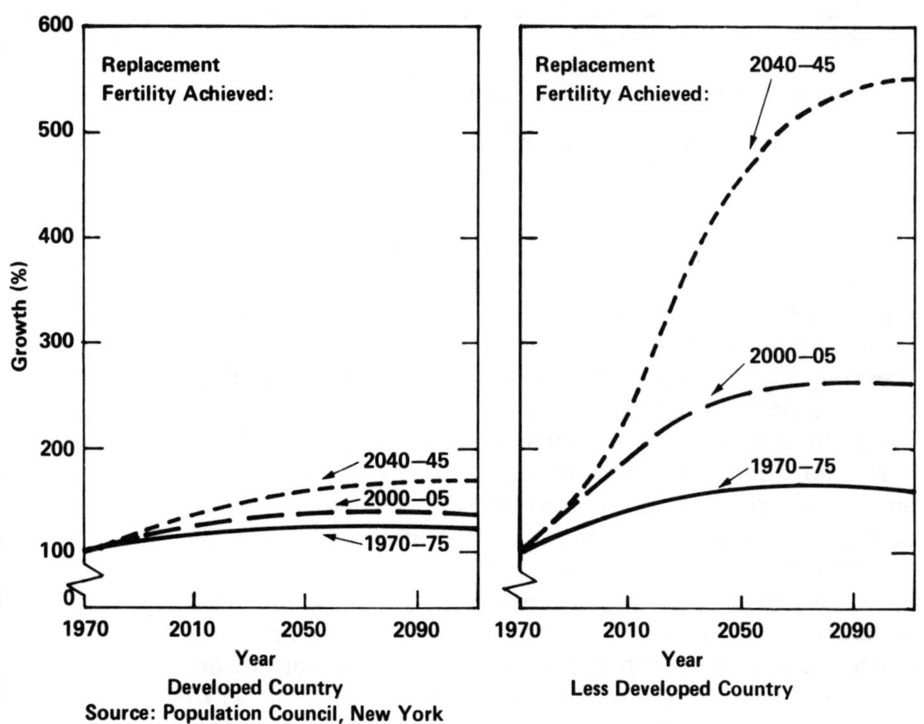

FIGURE 2. Population Growth and Fertility

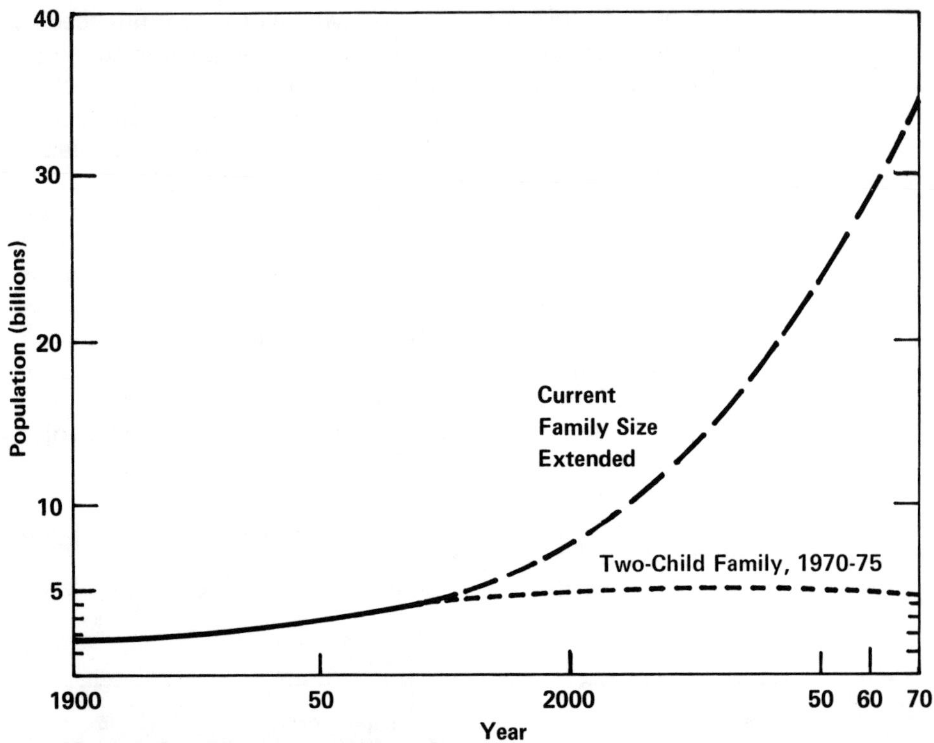

FIGURE 3. World Population Growth Projections

assuming average global energy consumption of 2, 3, or 5 kW per capita gives 24, 36, or 64 TW as the corresponding global energy production levels (Figure 4). A simple extrapolation of secular trends in per capita increases of ~0.2 percent per year leads to the figure of 3 kW per capita, and a break in such secular trends as intended by the New Economic Order might result in a figure of 5 kW per capita. These figures are low compared with projections made 10 years ago. Even today it is not at all unreasonable to assume an average per capita demand of 20 kW,[6] which would correspond to a global energy production of 260 TW—35 times greater than today's value of 7.5 TW.

Another factor to be considered is the fundamental difference in nature between present-day energy sources and conceivable future energy sources, such as nuclear fission, bulk solar power, or modern, large-scale use of coal, all of which require conversion into a versatile form of secondary energy. Figure 5 illustrates the widening gap between such primary and secondary energy forms. Today only electricity production is accompanied by such large concentrations of waste heat; when the nonelectrical sector also experiences such conversion losses—for instance in the gasification of coal—the per capita consumption of primary energy could increase accordingly.

Further, a population of 12 billion would mean changes in life style in various parts of the world. For example, feeding a world population of this size might require an upgrading of water quality, and especially large-scale water desalination, which would increase world demand for energy

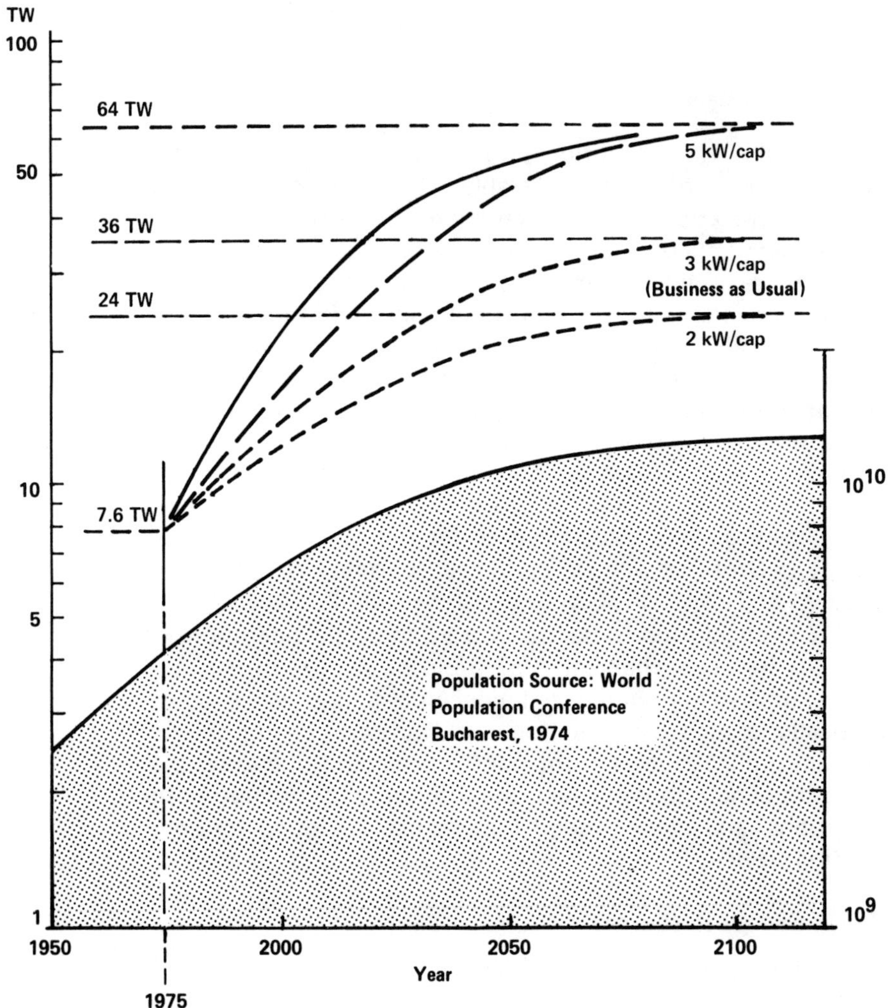

FIGURE 4. Global Energy Scenarios

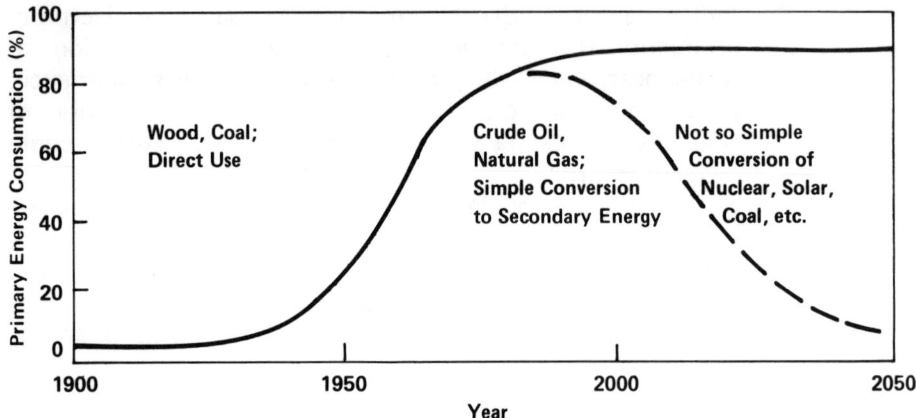

FIGURE 5. A Gap Between Primary and Secondary Energy in Federal Republic of Germany

significantly. If in addition to the 14.5 million square kilometers (km^2) of land now arable, 20 million km^2 in areas that are now arid were to be cultivated, requiring 200 centimeters per km^2 of water per year, 40,000 km^3 more water would be required per year. (For comparison, the total water runoff from the continents to the oceans is 35,000 km^3 per year.) At 50 kilowatt hours (kWh)/m^3, annual energy requirements for desalination alone would be 240 TW. Consideration of changes in life style 50 years hence must also focus on the question of urbanization. So far as the pronounced trend toward urbanization continues (Figure 6), concentration of population leads to high energy consumption densities. This is of relevance for the size of the energy supply network as well as for the appropriate kind of energy supply, and we will touch on these implications below.

The conclusion to be drawn from these considerations is not, for instance, that 240 TW would be required for water desalination. One should not conclude that 5 kW is automatically a high estimate for worldwide per capita energy consumption 50 years from now. Nor should one draw the opposite conclusions. In fact, the future is wide open in both directions, and equally good reasons can be found for contrasting views.

World Energy Resources

Of the world energy demand of 7.5 TW, oil and gas in 1975 supplied 5.5 TW or 73 percent. This figure includes the Soviet Union and Eastern European countries, but it does not include China. The market shares for oil and gas have been increasing over the past 100 years (Figure 7), the increase reflecting not only cheap production but easy transportation and storage and versatile end use. These advantages have led to the development of a supply system that is worldwide and thus to that extent centralized. Forty percent of the oil and gas produced in the noncommunist world, or enough to produce 1.7 TW, goes into trade.

The infrastructure that has developed reflects the availability of cheap oil and gas, and once such an infrastructure is in place, it tends in turn to dominate the outlook for reserves of fossil fuels in general, tending to depress estimates of coal reserves and inflate those of oil and gas reserves. Since reserves are a subset of resources and since estimates of resources have to make reference to price, to implications for groundwater, to effects on the atmosphere, and to many other such elements, it is not surprising that there are contrasting views on the future availability of resources.

As there is a long history of studies of coal resources and reserves (Figure 8), we will begin by considering this source of energy. Figure 8 illustrates projected coal production as a function of time by means the Hubbert method.[7] At the Detroit World Energy Conference of 1974 the total geological resources were estimated to be 10.7×10^{12} tons.[8] At the International Institute for Applied Systems Analysis

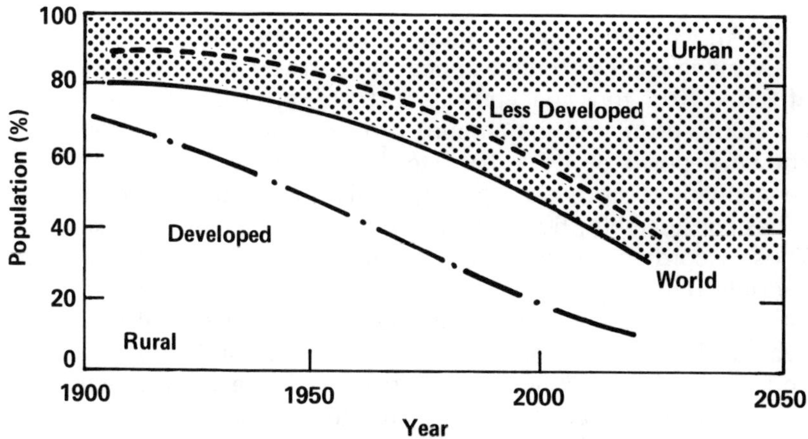

FIGURE 6. Urbanization Trends

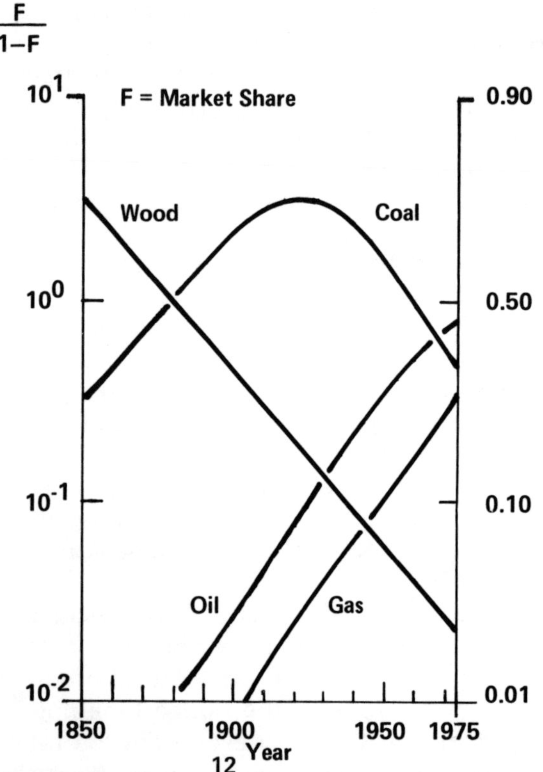

Source: Marchetti

FIGURE 7. World Energy Market Penetrations

(IIASA), Grenon[9] revised these data, taking into account the varying energy content of coal, and arrived at an estimate of 8.4×10^{12} tons coal equivalent (tce). Fettweis, after careful study, estimated the amount of coal available in already known deposits at 1.7×10^{12} tce.[9] A recovery factor must be applied to Fettweis's figure; while the recovery factor varies considerably, 50 percent is a reasonable global average. Thus, according to his data, ultimately recoverable reserves are only 0.85×10^{12} tce, or 10 percent of Grenon's revised estimate. Use of Hubbert's method then indicates that perhaps 7 TW per year could be produced around 2050. If improvements in technology are assumed, however, as much as 15 TW could be produced, and this factor-of-two difference, along with the differences in resource and reserve estimates, once again offers grounds for contrasting views of the future.

Fettweis then applied the same approach to the evalua-

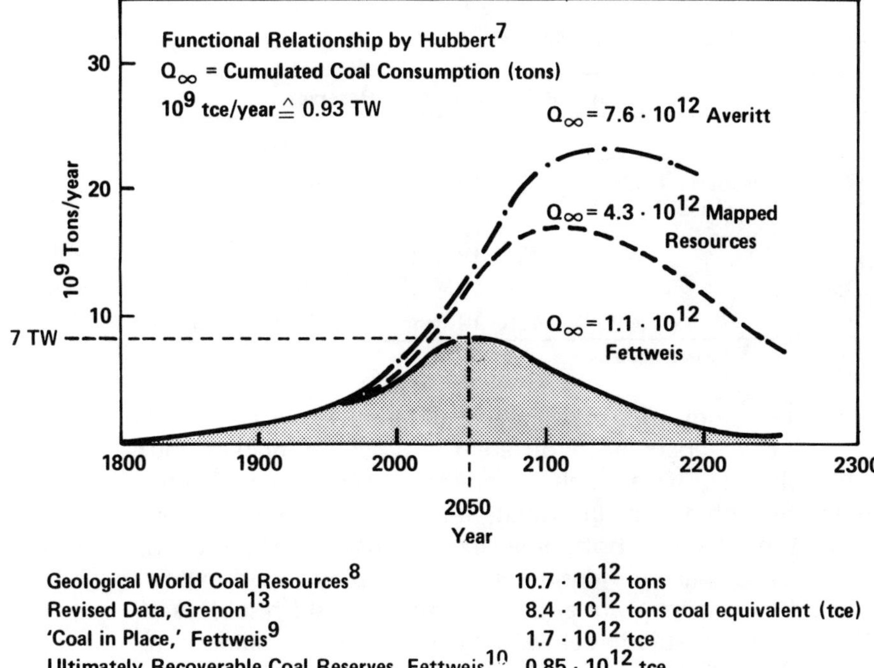

FIGURE 8. Coal Production and Reserves

tion of oil and gas reserves as envisaged by earlier sources (Table 1). He pointed out that while coal has always been understood in terms of total geological amounts, this has not been the case for oil and gas, where the search was only for reserves adequate for a certain time period. It is significant, therefore, that Fettweis tried to apply a single format to coal and to oil and gas (Table 2). He suggests 0.34×10^{12} tce for oil and 0.22 tce for gas as realistic estimates, and 0.67×10^{12} tce (oil) and 0.43×10^{12} tce (gas) as optimistic. His optimistic total for oil, gas, and coal is 2600 TW-years; the realistic total is 1300 TW-years (Table 2). The implications for getting hold of these amounts of energy indeed give room for contrasting views.

It is equally important to look at resources of uranium, thorium, lithium, and deuterium. At IIASA we are about to complete a study that compares the fast fission breeder and the fusion breeder, and the data reported here (Table 3) are from this study.[11] If burned in light-water reactors, low-cost uranium does not add very much to world fuel reserves, but breeding adds another dimension. There are contrasting views on the fast fission breeder and the fusion breeder, and

TABLE 1
Oil and Gas Reserves*

		Ultimately Recoverable Reserves	
		Amount	**Reference**
Oil	Recovery Factor 40%	$235 \cdot 10^9$ tons	Moody Tokyo, 1975
		$467 \cdot 10^9$ tons	Weeks, 1971
Gas	– –	$205 \cdot 10^{12} m^3$	Weeks, 1971
		$400 \cdot 10^{12} m^3$	U.S.S.R. Estimate

*Data selected and evaluated by Fettweis [10]

the IIASA study tries to evaluate their relative merits.

The breeders have different physical principles, but otherwise they have much in common. In the case of fission, uranium is bred into plutonium; in the case of fusion, lithium is bred into tritium. Both breeders make use of fast neutrons, and their engineering designs tend to employ liquid metals. Their fuel resources are also very similar. World resources of lithium and uranium below $100 per kilogram amount to about 200,000 TW-years each. Because the raw material share in the busbar total cost is less than 1 percent, the cost of raw material is essentially irrelevant for both breeders and price increases would have very little effect. This led A. Weinberg to observe in 1959 that if such breeders come into use, it will be possible to burn the rocks and the seas.[12] The breeder reactors offer the promise of hundreds of millions of TW-years.

Since 1974 the French 250-MWe Phenix reactor has been on the grid and is operating smoothly with an availability of 85 percent. The Soviet and British prototype

TABLE 2
Ultimately Recoverable Reserves of Fossil Fuels

	Estimates (10^{12} tce)	
	Realistic	**Optimistic**
Oil	0.34	0.67
Gas	0.22	0.43
Coal	0.85	1.70
Total	1.41	2.80

Source: Fettweis [10]

TABLE 3
Fuel Reserves and Consumption Rates

			Energy Content (TW-Year)
Fuel	World uranium*	To $ 33/kg	62
	World oil		400
	World ultimately recoverable gas, oil, and coal		2,600
	U.S. lithium		7,700
	U.S. uranium*	To $ 250/kg	22,000
	World lithium	To $ 60/kg	175,000
	World uranium	To $ 100/kg	200,000
Potential Fuel	Uranium in oceans*		$6 \cdot 10^6$
	Uranium on continental crust to 1 km*		$600 \cdot 10^6$
	Lithium in oceans	(25,000 kWh/g)	$700 \cdot 10^6$
	Deuterium in oceans		$500{,}000 \cdot 10^6$
Consumption Rates	World (1975)		7.5
	U.S. energy for electricity (1975)		0.6
	Annual, for 12 billion people at 5 kW/Capita		60

*For use in light- water reactors.
Source: Haefele et al.[11]

breeder reactors are now approaching full load, and construction of the German fast breeder prototype is far advanced. The fast breeder reactor is the only power generation device that can provide a virtually unlimited supply of energy with today's technology and on a scale that is industrially significant. There are contrasting views on the significance of this very fact.

Both breeders must face the problem of radiation damage. The problem is more severe in fusion breeders, where parts of the structural material must constantly be replaced. If this is done on a once-through basis, the limitations imposed might be more severe than those imposed by limitations of lithium supply (Table 4). The requirements for beryllium, for example, may far outstrip the reserves. Related to the problem of radiation damage is the problem of neutron-induced activation and the consequent problem of radioactive waste disposal. Even in this respect there are similarities between the fission and the fusion breeder. For more details we refer to the forthcoming IIASA study.[11]

Breeders, fission or fusion, are not the only way of providing practically unlimited amounts of energy, assuming

TABLE 4
Maximum Materials Requirements for D-T Fusion Reactors

Element	Maximum Requirement (T/Mwe)	Makeup (T/Mwe Year)	World Reserves (10^6 Tons, Present Prices)	Maximum Energy Depending on Resource Limitation (TWe Year)
Aluminum	2.42	0	2,000	7×10^3
Beryllium	0.43	0.011	0.04	1.6×10^3
Chromium	7.92	0.088	370	1×10^6
Niobium	1.93	0.193	7	3×10^4
Iron	777.82	0.304	276,000	1×10^8
Molybdenum	4.88	0.069	13	6×10^4
Nickel	5.91	0.069	24	9×10^4

the necessary technological advances are made. Solar energy is also nearly unlimited. One may also want to add the dry geothermal option, whose potential, though not unlimited, is still very large.

In relation to the sources of energy discussed in this section, energy conservation can be seen as a medium-range tactical requirement, not as a long-range strategic goal.

Spatial Patterns of Energy Demand and Centralization

About 120 years ago, in the days of James Watt, the size of an electric generator was about 10 kW. Today's generators are 1 million kW, five orders of magnitude larger. This gives a doubling time of roughly 7 years. This crude evaluation is essentially confirmed when more specific data are employed. Figure 9 shows the evolution of generator sizes in the United States from 1900 onward and in the Federal Republic of Germany from 1950 onward, reflecting doubling times of 8 to 10 years. If the load of the grid had remained constant, increasing centralization would have been the result, but the growth of the size of the generator was accompanied by improvements in the technology of electricity transportation. As a consequence, the distances over which electricity could be transported grew; this process can be quantified easily by considering the maximum transmission voltages, which show a doubling time of 23 years (Figure 10).

As the size of the area serviced grew, so too did the consumption per square meter, and it has been observed that the ratio of the size of the generator to the size of the load

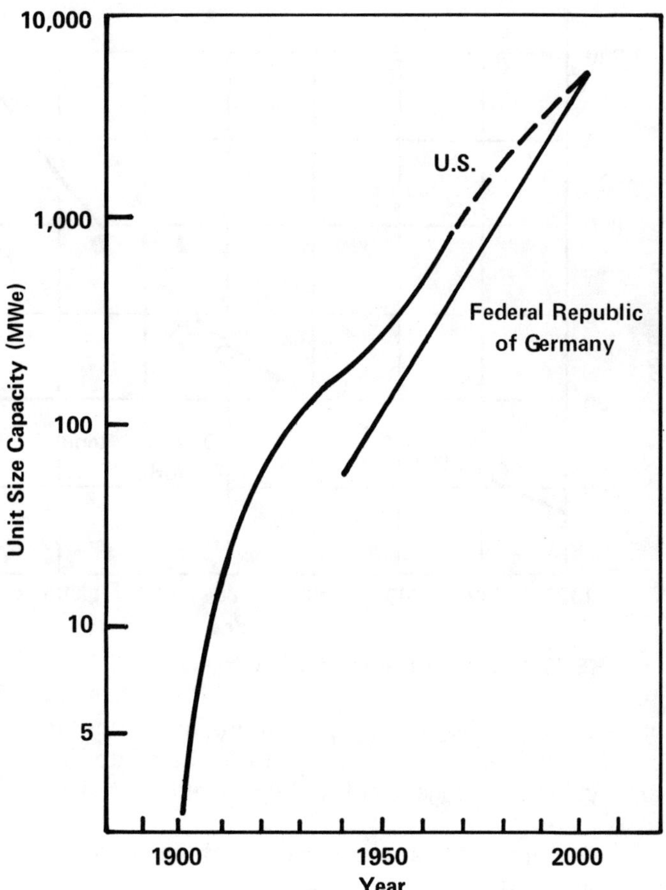

FIGURE 9. Trend of Unit Size in Turbine Generators

of the grid has remained remarkably constant, at about 10 percent.[12] Upon reflection, this is not too surprising, as there must be an optimum that reflects opposite trends. On the one hand, the law of scale drives in the direction of ever-increasing unit size. On the other hand, the demand for redundancy drives in the opposite direction. So far, the result is the constant ratio observed by Marchetti.[12] If the doubling time of the generator size were the same as the doubling time of electricity consumption, then there would be no increase in centralization. To the extent that the doubling time of the generator size tends to be shorter, there is increasing centralization. In other words, generator doubling times smaller than 10 years imply increasing centralization. The existence of widely interconnected grids does not

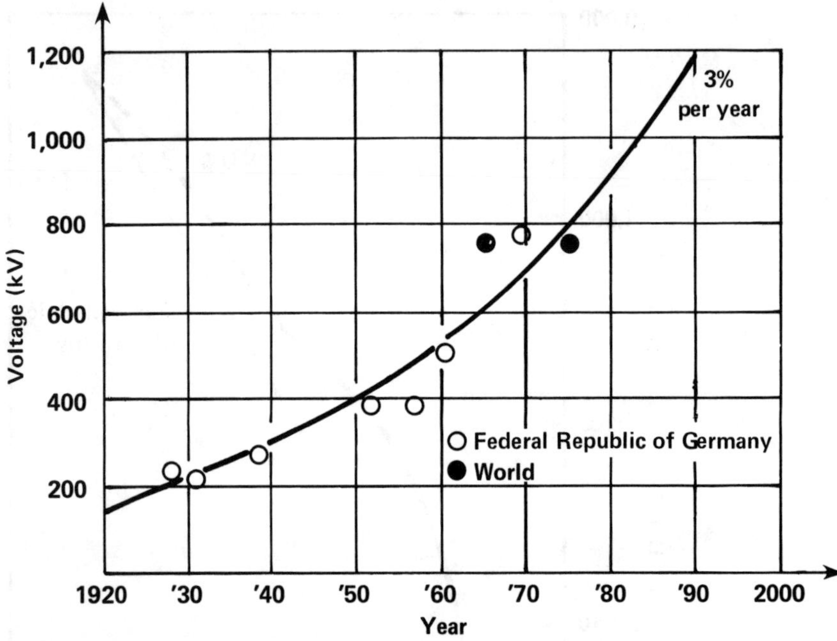

FIGURE 10. Maximum Transmission Voltages

undermine the argument—the interconnections are for the exchange of peak load and to ensure the reliability of power supply. The average electric kilowatt-hour travels only 50 to 100 km.

This line of reasoning leads us to the consideration of the density of total energy consumption. In order to put such energy consumption densities into perspective, a look at population densities is required. The figures given in Table 5 for the Federal Republic of Germany and for India may serve as examples. The average population densities in these two countries are not significantly different, but the average energy densities do differ. This gross understanding may invite one to consider soft options based on renewable low-supply-density sources such as the sun and wind. A closer look, however, reveals a situation quite unfavorable for such options. The energy densities in urban areas turn out to be the same in the two countries, while the urban population density is much higher for India. If the average inhabitant of urban areas in India were to improve his standard of living and thereby also his energy consumption, the result might be considerably higher energy densities than those that characterize the urban areas of developed countries. Therefore, the potential for soft options (locally available energy) may in fact be lower—not higher—in the urban areas of the third world.

For rural areas, the population densities of the two countries are again comparable. In both cases the energy consumption density is not much lower than the average. However, a difference in settlement pattern must be borne in mind: in the case of the Federal Republic of Germany, "rural" does refer to both farms and small towns. By contrast, in India energy consumption in the rural areas is exceedingly low and in addition more dispersed. Adoption of soft options should be accompanied by substantial increases in energy density in rural areas like India's. But the argument goes further.

J. K. Parikh and K. S. Parikh have analyzed at IIASA the potential of biogas for India.[13] Even with such low power densities, the use of biogas requires centralization and thus the establishment of a kind of unit cell. At the present low level of consumption, this unit cell is accordingly small. The Parikhs consider centralized supplies for villages ranging from 200 to 1000 people. While this may not seem to be hard technology in terms of industrialized countries, the earlier consideration of the size of turbine generators and their related grids indicates that as consumption increases, so will the size of economic unit cells.

To examine in some detail an urban area in a developed country, let us take Hannover in the Federal Republic of Germany. For Hannover, we have plotted not only the distribution of the density of the electrical load but also the distribution of the total demand for power. Figure 11 gives peak loads, not yearly averages as discussed so far. The ex-

TABLE 5
Settlement and Consumption Densities

	Population Density (Per square kilometer)			Energy Consumption Density (Watts per square meter)		
	Average	Urban*	Rural†	Average	Urban*	Rural†
Federal Republic of Germany	245	1,500‡	150	1.2	7.5	0.75
India	168	~12,000§	135	0.10	7.4	0.08

* Conurbations.
† Farms and small towns.
‡ 45% of population in these areas.
§ 69% of population in these areas.

tremely high densities in the center of the city are striking: density there is as high as 240 W/m^2. In addition, if present trends continue, these densities will increase with time.

The technological features discussed above with reference to the total grid load and the size of the electric generator have their counterparts in the case of energy distribution in the centers of cities. Here, the average power size of the final consumer is a decisive factor. Figure 12 shows the costs of such energy distribution. Once again the absolute numbers are striking. It would be hard to accept distribution costs in excess of $2/million Btu. Of course, each case has to be evaluated separately, but it is worth noting that for the Federal Republic of Germany, for example, local district heating requires densities far greater than 35 W/m^2. The fairly sharp slopes of the cost curves for energy distribution

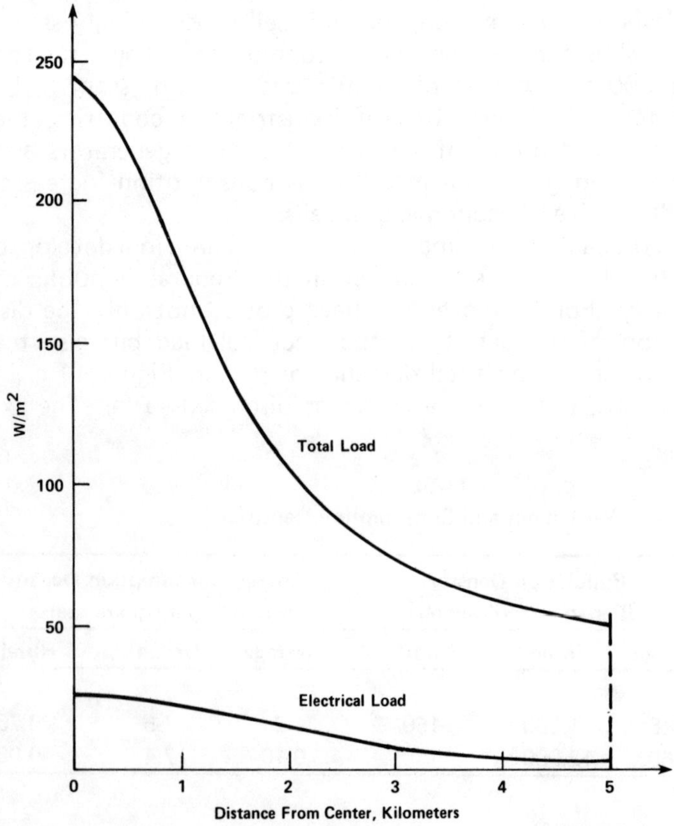

FIGURE 11. Peak Load Densities, Hannover, Federal Republic of Germany, 1973

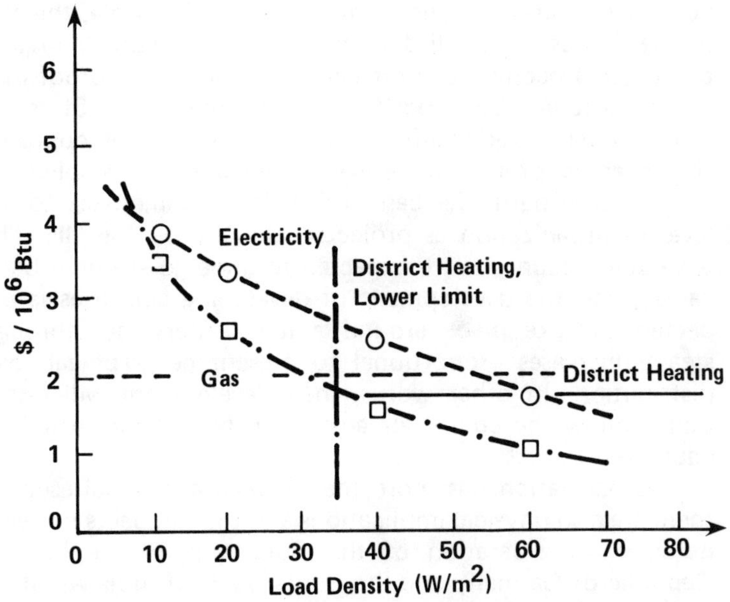

FIGURE 12. Distribution Costs of Urban Grids, Hannover, Federal Republic of Germany, 1973

at low densities make even the Ruhr conurbation a disconnected set of supply areas.

China offers some interesting departures from the trends described so far. Here, it may be that the demands of the political system will force development in a different direction. It has been observed that the Chinese seem to be aiming for unit cells of 300,000 people, which are expected to become self-sufficient; this is a purposeful attempt at decentralization. Marchetti has translated this general problem into the problem of the size of generators and grids. The average Chinese, according to UN data, consumes 150 kWh per year of electric energy, the corresponding figure for the United States is about 10,000 kWh, about 65, or $\sim 2^6$, times more. In the United States the doubling of the electric load took place every 10 years, so six doublings would bring us back to 1915. The population of the United States in 1915 was about half of that of 1975, so 150 kWh was the per capita consumption 10 years earlier, about 1905. If generator sizes double every 7 years, as reported above, in 70 years generator size doubles 10 times, or 1 Gigawatt (GW)/$2^{10} \cong 1$ MW. If the size of the generator is about 10 percent of the grid load, the total size of a grid would then be 10 MW. At 150 kWh per person per year, this relates to 300,000 people, the size of the Chinese unit cell described above. The reported figure of 300,000 people per unit cell in China is not proof of a course different from that followed by the United States or other industrialized countries. Nor is it proof of the opposite; the meaning of the Chinese experience is still ambiguous.

Leaving particular cases aside, let us come back to the level of urbanization as projected globally by the UN. The developing countries are expected to undergo strong urbanization, and the development of developing countries is expected to take place around a few centers, not through steady increases proportional to present geographical load distribution. In other words, their development will essentially follow the course already taken by the industrialized countries.

Urbanization has more than human and social aspects; there are also physical implications and their impacts on technologies. Consideration of the Ruhr area of the Federal Republic of Germany and the Hooghlyside of India will illustrate the truth of this observation (Figure 13). The Ruhr area covers 1200 km^2, but consists of a number of distinct urbanization centers (Verdichtungskerne), the largest of which are about 10 km in diameter. The Hooghlyside of India has an area of 570 km^2 and is an agglomeration of urbanization centers. A significant difference comes from the slum areas,

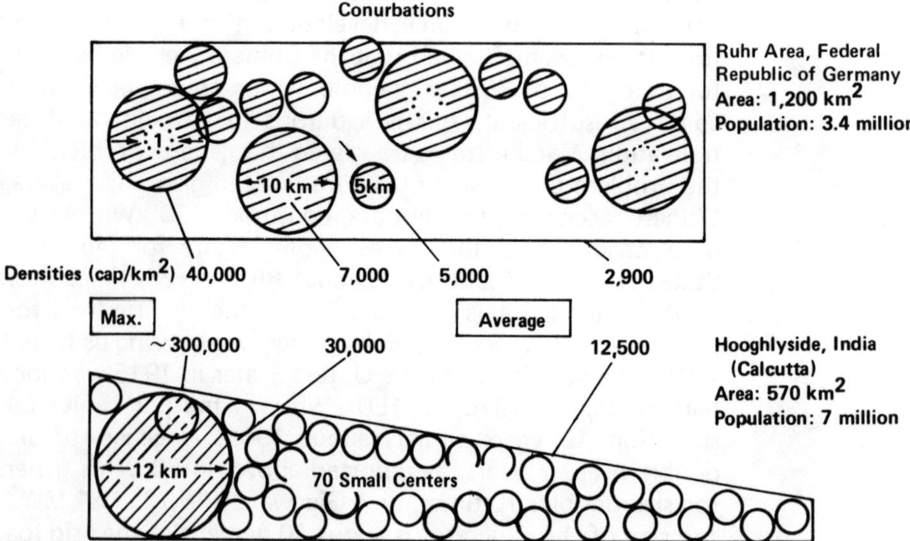

FIGURE 13. Selected Metropolitan Areas of West Germany and India

which, however, appear as a kind of singular point. Speaking in terms of energy technology, the situation is much the same.

In the past, urbanization and technology have matched each other closely. There is no doubt that centralization of energy services is first of all a reflection of the much more general phenomenon of urbanization. Whether urbanization drives technology or technology urbanization, however, is an open question. It seems to us that the evidence points to the systems character of energy technology, and this very systems character makes it impossible to separate the two elements: they are two sides of the same coin.

Let us now look at the densities of primary energy supply. At IIASA Grenon has considered the areas required for operation of a 1-Gwe plant powered by coal, by solar power, and by a light-water reactor (Table 6).[9] Strip mining of coal

TABLE 6
Fuel Supply Densities For 1-GWe Power Plants

Fuel	Attribute	Specifications	Area (km^2) For 1 GWe	Supply Density (Wth/m^2)
Coal*	Strip Mine	2-m seam	25	≈120
		10-m seam	5	600
Solar†	Tower	4 kWh/m^2/Day		
	Concept	$\eta = 0.2$	30	100
Nuclear*	LWR—U Shale	2-m seam	37	80
		10-m seam	7.5	400
	Fast Breeder Reactor—			
	U Shale	2-m seam	0.4	≈7,500

* Time scope 25 years.
† Time scope unclear.
Source: Grenon [9]

from a seam 10 m thick requires the temporary use of 5 km^2 per 25 years. An LWR that uses uranium-shale from 10-m seams requires 7.5 km^2 every 25 years, a similar figure. If both must make use of 2-m seams, the corresponding figures are 25 km^3 for coal and 37 km^2 for uranium. The use of such areas is temporary; land reclamation is possible and has been demonstrated on a large scale and it will probably become mandatory when the extraction of coal and uranium for large-scale uses becomes a necessity. For solar power the figure for the required area is very close to those reported so far: 30 km^2 for the middle latitudes. Unlike coal or uranium, however, use of land for solar harvesting does not require a new site after 25 years. Relating 1 GWe to roughly 3 GWth and using a figure of 30 km^2 for all cases leads to supply densities of the order of 100 Wth/m^2. The figure for the fast breeder, by contrast, is in the neighborhood of 10 $kWth/m^2$.

(It should be kept in mind that these figures refer to futuristic scenarios where 2-m seams must be exploited or solar power used for truly large-scale production of energy.)

Today's supply densities for oil vary significantly. A figure of 1 kW/m^2 may be typical for an upper range for LWR power; with high grade ores densities are at 3 kW/m^2. Such figures suggest that the ratio of supply density to consumption density that we enjoy today could continue in the future only if there is large-scale use of fast breeders; otherwise, the areas used for supply of primary energy must be increased by a factor of 30 to 100. While this may be of little importance worldwide, the effects in terms of regional development and urbanization probably will be significant.

Wind power is also of some interest. Bent Sørensen, for example, has examined its applicability in Denmark.[14] He considers vertical areas in the interior of Denmark that give 300 watts of mechanical energy per square meter. ($Wmech/m^2$). Other authors, heating less windy areas, employ a figure of only 100 $Wmech/m^2$. Sørensen assumes 50-m-high windmills and considers 2-km downwind strips for recovering the wind profile. If we relate 100 Wmech to 300 Wth, this leads to 300/40 = 7.5 W/m^2. While this is only 7.5 percent of the 100 W/m^2 supply density for coal, LWR, and solar power, the effects on the 2 km between rows of windmills are certainly not comparable to the effects of strip mining or solar power generation. The impact on land use patterns of extended rows of 50-m-high windmills every 2 km is unknown.

A more readily addressed question is that of the limited amount of available wind power. Of a global total of 2000 TW, roughly 50 percent will be across land, and 10^{-3} might be available to windmills. This would result in about 1 TW as an upper limit. One should realize that the harvesting of this 1 TW of wind power would require an all-out effort in terms of manpower, material, and infrastructure for energy collection and transportation, thereby destroying the original intent of making use of wind power for reasons of simplicity and ease. In practice, then, the share for wind power might be significantly lower than 1 TW. Wind power seems applicable only in special situations, and Sørensen suggests that Denmark offers such a special situation.

Energy Storage and Transportation

So far, we have dealt essentially with energy, not with power, which is the distribution of energy over time. At present, energy transportation relates a pattern of load variations to an energy supply that by its very nature could be constant. Indeed, adaptation of a given load curve to a set of power stations with different ratios between fuel and capital costs is the essence of supply system optimization. The load variations are daily, weekly and seasonal. The greater the distance energy can be transported, the larger is the smoothing effect on the resulting load curve. To that extent energy transportation, in the context of local distribution, substitutes for local energy storage that would otherwise be required for smoothing a load curve. Besides that, energy transportation allows redundancy and permits taking advantage of the law of scale.

In the case of solar energy, wind, and similar renewable resources variations in time and space must be taken into account. In there we have variations of the load and the source. In the time dimension solar and wind power show daily, weekly, (or monthly), and seasonal fluctuations. It is obvious that daily storage of such energy is necessary; there are no contrasting views on that. The feasibility of monthly and especially seasonal storage, however, is a greater problem. Unfortunately, the related technology is not as advanced or sophisticated as, for instance, the technology of fusion or fission, and this lack of sophistication is reflected in present cost estimates. At the Conference of the European Physical Society in Bucharest in September 1975, Schröder gave a careful evaluation of thermal energy storage (Table 7).[15] He considered the heating of a 100-m^2 apartment with a maximum of 100 W/m^2—a low figure, implying good insulation. He further assumed that heating is required for 2000 hours per year, for a total of 2×10^4 kWh per year. If daily storage is required, the cost is 0.32 mills/kWh. Monthly storage is 3.2 mills/kWh, and yearly storage 26 mills/kWh. Clearly, daily fluctuations could be bridged by storage, but not yearly fluctuations.

We now consider spatial fluctuations in the energy supply. The size of the shadows of clouds and the size of

TABLE 7
Economics of Optimal Heat Storage for Homes*

Operating Cycle	1 Day	30 Days	1 Year	
Total Throughput Per Year	10,000	15,000	20,000	kWh
Size of Storage Heater	70	2,000	20,000	kWh
Capital Costs† and Energy Delivered	0.32	3.2	26	mills/kWh

*100-m^2 apartment.
† 10% interest rate.
Source: Schroeder [15]

regions of low atmospheric pressure are typical of the distances that must be taken into account, and here energy transportation may again help to soften the requirements for local energy storage. This may already be significant for the monthly fluctuations, and it is definitely significant for the yearly cycles. At IIASA Partl has recently evaluated transportation of energy in the 10 GWe domain[16] over distances of 4000 km (Figure 14). Such transportation should be able to compensate for variations in energy supply patterns. Taking the best estimates for the related total costs (including all auxiliary equipment, transformer stations, and so on), we arrive at 6.5 mills/kWh for a distance of 4000 km. This is definitely lower than the costs for yearly energy storage and points to the interplay of energy storage and transportation. One must realize, of course, that if such arrangements for transportation are adopted, solar and wind power become a hard technology, and this prospect gives rise to contrasting views.

Capital Stocks, Investments, and LDC's

One of the most complex aspects of technological strategies for the next 30 to 50 years is the problem of capital stocks, investments, and the interrelation between developed and less developed countries. Specifically, the interplay between small and big, labor- and capital-intensive energy technologies must be seen in this context. Earlier in this paper we referred to the New Economic Order, of which the OPEC cartel may be seen as a part. One way of dealing with these problems is large-scale modeling of the kind being done by the Pestel–Mesarovic groups at Case Western Reserve University and the Technische Universität of Hannover. Here we present a minimodel that was recently conceived and studied at IIASA by W. Haefele, R. Buerk, and W. Sassin. Its purpose is not to produce numbers that are actually relevant, rather, it is meant to help in conceptualizing and understanding the complex relations.

To this end we consider the well-known balance between production and consumption of a gross national or regional product *Y*. If *K* refers to the capital stock, *P* to the population, and *c* to the per capita consumption, and if δ is the depreciation rate, we have the following simple equation:

$$Y(K,P) = \dot{K} + \delta \cdot K + cP. \tag{1}$$

For our purposes we assume *Y* to follow a Cobb-Douglas function. We assume

$$Y = A \cdot K^{\alpha} \cdot P^{1-\alpha}, \; 0 < \alpha < 1. \tag{2}$$

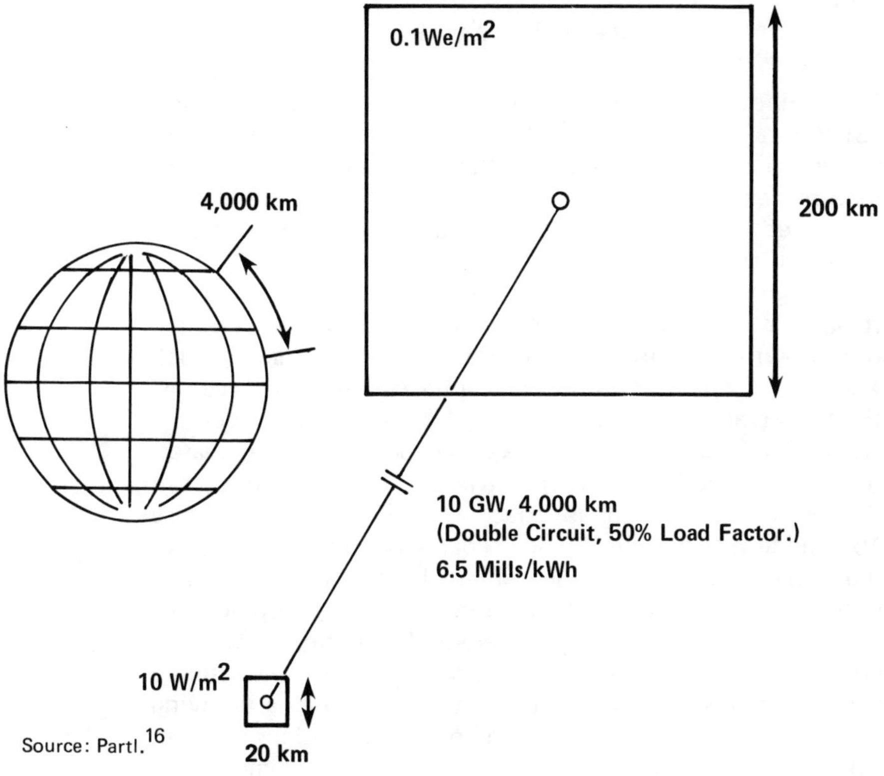

Source: Partl.[16]

FIGURE 14. Compensation of Energy Supply Patterns by Transportation

In (2) we implicitly assume the ratio between labor and population to be constant—that is, we eliminate from consideration the problems of education and employment. We further assume exponential growth of the population (as an approximation to reality):

$$\frac{P}{P} = \lambda. \tag{3}$$

If we now introduce k as the capital stock per capita, that is,

$$k = \frac{K}{P}, \tag{4}$$

we have the following simple relation

$$\dot{k} = AK^{\alpha} - (\lambda + \delta)k - c. \tag{5}$$

In (5) we have assumed c to be constant, or, more precisely, to be a "slow variable."

Let us consider the term X:

$$X = Ak^{\alpha} - (\lambda + \delta)k. \tag{6}$$

X describes the net production of a Gross National Product (GNP), taking into account depreciation and population growth. The latter comes into the picture as each newborn requires an infrastructure, or, in other words, his capital stock per capita. Equation (5) in turn equates X with consumption and the growth of capital stock per capita.

Figure 15 shows the character of X as a function of k. It passes through a maximum when depreciation and population growth rate surpass production. For appropriately small values of the consumption per capita c there are three distinct intervals for k. For $k < k_s$ or $k > k_a$, the capital stock per capita decreases: for $k_s < k < k_a$ the capital stock increases. In fact, k_s functions as a separatrix and k_a as an attractor in the one-dimensional phase space k, as illustrated in Figure 16. The terms "attractor" and "separatrix" point to a methodological background whose richness becomes visible only in multidimensional cases. The case in point is the topology of sets of nonlinear ordinary differential equations. While at IIASA Holling applied such a methodology in conducting a major systems analysis that dealt with pesticide spraying strategies in New Brunswick.[17] If many dimensions are considered, the regions $k < k_s$ and $k > k_s$ of the one-dimensional case become domains separated by interfaces. Attractors

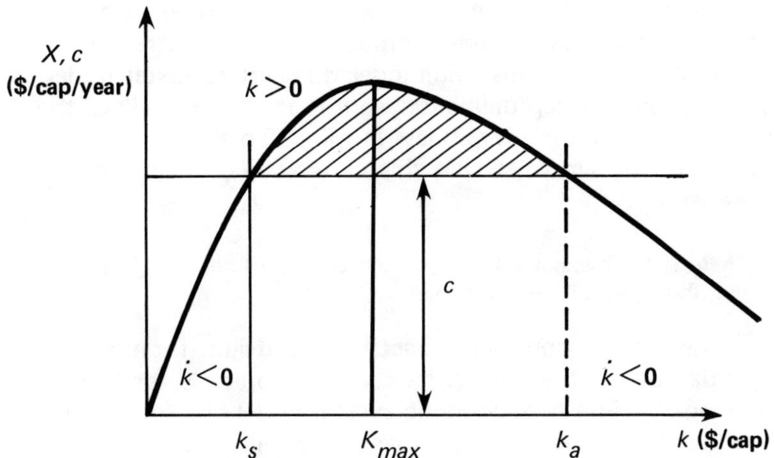

FIGURE 15. Net National Production as a Function of Capital Stock Per Capita

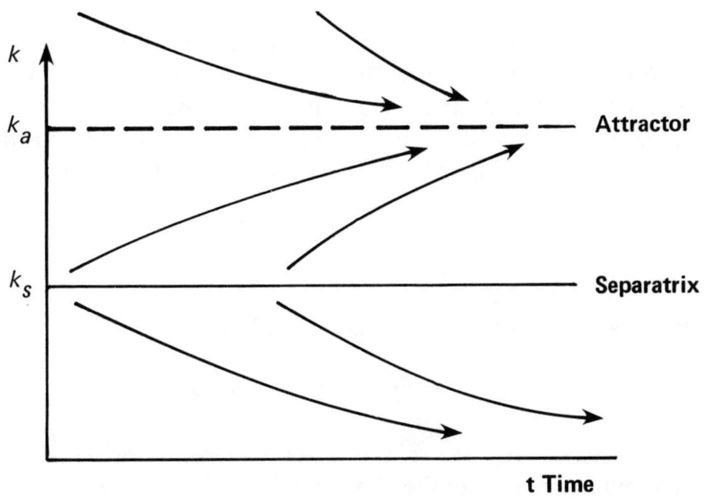

FIGURE 16. Attractor k_a and Separatrix k_s in One-Dimensional Phase Space k

become even lower-dimensional manifolds and represent the long-term tendencies of the trajectories that they attract.[18] In that context, Holling coined the term "resilience," by which he meant the capability of a system to remain in one region or basin even after finite impacts from the outside.[19]

In our model the region $k > k_s$ turns out to be a stability region that allows for the approach of k_a, the attractor. As mentioned before, this minimodel is meant to foster understanding and conceptualization, so it is important to keep the underlying topological richness in mind. To that extent, the model is not simple. Given this context, we make the *first* observation:

> There is the distinct value k_s, which acts as a threshold. Below that threshold it is no virtue to be small.

Let us now consider impacts from outside—for example, fluctuations of the per capita consumption *c*. If for a given moment in time the economic system under consideration is at *k*, with $k_{max} < k < k_a$, increases of *c* stabilize again at somewhat lower values of *k*. But if $k_s < k < k_{max}$, increases of *c* may well throw *k* across the separatrix k_s into the domain of extinction if these increases prevail long enough. This leads to the *second* observation:

> It is a resilient policy to have k beyond k_{max}.

Our minimodel can handle numbers. It should be borne in mind however, that such numbers are only a quantitative expression of a qualitative situation. A primary consideration is the value of α in developed and less developed countries. Consistent with (2) and the notation (4), let us consider the GNP per capita *y:*

$$y = AK^{\alpha}. \tag{7}$$

It is then interesting to obtain the following relation:

$$\frac{\delta y}{\delta k} = \alpha \frac{y}{k} = \alpha \frac{Y}{K}. \tag{8}$$

$\frac{Y}{K}$ is the inverse of the capital output ratio and $\frac{\delta y}{\delta k}$ is the shadow interest rate. We therefore have

$$\alpha = (\text{shadow interest rate}) \times (\text{capital output ratio}). \tag{9}$$

It is for this relation that less developed countries have a high α and developed countries have a low α. We now come to two sets of numbers that are somehow representative. The first set is indicative for developed countries, the second set is indicative for an average of Southeast Asian countries

TABLE 8
Data For the Minimodel

		Developed Countries	Less Developed Countries
$y = Y/P$	(1973)	\$2,270/cap/year	\$120/cap/year
k	(1973)	\$7,240/cap	\$250/cap
a		0.12	0.80
$A = y \cdot k^{-a}$		783	1.44
λ	(Per year)	0.008	0.028
δ	(Per year)	0.105	0.06
c Including Government		$\overline{c}$ = \$1,691/cap	$\overline{c}$ = \$105/cap
c Excluding Government		$\underline{c}$ = \$1,336/cap	$\underline{c}$ = \$91/cap

(Table 8). It is interesting to see that δ is in both cases larger than λ (depreciation overshadows growth rate) and that the two α's are so far apart. This has serious implications, which are shown in the semiqualitative comparison of developing and less developed countries (Figure 17). The salient point is that the high α's of the LDC's tend to push the maximum to extremely high values and the LDC's are, accordingly, working uphill far below the maximum. This brings us to the *third* observation:

> Developing countries do suffer from not being in the domain of high resiliency. Accordingly, whether governments invest in the capital stock can make all the difference in being below or above the separatrix k_S.

The theme of this volume is soft versus hard technologies, small versus big. The considerations presented here make these issues more subtle. This leads us to the *fourth* observation:

> The concern must be to keep the value of the capital elasticity coefficients low and thus ensure resilience. A more detailed analysis should be done to see whether the introduction of, for instance, solar or wind power in a given economy tends to increase or to decrease the capital output ratio and thus the capital elasticity.

We also feel that considerations of the sort included in this

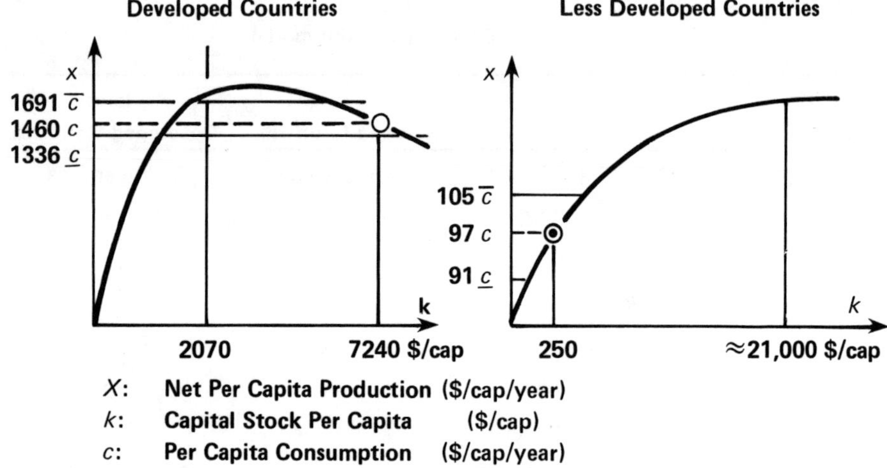

FIGURE 17. Comparing Developed and Less Developed Countries

minimodel provide us with a yardstick; this leads to the *fifth* observation:

> Small or big has to be seen in the context of the economic relations explained here. The values of k_s for the separatrix and k_{max} provide yardsticks against which technological labels like big or small can be evaluated.

This requires, when pursued in greater detail, models that are at least somewhat more complex. At IIASA Haefele and others are engaged in work that seems to permit explicit judgements on the feedback that investments in various kinds of technologies have on the overall economy.

There are further lessons to be gained from the model. In the crude language of the minimodel, expensive new technologies can to some extent be expressed by introducing high values for the assumed depreciation rate. If the assumed depreciation rate is higher than the actual one, this permits the formation of extra capital that may be required by new and expensive technologies. High values of δ (more precisely, of $\delta + \lambda$) tend to bring the maximum down. The region of stability is narrowed when the per capita consumption stays the same while k_s and the region of resiliency increase because k_a and k_{max} are brought closer together. This leads to the *sixth* observation:

For developed countries that already have a fairly low value of a large investments are important lest their region of stability and resilience become unduly small.

This model should not be overrated. The conceptual problems, let alone the actually relevant figures, are more complex than those seen by the model. But it may be a step toward placing the contrasting views on small and big, soft and hard technologies into a more appropriate perspective.

REFERENCES

1. Energy Conservation Study. A report to the National Science Foundation, Office of Energy R&D Policy. American Physical Society, New York City. 1974.

2. American Physical Society Study Group on Technological Aspects of Efficient Energy Utilization. Efficient Use of Energy. Physics Today, 28: 8, 1975, A-H.

3. Hafele, W., et al. Second Status Report of the IIASA Project on Energy Systems. RR-76-1. International Institute for Applied Systems Analysis, Laxenburg, Austria. 1976.

4. Häfele, W., and W. Sassin. Energy Strategies. Energy, 1 (2): 147-163, 1976 and RR-76-8. International Institute for Applied Systems Analysis, Laxenburg, Austria. 1976.

5. Häfele, W., and A. S. Manne. Strategies for a Transition from Fossil to Nuclear Fuels. Energy Policy, 3: 3-23, 1975 and RR-74-7. International Institute for Applied Systems Analysis, Laxenburg, Austria. 1974.

6. Weinberg, A. M., and R. P. Hammond. Global Effects of Increased Use of Energy. In: Peaceful Uses of Atomic Energy, V. 1. Proceedings of the Fourth International Conference, Geneva, 1971. International Atomic Energy Agency, Vienna. 1971. p. 171.

7. Hubbert, M. K. The Energy Resources of the Earth. Scientific American, 225: (3) 61, 1971.

8. Goeller, H. E., ed. Survey of Energy Resources 1974. Proceedings of the Ninth World Energy Conference, Detroit, 1974. World Energy Conference, London. 1975.

9. Grenon, M. Coal: Resources and Constraints. In: W. Häfele et al. Second Status Report of the IIASA Project on Energy Systems. RR-76-1. International Institute for Applied Systems Analysis, Laxenburg, Austria. 1976.

10. Fettweis, G. B. Weltkohlenvorräte: Eine vergleichende Analyse ihrer Erfassung und Bewertung. Bergbau-Rohstoffe-Energie, 12. Verlag Glückauf GmbH, Essen, FRG. 1976.

11. Häfele, W., et al. Report of the Fission Fusion Task Force. Forthcoming publication of the International Institute for Applied Systems Analysis, Laxenburg, Austria.

12. Marchetti, C. Private communication. International Institute for Applied Systems Analysis, Laxenburg, Austria, manuscript in preparation.

13. Parikh, J. K., and K. S. Parikh. Potential of Biogas Plants for Developing Countries and How to Realize It. Presented at Seminar on Microbiological Energy Conversion, Göttingen, FRG, Oct. 4-8, 1976. Proceedings to be published by E. Goltze KG, Göttingen.

14. Sørensen, B. Energy and Resources. Science, 189, 4199 (1975), 255-260.

15. Schröder, J. Thermal Energy Storage. In: Energy and Physics. Proceedings of the Third International Conference of the European Physical Society, Bucharest, Rumania, 1975. European Physical Society, Geneva, Switzerland. 1976.

16. Partl, R. Glacier Power from Greenland. Forthcoming publication of the International Institute for Applied Systems Analysis, Laxenburg, Austria.

17. Holling, C. S. Project Status Report Ecology/Environment. SR-74-2. International Institute for Applied Systems Analysis, Laxenburg, Austria. 1974.

18. Grümm, H. R. Resilience and Related Mathematical Concepts. In: IIASA Conference '76, V. 2. International Institute for Applied Systems Analysis, Laxenburg, Austria. 1976.

19. Holling, C. S. Resilience and Stability of Ecological Systems. RR-73-3. International Institute for Applied Systems Analysis, Laxenburg, Austria. 1973.

Thomas H. Lee
General Electric Company
Fairfield, Connecticut

Thomas H. Lee is manager of the Strategic Planning Operation of General Electric Company's Power Generation Group. He joined General Electric in 1950. In 1955 he started the company's vacuum interrupter development program. In 1959 he was appointed manager of engineering research at the Philadelphia Laboratory Operation, and in 1967 manager of the entire Laboratory Operation. In 1971 he became manager of the Power Delivery Group's Technical Resources Operation, with responsibility for all of General Electric's research and development work in the transmission and distribution business.

Mr. Lee was responsible for the successful development of practical vacuum interrupters, which have made circuit breakers using these elements a reality. Under his direction, the first HVDC conversion system using solid-state technology was developed. His many contributions to physics and electrical engineering span the fields of physical electronics, such as electron emission, arc cathode spots, the mechanism of electrical breakdown of high-temperature gases, transient circuit analysis, and development of high power switching and protective devices. In 1974, he was appointed to his present position.

Mr. Lee has been Adjunct Professor at Rensselaer Polytechnic Institute and lecturer in physics at the University of Pennsylvania. He is a Fellow of the Institute of Electrical and Electronic Engineers and president of its Power Engineering Society, a member of the American Vacuum Society, and a member of the American Physical Society. In 1975 he was elected to membership in the National Academy of Engineering.

Mr. Lee's bachelor's degree is from China's National Chiao-Tung University, his master's degree is from Union College and his doctorate was earned at Rensselaer Polytechnic Institute.

In 1975 he published the book Physics and Engineering of High Power Switching Devices. *Mr. Lee is the holder of 30 U.S. patents and the author of many technical papers.*

The Case for Evolutionary Optimization

Introduction

I had two serious reservations about participating in this conference. The first can perhaps be best expressed by quoting from an editorial I wrote last summer for the *Vineyard Gazette,* the newspaper published on the island of Martha's Vineyard:

> Almost two years have passed since our country recognized the unprecedented energy crisis. Despite Project Independence and much legislation, we still do not have a national energy policy. Instead we have a great national debate: on the balance between economic growth and environmental protection, on the desirability of nuclear energy, and on the potential of unproven technologies such as solar energy and fusion.
>
> A debate is not unlike sports, where each side has its strengths and weaknesses, and each side wants to win. But, in order to be sure that the contest is fair, both sides must play the same game and be bound by the same set of rules. Pitting a team of football players against a team of basketball players on a tennis court, no one can hope to prove anything. But this is exactly how we are conducting the debate on energy. If the pronuclear people are the football players and the antinuclear people are the basketball players, the tennis court is the political arena.
>
> The pronuclear people would like to use statistics and engineering reasoning to support their argument . . . the antinuclear people prefer to play a different game. They prefer to hit hard on the unknowns, to play the game on psychological grounds, to create the impression that scientific and engineering societies are polarized by pitting Nobel laureates against each other.
>
> So different games are being played on this tennis court. There is no hope that a meaningful winner will emerge unless the management of the tennis courts insists that the same game be played and the same set of rules be followed.

My first reservation, therefore, was: Are we really going to have a sensible dialogue or is everyone going to set his own rules for the game? Are some of us to be merely critics dissecting the weaknesses in a policy but free of the responsibility of considering alternative programs? And are these disinterested observers to be opposed by a group of administrators and strategists enmeshed in the dirty work of making heaters heat and often accused of losing sight of the forest for the trees?

I hope not. There is a role in society for pure managers and pure critics, and their interaction can be fruitful, even when bitter. But today, in America, the line between the two has been blurred. There are no pure outsiders and insiders in the energy game anymore. We are at the point where gaining something here means giving up something there, things that affect every one of us. All of us are facing more than yes-no decisions. We have choices to make.

My second reservation was: Will this conference become a debate between advocates of simple solutions? Searching for simple solutions is part of human nature. That desire gave us Newton's Laws and Maxwell's Equations. But, I also believe that this same desire has given us wars and the Luddites.

The energy problem is a real life problem. There is no simple unique solution. Our task is to find an optimum solution in the face of many constraints. To do that, we must examine all the problems in great depth. For that I welcome the theme of this conference: Scale. I would like to go one step further and confine myself to the question of scale in electrification.

First, I want to list a set of assumptions that is the basis for the thesis of evolutionary optimization:

- Future energy strategy will be decided by a democratic process just as it has been in the past, at least in the case of electrification.
- The majority of the public does not believe that our life styles must be drastically changed to solve the energy problem.
- Economics will remain as a major factor in choosing energy alternatives.
- The public will continue to demand reliable energy supplies.
- There can be no debate on the need of conservation, although how much can be conserved between now and the end of the century is very much a subject for debate. The question to be discussed here is: What is the optimum way to supply the electricity still needed after all reasonable efforts are made to conserve?

And, finally, a most important point, if not an assumption: Numbers generated from what has been learned in the past from many years of hard work in research, development, design, and production represent "hard" numbers. They are reliable. Future predictions based on "hard" numbers have far fewer uncertainties as compared with "soft" numbers based on unproven technology and untried systems. I hope that in making choices on energy strategy, due recognition will be given to the hardness of the numbers.

It is possible to debate the validity and sensitivity of every one of these assumptions. Among them, the first one—electrification was developed via a democratic process—is probably the most controversial. Too often, the accusation has been made that the electrical industry was responsible for the high rate of growth of electricity consumption because of its advertising, its promotional activities, rate structures designed to encourage more usage, and introduction of inefficient appliances.

To analyze the validity of this accusation, we must first understand how per capita energy consumption has changed in the past 100 years (Figure 1).[1] In the period 1870-1940, the per capita energy consumption was growing at a very slow rate, 0.5 percent per year. From 1940 on, the rate increased drastically. In the decade 1960-1970, the rate was 2.7 percent. This rapid increase in per capita energy consumption was due to the following factors:

- A shift in the working population from the agricultural to the industrial sector.
- More women entering the labor market.
- More energy used per unit of goods and services produced.

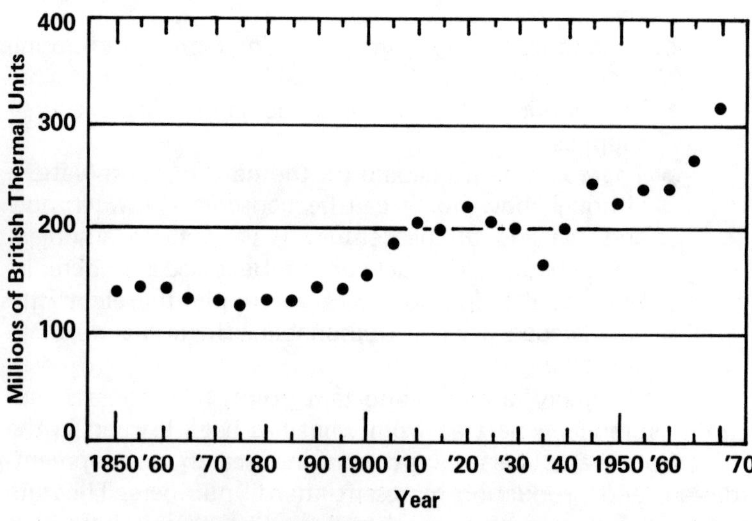

Source: Fisher, J.C. Energy Crises in Perspective. John Wiley and Sons, Inc., New York. 1974.

Figure 1. Per Capita Energy Consumption in United States, 1850-1970

- Increased affluence in our society, such as television, air conditioning, and second automobiles.

In addition, our population was growing at an average rate of 1.3 percent. The combination of all these factors resulted in the average annual growth rate of approximately 4 percent for total energy consumption.

In the meantime, electricity consumption was growing much faster—more than 7 percent a year. It is true that there was much competition among different kinds of energy—oil, gas, and electricity—and the electrical industry did its share of advertising and promotion. But before attributing this high growth rate to promotional efforts, let us look a little deeper into the situation. Examination of electricity consumption by three categories of end use—totally electrified, partially electrified, and unelectrified—permits us to draw several interesting conclusions (Table 1). First, for the matured end uses such as electrical drives, the rate of growth of electricity consumption matched closely that of Gross National Product (GNP). This match is not surprising. No matter how good the promotional effort, no factory would use more motors, more machine tools, or more electricity than it needed to make its products. A recent survey of over 100 industrial organizations by National Economic Research Associates revealed that after the initial conservation effort triggered by the oil embargo, there were very few short-term opportunities for conservation other than lighting and air conditioning. Surely, in the long run there will be more efficient processes or more efficient drives. But that takes time, and economics will play a key role in deciding whether and when this will happen.

The most mature end use is cooking. Here, the total energy consumption grew at almost the same rate as the population growth. None of us uses much more energy than our ancestors for cooking. This is an outstanding example of saturation. Electricity for cooking increased faster than total energy principally because it is clean and it is convenient. Although more expensive than gas, it is still low cost, and people chose convenience over economics.

The real reasons electricity consumption grew at a faster rate than total energy are not promotional efforts but are:

- Technological advances that offered people more enjoyment in life; television and air conditioning are two outstanding examples.
- Substitution for other fuels, such as in space heating.

The first reason is far more significant than the second,

TABLE 1

Electrification End Uses in the United States

End Uses	Percent Electric, 1963	Consumption, 1963 Percent of United States: Total Energy	Electricity	Annual Growth in Consumption, 1960-1968: Total Energy	Energy for Electricity
		(percent)		(percent)	
Electrified					
Industrial drive	~ 100	10.3	39.7	4.4	4.4
Refrigeration	~ 100	3.0	11.6	5.6	6.0
Electrolytic processes	100	1.5	5.8	3.7	3.7
Air conditioning	96	1.5	5.6	11.3	11.1
Television	100	0.7	2.6	9.6	9.6
Lighting and other	~ 100	5.0	19.1	9.4	9.4
	~ 100	22.0	84.4	6.0	6.0
Partially electrified					
Clothes drying	70	0.4	1.0	10.2	9.9
Cooking	40	1.4	2.1	2.0	3.0
Water heating	38	4.2	6.2	4.1	3.4
Direct heat	6	11.0	2.6	2.8	5.8
Space heating	5	20.8	3.3	4.1	24.0
	11	37.8	15.2	3.6	6.6
Unelectrified					
Transportation	——	25.6	0.4	4.1	——
Process steam	——	14.6	——	3.8	——
	——	40.2	0.4	4.0	--
All end uses	25	100.0	100.0	4.3	6.1

~ Indicates approximate percentage.

at least in the past 20 years. The public had a real role in deciding whether to consume electricity for these end uses. They had to decide if television and air conditioning were what they wanted in life and whether they were willing to pay the costs. And the public didn't make these decisions simply and naively. They were very critical when it came to economics. Air conditioning did not become popular until the cost came down to where the public was willing to buy. I am sure that if the energy cost of air conditioners becomes too high, the public will decide to curtail their use. *This is what I mean by a democratic process in energy strategies.*

A final point can be made regarding end uses: There are several that are not completely electrified, and they represent a significant fraction of total energy consumption. These are direct heat, process steam, space heating, and transportation. Practically all of them use oil and natural gas, two rapidly depleting energy sources. If we wish to solve our energy problems, the proper question to ask is: *How can we satisfy these end uses with fuels other than natural gas and oil?* But that is not the question we hear. Instead, we hear:

- Suggestions that energy problems can be solved by conservation alone. Are the American people going to be asked to change their life styles drastically? Do they deserve to know the alternatives before they make the choice?
- Attacks on the electrical industry on the ground that it is too big, selfish, bureaucratic, and inefficient. Have energy analysts demonstrated to the people ways to generate electricity more efficiently? Isn't this an alternative that analysts have an obligation to discuss?
- Attacks on every proven option in electricity generation: coal on environmental reasons, nuclear on safety and waste, hydro on ecological grounds, oil on economics, gas on supply. If all the attacks are successful, what will life be like in the United States? If the analysts believe in democratic processes, don't they have an obligation to tell the people?

Many suggestions have been made on how we can eventually depend on renewable energy sources. Even the strongest supporters of these suggestions admit that these are long-term options. How do we go from here to utopia? I have not yet seen a single responsible and convincing plan that shows the transition being made without using the two abundant indigenous energy sources: coal and uranium. Electricity is the only way to use uranium and is one of the best ways to use coal.

If there is agreement that electricity is needed to solve the energy problem, then the question of scale becomes important indeed. How can electricity be generated in the most economic way? Many parameters control the economics of electric power generation, but for the question of scale, the key parameters are economy of scale in plant construction and capacity factor of operating plants.

The Question of Economy of Scale

We will examine the question of economy of scale in three steps: components, plant construction cost, and optimization of system economics.

Components

An electrical system consists of many components: a boiler or a nuclear steam supply system, steam turbines, gas turbines, generators, transformers, station auxiliary equipment, circuit breakers, transmission lines, protective relays, lightning arresters, distribution equipment, meters, and entrance devices in user locations. Some have no significant impact on total system costs. I will therefore confine my discussion to the following major items: steam turbine generators, transformers, high-voltage direct-current (HVDC) conversion systems, and transmission lines. I have omitted boilers because I have not worked with them personally and like to avoid a discussion based on second-hand knowledge.

To discuss the economies of scale of components, one must make sure that components of different ratings are of the same degree of maturity in their stage of development. This is not easy. The famous learning curve concept inherently will make the smaller ones cheaper because larger ones are always developed later. But to be on the conservative side, I choose to overlook the learning curve effect and to inventory today's costs as a function of size for much of the equipment built for the electrical power system. In the discussion, the D-factor refers to the percentage reduction in cost per unit output when the rating of the equipment is doubled. This is a convenient way to express the economy of scale, although, as we shall see, the D-factor is usually not a constant, but rather a function of size.

Figure 2 shows a normalized curve of the cost of steam turbine generators as a function of size. The curve is normalized, not only because the actual cost figure is proprietary but also because it varies with time and location. Therefore, a normalized curve is much more meaningful. The D-factor in the range between 350 and 700 megawatts (mw) is 21 percent, but drops to 15 percent in the range of 500 to 1,000 mw—i.e., in dollars per kilowatt (kw), the cost of a 1,000-mw turbine is 15 percent lower than the cost of a 500-mw turbine of comparable design.

Figure 3 shows a similar curve for power transformers at

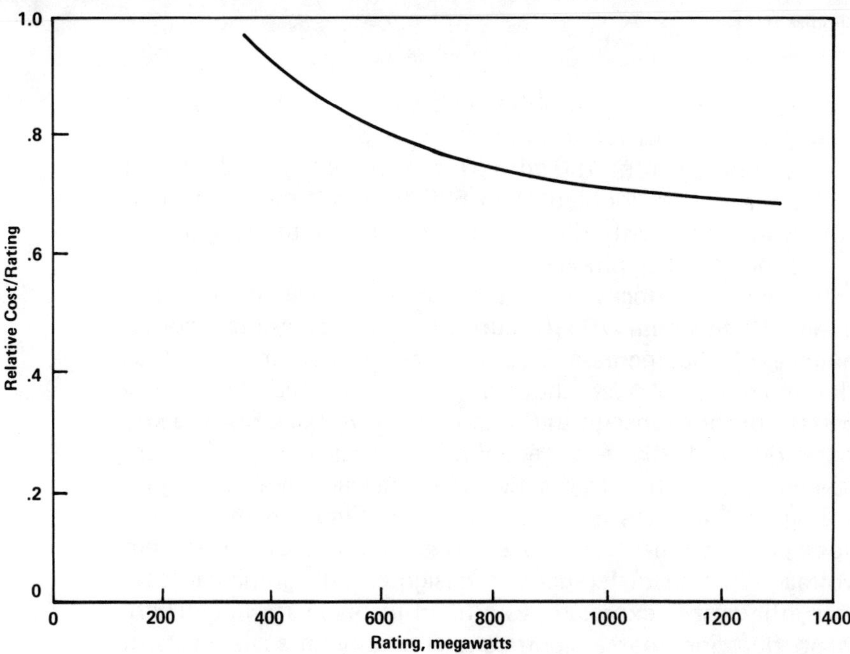

Figure 2. Cost vs. Rating of Fossil-Fired Steam Turbine Generators

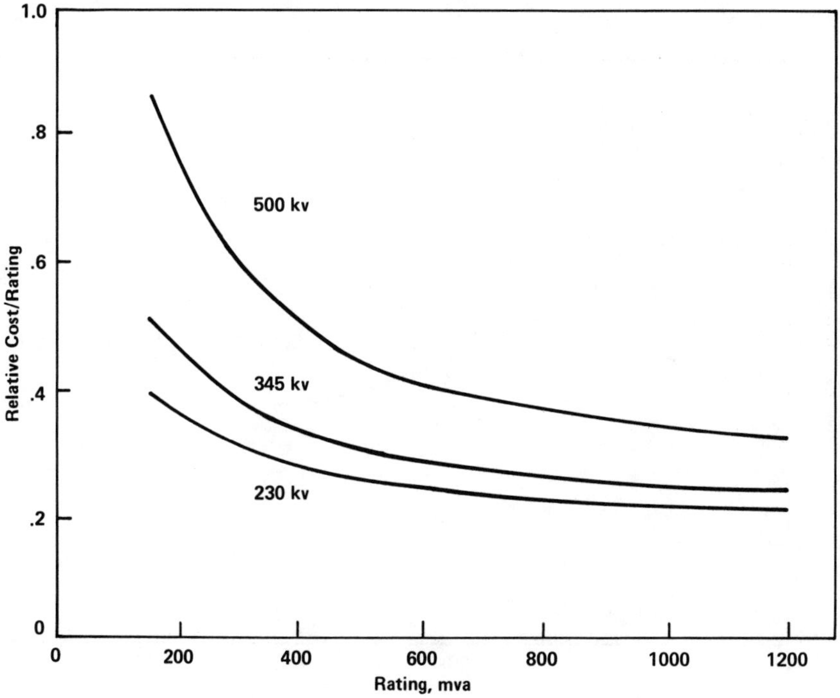

Figure 3. Cost vs. Rating of Power Transformers

different voltages. Take 500-kilovolt (kv) transformers as an example. The cost reduction per kva is 37 percent when the size is doubled from 150 to 300 mva but is only 23 percent when the size is doubled from 500 to 1,000 mva. A similar situation exists with HVDC conversion systems (Figure 4).

Since the beginning of the electrical industry, maximum transmission voltages in North America have grown from about 10 to about 765 (Figure 5). This increase is a consequence of the increase in plant sizes and of the fact that the maximum power that can be transmitted is approximately proportional to the square of the voltage. But the key question is: Is there an economy of scale—i.e., is it more economical to transmit a kw of electrical power at higher voltages? This question is answered in Figure 6, where the cost per mw-mile is expressed as a function of the system voltage. Due to differences in design and geographical location, the cost is expressed as a band instead of a line. There is no question that a significant economy of scale exists in transmission systems.

Now, does this mean that we should continue to push up transmission voltages beyond 765 kv or 1,100 kv? No one

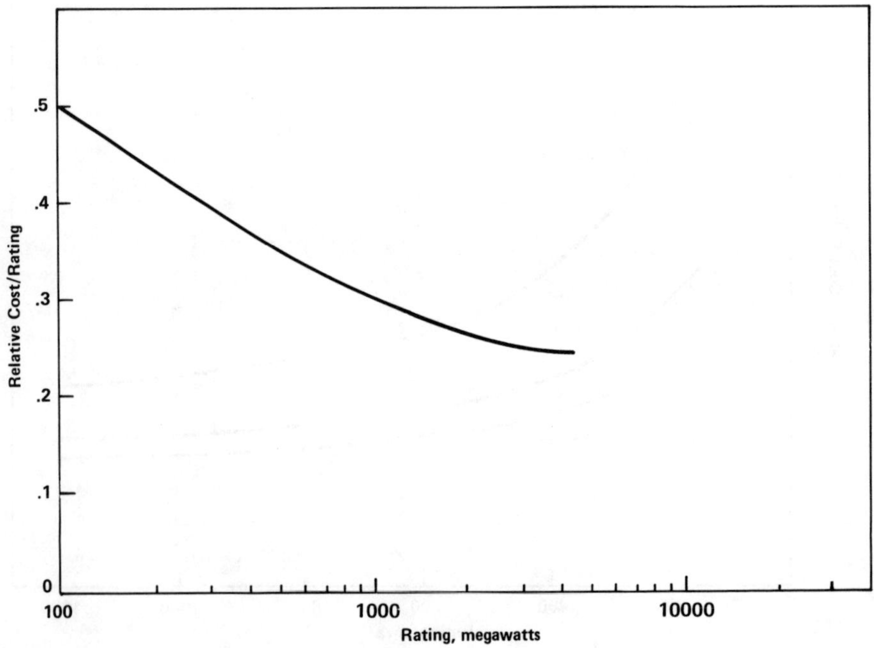

Figure 4. Cost vs. Rating of HVDC Conversion Equipment

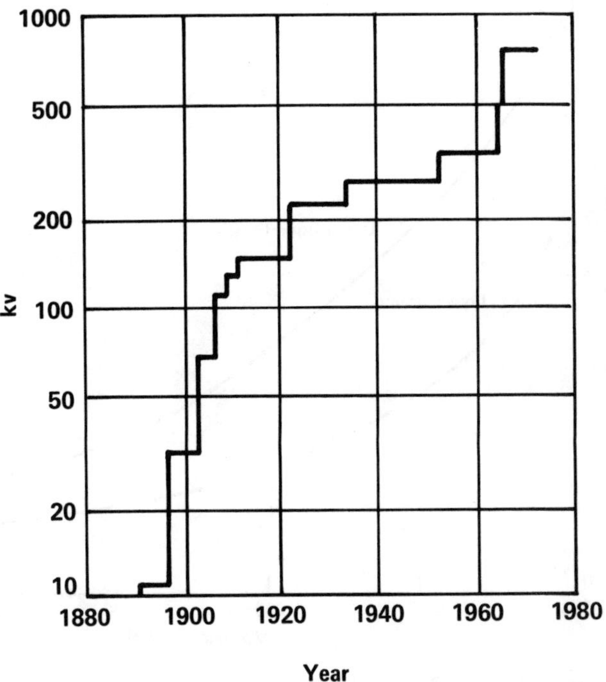

Figure 5. Maximum Transmission Voltage in North America, 1880-1975

can give a definite answer at this time. History tells us that as we move to higher voltage levels, new technical problems may surface. Below 345 kv, lightning used to be the controlling factor for insulation design. At 500 kv, switching surge took over that role. At 765 kv, we found a new problem—audible noise—and at 1,100 kv, another—electrostatic induction. We do not know at this time what problem will appear at voltages higher than 1,500 kv. On the other hand, history also shows that as these problems were discovered, solutions were found to preserve the economy of scale. For example, addition of a relatively inexpensive resistance and switch in 500-kv circuit breakers preserved the economic attractiveness of 500-kv transmission. Whether this trend will continue, no one can tell. But unless economics shows that higher voltage is more beneficial, I don't believe that anyone will move to higher transmission voltages just for the sake of change.

In summary, the actual costs of the major components that make up the electric power system clearly demonstrate a significant economy of scale. This, of course, has been known to the industry for some time.

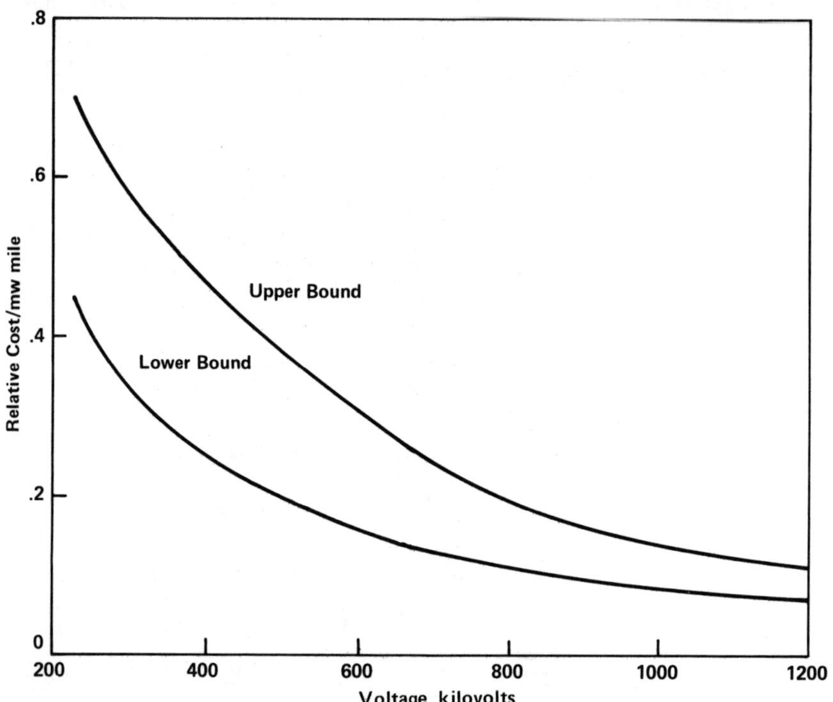

Figure 6. Cost vs. Rating of Transmission Lines (excluding right-of-way)

Plant construction cost

The economy of scale of plant construction cost is a much more complex question. Plant designs vary from one utility to another. Some plants are completely enclosed, while others are only partially enclosed. Labor productivity varies from one region to another. The first plant on a site includes the cost of the site, while the second one on the same site does not. Practices of private and public utilities can also be quite different, with the latter enjoying the benefit of tax-free, low-interest money. Unless the quantitative effects of these factors are sorted out, it is easy to draw wrong conclusions.

Nuclear plant costs are more complex than fossil fuel plant costs. Because standardization has not been a way of life in the nuclear industry, the design of two plants of comparable rating can be very different. Table 2 illustrates such an example where two plants of comparable rating used vastly different amounts of materials.[2] In addition to all the problems just described, continuously changing regulatory requirements, inherently long construction cycles, and unpredictable intervention introduce many other costs. As Table 3 shows, so much of the total plant cost is in interest,

TABLE 2

Comparison of Bulk Materials Used in Two Nuclear Power Plants

Item	Description	Unit	Plant 1	Plant 2	% Change, 1 vs. 2
1.	Plant size (net rating)	Megawatts	840	820	––
2.	Building volume*	Cubic feet/kilowatt	9.0	16.7	+ 85
	Actual (millions)	Cubic feet	7,600	13,665	
3.	Structural concrete (except cooling towers)	Cubic yards/kilowatt	0.08	0.22	+275
	Actual (thousands)	Cubic yards	67.0	176.0	
4.	Structural steel	Pounds/kilowatt tons	8.3	8.5	+ 2.4
	Actual (thousands)		3,500	3,500	
5.	Conduit (metal and nonmetal)	Linear feet/kilowatt	0.36	0.74	+100
	Actual (thousands)	Linear feet	305.0	610.0	
6.	Tray	Linear feet/kilowatt	0.02	0.086	+333
	Actual (thousands)	Linear feet	16.7	70.5	
7.	Piping (all sizes)	Linear feet/kilowatt	0.21	0.25	+ 19
	Actual (thousands)	Linear feet	176.0	205.0	
8.	Cable and wire	Linear feet/kilowatt	2.24	4.64	+107
	Actual (millions)	Linear feet	1,880	3,800	
9.	Manhours (manual and nonmanual)	––	5.3	19.5	+375
	Actual (millions)		4,420	16,200	
10.	Capital cost	$/kilowatt	285	495	+ 73
	Actual (millions)	$	238	406	

* Does not include miscellaneous buildings such as storeroom, screen house, water treatment, or service water structures.

Source: Palmeter, S.B. Information Exchange on Power Plant Construction. Paper given at joint meeting of International and American Nuclear Society, Mexico City, Sept. 29, 1975.

escalation, and indirect costs, that the question of economy of scale can no longer be answered by just looking at the direct materials and labor.

The ideal conditions to study the economy of scale of nuclear plants would be standardized design, stable regulatory requirements, optimized construction techniques, and large number of samples.

We are not blessed with any of these conditions. Short of that, what analysts have been doing is to study the costs of all the plants that have been built and arrive at some sta-

TABLE 3

Analysis of Power Plant Costs ($/kw)

	Coal	Oil	Nuclear
Equipment and construction costs	325	275	325
Indirect costs	90	75	120
Interest	170	165	230
Escalation	190	180	280
Total	775	695	955

tistical conclusions. How meaningful is the conclusion, I do not know. Does it represent what is theoretically possible? Most likely not!

In the case of fossil fuel plants, the situation is considerably better. At least here we are blessed with a more mature technology and a much larger population. Therefore, we decided to make a detailed study of the construction cost of fossil plants.[5] We looked at 305 plants built in the period 1960-1972, taking the cost information from the Federal Power Commission.[6] By necessity, the total plant cost is the summation of period dollars. It is difficult to sort out the effects of inflation. Fortunately, economic conditions during that period were relatively stable, and the construction cycle was not too long. Therefore, the effects of inflation and interest did not prevent drawing meaningful conclusions.

In spite of these favorable conditions, plotting all the costs of these 305 plants as a function of size on a single sheet of paper yields confusing results (Figure 7). However, it is quite obvious that some economy of scale existed below 200 mw, which is by no means a small plant. A single 200-mw plant can serve approximately 100,000 people in the United States, or 10 percent of a country like Egypt. But what about plants larger than 200 mw? To make some sense out of the maze of scattered points, we must separate out the effects of many of the factors mentioned before. The first critical parameter that must be normalized is time (year of operation). This was done in the following way: The average cost of all units completed during the same year was computed and divided by the overall average cost of all units to yield an index for that year. Then the cost of every unit completed in that year was divided by this index to yield a "time-independent" cost for that unit.

After all units were normalized for time, the entire

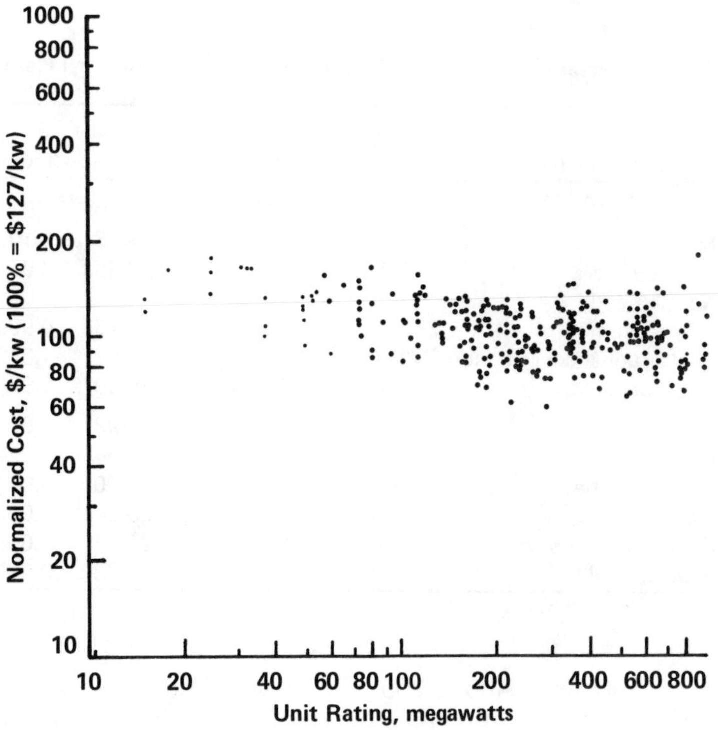

Figure 7. Plant Construction Cost of 305 Fossil Fuel Plants vs. Size

data base was placed in a retrievable computer format that included parameters for each unit that might affect capital costs (first unit, conventional boilers, supercritical boilers, conventional construction, outdoor boilers, etc.). Then each characteristic was analyzed for its effect on the average capital cost. For example, all units with outdoor boilers were isolated, and the average capital cost in dollars/kw was calculated. The ratio of this average capital cost to the average capital cost of all 305 units became an index of how significant that parameter might be. Each characteristic was analyzed this way, and then all indexes with ± 2.5 percent impact were judged significant. The significant ones are first or nonfirst unit, outdoor construction, and ownership (Table 4).

The capital cost of each unit that had any or all of these significant characteristics is then normalized by the appropriate indexes to allow the plot of a normalized curve of cost versus size (Figure 8).

It is particularly interesting that the curve shows a rise

TABLE 4

Index for Parameters Affecting Capital Costs of 305 Fossil Fuel Plants

Parameter	Av. Size,	No. of Units	Capital Cost,	Index
National	343	305	127.46	1.0
First units	381	80	143.47	1.126
Second units	354	76	122.86	0.964
Nonfirst and nonsecond units	317	149	119.73	0.939
Conventional construction	347	202	128.97	1.012
Outdoor construction	334	103	124.39	0.976
Coal-fired	363	249	126.98	0.996
Dual-fuel	251	56	130.54	1.024
Private utilities	366	258	126.20	0.99
Public utilities	215	47	139.24	1.092
Conventional boilers	258	237	127.87	1.003
Supercritical boilers	640	68	126.88	0.995

after 200 mw. Above 200 mw, steam conditions typically change from 2,400 pounds per square inch (psi) and 1,000°F to the supercritical conditions of 3,500 psi and 1,000°F. Between 200 mw and 500 mw, there is a mixture of supercritical systems and the lower pressure systems. If the supercritical systems are isolated from the rest, cost beyond 600 mw continues to drop. Figure 9 shows that the D-factor is about 18 percent for supercritical systems.

As another approach to the effect of time, the GNP deflator was applied to the plant costs on a uniform basis (Figure 10). However, since the spending on a plant construction is never uniform, we also tried to apply a GNP deflator with a 2-year lead time from the commercial operation date. The resultant curve shows that after adjusting for inflation, plant costs continue to decrease with increasing size.

Many papers have been written recently to point out that capital cost is not the only factor that affects the total cost of power plants. Capacity factor is equally important. This is certainly true, and we will discuss the question of capacity factor later.

Optimization of systems economics

Plant construction cost is only one term in the equation for the economics of electrical generation. For the public, the ultimate measurement is the price of electricity. Therefore, the optimum size of a generating unit must be decided by its total operating cost. In this section, we report the results of a study on the parameters that affect the total operating cost of a generating plant.

Any optimization problem has a set of constraints. The key constraints for the problem treated here are reliability and fuel availability. In the United States, the criterion for system reliability has been established for some time. It is measured in number of days in a given period of time that the system capacity fails to meet the load demand. Of two widely accepted criteria—1 day in 10 years or 1 day in 5 years—we use the latter.

On fuel availability, we assume that:

- Oil will not be used for base load generation.
- There will be sufficient coal for power generation if there is no enduring nuclear moratorium. To provide this coal will require major capital and urgent effort from the coal and railroad industries.
- Synthetic fuel from coal will not be available until late 1980's, and even then not in significant quantities.

We selected two power pools, a very large one and a

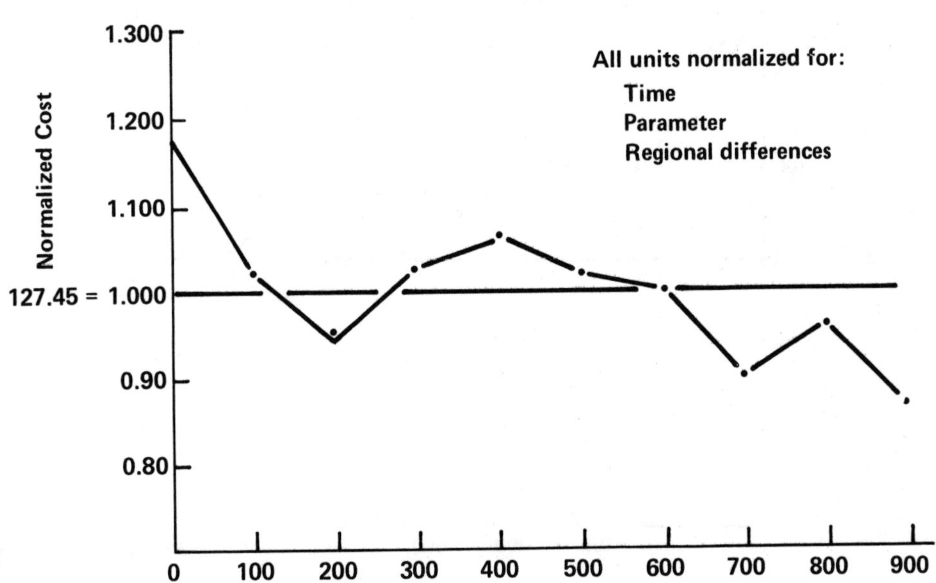

Figure 8. Normalized cost of 305 Fossil Fuel Power Plants vs. Size

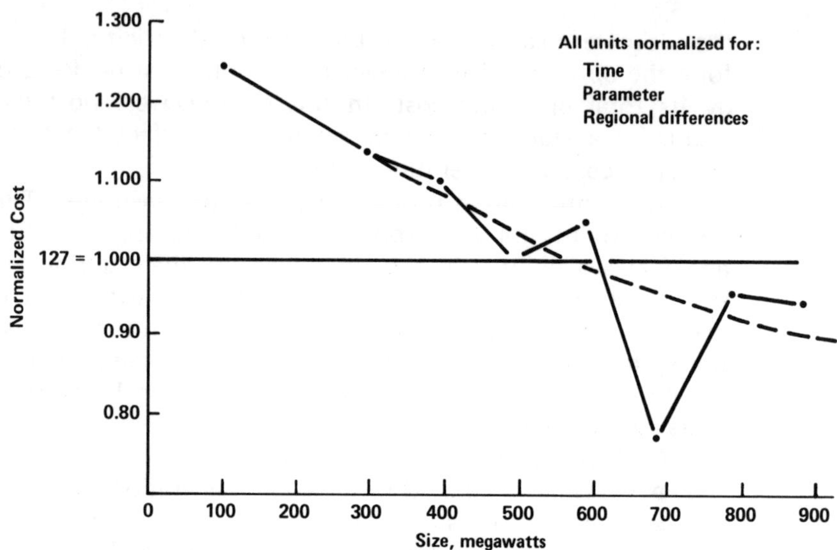

Figure 9. Normalized Cost for Supercritical Units of 305 Fossil Fuel Power Plants vs. Size

All units normalized for:
GNP deflator 1960 = 1.0
Regional differences
Parametric differences
--- GNP deflator lagged 2 years
—— GNP deflator

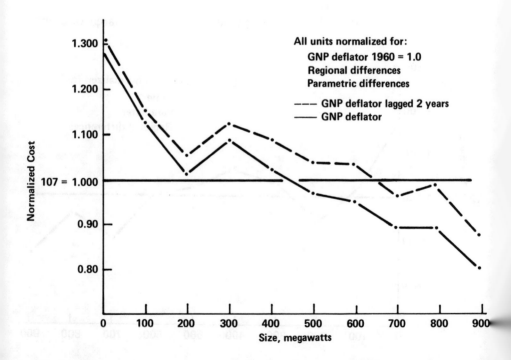

Figure 10. Normalized Cost (including inflation) of 305 Fossil Fuel Power Plants vs. Size

small one. For both, we assume that the electrical load growth in the next 15 years will be at an average annual rate of 5.7 percent. The results for both pools are almost identical. We will therefore show only the result for the large one.

The first step in this study is to compute the optimum generation mix that those two power pools must add to their systems in order to satisfy the demand. Then, we asked the question: Given the generation mix, what is the optimum size of the nuclear power plants?

We used three computer programs for this problem. The first is the optimum generation planning program, which was developed a number of years ago to facilitate the planning of the most economical mix for generation additions, given a forecasted rate of load growth. There is a very comprehensive data base in this program, including the characteristics of all turbine generators above a certain size in the United States and the daily and seasonal load profiles for all regions in the country.

In the second program, the load factor program, we divide all electrical load into nine end uses (Figure 11). Each has a typical daily and seasonal load profile. Included in the forecast of annual load growth rate of 5.7 percent are the degrees of penetration of all major appliances in both commercial and residential sectors. Combining this forecast with the load factor program, we can compute the future load factor as a function of time.

The financial program predicts total revenue, capital requirements, and price of electricity from a set of assumptions on load growth, fuel prices, construction costs, and plant capacity factor (Table 5).

TABLE 5

Assumptions for Financial Program

1981 Plant Costs,	$/Kwh	Plant Availability, %	Plant Cost Escalation, %/Year	1974 Fuel Costs, Cents/Million Btu	Fuel Escalation, %/Year	Heat Rate, Btu
Nuclear	800	68	6.2	0.40	6.1	10,400
Coal	670	73	6.2	0.62	6.2	8.900
Combined cycle	400	82	5.3	2.07	5.0	8,800
Gas turbine	180	88	5.3	2.07	5.0	12,500

Load growth: 5.5%, 1975-90.
Load factor: 2 points improvement.
Probability of loss-of-load: 1 day in 5 years.

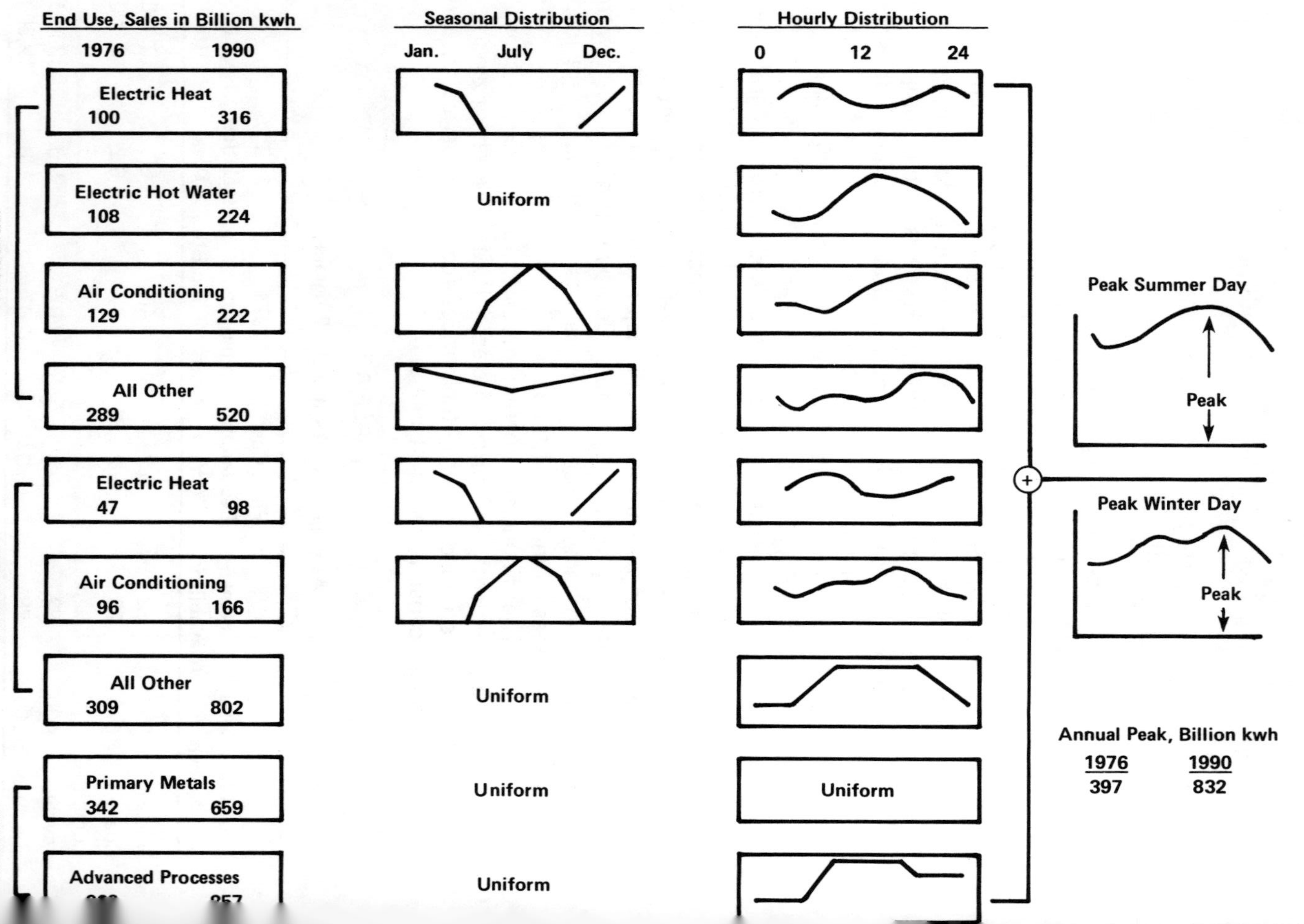

Figure 11. End Uses for Load Factor Program

Using these three programs, we compute the annual cost of operating a nuclear plant of a given size. This cost includes not only the capital cost, depreciation, and interest, but also fuel, operation, and maintenance costs. To test the sensitivity of the two key parameters controlling the cost, the D-factor and capacitor factor, we use several sets of numbers for any given plant size. The results (Figure 12) permit drawing several interesting conclusions:

- If availability does not decrease with increasing plant size, one should push for larger plants forever.
- If there is no economy of scale and availability does decrease with increasing plant size, then one should build the smallest plant.
- There is an optimum plant size—somewhere between 600 and 1,000 mw—for intermediate cases where

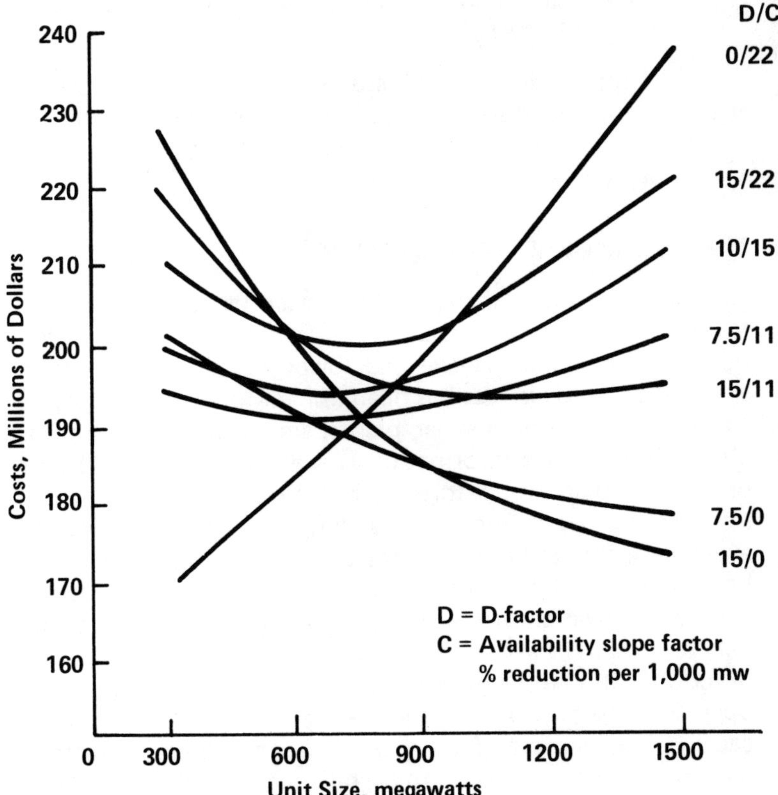

Figure 12. Annual Costs of Operating Nuclear Power Plant vs. Unit Size (costs based on an equivalent 1,300-mw unit)

availability does decrease with plant size and there is a reasonable economy of scale.

Although we do not know these two numbers with great accuracy, we do know their general behavior. Therefore, the intermediate case is much closer to the real world than the other two.

Our analysis also produced some conclusions on capital requirements and price of electricity:

- The capital requirements of an optimum-sized plant and other sizes differ significantly. Choosing the optimum size or not doing so can mean significant differences in capital requirements.
- However, in the next 15 years, the price of electricity is not affected as much by the choice as the capital requirements for one simple reason: a 500,000-mw system is already in place in this country; therefore, only fuel prices can change the price of electricity in a very short time.

And finally, for the case studied, the price of electricity need not increase faster than inflation to ensure the financial health of the utility industry and its ability to raise the needed capital.

The Importance of Capacity Factor

Recently, several papers have been written on the performance of power plants as a function of size. They pointed out that larger plants have had lower capacity factors than smaller plants. This fact has been known for fossil fuel plants for some time. For nuclear plants, similar indications exist, although much less information is available. The conclusions on nuclear plants therefore do not have as much statistical significance. Even more important than the quantity of available information is the fact that nuclear technology has been moving at a very rapid rate. On the average, the size of light water reactors has been increasing at the rate of 75 mw per year in the past 15 years. The new plants under design or construction do not have the benefit of feedback from operating plants with an earlier design. Many of the problems causing low capacity factors are generic in nature and not necessarily size related. Any statistical conclusion drawn from the available information may not represent what can happen in the future.

But, for fossil fuel plants, there is enough information

to indicate that the capacity factor decreases with increasing size. It is therefore logical to ask what is the optimum size for future additions.

From the analysis of 305 fossil plants, we found that below 200 mw, economy of scale is so great that the slightly higher capacity factor is inadequate to make smaller units economically attractive. The question remaining is: What is the optimum plant size beyond 200 mw? This question can be divided into two parts:

- Is economy of scale going to improve in the future?
- Is the relationship between capacity factor and size going to change?

Unfortunately, we cannot answer either question quantitatively. Our discussions therefore must be qualitative in nature and do involve subjective opinions. We have seen that for components, there is a definite economy of scale, although the D-factor does decrease as size increases. For the construction cost, the question is more complex. The size of the plant may have a significant impact on labor productivity. People experienced in construction often express the opinion that labor productivity on a construction site has a tendency to drop if there are more than 1,000 people working simultaneously in a relatively concentrated area such as a power plant. This, of course, is understandable because people get in each other's way.

Many things can be done to improve labor productivity. Standardization is one. For identical plants, one needs much less field supervision; much more can be done in factories; much better construction procedures and management can be put in place. Modular construction in the factory will reduce the amount of work needed in the field. I often wonder if a full size mockup were made (as is done for airplanes) to develop detail layouts and assembly procedures, how much we could improve the productivity on site and thereby enjoy the inherent economy of scale in the components of a power plant.

The second question is equally difficult to answer. We do not yet know all the reasons that cause larger machines to have lower capacity factors. As a general principle, it is true that if the design is identical, and if large sizes are built by using more components of the same kind, then larger machines should have lower capacity factors. For example, a six-flow turbine uses three times more low-pressure components than does a two-flow turbine. Therefore, if the low-pressure components are the main cause of forced outages, a turbine three times the size may have a forced outage rate three times higher. But are the low-pressure components the main cause of outages?

A larger nuclear reactor has more fuel bundles in proportion to its rating. Therefore, if fuel is the key contributor to unavailability, larger reactors would have lower availability than smaller ones. But is fuel the key cause of unavailability?

The information needed to answer these questions is available, but it is scattered, not collected on a basis to facilitate analysis of this kind. I believe that we should make an effort to study these questions. We must divide unavailability into generic and size-related causes. Once we have done that, we can challenge the engineers to solve the generic problems.

In spite of our limited knowledge, there is enough evidence to indicate that there is an optimum size because:

- Economy of scale is not constant; it decreases with increasing size.
- There are some size-related causes of unavailability.

As we understand the behavior of these two parameters better, we will be able to narrow down the range for the optimum size in an evolutionary way. In the meantime, all evidence points to the conclusions that "really small may not be beautiful" and "really huge may not be necessary."

On the Price of Electricity

A key reason for the recent attack on the electric utility industry is the rapid increase of price of electricity since 1973. However, on a constant dollar basis, the price had been dropping for many years (Figure 13). Let us examine capital cost, fuel cost, and operational and maintenance cost, the three components that determine the price of electricity (Table 6). The price across the nation has gone up 45 percent in the past 2 years. In the same 2 years, oil prices have gone up by 259 percent, coal by 195 percent, gas by 209 percent, and uranium by 400 percent. Adjusted for the mix in our electrical system, the increase in fuel cost is about 217 percent. This fact alone can explain the increase in the price of electricity. Why blame the electrical industry for something it cannot control?

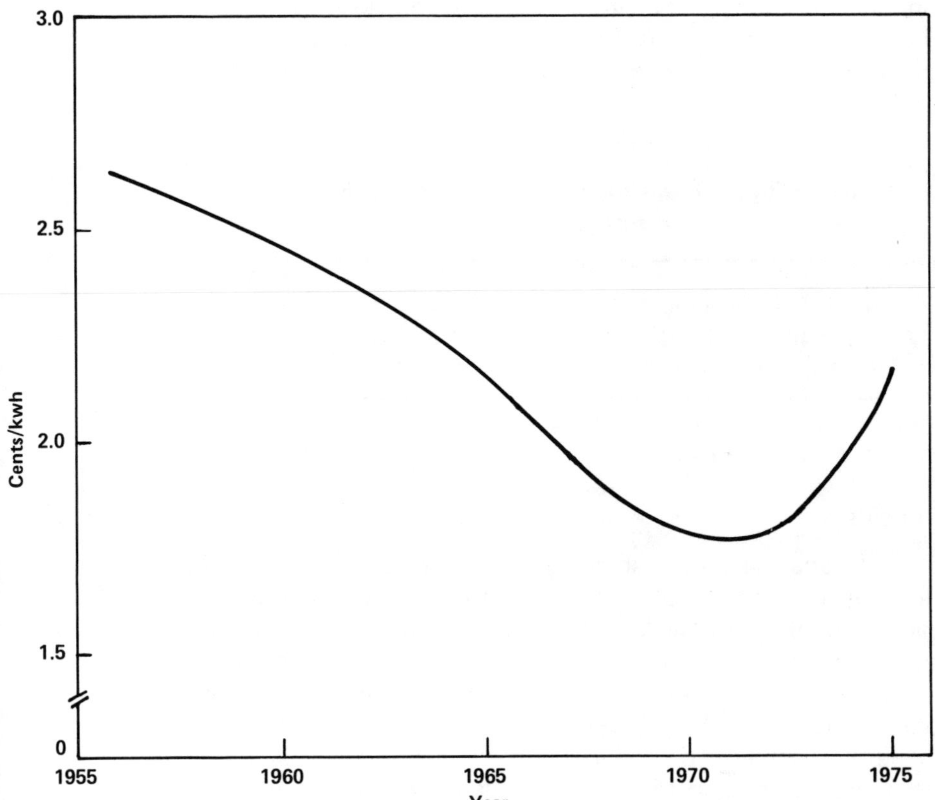

Figure 13. Average Price of Electricity in United States, 1955-1975 (constant 1972 dollars)

What about the effect of scale on the price of electricity? Table 7 shows prices in three locations: Pacific Gas & Electric Co. in California, one of the largest electrical systems; United Illuminating Co. in Connecticut, a relatively small system; and Block Island Power Co. in Rhode Island,

TABLE 6

Percentage Breakdown of Total Electrical Costs In United States, 1965-1975

Year	Fuel	O & M	Capital	Total Costs
1965	19	32	49	100
1973	31	28	41	100
1974	43	21	36	100
1975 (Est.)	45	19	36	100

one of the smallest systems. The only conclusion one can draw from this table is that Block Island's beauty cannot possibly be due to its electricity prices.

TABLE 7

Approximate Price of Electricity in Selected Plants, 1976 (Cents/kwh)

Pacific Gas & Electric Co.	2.8
United Illuminating Co.	5
Block Island Power Co.	10-11

Conclusion

My analysis clearly shows that there is economy of scale in both components and plant construction. By taking advantage of this through centralization, the industry has been able to continually reduce the price of electricity in the past 20 years. The increases in the past 2 years are principally due to fuel cost, which is beyond the control of the electric power companies. The analysis also shows that while we may have already reached the optimum plant size for the present technology, there is no reason to think that decentralizing the electric power systems will be beneficial to the public.

In the past decade, we have witnessed a great deal of turbulence in the United States: unrest in schools, mistrust of government, rising crime rates, declining moral standards. These are disturbing trends. There must be underlying reasons. We must search for these reasons and reverse the trends. Are the trends caused by the declining influence of religion, by failures in humanities education, by the behavior of our leaders, or by a combination of all these factors? These are very complex questions and deserve careful study. But to blame the problems in our society on "violent technology," as proposed by the advocates of "Small Is Beautiful," is like blaming the high electricity prices in the past 2 years on the "large, inefficient and bureaucratic power systems."

This type of approach reminds me of a story:

A man was bending over under a street light as if searching for something. A friend of his approached him and asked, "Have you lost something?" The man answered, "Yes, I lost my wallet." His friend said, "Let me help you. Do you know roughly where you lost it?" The man said, "I think it is somewhere in the dark." His friend asked, "Why are you looking here then?" He answered, "This is the only place I can see."

REFERENCES

1. Fisher, J. C. Energy Crises in Perspective. John Wiley & Sons, Inc., New York. 1974.

2. Palmeter, S. B. Information Exchange on Nuclear Power Plant Construction. Paper given at joint meeting of International and American Nuclear Society, Mexico City, Sept. 29, 1975.

Alvin M. Weinberg
Institute for Energy Analysis
Oak Ridge Associated Universities
Oak Ridge, Tennessee

Alvin M. Weinberg is director of the Institute for Energy Analysis, which he was instrumental in establishing at Oak Ridge Associated Universities in January 1974. After serving briefly as director of the Institute, Mr. Weinberg became director of the Federal Energy Administration's Office of Energy Research and Development; he returned to the Institute in July 1975.

For more than a quarter century Mr. Weinberg was director of the Oak Ridge National Laboratory (ORNL), one of the world's great scientific and technological institutions. He was also among the first members of the University of Chicago's war-time Metallurgical Laboratory. He joined ORNL in 1945 where he served as director of the Physics Division (1947-48), as research director (1948-1955) and as director (1955-1973).

For his role in the development of nuclear reactors, Weinberg shared the Atoms for Peace Award in 1960 and was one of the first recipients of the E. O. Lawrence Memorial Award. In 1966 he received The University of Chicago Alumni Medal for ". . . pioneering contributions to the application of science and technology to the service of mankind"; in 1971, the ANS Chernick Memorial Award; and in 1975 for "wide-ranging scientific contributions—especially in energy technology and particularly for contributions to the development of peaceful uses of nuclear energy," he was awarded the first Heinrich Hertz Prize on the 150th anniversary of the University of Karlsruhe.

Weinberg has articulated and clearly formulated the role of the national laboratory in our modern society. He has also contributed to the formulation of public policy concerning the relationship between science and government, as well as the debate on issues associated with energy policy and the relation between technology and society.

Weinberg received the S.B., S.M., and Ph.D. in physics at the University of Chicago. He is a member of the National Academy of Sciences and the National Academy of Engineering.

Can We Do Without Uranium?

Uranium epitomizes one extreme in energy strategy: It is heavily centralized, demands sophisticated "hard" technology, its output is primarily electricity. I avoid for the moment the question of whether technologies of this character are intrinsically bad; instead, I ask whether we can do without uranium, what is gained, what is lost if we reject fission—or to put it more dramatically, what man would have done for energy had he evolved 2 billion years later, at which time all of the uranium-235 would have disappeared. I shall examine the matter both in the relatively short run, say from now to 2010, and in the extremely long run, when we can no longer depend on fossil fuels—either because they have been exhausted, or because they have been proscribed because their combustion adds too much carbon dioxide to the atmosphere.

The Short Run: to 2010

For the short run, I draw on the results of a study we have just completed at the Institute for Energy Analysis: "Economic and Environmental Implications of a U.S. Nuclear Moratorium—1985-2010."[1] This study attempts to clarify the technical and economic issues that underlie much of the great debate on fission energy that now grips many Western countries.

We assumed the nuclear moratorium goes into effect in 1985. Reactors on line or under construction by then are allowed to run their course, but no new reactors are built after 1985. We tried to estimate economic and environmental consequences of such a moratorium.

As a first step, we estimated the future course of economic growth in the United States. Our main conclusion is that Gross National Product (GNP), measured in constant dollars, is unlikely to continue to rise at as rapid a rate as it has in the past decade. At least four factors account for this result:

- Fertility rate has been falling precipitously and now stands at 1.8.
- Population by 2000 is therefore likely to be around 250 million, some 100 million lower than was projected 10 years ago.
- Labor productivity will increase at a rate ranging from 1.7 percent per year to 2.4 percent per year—a rate somewhat below the average (2.27 percent)

from 1950 to 1965, but above the average (0.9 percent) from 1965 to 1975.

- Various structural changes—such as increase in transfer payments, increased ratio of capital expenditure to GNP, as well as increased cost of energy—point toward a lower growth rate of GNP.

We have estimated a range of plausible energy demands based on our estimates of GNP and population, and from an analysis of the likely improvements in energy efficiency in each of the sectors of the economy. Two main results stand out:

- Total energy demand in 2000 ranges from a low of 101 quadrillion British thermal units (Btu), or quads, to a high of 126 quads. These values are much lower than those projected by almost all other studies; the notable exception is the Ford Foundation-sponsored Energy Policy Project, which projects 100 quads in the "zero energy growth" scenario, 125 quads in the "technical fix" scenario. However, the Ford zero energy growth is normative—it represents what the project believed *ought* to happen; in contrast, our projections are nonnormative—they represent simply what we believe is *likely* to happen.
- Though total energy demand is considerably below most other estimates, we estimate that demand for electricity will increase sharply—from 20 quads (approximately 28 percent of present demand) to about 47 to 64 quads in 2000 (approximately 50 percent demand). This shift toward electricity is accounted for largely by the much higher rise in price of oil and gas as compared to coal and nuclear (which are used for electricity) and by our belief that new electrical devices, especially heat pumps, will penetrate the market by then faster than new supplies of oil and synthetic gas. These two factors suggest that a shift to coal and electricity is more plausible than the opposite trend. As for transport, we assume the mandated improvements in energy efficiency for our auto fleet will be nearly achieved.

The year 2000 might find us then with some 47 to 64 quads of energy going into electricity. We would estimate that almost all of this would be generated in large central plants–the equivalent of some 800 plants of 1,000 megawatts (mw) each (in the low case). In the absence of a moratorium, rather more than half these plants would be nuclear;

in the presence of a moratorium, only 25 percent would be nuclear. We do not see how, between now and 2000, the "soft" modes–solar and its children, or geothermal–can contribute more than a percent or so of the total except for passive heating and cooling of buildings.

The economic consequences of the moratorium depend mainly on the relative cost of generation of electricity from coal and from Light Water Reactors (LWR's). Our analysis suggests that in most parts of the country, the cost of electricity from coal-fired plants will be somewhat higher than from LWR's–but the margin is generally not more than about 10 to 20 percent. In New England and California, the margin favoring nuclear is considerably greater; in the Rocky Mountain region, the situation is reversed. Thus nuclear energy vis-à-vis coal seems to have reverted to its historical position: less expensive than coal in some parts of the country, more expensive in others. The euphoria of extremely low-cost nuclear electricity (a euphoria to which I contributed at the time of the Oyster Creek Nuclear Power Plant) has been superseded by the less euphoric actual turn of events.

The direct economic impact of a moratorium ranges from nil in those parts of the country where coal is cheaper than nuclear, to quite heavy in New England and California, where the reverse is the case. As for the *overall* cost to the economy, we estimate that, averaged over the entire country, we would have to pay more for electricity if we were denied the nuclear option. However, the total burden in no case is as much as about 1 percent of the cumulative GNP, and in most cases it is less than ½ percent of the cumulative GNP. One concludes then that the direct cost of the nuclear moratorium we postulate is bearable overall, but might be very awkward in some parts of the United States, particularly New England.

The primary effect of the moratorium is the immense pressure it puts on coal. There seems to be no alternative to burning much more coal in the next 30 years. Even without the moratorium, we might need as much as 2.5 billion tons of coal per year by 2010; with the moratorium, this number approaches 5 billion tons per year–about an eightfold increase over what we now use. When one considers that our coal production in the years since the embargo of 1973 has increased by only about 100 million tons per year in a 3-year period, it seems difficult to contemplate an increase of coal production of the magnitude we estimate. Thus when

I assert that a nuclear moratorium would not be unbearable, I am making the large additional assumption that a vast increase in coal is credible.

As for the environment, on the whole we found these environmental impacts that would *certainly* occur to be larger with the moratorium than without. More land would be disturbed, more noxious fumes would be emitted, more carbon dioxide would be added to the atmosphere. On the other hand, total emissions of sulfur dioxide, carbon monoxide, nitrogen oxides, and particulates would, surprisingly, be less than, or at least comparable to, our present emission levels, provided that the New Source Performance Standards promulgated by the U.S. Environmental Protection Agency are enforced.

As for presumptive insults to the environment—insults that would result from malfunction of the system, or even from routine emissions such as carbon dioxide whose magnitude can be estimated but whose consequences are clouded in doubt—we can only quote numbers, hardly draw conclusions. The main presumptive consequence of a nuclear scenario is an uncontained core meltdown: With the moratorium, the expected number of such meltdowns is 0.2; without the moratorium, it is 0.6 during the period to 2010. These estimates are based on the Rasmussen report on nuclear reactor safety,[2] despite Rasmussen's injunction *not* to apply his results to more than 100 LWR's of current type.

As for carbon dioxide, a U.S. moratorium would increase atmospheric levels, but hardly enough to make much difference in the period under consideration. On the other hand, should a U.S. moratorium lead to a global moratorium on nuclear energy, the carbon dioxide level, which now stands at about 330 parts per million (ppm) (10 percent higher than the preindustrial level), could significantly hasten the time when the atmospheric concentration doubles. Even before this point is reached, climate changes are likely to occur that might alter our patterns of food production.

Our findings then are likely to support both those who dislike and those who like nuclear energy. For the former, we point to a lower overall energy use and relatively bearable impact on GNP of a moratorium. For the latter, we point to large regional economic impacts and what to us appears to be an incredibly large demand for coal caused by a moratorium.

Our findings on the whole give little consolation to those who urge decentralization. We do project a trend toward electricity, and during the rather short time span—to

2010–it is unclear how this electricity can be supplied except by large central power plants. The projected amount–some 50 to 60 quads by 2000–simply seems too large to be supplied by other than large-scale generating plants, whether nuclear or coal-fired.

An Asymptotic World

I turn now to the asymptotic situation, when fossil fuel is gone or has been foresworn because of the carbon dioxide problem. I shall try to visualize and compare a future based on the sun with an alternative future based on uranium or thorium breeders.

To evaluate these two alternatives, I shall consider an ultimate world in which the great economic discrepancies between poor and rich have been eliminated. Robert Heilbroner's "wars of redistribution"[3] will have been avoided, and all people will have reached a living standard comparable to that of Western Europe. I choose such a scenario because it brings out most clearly what may be the essential choice: between a stable world in which all have a relatively large per capita energy but which places great pressure on the environment, and an unstable world in which the average per capita demand is very low (about 50 million Btu per person) but the environmental pressures are much smaller.

I shall assume F. Niehaus' asymptotic world energy demand[4]–2×10^{18} Btu (or 2000 quads)–reached in about 100 years, compared to 220 quads today (Figure 1). This corresponds to about 280 million Btu per person for a world of 7.5 billion people or 140 million Btu per person for a world of 15 billion. The latter per capita energy demand corresponds to the current West German demand, and is somewhat less than half the U.S. level.

Our present age of fossil fuel obviously will end rather quickly once this demand is reached. Oil and gas–about 30,000 quads–would last but a few years. The estimated 8×10^{12} tons of coal (assuming all the energy comes from coal) would be used up in about 100 years. Estimates of the total recoverable reserve of shale oil are most uncertain; I shall use the figure of about 100,000 quads given to me by G. Marland of the Institute for Energy Analysis.[5] This adds another 50 or so years to the time before the fossil fuels are exhausted.

The carbon dioxide added to the atmosphere might end the age of fossil fuel before the fuels are exhausted. About

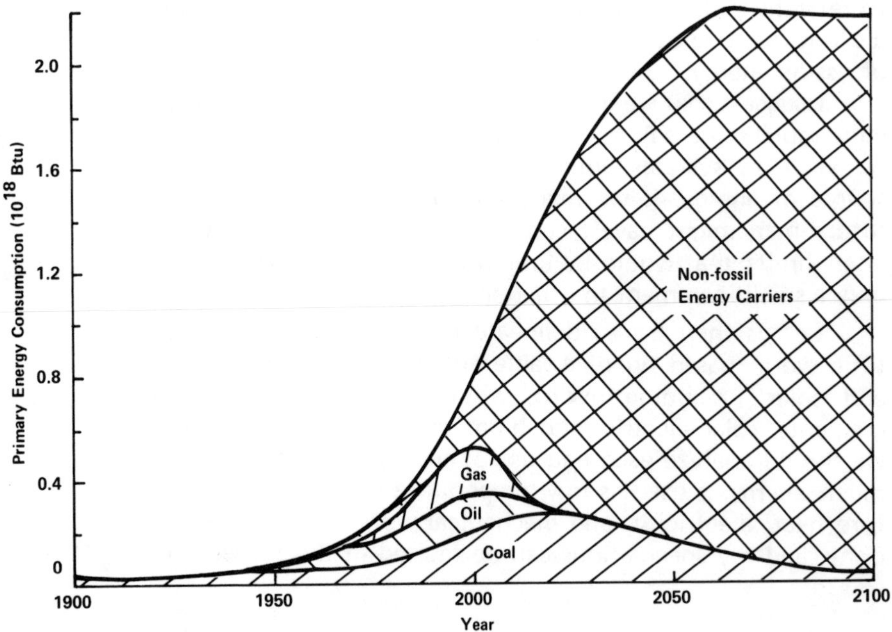

FIGURE 1. Projection of Primary Energy Consumption (After Niehaus)

half of the manmade carbon dioxide seems to remain in the atmosphere. Its concentration in the atmosphere is rising at a rate of about ½ ppm per year, and is now some 10 percent greater than it was in the preindustrial era (Figure 2). It has been suggested that if 20 percent of the estimated fossil resource of approximately 300,000 quads is burned, the concentration of carbon dioxide in the atmosphere would double; this might lead to unacceptable heating of the globe. It is conceivable that we shall have to shift to nonfossil energy sources much sooner than one would estimate from the projected depletion of coal resources—say, by the middle of the next century. The issue of the sun and uranium then might become nonacademic within some of our lifetimes.

I propose to examine the full implications of dependence on fission and on solar energy in this asymptotic world. In the early days of fission, we generally ignored its very long-term implications. The systems problems that plague fission now that it is widely deployed—safety, public acceptance, wastes, transport of radioactivity—somehow did not seem very important earlier, when it was small and was perhaps not taken seriously. (I remember a colleague on the President's Science Advisory Committee who, in 1960, used to refer to fission as a "solution looking for a problem.") We did not, so to speak, face the full implications of the success of fission energy.

I suggest that we ought not fall into this same trap as we contemplate the sun as the base of our energy system. Can we visualize systems limits if solar energy were our main source of energy—if we really had to face the hypothetical future man might have faced had he evolved 2 billion years later—limits that would be unimportant if solar energy were only a small increment to other energy systems?

Let us then try to delineate in more detail an asymptotic world based on renewable energy sources: geothermal and solar (including hydro, wind, waves, ocean thermal gradients, solar electric, and biomass). To do this properly, we should analyze each end use of energy, and estimate how much energy is used as low-temperature heat, high-temperature heat, electricity, and mechanical work. This I have not done, and my speculations can be faulted in this respect. In-

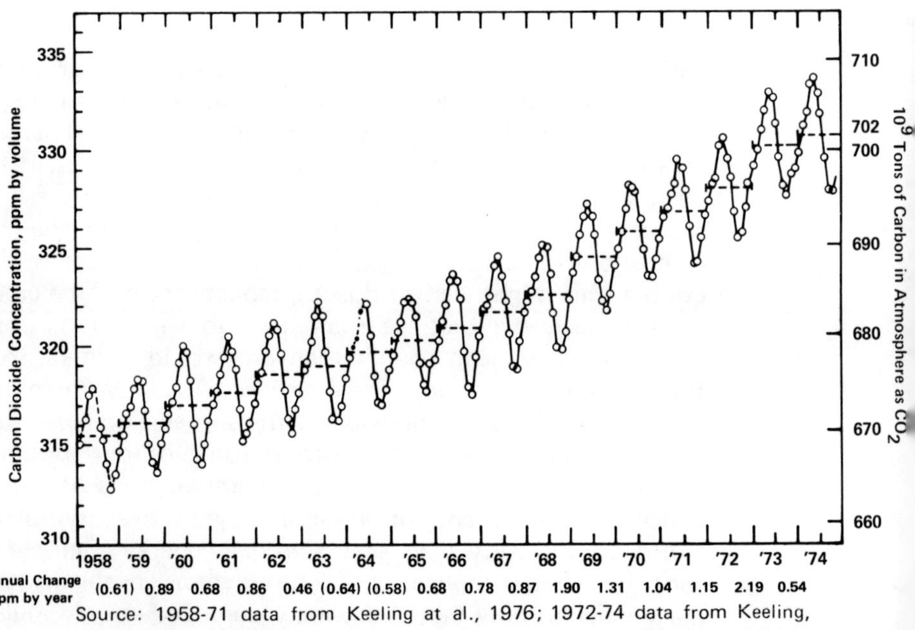

FIGURE 2. Atmospheric Carbon Dioxide Concentration at Mauna Loa Observatory (1958-71 data from Keeling et al., 1976; 1972-74 data from Keeling, private communication)

stead, I have lumped together all heat, regardless of temperature, and have done the same for electricity (Table 1). I have taken the present U.S. breakdown of end use demands and assumed this same pattern for the asymptotic future. This I call Case A: Transport is based on liquid fuels derived from biomass, and, at least for a fairly long time, from coal. I consider also Case B, in which transport is based on electricity: battery-driven cars, or electric trains, or conceivably, hydrogen-powered fuel cells of very high efficiency, the hydrogen being generated electrically. In determining how much heat goes into electricity, I have assumed a conversion efficiency of 10,000 Btu per kilowatt-hour (kwh).

Let us now consider how much heat and electricity man can plausibly derive from each of the renewable resources.

TABLE 1

Asymptotic World Annual Energy Demand

	1,000 quads/year
Household (22%)	0.44
Commercial (13%)	0.26
Transport (26%)	0.52
Industrial (39%)	0.78
Total heat input	2.00

	Case A (fluid transport)	**Case B (electric transport)**
Electricity	68×10^{12} kwh	118×10^{12} kwh
Heat not used for electricity	1.32×10^{3} quads	0.8×10^{3} quads

Geothermal

Although the geothermal energy stored in the rocks down to 10 kilometers has been estimated to be as high as several million quads, it is all but impossible at this time to estimate how much can be usefully recovered. However, since we are speaking of an asymptotic future, we can no longer *mine* the accumulated heat in the rocks; instead, we shall have to depend on the constant geothermal gradient. This amounts to 200 quads for world land areas—about man's total energy demand at present. Since so much of this heat is at very low temperature, and much of it is in parts of the world where no one lives, it seems fair to assume that no more than, say, 10 percent of it can be utilized as electricity at an efficiency of, say, 30 percent. This amounts to no more than 2×10^{12} kwh of geothermal electricity worldwide in the steady state (Table 2). We also assign a total of 10 quads of geothermal energy as heat.

TABLE 2

Ultimate Contributions to Asymptotic World Annual Energy Demand From Hydro, Geothermal, Wind, and Sun

	Electricity (kwh/year)	Heat (quads/year)
Hydro	10×10^{12}	–
Geothermal	2×10^{12}	10
Wind	0.8×10^{12}	–
Other (waves, tides)	1×10^{12}	–
Total	14×10^{12}	10
Needed from sun		
Case A (liquid transport)	$\sim 50 \times 10^{12}$	~1,300
Case B (electric transport)	$\sim 100 \times 10^{12}$	~ 800

Hydro

The ultimate world capacity for hydro we shall set at 10×10^{12} kwh. This is about 30 times the present total installed hydroelectricity.

Wind

Thirring[6] quotes Putnam for the total ultimate wind energy as 0.8×10^{12} kwh, or about 8 percent of the ultimate hydro capacity. To this, we probably ought to add wind for sailing ships, which might ply the oceans if we really must depend on the sun; this contribution, however, is surely small.

Waves and Tides

Wave energy may be a larger ultimate source than we had once believed; nevertheless, it is hard to imagine so dilute a source contributing substantially. Similarly, we would expect tidal power in aggregate to be very small. We rather aribitrarily place the combined contribution of waves and tides at no more than 1×10^{12} kwh.

Sun

The demand for electricity from the sun varies between 50 and 100 x 10^{12} kwh per year in the two cases; for heat, between 1,300 and 800 quads per year. At present, about 25 percent of our total energy in the United States goes for space and water heating. If the same fraction ultimately went for these purposes throughout the world, this would amount to about 500 quads. Let us further assume that *all* of this heat is provided directly by the sun; or alternatively, that better methods of insulation reduce the demand so that the entire space and water heating load can be handled directly by the sun. The remaining demand would have to be met either from biomass or solar electricity. Thus our hypothetical world displaced in time by 2 billion years would face the task of drawing between 300 quads and 800 quads from the sun as biomass; and from 50 to 100 x 10^{12} kwh as electrical energy. What are the prospects for achieving these outputs?

The average solar insolation in the Southeast United States is about 560,000 Btu per square foot per year—i.e., 0.2 kw per square meter (Table 3). If this is converted to electricity at 18 percent efficiency (a theoretical value for solar cells), we can extract roughly 300 kwh/m^2/year from the sun. Let us assume the sun's energy is converted into biomass at, say, 0.7 percent conversion efficiency; this corresponds to about 10 tons dry weight per acre per year, 7,500 Btu per pound dry weight, and is five times the global average efficiency of 0.13 percent. On this assumption, the energy obtained by burning biomass is 3.6 x 10^4 Btu/m^2/year—i.e., 10,000 square miles per quad of heat per year. Note that if the biomass is converted to electricity at 30

TABLE 3

Production of 800 Quads/Year Via Biomass

Average solar insolation (Southeast U.S.)	0.2 kw/m^2
Conversion of solar insolation to electricity, 18% efficiency	300 kwh/m^2/year
Conversion of solar insolation to biomass, 0.7% efficiency	3.6 x 10^4 kj/m^2/year
Conversion of biomass to electricity, 30% efficiency	3 kwh/m^2/year
Land requirement	8 million square miles

percent efficiency, we arrive at 3 kwh/m^2/year, about 10 times less than the efficiency of electrical conversion assumed for photocells.

We now examine limits on biomass and solar electricity in more detail.

Biomass

To get 800 quads per year from biomass would require about 8 million square miles—roughly 1/6 the total land area of the earth. Thus the high biomass scenario seems implausible. Even to supply the 300 quads in Case B (electric transport) requires 3 million square miles—a very formidable demand.

It would seem that biomass simply cannot provide the basis for the abundant energy future I visualize unless the effective photosynthetic yields can be increased much above the 0.7 percent I have assumed, or unless really large-scale farming of the sea (say for kelp) becomes feasible. Several possibilities suggest themselves: from improving crop management so as to harvest year in and year out those plants that in special situations now yield much more than 0.7 percent, to genetic engineering that might increase the effective photosynthetic efficiency, say, fivefold. I have no idea whether photosynthetic efficiency five times higher than the present average is achievable—whether, say, this is more likely than the development of practical controlled thermonuclear fusion. These estimates merely suggest how important such an achievement would be, and suggest possibly vital directions for future genetic research.

Solar electric systems

The yearly demand for solar electricity (between 50 and 100 x 10^{12} kwh) could be met, in principle, by photovoltaic arrays (PV), by power towers (PT), or by ocean thermal energy converters (OTEC). The first two are intermittent, the last is not. If these intermittent systems are small and are backed up by firm power from a grid, they would need little storage; if they stand alone, or if the total demand exceeds what can be met by reliable backup, these systems would need large amounts of storage—say 6 to 12 days. *Electrical* storage is much more expensive than is *heat* storage; hence, a priori, we would expect the PV system with full electric storage to be more expensive than the PT, which uses heat storage.

A few numbers illustrate the point. If a PV system, possibly with a light condensing system, can be installed for \$10/square foot (ft^2) without storage (this is 15 times cheaper than the present cost of photovoltaic silicon surfaces), then at our average output of 30 kwh/ft^2/year, the capital cost of the system is about 33 cents/kwh/year; at 20 percent fixed charges, this comes to about 7 cents/kwh; at 10 percent fixed charge, 3.5 cents per kwh. If the system were supplied with 6 days' storage and the batteries cost, with one replacement, \$40/kwh, we would add 66 cents/kwh/year to the capital costs (Table 4). The total cost of firm electricity would come to 20 cents/kwh and 10 cents/kwh at 20 percent and 10 percent fixed charges, respectively. Actually, even these may be underestimates for a full solar system, since we have not taken into account the variation in solar flux between winter and summer. This is about a factor of two to three, depending on the latitude. Thus to provide *firm* power, winter as well as summer, might require three times the capital investment in collectors, though not in storage. The storage for the PT system is much cheaper, though it is too early to say whether the PT or PV system itself is the cheaper. Thus if a large PT can be installed complete for as little as \$10/ft^2, we might achieve solar electricity at 20 percent fixed charges for, say 10 cents/kwh, but this still does not take into account the winter/summer variation. *Firm* power winter as well as summer might cost at least twice as much.

The total land required in the 100 x 10^{12} kwh/year scenario is about 80,000 square miles. The capital outlay, at 100 cents/kwh/year (including storage for the PV system), would be \$100 x 10^{12}. The annual per capita income

TABLE 4

Production of 100 x 10^{12} kwh/Year Via Solar Electricity

Solar electricity density, 18% efficiency	300 kwh/m^2/year
Cost of PV installed, 6-day storage	\$300/m^2
Capital cost	100 cents/kwh/year
Cost of electricity:	
@ 20% fixed charge	20 cents/kwh
@ 10% fixed charge	10 cents/kwh
Total capital cost	~\$100 x 10^{12}
Gross world product	~\$ 75 x 10^{12}

at that time would be equivalent, say, to the West German average of $5,000 per person per year. Thus the Gross World Product (GWP) would come to $75 x 10^{12} per year. A world electrical system whose capital cost is, say, 1.3 times the GWP may be acceptable, since the present U.S. electrical system, if it were to be duplicated, would cost about $500 billion, or 40 percent of our GNP.

One possibility that has perhaps received insufficient attention is OTEC. We have modified Zener's estimate,[7] and find that if the ocean surface temperature were reduced by 1°C from 20°N to 20°S latitude, some 100 x 10^{12} kwh conceivably could be obtained at a cost of perhaps 5 cents/kwh (20 percent fixed charge). However, if OTEC were deployed on so enormous a scale, the amount of water evaporated from the ocean would be reduced significantly, and this might induce serious changes in the climate.

To summarize, it would appear that the high solar electric scenario seems to be very expensive; the high biomass scenario seems to use too much land; the high OTEC scenario seems to imply serious climatic changes. An all-solar future is almost surely a low-energy future, unless man is prepared to pay a much larger share of his total income for energy than he now pays.

An Ultimate Future Based on Breeders

Let us now see what would be involved in providing the electric transport scenario with nuclear energy—i.e., 100 x 10^{12} kwh for direct electricity and transport and 300 quads for all other purposes except space and water heating, which we still assign to the sun. We assume the "all other purposes" will be met by hydrogen generated electrolytically, rather than by biomass as we did in the previous scenario. At 70 percent efficiency of conversion from electricity to hydrogen, 300 quads of hydrogen require 125 x 10^{12} kwh. (This number might in effect be halved if thermochemical splitting of water at 60 percent efficiency could be achieved.) Thus our total breeder system must supply about 225 x 10^{12} kwh of electricity each year (Table 5). In the asymptotic era, we assume that each breeder produces 5,000 mw for 7,000 hours, or 35 billion kwh of electricity per year. Thus the asymptotic nuclear world would be powered by about 7,000 enormous breeders, each producing 5,000 mw of electricity at 80 percent capacity factor, and about half of them converting the electricity into hydrogen or other liquid fuel. Is

TABLE 5

Production of 225 x 10^{12} kw/Year Via Nuclear Breeder System

Direct electricity and transport	100 x 10^{12} kwh/year
Electricity for "all other purposes"	125 x 10^{12} kwh/year
Total electricity	225 x 10^{12} kwh/year

Number of reactors	7,000	Cost of electricity:	
		@ 20% fixed charge	5 cents/kwh
Size of reactor	5,000 mw	@ 10% fixed charge	3 cents/kwh
Cost/kw	$1,500	Cost of hydrogen/million Btu:	
		@ 20% fixed charge	$20
Capital cost of system	$50 x 10^{12}	@ 10% fixed charge	$10

Number of reactors	7,000
Number of sites	1,500
Number of reactors built/year	~ 150
Uranium required	~40,000 tons/year
Pu inventory	100,000 tons
Excess Pu produced per day	10 tons
Accident rate @ .5 x 10^{-4}/reactor/year	0.3/year
High-level wastes produced	~2 million cubic feet/year
High-level waste burial land	15 square miles/year

such an energy system at all plausible? Let us examine various possible limits to such a system.

Cost

We shall assume the breeder system, together with its hydrogen generating plant, costs 50 percent more than present-day reactors–say $1,500/kw. The capital cost of the whole system would come to about $50 x 10^{12}–about half the cost of the solar electric system with electric transport–yet the nuclear system takes care of essentially all of the society's energy (except for space heating), whereas the solar electric system met only the demand for direct electricity and transport.

At $1,500/kw, the capital cost is about 21 cents/kwh/year. With fixed charges at 20 percent, and operating and fuel costs of 1 cent/kwh, this leads to electricity at 5 cents/kwh; at 10 percent, to 3 cents/kwh. Hydrogen from the system would cost roughly $10 to $20/million Btu, i.e., five to ten times present costs of fluid fuel. We estimate the yearly world expenditure on all energy in the high scenario to be about 15 x 10^{12} at 20 percent fixed charge, 10 x 10^{12} at 10 percent fixed charge—that is, 20 percent and 15 percent of GWP respectively.

Siting

To site 7,000 reactors, each producing 5,000 mw, is a formidable task. It seems clear that cluster siting will be adopted by then—perhaps five reactors at each site. About 1,500 sites would be needed. If each site occupied 40 square miles, the entire system would require 60,000 square miles. In the United States, assuming an asymptotic population of 300 million and that everything scales according to population, we would need about 50 sites.

Rate of building

If each reactor lasts 50 years, 150 reactors would be built each year. The total work force on the site, at say, 5,000 per reactor, would be close to 1 million. This number probably would be at least trebled if we count workers at component factories.

Uranium requirement

Each breeder "burns" about 15 kilograms of uranium per day. To keep the entire system going would require about 40,000 tons of uranium per year. This demand could be met only by "burning the rocks"—i.e., extracting the 12 ppm or so of uranium and thorium from the granitic rocks, or by extracting uranium from sea water.

Plutonium inventory

Each reactor and supporting chemical plant will contain about 25 tons of plutonium. The total system would contain about 75,000 tons of plutonium. If we assume a breeding ratio of 1.1 for the entire system, we estimate 10 tons of excess plutonium will be produced each day.

Accident rate

We have no real estimates of accident probabilities for Liquid Metal Fast Breeder Reactors (LMFBR's). The Rasmussen estimate (one in 20,000 per reactor year with an uncertainty of five either way) would lead to a meltdown every 3 years. It seems clear that this is an unacceptable rate; an accident rate at least 10 times lower, and possibly 100 times lower would be needed if the system is to be acceptable.

Waste disposal

Each 5,000-mw LMFBR produces about 275 cubic feet of high-level solidified waste per year, contained in about 50 steel cans. According to present plans, these would occupy about 1.5 acres of burial space. Thus the entire system of 7,000 reactors would require about 15 square miles of burial space per year. After 1,000 years, 15,000 square miles will have been used up; by that time, the radioactivity in the high-level wastes will have decayed sufficiently to allow fresh wastes to be layered over the older wastes. Thus the 15,000 square miles devoted to high-level wastes might be usable for much longer than 1,000 years.

To summarize, although we cannot identify physical limits that make a world of 7,000 large LMFBR's impossible, one would have to concede that the demands on the technology would be formidable. Two issues appear to me to predominate: First, the acceptable accident rate will have to be much lower than the Rasmussen report suggests. If one uncontained core meltdown per 100 years is acceptable (and we have no way of knowing what an acceptable rate really is), then the probability of such an accident will have to be reduced to about 1 in 1 million per reactor per year. This is the design goal for the LMFBR project in the United States. Second, a nuclear world such as we envisage will have long since had to make peace with plutonium. Ten *tons* of plutonium per day is mind-boggling. It is hard to conceive of the enterprise being conducted except in well-defined, permanent sites, and under the supervision of a special cadre—perhaps a kind of nuclear United Nations.

Thus we can hardly escape the impression that the price nuclear energy demands, if it is indeed to become the dominant energy system, may be an attention to detail, and a dedication of the nuclear cadre that goes much beyond what other technologies have demanded. It is only when one projects to an asymptotic nuclear future such as we have attempted that one recognizes the magnitude of the social problem posed by this particular technology.

Do Energy Modes Determine the Society's Structure?

The Niehaus scenario on which this paper is based makes us cornucopians. To us a stable world in which the discrepancy between the energy consumption of rich and poor remains 30-fold or even 50-fold seems impossible. If one concedes this proposition, then the rest of the cornucopian argument follows: the necessity for a world per capita average energy of, say, 140×10^6 Btu, and therefore the necessity for an energy source that is largely centralized, largely nonsolar (unless OTEC succeeds), and heavily electric—in short, the probable necessity of fission.

Wherein can this cornucopian argument break down? We see three possibilities—that our estimate of 140×10^5 Btu per year per capita is too high, that the world population of 15 billion is too high, or that a solar-based system like OTEC proves to be as cheap as its proponents claim. Considering the additional energy that will be required to extract lower grade resources—perhaps 30 percent more than we now use per ton of average metal—the 140×10^6 Btu per capita in fact corresponds to an energy budget only one-third the present U.S. budget. Granted that we can and will conserve energy in many ways—by cogeneration (which is possible only with decentralized systems), by using solar heat, by insulating our houses—it seems to us very difficult to imagine, say, a tenfold rather than a threefold reduction in energy use without a drastic lowering of the living standard.

The argument for decentralization was introduced originally because decentralized systems make cogeneration possible and this saves energy; but the matter now has taken on deeper social and political significance. What is at issue is the ideal of the good society: Can we have a good society in which the primary means of production, in this case energy, are in the hands of a small, remote elite? Is true human freedom possible in a society so technologically centralized? Is such a society resilient against disaster, either social or technological? In short, can man be both free and the slave of the centralized technologies of energy generation? To us this issue has two dimensions. First is the relation of centralized electricity to the basic structure of the society and to human freedom. One would think that a far greater determinant of the structure of our society, and a correspondingly greater threat to human freedom, lies in the remote and centralized control of the means of communication and the avenues of information than in the control of means of production of electricity. Electricity, after all, is impersonal and carries little semantic connotation. Television, on the other hand, intrudes on our personalities far more directly and pervasively. In some of the stormy debate, centralized electricity has been made a whipping boy for other, possibly more legitimate, concerns about our society's trend toward centralization.

But we would turn the question around. Can human freedom prevail in a society that is short of energy? Energy provides choices—perhaps the most important choice is the use of time (a point stressed by D. Spreng[8]). When neo-Malthusians insist that energy (and the environment) is limited, they ignore the other great human resource that is limited—time. There seems to be something fundamental in man's readiness to exchange energy for time—not simply, one supposes, because as primates we are lazy, but also because we are playful. Only with time can we be playful, and indeed, from our playfulness, develop the skills and arts and works that represent the fulfillment of our human potential.

We are caught in a powerful dilemma. To be fully human, we probably need 140×10^6 Btu per person; but to achieve this energy expenditure at a population of 15 billion, we almost surely will need energy systems, probably nuclear systems, that demand more of us than some think we are capable as humans of doing; social stability, technical ingenuity. Is this a resolvable dilemma?

No one can really say, since we speak here of a world in the distant future—at least a world that no longer burns fossil fuel. Two courses seem to be necessary. The first is to keep the population, say, to 5 billion. In that case, our world energy system demands only about 600 quads, an amount that probably can be supplied by the benign modes, plus a little nuclear. But we may not succeed in keeping the population down. We may have little alternative to perfecting the nuclear system. In that case, we probably will require every energy source, and we shall be confronted with the hazards and limits posed by each of them. In particular, we may have to learn to live with many thousands of breeder reactors. Admittedly this is a formidable task that will almost surely require large reorganization of our world. But such reorganization may be presaged in our attempts to deal with the other shorter-range issues posed by the long-term energy dilemma. If within 20 years it becomes clear that the carbon dioxide burden will become intolerable unless we greatly reduce the burning of fossil fuels, do we have any credible social mechanism for enforcing a ban on fossil fuel? Will this not pose a dilemma as powerful as the one we face in the deployment of 7,000 or even 2,500 breeders, or in the necessity to keep the world's population at, say, 5 billion? Yet these issues must be confronted and dealt with. It is by no means idle to speculate on the character of these future worlds even now—we may have much less time to deal with the dilemmas they present than people customarily realize.

REFERENCES

1. Whittle, Charles E., et al. Economic and Environmental Implications of a U.S. Nuclear Moratorium, 1985-2010. ORAU/IEA-76-4. Institute for Energy Analysis, Oak Ridge Associated Universities, Oak Ridge, Tenn. September 1976.

2. Rasmussen, Norman C. Reactor Safety Study: An Assessment of Accident Risks in U.S. Commercial Nuclear Power Plants. Report of U.S. Atomic Energy Commission Task Force. WASH-1400 (NUREG 75/014). National Technical Information Service, Washington, D.C. 1975.

3. Heilbroner, Robert L. An Inquiry into the Human Prospect. W. W. Norton & Co. Inc., New York. 1974.

4. Niehaus, Friedrich. A Nonlinear Eight Level Tandem Model to Calculate the Future CO_2 and C-14 Burden to the Atmosphere. Research Memorandum RM-76-35. International Institute for Applied Systems Analysis, Laxenburg, Austria. April 1976.

5. Marland, G. Private communication.

6. Thirring, Hans. Energy for Man: From Windmills to Nuclear Power. Harper & Row, Publishers, New York. 1976. p. 281.

7. Zener, Clarence. Solar Sea Power. Physics Today, January 1973. p. 48-52.

8. Spreng, Daniel T. Useful Questions in Energy Accounting. ORAU/IEA (M) 76-7. Institute for Energy Analysis, Oak Ridge Associated Universities, Oak Ridge, Tenn. June 1976.

Sam H. Schurr and Joel Darmstadter
Resources for the Future, Inc.
Washington, D.C.

Prior to his return in January 1976 to Resources for the Future, Inc., as Senior Fellow and co-director of the Energy and Materials Division, Sam H. Schurr was director of the Energy Systems, Environment and Conservation Division of the Electric Power Research Institute (EPRI), in Palo Alto, California.

Mr. Schurr had been director of the Energy and Minerals Program at Resources for the Future, Inc., in Washington, D.C., for 19 years.

Mr. Schurr has been a staff member of the National Bureau of Economic Research, the Cowles Commission for Research in Economics at the University of Chicago, and the Economics Division of the Rand Corporation. He has served with the U.S. Government in the War Production Board, the Office of Strategic Services, the Department of State, the Bureau of Labor Statistics, and the Bureau of Mines, where he served as chief economist.

Mr. Schurr earned his B.A. and M.A. degrees in economics at Rutgers University and did additional graduate work in economics at Columbia University.

Joel Darmstadter, a native of Germany, was educated at George Washington University and the New School for Social Research. He has worked as an economist with the National Planning Association in Washington, D.C. Since 1966 he has been employed by Resources for the Future as an economist and fellow. He is currently involved with such research projects as energy use relative to income levels and national energy strategies. He has written many articles and books dealing with national and international energy and economic questions. Mr. Darmstadter is a member of the Working Group on Growth, Global Commons and Environment, 1980s Project, Council on Foreign Relations; Demand/Conservation Panel, Committee on Nuclear and Alternative Energy Systems and the Board on Energy Studies, National Academy of Sciences; and the editorial board of Annual Review of Energy.

Some Observations on Energy and Economic Growth

The subject of energy and economic growth, which until just a few years ago was a matter of interest only to energy specialists, has recently moved from that limited sphere into the realm of broader public concern and even political debate. This concern and the associated political debate take the form of questioning the relevance for the future of the historical relationships that have prevailed between the growth of industrial economies and the amounts of energy they have consumed. Numerous voices argue strenuously that the future economic growth of advanced countries need not require as much energy as projections of past energy/economic growth relationships into the future suggest. However, other strong voices reply that to reduce the growth in energy consumption to levels well below those experienced in the past would lead to a significant downturn in the rate of economic growth in the future, rising unemployment, and sharply reduced standards of living.

Those who argue that the past is a false guide to the needs of the future offer a number of arguments in support of their position. One argument is that comparative energy prices will be higher in the future than in the past, leading consumers of energy to economize in its use, whether this be in the home, in commerce, in industrial processes, or in other applications. In addition to these voluntary responses, it is argued that as a matter of public policy it is essential and possible to adopt regulations and to introduce incentives that will lead to the more efficient use of energy.

There is also a body of opinion that goes beyond arguing for ways and means of achieving more efficient use of energy, to argue a more extreme position, namely that the life styles (a term not yet precisely defined) that have evolved in highly industrialized countries cannot and should not be sustained in the future for a variety of reasons. In other words, that sharp changes are needed in the evolutionary path of economic development away from the types of goods and services previously consumed and toward a simpler or more austere basket of goods; not only, according to this point of view, will such changes economize on energy, but they will save society from a number of serious dangers, perhaps even collapse, that would be implied if past trends unfolded into the indefinite future.

Yet, despite the growing interest in the overall subject of energy/economic relationships and its emergence into the front rank of policy concerns within industrialized countries, there is a striking absence of solid information on the relationship between energy and economic growth in the past and on the more tenuous question of its prospects for the future. This is a matter of serious concern, for it could result in planning for the future without a solid basis in fact, which could, in turn, lead to unexpected, harmful consequences.

History: The Broad Trends

What are the facts about the past? Some years ago, Resources for the Future published a detailed quantitative study of the interrelationships between energy and economic growth in the United States over a long period beginning in the 19th century.[1] In that study, we uncovered some interesting facts about the contrasting long-run trends in the movement of energy and Gross National Product (GNP) in the United States, to wit: Between the latter half of the 19th century and the first decade of the 20th century, energy consumption in the United States increased at a faster rate than GNP, while following the end of World War I and lasting until the early 1940's, the growth in energy consumption had generally been at a slower rate than the growth in GNP (Figure 1).[2] Note that these findings do not support the widely held belief that energy and GNP have grown at essentially the same rate in advanced industrial economies. Not only did they not grow at the same rate, but their comparative rates show divergence in different directions depending upon the particular period of U.S. economic history being covered.

This set of statistical findings was subjected to a deeper probing in order to explain the divergent trends for the different time periods. We found that a major factor explaining the rise in energy relative to GNP during the earlier period, and its decline relative to GNP in the later period, was the changing structure of output of the U.S. economy. Whereas prior to World War I, development of heavy industry was the dominant element in national economic growth, in the period following the 1920's, lighter manufacturing and the broad service component of national output were growing rapidly. The more intensive energy consumption relative to national output was thus associated with heavy industry, while the comparative decline was associated with the less energy intensive service sector and light manufacturing.

Electrification

In addition to the change in the composition of national

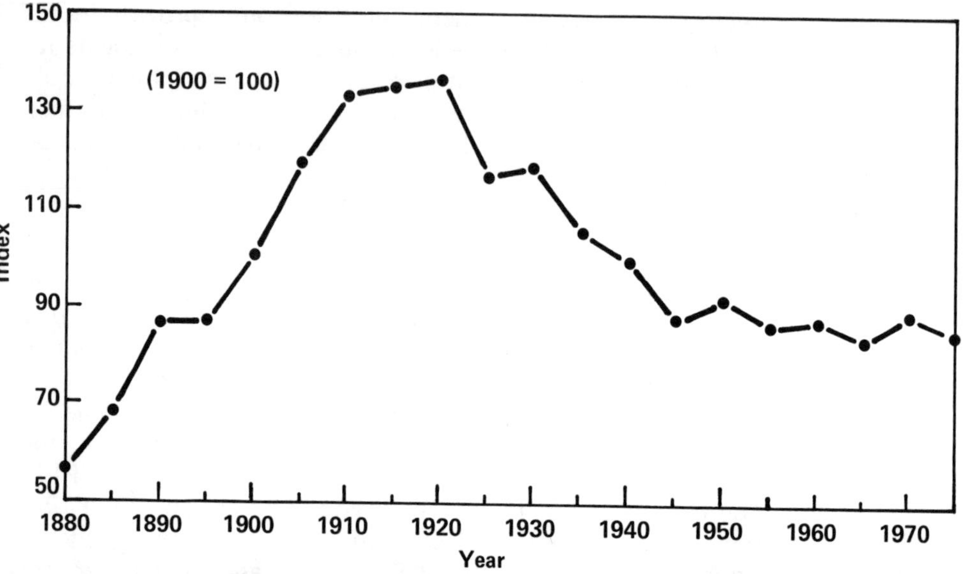

Source: 1880-1950: Schurr, Sam H., and Bruce C. Netschert. Energy in the American Economy, 1850-1975. Johns Hopkins Press, Baltimore. 1960. p. 524-525.

1955-1975: Alterman, Jack. Study (in progress) for Bureau of Economic Analysis, U.S. Department of Commerce. Alterman data adjusted by linking to series used for period 1880-1950. 1975 estimated.

FIGURE 1. Index of Energy Consumption Per Unit of Gross National Product, United States, 1880-1975

output, another major influence during the period of declining energy consumption relative to GNP was the change in the composition of energy output. Of particular importance was the far more rapid growth of electricity than of other elements within the energy total. From 1920 to 1955, electricity grew at a rate some five times faster than that of all other energy.

Two subtle points illustrated by the impact of electrification reveal major aspects of energy/economic relationships that are frequently overlooked. They are, however, essential to a proper understanding of the role of energy in economic growth.

One point concerns the question of thermal efficiency, as compared to the economic efficiency, of energy use. There is a school of thought, which has received much attention, that is concerned with what is called "net energy use." In the net energy approach, major attention is paid to energy balances: how much energy is put in compared to how much useful energy is obtained. This is in many cases a useful exercise, but it can convey a partial and, therefore, misleading impression. Thus, in net energy terms, electricity might be regarded as an undesirable energy form because it requires several British thermal units (Btu) of fossil fuel to produce 1 Btu of electricity. In terms of thermal efficiency, electricity is, by definition, less efficient than, say, the direct use of heat. However, in economic efficiency terms, which we would argue is the decisive factor to consider, electricity has been a very desirable energy form in the bulk of its applications. Its unique characteristics have permitted tasks to be performed in altogether different ways than if fuel had to be used directly as a source of energy.

The impact of electrification in industrial processes is the clearest case in point. A significant but not well recognized aspect of electrification was in its effect on the overall productive efficiency of the economy, particularly in the manufacturing sector. Historical examination of the organization of production within manufacturing shows that the growth of electrification permitted the layout and organization of productive processes within the factory in a manner that was altogether impossible when factories were powered by prime movers with shafts and belting carrying mechanical power to the various points of use. Electricity, which made possible the use of electric motors to which the power was delivered by wire, paved the way for a major reorganization of the sequence and layout of production. The new arrangements were more in keeping with the logic of the productive process than were the more rigid locational requirements imposed by a system of shafts and belting (e.g., locating heavy energy uses close to the prime mover). This was a factor of enormous importance in the growth of manufacturing productivity and thereby in the productivity of the total economy.

Beyond its key part in the electrification of industrial processes, electricity also has been of unique importance in communications, automatic controls of various kinds, and in the performance of numerous household tasks, to cite but a few examples. The essentially different applications made possible by electricity have greatly multiplied the efficiency with which labor and capital are employed, and have thereby enhanced the overall productivity of economic processes. Thermal efficiency considerations notwithstanding, the economic efficiency of electrification has been of outstanding significance.

The second point, related to the first, is that because of

electrification not only was the overall productivity of economic processes enhanced, resulting in the greater output of goods and services per unit of labor and capital employed, but, more subtly, so was the productivity of energy use itself enhanced, resulting in a decline in the amount of raw energy required per unit of national output. In this sense, despite the heat losses involved in its generation, electricity has also enhanced the productivity of energy use, as measured by the ratio of GNP to energy consumption.

This is not to suggest that greater thermal efficiency in the generation of electricity has not been significant in the historical record. Indeed, there have been vast improvements in the efficiency with which electricity has been converted from fossil fuels. And there are undoubtedly opportunities for achieving still greater thermal efficiency in the future. Nor do we mean to suggest that electricity is an appropriate energy form for all types of use; it is important to match all energy forms to the uses in which they can best be applied. But what's best cannot be decided on narrow thermal efficiency grounds.

Liquid fuels

In an analogous fashion, it may be reasoned that the internal combustion engine, powered by liquid fuels, permitted the substantial mechanization of agriculture, which played so great a part in the rising productivity of the American economy. Similarly, the growth of truck transportation made it possible for industry to move away from sites dictated by the location of railroad facilities or waterways. Again it is worth noting that it is not through its thermal efficiency characteristics, but in its broader economic impacts, that the internal combustion engine has left its imprint. (To be sure, substantial negative effects have resulted from the automobile, but it is likely that in the future technical and institutional improvements will serve to reduce such impacts.)

Thus, the change in the composition of energy output towards the more flexible forms of electricity and gasoline made possible shifts in production techniques and locations within industry, agriculture, and transportation, shifts that greatly enhanced the growth of national output and productivity. In enhancing the growth of productivity characterizing the overall economy—that is, the efficiency with which labor and capital are employed—the changes in the composition of energy output have also enhanced the efficiency with which energy itself has been employed as a factor of production; that is to say, over much of the time, energy consumption per unit of national output has persistently declined.

Energy and employment

What this past course of development has signified for the growth of jobs is a point worth pondering a bit more explicitly. The interrelationship of energy consumption and employment in the process of economic growth displays certain distinct and intuitive features in a macroeconomic and long-term historical context. In a disaggregated and shorter-term perspective, the picture is more elusive. The self-evident proposition that labor productivity is sensitive to the availability of energy-using capital equipment as a complementary productive input shows up in the close historical movement of energy per worker and output per worker (Figure 2). Of course, the paths of these two series are not proportionate. Indeed, the more rapid rise after World War I in output per worker than in energy per worker points to the innovative characteristics of industrial mechanization and electrification already commented upon in connection with the declining energy/output ratio that dates from around this time.

But one cannot draw the inference from these long-term trends that any deceleration in energy use, quite apart from the obvious reduction stemming from increased efficiency, necessarily jeopardizes the prospects for future productivity growth. At least 40 percent of U.S. energy consumption represents as much the "proceeds" of income growth (which persons deploy on such things as passenger transportation and household fuels and power) as it does the springboard for growth through its role in the productive process (Figure 3). In other words, no more than 60 percent of yearly energy use goes to the business sector—industry, freight transportation, agriculture, and commercial enterprises.

For the above reason—and also because of the widely different distribution of energy consumption, on the one hand, and jobs on the other, between the goods-producing sector and the service-producing sector—it seems to us questionable to assert a rigid linkage between energy consumption and employment. Clearly, it takes less energy, by far, to support a job in the service sector than in the goods sector. And it is the service sector in which the more rapid

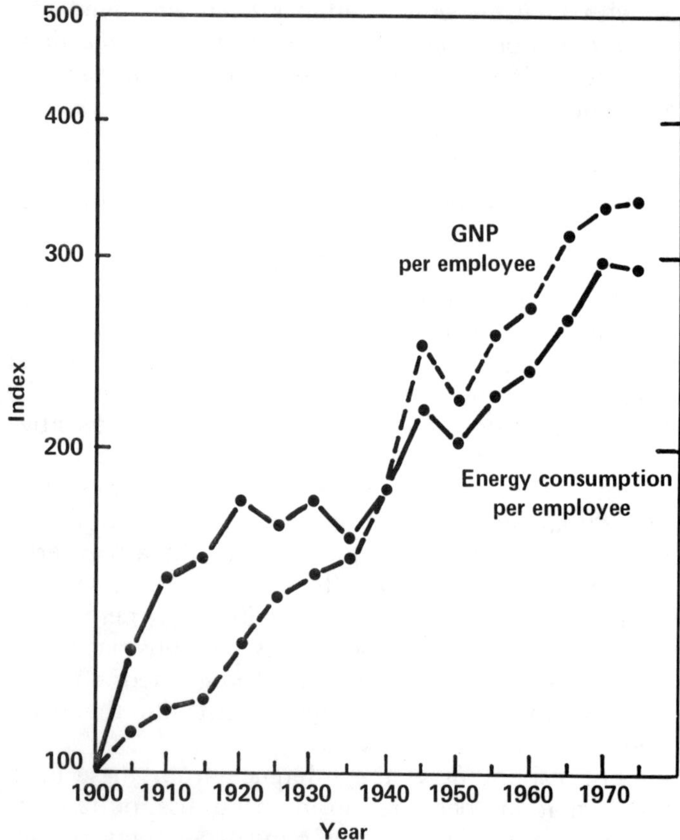

Source: Historical Statistics of the United States, Colonial Times to 1970. U.S. Department of Commerce, Bureau of the Census. U.S. Government Printing Office, Washington, D.C. 1975. and: Council of Economic Advisers. Economic Report of the President.

FIGURE 2. Indexes of Energy Per Employee and Gross National Product Per Employee, United States, 1900-1975

growth in employment is taking place (Figure 4).

We cite these instances to make the point that an understanding of the relationship among energy, economic growth, and employment requires a deep probing of the structure of the economy and the structure of energy output. In these changes in structure, one begins to see more clearly the two-way relationship between energy and economic growth: energy consumption as it is affected by the composition (and not just the level) of national output and employment, and the level of efficiency with which the national output is produced as it is affected by changes in the composition of energy output. Both are of critical importance to a proper appreciation of energy's relationship to economic growth.

Assessing the Future

Of what value are these findings for assessing the future? One factor that has been highlighted as an explanation of the past experience is that of structural change. Now, if as many believe, the economy of the United States and other advanced countries will in the future be shifting ever more heavily in the direction of services, there is some reason, using history as our teacher, for believing that energy relative to GNP will continue to be on a declining trend and perhaps at an accelerating rate of decline. This would be the energy

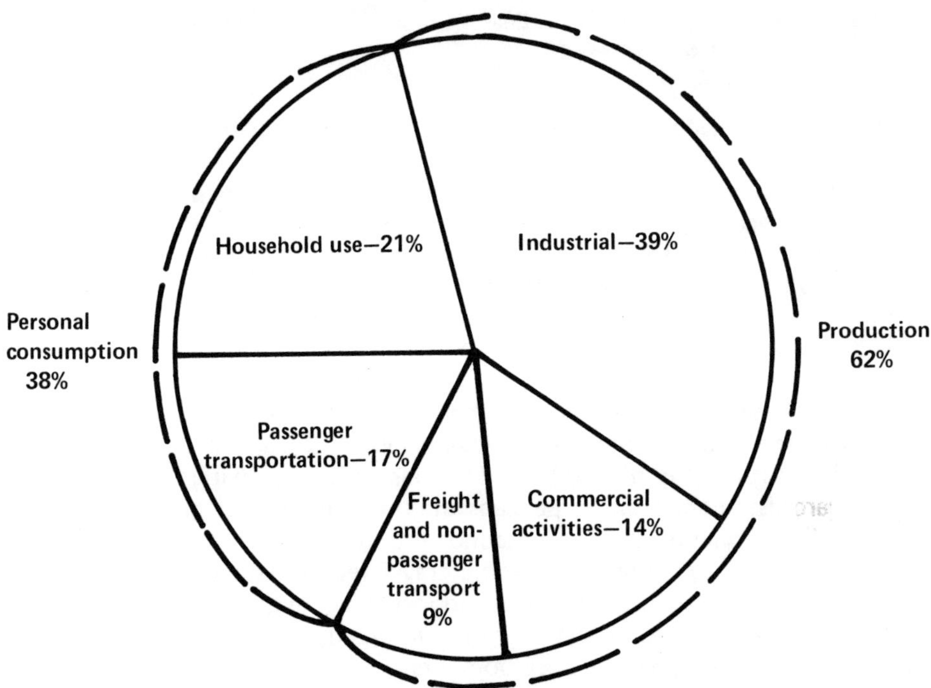

Source: Based on U.S. Bureau of Mines news release, April 5, 1976, which combines the residential-commercial sectors and the two transportation components. The greater detail shown here is an update by Resources for the Future of data for 1972 developed by the Federal Energy Administration.

FIGURE 3. Energy Used In Productive Activity vs. Energy Used In Personal Consumption, United States, 1975

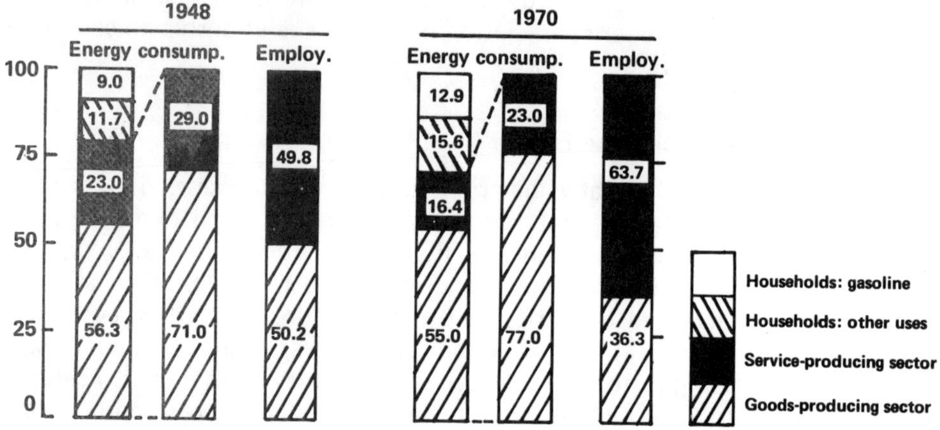

Sources: Alterman, Jack. Study (in progress) for Bureau of Economic Analysis, U.S. Department of Commerce, and Motor Vehicle Manufacturers Association of the U.S., Inc.

FIGURE 4. Percentage Distribution of Energy Consumption and Employment, United States, 1948 and 1970

counterpart of what has been called the "postindustrial economy."

However, a word of caution is called for. Services are a highly heterogeneous category of activities, and some of them may turn out to be quite energy intensive. Consider, for example, leisure-time activities, which will in the future account for increasingly large percentages of the personal services consumers will demand. It is not unusual for people to travel great distances by airplane or automobile to have a skiing weekend, or to engage in other types of leisure activity requiring substantial travel. This is obviously a high-energy-intensity form of services. There is also a growing trend towards second homes, perhaps the future counterpart of the earlier phenomenon of second and third cars. The construction of such homes, and the travel required to go from the city residence to the weekend residence, may again both turn out to be comparatively high in energy intensity. We mention these types of developments only to make the point that one should not too easily fall into the trap of believing that the growth of nonindustrial activities in the future will necessarily be associated with lower intensities of energy use.

One bit of insight into unfolding energy-consumption patterns is provided by a picture of the energy-using characteristics of different income groups. If the share of household budgets devoted to energy fell as families entered higher income-size classes, one would be tempted to suggest that, whatever else occurred, rising affluence implied less-than-proportionate increase in energy demand. In fact, it has been estimated that energy consumption does rise pretty much in line with income as households pass through successively higher income brackets (Figure 5). True, relative utilization of *direct* energy—motor gasoline, residential fuels and power—drop off as income grows. But this is apparently more than offset by the rise in the *indirect* energy embodied in purchases of nonenergy goods and services. An unanswerable question, of course, is what will happen to the spending patterns of those currently in the high income brackets. That aspect of future behavior involves pure conjecture.

Another factor that needs to be stressed in interpreting the historical record is that in the long period examined in the study made by Resources for the Future, energy prices were falling relative to the prices of other productive factors.

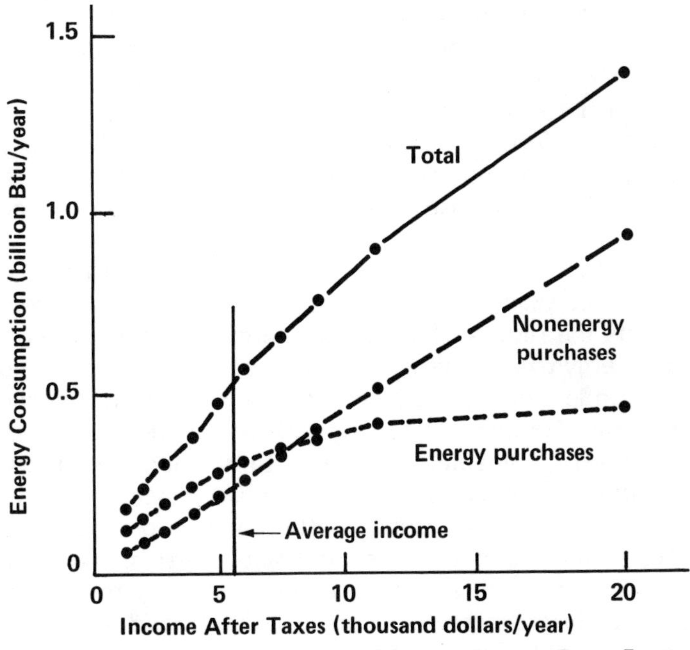

Source: Calculation by R. Herendeen. Shown in: Hannon, Bruce. Energy, Growth and Altruism. Mitchell Award Essay. University of Illinois, Center for Advanced Computation. 1975.

FIGURE 5. Direct, Indirect, and Total Family Energy Consumption vs. Family Income, United States, Early 1960's

Thus, price movements served to favor the substitution of energy for other factors of production. In the future, however, energy prices are expected to be rising relative to the prices of other factors. It may be correct, therefore, to expect that the use of energy will decline relative to that of other factors of production. Clearly, the whole question of energy demand and its response to price change, both up and down, is one which requires intensive investigation.

It is also reasonable to expect that the dampening effects of rising energy prices on the growth of demand will be abetted by advances that will be achieved in energy use technology. There are undoubtedly a number of energy-saving developments, not necessarily easy to foresee in detail, that will flow from technological response to higher prices of energy and to institutional and other reforms designed to conserve. Just as declining energy prices in the past encouraged the growth of energy-using machines and appliances, so price rises in the future should result in development and use of energy-saving equipment of various kinds.

Insights from international comparisons

If we are to achieve a dependable basis for projecting energy needs relative to future economic growth, we sorely need a better understanding of the combined effects on the growth of future energy consumption of structural changes in the economy, of relative price increases in energy, and of changes in energy-use technology in response to rising energy prices. Unfortunately, the historic record for any particular country may be quite inadequate to this task. In the United States, for example, the historical statistical record essentially depicts the results of declining energy prices and the growth of energy-using, rather than energy-saving, technologies.

Under these circumstances, comparative analyses among countries having contrasting energy characteristics should prove very valuable. Compared to the United States, the countries of Western Europe have, in general, experienced high energy prices. Consequently, comparisons among countries of the use of energy should help in assessing how energy consumption will respond to differences in energy prices, and also to differences in economic structure and in the energy-using technologies employed.

This complex issue can be illuminated by examining the comparative international studies designed to explain why energy consumed in relation to gross domestic product was, for example, about 70 percent of the U.S. level in West ern Germany and Sweden; about half the U.S. level in France; and about 90 percent the U.S. level in the Netherlands (Figure 6). An ongoing effort of this kind at Resources for the Future (sponsored by the Electric Power Research Institute) points to a number of important intercountry contrasts in energy consumption that contribute to the variability of these energy/GDP ratios.

Comparative transport practices are a key reason for the higher U.S. ratio. Not only are American passenger cars about 50 percent more energy intensive (in terms of passenger miles) than European cars. Relative to given income levels, Americans also drive a lot more than Europeans. Indeed, this factor is quantitatively more important than automotive efficiency in explaining the far greater amount of energy devoted to transportation in the United States compared to Western Europe, relative to income. A third contributory element is the proportionately greater share of (more efficient) public transport modes in the foreign energy mix. The RFF study found these differences to be a function not

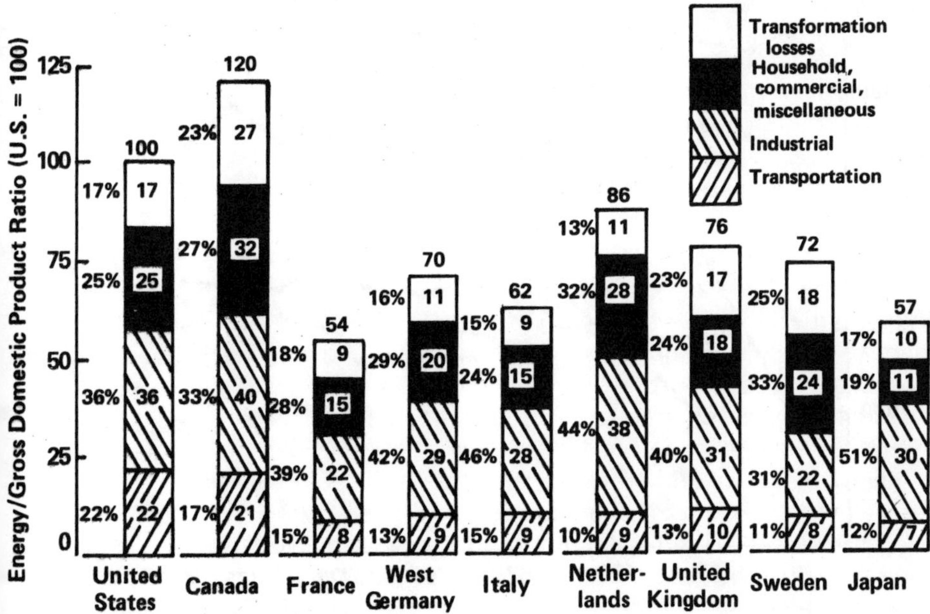

Source: Darmstadter, Joel, Joy Dunkerley, and Jack Alterman. How Industrial Societies Use Energy: A Comparative Analysis. A Resources for the Future study sponsored by the Electric Power Research Institute.

FIGURE 6. Energy/Gross Domestic Product Ratios and Sectoral Composition of Energy Consumption, Selected Countries, 1972

merely of the higher prices of acquiring and operating cars abroad—although that is certainly a critical element. Such aspects as urban density differentials and public-policy measures resulting in highly subsidized public transport also loom large.

Freight transport also contributes to the higher U.S. energy/GNP ratio. But, this comes about exclusively by virtue of the high volume of traffic (relative to GDP) that is generated in the United States compared to the grouping of European countries that we analyzed alongside. Indeed, the U.S. freight modal mix is, more than Western Europe's, oriented to such energy-saving forms as rail, pipelines, and waterborne traffic. If one argues that size of country and long-distance haulage of bulk commodities (such as ores, grains, and coal) are inherent characteristics of U.S. economic structure and geography, we have here a case where a rela-

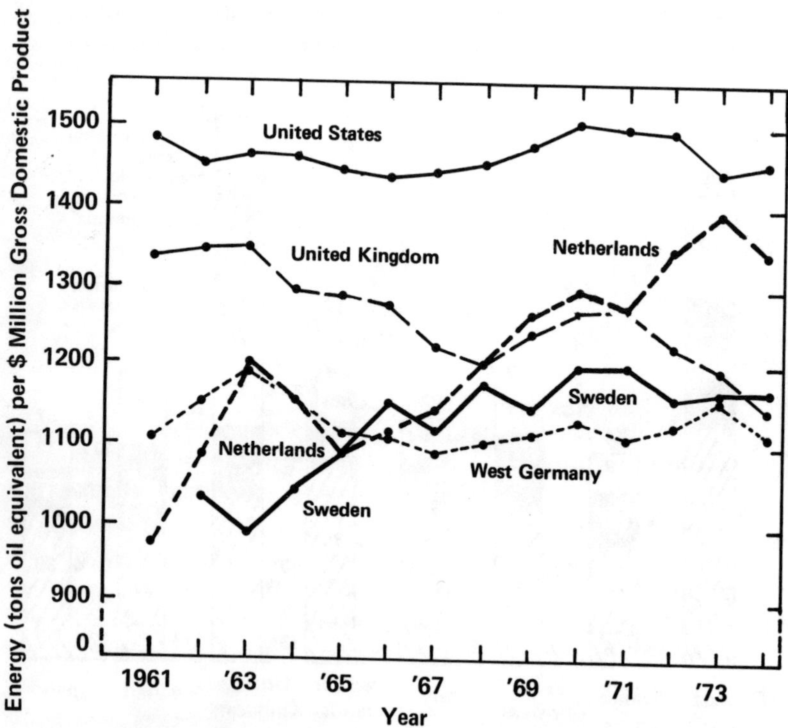

Source: Darmstadter, Joel, Joy Dunkerley, and Jack Alterman. How Industrial Societies Use Energy: A Comparative Analysis. A Resources for the Future study sponsored by the Electric Power Research Institute.

FIGURE 7. Energy Per Million Dollars of Gross Domestic Product, Selected Countries, 1961-1974

tively high energy/GNP ratio in no obvious way reflects comparative energy "inefficiency."

The industrial contribution to higher U.S. energy/GNP ratios comes about from more energy intensive processes in a number of manufacturing segments. That is, "structure" is not the reason—indeed, if industrial value-added were as high a proportion of national GNP as it is in a number of foreign countries, U.S. energy consumption would be even higher. Of course, energy represents only one component of industrial costs. The RFF study was not able to explore such complementary inputs as labor, capital, and other natural resources. Since input proportions can frequently be varied at a given level of output, differences in energy intensities—especially minor differences—do not tell us anything about comparative *overall* efficiency. The whole question of how energy intensity relates to economic efficiency has not yet received much attention, particularly of an empirical sort.

In the residential sector, higher U.S. energy consumption (relative to income) seems to stem, in large part, from the prevalence of larger, single-family houses and from higher temperatures for heating.

Another point is worth citing from the multicountry RFF study. The share of primary energy delivered in the form of electricity is not appreciably higher in the United States than elsewhere. In fact, it falls below the share of a number of other countries. Thus, the higher U.S. energy/GNP ratio is not due to disproportionate reliance on electricity in its energy mix.

Public policy and social attitudes

The possible effects of structural changes and of energy-price movements on future relationships between energy and economic growth lend themselves to quantitative investigation and are certain to receive more research attention in the future than in the past. However, a range of other considerations will need to be taken into account.

To turn first to public policy: There appears to be a growing tendency to want to mandate energy conservation, or to encourage it through the use of various types of incentives. Such regulatory and institutional factors are bound to have a lasting impact in altering the relationship between energy and economic growth, but the magnitude of the change is difficult to foresee.

For example, the law requires that automobiles be built in the future to achieve a certain average mileage per gallon, or laws may be passed to make it easier for people to invest in insulation for their homes. Such actions will have a substantial impact. However, there is some reason for wondering whether such legislatively enacted provisions will achieve their goals if the objective conditions of energy supply, essentially energy price and availability, do not serve to provide the justification for—or worse, act, as they now do, at cross-purposes with—proceeding in the direction of greater efficiency in energy utilization. In other words, regulations may not succeed if the underlying economic motivations, as reflected in price, do not provide a continuing spur. Nevertheless, in looking to the future it will be important to try to identify and measure the possible effects of regulatory and institutional intervention in altering relationships between energy growth and national output.

In addition, it is necessary to consider the matter of how changes in social attitudes may affect energy/economic growth relationships in the future. This subject is attracting an increasing amount of attention in the United States where it generally goes under the name of "changing life styles." It is difficult to describe what this general heading covers because the term means different things to different people.

As viewed by those who take an optimistic view of its potentialities, life style changes could, for example, lead to sharp departures from existing patterns of urban-rural population distribution towards modes of settlement that avoid the great distances presently separating places of work from peoples' homes, a striking characteristic of life in the United States. This goal might be accomplished either by decentralizing commerce and industry out of metropolitan centers, or, alternatively, by attracting more people into large cities where they would dwell in apartment buildings and use mass transportation to get from home to work. Developments of this sort—in either direction—could save greatly on the amount of energy consumed by transportation, and perhaps in other uses, such as the heating and cooling of residences.

A more extreme view of the potentialities of life style changes foresees a return to patterns of life having a closer tie to nature, with people doing for themselves many things that are presently performed by the market economy. The assumption is that the more self-sustaining ways of life would be far more sparing in their energy needs than is the case with the expansive, market-organized style of living. Woven into all of these considerations of life style change there is also the belief that in the future society can (or will) move in the direction of wanting a simpler basket of goods and services than has evolved in the high-consumption economies of the industrialized Western world.

Each of us is entitled to his own estimate of how far such developments may carry because as yet, there is no experience to refer to. Nor have there been any solid studies of the feasibility of achieving such a turnaround in social attitudes and in ways of living, and of the energy and materials implications of such changes if they were to be achieved.

Uncertainty concerning the future

So far, we have touched briefly on the changing structure of the economy and the changing composition of energy outputs, energy price developments, conservation policies, and the class of factors that goes under the name of life styles. These matters need to be considered in understanding the relationships between energy and economic growth that have characterized developments to date, and that throw light on what to expect in the future.

But while research and analysis proceed on these subjects, decisions must be made in respect to future planning for energy supply. These decisions are complicated by the uncertainties in estimating future energy needs that are either implicit in the underlying changes that are certain to occur—such as rising prices for energy or changes in economic structure; or in changes that many would like to see happen—such as public policies designed to achieve the conservation of energy, or changes in life styles being urged by many groups concerned not just with energy matters but with the quality of life generally.

It is necessary to pay heed to the discontinuities in basic relationships that may be in prospect, but it is also necessary to understand that changes, particularly those that are fundamental in nature, cannot ordinarily be achieved with great speed. Historical factors embedded in energy/economic relationships that have already been experienced are the product of a long evolutionary process. Although far from immutable, they need to be regarded as possessing a degree of continuity that may be hard to alter.

The Dynamic Thrust of Energy Technology

There is, finally, one additional perspective worth noting. Energy developments have in the past been a dynamic influence in the economic growth and development of industrial economies. Many of the fundamental features of contemporary Western society have had their origins, at least in part, in developments in the field of energy supply technology. One need only think of a change in energy supply technology, such as the emergence of coal in the 19th century (and earlier) as a replacement for fuelwood, to recognize the significance of this one energy supply shift for the growth of industrialization in Europe and the United States. For it was the transition from wood to coal that made possible the unimpeded growth of the iron and steel industry and the rapid expansion of railroad transportation. In similar fashion, the subsequent development of electrification in the 20th century has been of strategic importance. It supported the growth of productivity that led to higher living standards in the industrialized countries and also permitted introduction of new technologies into the home—electric lighting and various electric appliances, for example—that have fundamentally altered the comfort and convenience of the mass of the population. The more recent growth of liquid fuels has brought with it the use of automobiles, fundamentally altering the patterns of life of ordinary people by giving them far greater mobility.

It is not necessary to present a complete catalog of all of the basic social and economic changes that have depended upon, and indeed been produced by, developments in the field of energy supply technology. The point is that energy technology has been a dynamic element of great importance to American economic and social development. To be sure, numerous effects that have been produced are negative rather than beneficial in nature. There are air and water pollution, a blighted landscape, and deleterious effects of various kinds that are by-products of energy development. Many of these social costs can be substantially reduced through improved technologies and through institutional and regulatory reforms of various kinds. It is essential that we bend every effort to achieve environmentally benign energy technologies, but it is essential also that we not lose sight of the benefits in terms of improved living standards, decreased drudgery at work and at home, and greater mobility of all kinds—not just physical but also societal—that are the result of the dynamic thrust given by energy technology to economic and social conditions in the industrialized world.

REFERENCES

1. Schurr, Sam H., Bruce C. Netschert, et al., Energy in the American Economy, 1850-1975: Its History and Prospects (Baltimore: The Johns Hopkins Press for Resources for the Future, 1960).

2. During the more recent past the ratio between energy consumption and GNP has shown considerable short-term fluctuation, but no persistent secular trend is discernible. Some of these short-period movements, in particular the rise in energy consumption relative to GNP in the late 1960's (subsequently reversed), attracted much attention and concern.